STUDENT'S
SHOP REFERENCE
HANDBOOK

STUDENT'S SHOP REFERENCE HANDBOOK

Compiled by
Edward G. Hoffman

INDUSTRIAL PRESS INC.
New York

Library of Congress Cataloging in Publication Data
Main entry under title:

Student's shop reference handbook.

   Includes index.
   1. Machine-shop practice—Handbooks, manuals, etc.
I. Hoffman, Edward G.
TJ1165.S845   1985   670.42'3   85-8200
ISBN 0-8311-1161-5

INDUSTRIAL PRESS INC.
200 Madison Avenue
New York, N.Y. 10157

FIRST EDITION

STUDENT'S SHOP REFERENCE HANDBOOK

Printed and bound by Maple-Vail Book Manufacturing Group, Binghamton, New York.

6   8   7

PREFACE

THE AIM OF THE EDITOR in compiling the *Student's Shop Reference Handbook* has been to provide an affordable and dependable handbook for students in the machine shop, the tool room, and the drafting room. The material included has been taken from authoritative sources and is presented in as clear, accurate, and easy-to-follow form as possible. In it the reader will find a wide range of useful formulas and data together with an extensive text.

Data and information from many American National Standards Institute (ANSI) Standards will be found in this book and have been extracted with permission of the publisher, the American Society of Mechanical Engineers, United Engineering Center, 345 East 47 Street, New York, New York 10017. These Standards are revised periodically; ASME should be contacted for information concerning the current edition of any particular document.

The eight sections of this handbook cover those areas of interest commonly encountered by machinists, toolmakers, diemakers, draftsmen, and other shop and manufacturing personnel. From the ability to understand and use shop mathematics to the reading and interpreting of shop drawings, the editor's intent is to provide the information and know-how that students will need as they prepare themselves for jobs in the metalworking industries.

The section on shop mathematics covers those aspects of applied mathematics that are needed on the job. From basic conversions between the inch and metric systems to the solving of problems involving arithmetic, geometry, algebra, and trigonometry, all of the essential information, formulas, and tables needed to make accurate calculations are provided.

The section covering inspection deals with the proper use of measuring tools and the understanding of the application of tolerances, allowances, and fits.

The sections covering machine elements describe the sizes, forms, and dimensions of standard keys, keyseats, tapers, screw threads, and gears.

The machining methods and materials sections specify the recommended speeds and feeds for various kinds of machining operations on different materials and the types and compositions of metals commonly used in machine construction.

The final section outlines the standard methods of presentation and the conventions used in preparing engineering drawings.

An extensive index has been prepared to enable the user to quickly and conveniently find the information and data that he or she requires.

Suggestions and comments concerning this Handbook are welcome.

CONTENTS

Preface	v
Mathematics	1
Inspection	151
Tapers and Keys	199
Screw Thread Systems	229
Common Hardware	278
Gears	290
Machining Methods	320
Materials	416
Engineering Drawings	479
Index	517

STUDENT'S
SHOP REFERENCE
HANDBOOK

Metric Conversion Factors

(Symbols of SI units, multiples and submultiples are
given in parentheses in the right-hand column)

Multiply	By	To Obtain
LENGTH		
centimetre	0.03280840	foot
centimetre	0.3937008	inch
fathom	1.8288*	metre (m)
foot	0.3048*	metre (m)
foot	30.48*	centimetre (cm)
foot	304.8*	millimetre (mm)
inch	0.0254*	metre (m)
inch	2.54*	centimetre (cm)
inch	25.4*	millimetre (mm)
kilometre	0.6213712	mile [U. S. statute]
metre	39.37008	inch
metre	0.5468066	fathom
metre	3.280840	foot
metre	0.1988388	rod
metre	1.093613	yard
metre	0.0006213712	mile [U. S. statute]
microinch	0.0254*	micrometre [micron] (μm)
micrometre [micron]	39.37008	microinch
mile [U. S. statute]	1609.344*	metre (m)
mile [U. S. statute]	1.609344*	kilometre (km)
millimetre	0.003280840	foot
millimetre	0.03937008	inch
rod	5.0292*	metre (m)
yard	0.9144*	metre (m)
AREA		
acre	4046.856	metre^2 (m^2)
acre	0.4046856	hectare
centimetre^2	0.1550003	inch^2
centimetre^2	0.001076391	foot^2
foot^2	0.09290304*	metre^2 (m^2)
foot^2	929.0304*	centimetre^2 (cm^2)
foot^2	92,903.04*	millimetre^2 (mm^2)
hectare	2.471054	acre
inch^2	645.16*	millimetre^2 (mm^2)
inch^2	6.4516*	centimetre^2 (cm^2)
inch^2	0.00064516*	metre^2 (m^2)
metre^2	1550.003	inch^2
metre^2	10.763910	foot^2
metre^2	1.195990	yard^2
metre^2	0.0002471054	acre
millimetre^2	0.00001076391	foot^2
millimetre^2	0.001550003	inch^2
yard^2	0.8361274	metre^2 (m^2)

* Where an asterisk is shown, the figure is exact.

Metric Conversion Factors (*Continued*)

Multiply	By	To Obtain
VOLUME (including CAPACITY)		
centimetre³	0.06102376	inch³
foot³	0.02831685	metre³ (m³)
foot³	28.31685	litre
gallon [U. K. liquid]	0.004546092	metre³ (m³)
gallon [U. K. liquid]	4.546092	litre
gallon [U. S. liquid]	0.003785412	metre³ (m³)
gallon [U. S. liquid]	3.785412	litre
inch³	16,387.06	millimetre³ (mm³)
inch³	16.38706	centimetre³ (cm³)
inch³	0.00001638706	metre³ (m³)
litre	0.001*	metre³ (m³)
litre	0.2199692	gallon [U. K. liquid]
litre	0.2641720	gallon [U. S. liquid]
litre	0.03531466	foot³
metre³	219.9692	gallon [U. K. liquid]
metre³	264.1720	gallon [U. S. liquid]
metre³	35.31466	foot³
metre³	1.307951	yard³
metre³	1000.*	litre
metre³	61,023.76	inch³
millimetre³	0.00006102376	inch³
yard³	0.7645549	metre³ (m³)
VELOCITY, ACCELERATION, and FLOW		
centimetre/second	1.968504	foot/minute
centimetre/second	0.03280840	foot/second
centimetre/minute	0.3937008	inch/minute
foot/hour	0.00008466667	metre/second (m/s)
foot/hour	0.00508*	metre/minute
foot/hour	0.3048*	metre/hour
foot/minute	0.508*	centimetre/second
foot/minute	18.288*	metre/hour
foot/minute	0.3048*	metre/minute
foot/minute	0.00508*	metre/second (m/s)
foot/second	30.48*	centimetre/second
foot/second	18.288*	metre/minute
foot/second	0.3048*	metre/second (m/s)
foot/second²	0.3048*	metre/second² (m/s²)
foot³/minute	28.31685	litre/minute
foot³/minute	0.0004719474	metre³/second (m³/s)
gallon [U. S. liquid]/min.	0.003785412	metre³/minute
gallon [U. S. liquid]/min.	0.00006309020	metre³/second (m³/s)
gallon [U. S. liquid]/min.	0.06309020	litre/second
gallon [U. S. liquid]/min.	3.785412	litre/minute
gallon [U. K. liquid]/min.	0.004546092	metre³/minute
gallon [U. K. liquid]/min.	0.00007576820	metre³/second (m³/s)
inch/minute	25.4*	millimetre/minute
inch/minute	2.54*	centimetre/minute
inch/minute	0.0254*	metre/minute
inch/second²	0.0254*	metre/second² (m/s²)

* Where an asterisk is shown, the figure is exact.

Metric Conversion Factors (*Continued*)

Multiply	By	To Obtain
VELOCITY, ACCELERATION, and FLOW (*Continued*)		
kilometre/hour	0.6213712	mile/hour [U. S. statute]
litre/minute	0.03531466	foot^3/minute
litre/minute	0.2641720	gallon [U. S. liquid]/minute
litre/second	15.85032	gallon [U. S. liquid]/minute
mile/hour	1.609344*	kilometre/hour
millimetre/minute	0.03937008	inch/minute
metre/second	11,811.02	foot/hour
metre/second	196.8504	foot/minute
metre/second	3.280840	foot/second
metre/second^2	3.280840	foot/second^2
metre/second^2	39.37008	inch/second^2
metre/minute	3.280840	foot/minute
metre/minute	0.05468067	foot/second
metre/minute	39.37008	inch/minute
metre/hour	3.280840	foot/hour
metre/hour	0.05468067	foot/minute
metre^3/second	2118.880	foot^3/minute
metre^3/second	13,198.15	gallon [U. K. liquid]/minute
metre^3/second	15,850.32	gallon [U. S. liquid]/minute
metre^3/minute	219.9692	gallon [U. K. liquid]/minute
metre^3/minute	264.1720	gallon [U. S. liquid]/minute
MASS and DENSITY		
grain [1/7000 lb avoirdupois]	0.06479891	gram (g)
gram	15.43236	grain
gram	0.001*	kilogram (kg)
gram	0.03527397	ounce [avoirdupois]
gram	0.03215074	ounce [troy]
gram/centimetre^3	0.03612730	pound/inch^3
hundredweight [long]	50.80235	kilogram (kg)
hundredweight [short]	45.35924	kilogram (kg)
kilogram	1000.*	gram (g)
kilogram	35.27397	ounce [avoirdupois]
kilogram	32.15074	ounce [troy]
kilogram	2.204622	pound [avoirdupois]
kilogram	0.06852178	slug
kilogram	0.0009842064	ton [long]
kilogram	0.001102311	ton [short]
kilogram	0.001*	ton [metric]
kilogram	0.001*	tonne
kilogram	0.01968413	hundredweight [long]
kilogram	0.02204622	hundredweight [short]
kilogram/metre^3	0.06242797	pound/foot^3
kilogram/metre^3	0.01002242	pound/gallon [U. K. liquid]
kilogram/metre^3	0.008345406	pound/gallon [U. S. liquid]
ounce [avoirdupois]	28.34952	gram (g)
ounce [avoirdupois]	0.02834952	kilogram (kg)

* Where an asterisk is shown, the figure is exact.

Metric Conversion Factors (*Continued*)

Multiply	By	To Obtain
MASS and DENSITY (*Continued*)		
ounce [troy]	31.10348	gram (g)
ounce [troy]	0.03110348	kilogram (kg)
pound [avoirdupois]	0.4535924	kilogram (kg)
pound/foot^3	16.01846	kilogram/metre^3 (kg/m^3)
pound/inch^3	27.67990	gram/centimetre^3 (g/cm^3)
pound/gal [U. S. liquid]	119.8264	kilogram/metre^3 (kg/m^3)
pound/gal [U. K. liquid]	99.77633	kilogram/metre^3 (kg/m^3)
slug	14.59390	kilogram (kg)
ton [long 2240 lb]	1016.047	kilogram (kg)
ton [short 2000 lb]	907.1847	kilogram (kg)
ton [metric]	1000.*	kilogram (kg)
tonne	1000.*	kilogram (kg)
FORCE and FORCE/LENGTH		
dyne	0.00001*	newton (N)
kilogram-force	9.806650*	newton (N)
kilopond	9.806650*	newton (N)
newton	0.1019716	kilogram-force
newton	0.1019716	kilopond
newton	0.2248089	pound-force
newton	100,000.*	dyne
newton	7.23301	poundal
newton	3.596942	ounce-force
newton/metre	0.005710148	pound/inch
newton/metre	0.06852178	pound/foot
ounce-force	0.2780139	newton (N)
pound-force	4.448222	newton (N)
poundal	0.1382550	newton (N)
pound/inch	175.1268	newton/metre (N/m)
pound/foot	14.59390	newton/metre (N/m)
BENDING MOMENT or TORQUE		
dyne-centimetre	0.0000001*	newton-metre (N · m)
kilogram-metre	9.806650*	newton-metre (N · m)
ounce-inch	7.061552	newton-millimetre
ounce-inch	0.007061552	newton-metre (N · m)
newton-metre	0.7375621	pound-foot
newton-metre	10,000,000.*	dyne-centimetre
newton-metre	0.1019716	kilogram-metre
newton-metre	141.6119	ounce-inch
newton-millimetre	0.1416119	ounce-inch
pound-foot	1.355818	newton-metre (N · m)

* Where an asterisk is shown, the figure is exact.

Metric Conversion Factors (*Continued*)

Multiply	By	To Obtain
MOMENT OF INERTIA and SECTION MODULUS		
moment of inertia [kg · m²] moment of inertia [kg · m²]	23.73036 3417.171	pound-foot² pound-inch²
moment of inertia [lb · ft²]	0.04214011	kilogram-metre² (kg · m²)
moment of inertia [lb · inch²]	0.0002926397	kilogram-metre² (kg · m²)
moment of section [foot⁴]	0.008630975	metre⁴ (m⁴)
moment of section [inch⁴]	41.62314	centimetre⁴
moment of section [metre⁴]	115.8618	foot⁴
moment of section [centimetre⁴]	0.02402510	inch⁴
section modulus [foot³]	0.02831685	metre³ (m³)
section modulus [inch³]	0.00001638706	metre³ (m³)
section modulus [metre³] section modulus [metre³]	35.31466 61,023.76	foot³ inch³
MOMENTUM		
kilogram-metre/second kilogram-metre/second	7.233011 86.79614	pound-foot/second pound-inch/second
pound-foot/second	0.1382550	kilogram-metre/second (kg · m/s)
pound-inch/second	0.01152125	kilogram-metre/second (kg · m/s)
PRESSURE and STRESS		
atmosphere [14.6959 lb/inch²]	101,325.	pascal (Pa)
bar bar bar	100,000.* 14.50377 100,000.*	pascal (Pa) pound/inch² newton/metre² (N/m²)
hectobar	0.6474898	ton [long]/inch²
kilogram/centimetre²	14.22334	pound/inch²
kilogram/metre² kilogram/metre² kilogram/metre²	9.806650* 9.806650* 0.2048161	newton/metre² (N/m²) pascal (Pa) pound/foot²
kilonewton/metre²	0.1450377	pound/inch²
newton/centimetre²	1.450377	pound/inch²
newton/metre² newton/metre² newton/metre² newton/metre²	0.00001* 1.0* 0.0001450377 0.1019716	bar pascal (Pa) pound/inch² kilogram/metre²
newton/millimetre²	145.0377	pound/inch²
pascal pascal pascal pascal pascal pascal	0.00000986923 0.00001* 0.1019716 1.0* 0.02088543 0.0001450377	atmosphere bar kilogram/metre² newton/metre² (N/m²) pound/foot² pound/inch²

* Where an asterisk is shown, the figure is exact.

Metric Conversion Factors (*Continued*)

Multiply	By	To Obtain
PRESSURE and STRESS (*Continued*)		
pound/foot^2	4.882429	kilogram/metre^2
pound/foot^2	47.88026	pascal (Pa)
pound/inch^2	0.06894757	bar
pound/inch^2	0.07030697	kilogram/centimetre^2
pound/inch^2	0.6894757	newton/centimetre^2
pound/inch^2	6.894757	kilonewton/metre^2
pound/inch^2	6894.757	newton/metre^2 (N/m^2)
pound/inch^2	0.006894757	newton/millimetre^2 (N/mm^2)
pound/inch^2	6894.757	pascal (Pa)
ton [long]/inch^2	1.544426	hectobar
ENERGY and WORK		
Btu [International Table]	1055.056	joule (J)
Btu [mean]	1055.87	joule (J)
calorie [mean]	4.19002	joule (J)
foot-pound	1.355818	joule (J)
foot-poundal	0.04214011	joule (J)
joule	0.0009478170	Btu [International Table]
joule	0.0009470863	Btu [mean]
joule	0.2386623	calorie [mean]
joule	0.7375621	foot-pound
joule	23.73036	foot-poundal
joule	0.9998180	joule [International U. S.]
joule	0.9999830	joule [U. S. legal, 1948]
joule [International U. S.]	1.000182	joule (J)
joule [U. S. legal, 1948]	1.000017	joule (J)
joule	.0002777778	watt-hour
watt-hour	3600. *	joule (J)
POWER		
Btu [International Table]/hour	0.2930711	watt (W)
foot-pound/hour	0.0003766161	watt (W)
foot-pound/minute	0.02259697	watt (W)
horsepower [550 ft-lb/s]	0.7456999	kilowatt (kW)
horsepower [550 ft-lb/s]	745.6999	watt (W)
horsepower [electric]	746. *	watt (W)
horsepower [metric]	735.499	watt (W)
horsepower [U. K.]	745.70	watt (W)
kilowatt	1.341022	horsepower [550 ft-lb/s]
watt	2655.224	foot-pound/hour
watt	44.25372	foot-pound/minute
watt	0.001341022	horsepower [550 ft-lb/s]
watt	0.001340483	horsepower [electric]
watt	0.001359621	horsepower [metric]
watt	0.001341022	horsepower [U. K.]
watt	3.412141	Btu [International Table]/hour

* Where an asterisk is shown, the figure is exact.

Metric Conversion Factors (*Concluded*)

Multiply	By	To Obtain
VISCOSITY		
centipoise	0.001*	pascal-second (Pa · s)
centistoke	0.000001*	metre²/second (m²/s)
metre²/second	1,000,000.*	centistoke
metre²/second	10,000.*	stoke
pascal-second	1000.*	centipoise
pascal-second	10.*	poise
poise	0.1*	pascal-second (Pa · s)
stoke	0.0001*	metre²/second (m²/s)
TEMPERATURE		

To Convert From	To	Use Formula
temperature Celsius, t_C	temperature Kelvin, t_K	$t_K = t_C + 273.15$
temperature Fahrenheit, t_F	temperature Kelvin, t_K	$t_K = (t_F + 459.67)/1.8$
temperature Celsius, t_C	temperature Fahrenheit, t_F	$t_F = 1.8\,t_C + 32$
temperature Fahrenheit, t_F	temperature Celsius, t_C	$t_C = (t_F - 32)/1.8$
temperature Kelvin, t_K	temperature Celsius, t_C	$t_C = t_K - 273.15$
temperature Kelvin, t_K	temperature Fahrenheit, t_F	$t_F = 1.8\,t_K - 459.67$
temperature Kelvin, t_K	temperature Rankine, t_R	$t_R = 9/5\,t_K$
temperature Rankine, t_R	temperature Kelvin, t_K	$t_K = 5/9\,t_R$

* Where an asterisk is shown, the figure is exact.

Miscellaneous Conversion Factors
(English Units)

Multiply	By	To Obtain
atmospheres	29.92	inches of mercury (32 deg. F.)
atmospheres	14.70	pounds/inch²
British thermal units/hour	12.96	foot-pounds/minute
circular mils	0.7854	square mils
feet of water (60 deg. F.)	0.8843	inches of mercury (60 deg. F.)
feet of water (60 deg. F.)	0.4331	pounds/inch²
feet/minute	0.01136	miles/hour
foot-pounds/second	0.07716	British thermal units/minute
gallons (U.S.) of water (60 deg. F.)	8.337	pounds of water (60 deg. F.)
gallons (U.S.)/second	8.021	feet³/minute
inches of mercury (32 deg. F.)	0.03342	atmospheres
inches of mercury (60 deg. F.)	1.131	feet of water (60 deg. F.)
inches of mercury (60 deg. F.)	0.4898	pounds/inch²
inches of water (60 deg. F.)	0.03609	pounds/inch²
knots (International)	1.151	miles (statute)/hour
miles/hour	88	feet/minute
miles (statute)/hour	0.8690	knots (International)
ounces (avoirdupois)	0.9115	ounces (troy)
ounces (troy)	1.097	ounces (avoirdupois)
ounces (troy)	0.06857	pounds (avoirdupois)
pounds (avoirdupois)	14.58	ounces (troy)
pounds of water (60 deg. F.)	0.01603	feet³
pounds of water (60 deg. F.)	0.1199	gallons (U.S.)
pounds/inch²	0.06805	atmospheres
pounds/inch²	2.309	feet of water (60 deg. F.)
pounds/inch²	2.042	inches of mercury (60 deg. F.)
pounds/inch²	27.71	inches of water (60 deg. F.)
square mils	1.273	circular mils

MATHEMATICS

Fractions of One Inch
Converted to Decimal Inches
and Millimeters

Fraction	Decimal	mm	Fraction	Decimal	mm
1/64	0.015 625	0.396 9	33/64	0.515 625	13.096 9
1/32	0.031 25	0.793 8	17/32	0.531 25	13.493 8
3/64	0.046 875	1.190 6	35/64	0.546 875	13.890 6
1/16	0.062 5	1.587 5	9/16	0.562 5	14.287 5
5/64	0.078 125	1.984 4	37/64	0.578 125	14.684 4
3/32	0.093 75	2.381 3	19/32	0.593 75	15.081 3
7/64	0.109 375	2.778 1	39/64	0.609 375	15.478 1
1/8	0.125	3.175	5/8	0.625	15.875
9/64	0.140 625	3.571 9	41/64	0.640 625	16.271 9
5/32	0.156 25	3.968 8	21/32	0.656 25	16.668 8
11/64	0.171 875	4.365 6	43/64	0.671 875	17.065 6
3/16	0.187 5	4.762 5	11/16	0.687 5	17.462 5
13/64	0.203 125	5.159 4	45/64	0.703 125	17.859 4
7/32	0.218 75	5.556 3	23/32	0.718 75	18.256 3
15/64	0.234 375	5.953 1	47/64	0.734 375	18.653 1
1/4	0.25	6.35	3/4	0.75	19.05
17/64	0.265 625	6.746 9	49/64	0.765 625	19.446 9
9/32	0.281 25	7.143 8	25/32	0.781 25	19.843 8
19/64	0.296 875	7.540 6	51/64	0.796 875	20.240 6
5/16	0.312 5	7.937 5	13/16	0.812 5	20.637 5
21/64	0.328 125	8.334 4	53/64	0.828 125	21.034 4
11/32	0.343 75	8.731 3	27/32	0.843 75	21.431 3
23/64	0.359 375	9.128 1	55/64	0.859 375	21.828 1
3/8	0.375	9.525	7/8	0.875	22.225
25/64	0.390 625	9.921 9	57/64	0.890 625	22.621 9
13/32	0.406 25	10.318 8	29/32	0.906 25	23.018 8
27/64	0.421 875	10.715 6	59/64	0.921 875	23.415 6
7/16	0.437 5	11.112 5	15/16	0.937 5	23.812 5
29/64	0.453 125	11.509 4	61/64	0.953 125	24.209 4
15/32	0.468 75	11.906 3	31/32	0.968 75	24.606 3
31/64	0.484 375	12.303 1	63/64	0.984 375	25.003 1
1/2	0.5	12.7	1	1	25.4

Use of Conversion Tables. — On this and following pages tables are given which permit conversion from English to metric units and vice versa over a wide range of values. Where the desired value cannot be obtained directly from these tables, a simple addition of two or more values taken directly from the table will suffice as shown in the following examples:

Example 1: Find the millimetre equivalent of 0.4476 inch.

$$
\begin{aligned}
.4 \quad \text{in.} &= 10.16000 \text{ mm} \\
.04 \quad \text{in.} &= 1.01600 \text{ mm} \\
.007 \quad \text{in.} &= .17780 \text{ mm} \\
.0006 \text{ in.} &= .01524 \text{ mm} \\
\hline
.4476 \text{ in.} &= 11.36904 \text{ mm}
\end{aligned}
$$

Example 2: Find the inch equivalent of 84.9 mm.

$$
\begin{aligned}
80. \quad \text{mm} &= 3.14961 \text{ in.} \\
4. \quad \text{mm} &= 0.15748 \text{ in.} \\
0.9 \text{ mm} &= 0.03543 \text{ in.} \\
\hline
84.9 \text{ mm} &= 3.34252 \text{ in.}
\end{aligned}
$$

Inch—Millimetre and Inch—Centimetre Conversion Table*
(Based on 1 inch = 25.4 millimetres, exactly)

INCHES TO MILLIMETRES

in.	mm	in.	mm	in.	mm	in.	mm	in.	mm	in.	mm
10	254.00000	1	25.40000	.1	2.54000	.01	.25400	.001	.02540	.0001	.00254
20	508.00000	2	50.80000	.2	5.08000	.02	.50800	.002	.05080	.0002	.00508
30	762.00000	3	76.20000	.3	7.62000	.03	.76200	.003	.07620	.0003	.00762
40	1,016.00000	4	101.60000	.4	10.16000	.04	1.01600	.004	.10160	.0004	.01016
50	1,270.00000	5	127.00000	.5	12.70000	.05	1.27000	.005	.12700	.0005	.01270
60	1,524.00000	6	152.40000	.6	15.24000	.06	1.52400	.006	.15240	.0006	.01524
70	1,778.00000	7	177.80000	.7	17.78000	.07	1.77800	.007	.17780	.0007	.01778
80	2,032.00000	8	203.20000	.8	20.32000	.08	2.03200	.008	.20320	.0008	.02032
90	2,286.00000	9	228.60000	.9	22.86000	.09	2.28600	.009	.22860	.0009	.02286
100	2,540.00000	10	254.00000	1.0	25.40000	.10	2.54000	.010	.25400	.0010	.02540

MILLIMETRES TO INCHES

mm	in.	mm	in.	mm.	in.	mm	in.	mm	in.	mm	in.
100	3.93701	10	.39370	1	.03937	.1	.00394	.01	.00039	.001	.00004
200	7.87402	20	.78740	2	.07874	.2	.00787	.02	.00079	.002	.00008
300	11.81102	30	1.18110	3	.11811	.3	.01181	.03	.00118	.003	.00012
400	15.74803	40	1.57480	4	.15748	.4	.01575	.04	.00157	.004	.00016
500	19.68504	50	1.96850	5	.19685	.5	.01969	.05	.00197	.005	.00020
600	23.62205	60	2.36220	6	.23622	.6	.02362	.06	.00236	.006	.00024
700	27.55906	70	2.75591	.7	.27559	.7	.02756	.07	.00276	.007	.00028
800	31.49606	80	3.14961	8	.31496	.8	.03150	.08	.00315	.008	.00031
900	35.43307	90	3.54331	9	.35433	.9	.03543	.09	.00354	.009	.00035
1,000	39.37008	100	3.93701	10	.39370	1.0	.03937	.10	.00394	.010	.00039

* For inches to centimetres, shift decimal point in mm column one place to left and read centimetres, thus:

40 in. = 1016 mm = 101.6 cm

For centimetres to inches, shift decimal point of centimetre value one place to right and enter mm column, thus:

70 cm = 700 mm = 27.55906 inches

Decimals of an Inch to Millimeters
(Based on 1 inch = 25.4 millimeters, exactly)

Inches	0.000	0.001	0.002	0.003	0.004	0.005	0.006	0.007	0.008	0.009
					Millimeters					
0.000	...	0.0254	0.0508	0.0762	0.1016	0.1270	0.1524	0.1778	0.2032	0.2286
0.010	0.2540	0.2794	0.3048	0.3302	0.3556	0.3810	0.4064	0.4318	0.4572	0.4826
0.020	0.5080	0.5334	0.5588	0.5842	0.6096	0.6350	0.6604	0.6858	0.7112	0.7366
0.030	0.7620	0.7874	0.8128	0.8382	0.8636	0.8890	0.9144	0.9398	0.9652	0.9906
0.040	1.0160	1.0414	1.0668	1.0922	1.1176	1.1430	1.1684	1.1938	1.2192	1.2446
0.050	1.2700	1.2954	1.3208	1.3462	1.3716	1.3970	1.4224	1.4478	1.4732	1.4986
0.060	1.5240	1.5494	1.5748	1.6002	1.6256	1.6510	1.6764	1.7018	1.7272	1.7526
0.070	1.7780	1.8034	1.8288	1.8542	1.8796	1.9050	1.9304	1.9558	1.9812	2.0066
0.080	2.0320	2.0574	2.0828	2.1082	2.1336	2.1590	2.1844	2.2098	2.2352	2.2606
0.090	2.2860	2.3114	2.3368	2.3622	2.3876	2.4130	2.4384	2.4638	2.4892	2.5146
0.100	2.5400	2.5654	2.5908	2.6162	2.6416	2.6670	2.6924	2.7178	2.7432	2.7686
0.110	2.7940	2.8194	2.8448	2.8702	2.8956	2.9210	2.9464	2.9718	2.9972	3.0226
0.120	3.0480	3.0734	3.0988	3.1242	3.1496	3.1750	3.2004	3.2258	3.2512	3.2766
0.130	3.3020	3.3274	3.3528	3.3782	3.4036	3.4290	3.4544	3.4798	3.5052	3.5306
0.140	3.5560	3.5814	3.6068	3.6322	3.6576	3.6830	3.7084	3.7338	3.7592	3.7846
0.150	3.8100	3.8354	3.8608	3.8862	3.9116	3.9370	3.9624	3.9878	4.0132	4.0386
0.160	4.0640	4.0894	4.1148	4.1402	4.1656	4.1910	4.2164	4.2418	4.2672	4.2926
0.170	4.3180	4.3434	4.3688	4.3942	4.4196	4.4450	4.4704	4.4958	4.5212	4.5466
0.180	4.5720	4.5974	4.6228	4.6482	4.6736	4.6990	4.7244	4.7498	4.7752	4.8006
0.190	4.8260	4.8514	4.8768	4.9022	4.9276	4.9530	4.9784	5.0038	5.0292	5.0546
0.200	5.0800	5.1054	5.1308	5.1562	5.1816	5.2070	5.2324	5.2578	5.2832	5.3086
0.210	5.3340	5.3594	5.3848	5.4102	5.4356	5.4610	5.4864	5.5118	5.5372	5.5626
0.220	5.5880	5.6134	5.6388	5.6642	5.6896	5.7150	5.7404	5.7658	5.7912	5.8166
0.230	5.8420	5.8674	5.8928	5.9182	5.9436	5.9690	5.9944	6.0198	6.0452	6.0706
0.240	6.0960	6.1214	6.1468	6.1722	6.1976	6.2230	6.2484	6.2738	6.2992	6.3246
0.250	6.3500	6.3754	6.4008	6.4262	6.4516	6.4770	6.5024	6.5278	6.5532	6.5786
0.260	6.6040	6.6294	6.6548	6.6802	6.7056	6.7310	6.7564	6.7818	6.8072	6.8326
0.270	6.8580	6.8834	6.9088	6.9342	6.9596	6.9850	7.0104	7.0358	7.0612	7.0866
0.280	7.1120	7.1374	7.1628	7.1882	7.2136	7.2390	7.2644	7.2898	7.3152	7.3406
0.290	7.3660	7.3914	7.4168	7.4422	7.4676	7.4930	7.5184	7.5438	7.5692	7.5946
0.300	7.6200	7.6454	7.6708	7.6962	7.7216	7.7470	7.7724	7.7978	7.8232	7.8486
0.310	7.8740	7.8994	7.9248	7.9502	7.9756	8.0010	8.0264	8.0518	8.0772	8.1026
0.320	8.1280	8.1534	8.1788	8.2042	8.2296	8.2550	8.2804	8.3058	8.3312	8.3566
0.330	8.3820	8.4074	8.4328	8.4582	8.4836	8.5090	8.5344	8.5598	8.5852	8.6106
0.340	8.6360	8.6614	8.6868	8.7122	8.7376	8.7630	8.7884	8.8138	8.8392	8.8646
0.350	8.8900	8.9154	8.9408	8.9662	8.9916	9.0170	9.0424	9.0678	9.0932	9.1186
0.360	9.1440	9.1694	9.1948	9.2202	9.2456	9.2710	9.2964	9.3218	9.3472	9.3726
0.370	9.3980	9.4234	9.4488	9.4742	9.4996	9.5250	9.5504	9.5758	9.6012	9.6266
0.380	9.6520	9.6774	9.7028	9.7282	9.7536	9.7790	9.8044	9.8298	9.8552	9.8806
0.390	9.9060	9.9314	9.9568	9.9822	10.0076	10.0330	10.0584	10.0838	10.1092	10.1346
0.400	10.1600	10.1854	10.2108	10.2362	10.2616	10.2870	10.3124	10.3378	10.3632	10.3886
0.410	10.4140	10.4394	10.4648	10.4902	10.5156	10.5410	10.5664	10.5918	10.6172	10.6426
0.420	10.6680	10.6934	10.7188	10.7442	10.7696	10.7950	10.8204	10.8458	10.8712	10.8966
0.430	10.9220	10.9474	10.9728	10.9982	11.0236	11.0490	11.0744	11.0998	11.1252	11.1506
0.440	11.1760	11.2014	11.2268	11.2522	11.2776	11.3030	11.3284	11.3538	11.3792	11.4046
0.450	11.4300	11.4554	11.4808	11.5062	11.5316	11.5570	11.5824	11.6078	11.6332	11.6586
0.460	11.6840	11.7094	11.7348	11.7602	11.7856	11.8110	11.8364	11.8618	11.8872	11.9126
0.470	11.9380	11.9634	11.9888	12.0142	12.0396	12.0650	12.0904	12.1158	12.1412	12.1666
0.480	12.1920	12.2174	12.2428	12.2682	12.2936	12.3190	12.3444	12.3698	12.3952	12.4206
0.490	12.4460	12.4714	12.4968	12.5222	12.5476	12.5730	12.5984	12.6238	12.6492	12.6746

Use previous table to obtain whole inch equivalents to add to decimal equivalents above. All values given in this table are exact; figures to the right of the last place figures are all zeros.

Decimals of an Inch to Millimeters
(Based on 1 inch = 25.4 millimeters, exactly)

Inches	0.000	0.001	0.002	0.003	0.004	0.005	0.006	0.007	0.008	0.009
					Millimeters					
0.500	12.7000	12.7254	12.7508	12.7762	12.8016	12.8270	12.8524	12.8778	12.9032	12.9286
0.510	12.9540	12.9794	13.0048	13.0302	13.0556	13.0810	13.1064	13.1318	13.1572	13.1826
0.520	13.2080	13.2334	13.2588	13.2842	13.3096	13.3350	13.3604	13.3858	13.4112	13.4366
0.530	13.4620	13.4874	13.5128	13.5382	13.5636	13.5890	13.6144	13.6398	13.6652	13.6906
0.540	13.7160	13.7414	13.7668	13.7922	13.8176	13.8430	13.8684	13.8938	13.9192	13.9446
0.550	13.9700	13.9954	14.0208	14.0462	14.0716	14.0970	14.1224	14.1478	14.1732	14.1986
0.560	14.2240	14.2494	14.2748	14.3002	14.3256	14.3510	14.3764	14.4018	14.4272	14.4526
0.570	14.4780	14.5034	14.5288	14.5542	14.5796	14.6050	14.6304	14.6558	14.6812	14.7066
0.580	14.7320	14.7574	14.7828	14.8082	14.8336	14.8590	14.8844	14.9098	14.9352	14.9606
0.590	14.9860	15.0114	15.0368	15.0622	15.0876	15.1130	15.1384	15.1638	15.1892	15.2146
0.600	15.2400	15.2654	15.2908	15.3162	15.3416	15.3670	15.3924	15.4178	15.4432	15.4686
0.610	15.4940	15.5194	15.5448	15.5702	15.5956	15.6210	15.6464	15.6718	15.6972	15.7226
0.620	15.7480	15.7734	15.7988	15.8242	15.8496	15.8750	15.9004	15.9258	15.9512	15.9766
0.630	16.0020	16.0274	16.0528	16.0782	16.1036	16.1290	16.1544	16.1798	16.2052	16.2306
0.640	16.2560	16.2814	16.3068	16.3322	16.3576	16.3830	16.4084	16.4338	16.4592	16.4846
0.650	16.5100	16.5354	16.5608	16.5862	16.6116	16.6370	16.6624	16.6878	16.7132	16.7386
0.660	16.7640	16.7894	16.8148	16.8402	16.8656	16.8910	16.9164	16.9418	16.9672	16.9926
0.670	17.0180	17.0434	17.0688	17.0942	17.1196	17.1450	17.1704	17.1958	17.2212	17.2466
0.680	17.2720	17.2974	17.3228	17.3482	17.3736	17.3990	17.4244	17.4498	17.4752	17.5006
0.690	17.5260	17.5514	17.5768	17.6022	17.6276	17.6530	17.6784	17.7038	17.7292	17.7546
0.700	17.7800	17.8054	17.8308	17.8562	17.8816	17.9070	17.9324	17.9578	17.9832	18.0086
0.710	18.0340	18.0594	18.0848	18.1102	18.1356	18.1610	18.1864	18.2118	18.2372	18.2626
0.720	18.2880	18.3134	18.3388	18.3642	18.3896	18.4150	18.4404	18.4658	18.4912	18.5166
0.730	18.5420	18.5674	18.5928	18.6182	18.6436	18.6690	18.6944	18.7198	18.7452	18.7706
0.740	18.7960	18.8214	18.8468	18.8722	18.8976	18.9230	18.9484	18.9738	18.9992	19.0246
0.750	19.0500	19.0754	19.1008	19.1262	19.1516	19.1770	19.2024	19.2278	19.2532	19.2786
0.760	19.3040	19.3294	19.3548	19.3802	19.4056	19.4310	19.4564	19.4818	19.5072	19.5326
0.770	19.5580	19.5834	19.6088	19.6342	19.6596	19.6850	19.7104	19.7358	19.7612	19.7866
0.780	19.8120	19.8374	19.8628	19.8882	19.9136	19.9390	19.9644	19.9898	20.0152	20.0406
0.790	20.0660	20.0914	20.1168	20.1422	20.1676	20.1930	20.2184	20.2438	20.2692	20.2946
0.800	20.3200	20.3454	20.3708	20.3962	20.4216	20.4470	20.4724	20.4978	20.5232	20.5486
0.810	20.5740	20.5994	20.6248	20.6502	20.6756	20.7010	20.7264	20.7518	20.7772	20.8026
0.820	20.8280	20.8534	20.8788	20.9042	20.9296	20.9550	20.9804	21.0058	21.0312	21.0566
0.830	21.0820	21.1074	21.1328	21.1582	21.1836	21.2090	21.2344	21.2598	21.2852	21.3106
0.840	21.3360	21.3614	21.3868	21.4122	21.4376	21.4630	21.4884	21.5138	21.5392	21.5646
0.850	21.5900	21.6154	21.6408	21.6662	21.6916	21.7170	21.7424	21.7678	21.7932	21.8186
0.860	21.8440	21.8694	21.8948	21.9202	21.9456	21.9710	21.9964	22.0218	22.0472	22.0726
0.870	22.0980	22.1234	22.1488	22.1742	22.1996	22.2250	22.2504	22.2758	22.3012	22.3266
0.880	22.3520	22.3774	22.4028	22.4282	22.4536	22.4790	22.5044	22.5298	22.5552	22.5806
0.890	22.6060	22.6314	22.6568	22.6822	22.7076	22.7330	22.7584	22.7838	22.8092	22.8346
0.900	22.8600	22.8854	22.9108	22.9362	22.9616	22.9870	23.0124	23.0378	23.0632	23.0886
0.910	23.1140	23.1394	23.1648	23.1902	23.2156	23.2410	23.2664	23.2918	23.3172	23.3426
0.920	23.3680	23.3934	23.4188	23.4442	23.4696	23.4950	23.5204	23.5458	23.5712	23.5966
0.930	23.6220	23.6474	23.6728	23.6982	23.7236	23.7490	23.7744	23.7998	23.8252	23.8506
0.940	23.8760	23.9014	23.9268	23.9522	23.9776	24.0030	24.0284	24.0538	24.0792	24.1046
0.950	24.1300	24.1554	24.1808	24.2062	24.2316	24.2570	24.2824	24.3078	24.3332	24.3586
0.960	24.3840	24.4094	24.4348	24.4602	24.4856	24.5110	24.5364	24.5618	24.5872	24.6126
0.970	24.6380	24.6634	24.6888	24.7142	24.7396	24.7650	24.7904	24.8158	24.8412	24.8666
0.980	24.8920	24.9174	24.9428	24.9682	24.9936	25.0190	25.0444	25.0698	25.0952	25.1206
0.990	25.1460	25.1714	25.1968	25.2222	25.2476	25.2730	25.2984	25.3238	25.3492	25.3746
1.000	25.4000	...	..	...	...	...	...	...	...	...

Use previous table to obtain whole inch equivalents to add to decimal equivalents above. All values given in this table are exact; figures to the right of the last place figures are all zeros.

MATHEMATICS

Millimeters to Inches
(Based on 1 inch = 25.4 millimeters, exactly)

Milli-meters	0	1	2	3	4	5	6	7	8	9
					Inches					
0	...	0.03937	0.07874	0.11811	0.15748	0.19685	0.23622	0.27559	0.31496	0.35433
10	0.39370	0.43307	0.47244	0.51181	0.55118	0.59055	0.62992	0.66929	0.70866	0.74803
20	0.78740	0.82677	0.86614	0.90551	0.94488	0.98425	1.02362	1.06299	1.10236	1.14173
30	1.18110	1.22047	1.25984	1.29921	1.33858	1.37795	1.41732	1.45669	1.49606	1.53543
40	1.57480	1.61417	1.65354	1.69291	1.73228	1.77165	1.81102	1.85039	1.88976	1.92913
50	1.96850	2.00787	2.04724	2.08661	2.12598	2.16535	2.20472	2.24409	2.28346	2.32283
60	2.36220	2.40157	2.44094	2.48031	2.51969	2.55906	2.59843	2.63780	2.67717	2.71654
70	2.75591	2.79528	2.83465	2.87402	2.91339	2.95276	2.99213	3.03150	3.07087	3.11024
80	3.14961	3.18898	3.22835	3.26772	3.30709	3.34646	3.38583	3.42520	3.46457	3.50394
90	3.54331	3.58268	3.62205	3.66142	3.70079	3.74016	3.77953	3.81890	3.85827	3.89764
100	3.93701	3.97638	4.01575	4.05512	4.09449	4.13386	4.17323	4.21260	4.25197	4.29134
110	4.33071	4.37008	4.40945	4.44882	4.48819	4.52756	4.56693	4.60630	4.64567	4.68504
120	4.72441	4.76378	4.80315	4.84252	4.88189	4.92126	4.96063	5.00000	5.03937	5.07874
130	5.11811	5.15748	5.19685	5.23622	5.27559	5.31496	5.35433	5.39370	5.43307	5.47244
140	5.51181	5.55118	5.59055	5.62992	5.66929	5.70866	5.74803	5.78740	5.82677	5.86614
150	5.90551	5.94488	5.98425	6.02362	6.06299	6.10236	6.14173	6.18110	6.22047	6.25984
160	6.29921	6.33858	6.37795	6.41732	6.45669	6.49606	6.53543	6.57480	6.61417	6.65354
170	6.69291	6.73228	6.77165	6.81102	6.85039	6.88976	6.92913	6.96850	7.00787	7.04724
180	7.08661	7.12598	7.16535	7.20472	7.24409	7.28346	7.32283	7.36220	7.40157	7.44094
190	7.48031	7.51969	7.55906	7.59843	7.63780	7.67717	7.71654	7.75591	7.79528	7.83465
200	7.87402	7.91339	7.95276	7.99213	8.03150	8.07087	8.11024	8.14961	8.18898	8.22835
210	8.26772	8.30709	8.34646	8.38583	8.42520	8.46457	8.50394	8.54331	8.58268	8.62205
220	8.66142	8.70079	8.74016	8.77953	8.81890	8.85827	8.89764	8.93701	8.97638	9.01575
230	9.05512	9.09449	9.13386	9.17323	9.21260	9.25197	9.29134	9.33071	9.37008	9.40945
240	9.44882	9.48819	9.52756	9.56693	9.60630	9.64567	9.68504	9.72441	9.76378	9.80315
250	9.84252	9.88189	9.92126	9.96063	10.0000	10.0394	10.0787	10.1181	10.1575	10.1969
260	10.2362	10.2756	10.3150	10.3543	10.3937	10.4331	10.4724	10.5118	10.5512	10.5906
270	10.6299	10.6693	10.7087	10.7480	10.7874	10.8268	10.8661	10.9055	10.9449	10.9843
280	11.0236	11.0630	11.1024	11.1417	11.1811	11.2205	11.2598	11.2992	11.3386	11.3780
290	11.4173	11.4567	11.4961	11.5354	11.5748	11.6142	11.6535	11.6929	11.7323	11.7717
300	11.8110	11.8504	11.8898	11.9291	11.9685	12.0079	12.0472	12.0866	12.1260	12.1654
310	12.2047	12.2441	12.2835	12.3228	12.3622	12.4016	12.4409	12.4803	12.5197	12.5591
320	12.5984	12.6378	12.6772	12.7165	12.7559	12.7953	12.8346	12.8740	12.9134	12.9528
330	12.9921	13.0315	13.0709	13.1102	13.1496	13.1890	13.2283	13.2677	13.3071	13.3465
340	13.3858	13.4252	13.4646	13.5039	13.5433	13.5827	13.6220	13.6614	13.7008	13.7402
350	13.7795	13.8189	13.8583	13.8976	13.9370	13.9764	14.0157	14.0551	14.0945	14.1339
360	14.1732	14.2126	14.2520	14.2913	14.3307	14.3701	14.4094	14.4488	14.4882	14.5276
370	14.5669	14.6063	14.6457	14.6850	14.7244	14.7638	14.8031	14.8425	14.8819	14.9213
380	14.9606	15.0000	15.0394	15.0787	15.1181	15.1575	15.1969	15.2362	15.2756	15.3150
390	15.3543	15.3937	15.4331	15.4724	15.5118	15.5512	15.5906	15.6299	15.6693	15.7087
400	15.7480	15.7874	15.8268	15.8661	15.9055	15.9449	15.9843	16.0236	16.0630	16.1024
410	16.1417	16.1811	16.2205	16.2598	16.2992	16.3386	16.3780	16.4173	16.4567	16.4961
420	16.5354	16.5748	16.6142	16.6535	16.6929	16.7323	16.7717	16.8110	16.8504	16.8898
430	16.9291	16.9685	17.0079	17.0472	17.0866	17.1260	17.1654	17.2047	17.2441	17.2835
440	17.3228	17.3622	17.4016	17.4409	17.4803	17.5197	17.5591	17.5984	17.6378	17.6772
450	17.7165	17.7559	17.7953	17.8346	17.8740	17.9134	17.9528	17.9921	18.0315	18.0709
460	18.1102	18.1496	18.1890	18.2283	18.2677	18.3071	18.3465	18.3858	18.4252	18.4646
470	18.5039	18.5433	18.5827	18.6220	18.6614	18.7008	18.7402	18.7795	18.8189	18.8583
480	18.8976	18.9370	18.9764	19.0157	19.0551	19.0945	19.1339	19.1732	19.2126	19.2520
490	19.2913	19.3307	19.3701	19.4094	19.4488	19.4882	19.5276	19.5669	19.6063	19.6457

Millimeters to Inches
(Based on 1 inch = 25.4 millimeters, exactly)

Milli-meters	0	1	2	3	4	5	6	7	8	9
						Inches					
500	19.6850	19.7244	19.7638	19.8031	19.8425	19.8819	19.9213	19.9606	20.0000	20.0394
510	20.0787	20.1181	20.1575	20.1969	20.2362	20.2756	20.3150	20.3543	20.3937	20.4331
520	20.4724	20.5118	20.5512	20.5906	20.6299	20.6693	20.7087	20.7480	20.7874	20.8268
530	20.8661	20.9055	20.9449	20.9843	21.0236	21.0630	21.1024	21.1417	21.1811	21.2205
540	21.2598	21.2992	21.3386	21.3780	21.4173	21.4567	21.4961	21.5354	21.5748	21.6142
550	21.6535	21.6929	21.7323	21.7717	21.8110	21.8504	21.8898	21.9291	21.9685	22.0079
560	22.0472	22.0866	22.1260	22.1654	22.2047	22.2441	22.2835	22.3228	22.3622	22.4016
570	22.4409	22.4803	22.5197	22.5591	22.5984	22.6378	22.6772	22.7165	22.7559	22.7953
580	22.8346	22.8740	22.9134	22.9528	22.9921	23.0315	23.0709	23.1102	23.1496	23.1890
590	23.2283	23.2677	23.3071	23.3465	23.3858	23.4252	23.4646	23.5039	23.5433	23.5827
600	23.6220	23.6614	23.7008	23.7402	23.7795	23.8189	23.8583	23.8976	23.9370	23.9764
610	24.0157	24.0551	24.0945	24.1339	24.1732	24.2126	24.2520	24.2913	24.3307	24.3701
620	24.4094	24.4488	24.4882	24.5276	24.5669	24.6063	24.6457	24.6850	24.7244	24.7638
630	24.8031	24.8425	24.8819	24.9213	24.9606	25.0000	25.0394	25.0787	25.1181	25.1575
640	25.1969	25.2362	25.2756	25.3150	25.3543	25.3937	25.4331	25.4724	25.5118	25.5512
650	25.5906	25.6299	25.6693	25.7087	25.7480	25.7874	25.8268	25.8661	25.9055	25.9449
660	25.9843	26.0236	26.0630	26.1024	26.1417	26.1811	26.2205	26.2598	26.2992	26.3386
670	26.3780	26.4173	26.4567	26.4961	26.5354	26.5748	26.6142	26.6535	26.6929	26.7323
680	26.7717	26.8110	26.8504	26.8898	26.9291	26.9685	27.0079	27.0472	27.0866	27.1260
690	27.1654	27.2047	27.2441	27.2835	27.3228	27.3622	27.4016	27.4409	27.4803	27.5197
700	27.5591	27.5984	27.6378	27.6772	27.7165	27.7559	27.7953	27.8346	27.8740	27.9134
710	27.9528	27.9921	28.0315	28.0709	28.1102	28.1496	28.1890	28.2283	28.2677	28.3071
720	28.3465	28.3858	28.4252	28.4646	28.5039	28.5433	28.5827	28.6220	28.6614	28.7008
730	28.7402	28.7795	28.8189	28.8583	28.8976	28.9370	28.9764	29.0157	29.0551	29.0945
740	29.1339	29.1732	29.2126	29.2520	29.2913	29.3307	29.3701	29.4094	29.4488	29.4882
750	29.5276	29.5669	29.6063	29.6457	29.6850	29.7244	29.7638	29.8031	29.8425	29.8819
760	29.9213	29.9606	30.0000	30.0394	30.0787	30.1181	30.1575	30.1969	30.2362	30.2756
770	30.3150	30.3543	30.3937	30.4331	30.4724	30.5118	30.5512	30.5906	30.6299	30.6693
780	30.7087	30.7480	30.7874	30.8268	30.8661	30.9055	30.9449	30.9843	31.0236	31.0630
790	31.1024	31.1417	31.1811	31.2205	31.2598	31.2992	31.3386	31.3780	31.4173	31.4567
800	31.4961	31.5354	31.5748	31.6142	31.6535	31.6929	31.7323	31.7717	31.8110	31.8504
810	31.8898	31.9291	31.9685	32.0079	32.0472	32.0866	32.1260	32.1654	32.2047	32.2441
820	32.2835	32.3228	32.3622	32.4016	32.4409	32.4803	32.5197	32.5591	32.5984	32.6378
830	32.6772	32.7165	32.7559	32.7953	32.8346	32.8740	32.9134	32.9528	32.9921	33.0315
840	33.0709	33.1102	33.1496	33.1890	33.2283	33.2677	33.3071	33.3465	33.3858	33.4252
850	33.4646	33.5039	33.5433	33.5827	33.6220	33.6614	33.7008	33.7402	33.7795	33.8189
860	33.8583	33.8976	33.9370	33.9764	34.0157	34.0551	34.0945	34.1339	34.1732	34.2126
870	34.2520	34.2913	34.3307	34.3701	34.4094	34.4488	34.4882	34.5276	34.5669	34.6063
880	34.6457	34.6850	34.7244	34.7638	34.8031	34.8425	34.8819	34.9213	34.9606	35.0000
890	35.0394	35.0787	35.1181	35.1575	35.1969	35.2362	35.2756	35.3150	35.3543	35.3937
900	35.4331	35.4724	35.5118	35.5512	35.5906	35.6299	35.6693	35.7087	35.7480	35.7874
910	35.8268	35.8661	35.9055	35.9449	35.9843	36.0236	36.0630	36.1024	36.1417	36.1811
920	36.2205	36.2598	36.2992	36.3386	36.3780	36.4173	36.4567	36.4961	36.5354	36.5748
930	36.6142	36.6535	36.6929	36.7323	36.7717	36.8110	36.8504	36.8898	36.9291	36.9685
940	37.0079	37.0472	37.0866	37.1260	37.1654	37.2047	37.2441	37.2835	37.3228	37.3622
950	37.4016	37.4409	37.4803	37.5197	37.5591	37.5984	37.6378	37.6772	37.7165	37.7559
960	37.7953	37.8346	37.8740	37.9134	37.9528	37.9921	38.0315	38.0709	38.1102	38.1496
970	38.1890	38.2283	38.2677	38.3071	38.3465	38.3858	38.4252	38.4646	38.5039	38.5433
980	38.5827	38.6220	38.6614	38.7008	38.7402	38.7795	38.8189	38.8583	38.8976	38.9370
990	38.9764	39.0157	39.0551	39.0945	39.1339	39.1732	39.2126	39.2520	39.2913	39.3307
1000	39.3701	...	...	...	...	...	...	...	...	...

Fractional Inch—Millimetre and Foot—Millimetre Conversion Tables

(Based on 1 inch = 25.4 millimetres, exactly)

FRACTIONAL INCH TO MILLIMETRES

in.	mm	in.	mm	in.	mm	in.	mm
1/64	0.397	17/64	6.747	33/64	13.097	49/64	19.447
1/32	0.794	9/32	7.144	17/32	13.494	25/32	19.844
3/64	1.191	19/64	7.541	35/64	13.891	51/64	20.241
1/16	1.588	5/16	7.938	9/16	14.288	13/16	20.638
5/64	1.984	21/64	8.334	37/64	14.684	53/64	21.034
3/32	2.381	11/32	8.731	19/32	15.081	27/32	21.431
7/64	2.778	23/64	9.128	39/64	15.478	55/64	21.828
1/8	3.175	3/8	9.525	5/8	15.875	7/8	22.225
9/64	3.572	25/64	9.922	41/64	16.272	57/64	22.622
5/32	3.969	13/32	10.319	21/32	16.669	29/32	23.019
11/64	4.366	27/64	10.716	43/64	17.066	59/64	23.416
3/16	4.762	7/16	11.112	11/16	17.462	15/16	23.812
13/64	5.159	29/64	11.509	45/64	17.859	61/64	24.209
7/32	5.556	15/32	11.906	23/32	18.256	31/32	24.606
15/64	5.953	31/64	12.303	47/64	18.653	63/64	25.003
1/4	6.350	1/2	12.700	3/4	19.050	1	25.400

INCHES TO MILLIMETRES

in.	mm	in.	mm	in.	mm	in.	mm	in.	mm	in.	mm
1	25.4	3	76.2	5	127.0	7	177.8	9	228.6	11	279.4
2	50.8	4	101.6	6	152.4	8	203.2	10	254.0	12	304.8

FEET TO MILLIMETRES

ft	mm	ft	mm	ft	mm	ft	mm	ft	mm
100	30,480	10	3,048	1	304.8	0.1	30.48	0.01	3.048
200	60,960	20	6,096	2	609.6	0.2	60.96	0.02	6.096
300	91,440	30	9,144	3	914.4	0.3	91.44	0.03	9.144
400	121,920	40	12,192	4	1,219.2	0.4	121.92	0.04	12.192
500	152,400	50	15,240	5	1,524.0	0.5	152.40	0.05	15.240
600	182,880	60	18,288	6	1,828.8	0.6	182.88	0.06	18.288
700	213,360	70	21,336	7	2,133.6	0.7	213.36	0.07	21.336
800	243,840	80	24,384	8	2,438.4	0.8	243.84	0.08	24.384
900	274,320	90	27,432	9	2,743.2	0.9	274.32	0.09	27.432
1,000	304,800	100	30,480	10	3,048.0	1.0	304.80	0.10	30.480

Example 1: Find millimetre equivalent of 293 feet, 5⁴⁷/₆₄ inches.

$$
\begin{aligned}
200 \text{ ft} &= 60,960. &&\text{mm}\\
90 \text{ ft} &= 27,432. &&\text{mm}\\
3 \text{ ft} &= 914.4 &&\text{mm}\\
5 \text{ in.} &= 127.0 &&\text{mm}\\
4\tfrac{7}{64} \text{ in.} &= 18.653 &&\text{mm}\\
\hline
293 \text{ ft, } 5\tfrac{47}{64} \text{ in.} &= 89,452.053 &&\text{mm}
\end{aligned}
$$

Example 2: Find millimetre equivalent of 71.86 feet.

$$
\begin{aligned}
70. \quad \text{ft} &= 21,336. &&\text{mm}\\
1. \quad \text{ft} &= 304.8 &&\text{mm}\\
.80 \text{ ft} &= 243.84 &&\text{mm}\\
.06 \text{ ft} &= 18.288 &&\text{mm}\\
\hline
71.86 \text{ ft} &= 21,902.928 &&\text{mm}
\end{aligned}
$$

Microinches to Micrometers (microns)
(Based on 1 microinch = 0.0254 micrometers, exactly)

Micro-inches	0	1	2	3	4	5	6	7	8	9
	Micrometers (microns)									
0	.....	0.025	0.051	0.076	0.102	0.127	0.152	0.178	0.203	0.229
10	0.254	0.279	0.305	0.330	0.356	0.381	0.406	0.432	0.457	0.483
20	0.508	0.533	0.559	0.584	0.610	0.635	0.660	0.686	0.711	0.737
30	0.762	0.787	0.813	0.838	0.864	0.889	0.914	0.940	0.965	0.991
40	1.016	1.041	1.067	1.092	1.118	1.143	1.168	1.194	1.219	1.245
50	1.270	1.295	1.321	1.346	1.372	1.397	1.422	1.448	1.473	1.499
60	1.524	1.549	1.575	1.600	1.626	1.651	1.676	1.702	1.727	1.753
70	1.778	1.803	1.829	1.854	1.880	1.905	1.930	1.956	1.981	2.007
80	2.032	2.057	2.083	2.108	2.134	2.159	2.184	2.210	2.235	2.261
90	2.286	2.311	2.337	2.362	2.388	2.413	2.438	2.464	2.489	2.515
100	2.540	2.565	2.591	2.616	2.642	2.667	2.692	2.718	2.743	2.769
110	2.794	2.819	2.845	2.870	2.896	2.921	2.946	2.972	2.997	3.023
120	3.048	3.073	3.099	3.124	3.150	3.175	3.200	3.226	3.251	3.277
130	3.302	3.327	3.353	3.378	3.404	3.429	3.454	3.480	3.505	3.531
140	3.556	3.581	3.607	3.632	3.658	3.683	3.708	3.734	3.759	3.785
150	3.810	3.835	3.861	3.886	3.912	3.937	3.962	3.988	4.013	4.039
160	4.064	4.089	4.115	4.140	4.166	4.191	4.216	4.242	4.267	4.293
170	4.318	4.343	4.369	4.394	4.420	4.445	4.470	4.496	4.521	4.547
180	4.572	4.597	4.623	4.648	4.674	4.699	4.724	4.750	4.775	4.801
190	4.826	4.851	4.877	4.902	4.928	4.953	4.978	5.004	5.029	5.055
200	5.080	5.105	5.131	5.156	5.182	5.207	5.232	5.258	5.283	5.309
210	5.334	5.359	5.385	5.410	5.436	5.461	5.486	5.512	5.537	5.563
220	5.588	5.613	5.639	5.664	5.690	5.715	5.740	5.766	5.791	5.817
230	5.842	5.867	5.893	5.918	5.944	5.969	5.994	6.020	6.045	6.071
240	6.096	6.121	6.147	6.172	6.198	6.223	6.248	6.274	6.299	6.325
250	6.350	6.375	6.401	6.426	6.452	6.477	6.502	6.528	6.553	6.579
260	6.604	6.629	6.655	6.680	6.706	6.731	6.756	6.782	6.807	6.833
270	6.858	6.883	6.909	6.934	6.960	6.985	7.010	7.036	7.061	7.087
280	7.112	7.137	7.163	7.188	7.214	7.239	7.264	7.290	7.315	7.341
290	7.366	7.391	7.417	7.442	7.468	7.493	7.518	7.544	7.569	7.595

The following short table permits conversion of microinches to micrometers for ranges higher than in the main table given above. Appropriate quantities chosen from both tables are simply added to obtain the higher converted value:

μin.	μm	μin.	μm	μin.	μm	μin.	μm	μin.	μm
300	7.620	900	22.860	1500	38.100	2100	53.340	2700	68.580
600	15.240	1200	30.480	1800	45.720	2400	60.960	3000	76.200

Example: Convert 1375 μin. to μm:

From above table: 1200 μin. = 30.480 μm
From main table: 175 μin. = 4.445 μm

    1375 μin. = 34.925 μm

Micrometers (microns) to Microinches — 1
(Based on 1 microinch = 0.0254 micrometers, exactly)

Micrometers (microns)	0	0.01	0.02	0.03	0.04	0.05	0.06	0.07	0.08	0.09
	Microinches									
0	.....	0.4	0.8	1.2	1.6	2.0	2.4	2.8	3.1	3.5
0.10	3.9	4.3	4.7	5.1	5.5	5.9	6.3	6.7	7.1	7.5
0.20	7.9	8.3	8.7	9.1	9.4	9.8	10.2	10.6	11.0	11.4
0.30	11.8	12.2	12.6	13.0	13.4	13.8	14.2	14.6	15.0	15.4
0.40	15.7	16.1	16.5	16.9	17.3	17.7	18.1	18.5	18.9	19.3

Micrometers (microns) to Microinches — 2

Micro-meters (microns)	0	0.01	0.02	0.03	0.04	0.05	0.06	0.07	0.08	0.09
	Microinches									
0.50	19.7	20.1	20.5	20.9	21.3	21.7	22.0	22.4	22.8	23.2
0.60	23.6	24.0	24.4	24.8	25.2	25.6	26.0	26.4	26.8	27.2
0.70	27.6	28.0	28.3	28.7	29.1	29.5	29.9	30.3	30.7	31.1
0.80	31.5	31.9	32.3	32.7	33.1	33.5	33.9	34.3	34.6	35.0
0.90	35.4	35.8	36.2	36.6	37.0	37.4	37.8	38.2	38.6	39.0
1.00	39.4	39.8	40.2	40.6	40.9	41.3	41.7	42.1	42.5	42.9
1.10	43.3	43.7	44.1	44.5	44.9	45.3	45.7	46.1	46.5	46.9
1.20	47.2	47.6	48.0	48.4	48.8	49.2	49.6	50.0	50.4	50.8
1.30	51.2	51.6	52.0	52.4	52.8	53.1	53.5	53.9	54.3	54.7
1.40	55.1	55.5	55.9	56.3	56.7	57.1	57.5	57.9	58.3	58.7
1.50	59.1	59.4	59.8	60.2	60.6	61.0	61.4	61.8	62.2	62.6
1.60	63.0	63.4	63.8	64.2	64.6	65.0	65.4	65.7	66.1	66.5
1.70	66.9	67.3	67.7	68.1	68.5	68.9	69.3	69.7	70.1	70.5
1.80	70.9	71.3	71.7	72.0	72.4	72.8	73.2	73.6	74.0	74.4
1.90	74.8	75.2	75.6	76.0	76.4	76.8	77.2	77.6	78.0	78.3
2.00	78.7	79.1	79.5	79.9	80.3	80.7	81.1	81.5	81.9	82.3
2.10	82.7	83.1	83.5	83.9	84.3	84.6	85.0	85.4	85.8	86.2
2.20	86.6	87.0	87.4	87.8	88.2	88.6	89.0	89.4	89.8	90.2
2.30	90.6	90.9	91.3	91.7	92.1	92.5	92.9	93.3	93.7	94.1
2.40	94.5	94.9	95.3	95.7	96.1	96.5	96.9	97.2	97.6	98.0
2.50	98.4	98.8	99.2	99.6	100.0	100.4	100.8	101.2	101.6	102.0
2.60	102.4	102.8	103.1	103.5	103.9	104.3	104.7	105.1	105.5	105.9
2.70	106.3	106.7	107.1	107.5	107.9	108.3	108.7	109.1	109.4	109.8
2.80	110.2	110.6	111.0	111.4	111.8	112.2	112.6	113.0	113.4	113.8
2.90	114.2	114.6	115.0	115.4	115.7	116.1	116.5	116.9	117.3	117.7
3.00	118.1	118.5	118.9	119.3	119.7	120.1	120.5	120.9	121.3	121.7
3.10	122.0	122.4	122.8	123.2	123.6	124.0	124.4	124.8	125.2	125.6
3.20	126.0	126.4	126.8	127.2	127.6	128.0	128.3	128.7	129.1	129.5
3.30	129.9	130.3	130.7	131.1	131.5	131.9	132.3	132.7	133.1	133.5
3.40	133.9	134.3	134.6	135.0	135.4	135.8	136.2	136.6	137.0	137.4
3.50	137.8	138.2	138.6	139.0	139.4	139.8	140.2	140.6	140.9	141.3
3.60	141.7	142.1	142.5	142.9	143.3	143.7	144.1	144.5	144.9	145.3
3.70	145.7	146.1	146.5	146.9	147.2	147.6	148.0	148.4	148.8	149.2
3.80	149.6	150.0	150.4	150.8	151.2	151.6	152.0	152.4	152.8	153.1
3.90	153.5	153.9	154.3	154.7	155.1	155.5	155.9	156.3	156.7	157.1
4.00	157.5	157.9	158.3	158.7	159.1	159.4	159.8	160.2	160.6	161.0
4.10	161.4	161.8	162.2	162.6	163.0	163.4	163.8	164.2	164.6	165.0
4.20	165.4	165.7	166.1	166.5	166.9	167.3	167.7	168.1	168.5	168.9
4.30	169.3	169.7	170.1	170.5	170.9	171.3	171.7	172.0	172.4	172.8
4.40	173.2	173.6	174.0	174.4	174.8	175.2	175.6	176.0	176.4	176.8
4.50	177.2	177.6	178.0	178.3	178.7	179.1	179.5	179.9	180.3	180.7
4.60	181.1	181.5	181.9	182.3	182.7	183.1	183.5	183.9	184.3	184.6
4.70	185.0	185.4	185.8	186.2	186.6	187.0	187.4	187.8	188.2	188.6
4.80	189.0	189.4	189.8	190.2	190.6	190.9	191.3	191.7	192.1	192.5
4.90	192.9	193.3	193.7	194.1	194.5	194.9	195.3	195.7	196.1	196.5
5.00	196.9	197.2	197.6	198.0	198.4	198.8	199.2	199.6	200.0	200.4

The table given below can be used with the preceding main table to obtain higher converted values, simply by adding appropriate quantities chosen from each table:

μm	μin.	μm	μin.	μm	μin.	μm	μin.	μm	μin.
10	393.7	20	787.4	30	1,181.1	40	1,574.8	50	1,968.5
15	590.6	25	984.3	35	1,378.0	45	1,771.7	55	2,165.4

Example: Convert 23.55 μm to μin.:
From above table: 20.00 μm = 787.4 μin.
From main table: 3.55 μm = 139.8 μin.
23.55 μm = 927.2 μin.

TEMPERATURE CONVERSION

Thermometer Scales. — There are two thermometer scales in general use: the Fahrenheit (F), which is used in the United States and in other countries still using the English system of units, and the Celsius (C) or Centrigrade used throughout the rest of the world.

In the Fahrenheit thermometer, the freezing point of water is marked at 32 degrees on the scale and the boiling point, at atmospheric pressure, at 212 degrees. The distance between these two points is divided into 180 degrees. On the Celsius scale, the freezing point of water is at 0 degrees and the boiling point at 100 degrees. The following formulas may be used for converting temperatures given on any one of the scales to the other scale:

$$\text{Degrees Fahrenheit} = \frac{9 \times \text{degrees C}}{5} + 32$$

$$\text{Degrees Celsius} = \frac{5 \times (\text{degrees F} - 32)}{9}$$

Tables appear on the pages which follow that can be used to convert degrees Celsius into degrees Fahrenheit or vice versa. In the event that the conversions are not covered in the tables use those applicable portions of the formulas given above for converting.

Absolute Temperature and Absolute Zero. — A point has been determined on the thermometer scale, by theoretical considerations, which is called the absolute zero and beyond which a further decrease in temperature is inconceivable. This point is located at -273.2 degrees Celsius or -459.7 degrees F. A temperature reckoned from this point, instead of from the zero on the ordinary thermometers, is called absolute temperature. Absolute temperature in degrees C is known as "degrees Kelvin" or the "Kelvin scale" (K) and absolute temperature in degrees F is known as "degrees Rankine" or the "Rankine scale" (R).

$$\text{Degrees Kelvin} = \text{degrees C} + 273.2$$

$$\text{Degrees Rankine} = \text{degrees F} + 459.7$$

Measures of the Quantity of Thermal Energy. — The unit of quantity of thermal energy used in the United States is the British thermal unit, which is the quantity of heat or thermal energy required to raise the temperature of one pound of pure water one degree F. (American National Standard abbreviation, Btu; conventional British symbol, B.Th.U.) The French thermal unit or *kilogram calorie*, is the quantity of heat or thermal energy required to raise the temperature of one kilogram of pure water one degree C. One kilogram calorie = 3.968 British thermal units = 1000 gram calories. The number of foot-pounds of mechanical energy equivalent to one British thermal unit is called the *mechanical equivalent of heat*, and equals 778 foot-pounds.

In the modern metric or SI system of units, the unit for thermal energy is the *joule* (J); a commonly used multiple being the kilojoule (kJ) or 1000 joules. One kilojoule = 0.9478 Btu. Also in the SI System, the *watt* (W), equal to joule per second (J/s), is used for power, where one watt = 3.412 Btu per hour.

Fahrenheit — Celsius (Centigrade) Conversion. — A simple way to convert a Fahrenheit temperature reading into a Celsius temperature reading or vice versa

is to enter the accompanying table in the center or boldface column of figures. These figures refer to the temperature in either Fahrenheit of Celsius degrees. If it is desired to convert from Fahrenheit to Celsius degrees, consider the center column as a table of Fahrenheit temperatures and read the corresponding Celsius temperature in the column at the left. If it is desired to conve₁t from Celsius to Fahrenheit degrees, consider the center column as a table of Celsius values, and read the corresponding Fahrenheit temperature on the right.

Interpolation Factors

deg C		deg F	deg C		deg F
0.56	1	1.8	3.33	6	10.8
1.11	2	3.6	3.89	7	12.6
1.67	3	5.4	4.44	8	14.4
2.22	4	7.2	5.00	9	16.2
2.78	5	9.0	5.56	10	18.0

Interpolation factors are given for use with that portion of the table in which the center column advances in increments of 10. To illustrate, suppose it is desired to find the Fahrenheit equivalent of 314 degrees C. The equivalent of 310 degrees C , found in the body of the main table, is seen to be 590.0 degrees F. The Fahrenheit equivalent of a 4-degree C difference is seen to be 7.2, as read in the table of interpolating factors. The answer is the sum or 597.2 degrees F.

Fahrenheit — Celsius (Centigrade) Conversion Table

deg C	deg F.		deg C		deg F	deg C		deg F	deg C		deg F
-273	-459.4	...	-101	-150	-238	-8.3	17	62.6	9.4	49	120.2
-268	-450	...	-96	-140	-220	-7.8	18	64.4	10.0	50	122.0
-262	-440	...	-90	-130	-202	-7.2	19	66.2	10.6	51	123.8
-257	-430	...	-84	-120	-184	-6.7	20	68.0	11.1	52	125.6
-251	-420	...	-79	-110	-166	-6.1	21	69.8	11.7	53	127.4
-246	-410	...	-73	-100	-148	-5.6	22	71.6	12.2	54	129.2
-240	-400	...	-68	-90	-130	-5.0	23	73.4	12.8	55	131.0
-234	-390	...	-62	-80	-112	-4.4	24	75.2	13.3	56	132.8
-229	-380	...	-57	-70	-94	-3.9	25	77.0	13.9	57	134.6
-223	-370	...	-51	-60	-76	-3.3	26	78.8	14.4	58	136.4
-218	-360	...	-46	-50	-58	-2.8	27	80.6	15.0	59	138.2
-212	-350	...	-40	-40	-40	-2.2	28	82.4	15.6	60	140.0
-207	-340	...	-34	-30	-22	-1.7	29	84.2	16.1	61	141.8
-201	-330	...	-29	-20	-4	-1.1	30	86.0	16.7	62	143.6
-196	-320	...	-23	-10	14	-0.6	31	87.8	17.2	63	145.4
-190	-310	...	-17.8	0	32-	0—	32	89.6	17.8	64	147.2
-184	-300	...	-17.2	1	33.8	0.6	33	91.4	18.3	65	149.0
-179	-290	...	-16.7	2	35.6	1.1	34	93.2	18.9	66	150.8
-173	-280	...	-16.1	3	37.4	1.7	35	95.0	19.4	67	152.6
-169	-273	-459.4	-15.6	4	39.2	2.2	36	96.8	20.0	68	154.4
-168	-270	-454	-15.0	5	41.0	2.7	37	98.6	20.6	69	156.2
-162	-260	-436	-14.4	6	42.8	3.3	38	100.4	21.1	70	158.0
-157	-250	-418	-13.9	7	44.6	3.9	39	102.2	21.7	71	159.8
-151	-240	-400	-13.3	8	46.4	4.4	40	104.0	22.2	72	161.6
-146	-230	-382	-12.8	9	48.2	5.0	41	105.8	22.8	73	163.4
-140	-220	-364	-12.2	10	50.0	5.6	42	107.6	23.3	74	165.2
-134	-210	-346	-11.7	11	51.8	6.1	43	109.4	23.9	75	167.0
-129	-200	-328	-11.1	12	53.6	6.7	44	111.2	24.4	76	168.8
-123	-190	-310	-10.6	13	55.4	7.2	45	113.0	25.0	77	170.6
-118	-180	-292	-10.0	14	57.2	7.8	46	114.8	25.6	78	172.4
-112	-170	-274	-9.4	15	59.0	8.3	47	116.6	26.1	79	174.2
-107	-160	-256	-8.9	16	60.8	8.9	48	118.4	26.7	80	176.0

Fahrenheit — Celsius (Centigrade) Conversion Table (*Continued*)

deg C		deg F	deg C		deg F	deg C		deg F	deg C		deg F
27.2	81	177.8	58.3	137	278.6	89.4	193	379.4	304.4	580	1076
27.8	82	179.6	58.9	138	280.4	90.0	194	381.2	310.0	590	1094
28.3	83	181.4	59.4	139	282.2	90.6	195	383.0	315.6	600	1112
28.9	84	183.2	60.0	140	284.0	91.1	196	384.8	321.1	610	1130
29.4	85	185.0	60.6	141	285.8	91.7	197	386.6	326.7	620	1148
30.0	86	186.8	61.1	142	287.6	92.2	198	388.4	332.2	630	1166
30.6	87	188.6	61.7	143	289.4	92.8	199	390.2	337.8	640	1184
31.1	88	190.4	62.2	144	291.2	93.3	200	392.0	343.3	650	1202
31.7	89	192.2	62.8	145	293.0	93.9	201	393.8	348.9	660	1220
32.2	90	194.0	63.3	146	294.8	94.4	202	395.6	354.4	670	1238
32.8	91	195.8	63.9	147	296.6	95.0	203	397.4	360.0	680	1256
33.3	92	197.6	64.4	148	298.4	95.6	204	399.2	365.6	690	1274
33.9	93	199.4	65.0	149	300.2	96.1	205	401.0	371.1	700	1292
34.4	94	201.2	65.6	150	302.0	96.7	206	402.8	376.7	710	1310
35.0	95	203.0	66.1	151	303.8	97.2	207	404.6	382.2	720	1328
35.6	96	204.8	66.7	152	305.6	97.8	208	406.4	387.8	730	1346
36.1	97	206.6	67.2	153	307.4	98.3	209	408.2	393.3	740	1364
36.7	98	208.4	67.8	154	309.2	98.9	210	410.0	398.9	750	1382
37.2	99	210.2	68.3	155	311.0	99.4	211	411.8	404.4	760	1400
37.8	100	212.0	68.9	156	312.8	100.0	212	413.6	410.0	770	1418
38.3	101	213.8	69.4	157	314.6	104.4	220	428.0	415.6	780	1436
38.9	102	215.6	70.0	158	316.4	110.0	230	446.0	421.1	790	1454
39.4	103	217.4	70.6	159	318.2	115.6	240	464.0	426.7	800	1472
40.0	104	219.2	71.1	160	320.0	121.1	250	482.0	432.2	810	1490
40.6	105	221.0	71.7	161	321.8	126.7	260	500.0	437.8	820	1508
41.1	106	222.8	72.2	162	323.6	132.2	270	518.0	443.3	830	1526
41.7	107	224.6	72.8	163	325.4	137.8	280	536.0	448.9	840	1544
42.2	108	226.4	73.3	164	327.2	143.3	290	554.0	454.4	850	1562
42.8	109	228.2	73.9	165	329.0	148.9	300	572.0	460.0	860	1580
43.3	110	230.0	74.4	166	330.8	154.4	310	590.0	465.6	870	1598
43.9	111	231.8	75.0	167	332.6	160.0	320	608.0	471.1	880	1616
44.4	112	233.6	75.6	168	334.4	165.6	330	626.0	476.7	890	1634
45.0	113	235.4	76.1	169	336.2	171.1	340	644.0	482.2	900	1652
45.6	114	237.2	76.7	170	338.0	176.7	350	662.0	487.8	910	1670
46.1	115	239.0	77.2	171	339.8	182.2	360	680.0	493.3	920	1688
46.7	116	240.8	77.8	172	341.6	187.8	370	698.0	498.9	930	1706
47.2	117	242.6	78.3	173	343.4	193.3	380	716.0	504.4	940	1724
47.8	118	244.4	78.9	174	345.2	198.9	390	734.0	510.0	950	1742
48.3	119	246.2	79.4	175	347.0	204.4	400	752.0	515.6	960	1760
48.9	120	248.0	80.0	176	348.8	210	410	770.0	521.1	970	1778
49.4	121	249.8	80.6	177	350.6	215.6	420	788	526.7	980	1796
50.0	122	251.6	81.1	178	352.4	221.1	430	806	532.2	990	1814
50.6	123	253.4	81.7	179	354.2	226.7	440	824	537.8	1000	1832
51.1	124	255.2	82.2	180	356.0	232.2	450	842	565.6	1050	1922
51.7	125	257.0	82.8	181	357.8	237.8	460	860	593.3	1100	2012
52.2	126	258.8	83.3	182	359.6	243.3	470	878	621.1	1150	2102
52.8	127	260.6	83.9	183	361.4	248.9	480	896	648.9	1200	2192
53.3	128	262.4	84.4	184	363.2	254.4	490	914	676.7	1250	2282
53.9	129	264.2	85.0	185	365.0	260.0	500	932	704.4	1300	2372
54.4	130	266.0	85.6	186	366.8	265.6	510	950	732.2	1350	2462
55.0	131	267.8	86.1	187	368.6	271.1	520	968	760.0	1400	2552
55.6	132	269.6	86.7	188	370.4	276.7	530	986	787.8	1450	2642
56.1	133	271.4	87.2	189	372.2	282.2	540	1004	815.6	1500	2732
56.7	134	273.2	87.8	190	374.0	287.8	550	1022	1093.9	2000	3632
57.2	135	275.0	88.3	191	375.8	293.3	560	1040	1648.9	3000	5432
57.8	136	276.8	88.9	192	377.6	298.9	570	1058	2760.0	5000	9032

Above 1000 in the center column, the table increases in increments of 50. To convert 1462 degrees F to Celsius, for instance, add to the Celsius equivalent of 1400 degrees F ten times the interpolation factor for 6 and the interpolation factor for 2 or 760.0 + 33.3 + 1.11, which equals 794.4.

ARITHMETIC*

Common Fractions. — Common fractions consist of two basic parts, a denominator, or bottom number, and a numerator, or top number. The denominator shows how many parts the whole unit has been divided into. The numerator indicates the number of parts of the whole that are being considered. A fraction having a value of $\frac{5}{32}$, means the whole unit has been divided into 32 equal parts and 5 of these parts are considered in the value of the fraction.

The following are the basic facts, rules, and definitions concerning common fractions.

1. A proper fraction is a common fraction having a numerator smaller than its denominator.

Examples: $\frac{1}{4}$, $\frac{1}{2}$, $\frac{3}{4}$, $\frac{15}{16}$, $\frac{23}{32}$, and $\frac{47}{64}$

2. An improper fraction is a common fraction having a numerator larger than its denominator

Examples: $\frac{3}{2}$, $\frac{5}{4}$, $\frac{10}{8}$, $\frac{19}{16}$, $\frac{45}{32}$, and $\frac{84}{64}$

3. A common fraction having the same numerator and denominator is equal to 1.

Examples: $\frac{2}{2}$, $\frac{4}{4}$, $\frac{8}{8}$, $\frac{16}{16}$, $\frac{32}{32}$, and $\frac{64}{64}$ all equal 1

4. A reducible fraction is a common fraction that can be reduced to lower terms.

Examples: $\frac{2}{4}$, $\frac{6}{8}$, $\frac{12}{16}$, $\frac{28}{32}$, and $\frac{62}{64}$

5. To reduce a common fraction to lower terms, divide both the numerator and the denominator by the same number.

Examples: $\frac{24}{32}$: $\frac{24}{32} \div \frac{8}{8} = \frac{3}{8}$
$$\text{or}$$
$\frac{6}{8}$: $\frac{6}{8} \div \frac{2}{2} = \frac{3}{4}$

6. A fraction may also be converted to higher terms by multiplying the numerator and denominator by the same number.

Examples: $\frac{1}{4}$ in 16ths $= \frac{1}{4} \times \frac{4}{4} = \frac{4}{16}$
$$\text{and}$$
$\frac{3}{8}$ in 32nds $= \frac{3}{8} \times \frac{4}{4} = \frac{12}{32}$

7. A least common denominator is the smallest denominator value that is evenly divisible by the other denominator values in the problem.

Examples: $\frac{1}{2} + \frac{1}{4} + \frac{3}{8}$: The least common denominator is 8

8. A mixed number is a combination of a whole number and a common fraction.

Examples: $2\frac{1}{2}$, $1\frac{7}{8}$, $3\frac{15}{16}$, and $1\frac{9}{32}$

9. To convert mixed numbers to improper fractions, multiply the whole number by the denominator and add the numerator to obtain the new numerator. The denominator remains the same.

Examples: $2\frac{1}{2}$: $2 \times 2 + 1 = \frac{5}{2}$
$$\text{or}$$
$3\frac{7}{16}$: $3 \times 16 + 7 = \frac{55}{16}$

*Portions of this section have been reprinted from *Print Reading for Industry,* by permission of South-Western Publishing Co., Cincinnati, OH.

10. To convert a whole number to an improper fraction, place the whole number over 1.

Examples: $\qquad 4 = \frac{4}{1}$, or $3 = \frac{3}{1}$

11. To change a whole number to a common fraction with a specific denominator value, convert the whole number to a fraction and multiply the numerator and denominator by the desired denominator value.

Examples: $\qquad 4$ in 16ths $= \frac{4}{1} \times \frac{16}{16} = \frac{64}{16}$
or
3 in 32nds $= \frac{3}{1} \times \frac{32}{32} = \frac{96}{32}$

12. To convert an improper fraction to a mixed number, divide the numerator by the denominator and reduce the remaining fraction to its lowest terms.

Examples: $\qquad \frac{17}{8} = 17 \div 8 = 2\frac{1}{8}$
or
$\frac{26}{16} = 26 \div 16 = 1\frac{10}{16} = 1\frac{5}{8}$

Adding Fractions and Mixed Numbers: To add common fractions:

1. Find and convert to the least common denominator.

2. Add the numerators.

3. Convert the answer to a mixed number, if necessary.

4. Reduce the fraction to its lowest terms.

Example: $\qquad \frac{1}{4} + \frac{3}{16} + \frac{7}{8} = \begin{array}{l} \frac{1}{4} = \frac{4}{16} \\ \frac{3}{16} = \frac{3}{16} \\ \frac{7}{8} = \frac{14}{16} \\ \hline \frac{21}{16} = 1\frac{5}{16} \end{array}$

To add mixed numbers:

1. Find and convert to the least common denominator.

2. Add the numerators.

3. Add the whole numbers.

4. Reduce the answer to its lowest terms.

Example: $\qquad 2\frac{1}{2} + 4\frac{1}{4} + 1\frac{15}{32} = \begin{array}{l} 2\frac{1}{2} = 2\frac{16}{32} \\ 4\frac{1}{4} = 4\frac{8}{32} \\ 1\frac{15}{32} = 1\frac{15}{32} \\ \hline 7\frac{39}{32} = 8\frac{7}{32} \end{array}$

Subtracting Fractions and Mixed Numbers: To subtract common fractions:

1. Find and convert to the least common denominator.

2. Subtract the numerators.

3. Reduce the answer to its lowest terms.

Example: $\qquad \frac{15}{16} - \frac{7}{32} = \begin{array}{l} \frac{15}{16} = \frac{30}{32} \\ -\frac{7}{32} = -\frac{7}{32} \\ \hline \frac{23}{32} \end{array}$

To subtract mixed numbers:

1. Find and convert to the least common denominator.

2. Subtract the numerators.

3. Subtract the whole numbers.

4. Reduce the answer to its lowest terms.

Example:
$$2\tfrac{3}{8} - 1\tfrac{1}{16} = \quad 2\tfrac{3}{8} = \quad 2\tfrac{6}{16}$$
$$-1\tfrac{1}{16} = -1\tfrac{1}{16}$$
$$\overline{\qquad 1\tfrac{5}{16}}$$

Multiplying Fractions and Mixed Numbers: To multiply common fractions:

1. Multiply the numerators.

2. Multiply the denominators.

3. Convert improper fractions to mixed numbers, if necessary.

Example:
$$\tfrac{3}{4} \times \tfrac{7}{16} = \frac{3 \times 7}{4 \times 16} = \frac{21}{64}$$

To multiply mixed numbers:

1. Convert the mixed numbers to improper fractions.

2. Multiply the numerators.

3. Multiply the denominators.

4. Convert improper fractions to mixed numbers, if necessary.

Example:
$$2\tfrac{1}{4} \times 3\tfrac{1}{2} = \frac{9 \times 7}{4 \times 2} = \frac{63}{8}$$
$$\tfrac{63}{8} = 7\tfrac{7}{8}$$

Dividing Fractions and Mixed Numbers: To divide common fractions:

1. Write the fractions to be divided.

2. Invert (switch) the numerator and denominator in the dividing fraction.

3. Multiply the numerators and denominators.

4. Convert improper fractions to mixed numbers, if necessary.

Example:
$$\tfrac{3}{4} \div \tfrac{1}{2} = \frac{3 \times 2}{4 \times 1} = \frac{6}{4}$$
$$\tfrac{6}{4} = 1\tfrac{1}{2}$$

To divide mixed numbers:

1. Convert the mixed numbers to improper fractions.

2. Write the improper fraction to be divided.

3. Invert (switch) the numerator and denominator in the dividing fraction.

4. Multiplying numerators and denominators.

5. Convert improper fractions to mixed numbers, if necessary.

Example:
$$2\tfrac{1}{2} \div 1\tfrac{7}{8} = \frac{5 \times 8}{2 \times 15} = \frac{40}{30}$$
$$\tfrac{40}{30} = 1\tfrac{1}{3}$$

Decimal Fractions.—Decimal fractions are fractional parts of a whole unit, which have implied denominators that are multiples of 10. A decimal fraction of .1 has a value of 1/10th, .01 has a value of 1/100th, and .001 has a value of 1/1000th. As the number of decimal place values increases, the value of the decimal number changes by a multiple of 10. A single number placed to the right of a decimal point has a value expressed in tenths; two numbers to the right of a decimal point have a value expressed in hundredths; three numbers to the right have a value expressed in thousandths; and four numbers are expressed in ten-thousandths. Since the denominator is implied, the number of decimal places in the numerator indicates the value of the decimal fraction. So a decimal fraction expressed as a .125 means the whole unit has been divided into 1000 parts and 125 of these parts are considered in the value of the decimal fraction.

In industry, most decimal fractions are expressed in terms of thousandths rather than tenths or hundredths. So a decimal fraction of .2 is expressed as 200 thousandths, not 2 tenths, and a value of .75 is expressed as 750 thousandths, rather than 75 hundredths. In the case of four place decimals, the values are expressed in terms of ten-thousandths. So a value of .1875 is expressed as 1 thousand 8 hundred and 75 ten-thousandths. When whole numbers and decimal fractions are used together, whole units are shown to the left of a decimal point, while fractional parts of a whole unit are shown to the right.

Example: 10.125
                     Whole Fractional
                     Units Units

Adding Decimal Fractions: To add decimal fractions:

1. Write the problem with all decimal points aligned vertically.

2. Add the numbers as whole number values.

3. Insert the decimal point in the same vertical column in the answer.

Examples: .125   or   1.750
                     1.0625   .875
                     2.50   .125
                     .1875   2.0005
                     _____   _____
                     3.8750   4.7505

Subtracting Decimal Fractions: To subtract decimal fractions:

1. Write the problem with all decimal points aligned vertically.

2. Subtract the numbers as whole number values.

3. Insert the decimal point in the same vertical column in the answer.

Examples: 1.750   or   2.625
                   −   .250   − 1.125
                     _____   _____
                     1.500   1.500

Multiplying Decimal Fractions: To multiply decimal fractions:

1. Write the problem with the decimal points aligned.

2. Multiply the values as whole numbers.

3. Count the number of decimal places in both multiplied values.

4. Counting from right to left in the answer, insert the decimal point so the number of decimal places in the answer equals the total number of decimal places in the numbers multiplied.

Examples: .75 (four decimal   1.625 (six decimal
               × .25 places)   × .033 places)

        375   4875
        150   4875

         .1875 (four decimal places   .053625 (six decimal places
                in answer)   in answer)

Dividing Decimal Fractions: To divide decimal fractions:

1. Write the problem as whole number division by moving the decimal point of the dividing number to the extreme right (this will make the decimal fraction appear as a whole number). Note the number of places the decimal point was moved.

2. Move the decimal point in the divided number to the right the same number of places. Add zeros if necessary.

3. Divide as a whole number problem.

4. Locate the decimal point in the answer directly above the new position in the divided number.

5. Continue the division until the desired number of decimal places in the answer is reached. Add zeros if necessary.

Examples: 2.75 ÷ .625 =

$$
\begin{array}{r}
4.4 \\
.625.\overline{)2.750.0} \\
2500 \\
\hline
2500 \\
2500 \\
\hline
0
\end{array}
$$

or

                           5.6682 ÷ 2.557 =

$$
\begin{array}{r}
2.216738 \\
2.557.\overline{)5.668.200000} \\
5\,114 \\
\hline
5542 \\
5114 \\
\hline
4280 \\
2557 \\
\hline
17230 \\
15342 \\
\hline
18880 \\
17899 \\
\hline
9810 \\
7671 \\
\hline
21390 \\
20456 \\
\hline
934
\end{array}
$$

Converting Common and Decimal Fractions.—Conversions between common fractions and decimal fractions may be accomplished by using a decimal equivalent chart (see page 8) or by calculating the equivalent values.

Converting Common Fractions To Decimal Fractions.—To convert common fractions to decimal fractions:

1. Write the common fraction.

2. Divide the numerator by the denominator.

3. Continue the division until there is no remainder or until the desired number of decimal places in the answer is reached (normally, five places and round off to four).

Examples:

$$\frac{1}{2} = 2\overline{)\begin{array}{r} .5 \\ 1.0 \\ \underline{1\ 0} \\ 0 \end{array}}$$

or

$$\frac{1}{8} = 8\overline{)\begin{array}{r} .125 \\ 1.000 \\ \underline{8} \\ 20 \\ \underline{16} \\ 40 \\ \underline{40} \\ 0 \\ 0 \end{array}}$$

To convert decimal fractions to common fractions:

1. Write the decimal fraction as a common fraction. Remember to use the proper number of zeros in the denominator.

2. Reduce this common fraction to its lowest terms.

Examples:

$$.5 = \frac{5}{10} = \frac{1}{2}$$

or

$$.125 = \frac{125}{1000} = \frac{1}{8}$$

Principal Algebraic Expressions and Formulas

$$a \times a = aa = a^2$$
$$a \times a \times a = aaa = a^3$$
$$a \times b = ab$$
$$a^2b^2 = (ab)^2$$
$$a^2a^3 = a^{2+3} = a^5$$
$$a^4 \div a^3 = a^{4-3} = a$$
$$a^0 = 1$$
$$a^2 - b^2 = (a+b)(a-b)$$
$$(a+b)^2 = a^2 + 2ab + b^2$$
$$(a-b)^2 = a^2 - 2ab + b^2$$

$$\frac{a^3}{b^3} = \left(\frac{a}{b}\right)^3$$
$$\frac{1}{a^3} = \left(\frac{1}{a}\right)^3 = a^{-3}$$
$$(a^2)^3 = a^{2 \times 3} = (a^3)^2 = a^6$$
$$a^3 + b^3 = (a+b)(a^2 - ab + b^2)$$
$$a^3 - b^3 = (a-b)(a^2 + ab + b^2)$$
$$(a+b)^3 = a^3 + 3a^2b + 3ab^2 + b^3$$
$$(a-b)^3 = a^3 - 3a^2b + 3ab^2 - b^3$$

$$\sqrt{a} \times \sqrt{a} = a$$
$$\sqrt[3]{a} \times \sqrt[3]{a} \times \sqrt[3]{a} = a$$
$$(\sqrt[3]{a})^3 = a$$
$$\sqrt[3]{a^2} = (\sqrt[3]{a})^2 = a^{2/3}$$
$$\sqrt[4]{\sqrt[3]{a}} = \sqrt[4 \times 3]{a} = \sqrt[3]{\sqrt[4]{a}}$$

$$\sqrt[3]{ab} = \sqrt[3]{a} \times \sqrt[3]{b}$$
$$\sqrt[3]{\frac{a}{b}} = \frac{\sqrt[3]{a}}{\sqrt[3]{b}}$$
$$\sqrt[3]{\frac{1}{a}} = \frac{1}{\sqrt[3]{a}} = a^{-1/3}$$

$$\sqrt{a} + \sqrt{b} = \sqrt{a + b + 2\sqrt{ab}}$$

When
$$a \times b = x, \quad \text{then} \quad \log a + \log b = \log x$$
$$a \div b = x, \quad \text{then} \quad \log a - \log b = \log x$$
$$a^3 = x, \quad \text{then} \quad 3 \log a = \log x$$
$$\sqrt[3]{a} = x, \quad \text{then} \quad \frac{\log a}{3} = \log x$$

Equations

An equation is a statement of equality between two expressions, as $5x = 105$. The unknown quantity in an equation is generally designated by the letter x. If there is more than one unknown quantity, the others are designated by letters also selected at the end of the alphabet, as y, z, u, t, etc.

An equation of the first degree is one which contains the unknown quantity only in the first power, as $3x = 9$. A quadratic equation is one which contains the unknown quantity in the second, but no higher, power, as $x^2 + 3x = 10$.

Solving Equations of the First Degree with One Unknown. — Transpose all the terms containing the unknown x to one side of the equals sign, and all the other terms to the other side. Combine and simplify the expressions as far as possible, and divide both sides by the coefficient of the unknown x. (See the rules given for transposition of formulas.)

Example:
$$22x - 11 = 15x \times 10$$
$$22x - 15x = 10 + 11$$
$$7x = 21$$
$$x = 3$$

Solution of Equations of the First Degree with Two Unknowns. — The form of the simplified equations is:

$$ax + by = c$$
$$a_1x + b_1y = c_1$$

Then,

$$x = \frac{cb_1 - c_1b}{ab_1 - a_1b} \qquad\qquad y = \frac{ac_1 - a_1c}{ab_1 - a_1b}$$

Example:

$$3x + 4y = 17$$
$$5x - 2y = 11$$
$$x = \frac{17 \times (-2) - 11 \times 4}{3 \times (-2) - 5 \times 4} = \frac{-34 - 44}{-6 - 20} = \frac{-78}{-26} = 3.$$

The value of y can now be most easily found by inserting the value of x in one of the equations:

$$5 \times 3 - 2y = 11; \quad 2y = 15 - 11 = 4; \quad y = 2.$$

Solution of Quadratic Equations with One Unknown. — If the form of the equation is $ax^2 + bx + c = 0$, then

$$x = \frac{-b \pm \sqrt{b^2 - 4ac}}{2a}$$

Example: Given the equation, $1x^2 + 6x + 5 = 0$, then $a = 1$, $b = 6$, and $c = 5$.

$$x = \frac{-6 \pm \sqrt{6^2 - 4 \times 1 \times 5}}{2 \times 1} = \frac{(-6) + 4}{2} = -1; \quad \text{or} \quad \frac{(-6) - 4}{2} = -5$$

If the form of the equation is $ax^2 + bx = c$, then

$$x = \frac{-b \pm \sqrt{b^2 + 4ac}}{2a}$$

Example: A right-angle triangle has a hypotenuse 5 inches long and one side which is one inch longer than the other; find the lengths of the two sides.

Let $x =$ one side and $x + 1 =$ other side; then $x^2 + (x + 1)^2 = 5^2$ or $x^2 + x^2 + 2x + 1 = 25$; or $2x^2 + 2x = 24$; or $x^2 + x = 12$. Now referring to the basic formula, $ax^2 + bx = c$, we find, in this case, that $a = 1$; $b = 1$, and $c = 12$; hence

$$x = \frac{-1 \pm \sqrt{1 + 4 \times 1 \times 12}}{2 \times 1} = \frac{(-1) + 7}{2} = 3 \quad \text{or} \quad x = \frac{(-1) - 7}{2} = -4$$

Since the positive value (3) would apply in this case, the lengths of the two sides are $x = 3$ inches and $x + 1 = 4$ inches.

Cubic Equations. — If the given equation has the form: $x^3 + ax + b = 0$, then

$$x = \left(-\frac{b}{2} + \sqrt{\frac{a^3}{27} + \frac{b^2}{4}} \right)^{1/3} + \left(-\frac{b}{2} - \sqrt{\frac{a^3}{27} + \frac{b^2}{4}} \right)^{1/3}$$

The equation $x^3 + px^2 + qx + r = 0$, may be reduced to the form $x_1^3 + ax_1 + b = 0$ by substituting $x_1 - \dfrac{p}{3}$ for x in the given equation.

Rearrangement and Transposition of Terms in Formulas

A formula is a rule for a calculation expressed by using letters and signs instead of writing out the rule in words; by this means it is possible to condense, in a very small space, the essentials of long and cumbersome rules. The letters used in formulas simply stand in place of the figures which are to be substituted when solving a specific problem.

As an example, the formula for the horsepower transmitted by belting may be written:

$$P = \frac{SVW}{33,000}$$

in which
P = horsepower transmitted;
S = working stress of belt per inch of width, in pounds;
V = velocity of belt in feet per minute;
W = width of belt in inches.

If the working stress S, the velocity V, and the width W are known, the horsepower can be found directly from this formula by inserting the given values. Assume $S = 33$; $V = 600$; and $W = 5$. Then:

$$P = \frac{33 \times 600 \times 5}{33,000} = 3.$$

Assume, however, that the horsepower P, the stress S, and the velocity V are known, and that the width of belt, W, is to be found. The formula must then be rearranged so that the symbol W will be on one side of the equals sign and all the known quantities on the other. The rearranged formula is as follows:

$$\frac{P \times 33,000}{SV} = W.$$

The quantities (S and V) that were in the numerator on the right side of the equals sign are moved to the denominator on the left side, and "33,000" which was in the denominator on the right side of the equals sign is moved to the numerator on the other side. Symbols which are not part of a fraction, like "P" in the formula first given, are to be considered as being numerators (having the denominator 1).

Thus, any formula of the form $A = \dfrac{B}{C}$ can be rearranged as below:

$$A \times C = B, \quad \text{and} \quad C = \frac{B}{A}$$

Suppose a formula to be of the form:

$$A = \frac{B \times C}{D}$$

Then:
$$D = \frac{B \times C}{A}; \quad \frac{A \times D}{C} = B; \quad \frac{A \times D}{B} = C.$$

The method given is only directly applicable when all the quantities in the numerator or denominator are standing independently or are *factors of a product*. If connected by $+$ or $-$ signs, the entire numerator or denominator must be moved as a unit, thus,

Given:
$$\frac{B + C}{A} = \frac{D + E}{F}, \quad \text{to solve for } F$$

then
$$\frac{F}{A} = \frac{D + E}{B + C}$$

and
$$F = \frac{A(D + E)}{B + C}$$

A quantity preceded by a $+$ or $-$ sign can be transposed to the opposite side of the equals sign by changing its sign; if the sign is $+$, change it to $-$ on the other side; if it is $-$, change it to $+$. This is called *transposition* of terms.

Example:

$$B + C = A - D; \text{ then } B + C + D = A; B = A - D - C;$$
$$C = A - D - B;$$

Order of Performing Arithmetic Operations

When several numbers or quantities in a formula are connected by signs indicating that additions, subtractions, multiplications, or divisions are to be made, the multiplications and divisions should be carried out first, in the order in which they appear, before the additions or subtractions are performed.

Examples:

$$10 + 26 \times 7 - 2 = 10 + 182 - 2 = 190.$$
$$18 \div 6 + 15 \times 3 = 3 + 45 = 48.$$
$$12 + 14 \div 2 - 4 = 12 + 7 - 4 = 15.$$

When it is required that certain additions and subtractions should precede multiplications and divisions, use is made of parentheses ( ) and brackets [ ]. These indicate that the calculation inside the parentheses or brackets should be carried out complete by itself before the remaining calculations are commenced. If one bracket is placed inside of another, the one inside is first calculated.

Examples:

$$(6 - 2) \times 5 + 8 = 4 \times 5 + 8 = 20 + 8 = 28.$$
$$6 \times (4 + 7) \div 22 = 6 \times 11 \div 22 = 66 \div 22 = 3.$$
$$2 + [10 \times 6(8 + 2) - 4] \times 2 = 2 + [10 \times 6 \times 10 - 4] \times 2$$
$$= 2 + [600 - 4] \times 2 = 2 + 596 \times 2 = 2 + 1192 = 1194.$$

The parentheses are considered as a sign of multiplication; for example, $6 (8 + 2)$ $= 6 \times (8 + 2)$.

The line or bar between the numerator and denominator in a fractional expression is to be considered as a division sign. For example,

$$\frac{12 + 16 + 22}{10} = (12 + 16 + 22) \div 10 = 50 \div 10 = 5.$$

In formulas the multiplication sign ($\times$) is often left out between symbols or letters, the values of which are to be multiplied. Thus

$$AB = A \times B, \quad \text{and} \quad \frac{ABC}{D} = (A \times B \times C) \div D$$

Imaginary Quantities

An imaginary quantity is an even root of a negative number, such as $\sqrt{-1}$, $\sqrt{-a^2}$. The square root of -1 is the simplest imaginary quantity. It is usually denoted by i. In electrical work, where imaginary quantities appear frequently, the square root of -1 is denoted by j, since i is used to represent current flow.

The following examples illustrate the use of i in mathematical operations:

$$\sqrt{-9} = \sqrt{(9)(-1)} = \sqrt{9}\sqrt{-1} = 3i$$
$$\sqrt{-25} = \sqrt{(25)(-1)} = \sqrt{25}\sqrt{-1} = 5i$$
$$(6i)^2 = 6^2 \times i^2 = (6 \times 6)(i \times i)$$
$$= 36(\sqrt{-1} \times \sqrt{-1}) = 36(-1) = -36$$
$$\sqrt{-b^2} = \sqrt{(b^2)(-1)} = \sqrt{b^2}\sqrt{-1} = bi$$
$$(bi)^2 = b^2 \times i^2 = (b^2)(-1) = -b^2$$

Note that $i^2 = -1$, $i^3 = -\sqrt{-1} = -i$, $i^4 = 1$, $i^5 = i$, and so on, in a repeating pattern of i, -1, $-i$, 1, etc.

Squares and Cubes of Numbers from 1/32 to 100

Advancing by 32nds to 2; from 2 to 10 by 16ths; from 10 to 100 by 8ths

No.	Square	Cube	No.	Square	Cube	No.	Square	Cube
1/32	0.000977	0.00031	1 17/32	2.344727	3.590363	4	16.0000	64.0000
1/16	0.003906	0.000244	9/16	2.441406	3.814697	1/16	16.5039	67.0471
3/32	0.008789	0.000824	19/32	2.540039	4.048187	1/8	17.0156	70.1895
1/8	0.015625	0.001953	5/8	2.640625	4.291016	3/16	17.5352	73.4285
5/32	0.024414	0.003815	21/32	2.743164	4.543365	1/4	18.0625	76.7656
3/16	0.035156	0.006592	11/16	2.847656	4.805420	5/16	18.5977	80.2024
7/32	0.047852	0.010468	23/32	2.954102	5.077362	3/8	19.1406	83.7402
1/4	0.062500	0.015625	3/4	3.062500	5.359375	7/16	19.6914	87.3806
9/32	0.079102	0.022247	25/32	3.172852	5.651642	1/2	20.2500	91.1250
5/16	0.097656	0.030518	13/16	3.285156	5.954346	9/16	20.8164	94.9749
11/32	0.118164	0.040619	27/32	3.399414	6.267670	5/8	21.3906	98.9316
3/8	0.140625	0.052734	7/8	3.515625	6.591797	11/16	21.9727	102.9968
13/32	0.165039	0.067047	29/32	3.633789	6.926910	3/4	22.5625	107.1719
7/16	0.191406	0.083740	15/16	3.753906	7.273193	13/16	23.1602	111.4583
15/32	0.219727	0.102997	31/32	3.875977	7.630829	7/8	23.7656	115.8574
1/2	0.250000	0.125000	2	4.00000	8.00000	15/16	24.3789	120.3708
17/32	0.282227	0.149933	1/32	4.12598	8.38089	5	25.0000	125.0000
9/16	0.316406	0.177979	1/16	4.25391	8.77368	1/16	25.6289	129.7463
19/32	0.352539	0.209320	1/8	4.51563	9.59570	1/8	26.2656	134.6113
5/8	0.390625	0.244141	3/16	4.78516	10.46753	3/16	26.9102	139.5964
21/32	0.430664	0.282623	1/4	5.06250	11.39063	1/4	27.5625	144.7031
11/16	0.472656	0.324951	5/16	5.34766	12.36646	5/16	28.2227	149.9329
23/32	0.516602	0.371307	3/8	5.64063	13.39648	3/8	28.8906	155.2871
3/4	0.562500	0.421875	7/16	5.94141	14.48218	7/16	29.5664	160.7673
25/32	0.610352	0.476837	1/2	6.25000	15.62500	1/2	30.2500	166.3750
13/16	0.660156	0.536377	9/16	6.56641	16.82642	9/16	30.9414	172.1116
27/32	0.711914	0.600677	5/8	6.89063	18.08789	5/8	31.6406	177.9785
7/8	0.765625	0.669922	11/16	7.22266	19.41089	11/16	32.3477	183.9773
29/32	0.821289	0.744293	3/4	7.56250	20.79688	3/4	33.0625	190.1094
15/16	0.878906	0.823975	13/16	7.91016	22.24731	13/16	33.7852	196.3762
31/32	0.938477	0.909149	7/8	8.26563	23.76367	7/8	34.5156	202.7793
1	1.000000	1.000000	15/16	8.62891	25.34741	15/16	35.2539	209.3201
1/32	1.063477	1.096710	3	9.00000	27.00000	6	36.0000	216.0000
1/16	1.128906	1.199463	1/16	9.37891	28.72290	1/16	36.7539	222.8206
3/32	1.196289	1.308441	1/8	9.76563	30.51758	1/8	37.5156	229.7832
1/8	1.265625	1.423828	3/16	10.16016	32.38550	3/16	38.2852	236.8894
5/32	1.336914	1.545807	1/4	10.56250	34.32813	1/4	39.0625	244.1406
3/16	1.410156	1.674561	5/16	10.97266	36.34692	5/16	39.8477	251.5383
7/32	1.485352	1.810272	3/8	11.39063	38.44336	3/8	40.6406	259.0840
1/4	1.562500	1.953125	7/16	11.81641	40.61890	7/16	41.4414	266.7791
9/32	1.641602	2.103302	1/2	12.25000	42.87500	1/2	42.2500	274.6250
5/16	1.722656	2.260986	9/16	12.69141	45.21313	9/16	43.0664	282.6233
11/32	1.805664	2.426361	5/8	13.14063	47.63477	5/8	43.8906	290.7754
3/8	1.890625	2.599609	11/16	13.59766	50.14136	11/16	44.7227	299.0828
13/32	1.977539	2.780914	3/4	14.06250	52.73438	3/4	45.5625	307.5469
7/16	2.066406	2.970459	13/16	14.53516	55.41528	13/16	46.4102	316.1692
15/32	2.157227	3.168427	7/8	15.01563	58.18555	7/8	47.2656	324.9512
1/2	2.250000	3.375000	15/16	15.50391	61.04663	15/16	48.1289	333.8943

Squares and Cubes of Numbers from 1/32 to 100 (Continued)

No.	Square	Cube	No.	Square	Cube	No.	Square	Cube
7	49.0000	343.0000	10	100.0000	1000.0000	16	256.0000	4096.000
1/16	49.8789	352.2698	1/8	102.5156	1037.9707	1/8	260.0156	4192.752
1/8	50.7656	361.7051	1/4	105.0625	1076.8906	1/4	264.0625	4291.016
3/16	51.6602	371.3074	3/8	107.6406	1116.7715	3/8	268.1406	4390.803
1/4	52.5625	381.0781	1/2	110.2500	1157.6250	1/2	272.2500	4492.125
5/16	53.4727	391.0188	5/8	112.8906	1199.4629	5/8	276.3906	4594.994
3/8	54.3906	401.1309	3/4	115.5625	1242.2969	3/4	280.5625	4699.422
7/16	55.3164	411.4158	7/8	118.2656	1286.1387	7/8	284.7656	4805.420
1/2	56.2500	421.8750	11	121.0000	1331.0000	17	289.0000	4913.000
9/16	57.1914	432.5100	1/8	123.7656	1376.8926	1/8	293.2656	5022.174
5/8	58.1406	443.3223	1/4	126.5625	1423.8281	1/4	297.5625	5132.953
11/16	59.0977	454.3132	3/8	129.3906	1471.8184	3/8	301.8906	5245.350
3/4	60.0625	465.4844	1/2	132.2500	1520.8750	1/2	306.2500	5359.375
13/16	61.0352	476.8372	5/8	135.1406	1571.0098	5/8	310.6406	5475.041
7/8	62.0156	488.3730	3/4	138.0625	1622.2344	3/4	315.0625	5592.359
15/16	63.0039	500.0935	7/8	141.0156	1674.5605	7/8	319.5156	5711.342
8	64.0000	512.0000	12	144.0000	1728.0000	18	324.0000	5832.000
1/16	65.0039	524.0940	1/8	147.0156	1782.5645	1/8	328.5156	5954.346
1/8	66.0156	536.3770	1/4	150.0625	1838.2656	1/4	333.0625	6078.391
3/16	67.0352	548.8503	3/8	153.1406	1895.1152	3/8	337.6406	6204.146
1/4	68.0625	561.5156	1/2	156.2500	1953.1250	1/2	342.2500	6331.625
5/16	69.0977	574.3743	5/8	159.3906	2012.3066	5/8	346.8906	6460.838
3/8	70.1406	587.4277	3/4	162.5625	2072.6719	3/4	351.5625	6591.797
7/16	71.1914	600.6775	7/8	165.7656	2134.2324	7/8	356.2656	6724.514
1/2	72.2500	614.1250	13	169.0000	2197.0000	19	361.0000	6859.000
9/16	73.3164	627.7717	1/8	172.2656	2260.9863	1/8	365.7656	6995.268
5/8	74.3906	641.6191	1/4	175.5625	2326.2031	1/4	370.5625	7133.328
11/16	75.4727	655.6687	3/8	178.8906	2392.6621	3/8	375.3906	7273.193
3/4	76.5625	669.9219	1/2	182.2500	2460.3750	1/2	380.2500	7414.875
13/16	77.6602	684.3801	5/8	185.6406	2529.3535	5/8	385.1406	7558.385
7/8	78.7656	699.0449	3/4	189.0625	2599.6094	3/4	390.0625	7703.734
15/16	79.8789	713.9177	7/8	192.5156	2671.1543	7/8	395.0156	7850.936
9	81.0000	729.0000	14	196.0000	2744.0000	20	400.0000	8000.000
1/16	82.1289	744.2932	1/8	199.5156	2818.1582	1/8	405.0156	8150.939
1/8	83.2656	759.7988	1/4	203.0625	2893.6406	1/4	410.0625	8303.766
3/16	84.4102	775.5183	3/8	206.6406	2970.4590	3/8	415.1406	8458.490
1/4	85.5625	791.4531	1/2	210.2500	3048.6250	1/2	420.2500	8615.125
5/16	86.7227	807.6047	5/8	213.8906	3128.1504	5/8	425.3906	8773.682
3/8	87.8906	823.9746	3/4	217.5625	3209.0469	3/4	430.5625	8934.172
7/16	89.0664	840.5642	7/8	221.2656	3291.3262	7/8	435.7656	9096.607
1/2	90.2500	857.3750	15	225.0000	3375.0000	21	441.0000	9261.000
9/16	91.4414	874.4804	1/8	228.7656	3460.0801	1/8	446.2656	9427.361
5/8	92.6406	891.6660	1/4	232.5625	3546.5781	1/4	451.5625	9595.703
11/16	93.8477	909.1492	3/8	236.3906	3634.5059	3/8	456.8906	9766.037
3/4	95.0625	926.8594	1/2	240.2500	3723.8750	1/2	462.2500	9,938.375
13/16	96.2852	944.7981	5/8	244.1406	3814.6973	5/8	467.6406	10,112.729
7/8	97.5156	962.9668	3/4	248.0625	3906.9844	3/4	473.0625	10,289.109
15/16	98.7539	981.3669	7/8	252.0156	4000.7480	7/8	478.5156	10,467.529

Squares and Cubes of Numbers from 1/32 to 100 (Continued)

No.	Square	Cube	No.	Square	Cube	No.	Square	Cube
22	484.0000	10,648.000	28	784.000	21,952.000	34	1156.000	39,304.000
⅛	489.5156	10,830.533	⅛	791.016	22,247.314	⅛	1164.516	39,739.096
¼	495.0625	11,015.140	¼	798.063	22,545.266	¼	1173.063	40,177.391
⅜	500.6406	11,201.834	⅜	805.141	22,845.865	⅜	1181.641	40,618.896
½	506.2500	11,390.625	½	812.250	23,149.125	½	1190.250	41,063.625
⅝	511.8906	11,581.525	⅝	819.391	23,455.057	⅝	1198.891	41,511.588
¾	517.5625	11,774.547	¾	826.563	23,763.672	¾	1207.563	41,962.797
⅞	523.2656	11,969.701	⅞	833.766	24,074.982	⅞	1216.266	42,417.264
23	529.0000	12,167.000	29	841.000	24,389.000	35	1225.000	42,875.000
⅛	534.7656	12,366.455	⅛	848.266	24,705.736	⅛	1233.766	43,336.018
¼	540.5625	12,568.078	¼	855.563	25,025.203	¼	1242.563	43,800.328
⅜	546.3906	12,771.881	⅜	862.891	25,347.412	⅜	1251.391	44,267.943
½	552.2500	12,977.875	½	870.250	25,672.375	½	1260.250	44,738.875
⅝	558.1406	13,186.072	⅝	877.641	26,000.104	⅝	1269.141	45,213.135
¾	564.0625	13,396.484	¾	885.063	26,330.609	¾	1278.063	45,690.734
⅞	570.0156	13,609.123	⅞	892.516	26,663.904	⅞	1287.016	46,171.686
24	576.0000	13,824.000	30	900.000	27,000.000	36	1296.000	46,656.000
⅛	582.0156	14,041.127	⅛	907.516	27,338.908	⅛	1305.016	47,143.689
¼	588.0625	14,260.516	¼	915.063	27,680.641	¼	1314.063	47,634.766
⅜	594.1406	14,482.178	⅜	922.641	28,025.209	⅜	1323.141	48,129.240
½	600.2500	14,706.125	½	930.250	28,372.625	½	1332.250	48,627.125
⅝	606.3906	14,932.369	⅝	937.891	28,722.900	⅝	1341.391	49,128.432
¾	612.5625	15,160.922	¾	945.563	29,076.047	¾	1350.563	49,633.172
⅞	618.7656	15,391.795	⅞	953.266	29,432.076	⅞	1359.766	50,141.357
25	625.0000	15,625.000	31	961.000	29,791.000	37	1369.000	50,653.000
⅛	631.2656	15,860.549	⅛	968.766	30,152.830	⅛	1378.266	51,168.111
¼	637.5625	16,098.453	¼	976.563	30,517.578	¼	1387.563	51,686.703
⅜	643.8906	16,338.725	⅜	984.391	30,885.256	⅜	1396.891	52,208.787
½	650.2500	16,581.375	½	992.250	31,255.875	½	1406.250	52,734.375
⅝	656.6406	16,826.416	⅝	1000.141	31,629.447	⅝	1415.641	53,263.479
¾	663.0625	17,073.859	¾	1008.063	32,005.984	¾	1425.063	53,796.109
⅞	669.5156	17,323.717	⅞	1016.016	32,385.498	⅞	1434.516	54,332.279
26	676.0000	17,576.000	32	1024.000	32,768.000	38	1444.000	54,872.000
⅛	682.5156	17,830.721	⅛	1032.016	33,153.502	⅛	1453.516	55,415.283
¼	689.0625	18,087.891	¼	1040.063	33,542.016	¼	1463.063	55,962.141
⅜	695.6406	18,347.521	⅜	1048.141	33,933.553	⅜	1472.641	56,512.584
½	702.2500	18,609.625	½	1056.250	34,328.125	½	1482.250	57,066.625
⅝	708.8906	18,874.213	⅝	1064.391	34,725.744	⅝	1491.891	57,624.275
¾	715.5625	19,141.297	¾	1072.563	35,126.422	¾	1501.563	58,185.547
⅞	722.2656	19,410.889	⅞	1080.766	35,530.170	⅞	1511.266	58,750.451
27	729.0000	19,683.000	33	1089.000	35,937.000	39	1521.000	59,319.000
⅛	735.7656	19,957.643	⅛	1097.266	36,346.924	⅛	1530.766	59,891.205
¼	742.5625	20,234.828	¼	1105.563	36,759.953	¼	1540.563	60,467.078
⅜	749.3906	20,514.568	⅜	1113.891	37,176.100	⅜	1550.391	61,046.631
½	756.2500	20,796.875	½	1122.250	37,595.375	½	1560.250	61,629.875
⅝	763.1406	21,081.760	⅝	1130.641	38,017.791	⅝	1570.141	62,216.822
¾	770.0625	21,369.234	¾	1139.063	38,443.359	¾	1580.063	62,807.484
⅞	777.0156	21,659.311	⅞	1147.516	38,872.092	⅞	1590.016	63,401.873

Squares and Cubes of Numbers from 1/32 to 100 (Continued)

No.	Square	Cube	No.	Square	Cube	No.	Square	Cube
40	1600.000	64,000.000	46	2116.000	97,336.00	52	2704.000	140,608.00
1/8	1610.016	64,601.877	1/8	2127.516	98,131.66	1/8	2717.016	141,624.44
1/4	1620.063	65,207.516	1/4	2139.063	98,931.64	1/4	2730.063	142,645.77
3/8	1630.141	65,816.928	3/8	2150.641	99,735.96	3/8	2743.141	143,671.99
1/2	1640.250	66,430.125	1/2	2162.250	100,544.63	1/2	2756.250	144,703.13
5/8	1650.391	67,047.119	5/8	2173.891	101,357.65	5/8	2769.391	145,739.18
3/4	1660.563	67,667.922	3/4	2185.563	102,175.05	3/4	2782.563	146,780.17
7/8	1670.766	68,292.545	7/8	2197.266	102,996.83	7/8	2795.766	147,826.11
41	1681.000	68,921.000	47	2209.000	103,823.00	53	2809.000	148,877.00
1/8	1691.266	69,553.299	1/8	2220.766	104,653.58	1/8	2822.266	149,932.86
1/4	1701.563	70,189.453	1/4	2232.563	105,488.58	1/4	2835.563	150,993.70
3/8	1711.891	70,829.475	3/8	2244.391	106,328.01	3/8	2848.891	152,059.54
1/2	1722.250	71,473.375	1/2	2256.250	107,171.88	1/2	2862.250	153,130.38
5/8	1732.641	72,121.166	5/8	2268.141	108,020.20	5/8	2875.641	154,206.23
3/4	1743.063	72,772.859	3/4	2280.063	108,872.98	3/4	2889.063	155,287.11
7/8	1753.516	73,428.467	7/8	2292.016	109,730.25	7/8	2902.516	156,373.03
42	1764.000	74,088.000	48	2304.000	110,592.00	54	2916.000	157,464.00
1/8	1774.516	74,751.471	1/8	2316.016	111,458.25	1/8	2929.516	158,560.03
1/4	1785.063	75,418.891	1/4	2328.063	112,329.02	1/4	2943.063	159,661.14
3/8	1795.641	76,090.271	3/8	2340.141	113,204.30	3/8	2956.641	160,767.33
1/2	1806.250	76,765.625	1/2	2352.250	114,084.13	1/2	2970.250	161,878.63
5/8	1816.891	77,444.963	5/8	2364.391	114,968.49	5/8	2983.891	162,995.03
3/4	1827.563	78,128.297	3/4	2376.563	115,857.42	3/4	2997.563	164,116.55
7/8	1838.266	78,815.639	7/8	2388.766	116,750.92	7/8	3011.266	165,243.20
43	1849.000	79,507.000	49	2401.000	117,649.00	55	3025.000	166,375.00
1/8	1859.766	80,202.393	1/8	2413.266	118,551.67	1/8	3038.766	167,511.96
1/4	1870.563	80,901.828	1/4	2425.563	119,458.95	1/4	3052.563	168,654.08
3/8	1881.391	81,605.318	3/8	2437.891	120,370.85	3/8	3066.391	169,801.38
1/2	1892.250	82,312.875	1/2	2450.250	121,287.38	1/2	3080.250	170,953.88
5/8	1903.141	83,024.510	5/8	2462.641	122,208.54	5/8	3094.141	172,111.57
3/4	1914.063	83,740.234	3/4	2475.063	123,134.36	3/4	3108.063	173,274.48
7/8	1925.016	84,460.061	7/8	2487.516	124,064.84	7/8	3122.016	174,442.62
44	1936.000	85,184.000	50	2500.000	125,000.00	56	3136.000	175,616.00
1/8	1947.016	85,912.064	1/8	2512.516	125,939.85	1/8	3150.016	176,794.63
1/4	1958.063	86,644.266	1/4	2525.063	126,884.39	1/4	3164.063	177,978.52
3/8	1969.141	87,380.615	3/8	2537.641	127,833.65	3/8	3178.141	179,167.68
1/2	1980.250	88,121.125	1/2	2550.250	128,787.63	1/2	3192.250	180,362.13
5/8	1991.391	88,865.807	5/8	2562.891	129,746.34	5/8	3206.391	181,561.87
3/4	2002.563	89,614.672	3/4	2575.563	130,709.80	3/4	3220.563	182,766.92
7/8	2013.766	90,367.732	7/8	2588.266	131,678.01	7/8	3234.766	183,977.29
45	2025.000	91,125.000	51	2601.000	132,651.00	57	3249.000	185,193.00
1/8	2036.266	91,886.486	1/8	2613.766	133,628.77	1/8	3263.266	186,414.05
1/4	2047.563	92,652.203	1/4	2626.563	134,611.33	1/4	3277.563	187,640.45
3/8	2058.891	93,422.162	3/8	2639.391	135,598.69	3/8	3291.891	188,872.22
1/2	2070.250	94,196.375	1/2	2652.250	136,590.88	1/2	3306.250	190,109.38
5/8	2081.641	94,974.854	5/8	2665.141	137,587.88	5/8	3320.641	191,351.92
3/4	2093.063	95,757.609	3/4	2678.063	138,589.73	3/4	3335.063	192,599.86
7/8	2104.516	96,544.654	7/8	2691.016	139,596.44	7/8	3349.516	193,853.22

Squares and Cubes of Numbers from 1/32 to 100 (Continued)

No.	Square	Cube	No.	Square	Cube	No.	Square	Cube
58	3364.000	195,112.00	64	4096.000	262,144.00	70	4900.000	343,000.00
1/8	3378.516	196,376.22	1/8	4112.016	263,683.00	1/8	4917.516	344,840.78
1/4	3393.063	197,645.89	1/4	4128.063	265,228.02	1/4	4935.063	346,688.14
3/8	3407.641	198,921.02	3/8	4144.141	266,779.05	3/8	4952.641	348,542.08
1/2	3422.250	200,201.63	1/2	4160.250	268,336.13	1/2	4970.250	350,402.63
5/8	3436.891	201,487.71	5/8	4176.391	269,899.24	5/8	4987.891	352,269.78
3/4	3451.563	202,779.30	3/4	4192.563	271,468.42	3/4	5005.563	354,143.55
7/8	3466.266	204,076.39	7/8	4208.766	273,043.67	7/8	5023.266	356,023.95
59	3481.000	205,379.00	65	4225.000	274,625.00	71	5041.000	357,911.00
1/8	3495.766	206,687.14	1/8	4241.266	276,212.42	1/8	5058.766	359,804.71
1/4	3510.563	208,000.83	1/4	4257.563	277,805.95	1/4	5076.563	361,705.08
3/8	3525.391	209,320.07	3/8	4273.891	279,405.60	3/8	5094.391	363,612.13
1/2	3540.250	210,644.88	1/2	4290.250	281,011.38	1/2	5112.250	365,525.88
5/8	3555.141	211,975.26	5/8	4306.641	282,623.29	5/8	5130.141	367,446.32
3/4	3570.063	213,311.23	3/4	4323.063	284,241.36	3/4	5148.063	369,373.48
7/8	3585.016	214,652.81	7/8	4339.516	285,865.59	7/8	5166.016	371,307.37
60	3600.000	216,000.00	66	4356.000	287,496.00	72	5184.000	373,248.00
1/8	3615.016	217,352.81	1/8	4372.516	289,132.60	1/8	5202.016	375,195.38
1/4	3630.063	218,711.27	1/4	4389.063	290,775.39	1/4	5220.063	377,149.52
3/8	3645.141	220,075.37	3/8	4405.641	292,424.40	3/8	5238.141	379,110.43
1/2	3660.250	221,445.13	1/2	4422.250	294,079.63	1/2	5256.250	381,078.13
5/8	3675.391	222,820.56	5/8	4438.891	295,741.09	5/8	5274.391	383,052.62
3/4	3690.563	224,201.67	3/4	4455.563	297,408.80	3/4	5292.563	385,033.92
7/8	3705.766	225,588.48	7/8	4472.266	299,082.76	7/8	5310.766	387,022.04
61	3721.000	226,981.00	67	4489.000	300,763.00	73	5329.000	389,017.00
1/8	3736.266	228,379.24	1/8	4505.766	302,449.52	1/8	5347.266	391,018.80
1/4	3751.563	229,783.20	1/4	4522.563	304,142.33	1/4	5365.563	393,027.45
3/8	3766.891	231,192.91	3/8	4539.391	305,841.44	3/8	5383.891	395,042.97
1/2	3782.250	232,608.38	1/2	4556.250	307,546.88	1/2	5402.250	397,065.38
5/8	3797.641	234,029.60	5/8	4573.141	309,258.63	5/8	5420.641	399,094.67
3/4	3813.063	235,456.61	3/4	4590.063	310,976.73	3/4	5439.063	401,130.86
7/8	3828.516	236,889.40	7/8	4607.016	312,701.19	7/8	5457.516	403,173.97
62	3844.000	238,328.00	68	4624.000	314,432.00	74	5476.000	405,224.00
1/8	3859.516	239,772.41	1/8	4641.016	316,169.19	1/8	5494.516	407,280.97
1/4	3875.063	241,222.64	1/4	4658.063	317,912.77	1/4	5513.063	409,344.89
3/8	3890.641	242,678.71	3/8	4675.141	319,662.74	3/8	5531.641	411,415.77
1/2	3906.250	244,140.63	1/2	4692.250	321,419.13	1/2	5550.250	413,493.63
5/8	3921.891	245,608.40	5/8	4709.391	323,181.93	5/8	5568.891	415,578.46
3/4	3937.563	247,082.05	3/4	4726.563	324,951.17	3/4	5587.563	417,670.30
7/8	3953.266	248,561.58	7/8	4743.766	326,726.86	7/8	5606.266	419,769.14
63	3969.000	250,047.00	69	4761.000	328,509.00	75	5625.000	421,875.00
1/8	3984.766	251,538.33	1/8	4778.266	330,297.61	1/8	5643.766	423,987.89
1/4	4000.563	253,035.58	1/4	4795.563	332,092.70	1/4	5662.563	426,107.83
3/8	4016.391	254,538.76	3/8	4812.891	333,894.29	3/8	5681.391	428,234.82
1/2	4032.250	256,047.88	1/2	4830.250	335,702.38	1/2	5700.250	430,368.88
5/8	4048.141	257,562.95	5/8	4847.641	337,516.98	5/8	5719.141	432,510.01
3/4	4064.063	259,083.98	3/4	4865.063	339,338.11	3/4	5738.063	434,658.23
7/8	4080.016	260,611.00	7/8	4882.516	341,165.78	7/8	5757.016	436,813.56

Squares and Cubes of Numbers from 1/32 to 100 (Continued)

No.	Square	Cube	No.	Square	Cube	No.	Square	Cube
76	5776.000	438,976.00	82	6724.000	551,368.00	88	7744.000	681,472.00
1/8	5795.016	441,145.56	1/8	6744.516	553,893.35	1/8	7766.016	684,380.13
1/4	5814.063	443,322.27	1/4	6765.063	556,426.39	1/4	7788.063	687,296.52
3/8	5833.141	445,506.12	3/8	6785.641	558,967.15	3/8	7810.141	690,221.18
1/2	5852.250	447,697.13	1/2	6806.250	561,515.63	1/2	7832.250	693,154.13
5/8	5871.390	449,895.30	5/8	6826.891	564,071.84	5/8	7854.391	696,095.37
3/4	5890.563	452,100.67	3/4	6847.563	566,635.80	3/4	7876.563	699,044.92
7/8	5909.766	454,313.23	7/8	6868.266	569,207.51	7/8	7898.766	702,002.79
77	5929.000	456,533.00	83	6889.000	571,787.00	89	7921.000	704,969.00
1/8	5948.266	458,759.99	1/8	6909.766	574,374.27	1/8	7943.266	707,943.55
1/4	5967.563	460,994.20	1/4	6930.563	576,969.33	1/4	7965.563	710,926.45
3/8	5986.891	463,235.66	3/8	6951.391	579,572.19	3/8	7987.891	713,917.72
1/2	6006.250	465,484.38	1/2	6972.250	582,182.88	1/2	8010.250	716,917.38
5/8	6025.641	467,740.35	5/8	6993.141	584,801.38	5/8	8032.641	719,925.42
3/4	6045.063	470,003.61	3/4	7014.063	587,427.73	3/4	8055.063	722,941.86
7/8	6064.516	472,274.15	7/8	7035.016	590,061.94	7/8	8077.516	725,966.72
78	6084.000	474,552.00	84	7056.000	592,704.00	90	8100.000	729,000.00
1/8	6103.516	476,837.16	1/8	7077.016	595,353.94	1/8	8122.516	732,041.72
1/4	6123.063	479,129.64	1/4	7098.063	598,011.77	1/4	8145.063	735,091.89
3/8	6142.641	481,429.46	3/8	7119.141	600,677.49	3/8	8167.641	738,150.52
1/2	6162.250	483,736.63	1/2	7140.250	603,351.13	1/2	8190.250	741,217.63
5/8	6181.891	486,051.15	5/8	7161.391	606,032.68	5/8	8212.891	744,293.21
3/4	6201.563	488,373.05	3/4	7182.563	608,722.17	3/4	8235.563	747,377.30
7/8	6221.266	490,702.33	7/8	7203.766	611,419.61	7/8	8258.266	750,469.89
79	6241.000	493,039.00	85	7225.000	614,125.00	91	8281.000	753,571.00
1/8	6260.766	495,383.08	1/8	7246.266	616,838.36	1/8	8303.766	755,680.64
1/4	6280.563	497,734.58	1/4	7267.563	619,559.70	1/4	8325.563	759,798.83
3/8	6300.391	500,093.51	3/8	7288.891	622,289.04	3/8	8349.391	762,925.57
1/2	6320.250	502,459.88	1/2	7310.250	625,026.38	1/2	8372.250	766,060.88
5/8	6340.141	504,833.70	5/8	7331.641	627,771.73	5/8	8395.141	769,204.76
3/4	6360.063	507,214.98	3/4	7353.063	630,525.11	3/4	8418.063	772,357.23
7/8	6380.016	509,603.75	7/8	7374.516	633,286.53	7/8	8441.016	775,518.31
80	6400.000	512,000.00	86	7396.000	636,056.00	92	8464.000	778,688.00
1/8	6420.016	514,403.75	1/8	7417.516	638,833.53	1/8	8487.016	781,866.31
1/4	6440.063	516,815.02	1/4	7439.063	641,619.14	1/4	8510.063	785,053.27
3/8	6460.141	519,233.80	3/8	7460.641	644,412.83	3/8	8533.141	788,248.87
1/2	6480.250	521,660.13	1/2	7482.250	647,214.63	1/2	8556.250	791,453.13
5/8	6500.391	524,093.99	5/8	7503.891	650,024.53	5/8	8579.391	794,666.06
3/4	6520.563	526,535.42	3/4	7525.563	652,842.55	3/4	8602.563	797,887.67
7/8	6540.766	528,984.42	7/8	7547.266	655,668.70	7/8	8625.766	801,117.98
81	6561.000	531,441.00	87	7569.000	658,503.00	93	8649.000	804,357.00
1/8	6581.266	533,905.17	1/8	7590.766	661,345.46	1/8	8672.266	807,604.74
1/4	6601.563	536,376.95	1/4	7612.563	664,196.08	1/4	8695.563	810,861.20
3/8	6621.891	538,856.35	3/8	7634.390	667,054.88	3/8	8718.891	814,126.41
1/2	6642.250	541,343.38	1/2	7656.250	669,921.88	1/2	8742.250	817,400.38
5/8	6662.641	543,838.04	5/8	7678.141	672,797.07	5/8	8765.641	820,683.10
3/4	6683.063	546,340.36	3/4	7700.063	675,680.48	3/4	8789.063	823,974.61
7/8	6703.516	548,850.34	7/8	7722.016	678,572.12	7/8	8812.516	827,274.90

Squares and Cubes of Numbers from 1/32 to 100 (Continued)

No.	Square	Cube	No.	Square	Cube	No.	Square	Cube
94	8836.000	830,584.00	96	9216.000	884,736.00	98	9604.00	941,192.0
⅛	8859.516	833,901.91	⅛	9240.016	888,196.50	⅛	9628.52	944,798.1
¼	8883.063	837,228.64	¼	9264.063	891,666.02	¼	9653.06	948,413.4
⅜	8906.641	840,564.21	⅜	9288.141	895,144.55	⅜	9677.64	952,037.9
½	8930.250	843,908.63	½	9312.250	898,632.13	½	9702.25	955,671.6
⅝	8953.891	847,261.90	⅝	9336.391	902,128.74	⅝	9726.89	959,314.6
¾	8977.563	850,624.05	¾	9360.563	905,634.42	¾	9751.56	962,966.8
⅞	9001.266	853,995.08	⅞	9384.766	909,149.17	⅞	9776.27	966,628.3
95	9025.000	857,375.00	97	9409.000	912,673.00	99	9801.00	970,299.0
⅛	9048.766	860,763.83	⅛	9433.266	916,205.92	⅛	9825.77	973,979.0
¼	9072.563	864,161.58	¼	9457.563	919,747.95	¼	9850.56	977,668.3
⅜	9096.391	867,568.26	⅜	9481.891	923,299.10	⅜	9875.39	981,366.9
½	9120.250	870,983.88	½	9506.250	926,859.38	½	9900.25	985,074.9
⅝	9144.141	874,408.45	⅝	9530.641	930,428.79	⅝	9925.14	988,792.1
¾	9168.063	877,841.98	¾	9555.063	934,007.36	¾	9950.06	992,518.7
⅞	9192.016	881,284.50	⅞	9579.516	937,595.09	⅞	9975.02	996,254.7
						100	10,000.00	1,000,000.0

Table of Fractions of π = 3.14159265

a	$\frac{\pi}{a}$	a	$\frac{\pi}{a}$	a	$\frac{\pi}{a}$	a	$\frac{\pi}{a}$	a	$\frac{\pi}{a}$
1	3.14159	21	0.14960	41	0.07662	61	0.05150	81	0.03879
2	1.57080	22	0.14280	42	0.07480	62	0.05067	82	0.03831
3	1.04720	23	0.13659	43	0.07306	63	0.04987	83	0.03785
4	0.78540	24	0.13090	44	0.07140	64	0.04909	84	0.03740
5	0.62832	25	0.12566	45	0.06981	65	0.04833	85	0.03696
6	0.52360	26	0.12083	46	0.06830	66	0.04760	86	0.03653
7	0.44880	27	0.11636	47	0.06684	67	0.04689	87	0.03611
8	0.39270	28	0.11220	48	0.06545	68	0.04620	88	0.03570
9	0.34907	29	0.10833	49	0.06411	69	0.04553	89	0.03530
10	0.31416	30	0.10472	50	0.06283	70	0.04488	90	0.03491
11	0.28560	31	0.10134	51	0.06160	71	0.04425	91	0.03452
12	0.26180	32	0.09817	52	0.06042	72	0.04363	92	0.03415
13	0.24166	33	0.09520	53	0.05928	73	0.04304	93	0.03378
14	0.22440	34	0.09240	54	0.05818	74	0.04245	94	0.03342
15	0.20944	35	0.08976	55	0.05712	75	0.04189	95	0.03307
16	0.19635	36	0.08727	56	0.05610	76	0.04134	96	c.03272
17	0.18480	37	0.08491	57	0.05512	77	0.04080	97	0.03239
18	0.17453	38	0.08267	58	0.05417	78	0.04028	98	0.03206
19	0.16535	39	0.08055	59	0.05325	79	0.03977	99	0.03173
20	0.15708	40	0.07854	60	0.05236	80	0.03927	100	0.03142

Pi (π). — The ratio of the circumference of a circle to its diameter, which is represented by the Greek letter pi (π), is an incommensurable quantity. The value 3.1416 is accurate enough for ordinary purposes and the value 22/7 is convenient for rough calculations. The fractions of π given in the above table will be found convenient in certain calculations and also the values in the table of constants on page 37.

Table of Commonly Used Constants

Constant	Numerical Value	Logarithm	Constant	Numerical Value	Logarithm
π	3.141593	0.49715	Weight in pounds of:		
2π	6.283185	0.79818	Water column, $1''\times1''\times1$ ft.	0.4335	$\bar{1}$.63699
$\pi\div4$	0.785398	$\bar{1}$.89509	1 U.S. gallon of water, 39.1°F.	8.34	0.92117
π^2	9.869604	0.99430	1 cu. ft. of water, 39.1° F...	62.4245	1.79536
π^3	31.006277	1.49145	1 cu. in. of water, 39.1° F...	0.0361	$\bar{2}$.55751
$1\div\pi$	0.318310	$\bar{1}$.50285	1 cu. ft. of air, 32° F., atmos-		
$1\div\pi^2$	0.101321	$\bar{1}$.00570	pheric pressure	0.08073	$\bar{2}$.90703
$1\div\pi^3$	0.032252	$\bar{2}$.50855	Volume in cu. ft. of:		
$\sqrt{\pi}$	1.772454	0.24858	1 pound of water, 39.1° F...	0.01602	$\bar{2}$.20466
$\sqrt[3]{\pi}$	1.464592	0.16572	1 pound of air, 32° F., atmos-		
g	32.16	1.50732	pheric pressure	12.387	1.09297
g^2	1034.266	3.01463	Volume in gallons of 1 pound		
$2g$	64.32	1.80835	of water, 39.1° F.........	0.1199	$\bar{1}$.07882
$1\div2g$	0.01555	$\bar{2}$.19165	Volume in cu. in. of 1 pound of		
$\sqrt{2g}$	8.01998	0.90417	water, 39.1° F.	27.70	1.44248
$1\div\sqrt{g}$	0.17634	$\bar{1}$.24635	One cubic ft. in gallons	7.4805	0.87393
$\pi\div\sqrt{g}$	0.55398	$\bar{1}$.74350	Atmospheric pressure in		
e	2.71828	0.43429	pounds per sq. in........	14.696	1.16720

Useful Constants Multiplied and Divided by 1 to 10

Constant	Multiplied by:							
	2	3	4	5	6	7	8	9
0.7854	1.5708	2.3562	3.1416	3.9270	4.7124	5.4978	6.2832	7.0686
3.1416	6.2832	9.4248	12.566	15.708	18.850	21.991	25.133	28.274
14.7	29.4	44.1	58.8	73.5	88.2	102.9	117.6	132.3
32.16	64.32	96.48	128.64	160.80	192.96	225.12	257.28	289.44
64.32	128.64	192.96	257.28	321.60	385.92	450.24	514.56	578.88
144	288	432	576	720	864	1,008	1,152	1,296
778	1,556	2,334	3,112	3,890	4,668	5,446	6,224	7,002
1,728	3,456	5,184	6,912	8,640	10,368	12,096	13,824	15,552
33,000	66,000	99,000	132,000	165,000	198,000	231,000	264,000	297,000

Constant	Divided by:							
	2	3	4	5	6	7	8	9
0.7854	0.3927	0.2618	0.1964	0.1571	0.1309	0.1122	0.0982	0.0873
3.1416	1.5708	1.0472	0.7854	0.6283	0.5236	0.4488	0.3927	0.3491
14.7	7.350	4.900	3.675	2.940	2.450	2.100	1.838	1.633
32.16	16.080	10.720	8.040	6.432	5.360	4.594	4.020	3.573
64.32	32.160	21.440	16.080	12.864	10.720	9.189	8.040	7.147
144	72	48	36	28.800	24	20.571	18	16
778	389	259.33	194.50	155.60	129.67	111.14	97.25	86.44
1,728	864	576	432	345.60	288	246.86	216	192
33,000	16,500	11,000	8250	6600	5500	4714.3	4125	3666.7

Circumferences and Areas of Circles*

Diameter	Circumference	Area	Diameter	Circumference	Area	Diameter	Circumference	Area
1/64	0.0491	0.0002	2	6.2832	3.1416	5	15.7080	19.635
1/32	0.0982	0.0008	1/16	6.4795	3.3410	1/16	15.9043	20.129
1/16	0.1963	0.0031	1/8	6.6759	3.5466	1/8	16.1007	20.629
3/32	0.2945	0.0069	3/16	6.8722	3.7583	3/16	16.2970	21.135
1/8	0.3927	0.0123	1/4	7.0686	3.9761	1/4	16.4934	21.648
5/32	0.4909	0.0192	5/16	7.2649	4.2000	5/16	16.6897	22.166
3/16	0.5890	0.0276	3/8	7.4613	4.4301	3/8	16.8861	22.691
7/32	0.6872	0.0376	7/16	7.6576	4.6664	7/16	17.0824	23.221
1/4	0.7854	0.0491	1/2	7.8540	4.9087	1/2	17.2788	23.758
9/32	0.8836	0.0621	9/16	8.0503	5.1572	9/16	17.4751	24.301
5/16	0.9817	0.0767	5/8	8.2467	5.4119	5/8	17.6715	24.850
11/32	1.0799	0.0928	11/16	8.4430	5.6727	11/16	17.8678	25.406
3/8	1.1781	0.1104	3/4	8.6394	5.9396	3/4	18.0642	25.967
13/32	1.2763	0.1296	13/16	8.8357	6.2126	13/16	18.2605	26.535
7/16	1.3744	0.1503	7/8	9.0321	6.4918	7/8	18.4569	27.109
15/32	1.4726	0.1726	15/16	9.2284	6.7771	15/16	18.6532	27.688
1/2	1.5708	0.1963	3	9.4248	7.0686	6	18.8496	28.274
17/32	1.6690	0.2217	1/16	9.6211	7.3662	1/8	19.2423	29.465
9/16	1.7671	0.2485	1/8	9.8175	7.6699	1/4	19.6350	30.680
19/32	1.8653	0.2769	3/16	10.0138	7.9798	3/8	20.0277	31.919
5/8	1.9635	0.3068	1/4	10.2102	8.2958	1/2	20.4204	33.183
21/32	2.0617	0.3382	5/16	10.4065	8.6179	5/8	20.8131	34.472
11/16	2.1598	0.3712	3/8	10.6029	8.9462	3/4	21.2058	35.785
23/32	2.2580	0.4057	7/16	10.7992	9.2806	7/8	21.5984	37.122
3/4	2.3562	0.4418	1/2	10.9956	9.6211	7	21.9911	38.485
25/32	2.4544	0.4794	9/16	11.1919	9.9678	1/8	22.3838	39.871
13/16	2.5525	0.5185	5/8	11.3883	10.321	1/4	22.7765	41.282
27/32	2.6507	0.5591	11/16	11.5846	10.680	3/8	23.1692	42.718
7/8	2.7489	0.6013	3/4	11.7810	11.045	1/2	23.5619	44.179
29/32	2.8471	0.6450	13/16	11.9773	11.416	5/8	23.9546	45.664
15/16	2.9452	0.6903	7/8	12.1737	11.793	3/4	24.3473	47.173
31/32	3.0434	0.7371	15/16	12.3700	12.177	7/8	24.7400	48.707
1	3.1416	0.7854	4	12.5664	12.566	8	25.1327	50.265
1/16	3.3379	0.8866	1/16	12.7627	12.962	1/8	25.5254	51.849
1/8	3.5343	0.9940	1/8	12.9591	13.364	1/4	25.9181	53.456
3/16	3.7306	1.1075	3/16	13.1554	13.772	3/8	26.3108	55.088
1/4	3.9270	1.2272	1/4	13.3518	14.186	1/2	26.7035	56.745
5/16	4.1233	1.3530	5/16	13.5481	14.607	5/8	27.0962	58.426
3/8	4.3197	1.4849	3/8	13.7445	15.033	3/4	27.4889	60.132
7/16	4.5160	1.6230	7/16	13.9408	15.466	7/8	27.8816	61.862
1/2	4.7124	1.7671	1/2	14.1372	15.904	9	28.2743	63.617
9/16	4.9087	1.9175	9/16	14.3335	16.349	1/8	28.6670	65.397
5/8	5.1051	2.0739	5/8	14.5299	16.800	1/4	29.0597	67.201
11/16	5.3014	2.2365	11/16	14.7262	17.257	3/8	29.4524	69.029
3/4	5.4978	2.4053	3/4	14.9226	17.721	1/2	29.8451	70.882
13/16	5.6941	2.5802	13/16	15.1189	18.190	5/8	30.2378	72.760
7/8	5.8905	2.7612	7/8	15.3153	18.665	3/4	30.6305	74.662
15/16	6.0868	2.9483	15/16	15.5116	19.147	7/8	31.0232	76.589

* All the figures given in the tables on pages 38 through 49 can be used for English units and those without common fractions can be used for metric units.

Circumferences and Areas of Circles

Diameter	Circumference	Area	Diameter	Circumference	Area	Diameter	Circumference	Area
10	31.4159	78.540	16	50.2655	201.06	22	69.1150	380.13
⅛	31.8086	80.516	⅛	50.6582	204.22	⅛	69.5077	384.46
¼	32.2013	82.516	¼	51.0509	207.39	¼	69.9004	388.82
⅜	32.5940	84.541	⅜	51.4436	210.60	⅜	70.2931	393.20
½	32.9867	86.590	½	51.8363	213.82	½	70.6858	397.61
⅝	33.3794	88.664	⅝	52.2290	217.08	⅝	71.0785	402.04
¾	33.7721	90.763	¾	52.6217	220.35	¾	71.4712	406.49
⅞	34.1648	92.886	⅞	53.0144	223.65	⅞	71.8639	410.97
11	34.5575	95.033	17	53.4071	226.98	23	72.2566	415.48
⅛	34.9502	97.205	⅛	53.7998	230.33	⅛	72.6493	420.00
¼	35.3429	99.402	¼	54.1925	233.71	¼	73.0420	424.56
⅜	35.7356	101.62	⅜	54.5852	237.10	⅜	73.4347	429.13
½	36.1283	103.87	½	54.9779	240.53	½	73.8274	433.74
⅝	36.5210	106.14	⅝	55.3706	243.98	⅝	74.2201	438.36
¾	36.9137	108.43	¾	55.7633	247.45	¾	74.6128	443.01
⅞	37.3064	110.75	⅞	56.1560	250.95	⅞	75.0055	447.69
12	37.6991	113.10	18	56.5487	254.47	24	75.3982	452.39
⅛	38.0918	115.47	⅛	56.9414	258.02	⅛	75.7909	457.11
¼	38.4845	117.86	¼	57.3341	261.59	¼	76.1836	461.86
⅜	38.8772	120.28	⅜	57.7268	265.18	⅜	76.5763	466.64
½	39.2699	122.72	½	58.1195	268.80	½	76.9690	471.44
⅝	39.6626	125.19	⅝	58.5122	272.45	⅝	77.3617	476.26
¾	40.0553	127.68	¾	58.9049	276.12	¾	77.7544	481.11
⅞	40.4480	130.19	⅞	59.2976	279.81	⅞	78.1471	485.98
13	40.8407	132.73	19	59.6903	283.53	25	78.5398	490.87
⅛	41.2334	135.30	⅛	60.0830	287.27	⅛	78.9325	495.79
¼	41.6261	137.89	¼	60.4757	291.04	¼	79.3252	500.74
⅜	42.0188	140.50	⅜	60.8684	294.83	⅜	79.7179	505.71
½	42.4115	143.14	½	61.2611	298.65	½	80.1106	510.71
⅝	42.8042	145.80	⅝	61.6538	302.49	⅝	80.5033	515.72
¾	43.1969	148.49	¾	62.0465	306.35	¾	80.8960	520.77
⅞	43.5896	151.20	⅞	62.4392	310.24	⅞	81.2887	525.84
14	43.9823	153.94	20	62.8319	314.16	26	81.6814	530.93
⅛	44.3750	156.70	⅛	63.2246	318.10	⅛	82.0741	536.05
¼	44.7677	159.48	¼	63.6173	322.06	¼	82.4668	541.19
⅜	45.1604	162.30	⅜	64.0100	326.05	⅜	82.8595	546.35
½	45.5531	165.13	½	64.4026	330.06	½	83.2522	551.55
⅝	45.9458	167.99	⅝	64.7953	334.10	⅝	83.6449	556.76
¾	46.3385	170.87	¾	65.1880	338.16	¾	84.0376	562.00
⅞	46.7312	173.78	⅞	65.5807	342.25	⅞	84.4303	567.27
15	47.1239	176.71	21	65.9734	346.36	27	84.8230	572.56
⅛	47.5166	179.67	⅛	66.3661	350.50	⅛	85.2157	577.87
¼	47.9093	182.65	¼	66.7588	354.66	¼	85.6084	583.21
⅜	48.3020	185.66	⅜	67.1515	358.84	⅜	86.0011	588.57
½	48.6947	188.69	½	67.5442	363.05	½	86.3938	593.96
⅝	49.0874	191.75	⅝	67.9369	367.28	⅝	86.7865	599.37
¾	49.4801	194.83	¾	68.3296	371.54	¾	87.1792	604.81
⅞	49.8728	197.93	⅞	68.7223	375.83	⅞	87.5719	610.27

Circumferences and Areas of Circles

Diameter	Circumference	Area	Diameter	Circumference	Area	Diameter	Circumference	Area
28	87.9646	615.75	34	106.814	907.92	40	125.664	1256.6
⅛	88.3573	621.26	⅛	107.207	914.61	⅛	126.056	1264.5
¼	88.7500	626.80	¼	107.600	921.32	¼	126.449	1272.4
⅜	89.1427	632.36	⅜	107.992	928.06	⅜	126.842	1280.3
½	89.5354	637.94	½	108.385	934.82	½	127.235	1288.2
⅝	89.9281	643.55	⅝	108.778	941.61	⅝	127.627	1296.2
¾	90.3208	649.18	¾	109.170	948.42	¾	128.020	1304.2
⅞	90.7135	654.84	⅞	109.563	955.25	⅞	128.413	1312.2
29	91.1062	660.52	35	109.956	962.11	41	128.805	1320.3
⅛	91.4989	666.23	⅛	110.348	969.00	⅛	129.198	1328.3
¼	91.8916	671.96	¼	110.741	975.91	¼	129.591	1336.4
⅜	92.2843	677.71	⅜	111.134	982.84	⅜	129.983	1344.5
½	92.6770	683.49	½	111.527	989.80	½	130.376	1352.7
⅝	93.0697	689.30	⅝	111.919	996.78	⅝	130.769	1360.8
¾	93.4624	695.13	¾	112.312	1003.8	¾	131.161	1369.0
⅞	93.8551	700.98	⅞	112.705	1010.8	⅞	131.554	1377.2
30	94.2478	706.86	36	113.097	1017.9	42	131.947	1385.4
⅛	94.6405	712.76	⅛	113.490	1025.0	⅛	132.340	1393.7
¼	95.0332	718.69	¼	113.883	1032.1	¼	132.732	1402.0
⅜	95.4259	724.64	⅜	114.275	1039.2	⅜	133.125	1410.3
½	95.8186	730.62	½	114.668	1046.3	½	133.518	1418.6
⅝	96.2113	736.62	⅝	115.061	1053.5	⅝	133.910	1427.0
¾	96.6040	742.64	¾	115.454	1060.7	¾	134.303	1435.4
⅞	96.9967	748.69	⅞	115.846	1068.0	⅞	134.696	1443.8
31	97.3894	754.77	37	116.239	1075.2	43	135.088	1452.2
⅛	97.7821	760.87	⅛	116.632	1082.5	⅛	135.481	1460.7
¼	98.1748	766.99	¼	117.024	1089.8	¼	135.874	1469.1
⅜	98.5675	773.14	⅜	117.417	1097.1	⅜	136.267	1477.6
½	98.9602	779.31	½	117.810	1104.5	½	136.659	1486.2
⅝	99.3529	785.51	⅝	118.202	1111.8	⅝	137.052	1494.7
¾	99.7456	791.73	¾	118.595	1119.2	¾	137.445	1503.3
⅞	100.138	797.98	⅞	118.988	1126.7	⅞	137.837	1511.9
32	100.531	804.25	38	119.381	1134.1	44	138.230	1520.5
⅛	100.924	810.54	⅛	119.773	1141.6	⅛	138.623	1529.2
¼	101.316	816.86	¼	120.166	1149.1	¼	139.015	1537.9
⅜	101.709	823.21	⅜	120.559	1156.6	⅜	139.408	1546.6
½	102.102	829.58	½	120.951	1164.2	½	139.801	1555.3
⅝	102.494	835.97	⅝	121.344	1171.7	⅝	140.194	1564.0
¾	102.887	842.39	¾	121.737	1179.3	¾	140.586	1572.8
⅞	103.280	848.83	⅞	122.129	1186.9	⅞	140.979	1581.6
33	103.673	855.30	39	122.522	1194.6	45	141.372	1590.4
⅛	104.065	861.79	⅛	122.915	1202.3	⅛	141.764	1599.3
¼	104.458	868.31	¼	123.308	1210.0	¼	142.157	1608.2
⅜	104.851	874.85	⅜	123.700	1217.7	⅜	142.550	1617.0
½	105.243	881.41	½	124.093	1225.4	½	142.942	1626.0
⅝	105.636	888.00	⅝	124.486	1233.2	⅝	143.335	1634.9
¾	106.029	894.62	¾	124.878	1241.0	¾	143.728	1643.9
⅞	106.421	901.26	⅞	125.271	1248.8	⅞	144.121	1652.9

Circumferences and Areas of Circles

Diam-eter	Circum-ference	Area	Diam-eter	Circum-ference	Area	Diam-eter	Circum-ference	Area
46	144.513	1661.9	52	163.363	2123.7	58	182.212	2642.1
⅛	144.906	1670.9	⅛	163.756	2133.9	⅛	182.605	2653.5
¼	145.299	1680.0	¼	164.148	2144.2	¼	182.998	2664.9
⅜	145.691	1689.1	⅜	164.541	2154.5	⅜	183.390	2676.4
½	146.084	1698.2	½	164.934	2164.8	½	183.783	2687.8
⅝	146.477	1707.4	⅝	165.326	2175.1	⅝	184.176	2699.3
¾	146.869	1716.5	¾	165.719	2185.4	¾	184.569	2710.9
⅞	147.262	1725.7	⅞	166.112	2195.8	⅞	184.961	2722.4
47	147.655	1734.9	53	166.504	2206.2	59	185.354	2734.0
⅛	148.048	1744.2	⅛	166.897	2216.6	⅛	185.747	2745.6
¼	148.440	1753.5	¼	167.290	2227.0	¼	186.139	2757.2
⅜	148.833	1762.7	⅜	167.683	2237.5	⅜	186.532	2768.8
½	149.226	1772.1	½	168.075	2248.0	½	186.925	2780.5
⅝	149.618	1781.4	⅝	168.468	2258.5	⅝	187.317	2792.2
¾	150.011	1790.8	¾	168.861	2269.1	¾	187.710	2803.9
⅞	150.404	1800.1	⅞	169.253	2279.6	⅞	188.103	2815.7
48	150.796	1809.6	54	169.646	2290.2	60	188.496	2827.4
⅛	151.189	1819.0	⅛	170.039	2300.8	⅛	188.888	2839.2
¼	151.582	1828.5	¼	170.431	2311.5	¼	189.281	2851.0
⅜	151.975	1837.9	⅜	170.824	2322.1	⅜	189.674	2862.9
½	152.367	1847.5	½	171.217	2332.8	½	190.066	2874.8
⅝	152.760	1857.0	⅝	171.609	2343.5	⅝	190.459	2886.6
¾	153.153	1866.5	¾	172.002	2354.3	¾	190.852	2898.6
⅞	153.545	1876.1	⅞	172.395	2365.0	⅞	191.244	2910.5
49	153.938	1885.7	55	172.788	2375.8	61	191.637	2922.5
⅛	154.331	1895.4	⅛	173.180	2386.6	⅛	192.030	2934.5
¼	154.723	1905.0	¼	173.573	2397.5	¼	192.423	2946.5
⅜	155.116	1914.7	⅜	173.966	2408.3	⅜	192.815	2958.5
½	155.509	1924.4	½	174.358	2419.2	½	193.208	2970.6
⅝	155.902	1934.2	⅝	174.751	2430.1	⅝	193.601	2982.7
¾	156.294	1943.9	¾	175.144	2441.1	¾	193.993	2994.8
⅞	156.687	1953.7	⅞	175.536	2452.0	⅞	194.386	3006.9
50	157.080	1963.5	56	175.929	2463.0	62	194.779	3019.1
⅛	157.472	1973.3	⅛	176.322	2474.0	⅛	195.171	3031.3
¼	157.865	1983.2	¼	176.715	2485.0	¼	195.564	3043.5
⅜	158.258	1993.1	⅜	177.107	2496.1	⅜	195.957	3055.7
½	158.650	2003.0	½	177.500	2507.2	½	196.350	3068.0
⅝	159.043	2012.9	⅝	177.893	2518.3	⅝	196.742	3080.3
¾	159.436	2022.8	¾	178.285	2529.4	¾	197.135	3092.6
⅞	159.829	2032.8	⅞	178.678	2540.6	⅞	197.528	3104.9
51	160.221	2042.8	57	179.071	2551.8	63	197.920	3117.2
⅛	160.614	2052.8	⅛	179.463	2563.0	⅛	198.313	3129.6
¼	161.007	2062.9	¼	179.856	2574.2	¼	198.706	3142.0
⅜	161.399	2073.0	⅜	180.249	2585.4	⅜	199.098	3154.5
½	161.792	2083.1	½	180.642	2596.7	½	199.491	3166.9
⅝	162.185	2093.2	⅝	181.034	2608.0	⅝	199.884	3179.4
¾	162.577	2103.3	¾	181.427	2619.4	¾	200.277	3191.9
⅞	162.970	2113.5	⅞	181.820	2630.7	⅞	200.669	3204.4

Circumferences and Areas of Circles

Diameter	Circumference	Area	Diameter	Circumference	Area	Diameter	Circumference	Area
64	201.062	3217.0	70	219.911	3848.5	76	238.761	4536.5
⅛	201.455	3229.6	⅛	220.304	3862.2	⅛	239.154	4551.4
¼	201.847	3242.2	¼	220.697	3876.0	¼	239.546	4566.4
⅜	202.240	3254.8	⅜	221.090	3889.8	⅜	239.939	4581.3
½	202.633	3267.5	½	221.482	3903.6	½	240.332	4596.3
⅝	203.025	3280.1	⅝	221.875	3917.5	⅝	240.725	4611.3
¾	203.418	3292.8	¾	222.268	3931.4	¾	241.117	4626.4
⅞	203.811	3305.6	⅞	222.660	3945.3	⅞	241.510	4641.5
65	204.204	3318.3	71	223.053	3959.2	77	241.903	4656.6
⅛	204.596	3331.1	⅛	223.446	3973.1	⅛	242.295	4671.8
¼	204.989	3343.9	¼	223.838	3987.1	¼	242.688	4686.9
⅜	205.382	3356.7	⅜	224.231	4001.1	⅜	243.081	4702.1
½	205.774	3369.6	½	224.624	4015.2	½	243.473	4717.3
⅝	206.167	3382.4	⅝	225.017	4029.2	⅝	243.866	4732.5
¾	206.560	3395.3	¾	225.409	4043.3	¾	244.259	4747.8
⅞	206.952	3408.2	⅞	225.802	4057.4	⅞	244.652	4763.1
66	207.345	3421.2	72	226.195	4071.5	78	245.044	4778.4
⅛	207.738	3434.2	⅛	226.587	4085.7	⅛	245.437	4793.7
¼	208.131	3447.2	¼	226.980	4099.8	¼	245.830	4809.0
⅜	208.523	3460.2	⅜	227.373	4114.0	⅜	246.222	4824.4
½	208.916	3473.2	½	227.765	4128.2	½	246.615	4839.8
⅝	209.309	3486.3	⅝	228.158	4142.5	⅝	247.008	4855.2
¾	209.701	3499.4	¾	228.551	4156.8	¾	247.400	4870.7
⅞	210.094	3512.5	⅞	228.944	4171.1	⅞	247.793	4886.2
67	210.487	3525.7	73	229.336	4185.4	79	248.186	4901.7
⅛	210.879	3538.8	⅛	229.729	4199.7	⅛	248.579	4917.2
¼	211.272	3552.0	¼	230.122	4214.1	¼	248.971	4932.7
⅜	211.665	3565.2	⅜	230.514	4228.5	⅜	249.364	4948.3
½	212.058	3578.5	½	230.907	4242.9	½	249.757	4963.9
⅝	212.450	3591.7	⅝	231.300	4257.4	⅝	250.149	4979.5
¾	212.843	3605.0	¾	231.692	4271.8	¾	250.542	4995.2
⅞	213.236	3618.3	⅞	232.085	4286.3	⅞	250.935	5010.9
68	213.628	3631.7	74	232.478	4300.8	80	251.327	5026.5
⅛	214.021	3645.0	⅛	232.871	4315.4	⅛	251.720	5042.3
¼	214.414	3658.4	¼	233.263	4329.9	¼	252.113	5058.0
⅜	214.806	3671.8	⅜	233.656	4344.5	⅜	252.506	5073.8
½	215.199	3685.3	½	234.049	4359.2	½	252.898	5089.6
⅝	215.592	3698.7	⅝	234.441	4373.8	⅝	253.291	5105.4
¾	215.984	3712.2	¾	234.834	4388.5	¾	253.684	5121.2
⅞	216.377	3725.7	⅞	235.227	4403.1	⅞	254.076	5137.1
69	216.770	3739.3	75	235.619	4417.9	81	254.469	5153.0
⅛	217.163	3752.8	⅛	236.012	4432.6	⅛	254.862	5168.9
¼	217.555	3766.4	¼	236.405	4447.4	¼	255.254	5184.9
⅜	217.948	3780.0	⅜	236.798	4462.2	⅜	255.647	5200.8
½	218.341	3793.7	½	237.190	4477.0	½	256.040	5216.8
⅝	218.733	3807.3	⅝	237.583	4491.8	⅝	256.433	5232.8
¾	219.126	3821.0	¾	237.976	4506.7	¾	256.825	5248.9
⅞	219.519	3834.7	⅞	238.368	4521.5	⅞	257.218	5264.9

Circumferences and Areas of Circles

Diameter	Circumference	Area	Diameter	Circumference	Area	Diameter	Circumference	Area
82	257.611	5281.0	88	276.460	6082.1	94	295.310	6939.8
⅛	258.003	5297.1	⅛	276.853	6099.4	⅛	295.702	6958.2
¼	258.396	5313.3	¼	277.246	6116.7	¼	296.095	6976.7
⅜	258.789	5329.4	⅜	277.638	6134.1	⅜	296.488	6995.3
½	259.181	5345.6	½	278.031	6151.4	½	296.881	7013.8
⅝	259.574	5361.8	⅝	278.424	6168.8	⅝	297.273	7032.4
¾	259.967	5378.1	¾	278.816	6186.2	¾	297.666	7051.0
⅞	260.359	5394.3	⅞	279.209	6203.7	⅞	298.059	7069.6
83	260.752	5410.6	89	279.602	6221.1	95	298.451	7088.2
⅛	261.145	5426.9	⅛	279.994	6238.6	⅛	298.844	7106.9
¼	261.538	5443.3	¼	280.387	6256.1	¼	299.237	7125.6
⅜	261.930	5459.6	⅜	280.780	6273.7	⅜	299.629	7144.3
½	262.323	5476.0	½	281.173	6291.2	½	300.022	7163.0
⅝	262.716	5492.4	⅝	281.565	6308.8	⅝	300.415	7181.8
¾	263.108	5508.8	¾	281.958	6326.4	¾	300.807	7200.6
⅞	263.501	5525.3	⅞	282.351	6344.1	⅞	301.200	7219.4
84	263.894	5541.8	90	282.743	6361.7	96	301.593	7238.2
⅛	264.286	5558.3	⅛	283.136	6379.4	⅛	301.986	7257.1
¼	264.679	5574.8	¼	283.529	6397.1	¼	302.378	7276.0
⅜	265.072	5591.4	⅜	283.921	6414.8	⅜	302.771	7294.9
½	265.465	5607.9	½	284.314	6432.6	½	303.164	7313.8
⅝	265.857	5624.5	⅝	284.707	6450.4	⅝	303.556	7332.8
¾	266.250	5641.2	¾	285.100	6468.2	¾	303.949	7351.8
⅞	266.643	5657.8	⅞	285.492	6486.0	⅞	304.342	7370.8
85	267.035	5674.5	91	285.885	6503.9	97	304.734	7389.8
⅛	267.428	5691.2	⅛	286.278	6521.8	⅛	305.127	7408.9
¼	267.821	5707.9	¼	286.670	6539.7	¼	305.520	7428.0
⅜	268.213	5724.7	⅜	287.063	6557.6	⅜	305.913	7447.1
½	268.606	5741.5	½	287.456	6575.5	½	306.305	7466.2
⅝	268.999	5758.3	⅝	287.848	6593.5	⅝	306.698	7485.3
¾	269.392	5775.1	¾	288.241	6611.5	¾	307.091	7504.5
⅞	269.784	5791.9	⅞	288.634	6629.6	⅞	307.483	7523.7
86	270.177	5808.8	92	289.027	6647.6	98	307.876	7543.0
⅛	270.570	5825.7	⅛	289.419	6665.7	⅛	308.269	7562.2
¼	270.962	5842.6	¼	289.812	6683.8	¼	308.661	7581.5
⅜	271.355	5859.6	⅜	290.205	6701.9	⅜	309.054	7600.8
½	271.748	5876.5	½	290.597	6720.1	½	309.447	7620.1
⅝	272.140	5893.5	⅝	290.990	6738.2	⅝	309.840	7639.5
¾	272.533	5910.6	¾	291.383	6756.4	¾	310.232	7658.9
⅞	272.926	5927.6	⅞	291.775	6774.7	⅞	310.625	7678.3
87	273.319	5944.7	93	292.168	6792.9	99	311.018	7697.7
⅛	273.711	5961.8	⅛	292.561	6811.2	⅛	311.410	7717.1
¼	274.104	5978.9	¼	292.954	6829.5	¼	311.803	7736.6
⅜	274.497	5996.0	⅜	293.346	6847.8	⅜	312.196	7756.1
½	274.889	6013.2	½	293.739	6866.1	½	312.588	7775.6
⅝	275.282	6030.4	⅝	294.132	6884.5	⅝	312.981	7795.2
¾	275.675	6047.6	¾	294.524	6902.9	¾	313.374	7814.8
⅞	276.067	6064.9	⅞	294.917	6921.3	⅞	313.767	7834.4

Circumferences and Areas of Circles

Diameter	Circumference	Area	Diameter	Circumference	Area	Diameter	Circumference	Area
100	314.16	7,854.0	150	471.24	17,671.5	200	628.32	31,415.9
101	317.30	8,011.8	151	474.38	17,907.9	201	631.46	31,730.9
102	320.44	8,171.3	152	477.52	18,145.8	202	634.60	32,047.4
103	323.58	8,332.3	153	480.66	18,385.4	203	637.74	32,365.5
104	326.73	8,494.9	154	483.81	18,626.5	204	640.88	32,685.1
105	329.87	8,659.0	155	486.95	18,869.2	205	644.03	33,006.4
106	333.01	8,824.7	156	490.09	19,113.4	206	647.17	33,329.2
107	336.15	8,992.0	157	493.23	19,359.3	207	650.31	33,653.5
108	339.29	9,160.9	158	496.37	19,606.7	208	653.45	33,979.5
109	342.43	9,331.3	159	499.51	19,855.7	209	656.59	34,307.0
110	345.58	9,503.3	160	502.65	20,106.2	210	659.73	34,636.1
111	348.72	9,676.9	161	505.80	20,358.3	211	662.88	34,966.7
112	351.86	9,852.0	162	508.94	20,612.0	212	666.02	35,298.9
113	355.00	10,028.7	163	512.08	20,867.2	213	669.16	35,632.7
114	358.14	10,207.0	164	515.22	21,124.1	214	672.30	35,968.1
115	361.28	10,386.9	165	518.36	21,382.5	215	675.44	36,305.0
116	364.42	10,568.3	166	521.50	21,642.4	216	678.58	36,643.5
117	367.57	10,751.3	167	524.65	21,904.0	217	681.73	36,983.6
118	370.71	10,935.9	168	527.79	22,167.1	218	684.87	37,325.3
119	373.85	11,122.0	169	530.93	22,431.8	219	688.01	37,668.5
120	376.99	11,309.7	170	534.07	22,698.0	220	691.15	38,013.3
121	380.13	11,499.0	171	537.21	22,965.8	221	694.29	38,359.6
122	383.27	11,689.9	172	540.35	23,235.2	222	697.43	38,707.6
123	386.42	11,882.3	173	543.50	23,506.2	223	700.58	39,057.1
124	389.56	12,076.3	174	546.64	23,778.7	224	703.72	39,408.1
125	392.70	12,271.8	175	549.78	24,052.8	225	706.86	39,760.8
126	395.84	12,469.0	176	552.92	24,328.5	226	710.00	40,115.0
127	398.98	12,667.7	177	556.06	24,605.7	227	713.14	40,470.8
128	402.12	12,868.0	178	559.20	24,884.6	228	716.28	40,828.1
129	405.27	13,069.8	179	562.35	25,164.9	229	719.42	41,187.1
130	408.41	13,273.2	180	565.49	25,446.9	230	722.57	41,547.6
131	411.55	13,478.2	181	568.63	25,730.4	231	725.71	41,909.6
132	414.69	13,684.8	182	571.77	26,015.5	232	728.85	42,273.3
133	417.83	13,892.9	183	574.91	26,302.2	233	731.99	42,638.5
134	420.97	14,102.6	184	578.05	26,590.4	234	735.13	43,005.3
135	424.12	14,313.9	185	581.19	26,880.3	235	738.27	43,373.6
136	427.26	14,526.7	186	584.34	27,171.6	236	741.42	43,743.5
137	430.40	14,741.1	187	587.48	27,464.6	237	744.56	44,115.0
138	433.54	14,957.1	188	590.62	27,759.1	238	747.70	44,488.1
139	436.68	15,174.7	189	593.76	28,055.2	239	750.84	44,862.7
140	439.82	15,393.8	190	596.90	28,352.9	240	753.98	45,238.9
141	442.96	15,614.5	191	600.04	28,652.1	241	757.12	45,616.7
142	446.11	15,836.8	192	603.19	28,952.9	242	760.27	45,996.1
143	449.25	16,060.6	193	606.33	29,255.3	243	763.41	46,377.0
144	452.39	16,286.0	194	609.47	29,559.2	244	766.55	46,759.5
145	455.53	16,513.0	195	612.61	29,864.8	245	769.69	47,143.5
146	458.67	16,741.5	196	615.75	30,171.9	246	772.83	47,529.2
147	461.81	16,971.7	197	618.89	30,480.5	247	775.97	47,916.4
148	464.96	17,203.4	198	622.04	30,790.7	248	779.11	48,305.1
149	468.10	17,436.6	199	625.18	31,102.6	249	782.26	48,695.5

Circumferences and Areas of Circles

Diameter	Circumference	Area	Diameter	Circumference	Area	Diameter	Circumference	Area
250	785.40	49,087.4	300	942.48	70,685.8	350	1099.56	96,211.3
251	788.54	49,480.9	301	945.62	71,157.9	351	1102.70	96,761.8
252	791.68	49,875.9	302	948.76	71,631.5	352	1105.84	97,314.0
253	794.82	50,272.6	303	951.90	72,106.6	353	1108.98	97,867.7
254	797.96	50,670.7	304	955.04	72,583.4	354	1112.12	98,423.0
255	801.11	51,070.5	305	958.19	73,061.7	355	1115.27	98,979.8
256	804.25	51,471.9	306	961.33	73,541.5	356	1118.41	99,538.2
257	807.39	51,874.8	307	964.47	74,023.0	357	1121.55	100,098
258	810.53	52,279.2	308	967.61	74,506.0	358	1124.69	100,660
259	813.67	52,685.3	309	970.75	74,990.6	359	1127.83	101,223
260	816.81	53,092.9	310	973.89	75,476.8	360	1130.97	101,788
261	819.96	53,502.1	311	977.04	75,964.5	361	1134.11	102,354
262	823.10	53,912.9	312	980.18	76,453.8	362	1137.26	102,922
263	826.24	54,325.2	313	983.32	76,944.7	363	1140.40	103,491
264	829.38	54,739.1	314	986.46	77,437.1	364	1143.54	104,062
265	832.52	55,154.6	315	989.60	77,931.1	365	1146.68	104,635
266	835.66	55,571.6	316	992.74	78,426.7	366	1149.82	105,209
267	838.81	55,990.2	317	995.88	78,923.9	367	1152.96	105,784
268	841.95	56,410.4	318	999.03	79,422.6	368	1156.11	106,362
269	845.09	56,832.2	319	1002.17	79,922.9	369	1159.25	106,941
270	848.23	57,255.5	320	1005.31	80,424.8	370	1162.39	107,521
271	851.37	57,680.4	321	1008.45	80,928.2	371	1165.53	108,103
272	854.51	58,106.9	322	1011.59	81,433.2	372	1168.67	108,687
273	857.65	58,534.9	323	1014.73	81,939.8	373	1171.81	109,272
274	860.80	58,964.6	324	1017.88	82,448.0	374	1174.96	109,858
275	863.94	59,395.7	325	1021.02	82,957.7	375	1178.10	110,447
276	867.08	59,828.5	326	1024.16	83,469.0	376	1181.24	111,036
277	870.22	60,262.8	327	1027.30	83,981.8	377	1184.38	111,628
278	873.36	60,698.7	328	1030.44	84,496.3	378	1187.52	112,221
279	876.50	61,136.2	329	1033.58	85,012.3	379	1190.66	112,815
280	879.65	61,575.2	330	1036.73	85,529.9	380	1193.81	113,411
281	882.79	62,015.8	331	1039.87	86,049.0	381	1196.95	114,009
282	885.93	62,458.0	332	1043.01	86,569.7	382	1200.09	114,608
283	889.07	62,901.8	333	1046.15	87,092.0	383	1203.23	115,209
284	892.21	63,347.1	334	1049.29	87,615.9	384	1206.37	115,812
285	895.35	63,794.0	335	1052.43	88,141.3	385	1209.51	116,416
286	898.50	64,242.4	336	1055.58	88,668.3	386	1212.65	117,021
287	901.64	64,692.5	337	1058.72	89,196.9	387	1215.80	117,628
288	904.78	65,144.1	338	1061.86	89,727.0	388	1218.94	118,237
289	907.92	65,597.2	339	1065.00	90,258.7	389	1222.08	118,847
290	911.06	66,052.0	340	1068.14	90,792.0	390	1225.22	119,459
291	914.20	66,508.3	341	1071.28	91,326.9	391	1228.36	120,072
292	917.35	66,966.2	342	1074.42	91,863.3	392	1231.50	120,687
293	920.49	67,425.6	343	1077.57	92,401.3	393	1234.65	121,304
294	923.63	67,886.7	344	1080.71	92,940.9	394	1237.79	121,922
295	926.77	68,349.3	345	1083.85	93,482.0	395	1240.93	122,542
296	929.91	68,813.4	346	1086.99	94,024.7	396	1244.07	123,163
297	933.05	69,279.2	347	1090.13	94,569.0	397	1247.21	123,786
298	936.19	69,746.5	348	1093.27	95,114.9	398	1250.35	124,410
299	939.34	70,215.4	349	1096.42	95,662.3	399	1253.50	125,036

MATHEMATICS

Circumferences and Areas of Circles

Diameter	Circumference	Area	Diameter	Circumference	Area	Diameter	Circumference	Area
400	1256.64	125,664	450	1413.72	159,043	500	1570.80	196,350
401	1259.78	126,293	451	1416.86	159,751	501	1573.94	197,136
402	1262.92	126,923	452	1420.00	160,460	502	1577.08	197,923
403	1266.06	127,556	453	1423.14	161,171	503	1580.22	198,713
404	1269.20	128,190	454	1426.28	161,883	504	1583.36	199,504
405	1272.35	128,825	455	1429.42	162,597	505	1586.50	200,296
406	1275.49	129,462	456	1432.57	163,313	506	1589.65	201,090
407	1278.63	130,100	457	1435.71	164,030	507	1592.79	201,886
408	1281.77	130,741	458	1438.85	164,748	508	1595.93	202,683
409	1284.91	131,382	459	1441.99	165,468	509	1599.07	203,482
410	1288.05	132,025	460	1445.13	166,190	510	1602.21	204,282
411	1291.19	132,670	461	1448.27	166,914	511	1605.35	205,084
412	1294.34	133,317	462	1451.42	167,639	512	1608.50	205,887
413	1297.48	133,965	463	1454.56	168,365	513	1611.64	206,692
414	1300.62	134,614	464	1457.70	169,093	514	1614.78	207,499
415	1303.76	135,265	465	1460.84	169,823	515	1617.92	208,307
416	1306.90	135,918	466	1463.98	170,554	516	1621.06	209,117
417	1310.04	136,572	467	1467.12	171,287	517	1624.20	209,928
418	1313.19	137,228	468	1470.27	172,021	518	1627.34	210,741
419	1316.33	137,885	469	1473.41	172,757	519	1630.49	211,556
420	1319.47	138,544	470	1476.55	173,494	520	1633.63	212,372
421	1322.61	139,205	471	1479.69	174,234	521	1636.77	213,189
422	1325.75	139,867	472	1482.83	174,974	522	1639.91	214,008
423	1328.89	140,531	473	1485.97	175,716	523	1643.05	214,829
424	1332.04	141,196	474	1489.11	176,460	524	1646.19	215,651
425	1335.18	141,863	475	1492.26	177,205	525	1649.34	216,475
426	1338.32	142,531	476	1495.40	177,952	526	1652.48	217,301
427	1341.46	143,201	477	1498.54	178,701	527	1655.62	218,128
428	1344.60	143,872	478	1501.68	179,451	528	1658.76	218,956
429	1347.74	144,545	479	1504.82	180,203	529	1661.90	219,787
430	1350.88	145,220	480	1507.96	180,956	530	1665.04	220,618
431	1354.03	145,896	481	1511.11	181,711	531	1668.19	221,452
432	1357.17	146,574	482	1514.25	182,467	532	1671.33	222,287
433	1360.31	147,254	483	1517.39	183,225	533	1674.47	223,123
434	1363.45	147,934	484	1520.53	183,984	534	1677.61	223,961
435	1366.59	148,617	485	1523.67	184,745	535	1680.75	224,801
436	1369.73	149,301	486	1526.81	185,508	536	1683.89	225,642
437	1372.88	149,987	487	1529.96	186,272	537	1687.04	226,484
438	1376.02	150,674	488	1533.10	187,038	538	1690.18	227,329
439	1379.16	151,363	489	1536.24	187,805	539	1693.32	228,175
440	1382.30	152,053	490	1539.38	188,574	540	1696.46	229,022
441	1385.44	152,745	491	1542.52	189,345	541	1699.60	229,871
442	1388.58	153,439	492	1545.66	190,117	542	1702.74	230,722
443	1391.73	154,134	493	1548.81	190,890	543	1705.88	231,574
444	1394.87	154,830	494	1551.95	191,665	544	1709.03	232,428
445	1398.01	155,528	495	1555.09	192,442	545	1712.17	233,283
446	1401.15	156,228	496	1558.23	193,221	546	1715.31	234,140
447	1404.29	156,930	497	1561.37	194,000	547	1718.45	234,998
448	1407.43	157,633	498	1564.51	194,782	548	1721.59	235,858
449	1410.58	158,337	499	1567.65	195,565	549	1724.73	236,720

Circumferences and Areas of Circles

Diameter	Circumference	Area	Diameter	Circumference	Area	Diameter	Circumference	Area
550	1727.88	237,583	600	1884.96	282,743	650	2042.04	331,831
551	1731.02	238,448	601	1888.10	283,687	651	2045.18	332,853
552	1734.16	239,314	602	1891.24	284,631	652	2048.32	333,876
553	1737.30	240,182	603	1894.38	285,578	653	2051.46	334,901
554	1740.44	241,051	604	1897.52	286,526	654	2054.60	335,927
555	1743.58	241,922	605	1900.66	287,475	655	2057.74	336,955
556	1746.73	242,795	606	1903.81	288,426	656	2060.88	337,985
557	1749.87	243,669	607	1906.95	289,379	657	2064.03	339,016
558	1753.01	244,545	608	1910.09	290,333	658	2067.17	340,049
559	1756.15	245,422	609	1913.23	291,289	659	2070.31	341,083
560	1759.29	246,301	610	1916.37	292,247	660	2073.45	342,119
561	1762.43	247,181	611	1919.51	293,206	661	2076.59	343,157
562	1765.58	248,063	612	1922.65	294,166	662	2079.73	344,196
563	1768.72	248,947	613	1925.80	295,128	663	2082.88	345,237
564	1771.86	249,832	614	1928.94	296,092	664	2086.02	346,279
565	1775.00	250,719	615	1932.08	297,057	665	2089.16	347,323
566	1778.14	251,607	616	1935.22	298,024	666	2092.30	348,368
567	1781.28	252,497	617	1938.36	298,992	667	2095.44	349,415
568	1784.42	253,388	618	1941.50	299,962	668	2098.58	350,464
569	1787.57	254,281	619	1944.65	300,934	669	2101.73	351,514
570	1790.71	255,176	620	1947.79	301,907	670	2104.87	352,565
571	1793.85	256,072	621	1950.93	302,882	671	2108.01	353,618
572	1796.99	256,970	622	1954.07	303,858	672	2111.15	354,673
573	1800.13	257,869	623	1957.21	304,836	673	2114.29	355,730
574	1803.27	258,770	624	1960.35	305,815	674	2117.43	356,788
575	1806.42	259,672	625	1963.50	306,796	675	2120.58	357,847
576	1809.56	260,576	626	1966.64	307,779	676	2123.72	358,908
577	1812.70	261,482	627	1969.78	308,763	677	2126.86	359,971
578	1815.84	262,389	628	1972.92	309,748	678	2130.00	361,035
579	1818.98	263,298	629	1976.06	310,736	679	2133.14	362,101
580	1822.12	264,208	630	1979.20	311,725	680	2136.28	363,168
581	1825.27	265,120	631	1982.34	312,715	681	2139.42	364,237
582	1828.41	266,033	632	1985.49	313,707	682	2142.57	365,308
583	1831.55	266,948	633	1988.63	314,700	683	2145.71	366,380
584	1834.69	267,865	634	1991.77	315,696	684	2148.85	367,453
585	1837.83	268,783	635	1994.91	316,692	685	2151.99	368,528
586	1840.97	269,703	636	1998.05	317,690	686	2155.13	369,605
587	1844.11	270,624	637	2001.19	318,690	687	2158.27	370,684
588	1847.26	271,547	638	2004.34	319,692	688	2161.42	371,764
589	1850.40	272,471	639	2007.48	320,695	689	2164.56	372,845
590	1853.54	273,397	640	2010.62	321,699	690	2167.70	373,928
591	1856.68	274,325	641	2013.76	322,705	691	2170.84	375,013
592	1859.82	275,254	642	2016.90	323,713	692	2173.98	376,099
593	1862.96	276,184	643	2020.04	324,722	693	2177.12	377,187
594	1866.11	277,117	644	2023.19	325,733	694	2180.27	378,276
595	1869.25	278,051	645	2026.33	326,745	695	2183.41	379,367
596	1872.39	278,986	646	2029.47	327,759	696	2186.55	380,459
597	1875.53	279,923	647	2032.61	328,775	697	2189.69	381,553
598	1878.67	280,862	648	2035.75	329,792	698	2192.83	382,649
599	1881.81	281,802	649	2038.89	330,810	699	2195.97	383,746

MATHEMATICS

Circumferences and Areas of Circles

Diameter	Circumference	Area	Diameter	Circumference	Area	Diameter	Circumference	Area
700	2199.11	384,845	750	2356.19	441,786	800	2513.27	502,655
701	2202.26	385,945	751	2359.34	442,965	801	2516.42	503,912
702	2205.40	387,047	752	2362.48	444,146	802	2519.56	505,171
703	2208.54	388,151	753	2365.62	445,328	803	2522.70	506,432
704	2211.68	389,256	754	2368.76	446,511	804	2525.84	507,694
705	2214.82	390,363	755	2371.90	447,697	805	2528.98	508,958
706	2217.96	391,471	756	2375.04	448,883	806	2532.12	510,223
707	2221.11	392,580	757	2378.19	450,072	807	2535.27	511,490
708	2224.25	393,692	758	2381.33	451,262	808	2538.41	512,758
709	2227.39	394,805	759	2384.47	452,453	809	2541.55	514,028
710	2230.53	395,919	760	2387.61	453,646	810	2544.69	515,300
711	2233.67	397,035	761	2390.75	454,841	811	2547.83	516,573
712	2236.81	398,153	762	2393.89	456,037	812	2550.97	517,848
713	2239.96	399,272	763	2397.04	457,234	813	2554.11	519,124
714	2243.10	400,393	764	2400.18	458,434	814	2557.26	520,402
715	2246.24	401,515	765	2403.32	459,635	815	2560.40	521,681
716	2249.38	402,639	766	2406.46	460,837	816	2563.54	522,962
717	2252.52	403,765	767	2409.60	462,041	817	2566.68	524,245
718	2255.66	404,892	768	2412.74	463,247	818	2569.82	525,529
719	2258.81	406,020	769	2415.88	464,454	819	2572.96	526,814
720	2261.95	407,150	770	2419.03	465,663	820	2576.11	528,102
721	2265.09	408,282	771	2422.17	466,873	821	2579.25	529,391
722	2268.23	409,415	772	2425.31	468,085	822	2582.39	530,681
723	2271.37	410,550	773	2428.45	469,298	823	2585.53	531,973
724	2274.51	411,687	774	2431.59	470,513	824	2588.67	533,267
725	2277.65	412,825	775	2434.73	471,730	825	2591.81	534,562
726	2280.80	413,965	776	2437.88	472,948	826	2594.96	535,858
727	2283.94	415,106	777	2441.02	474,168	827	2598.10	537,157
728	2287.08	416,248	778	2444.16	475,389	828	2601.24	538,456
729	2290.22	417,393	779	2447.30	476,612	829	2604.38	539,758
730	2293.36	418,539	780	2450.44	477,836	830	2607.52	541,061
731	2296.50	419,686	781	2453.58	479,062	831	2610.66	542,365
732	2299.65	420,835	782	2456.73	480,290	832	2613.81	543,671
733	2302.79	421,986	783	2459.87	481,519	833	2616.95	544,979
734	2305.93	423,138	784	2463.01	482,750	834	2620.09	546,288
735	2309.07	424,292	785	2466.15	483,982	835	2623.23	547,599
736	2312.21	425,447	786	2469.29	485,216	836	2626.37	548,912
737	2315.35	426,604	787	2472.43	486,451	837	2629.51	550,226
738	2318.50	427,762	788	2475.58	487,688	838	2632.65	551,541
739	2321.64	428,922	789	2478.72	488,927	839	2635.80	552,858
740	2324.78	430,084	790	2481.86	490,167	840	2638.94	554,177
741	2327.92	431,247	791	2485.00	491,409	841	2642.08	555,497
742	2331.06	432,412	792	2488.14	492,652	842	2645.22	556,819
743	2334.20	433,578	793	2491.28	493,897	843	2648.36	558,142
744	2337.34	434,746	794	2494.42	495,143	844	2651.50	559,467
745	2340.49	435,916	795	2497.57	496,391	845	2654.65	560,794
746	2343.63	437,087	796	2500.71	497,641	846	2657.79	562,122
747	2346.77	438,259	797	2503.85	498,892	847	2660.93	563,452
748	2349.91	439,433	798	2506.99	500,145	848	2664.07	564,783
749	2353.05	440,609	799	2510.13	501,399	849	2667.21	566,116

Circumferences and Areas of Circles

Diameter	Circumference	Area	Diameter	Circumference	Area	Diameter	Circumference	Area
850	2670.35	567,450	900	2827.43	636,173	950	2984.51	708,822
851	2673.50	568,786	901	2830.57	637,587	951	2987.65	710,315
852	2676.64	570,124	902	2833.72	639,003	952	2990.80	711,809
853	2679.78	571,463	903	2836.86	640,421	953	2993.94	713,306
854	2682.92	572,803	904	2840.00	641,840	954	2997.08	714,803
855	2686.06	574,146	905	2843.14	643,261	955	3000.22	716,303
856	2689.20	575,490	906	2846.28	644,683	956	3003.36	717,804
857	2692.34	576,835	907	2849.42	646,107	957	3006.50	719,306
858	2695.49	578,182	908	2852.57	647,533	958	3009.65	720,810
859	2698.63	579,530	909	2855.71	648,960	959	3012.79	722,316
860	2701.77	580,880	910	2858.85	650,388	960	3015.93	723,823
861	2704.91	582,232	911	2861.99	651,818	961	3019.07	725,332
862	2708.05	583,585	912	2865.13	653,250	962	3022.21	726,842
863	2711.19	584,940	913	2868.27	654,684	963	3025.35	728,354
864	2714.34	586,297	914	2871.42	656,118	964	3028.50	729,867
865	2717.48	587,655	915	2874.56	657,555	965	3031.64	731,382
866	2720.62	589,014	916	2877.70	658,993	966	3034.78	732,899
867	2723.76	590,375	917	2880.84	660,433	967	3037.92	734,417
868	2726.90	591,738	918	2883.98	661,874	968	3041.06	735,937
869	2730.04	593,102	919	2887.12	663,317	969	3044.20	737,458
870	2733.19	594,468	920	2890.27	664,761	970	3047.34	738,981
871	2736.33	595,835	921	2893.41	666,207	971	3050.49	740,506
872	2739.47	597,204	922	2896.55	667,654	972	3053.63	742,032
873	2742.61	598,575	923	2899.69	669,103	973	3056.77	743,559
874	2745.75	599,947	924	2902.83	670,554	974	3059.91	745,088
875	2748.89	601,320	925	2905.97	672,006	975	3063.05	746,619
876	2752.04	602,696	926	2909.11	673,460	976	3066.19	748,151
877	2755.18	604,073	927	2912.26	674,915	977	3069.34	749,685
878	2758.32	605,451	928	2915.40	676,372	978	3072.48	751,221
879	2761.46	606,831	929	2918.54	677,831	979	3075.62	752,758
880	2764.60	608,212	930	2921.68	679,291	980	3078.76	754,296
881	2767.74	609,595	931	2924.82	680,752	981	3081.90	755,837
882	2770.88	610,980	932	2927.96	682,216	982	3085.04	757,378
883	2774.03	612,366	933	2931.11	683,680	983	3088.19	758,922
884	2777.17	613,754	934	2934.25	685,147	984	3091.33	760,466
885	2780.31	615,143	935	2937.39	686,615	985	3094.47	762,013
886	2783.45	616,534	936	2940.53	688,084	986	3097.61	763,561
887	2786.59	617,927	937	2943.67	689,555	987	3100.75	765,111
888	2789.73	619,321	938	2946.81	691,028	988	3103.89	766,662
889	2792.88	620,717	939	2949.96	692,502	989	3107.04	768,214
890	2796.02	622,114	940	2953.10	693,978	990	3110.18	769,769
891	2799.16	623,513	941	2956.24	695,455	991	3113.32	771,325
892	2802.30	624,913	942	2959.38	696,934	992	3116.46	772,882
893	2805.44	626,315	943	2962.52	698,415	993	3119.60	774,441
894	2808.58	627,718	944	2965.66	699,897	994	3122.74	776,002
895	2811.73	629,124	945	2968.81	701,380	995	3125.88	777,564
896	2814.87	630,530	946	2971.95	702,865	996	3129.03	779,128
897	2818.01	631,938	947	2975.09	704,352	997	3132.17	780,693
898	2821.15	633,348	948	2978.23	705,840	998	3135.31	782,260
899	2824.29	634,760	949	2981.37	707,330	999	3138.45	783,828

Exact and Approximate Formulas for Circular Segment Area. — The areas of circular segments given in the table, pages 73 and 74, are based on the exact formula: $A = \frac{1}{2}[rl - c(r - h)]$. This and other formulas for segments are given on page 69.

In many cases, notably in calculating the area of an arch in construction work, only the length of the chord c and the height of the segment h may be known or can be measured directly. In such cases approximate formulas for obtaining A directly in terms of c and h are useful since they eliminate the need to first calculate the radius r or the angle θ in finding the area as would be the case if the exact formula is used.

An approximate formula which gives an error of about 0.1 per cent or less for circular segments ranging almost up to a semi-circle is:

$$A = \frac{4h^2}{3} \sqrt{\frac{c^2}{4h^2} + 0.392} \tag{1}$$

An approximate formula which gives an error of about 0.1 per cent when the ratio of h to c is ⅓ or less is:

$$A = \frac{2ch}{3} + \frac{h^3}{2c} \tag{2}$$

An approximate formula which is more accurate than Formula (2) for segments close to a semi-circle, i.e. with a ratio of h to c of from 0.454 to 0.500, is:

$$A = \frac{h^3}{2c} + 0.6604ch \tag{3}$$

Lengths of Chords for Spacing off the Circumference of Circles

On the following pages are given tables of the lengths of chords for spacing off the circumference of circles. The object of these tables is to make possible the division of the periphery into a number of equal parts without trials with the dividers. The first table is calculated for circles having a diameter equal to 1. For circles of other diameters, the length of chord given in the table should be multiplied by the diameter of the circle. This first table may be used by tool-makers when setting "buttons" in circular formation. Assume that it is required to divide the periphery of a circle of 20 inches diameter into thirty-two equal parts. From the table the length of the chord is found to be 0.098017 inch, if the diameter of the circle were 1 inch. With a diameter of 20 inches the length of the chord for one division would be 20 × 0.098017 = 1.9603 inches. Another example in metric units: For a 100 millimeter diameter requiring 5 equal divisions, the length of the chord for one division would be 100 × 0.587785 = 58.7785 millimeters.

The two following pages give an additional table for the spacing off of circles, the table, in this case, being worked out for diameters from 1/16 inch to 14 inches. As an example, assume that it is required to divide a circle having a diameter of 6½ inches into seven equal parts. Find first, in the column headed "6" and in line with 7 divisions, the length of the chord for a 6-inch circle, which is 2.604 inches. Then find the length of the chord for a ½-inch diameter circle, 7 divisions, which is 0.217. The sum of these two values, 2.604 + 0.217 = 2.821 inches, is the length of the chord required for spacing off the circumference of a 6½-inch circle into seven equal divisions.

As another example, assume that it is required to divide a circle having a diameter of 9²³⁄₃₂ inches into 15 equal divisions. First find the length of the chord for a 9-inch circle, which is 1.871 inch. The length of the chord for a 2³⁄₃₂-inch circle can easily be estimated from the table by taking the value that is exactly between those given for 1/16 and ¾ inch. The value for 1/16 inch is 0.143, and for ¾ inch, 0.156. For 2³⁄₃₂, the value would be 0.150. Then, 1.871 + 0.150 = 2.021 inches.

Lengths of Chords for Spacing Off the Circumference of Circles with a Diameter Equal to 1

For circles of other diameters multiply length given in table by diameter of circle. (English or metric units)

No. of Spaces	Length of Chord	No. of Spaces	Length of Chord	No. of Spaces	Length of Chord	No. of Spaces	Length of Chord
3	0.866025	51	0.061561	99	0.031728	147	0.021370
4	0.707107	52	0.060378	100	0.031411	148	0.021225
5	0.587785	53	0.059241	101	0.031100	149	0.021083
6	0.500000	54	0.058145	102	0.030795	150	0.020942
7	0.433884	55	0.057089	103	0.030496	151	0.020804
8	0.382683	56	0.056070	104	0.030203	152	0.020667
9	0.342020	57	0.055088	105	0.029915	153	0.020532
10	0.309017	58	0.054139	106	0.029633	154	0.020399
11	0.281733	59	0.053222	107	0.029356	155	0.020267
12	0.258819	60	0.052336	108	0.029085	156	0.020137
13	0.239316	61	0.051479	109	0.028818	157	0.020009
14	0.222521	62	0.050649	110	0.028556	158	0.019882
15	0.207912	63	0.049846	111	0.028299	159	0.019757
16	0.195090	64	0.049068	112	0.028046	160	0.019634
17	0.183750	65	0.048313	113	0.027798	161	0.019512
18	0.173648	66	0.047582	114	0.027554	162	0.019391
19	0.164595	67	0.046872	115	0.027315	163	0.019272
20	0.156434	68	0.046183	116	0.027079	164	0.019155
21	0.149042	69	0.045515	117	0.026848	165	0.019039
22	0.142315	70	0.044865	118	0.026621	166	0.018924
23	0.136167	71	0.044233	119	0.026397	167	0.018811
24	0.130526	72	0.043619	120	0.026177	168	0.018699
25	0.125333	73	0.043022	121	0.025961	169	0.018588
26	0.120537	74	0.042441	122	0.025748	170	0.018479
27	0.116093	75	0.041876	123	0.025539	171	0.018371
28	0.111964	76	0.041325	124	0.025333	172	0.018264
29	0.108119	77	0.040789	125	0.025130	173	0.018158
30	0.104528	78	0.040266	126	0.024931	174	0.018054
31	0.101168	79	0.039757	127	0.024734	175	0.017951
32	0.098017	80	0.039260	128	0.024541	176	0.017849
33	0.095056	81	0.038775	129	0.024351	177	0.017748
34	0.092268	82	0.038303	130	0.024164	178	0.017648
35	0.089639	83	0.037841	131	0.023979	179	0.017550
36	0.087156	84	0.037391	132	0.023798	180	0.017452
37	0.084806	85	0.036951	133	0.023619	181	0.017356
38	0.082579	86	0.036522	134	0.023443	182	0.017261
39	0.080467	87	0.036102	135	0.023269	183	0.017166
40	0.078459	88	0.035692	136	0.023098	184	0.017073
41	0.076549	89	0.035291	137	0.022929	185	0.016981
42	0.074730	90	0.034899	138	0.022763	186	0.016889
43	0.072995	91	0.034516	139	0.022599	187	0.016799
44	0.071339	92	0.034141	140	0.022438	188	0.016710
45	0.069756	93	0.033774	141	0.022279	189	0.016621
46	0.068242	94	0.033415	142	0.022122	190	0.016534
47	0.066793	95	0.033063	143	0.021967	191	0.016447
48	0.065403	96	0.032719	144	0.021815	192	0.016362
49	0.064070	97	0.032382	145	0.021664	193	0.016277
50	0.062791	98	0.032052	146	0.021516	194	0.016193

Table for Spacing Off the Circumference of Circles
(see page 50 for explantory matter.)

Diameter of Circle to be Spaced Off — Length of Chord

No. of Divisions	Degrees in Arc	1/16	1/8	3/16	1/4	5/16	3/8	7/16	1/2	9/16	5/8	11/16	3/4	13/16	7/8	15/16
3	120	0.054	0.108	0.162	0.217	0.271	0.325	0.379	0.433	0.487	0.541	0.595	0.650	0.704	0.758	0.812
4	90	0.044	0.088	0.133	0.177	0.221	0.265	0.309	0.354	0.398	0.442	0.486	0.530	0.575	0.619	0.663
5	72	0.037	0.073	0.110	0.147	0.184	0.220	0.257	0.294	0.331	0.367	0.404	0.441	0.478	0.514	0.551
6	60	0.031	0.063	0.094	0.125	0.156	0.188	0.219	0.250	0.281	0.313	0.344	0.375	0.406	0.438	0.469
7	51 3/7	0.027	0.054	0.081	0.108	0.136	0.163	0.190	0.217	0.244	0.271	0.298	0.325	0.353	0.386	0.407
8	45	0.024	0.048	0.072	0.096	0.120	0.144	0.167	0.191	0.215	0.239	0.263	0.287	0.311	0.335	0.359
9	40	0.021	0.043	0.064	0.086	0.107	0.128	0.150	0.171	0.192	0.214	0.235	0.257	0.278	0.299	0.321
10	36	0.019	0.039	0.058	0.077	0.097	0.116	0.135	0.155	0.174	0.193	0.212	0.232	0.251	0.270	0.290
11	32 8/11	0.018	0.035	0.053	0.070	0.088	0.106	0.123	0.141	0.158	0.176	0.194	0.211	0.229	0.247	0.264
12	30	0.016	0.032	0.049	0.065	0.081	0.097	0.113	0.129	0.146	0.162	0.178	0.194	0.210	0.226	0.243
13	27 9/13	0.015	0.030	0.045	0.060	0.075	0.090	0.105	0.120	0.135	0.150	0.165	0.179	0.194	0.209	0.224
14	25 5/7	0.014	0.028	0.042	0.056	0.070	0.083	0.097	0.111	0.125	0.139	0.153	0.167	0.181	0.195	0.209
15	24	0.013	0.026	0.039	0.052	0.065	0.078	0.091	0.104	0.117	0.130	0.143	0.156	0.169	0.182	0.195
16	22 1/2	0.012	0.024	0.037	0.049	0.061	0.073	0.085	0.098	0.110	0.122	0.134	0.146	0.159	0.171	0.183
17	21 3/17	0.011	0.023	0.034	0.046	0.057	0.069	0.080	0.092	0.103	0.115	0.126	0.138	0.149	0.161	0.172
18	20	0.011	0.022	0.033	0.043	0.054	0.065	0.076	0.087	0.098	0.109	0.119	0.130	0.141	0.152	0.163
19	18 18/19	0.010	0.021	0.031	0.041	0.051	0.062	0.072	0.082	0.093	0.103	0.113	0.123	0.134	0.144	0.154
20	18	0.010	0.020	0.029	0.039	0.049	0.059	0.068	0.078	0.088	0.098	0.108	0.117	0.127	0.137	0.147
21	17 1/7	0.009	0.019	0.028	0.037	0.047	0.056	0.065	0.075	0.084	0.093	0.102	0.112	0.121	0.130	0.140
22	16 4/11	0.009	0.018	0.027	0.036	0.044	0.053	0.062	0.071	0.080	0.089	0.098	0.107	0.116	0.125	0.133
23	15 15/23	0.009	0.017	0.026	0.034	0.043	0.051	0.060	0.068	0.077	0.085	0.094	0.102	0.111	0.119	0.128
24	15	0.008	0.016	0.024	0.033	0.041	0.049	0.057	0.065	0.073	0.082	0.090	0.098	0.106	0.114	0.122
25	14 2/5	0.008	0.016	0.023	0.031	0.039	0.047	0.055	0.063	0.070	0.078	0.086	0.094	0.102	0.110	0.117
26	13 11/13	0.008	0.015	0.023	0.030	0.038	0.045	0.053	0.060	0.068	0.075	0.083	0.090	0.098	0.105	0.113
28	12 6/7	0.007	0.014	0.021	0.028	0.035	0.042	0.049	0.056	0.063	0.070	0.077	0.084	0.091	0.098	0.105
30	12	0.007	0.013	0.020	0.026	0.033	0.039	0.046	0.052	0.059	0.065	0.072	0.078	0.085	0.091	0.098
32	11 1/4	0.006	0.012	0.018	0.025	0.031	0.037	0.043	0.049	0.055	0.061	0.067	0.074	0.080	0.086	0.092

Table for Spacing Off the Circumference of Circles

No. of Divisions	Degrees in Arc	Diameter of Circle to be Spaced Off													
		Length of Chord													
		1	2	3	4	5	6	7	8	9	10	11	12	13	14
3	120	0.866	1.732	2.598	3.464	4.330	5.196	6.062	6.928	7.794	8.660	9.526	10.392	11.258	12.124
4	90	0.707	1.414	2.121	2.828	3.536	4.243	4.950	5.657	6.364	7.071	7.778	8.485	9.192	9.899
5	72	0.588	1.176	1.763	2.351	2.939	3.527	4.114	4.702	5.290	5.878	6.466	7.053	7.641	8.229
6	60	0.500	1.000	1.500	2.000	2.500	3.000	3.500	4.000	4.500	5.000	5.500	6.000	6.500	7.000
7	51 8⁄17	0.434	0.868	1.302	1.736	2.169	2.603	3.037	3.471	3.995	4.339	4.773	5.207	5.640	6.074
8	45	0.383	0.765	1.148	1.531	1.913	2.296	2.679	3.061	3.444	3.827	4.210	4.592	4.975	5.358
9	40	0.342	0.684	1.026	1.368	1.710	2.052	2.394	2.736	3.078	3.420	3.762	4.104	4.446	4.788
10	36	0.309	0.618	0.927	1.236	1.545	1.854	2.163	2.472	2.781	3.090	3.399	3.708	4.017	4.326
11	32 8⁄11	0.282	0.563	0.845	1.127	1.409	1.690	1.972	2.254	2.536	2.817	3.099	3.381	3.663	3.944
12	30	0.259	0.518	0.776	1.035	1.294	1.553	1.812	2.071	2.329	2.588	2.847	3.106	3.365	3.623
13	27 9⁄13	0.239	0.479	0.718	0.957	1.197	1.436	1.675	1.915	2.154	2.393	2.632	2.872	3.111	3.350
14	25 5⁄7	0.223	0.445	0.668	0.890	1.113	1.335	1.558	1.780	2.003	2.225	2.448	2.670	2.893	3.115
15	24	0.208	0.416	0.624	0.832	1.040	1.247	1.455	1.663	1.871	2.079	2.287	2.495	2.703	2.911
16	22 1⁄2	0.195	0.390	0.585	0.780	0.975	1.171	1.366	1.561	1.756	1.951	2.146	2.341	2.536	2.731
17	21 9⁄17	0.184	0.367	0.551	0.735	0.919	1.102	1.286	1.470	1.654	1.837	2.021	2.205	2.389	2.572
18	20	0.174	0.347	0.521	0.695	0.868	1.042	1.216	1.389	1.563	1.736	1.910	2.084	2.257	2.431
19	18 18⁄19	0.165	0.329	0.494	0.658	0.823	0.988	1.152	1.317	1.481	1.646	1.811	1.975	2.140	2.304
20	18	0.156	0.313	0.469	0.626	0.782	0.939	1.095	1.251	1.408	1.564	1.721	1.877	2.034	2.190
21	17 2⁄7	0.149	0.298	0.447	0.596	0.745	0.894	1.043	1.192	1.341	1.490	1.639	1.789	1.938	2.087
22	16 4⁄11	0.142	0.285	0.427	0.569	0.712	0.854	0.996	1.139	1.281	1.423	1.565	1.708	1.850	1.992
23	15 15⁄23	0.136	0.272	0.408	0.545	0.681	0.817	0.953	1.089	1.225	1.362	1.498	1.634	1.770	1.906
24	15	0.131	0.261	0.392	0.522	0.653	0.783	0.914	1.044	1.175	1.305	1.436	1.566	1.697	1.827
25	14 2⁄5	0.125	0.251	0.376	0.501	0.627	0.752	0.877	1.003	1.128	1.253	1.379	1.504	1.629	1.755
26	13 11⁄13	0.121	0.241	0.362	0.482	0.603	0.723	0.844	0.964	1.085	1.205	1.326	1.446	1.567	1.688
28	12 9⁄7	0.112	0.224	0.336	0.448	0.560	0.672	0.784	0.896	1.008	1.120	1.232	1.344	1.456	1.568
30	12	0.105	0.209	0.314	0.418	0.523	0.627	0.732	0.836	0.941	1.045	1.150	1.254	1.359	1.463
32	11 1⁄4	0.098	0.196	0.294	0.392	0.490	0.588	0.686	0.784	0.882	0.980	1.078	1.176	1.274	1.372

Table 1. Hole Coordinate Dimension Factors for Jig Boring — Type "A" Hole Circles
(English or metric units)

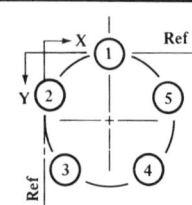

The diagram shows a type "A" circle for a 5-hole circle. Coordinates x, y are given in the table for hole circles of from 3 to 28 holes. Dimensions are for holes numbered in a counterclockwise direction (as shown). Dimensions given are based upon a hole circle of unit diameter. For a hole circle of, say, 3-inch or 3-centimeter diameter, multiply table values by 3.

3 Holes

Hole	x	y
1	0.50000	0.00000
2	0.06699	0.75000
3	0.93301	0.75000

4 Holes

Hole	x	y
1	0.50000	0.00000
2	0.00000	0.50000
3	0.50000	1.00000
4	1.00000	0.50000

5 Holes

Hole	x	y
1	0.50000	0.00000
2	0.02447	0.34549
3	0.20611	0.90451
4	0.79389	0.90451
5	0.97553	0.34549

6 Holes

Hole	x	y
1	0.50000	0.00000
2	0.06699	0.25000
3	0.06699	0.75000
4	0.50000	1.00000
5	0.93301	0.75000
6	0.93301	0.25000

7 Holes

Hole	x	y
1	0.50000	0.00000
2	0.10908	0.18826
3	0.01254	0.61126
4	0.28306	0.95048
5	0.71694	0.95048
6	0.98746	0.61126
7	0.89092	0.18826

8 Holes

Hole	x	y
1	0.50000	0.00000
2	0.14645	0.14645
3	0.00000	0.50000
4	0.14645	0.85355
5	0.50000	1.00000
6	0.85355	0.85355
7	1.00000	0.50000
8	0.85355	0.14645

9 Holes

Hole	x	y
1	0.50000	0.00000
2	0.17861	0.11698
3	0.00760	0.41318
4	0.06699	0.75000
5	0.32899	0.96985
6	0.67101	0.96985
7	0.93301	0.75000
8	0.99240	0.41318
9	0.82139	0.11698

10 Holes

Hole	x	y
1	0.50000	0.00000
2	0.20611	0.09549
3	0.02447	0.34549
4	0.02447	0.65451
5	0.20611	0.90451
6	0.50000	1.00000
7	0.79389	0.90451
8	0.97553	0.65451
9	0.97553	0.34549
10	0.79389	0.09549

11 Holes

Hole	x	y
1	0.50000	0.00000
2	0.22968	0.07937
3	0.04518	0.29229
4	0.00590	0.57116
5	0.12213	0.82743
6	0.35913	0.97975
7	0.64087	0.97975
8	0.87787	0.82743
9	0.99491	0.57116
10	0.95482	0.29229
11	0.77032	0.07937

12 Holes

Hole	x	y
1	0.50000	0.00000
2	0.25000	0.06699
3	0.06699	0.25000
4	0.00000	0.50000
5	0.06699	0.75000
6	0.25000	0.93301
7	0.50000	1.00000
8	0.75000	0.93301
9	0.93301	0.75000
10	1.00000	0.50000
11	0.93301	0.25000
12	0.75000	0.06699

13 Holes

Hole	x	y
1	0.50000	0.00000
2	0.26764	0.05727
3	0.08851	0.21597
4	0.00365	0.43973
5	0.03249	0.67730
6	0.16844	0.87426
7	0.38034	0.98547
8	0.61966	0.98547
9	0.83156	0.87426
10	0.96751	0.67730
11	0.99635	0.43973
12	0.91149	0.21597
13	0.73236	0.05727

14 Holes

Hole	x	y
1	0.50000	0.00000
2	0.28306	0.04952
3	0.10908	0.18826
4	0.01254	0.38874
5	0.01254	0.61126
6	0.10908	0.81174
7	0.28306	0.95048
8	0.50000	1.00000
9	0.71694	0.95048
10	0.89092	0.81174
11	0.98746	0.61126
12	0.98746	0.38874
13	0.89092	0.18826
14	0.71694	0.04952

15 Holes

Hole	x	y
1	0.50000	0.00000
2	0.29663	0.04323
3	0.12843	0.16543
4	0.02447	0.34549
5	0.00274	0.55228
6	0.06699	0.75000
7	0.20611	0.90451
8	0.39604	0.98907
9	0.60396	0.98907
10	0.79389	0.90451
11	0.93301	0.75000
12	0.99726	0.55228
13	0.97553	0.34549
14	0.87157	0.16543
15	0.70337	0.04323

16 Holes

Hole	x	y
1	0.50000	0.00000
2	0.30866	0.03806
3	0.14645	0.14645
4	0.03806	0.30866
5	0.00000	0.50000
6	0.03806	0.69134
7	0.14645	0.85355
8	0.30866	0.96194
9	0.50000	1.00000
10	0.69134	0.96194
11	0.85355	0.85355
12	0.96194	0.69134
13	1.00000	0.50000
14	0.96194	0.30866
15	0.85355	0.14645
16	0.69134	0.03806

17 Holes

Hole	x	y
1	0.50000	0.00000
2	0.31938	0.03376
3	0.16315	0.13050
4	0.05242	0.27713
5	0.00213	0.45387
6	0.01909	0.63683
7	0.10099	0.80132
8	0.23678	0.92511
9	0.40813	0.99149
10	0.59187	0.99149
11	0.76322	0.92511
12	0.89901	0.80132
13	0.98091	0.63683
14	0.99787	0.45387
15	0.94758	0.27713
16	0.83685	0.13050
17	0.68062	0.03376

18 Holes

Hole	x	y
1	0.50000	0.00000
2	0.32899	0.03015
3	0.17861	0.11698
4	0.06699	0.25000
5	0.00760	0.41318
6	0.00760	0.58682
7	0.06699	0.75000
8	0.17861	0.88302
9	0.32899	0.96985
10	0.50000	1.00000
11	0.67101	0.96985
12	0.82139	0.88302
13	0.93301	0.75000
14	0.99240	0.58682
15	0.99240	0.41318
16	0.93301	0.25000
17	0.82139	0.11698
18	0.67101	0.03015

19 Holes

Hole	x	y
1	0.50000	0.00000
2	0.33765	0.02709
3	0.19289	0.10543
4	0.08142	0.22653
5	0.01530	0.37726
6	0.00171	0.54129
7	0.04212	0.70085
..	....	

Table 1 (*Concluded*). **Hole Coordinate Dimension Factors for Jig Boring — Type "A"**
Hole Circles (English or metric units)

Column 1 (continuation — 19 Holes)

```
x8  0.13214      x9  0.26203      x10 0.41770
y8  0.83864      y9  0.93974      y10 0.99318
x11 0.58230      x12 0.73797      x13 0.86786
y11 0.99318      y12 0.93974      y13 0.83864
x14 0.95789      x15 0.99829      x16 0.98470
y14 0.70085      y15 0.54129      y16 0.37726
x17 0.91858      x18 0.80711      x19 0.66235
y17 0.22658      y18 0.10543      y19 0.02709

20 Holes
x1  0.50000      x2  0.34549      x3  0.20611
y1  0.00000      y2  0.02447      y3  0.09549
x4  0.09549      x5  0.02447      x6  0.00000
y4  0.20611      y5  0.34549      y6  0.50000
x7  0.02447      x8  0.09549      x9  0.20611
y7  0.65451      y8  0.79389      y9  0.90451
x10 0.34549      x11 0.50000      x12 0.65451
y10 0.97553      y11 1.00000      y12 0.97553
x13 0.79389      x14 0.90451      x15 0.97553
y13 0.90451      y14 0.79389      y15 0.65451
x16 1.00000      x17 0.97553      x18 0.90451
y16 0.50000      y17 0.34549      y18 0.20611
x19 0.79389      x20 0.65451
y19 0.09549      y20 0.02447
..  ....
..  ....
```

Column 2 (21 Holes, then 22 Holes)

```
21 Holes
x1  0.50000      x2  0.35262      x3  0.21834
y1  0.00000      y2  0.02221      y3  0.08688
x4  0.10908      x5  0.03456      x6  0.00140
y4  0.18826      y5  0.31733      y6  0.46263
x7  0.01254      x8  0.06699      x9  0.15991
y7  0.61126      y8  0.75000      y9  0.86653
x10 0.28306      x11 0.42548      x12 0.57452
y10 0.95048      y11 0.99442      y12 0.99442
x13 0.71694      x14 0.84009      x15 0.93301
y13 0.95048      y14 0.86653      y15 0.75000
x16 0.98746      x17 0.99860      x18 0.96544
y16 0.61126      y17 0.46263      y18 0.31733
x19 0.89092      x20 0.78166      x21 0.64738
y19 0.18826      y20 0.08688      y21 0.02221

22 Holes
x1  0.50000      x2  0.35913      x3  0.22968
y1  0.00000      y2  0.02025      y3  0.07937
x4  0.12213      x5  0.04518      x6  0.00509
y4  0.17257      y5  0.29229      y6  0.42884
x7  0.00509      x8  0.04518      x9  0.12213
y7  0.57116      y8  0.70771      y9  0.82743
x10 0.22968      x11 0.35913
y10 0.92063      y11 0.97975
```

Column 3 (22 Holes continued, then 23 Holes)

```
x12 0.50000      x13 0.64087      x14 0.77032
y12 1.00000      y13 0.97975      y14 0.92063
x15 0.87787      x16 0.95482      x17 0.99491
y15 0.82743      y16 0.70771      y17 0.57116
x18 0.99491      x19 0.95482      x20 0.87787
y18 0.42884      y19 0.29229      y20 0.17257
x21 0.77032      x22 0.64087
y21 0.07937      y22 0.02025

23 Holes
x1  0.50000      x2  0.36510      x3  0.24021
y1  0.00000      y2  0.01854      y3  0.07279
x4  0.13458      x5  0.05606      x6  0.01046
y4  0.15872      y5  0.26997      y6  0.39827
x7  0.00117      x8  0.02887      x9  0.09152
y7  0.53412      y8  0.66744      y9  0.78834
x10 0.18446      x11 0.30080      x12 0.43192
y10 0.88786      y11 0.95861      y12 0.99534
x13 0.56808      x14 0.69920      x15 0.81554
y13 0.99534      y14 0.95861      y15 0.88786
x16 0.90848      x17 0.97113      x18 0.99883
y16 0.78834      y17 0.66744      y18 0.53412
x19 0.98954      x20 0.94394      x21 0.86542
y19 0.39827      y20 0.26997      y21 0.15872
x22 0.75979
y22 0.07279
```

Column 4 (23 Holes last point, 24 Holes, then 25 Holes)

```
x23 0.63490
y23 0.01854

24 Holes
x1  0.50000      x2  0.37059      x3  0.25000
y1  0.00000      y2  0.01704      y3  0.06699
x4  0.14645      x5  0.06699      x6  0.01704
y4  0.14645      y5  0.25000      y6  0.37059
x7  0.00000      x8  0.01704      x9  0.06699
y7  0.50000      y8  0.62941      y9  0.75000
x10 0.14645      x11 0.25000      x12 0.37059
y10 0.85355      y11 0.93301      y12 0.98296
x13 0.50000      x14 0.62941      x15 0.75000
y13 1.00000      y14 0.98296      y15 0.93301
x16 0.85355      x17 0.93301      x18 0.98296
y16 0.85355      y17 0.75000      y18 0.62941
x19 1.00000      x20 0.98296      x21 0.93301
y19 0.50000      y20 0.37059      y21 0.25000
x22 0.85355      x23 0.75000      x24 0.62941
y22 0.14645      y23 0.06699      y24 0.01704

25 Holes
x1  0.50000      x2  0.37566      x3  0.25912
y1  0.00000      y2  0.01571      y3  0.06185
x4  0.15773      x5  0.07784      x6  0.02447
y4  0.13552      y5  0.23209      y6  0.34549
x7  0.00099
y7  0.46860
..  ....
```

Column 5 (25 Holes continued, then 26 Holes)

```
x8  0.00886      x9  0.04759      x10 0.11474
y8  0.59369      y9  0.71289      y10 0.81871
x11 0.20611      x12 0.31594      x13 0.43733
y11 0.90451      y12 0.96489      y13 0.99606
x14 0.56267      x15 0.68406      x16 0.79389
y14 0.99606      y15 0.96489      y16 0.90451
x17 0.88526      x18 0.95241      x19 0.99114
y17 0.81871      y18 0.71289      y19 0.59369
x20 0.99901      x21 0.97553      x22 0.92216
y20 0.46860      y21 0.34549      y22 0.23209
x23 0.84227      x24 0.74088      x25 0.62434
y23 0.13552      y24 0.06185      y25 0.01571

26 Holes
x1  0.50000      x2  0.38034      x3  0.26764
y1  0.00000      y2  0.01453      y3  0.05727
x4  0.16844      x5  0.08851      x6  0.03249
y4  0.12574      y5  0.21597      y6  0.32270
x7  0.00365      x8  0.00365      x9  0.03249
y7  0.43973      y8  0.56027      y9  0.67730
x10 0.08851      x11 0.16844      x12 0.26764
y10 0.78403      y11 0.87426      y12 0.94273
x13 0.38034      x14 0.50000      x15 0.61966
y13 0.98547      y14 1.00000      y15 0.98547
```

Column 6 (26 Holes continued, then 27 Holes)

```
x16 0.73236      x17 0.83156      x18 0.91149
y16 0.94273      y17 0.87426      y18 0.78403
x19 0.96751      x20 0.99635      x21 0.99635
y19 0.67730      y20 0.56027      y21 0.43973
x22 0.96751      x23 0.91149      x24 0.83156
y22 0.32270      y23 0.21597      y24 0.12574
x25 0.73236      x26 0.61966
y25 0.05727      y26 0.01453

27 Holes
x1  0.50000      x2  0.38469      x3  0.27560
y1  0.00000      y2  0.01348      y3  0.05318
x4  0.17861      x5  0.09894      x6  0.04089
y4  0.11698      y5  0.20142      y6  0.30196
x7  0.00760      x8  0.00085      x9  0.02101
y7  0.41318      y8  0.52907      y9  0.64340
x10 0.06699      x11 0.13631      x12 0.22525
y10 0.75000      y11 0.84312      y12 0.91774
x13 0.32899      x14 0.44195      x15 0.55805
y13 0.96985      y14 0.99662      y15 0.99662
x16 0.67101      x17 0.77475      x18 0.86369
y16 0.96985      y17 0.91774      y18 0.84312
x19 0.93301      x20 0.97899      x21 0.99915
y19 0.75000      y20 0.64340      y21 0.52907
x22 0.99240
y22 0.41318
```

Column 7 (27 Holes continued, then 28 Holes)

```
x23 0.95911      x24 0.90106      x25 0.82139
y23 0.30196      y24 0.20142      y25 0.11698
x26 0.72440      x27 0.61531
y26 0.05318      y27 0.01348

28 Holes
x1  0.50000      x2  0.38874      x3  0.28306
y1  0.00000      y2  0.01254      y3  0.04952
x4  0.18826      x5  0.10908      x6  0.04952
y4  0.10908      y5  0.18826      y6  0.28306
x7  0.01254      x8  0.00000      x9  0.01254
y7  0.38874      y8  0.50000      y9  0.61126
x10 0.04952      x11 0.10908      x12 0.18826
y10 0.71694      y11 0.81174      y12 0.89092
x13 0.28306      x14 0.38874      x15 0.50000
y13 0.95048      y14 0.98746      y15 1.00000
x16 0.61126      x17 0.71694      x18 0.81174
y16 0.98746      y17 0.95048      y18 0.89092
x19 0.89092      x20 0.95048      x21 0.98746
y19 0.81174      y20 0.71694      y21 0.61126
x22 1.00000      x23 0.98746      x24 0.95048
y22 0.50000      y23 0.38874      y24 0.28306
x25 0.89092      x26 0.81174      x27 0.71694
y25 0.18826      y26 0.10908      y27 0.04952
x28 0.61126
y28 0.01254
```

Table 2. Hole Coordinate Dimension Factors for Jig Boring — Type "B" Hole Circles
(English or metric units)

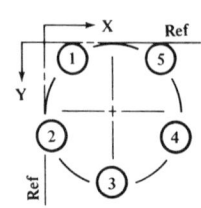

The diagram shows a type "B" circle for a 5-hole circle. Coordinates x, y are given in the table for hole circles of from 3 to 28 holes. Dimensions are for holes numbered in a counterclockwise direction (as shown). Dimensions given are based upon a hole circle of unit diameter. For a hole circle of, say, 3-inch or 3-centimeter diameter, multiply table values by 3.

3 Holes

No.	x	y
1	0.06699	0.25000
2	0.50000	1.00000
3	0.93301	0.25000

4 Holes

No.	x	y
1	0.14645	0.14645
2	0.14645	0.85355
3	0.85355	0.85355
4	0.85355	0.14645

5 Holes

No.	x	y
1	0.20611	0.09549
2	0.02447	0.65451
3	0.50000	1.00000
4	0.97553	0.65451
5	0.79389	0.09549

6 Holes

No.	x	y
1	0.25000	0.06699
2	0.00000	0.50000
3	0.25000	0.93301
4	0.75000	0.93301
5	1.00000	0.50000
6	0.75000	0.06699

7 Holes

No.	x	y
1	0.28306	0.04952
2	0.01254	0.38874
3	0.10908	0.81174
4	0.50000	1.00000
5	0.89092	0.81174
6	0.98746	0.38874
7	0.71694	0.04952

8 Holes

No.	x	y
1	0.30866	0.03806
2	0.03806	0.30866
3	0.03806	0.69134
4	0.30866	0.96194
5	0.69134	0.96194
6	0.96194	0.69134
7	0.96194	0.30866
8	0.69134	0.03806

9 Holes

No.	x	y
1	0.32899	0.03015
2	0.06699	0.25000
3	0.00760	0.58682
4	0.17861	0.88302
5	0.50000	1.00000
6	0.82139	0.88302
7	0.99240	0.58682
8	0.93301	0.25000
9	0.67101	0.03015

10 Holes

No.	x	y
1	0.34549	0.02447
2	0.09549	0.20611
3	0.00760	0.50000
4	0.09549	0.79389
5	0.34549	0.97553
6	0.65451	0.97553
7	0.90451	0.79389
8	1.00000	0.50000
9	0.90451	0.20611
10	0.65451	0.02447

11 Holes

No.	x	y
1	0.35913	0.02025
2	0.12213	0.17257
3	0.00509	0.42884
4	0.04518	0.70711
5	0.22968	0.92063
6	0.50000	1.00000
7	0.77032	0.92063
8	0.95482	0.70771
9	0.99491	0.42884
10	0.87787	0.17257
11	0.64087	0.02025

12 Holes

No.	x	y
1	0.37059	0.01704
2	0.14645	0.14645
3	0.01704	0.37059
4	0.01704	0.62941
5	0.14645	0.85355
6	0.37059	0.98296
7	0.62941	0.98296
8	0.85355	0.85355
9	0.98296	0.62941
10	0.98296	0.37059
11	0.85355	0.14645
12	0.62941	0.01704

13 Holes

No.	x	y
1	0.38034	0.01453
2	0.16844	0.12574
3	0.03249	0.32270
4	0.00365	0.56027
5	0.08851	0.78403
6	0.26764	0.94273
7	0.50000	1.00000
8	0.73236	0.94273
9	0.91149	0.78403
10	0.99635	0.56027
11	0.96751	0.32270
12	0.83156	0.12574
13	0.61966	0.01453

14 Holes

No.	x	y
1	0.38874	0.01254
2	0.18826	0.10908
3	0.04952	0.28306
4	0.00000	0.50000
5	0.04952	0.71694
6	0.18826	0.89092
7	0.38874	0.98746
8	0.61126	0.98746
9	0.81174	0.89092
10	0.95048	0.71694
11	1.00000	0.50000
12	0.95048	0.28306
13	0.81174	0.10908
14	0.61126	0.01254

15 Holes

No.	x	y
1	0.39604	0.01093
2	0.20611	0.09549
3	0.06699	0.25000
4	0.00274	0.44774
5	0.02447	0.65451
6	0.12843	0.83457
7	0.29663	0.95677
8	0.50000	1.00000
9	0.70337	0.95677
10	0.87157	0.83457
11	0.97553	0.65451
12	0.99726	0.44774
13	0.93301	0.25000
14	0.79389	0.09549
15	0.60396	0.01093

16 Holes

No.	x	y
1	0.40245	0.00961
2	0.22221	0.08427
3	0.08427	0.22221
4	0.00961	0.40245
5	0.00961	0.59755
6	0.08427	0.77779
7	0.23678	0.91573
8	0.40245	0.99039
9	0.59755	0.99039
10	0.77779	0.91573
11	0.91573	0.77779
12	0.99039	0.59755
13	0.99039	0.40245
14	0.91573	0.22221
15	0.77779	0.08427
16	0.59755	0.00961

17 Holes

No.	x	y
1	0.40813	0.00851
2	0.23678	0.07489
3	0.10099	0.19868
4	0.01909	0.36317
5	0.00213	0.54613
6	0.05242	0.72287
7	0.16315	0.86950
8	0.31938	0.96624
9	0.50000	1.00000
10	0.68062	0.96624
11	0.83685	0.86950
12	0.94758	0.72287
13	0.99787	0.54613
14	0.98091	0.36317
15	0.89901	0.19868
16	0.76322	0.07489
17	0.59187	0.00851

18 Holes

No.	x	y
1	0.41318	0.00760
2	0.25000	0.06699
3	0.11698	0.17861
4	0.03015	0.32899
5	0.00000	0.50000
6	0.03015	0.67101
7	0.11698	0.82139
8	0.25000	0.93301
9	0.41318	0.99240
10	0.58682	0.99240
11	0.75000	0.93301
12	0.88302	0.82139
13	0.96985	0.67101
14	1.00000	0.50000
15	0.96985	0.32899
16	0.88302	0.17861
17	0.75000	0.06699
18	0.58682	0.00760

19 Holes

No.	x	y
1	0.41770	0.00682
2	0.26203	0.06026
3	0.13214	0.16136
4	0.04211	0.29915
5	0.00171	0.45871
6	0.01530	0.62274
7	0.08142	0.77347
…	…	…

Table 2 (*Concluded*). **Hole Coordinate Dimension Factors for Jig Boring — Type "B"**
Hole Circles (English or metric units)

19 Holes (continued)

	x	y
8	0.19289	0.89457
9	0.33765	0.97291
10	0.50000	1.00000
11	0.66235	0.97291
12	0.80711	0.89457
13	0.91858	0.77347
14	0.98470	0.62274
15	0.99829	0.45871
16	0.95789	0.29915
17	0.86786	0.16136
18	0.73797	0.06026
19	0.58230	0.00682

20 Holes

	x	y
1	0.42178	0.00616
2	0.27300	0.05450
3	0.14645	0.14645
4	0.05450	0.27300
5	0.00616	0.42178
6	0.00616	0.57822
7	0.05450	0.72700
8	0.14645	0.85355
9	0.27300	0.94550
10	0.42178	0.99384
11	0.57822	0.99384
12	0.72700	0.94550
13	0.85355	0.85355
14	0.94550	0.72700
15	0.99384	0.57822
16	0.99384	0.42178
17	0.94550	0.27300
18	0.85355	0.14645
19	0.72700	0.05450
20	0.57822	0.00616

21 Holes

	x	y
1	0.42548	0.00558
2	0.28306	0.04952
3	0.15991	0.13347
4	0.06699	0.25000
5	0.01254	0.38874
6	0.00140	0.53737
7	0.03456	0.68267
8	0.10908	0.81174
9	0.21834	0.91312
10	0.35262	0.97779
11	0.50000	1.00000
12	0.64738	0.97779
13	0.78166	0.91312
14	0.89092	0.81174
15	0.96544	0.68267
16	0.98746	0.53737
17	0.98746	0.38874
18	0.93301	0.25000
19	0.84009	0.13347
20	0.71694	0.04952
21	0.57452	0.00558

22 Holes

	x	y
1	0.42884	0.00509
2	0.29229	0.04518
3	0.17257	0.12213
4	0.07937	0.22968
5	0.02025	0.35913
6	0.00000	0.50000
7	0.02025	0.64087
8	0.07937	0.77032
9	0.17257	0.87787
10	0.29229	0.95482
11	0.42884	0.99491
12	0.57116	0.99491
13	0.70771	0.95482
14	0.82743	0.87787
15	0.92063	0.77032
16	0.97975	0.64087
17	1.00000	0.50000
18	0.97975	0.35913
19	0.92063	0.22968
20	0.82743	0.12213
21	0.70771	0.04518
22	0.57116	0.00509

23 Holes

	x	y
1	0.43192	0.00466
2	0.30080	0.04139
3	0.18446	0.11214
4	0.09152	0.21166
5	0.02887	0.33256
6	0.00117	0.46588
7	0.01046	0.60173
8	0.05606	0.73003
9	0.13458	0.84128
10	0.24021	0.92721
11	0.36510	0.98146
12	0.50000	1.00000
13	0.63490	0.98146
14	0.75979	0.92721
15	0.86542	0.84128
16	0.94394	0.73003
17	0.98954	0.60173
18	0.99883	0.46588
19	0.97113	0.33256
20	0.90848	0.21166
21	0.81554	0.11214
22	0.69920	0.04139
23	0.56808	0.00466

24 Holes

	x	y
1	0.43474	0.00428
2	0.30866	0.03806
3	0.19562	0.10332
4	0.10332	0.19562
5	0.03806	0.30866
6	0.00428	0.43474
7	0.00428	0.56526
8	0.03806	0.69134
9	0.10332	0.80438
10	0.19562	0.89668
11	0.30866	0.96194
12	0.43474	0.99572
13	0.56526	0.99572
14	0.69134	0.96194
15	0.80438	0.89668
16	0.89668	0.80438
17	0.96194	0.69134
18	0.99572	0.56526
19	0.99572	0.43474
20	0.96194	0.30866
21	0.89668	0.19562
22	0.80438	0.10332
23	0.69134	0.03806
24	0.56526	0.00428

25 Holes

	x	y
1	0.43733	0.00394
2	0.31594	0.03511
3	0.20611	0.09549
4	0.11474	0.18129
5	0.04759	0.28711
6	0.00886	0.40631
7	0.00000	0.53140
8	0.02447	0.65451
9	0.07784	0.76791
10	0.15773	0.86448
11	0.25912	0.93815
12	0.37566	0.98429
13	0.50000	1.00000
14	0.62434	0.98429
15	0.74088	0.93815
16	0.84227	0.86448
17	0.92216	0.76791
18	0.97553	0.65451
19	0.99991	0.53140
20	0.99114	0.40631
21	0.95241	0.28711
22	0.88526	0.18129
23	0.79389	0.09549
24	0.68406	0.03511
25	0.56267	0.00394

26 Holes

	x	y
1	0.43973	0.00365
2	0.32270	0.03249
3	0.21597	0.08851
4	0.12574	0.16844
5	0.05727	0.26764
6	0.01453	0.38034
7	0.00000	0.50000
8	0.01453	0.61531
9	0.05727	0.72440
10	0.12574	0.83156
11	0.21597	0.91149
12	0.32270	0.96751
13	0.43973	0.99635
14	0.56027	0.99635
15	0.67730	0.96751
16	0.78403	0.91149
17	0.87426	0.83156
18	0.94273	0.73236
19	0.98547	0.61966
20	1.00000	0.50000
21	0.98547	0.38034
22	0.94273	0.26764
23	0.87426	0.16844
24	0.78403	0.08851
25	0.67730	0.03249
26	0.56027	0.00365

27 Holes

	x	y
1	0.44195	0.00338
2	0.32899	0.03015
3	0.22525	0.08226
4	0.13631	0.15688
5	0.06699	0.25000
6	0.02101	0.35660
7	0.00085	0.47093
8	0.00760	0.58682
9	0.04089	0.69804
10	0.09894	0.79858
11	0.17861	0.88302
12	0.27560	0.94682
13	0.38469	0.98652
14	0.50000	1.00000
15	0.61531	0.98652
16	0.72440	0.94682
17	0.82139	0.88302
18	0.90106	0.79858
19	0.95911	0.69804
20	0.99240	0.58682
21	0.99915	0.47093
22	0.97899	0.35660
23	0.93301	0.25000
24	0.86369	0.15688
25	0.77475	0.08226
26	0.67101	0.03015
27	0.55805	0.00338

28 Holes

	x	y
1	0.44402	0.00314
2	0.33486	0.02806
3	0.23398	0.07664
4	0.14645	0.14645
5	0.07664	0.23398
6	0.02806	0.33486
7	0.00314	0.44402
8	0.00314	0.55598
9	0.02806	0.66514
10	0.07664	0.76602
11	0.14645	0.85355
12	0.23398	0.92336
13	0.33486	0.97194
14	0.44402	0.99686
15	0.55598	0.99686
16	0.66514	0.97194
17	0.76602	0.92336
18	0.85355	0.85355
19	0.92336	0.76602
20	0.97194	0.66514
21	0.99686	0.55598
22	0.99686	0.44402
23	0.97194	0.33486
24	0.92336	0.23398
25	0.85355	0.14645
26	0.76602	0.07664
27	0.66514	0.02806
28	0.55598	0.00314

Table 3. Hole Coordinate Dimension Factors for Jig Boring — Type "A" Hole Circles, Central Coordinates (English or metric units)

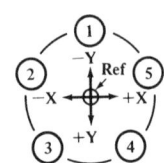

The diagram shows a type "A" circle for a 5-hole circle. Coordinates x, y are given in the table for hole circles of from 3 to 28 holes. Dimensions are for holes numbered in a counterclockwise direction (as shown). Dimensions given are based upon a hole circle of unit diameter. For a hole circle of, say, 3-inch or 3-centimeter diameter, multiply table values by 3.

3 Holes

	x	y
1	0.00000	−0.50000
2	−0.43301	+0.25000
3	+0.43301	+0.25000

4 Holes

	x	y
1	0.00000	−0.50000
2	−0.50000	0.00000
3	0.00000	+0.50000
4	+0.50000	0.00000

5 Holes

	x	y
1	0.00000	−0.50000
2	−0.47553	−0.15451
3	−0.29389	+0.40451
4	+0.29389	+0.40451
5	+0.47553	−0.15451

6 Holes

	x	y
1	0.00000	−0.50000
2	−0.43301	−0.25000
3	−0.43301	+0.25000
4	0.00000	+0.50000
5	+0.43301	+0.25000
6	+0.43301	−0.25000

7 Holes

	x	y
1	0.00000	−0.50000
2	−0.39092	−0.31174
3	−0.48746	+0.11126
4	−0.21694	+0.45048
5	+0.21694	+0.45048
6	+0.48746	+0.11126
7	+0.39092	−0.31174

8 Holes

	x	y
1	0.00000	−0.50000
2	−0.35355	−0.35355
3	−0.50000	0.00000
4	−0.35355	+0.35355
5	0.00000	+0.50000
6	+0.35355	+0.35355
7	+0.50000	0.00000
8	+0.35355	−0.35355

9 Holes

	x	y
1	0.00000	−0.50000
2	−0.32139	−0.38302
3	−0.49240	−0.08682
4	−0.43301	+0.25000
5	−0.17101	+0.46985
6	+0.17101	+0.46985
7	+0.43301	+0.25000
8	+0.49240	−0.08682
9	+0.32139	−0.38302

10 Holes

	x	y
1	0.00000	−0.50000
2	−0.29389	−0.40451
3	−0.47553	−0.15451
4	−0.47553	+0.15451
5	−0.29389	+0.40451
6	0.00000	+0.50000
7	+0.29389	+0.40451
8	+0.47553	+0.15451
9	+0.47553	−0.15451
10	+0.29389	−0.40451

11 Holes

	x	y
1	0.00000	−0.50000
2	−0.27032	−0.42063
3	−0.45482	−0.20771
4	−0.49491	+0.07116
5	−0.37787	+0.32743
6	−0.14087	+0.47975
7	+0.14087	+0.47975
8	+0.37787	+0.32743
9	+0.49491	+0.07116
10	+0.45482	−0.20771
11	+0.27032	−0.42063

12 Holes

	x	y
1	0.00000	−0.50000
2	−0.25000	−0.43301
3	−0.43301	−0.25000
4	−0.50000	0.00000
5	−0.43301	+0.25000
6	−0.25000	+0.43301
7	0.00000	+0.50000
8	+0.25000	+0.43301
9	+0.43301	+0.25000
10	+0.50000	0.00000
11	+0.43301	−0.25000
12	+0.25000	−0.43301

13 Holes

	x	y
1	0.00000	−0.50000
2	−0.23236	−0.44273
3	−0.41149	−0.28403
4	−0.49635	−0.06027
5	−0.46751	+0.17730
6	−0.33156	+0.37426
7	−0.11966	+0.48547
8	+0.11966	+0.48547
9	+0.33156	+0.37426
10	+0.46751	+0.17730
11	+0.49635	−0.06027
12	+0.41149	−0.28403
13	+0.23236	−0.44273

14 Holes

	x	y
1	0.00000	−0.50000
2	−0.21694	−0.45048
3	−0.39092	−0.31174
4	−0.48746	−0.11126
5	−0.48746	+0.11126
6	−0.39092	+0.31174
7	−0.21694	+0.45048
8	0.00000	+0.50000
9	+0.21694	+0.45048
10	+0.39092	+0.31174
11	+0.48746	+0.11126
12	+0.48746	−0.11126
13	+0.39092	−0.31174
14	+0.21694	−0.45048

15 Holes

	x	y
1	0.00000	−0.50000
2	−0.20337	−0.45677
3	−0.37157	−0.33457
4	−0.47553	−0.15451
5	−0.49726	+0.05226
6	−0.43301	+0.25000
7	−0.29389	+0.40451
8	−0.10396	+0.48907
9	+0.10396	+0.48907
10	+0.29389	+0.40451
11	+0.43301	+0.25000
12	+0.49726	+0.05226
13	+0.47553	−0.15451
14	+0.37157	−0.33457
15	+0.20337	−0.45677

16 Holes

	x	y
1	0.00000	−0.50000
2	−0.19134	−0.46194
3	−0.35355	−0.35355
4	−0.46194	−0.19134
5	−0.50000	0.00000
6	−0.46194	+0.19134
7	−0.35355	+0.35355
8	−0.19134	+0.46194
9	0.00000	+0.50000
10	+0.19134	+0.46194
11	+0.35355	+0.35355
12	+0.46194	+0.19134
13	+0.50000	0.00000
14	+0.46194	−0.19134
15	+0.35355	−0.35355
16	+0.19134	−0.46194

17 Holes

	x	y
1	0.00000	−0.50000
2	−0.18062	−0.46624
3	−0.33685	−0.36950
4	−0.44758	−0.22287
5	−0.49787	−0.04613
6	−0.48091	+0.13683
7	−0.39901	+0.30132
8	−0.26322	+0.42511
9	−0.09187	+0.49149
10	+0.09187	+0.49149
11	+0.26322	+0.42511
12	+0.39901	+0.30132
13	+0.48091	+0.13683
14	+0.49787	−0.04613
15	+0.44758	−0.22287
16	+0.33685	−0.36950
17	+0.18062	−0.46624

18 Holes

	x	y
1	0.00000	−0.50000
2	−0.17101	−0.46985
3	−0.32139	−0.38302
4	−0.43301	−0.25000
5	−0.49240	−0.08682
6	−0.49240	+0.08682
7	−0.43301	+0.25000
8	−0.32139	+0.38302
9	−0.17101	+0.46985
10	0.00000	+0.50000
11	+0.17101	+0.46985
12	+0.32139	+0.38302
13	+0.43301	+0.25000
14	+0.49240	+0.08682
15	+0.49240	−0.08682
16	+0.43301	−0.25000
17	+0.32139	−0.38302
18	+0.17101	−0.46985

19 Holes

	x	y
1	0.00000	−0.50000
2	−0.16235	−0.47291
3	−0.30711	−0.39457
4	−0.41858	−0.27347
5	−0.48470	−0.12274
6	−0.49829	+0.04129
7	−0.45789	+0.20085
..	....	....

Table 3 (*Concluded*). Hole Coordinate Dimension Factors for Jig Boring — Type "A" Hole Circles, Central Coordinates (English or metric units)

x8 -0.36786	21 Holes	x12 0.00000	x23 +0.13490	x8 -0.49114	x16 +0.23236	x23 +0.45911
y8 +0.33864	x1 0.00000	y12 +0.50000	y23 -0.48146	y8 +0.09369	y16 +0.44273	y23 -0.19804
x9 -0.23797	y1 -0.50000	x13 +0.14087	24 Holes	x9 -0.45241	x17 +0.33156	x24 +0.40106
y9 +0.43974	x2 -0.14738	y13 +0.47975	x1 0.00000	y9 +0.21289	y17 +0.37426	y24 -0.29858
x10 -0.08230	y2 -0.47779	x14 +0.27032	y1 -0.50000	x10 -0.38526	x18 +0.41149	x25 +0.32139
y10 +0.49318	x3 -0.28166	y14 +0.42063	x2 -0.12941	y10 +0.31871	y18 +0.28403	y25 -0.38302
x11 +0.08230	y3 -0.41312	x15 +0.37787	y2 -0.48296	x11 -0.29389	x19 +0.46751	x26 +0.22440
y11 +0.49318	x4 -0.39092	y15 +0.32743	x3 -0.25000	y11 +0.40451	y19 +0.17730	y26 -0.44682
x12 +0.23797	y4 -0.31174	x16 +0.45482	y3 -0.43301	x12 -0.18406	x20 +0.49635	x27 +0.11531
y12 +0.43974	x5 -0.46544	y16 +0.20771	x4 -0.35355	y12 +0.46489	y20 +0.06027	y27 -0.48652
x13 +0.36786	y5 -0.18267	x17 +0.49491	y4 -0.35355	x13 -0.06267	x21 +0.49635	
y13 +0.33864	x6 -0.49860	y17 +0.07116	x5 -0.43301	y13 +0.49606	y21 -0.06027	28 Holes
x14 +0.45789	y6 -0.03737	x18 +0.49491	y5 -0.25000	x14 +0.06267	x22 +0.46751	x1 0.00000
y14 +0.20085	x7 -0.48746	y18 -0.07116	x6 -0.48296	y14 +0.49606	y22 -0.17730	y1 -0.50000
x15 +0.49829	y7 +0.11126	x19 +0.45482	y6 -0.12941	x15 +0.18406	x23 +0.41149	x2 -0.11126
y15 +0.04129	x8 -0.43301	y19 -0.20771	x7 -0.50000	y15 +0.46489	y23 -0.28403	y2 -0.48746
x16 +0.48470	y8 +0.25000	x20 +0.37787	y7 0.00000	x16 +0.29389	x24 +0.33156	x3 -0.21694
y16 -0.12274	x9 -0.34009	y20 -0.32743	x8 -0.48296	y16 +0.40451	y24 -0.37426	y3 -0.45048
x17 +0.41858	y9 +0.36653	x21 +0.27032	y8 +0.12941	x17 +0.38526	x25 +0.23236	x4 -0.31174
y17 -0.27347	x10 -0.21694	y21 -0.42063	x9 -0.43301	y17 +0.31871	y25 -0.44273	y4 -0.39092
x18 +0.30711	y10 +0.45048	x22 +0.14087	y9 +0.25000	x18 +0.45241	x26 +0.11966	x5 -0.39092
y18 -0.39457	x11 -0.07452	y22 -0.47975	x10 -0.35355	y18 +0.21289	y26 -0.48547	y5 -0.31174
x19 +0.16235	y11 +0.49442		y10 +0.35355	x19 +0.49114		x6 -0.45048
y19 -0.47291	x12 +0.07452	23 Holes	x11 -0.25000	y19 +0.09369	27 Holes	y6 -0.21694
	y12 +0.49442	x1 0.00000	y11 +0.43301	x20 +0.49901	x1 0.00000	x7 -0.48746
20 Holes	x13 +0.21694	y1 -0.50000	x12 -0.12941	y20 -0.03140	y1 -0.50000	y7 -0.11126
x1 0.00000	y13 +0.45048	x2 -0.13490	y12 +0.48296	x21 +0.47553	x2 -0.11531	x8 -0.50000
y1 -0.50000	x14 +0.34009	y2 -0.48146	x13 0.00000	y21 -0.15451	y2 -0.48652	y8 0.00000
x2 -0.15451	y14 +0.36653	x3 -0.25979	y13 +0.50000	x22 +0.42216	x3 -0.22440	x9 -0.48746
y2 -0.47553	x15 +0.43301	y3 -0.42721	x14 +0.12941	y22 -0.26791	y3 -0.44682	y9 +0.11126
x3 -0.29389	y15 +0.25000	x4 -0.36542	y14 +0.48296	x23 +0.34227	x4 -0.32139	x10 -0.45048
y3 -0.40451	x16 +0.48746	y4 -0.34128	x15 +0.25000	y23 -0.36448	y4 -0.38302	y10 +0.21694
x4 -0.40451	y16 +0.11126	x5 -0.44394	y15 +0.43301	x24 +0.24088	x5 -0.40106	x11 -0.39092
y4 -0.29389	x17 +0.49860	y5 -0.22003	x16 +0.35355	y24 -0.43815	y5 -0.29858	y11 +0.31174
x5 -0.47553	y17 -0.03737	x6 -0.48954	y16 +0.35355	x25 +0.12434	x6 -0.45911	x12 -0.31174
y5 -0.15451	x18 +0.46544	y6 -0.10173	x17 +0.43301	y25 -0.48429	y6 -0.19804	y12 +0.39092
x6 -0.50000	y18 -0.18267	x7 -0.49883	y17 +0.25000		x7 -0.49240	x13 -0.21694
y6 0.00000	x19 +0.39092	y7 +0.03412	x18 +0.48296	26 Holes	y7 -0.08682	y13 +0.45048
x7 -0.47553	y19 -0.31174	x8 -0.47113	y18 +0.12941	x1 0.00000	x8 -0.49915	x14 -0.11126
y7 +0.15451	x20 +0.28166	y8 +0.16744	x19 +0.50000	y1 -0.50000	y8 +0.02907	y14 +0.48746
x8 -0.40451	y20 -0.41312	x9 -0.40848	y19 0.00000	x2 -0.11966	x9 -0.47899	x15 0.00000
y8 +0.29389	x21 +0.14738	y9 +0.28834	x20 +0.48296	y2 -0.48547	y9 +0.14340	y15 +0.50000
x9 -0.29389	y21 -0.47779	x10 -0.31554	y20 -0.12941	x3 -0.23236	x10 -0.43301	x16 +0.11126
y9 +0.40451		y10 +0.38786	x21 +0.43301	y3 -0.44273	y10 +0.25000	y16 +0.48746
x10 -0.15451	22 Holes	x11 -0.19920	y21 -0.25000	x4 -0.33156	x11 -0.36369	x17 +0.21694
y10 +0.47553	x1 0.00000	y11 +0.45861	x22 +0.35355	y4 -0.37426	y11 +0.34312	y17 +0.45048
x11 0.00000	y1 -0.50000	x12 -0.06808	y22 -0.35355	x5 -0.41149	x12 -0.27475	x18 +0.31174
y11 +0.50000	x2 -0.14087	y12 +0.49534	x23 +0.25000	y5 -0.28403	y12 +0.41774	y18 +0.39092
x12 +0.15451	y2 -0.47975	x13 +0.06808	y23 -0.43301	x6 -0.46751	x13 -0.17101	x19 +0.39092
y12 +0.47553	x3 -0.27032	y13 +0.49534	x24 +0.12941	y6 -0.17730	y13 +0.46985	y19 +0.31174
x13 +0.29389	y3 -0.42063	x14 +0.19920	y24 -0.48296	x7 -0.49635	x14 -0.05805	x20 +0.45048
y13 +0.40451	x4 -0.37787	y14 +0.45861		y7 -0.06027	y14 +0.49662	y20 +0.21694
x14 +0.40451	y4 -0.32743	x15 +0.31554	25 Holes	x8 -0.49635	x15 +0.05805	x21 +0.48746
y14 +0.29389	x5 -0.45482	y15 +0.38786	x1 0.00000	y8 +0.06027	y15 +0.49662	y21 +0.11126
x15 +0.47553	y5 -0.20771	x16 +0.40848	y1 -0.50000	x9 -0.46751	x16 +0.17101	x22 +0.50000
y15 +0.15451	x6 -0.49491	y16 +0.28834	x2 -0.12434	y9 +0.17730	y16 +0.46985	y22 0.00000
x16 +0.50000	y6 -0.07116	x17 +0.47113	y2 -0.48429	x10 -0.41149	x17 +0.27475	x23 +0.48746
y16 0.00000	x7 -0.49491	y17 +0.16744	x3 -0.24088	y10 +0.28403	y17 +0.41774	y23 -0.11126
x17 +0.47553	x8 -0.45482	x18 +0.49883	y3 -0.43815	x11 -0.33156	x18 +0.36369	x24 +0.45048
y17 -0.15451	y8 +0.20771	y18 +0.03412	x4 -0.34227	y11 +0.37426	y18 +0.34312	y24 -0.21694
x18 +0.40451	x9 -0.37787	y19 -0.10173	y4 -0.36448	x12 -0.23236	x19 +0.43301	x25 +0.39092
y18 -0.29389	y9 +0.32743	x20 -0.44394	x5 -0.42216	y12 +0.44273	y19 +0.25000	y25 -0.31174
x19 +0.29389	x10 -0.27032	y20 -0.22003	y5 -0.26791	x13 -0.11966	x20 +0.47899	x26 +0.31174
y19 -0.40451	y10 +0.42063	x21 +0.36542	x6 -0.47553	y13 +0.48547	y20 +0.14340	y26 -0.39092
x20 +0.15451	x11 -0.14087	y21 -0.34128	y6 -0.15451	x14 0.00000	x21 +0.49915	x27 +0.21694
y20 -0.47553	y11 +0.47975	x22 +0.25979	x7 -0.49901	y14 +0.50000	y21 +0.02907	y27 -0.45048
.. ....		y22 -0.42721	y7 -0.03140	x15 +0.11966	x22 +0.49240	x28 +0.11126
.. ....				y15 +0.48547	y22 -0.08682	y28 -0.48746

Table 4. Hole Coordinate Dimension Factors for Jig Boring — Type "B" Hole Circles
Central Coordinates (English or metric units)

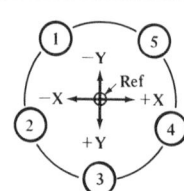

The diagram shows a type "B" circle for a 5-hole circle. Coordinates x, y are given in the table for hole circles of from 3 to 28 holes. Dimensions are for holes numbered in a counterclockwise direction (as shown). Dimensions given are based upon a hole circle of unit diameter. For a hole circle of, say, 3-inch or 3-centimeter diameter, multiply table values by 3.

3 Holes

Hole	x	y
1	−0.43301	−0.25000
2	0.00000	+0.50000
3	+0.43301	−0.25000

4 Holes

Hole	x	y
1	−0.35355	−0.35355
2	−0.35355	+0.35355
3	+0.35355	+0.35355
4	+0.35355	−0.35355

5 Holes

Hole	x	y
1	−0.29389	−0.40451
2	−0.47553	+0.15451
3	0.00000	+0.50000
4	+0.47553	+0.15451
5	+0.29389	−0.40451

6 Holes

Hole	x	y
1	−0.25000	−0.43301
2	−0.50000	0.00000
3	−0.25000	+0.43301
4	+0.25000	+0.43301
5	+0.50000	0.00000
6	+0.25000	−0.43301

7 Holes

Hole	x	y
1	−0.21694	−0.45048
2	−0.48746	−0.11126
3	−0.39092	+0.31174
4	0.00000	+0.50000
5	+0.39092	+0.31174
6	+0.48746	−0.11126
7	+0.21694	−0.45048

8 Holes

Hole	x	y
1	−0.19134	−0.46194
2	−0.46194	−0.19134
3	−0.46194	+0.19134
4	−0.19134	+0.46194
5	+0.19134	+0.46194
6	+0.46194	+0.19134
7	+0.46194	−0.19134
8	+0.19134	−0.46194

9 Holes

Hole	x	y
1	−0.17101	−0.46985
2	−0.43301	−0.25000
3	−0.49240	+0.08682
4	−0.32139	+0.38302
5	0.00000	+0.50000
6	+0.32139	+0.38302
7	+0.49240	+0.08682
8	+0.43301	−0.25000
9	+0.17101	−0.46985

10 Holes

Hole	x	y
1	−0.15451	−0.47553
2	−0.40451	−0.29389
3	−0.50000	0.00000
4	−0.40451	+0.29389
5	−0.15451	+0.47553
6	+0.15451	+0.47553
7	+0.40451	+0.29389
8	+0.50000	0.00000
9	+0.40451	−0.29389
10	+0.15451	−0.47553

11 Holes

Hole	x	y
1	−0.14087	−0.47975
2	−0.37787	−0.32743
3	−0.49491	−0.07116
4	−0.45482	+0.20771
5	−0.27032	+0.42063
6	0.00000	+0.50000
7	+0.27032	+0.42063
8	+0.45482	+0.20771
9	+0.49491	−0.07116
10	+0.37787	−0.32743
11	+0.14087	−0.47975

12 Holes

Hole	x	y
1	−0.12941	−0.48296
2	−0.35355	−0.35355
3	−0.48296	−0.12941
4	−0.48296	+0.12941
5	−0.35355	+0.35355
6	−0.12941	+0.48296
7	+0.12941	+0.48296
8	+0.35355	+0.35355
9	+0.48296	+0.12941
10	+0.48296	−0.12941
11	+0.35355	−0.35355
12	+0.12941	−0.48296

13 Holes

Hole	x	y
1	−0.11966	−0.48547
2	−0.33156	−0.37426
3	−0.46751	−0.17730
4	−0.49635	+0.06027
5	−0.41149	+0.28403
6	−0.23236	+0.44273
7	0.00000	+0.50000
8	+0.23236	+0.44273
9	+0.41149	+0.28403
10	+0.49635	+0.06027
11	+0.46751	−0.17730
12	+0.33156	−0.37426
13	+0.11966	−0.48547

14 Holes

Hole	x	y
1	−0.11126	−0.48746
2	−0.31174	−0.39092
3	−0.45048	−0.21694
4	−0.50000	0.00000
5	−0.45048	+0.21694
6	−0.31174	+0.39092
7	−0.11126	+0.48746
8	+0.11126	+0.48746
9	+0.31174	+0.39092
10	+0.45048	+0.21694
11	+0.50000	0.00000
12	+0.45048	−0.21694
13	+0.31174	−0.39092
14	+0.11126	−0.48746

15 Holes

Hole	x	y
1	−0.10396	−0.48907
2	−0.29389	−0.40451
3	−0.43301	−0.25000
4	−0.49726	−0.05226
5	−0.47553	+0.15451
6	−0.37157	+0.33457
7	−0.20337	+0.45677
8	0.00000	+0.50000
9	+0.20337	+0.45677
10	+0.37157	+0.33457
11	+0.47553	+0.15451
12	+0.49726	−0.05226
13	+0.43301	−0.25000
14	+0.29389	−0.40451
15	+0.10396	−0.48907

16 Holes

Hole	x	y
1	−0.09755	−0.49039
2	−0.27779	−0.41573
3	−0.41573	−0.27779
4	−0.49039	−0.09755
5	−0.49039	+0.09755
6	−0.41573	+0.27779
7	−0.27779	+0.41573
8	−0.09755	+0.49039
9	+0.09755	+0.49039
10	+0.27779	+0.41573
11	+0.41573	+0.27779
12	+0.49039	+0.09755
13	+0.49039	−0.09755
14	+0.41573	−0.27779
15	+0.27779	−0.41573
16	+0.09755	−0.49039

17 Holes

Hole	x	y
1	−0.09187	−0.49149
2	−0.26322	−0.42511
3	−0.39901	−0.30132
4	−0.48091	−0.13683
5	−0.49787	+0.04613
6	−0.44758	+0.22287
7	−0.33685	+0.36950
8	−0.18062	+0.46624
9	0.00000	+0.50000
10	+0.18062	+0.46624
11	+0.33685	+0.36950
12	+0.44758	+0.22287
13	+0.49787	+0.04613
14	+0.48091	−0.13683
15	+0.39901	−0.30132
16	+0.26322	−0.42511
17	+0.09187	−0.49149

18 Holes

Hole	x	y
1	−0.08682	−0.49240
2	−0.25000	−0.43301
3	−0.38302	−0.32139
4	−0.46985	−0.17101
5	−0.50000	0.00000
6	−0.46985	+0.17101
7	−0.38302	+0.32139
8	−0.25000	+0.43301
9	−0.08682	+0.49240
10	+0.08682	+0.49240
11	+0.25000	+0.43301
12	+0.38302	+0.32139
13	+0.46985	+0.17101
14	+0.50000	0.00000
15	+0.46985	−0.17101
16	+0.38302	−0.32139
17	+0.25000	−0.43301
18	+0.08682	−0.49240

19 Holes

Hole	x	y
1	−0.08230	−0.49318
2	−0.23797	−0.43974
3	−0.36786	−0.33864
4	−0.45789	−0.20085
5	−0.49829	−0.04129
6	−0.48470	+0.12274
7	−0.41858	+0.27347
…	……	……

Table 4 (*Concluded*). **Hole Coordinate Dimension Factors for Jig Boring — Type "B"**
Hole Circles, Central Coordinates (English or metric units)

19 Holes (continued)

i	x	y
8	−0.30711	+0.39457
9	−0.16235	+0.47291
10	0.00000	+0.50000
11	+0.16235	+0.47291
12	+0.30711	+0.39457
13	+0.41858	+0.27347
14	+0.48470	+0.12274
15	+0.49829	−0.04129
16	+0.45789	−0.20085
17	+0.36786	−0.33864
18	+0.23797	−0.43974
19	+0.08230	−0.49318

20 Holes

i	x	y
1	−0.07822	−0.49384
2	−0.22700	−0.44550
3	−0.35355	−0.35355
4	−0.44550	−0.22700
5	−0.49384	−0.07822
6	−0.49384	+0.07822
7	−0.44550	+0.22700
8	−0.35355	+0.35355
9	−0.22700	+0.44550
10	−0.07822	+0.49384
11	+0.07822	+0.49384
12	+0.22700	+0.44550
13	+0.35355	+0.35355
14	+0.44550	+0.22700
15	+0.49384	+0.07822
16	+0.49384	−0.07822
17	+0.44550	−0.22700
18	+0.35355	−0.35355
19	+0.22700	−0.44550
20	+0.07822	−0.49384

21 Holes

i	x	y
1	−0.07452	−0.49442
2	−0.21694	−0.45048
3	−0.34009	−0.36653
4	−0.43301	−0.25000
5	−0.48746	−0.11126
6	−0.49860	+0.03737
7	−0.46544	+0.18267
8	−0.39092	+0.31174
9	−0.28166	+0.41312
10	−0.14738	+0.47779
11	0.00000	+0.50000
12	+0.14738	+0.47779
13	+0.28166	+0.41312
14	+0.39092	+0.31174
15	+0.46544	+0.18267
16	+0.49860	+0.03737
17	+0.48746	−0.11126
18	+0.43301	−0.25000
19	+0.34009	−0.36653
20	+0.21694	−0.45048
21	+0.07452	−0.49442

22 Holes

i	x	y
1	−0.07116	−0.49491
2	−0.20771	−0.45482
3	−0.32743	−0.37787
4	−0.42063	−0.27032
5	−0.47975	−0.14087
6	−0.50000	0.00000
7	−0.47975	+0.14087
8	−0.42063	+0.27032
9	−0.32743	+0.37787
10	−0.20771	+0.45482
11	−0.07116	+0.49491
12	+0.07116	+0.49491
13	+0.20771	+0.45482
14	+0.32743	+0.37787
15	+0.42063	+0.27032
16	+0.47975	+0.14087
17	+0.50000	0.00000
18	+0.47975	−0.14087
19	+0.42063	−0.27032
20	+0.32743	−0.37787
21	+0.20771	−0.45482
22	+0.07116	−0.49491

23 Holes

i	x	y
1	−0.06808	−0.49534
2	−0.19920	−0.45861
3	−0.31554	−0.38786
4	−0.40848	−0.28834
5	−0.47113	−0.16744
6	−0.49883	−0.03412
7	−0.48954	+0.10173
8	−0.44394	+0.23003
9	−0.36542	+0.34128
10	−0.25979	+0.42721
11	−0.13490	+0.48146
12	0.00000	+0.50000
13	+0.13490	+0.48146
14	+0.25979	+0.42721
15	+0.36542	+0.34128
16	+0.44394	+0.23003
17	+0.48954	+0.10173
18	+0.49883	−0.03412
19	+0.47113	−0.16744
20	+0.40848	−0.28834
21	+0.31554	−0.38786
22	+0.19920	−0.45861
23	+0.06808	−0.49534

24 Holes

i	x	y
1	−0.06526	−0.49572
2	−0.19134	−0.46194
3	−0.30438	−0.39668
4	−0.39668	−0.30438
5	−0.46194	−0.19134
6	−0.49572	−0.06526
7	−0.49572	+0.06526
8	−0.46194	+0.19134
9	−0.39668	+0.30438
10	−0.30438	+0.39668
11	−0.19134	+0.46194
12	−0.06526	+0.49572
13	+0.06526	+0.49572
14	+0.19134	+0.46194
15	+0.30438	+0.39668
16	+0.39668	+0.30438
17	+0.46194	+0.19134
18	+0.49572	+0.06526
19	+0.49572	−0.06526
20	+0.46194	−0.19134
21	+0.39668	−0.30438
22	+0.30438	−0.39668
23	+0.19134	−0.46194
24	+0.06526	−0.49572

25 Holes

i	x	y
1	−0.06267	−0.49606
2	−0.18406	−0.46489
3	−0.29389	−0.40451
4	−0.38526	−0.31871
5	−0.45241	−0.21289
6	−0.49114	−0.09369
7	−0.49901	+0.03140
8	−0.47553	+0.15451
9	−0.42216	+0.26791
10	−0.34227	+0.36448
11	−0.24088	+0.43815
12	−0.12434	+0.48429
13	0.00000	+0.50000
14	+0.12434	+0.48429
15	+0.24088	+0.43815
16	+0.34227	+0.36448
17	+0.42216	+0.26791
18	+0.47553	+0.15451
19	+0.49901	+0.03140
20	+0.49114	−0.09369
21	+0.45241	−0.21289
22	+0.38526	−0.31871
23	+0.29389	−0.40451
24	+0.18406	−0.46489
25	+0.06267	−0.49606

26 Holes

i	x	y
1	−0.06027	−0.49635
2	−0.17730	−0.46751
3	−0.28403	−0.41149
4	−0.37426	−0.33156
5	−0.44273	−0.23236
6	−0.48547	−0.11966
7	−0.50000	0.00000
8	−0.48547	+0.11966
9	−0.44273	+0.23236
10	−0.37426	+0.33156
11	−0.28403	+0.41149
12	−0.17730	+0.46751
13	−0.06027	+0.49635
14	+0.06027	+0.49635
15	+0.17730	+0.46751
16	+0.28403	+0.41149
17	+0.37426	+0.33156
18	+0.44273	+0.23236
19	+0.48547	+0.11966
20	+0.50000	0.00000
21	+0.48547	−0.11966
22	+0.44273	−0.23236
23	+0.37426	−0.33156
24	+0.28403	−0.41149
25	+0.17730	−0.46751
26	+0.06027	−0.49635

27 Holes

i	x	y
1	−0.05805	−0.49662
2	−0.17101	−0.46985
3	−0.27475	−0.41774
4	−0.36369	−0.34312
5	−0.43301	−0.25000
6	−0.47899	−0.14340
7	−0.49915	−0.02907
8	−0.49240	+0.08682
9	−0.45911	+0.19804
10	−0.40106	+0.29858
11	−0.32139	+0.38302
12	−0.22440	+0.44682
13	−0.11531	+0.48652
14	0.00000	+0.50000
15	+0.11531	+0.48652
16	+0.22440	+0.44682
17	+0.32139	+0.38302
18	+0.40106	+0.29858
19	+0.45911	+0.19804
20	+0.49240	+0.08682
21	+0.49915	−0.02907
22	+0.47899	−0.14340
23	+0.43301	−0.25000
24	+0.36369	−0.34312
25	+0.27475	−0.41774
26	+0.17101	−0.46985
27	+0.05805	−0.49662

28 Holes

i	x	y
1	−0.05598	−0.49686
2	−0.16514	−0.47194
3	−0.26602	−0.42336
4	−0.35355	−0.35355
5	−0.42336	−0.26602
6	−0.47194	−0.16514
7	−0.49686	−0.05598
8	−0.49686	+0.05598
9	−0.47194	+0.16514
10	−0.42336	+0.26602
11	−0.35355	+0.35355
12	−0.26602	+0.42336
13	−0.16514	+0.47194
14	−0.05598	+0.49686
15	+0.05598	+0.49686
16	+0.16514	+0.47194
17	+0.26602	+0.42336
18	+0.35355	+0.35355
19	+0.42336	+0.26602
20	+0.47194	+0.16514
21	+0.49686	+0.05598
22	+0.49686	−0.05598
23	+0.47194	−0.16514
24	+0.42336	−0.26602
25	+0.35355	−0.35355
26	+0.26602	−0.42336
27	+0.16514	−0.47194
28	+0.05598	−0.49686

MATHEMATICS

Diameters of Circles and Sides of Squares of Equal Area

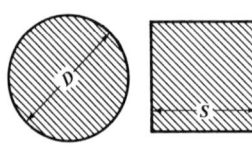

The table below will be found useful for determining the diameter of a circle of an area equal to that of a square, the side of which is known, or for determining the side of a square which has an area equal to that of a circle, the area or diameter of which is known. For example, if the diameter of a circle is 17½ inches, it is found from the table that the side of a square of the same area is 15.51 inches.

Diam. of Circle, D	Side of Square, S	Area of Circle or Square	Diam. of Circle, D	Side of Square, S	Area of Circle or Square	Diam. of Circle, D	Side of Square S	Area of Circle or Square
½	0.44	0.196	20½	18.17	330.06	40½	35.89	1288.25
1	0.89	0.785	21	18.61	346.36	41	36.34	1320.25
1½	1.33	1.767	21½	19.05	363.05	41½	36.78	1352.65
2	1.77	3.142	22	19.50	380.13	42	37.22	1385.44
2½	2.22	4.909	22½	19.94	397.61	42½	37.66	1418.63
3	2.66	7.069	23	20.38	415.48	43	38.11	1452.20
3½	3.10	9.621	23½	20.83	433.74	43½	38.55	1486.17
4	3.54	12.566	24	21.27	452.39	44	38.99	1520.53
4½	3.99	15.904	24½	21.71	471.44	44½	39.44	1555.28
5	4.43	19.635	25	22.16	490.87	45	39.88	1590.43
5½	4.87	23.758	25½	22.60	510.71	45½	40.32	1625.97
6	5.32	28.274	26	23.04	530.93	46	40.77	1661.90
6½	5.76	33.183	26½	23.49	551.55	46½	41.21	1698.23
7	6.20	38.485	27	23.93	572.56	47	41.65	1734.94
7½	6.65	44.179	27½	24.37	593.96	47½	42.10	1772.05
8	7.09	50.265	28	24.81	615.75	48	42.54	1809.56
8½	7.53	56.745	28½	25.26	637.94	48½	42.98	1847.45
9	7.98	63.617	29	25.70	660.52	49	43.43	1885.74
9½	8.42	70.882	29½	26.14	683.49	49½	43.87	1924.42
10	8.86	78.540	30	26.59	706.86	50	44.31	1963.50
10½	9.31	86.590	30½	27.03	730.62	50½	44.75	2002.96
11	9.75	95.033	31	27.47	754.77	51	45.20	2042.82
11½	10.19	103.87	31½	27.92	779.31	51½	45.64	2083.07
12	10.63	113.10	32	28.36	804.25	52	46.08	2123.72
12½	11.08	122.72	32½	28.80	829.58	52½	46.53	2164.75
13	11.52	132.73	33	29.25	855.30	53	46.97	2206.18
13½	11.96	143.14	33½	29.69	881.41	53½	47.41	2248.01
14	12.41	153.94	34	30.13	907.92	54	47.86	2290.22
14½	12.85	165.13	34½	30.57	934.82	54½	48.30	2332.83
15	13.29	176.71	35	31.02	962.11	55	48.74	2375.83
15½	13.74	188.69	35½	31.46	989.80	55½	49.19	2419.22
16	14.18	201.06	36	31.90	1017.88	56	49.63	2463.01
16½	14.62	213.82	36½	32.35	1046.35	56½	50.07	2507.19
17	15.07	226.98	37	32.79	1075.21	57	50.51	2551.76
17½	15.51	240.53	37½	33.23	1104.47	57½	50.96	2596.72
18	15.95	254.47	38	33.68	1134.11	58	51.40	2642.08
18½	16.40	268.80	38½	34.12	1164.16	58½	51.84	2687.83
19	16.84	283.53	39	34.56	1194.59	59	52.29	2733.97
19½	17.28	298.65	39½	35.01	1225.42	59½	52.73	2780.51
20	17.72	314.16	40	35.45	1256.64	60	53.17	2827.43

Distance Across Corners of Squares and Hexagons
(English and metric units)

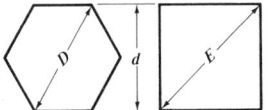

$$D = 1.154701\ d$$
$$E = 1.414214\ d$$

d	D	E	d	D	E	d	D	E	d	D	E
1/32	0.0361	0.0442	0.9	1.0392	1.2728	32	36.9504	45.2548	67	77.3649	94.7523
1/16	0.0722	0.0884	29/32	1.0464	1.2816	33	38.1051	46.6690	68	78.5196	96.1665
3/32	0.1083	0.1326	15/16	1.0825	1.3258	34	39.2598	48.0833	69	79.6743	97.5807
0.1	0.1155	0.1414	31/32	1.1186	1.3700	35	40.4145	49.4975	70	80.8290	98.9949
1/8	0.1443	0.1768	1.0	1.1547	1.4142	36	41.5692	50.9117	71	81.9837	100.409
5/32	0.1804	0.2210	2.0	2.3094	2.8284	37	42.7239	52.3259	72	83.1384	101.823
3/16	0.2165	0.2652	3.0	3.4641	4.2426	38	43.8786	53.7401	73	84.2931	103.238
0.2	0.2309	0.2828	4.0	4.6188	5.6569	39	45.0333	55.1543	74	85.4478	104.652
7/32	0.2526	0.3094	5.0	5.7735	7.0711	40	46.1880	56.5685	75	86.6025	106.066
1/4	0.2887	0.3536	6.0	6.9282	8.4853	41	47.3427	57.9828	76	87.7572	107.480
9/32	0.3248	0.3977	7.0	8.0829	9.8995	42	48.4974	59.3970	77	88.9119	108.894
0.3	0.3464	0.4243	8.0	9.2376	11.3137	43	49.6521	60.8112	78	90.0666	110.309
5/16	0.3608	0.4419	9.0	10.3923	12.7279	44	50.8068	62.2254	79	91.2213	111.723
11/32	0.3969	0.4861	10	11.5470	14.1421	45	51.9615	63.6396	80	92.3760	113.137
3/8	0.4330	0.5303	11	12.7017	15.5563	46	53.1162	65.0538	81	93.5307	114.551
0.4	0.4619	0.5657	12	13.8564	16.9706	47	54.2709	66.4680	82	94.6854	115.966
13/32	0.4691	0.5745	13	15.0111	18.3848	48	55.4256	67.8823	83	95.8401	117.380
7/16	0.5052	0.6187	14	16.1658	19.7990	49	56.5803	69.2965	84	96.9948	118.794
15/32	0.5413	0.6629	15	17.3205	21.2132	50	57.7350	70.7107	85	98.1495	120.208
0.5	0.5774	0.7071	16	18.4752	22.6274	51	58.8897	72.1249	86	99.3042	121.622
17/32	0.6134	0.7513	17	19.6299	24.0416	52	60.0444	73.5391	87	100.459	123.037
9/16	0.6495	0.7955	18	20.7846	25.4558	53	61.1991	74.9533	88	101.614	124.451
19/32	0.6856	0.8397	19	21.9393	26.8701	54	62.3538	76.3675	89	102.768	125.865
0.6	0.6928	0.8485	20	23.0940	28.2843	55	63.5085	77.7817	90	103.923	127.279
5/8	0.7217	0.8839	21	24.2487	29.6985	56	64.6632	79.1960	91	105.078	128.693
21/32	0.7578	0.9281	22	25.4034	31.1127	57	65.8179	80.6102	92	106.232	130.108
11/16	0.7939	0.9723	23	26.5581	32.5269	58	66.9726	82.0244	93	107.387	131.522
0.7	0.8083	0.9899	24	27.7128	33.9411	59	68.1273	83.4386	94	108.542	132.936
23/32	0.8299	1.0165	25	28.8675	35.3553	60	69.2820	84.8528	95	109.697	134.350
3/4	0.8660	1.0607	26	30.0222	36.7696	61	70.4367	86.2670	96	110.851	135.765
25/32	0.9021	1.1049	27	31.1769	38.1838	62	71.5914	87.6812	97	112.006	137.179
0.8	0.9238	1.1314	28	32.3316	39.5980	63	72.7461	89.0955	98	113.161	138.593
13/16	0.9382	1.1490	29	33.4863	41.0122	64	73.9008	90.5097	99	114.315	140.007
27/32	0.9743	1.1932	30	34.6410	42.4264	65	75.0555	91.9239	100	115.470	141.421
7/8	1.0104	1.2374	31	35.7957	43.8406	66	76.2102	93.3381	...	...	...

A desired value not given directly in the table can be obtained by the simple addition of two or more values taken directly from the table. Further values can be obtained by shifting the decimal point.

Example 1: Find D when $d = 2\,5/16$ inches. From the table, $2 = 2.3094$, and $5/16 = 0.3608$. Therefore, $D = 2.3094 + 0.3608 = 2.6702$ inches.

Example 2: Find E when $d = 20.25$ millimeters. From the table, $20 = 28.2843$; $0.2 = 0.2828$; and $0.05 = 0.0707$ (obtained by shifting the decimal point one place to the left at $d = 0.5$). Thus, $E = 28.2843 + 0.2828 + 0.0707 = 28.6378$ millimeters.

Formulas and Table for Regular Polygons
(English and metric units)

N = number of sides.
S = length of side.
R = radius of circumscribed circle.
r = radius of inscribed circle.
A = area of polygon.
$\alpha = 180° \div N$ = one-half center angle of one side.

Formulas:

$A = (N \times \cot \alpha \times S^2) \div 4$
$A = N \times \sin \alpha \times \cos \alpha \times R^2$
$A = N \times \tan \alpha \times r^2$
$r = R \times \cos \alpha$
$r = (S \times \cot \alpha) \div 2$
$r = \sqrt{(A \times \cot \alpha) \div N}$ *

$R = S \div (2 \sin \alpha)$
$R = r \div \cos \alpha$
$R = \sqrt{A} \div (N \sin \alpha \cos \alpha)$ *

$S = 2R \times \sin \alpha$
$S = 2r \times \tan \alpha$
$S = 2\sqrt{(A \times \tan \alpha) \div N}$ *

* These formulas may be used to calculate R, S, or r needed to provide a required area A.

Examples of Use of Table. (English and metric units)

A regular hexagon is inscribed in a circle of 6 inches diameter. Find the area and the radius of an inscribed circle. — Here $R = 3$. From the table, area $(A) = 2.5981 R^2 = 2.5981 \times 9 = 23.3829$ square inches. Radius of inscribed circle, $r = 0.866 R = 0.866 \times 3 = 2.598$ inches.

An octagon is inscribed in a circle of 100 millimeters diameter. Thus $R = 50$. Find the area and radius of an inscribed circle. From the table, $A = 2.8284 R^2 = 2.8284 \times 2500 = 7071$ mm² $= 70.7$ cm². Radius of inscribed circle, $r = 0.9239 R = 0.9239 \times 50 = 46.195$ mm.

Thirty-two bolts are to be equally spaced on the periphery of a bolt-circle, 16 inches in diameter. Find the chordal distance between the bolts. — Chordal distance equals the side (S) of a polygon with 32 sides. $R = 8$. Hence, $S = 0.196 R = 0.196 \times 8 = 1.568$ inch.

Sixteen bolts are to be equally spaced on the periphery of a bolt-circle, 250 millimeters diameter. Find the chordal distance between the bolts. — Chordal distance equals the side (S) of a polygon with 16 sides. $R = 125$. Thus, $S = 0.3902 R = 0.3902 \times 125 = 48.775$ millimeters.

No. of Sides	$A =$	$A =$	$A =$	$R =$	$R =$	$S =$	$S =$	$r =$	$r =$	No. of Sides
3	0.4330 S^2	1.2990 R^2	5.1962 r^2	0.5774 S	2.0000 r	1.7321 R	3.4641 r	0.5000 R	0.2887 S	3
4	1.0000 S^2	2.0000 R^2	4.0000 r^2	0.7071 S	1.4142 r	1.4142 R	2.0000 r	0.7071 R	0.5000 S	4
5	1.7205 S^2	2.3776 R^2	3.6327 r^2	0.8507 S	1.2361 r	1.1756 R	1.4531 r	0.8090 R	0.6882 S	5
6	2.5981 S^2	2.5981 R^2	3.4641 r^2	1.0000 S	1.1547 r	1.0000 R	1.1547 r	0.8660 R	0.8660 S	6
7	3.6339 S^2	2.7364 R^2	3.3710 r^2	1.1524 S	1.1099 r	0.8678 R	0.9631 r	0.9010 R	1.0383 S	7
8	4.8284 S^2	2.8284 R^2	3.3137 r^2	1.3066 S	1.0824 r	0.7654 R	0.8284 r	0.9239 R	1.2071 S	8
9	6.1818 S^2	2.8925 R^2	3.2757 r^2	1.4619 S	1.0642 r	0.6840 R	0.7279 r	0.9397 R	1.3737 S	9
10	7.6942 S^2	2.9389 R^2	3.2492 r^2	1.6180 S	1.0515 r	0.6180 R	0.6498 r	0.9511 R	1.5388 S	10
12	11.196 S^2	3.0000 R^2	3.2154 r^2	1.9319 S	1.0353 r	0.5176 R	0.5359 r	0.9659 R	1.8660 S	12
16	20.109 S^2	3.0615 R^2	3.1826 r^2	2.5629 S	1.0196 r	0.3902 R	0.3978 r	0.9868 R	2.5137 S	16
20	31.569 S^2	3.0902 R^2	3.1677 r^2	3.1962 S	1.0125 r	0.3129 R	0.3168 r	0.9877 R	3.1569 S	20
24	45.575 S^2	3.1058 R^2	3.1597 r^2	3.8306 S	1.0086 r	0.2611 R	0.2633 r	0.9914 R	3.7979 S	24
32	81.225 S^2	3.1214 R^2	3.1517 r^2	5.1011 S	1.0048 r	0.1960 R	0.1970 r	0.9952 R	5.0766 S	32
48	183.08 S^2	3.1326 R^2	3.1461 r^2	7.6449 S	1.0021 r	0.1308 R	0.1311 r	0.9979 R	7.6285 S	48
64	325.69 S^2	3.1365 R^2	3.1441 r^2	10.190 S	1.0012 r	0.0981 R	0.0983 r	0.9988 R	10.178 S	64

Areas and Dimensions of Plane Figures

In the following tables are given the areas of plane figures, together with other formulas relating to their dimensions and properties; the surfaces of solids; and the volumes of solids. The notation used in the formulas is, as far as possible, given in the illustration accompanying them; where this has not been possible, it is given at the beginning of each set of formulas.

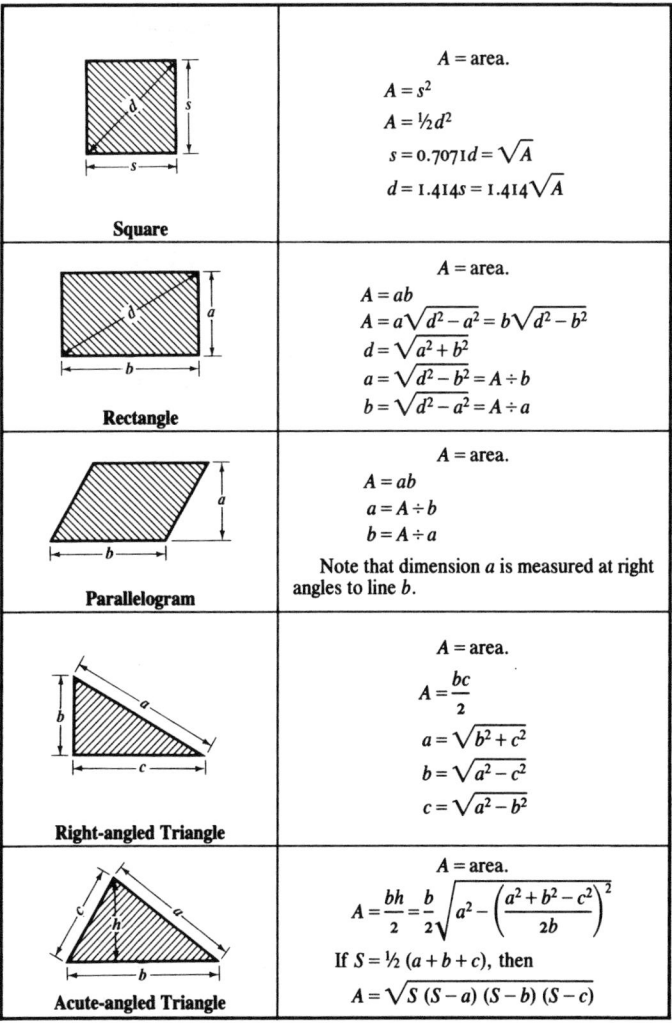

Square

A = area.

$A = s^2$

$A = \frac{1}{2} d^2$

$s = 0.7071 d = \sqrt{A}$

$d = 1.414 s = 1.414 \sqrt{A}$

Rectangle

A = area.

$A = ab$

$A = a\sqrt{d^2 - a^2} = b\sqrt{d^2 - b^2}$

$d = \sqrt{a^2 + b^2}$

$a = \sqrt{d^2 - b^2} = A \div b$

$b = \sqrt{d^2 - a^2} = A \div a$

Parallelogram

A = area.

$A = ab$

$a = A \div b$

$b = A \div a$

Note that dimension a is measured at right angles to line b.

Right-angled Triangle

A = area.

$A = \frac{bc}{2}$

$a = \sqrt{b^2 + c^2}$

$b = \sqrt{a^2 - c^2}$

$c = \sqrt{a^2 - b^2}$

Acute-angled Triangle

A = area.

$A = \frac{bh}{2} = \frac{b}{2}\sqrt{a^2 - \left(\frac{a^2 + b^2 - c^2}{2b}\right)^2}$

If $S = \frac{1}{2}(a + b + c)$, then

$A = \sqrt{S(S-a)(S-b)(S-c)}$

Examples of the Use of the Formulas (English and metric units)

Below are given examples, some in English and some in metric units, showing the use of the formulas on the opposite page. Each section corresponds to the opposite section on the previous page, and the illustration on that page should be referred to. The notation used in the illustrations is also used in the examples given.

Square. — Assume that the side s of a square is 15 inches. Find the area and the length of the diagonal.

$$\text{Area} = A = s^2 = 15^2 = 225 \text{ square inches.}$$

$$\text{Diagonal} = d = 1.414s = 1.414 \times 15 = 21.21 \text{ inches.}$$

The area of a square is 625 square inches. Find the length of the side s and the diagonal d.

$$s = \sqrt{A} = \sqrt{625} = 25 \text{ inches.}$$

$$d = 1.414\sqrt{A} = 1.414 \times 25 = 35.35 \text{ inches.}$$

Rectangle. — The side a of a rectangle is 12 centimeters, and the area 70.5 square centimeters. Find the length of the side b, and the diagonal d.

$$b = A \div a = 70.5 \div 12 = 5.875 \text{ centimeters.}$$

$$d = \sqrt{a^2 + b^2} = \sqrt{12^2 + 5.875^2} = \sqrt{178.516} = 13.361 \text{ centimeters.}$$

The sides of a rectangle are 30.5 and 11 centimeters long. Find the area.

$$\text{Area} = a \times b = 30.5 \times 11 = 335.5 \text{ square centimeters.}$$

Parallelogram. — The base b of a parallelogram is 16 feet. The height a is 5.5 feet. Find the area.

$$\text{Area} = A = a \times b = 5.5 \times 16 = 88 \text{ square feet.}$$

The area of a parallelogram is 12 square inches. The height is 1.5 inches. Find the length of the base b.

$$b = A \div a = 12 \div 1.5 = 8 \text{ inches.}$$

Right-angled Triangle. — The sides b and c in a right-angled triangle are 6 and 8 inches. Find side a and the area.

$$a = \sqrt{b^2 + c^2} = \sqrt{6^2 + 8^2} = \sqrt{36 + 64} = \sqrt{100} = 10 \text{ inches.}$$

$$A = \frac{b \times c}{2} = \frac{6 \times 8}{2} = \frac{48}{2} = 24 \text{ square inches.}$$

If $a = 10$ and $b = 6$ had been known, but not c, the latter would have been found as follows:

$$c = \sqrt{a^2 - b^2} = \sqrt{10^2 - 6^2} = \sqrt{100 - 36} = \sqrt{64} = 8 \text{ inches.}$$

Acute-angled Triangle. — If $a = 10$, $b = 9$, and $c = 8$ centimeters, what is the area of the triangle?

$$A = \frac{b}{2}\sqrt{a^2 - \left(\frac{a^2 + b^2 - c^2}{2b}\right)^2} = \frac{9}{2}\sqrt{10^2 - \left(\frac{10^2 + 9^2 - 8^2}{2 \times 9}\right)^2} = 4.5\sqrt{100 - \left(\frac{117}{18}\right)^2}$$

$$= 4.5\sqrt{100 - 42.25} = 4.5\sqrt{57.75} = 4.5 \times 7.60 = 34.20 \text{ square centimeters.}$$

Areas and Dimensions of Plane Figures

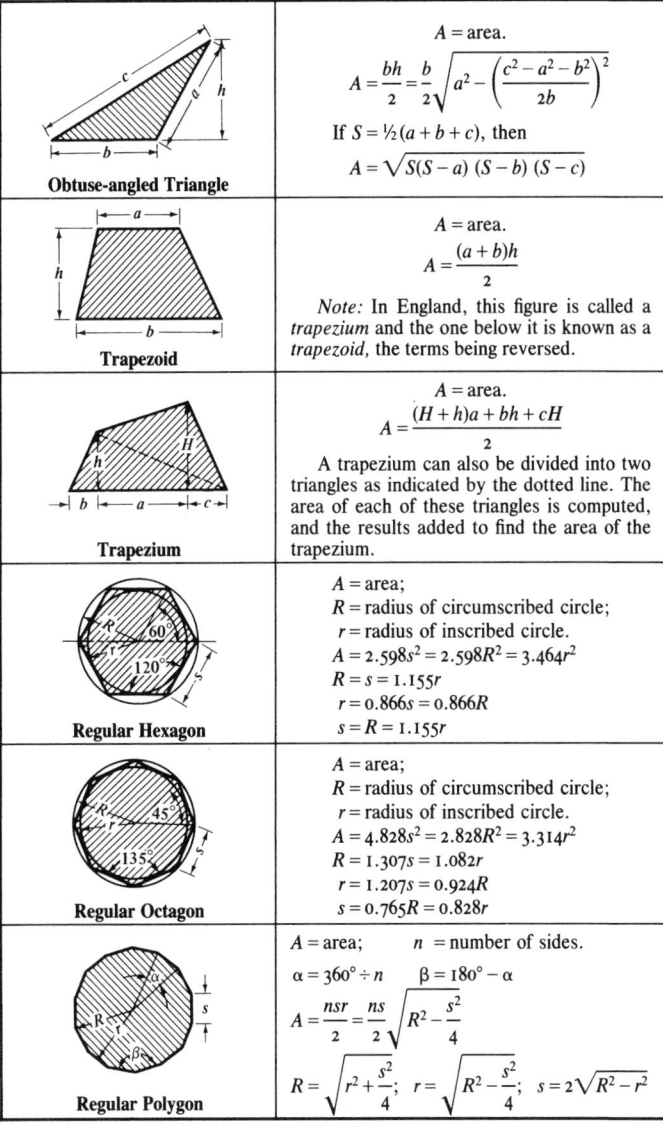

Obtuse-angled Triangle

A = area.

$$A = \frac{bh}{2} = \frac{b}{2}\sqrt{a^2 - \left(\frac{c^2 - a^2 - b^2}{2b}\right)^2}$$

If $S = \frac{1}{2}(a + b + c)$, then

$$A = \sqrt{S(S - a)(S - b)(S - c)}$$

Trapezoid

A = area.

$$A = \frac{(a + b)h}{2}$$

Note: In England, this figure is called a *trapezium* and the one below it is known as a *trapezoid*, the terms being reversed.

Trapezium

A = area.

$$A = \frac{(H + h)a + bh + cH}{2}$$

A trapezium can also be divided into two triangles as indicated by the dotted line. The area of each of these triangles is computed, and the results added to find the area of the trapezium.

Regular Hexagon

A = area;
R = radius of circumscribed circle;
r = radius of inscribed circle.
$A = 2.598s^2 = 2.598R^2 = 3.464r^2$
$R = s = 1.155r$
$r = 0.866s = 0.866R$
$s = R = 1.155r$

Regular Octagon

A = area;
R = radius of circumscribed circle;
r = radius of inscribed circle.
$A = 4.828s^2 = 2.828R^2 = 3.314r^2$
$R = 1.307s = 1.082r$
$r = 1.207s = 0.924R$
$s = 0.765R = 0.828r$

Regular Polygon

A = area; n = number of sides.

$\alpha = 360° \div n$ $\beta = 180° - \alpha$

$$A = \frac{nsr}{2} = \frac{ns}{2}\sqrt{R^2 - \frac{s^2}{4}}$$

$$R = \sqrt{r^2 + \frac{s^2}{4}}; \quad r = \sqrt{R^2 - \frac{s^2}{4}}; \quad s = 2\sqrt{R^2 - r^2}$$

Examples of the Use of the Formulas (English and metric units)

Obtuse-angled Triangle. — The side $a = 5$, side $b = 4$, and side $c = 8$ inches. Find the area.

$$S = \tfrac{1}{2}(a + b + c) = \tfrac{1}{2}(5 + 4 + 8) = \tfrac{1}{2} \times 17 = 8.5$$

$$A = \sqrt{S(S-a)(S-b)(S-c)} = \sqrt{8.5(8.5-5)(8.5-4)(8.5-8)}$$

$$= \sqrt{8.5 \times 3.5 \times 4.5 \times 0.5} = \sqrt{66.937} = 8.18 \text{ square inches.}$$

Trapezoid. — Side $a = 23$ meters, side $b = 32$ meters, and height $h = 12$ meters. Find the area.

$$A = \frac{(a+b)h}{2} = \frac{(23+32)12}{2} = \frac{55 \times 12}{2} = \frac{660}{2} = 330 \text{ square meters.}$$

Trapezium. — Let $a = 10$, $b = 2$, $c = 3$, $h = 8$, and $H = 12$ inches. Find the area.

$$A = \frac{(H+h)a + bh + cH}{2} = \frac{(12+8)10 + 2 \times 8 + 3 \times 12}{2}$$

$$= \frac{20 \times 10 + 16 + 36}{2} = \frac{252}{2} = 126 \text{ square inches.}$$

Regular Hexagon. — The side s of a regular hexagon is 40 millimeters. Find the area and the radius r of the inscribed circle.

$$A = 2.598 s^2 = 2.598 \times 40^2 = 2.598 \times 1600 = 4156.8 \text{ square millimeters.}$$
$$r = 0.866 s = 0.866 \times 40 = 34.64 \text{ millimeters.}$$

What is the length of the side of a hexagon that is described about a circle of 50 millimeters radius? — Here $r = 50$. Hence,

$$s = 1.155 r = 1.155 \times 50 = 57.75 \text{ millimeters.}$$

Regular Octagon. — Find the area and the length of the side of an octagon that is inscribed in a circle of 12 inches diameter.

Diameter of circumscribed circle = 12 inches; hence, $R = 6$ inches.

$$A = 2.828 R^2 = 2.828 \times 6^2 = 2.828 \times 36 = 101.81 \text{ square inches.}$$

$$s = 0.765 R = 0.765 \times 6 = 4.590 \text{ inches.}$$

Regular Polygon. — Find the area of a polygon having 12 sides, inscribed in a circle of 8 centimeters radius. The length of the side s is 4.141 centimeters.

$$A = \frac{ns}{2}\sqrt{R^2 - \frac{s^2}{4}} = \frac{12 \times 4.141}{2}\sqrt{8^2 - \frac{4.141^2}{4}} = 24.846\sqrt{59.713}$$

$$= 24.846 \times 7.727 = 191.98 \text{ square centimeters.}$$

Areas and Dimensions of Plane Figures

Circle	A = area; $\;C$ = circumference. $A = \pi r^2 = 3.1416r^2 = 0.7854d^2$ $C = 2\pi r = 6.2832r = 3.1416d$ $r = C \div 6.2832 = \sqrt{A \div 3.1416} = 0.564\sqrt{A}$ $d = C \div 3.1416 = \sqrt{A \div 0.7854} = 1.128\sqrt{A}$ Length of arc for center-angle of $1° = 0.008727d$ Length of arc for center-angle of $n° = 0.008727nd$
Circular Sector	A = area; $\;l$ = length of arc; $\;\alpha$ = angle, in degrees. $l = \dfrac{r \times \alpha \times 3.1416}{180} = 0.01745r\alpha = \dfrac{2A}{r}$ $A = \frac{1}{2}rl = 0.008727\alpha r^2$ $\alpha = \dfrac{57.296\,l}{r} \qquad r = \dfrac{2A}{l} = \dfrac{57.296\,l}{\alpha}$
Circular Segment	A = area; $\;l$ = length of arc; $\;\alpha$ = angle, in degrees. $c = 2\sqrt{h(2r - h)} \qquad A = \frac{1}{2}[rl - c(r - h)]^*$ $r = \dfrac{c^2 + 4h^2}{8h} \qquad\qquad l = 0.01745r\alpha$ $h = r - \frac{1}{2}\sqrt{4r^2 - c^2} \qquad \alpha = \dfrac{57.296\,l}{r}$ $h = r[1 - \cos(\alpha/2)] \qquad$ * See also p. 50.
Circular Ring	A = area. $A = \pi(R^2 - r^2) = 3.1416(R^2 - r^2)$ $\quad = 3.1416(R + r)\,(R - r)$ $\quad = 0.7854(D^2 - d^2) = 0.7854(D + d)\,(D - d)$
Circular Ring Sector	A = area; $\;\alpha$ = angle, in degrees. $A = \dfrac{\alpha\pi}{360}(R^2 - r^2) = 0.00873\alpha(R^2 - r^2)$ $\quad = \dfrac{\alpha\pi}{4 \times 360}(D^2 - d^2) = 0.00218\alpha(D^2 - d^2)$
Spandrel or Fillet	A = area. $A = r^2 - \dfrac{\pi r^2}{4} = 0.215r^2$ $\quad = 0.1075c^2$

Examples of the Use of the Formulas (English and metric units)

Circle. — Find the area A and circumference C of a circle with a diameter of $2\frac{3}{4}$ inches.

$A = 0.7854d^2 = 0.7854 \times 2.75^2 = 0.7854 \times 2.75 \times 2.75 = 5.9396$ square inches.
$C = 3.1416d = 3.1416 \times 2.75 = 8.6394$ inches.

The area of a circle is 16.8 square inches. Find its diameter.

$d = 1.128\sqrt{A} = 1.128\sqrt{16.8} = 1.128 \times 4.099 = 4.624$ inches.

Circular Sector. — The radius of a circle is 35 millimeters, and angle α of a sector of the circle is 60 degrees. Find the area of the sector and the length of arc l.

$A = 0.008727\alpha r^2 = 0.008727 \times 60 \times 35^2 = 0.5236 \times 35 \times 35$
$\qquad\qquad\qquad\qquad\qquad\qquad\quad = 641.41$ square millimeters
$\qquad\qquad\qquad\qquad\qquad\qquad\quad = 6.41$ square centimeters.
$l = 0.01745r\alpha = 0.01745 \times 35 \times 60 = 36.645$ millimeters.

Circular Segment. — The radius r of a circular segment is 60 inches and the height h is 8 inches. Find the length of the chord c.

$c = 2\sqrt{h(2r - h)} = 2\sqrt{8 \times (2 \times 60 - 8)} = 2\sqrt{896} = 2 \times 29.93 = 59.86$ inches.

If $c = 16$, and $h = 6$ inches, what is the radius of the circle of which the segment is a part?

$$r = \frac{c^2 + 4h^2}{8h} = \frac{16^2 + 4 \times 6^2}{8 \times 6} = \frac{256 + 144}{48} = \frac{400}{48} = 8\frac{1}{3} \text{ inches.}$$

Circular Ring. — Let the outside diameter $D = 12$ centimeters and the inside diameter $d = 8$ centimeters. Find area of ring.

$A = 0.7854\,(D^2 - d^2) = 0.7854\,(12^2 - 8^2) = 0.7854\,(144 - 64) = 0.7854 \times 80$
$\qquad = 62.83$ square centimeters.
By the alternative formula:

$A = 0.7854\,(D + d)\,(D - d) = 0.7854\,(12 + 8)\,(12 - 8) = 0.7854 \times 20 \times 4$
$\qquad = 62.83$ square centimeters.

Circular Ring Sector. — Find the area, if the outside radius $R = 5$ inches, the inside radius $r = 2$ inches, and $\alpha = 72$ degrees.

$\qquad\qquad A = 0.00873\alpha(R^2 - r^2) = 0.00873 \times 72(5^2 - 2^2)$
$\qquad\qquad\qquad = 0.6286(25 - 4) = 0.6286 \times 21 = 13.2$ square inches.

Spandrel or Fillet. — Find the area of a spandrel, the radius of which is 0.7 inch.

$\qquad\qquad A = 0.215r^2 = 0.215 \times 0.7^2 = 0.215 \times 0.7 \times 0.7 = 0.105$ square inch.

If chord c were given as 2.2 inches, what would be the area?

$\qquad\qquad A = 0.1075c^2 = 0.1075 \times 2.2^2 = 0.1075 \times 4.84 = 0.520$ square inch.

Areas and Dimensions of Plane Figures

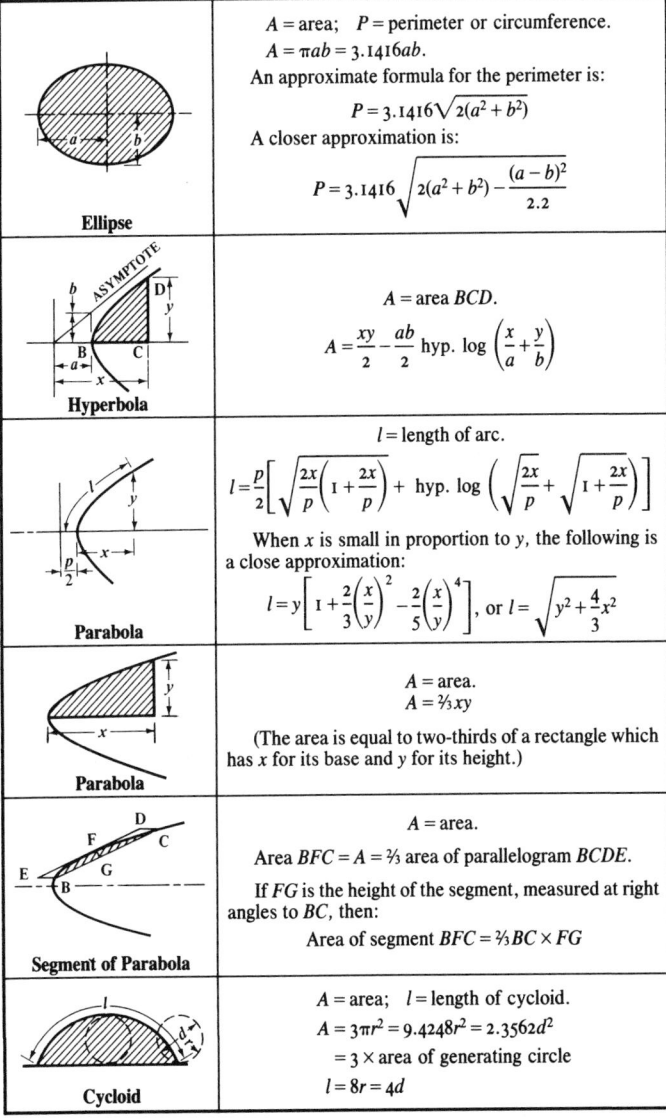

Ellipse

A = area; P = perimeter or circumference.
$A = \pi ab = 3.1416ab$.
An approximate formula for the perimeter is:

$$P = 3.1416\sqrt{2(a^2 + b^2)}$$

A closer approximation is:

$$P = 3.1416\sqrt{2(a^2 + b^2) - \frac{(a - b)^2}{2.2}}$$

Hyperbola

A = area BCD.

$$A = \frac{xy}{2} - \frac{ab}{2}\ \text{hyp. log}\left(\frac{x}{a} + \frac{y}{b}\right)$$

Parabola

l = length of arc.

$$l = \frac{p}{2}\left[\sqrt{\frac{2x}{p}\left(1 + \frac{2x}{p}\right)} + \text{hyp. log}\left(\sqrt{\frac{2x}{p}} + \sqrt{1 + \frac{2x}{p}}\right)\right]$$

When x is small in proportion to y, the following is a close approximation:

$$l = y\left[1 + \frac{2}{3}\left(\frac{x}{y}\right)^2 - \frac{2}{5}\left(\frac{x}{y}\right)^4\right], \text{ or } l = \sqrt{y^2 + \frac{4}{3}x^2}$$

Parabola

A = area.
$A = \tfrac{2}{3}xy$

(The area is equal to two-thirds of a rectangle which has x for its base and y for its height.)

Segment of Parabola

A = area.

Area $BFC = A = \tfrac{2}{3}$ area of parallelogram $BCDE$.

If FG is the height of the segment, measured at right angles to BC, then:

Area of segment $BFC = \tfrac{2}{3}BC \times FG$

Cycloid

A = area; l = length of cycloid.
$A = 3\pi r^2 = 9.4248r^2 = 2.3562d^2$
$\quad = 3 \times$ area of generating circle
$l = 8r = 4d$

Examples of the Use of the Formulas (English and metric units)

Ellipse. — The larger or major axis is 200 millimeters. The smaller or minor axis is 150 millimeters. Find the area and the approximate circumference. Here, then, $a = 100$, and $b = 75$.

$A = 3.1416ab = 3.1416 \times 100 \times 75 = 23,562$ square millimeters

$= 235.62$ square centimeters.

$P = 3.1416\sqrt{2(a^2 + b^2)} = 3.1416\sqrt{2\ (100^2 + 75^2)} = 3.1416\sqrt{2 \times 15,625}$

$= 3.1416\sqrt{31,250} = 3.1416 \times 176.78 = 555.37$ millimeters

$= 55.537$ centimeters.

Hyperbola. — The half-axes a and b are 3 and 2 inches, respectively. Find area shown shaded in illustration for $x = 8$ and $y = 5$.

Inserting the known values in the formula:

$A = \dfrac{8 \times 5}{2} - \dfrac{3 \times 2}{2} \times \text{hyp. log}\ \left(\dfrac{8}{3} + \dfrac{5}{2}\right) = 20 - 3 \times \text{hyp. log } 5.167$

$= 20 - 3 \times 1.6423 = 20 - 4.927 = 15.073$ square inches.

Parabola. — If $x = 2$ and $y = 24$ feet, what is the approximate length l of the parabolic curve?

$l = y\left[1 + \dfrac{2}{3}\left(\dfrac{x}{y}\right)^2 - \dfrac{2}{5}\left(\dfrac{x}{y}\right)^4\right] = 24\left[1 + \dfrac{2}{3}\left(\dfrac{2}{24}\right)^2 - \dfrac{2}{5}\left(\dfrac{2}{24}\right)^4\right]$

$= 24\left[1 + \dfrac{2}{3} \times \dfrac{1}{144} - \dfrac{2}{5} \times \dfrac{1}{20,736}\right] = 24 \times 1.0046 = 24.11$ feet.

Parabola. — Let the dimension x in the illustration be 15 centimeters, and y, 9 centimeters. Find the area of the shaded portion of the parabola.

$A = \tfrac{2}{3} \times xy = \tfrac{2}{3} \times 15 \times 9 = 10 \times 9 = 90$ square centimeters.

Segment of Parabola. — The length of the chord $BC = 19.5$ inches. The distance between lines BC and DE, measured at right angles to BC, is 2.25 inches. This is the height of the segment. Find the area.

Area $= A = \tfrac{2}{3}BC \times FG = \tfrac{2}{3} \times 19.5 \times 2.25 = 29.25$ square inches.

Cycloid. — The diameter of the generating circle of a cycloid is 6 inches. Find the length l of the cycloidal curve, and the area enclosed between the curve and the base line.

$l = 4d = 4 \times 6 = 24$ inches.

$A = 2.3562d^2 = 2.3562 \times 6^2 = 2.3562 \times 36 = 84.82$ square inches.

Segments of Circles for Radius = 1
(English or metric units)

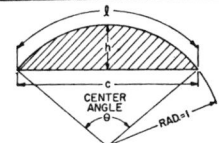

Length of arc, height of segment, length of chord, and area of segment for angles from 1 to 180 degrees and radius = 1. For other radii, multiply the values of l, h and c in the table by the given radius r, and the values for areas, by r^2, the square of the radius. The values in the tables can be used for English or metric units.

Center Angle θ, Degrees	l	h	c	Area of Segment A	Center Angle θ, Degrees	l	h	c	Area of Segment A
1	0.01745	0.00004	0.01745	0.00000	46	0.803	0.0795	0.781	0.04176
2	0.03491	0.00015	0.03490	0.00000	47	0.820	0.0829	0.797	0.04448
3	0.05236	0.00034	0.05235	0.00001	48	0.838	0.0865	0.813	0.04731
4	0.06981	0.00061	0.06980	0.00003	49	0.855	0.0900	0.829	0.05025
5	0.08727	0.00095	0.08724	0.00006	50	0.873	0.0937	0.845	0.05331
6	0.10472	0.00137	0.10467	0.00010	51	0.890	0.0974	0.861	0.05649
7	0.12217	0.00187	0.12210	0.00015	52	0.908	0.1012	0.877	0.05978
8	0.13963	0.00244	0.13951	0.00023	53	0.925	0.1051	0.892	0.06319
9	0.15708	0.00308	0.15692	0.00032	54	0.942	0.1090	0.908	0.06673
10	0.17453	0.00381	0.17431	0.00044	55	0.960	0.1130	0.923	0.07039
11	0.19199	0.00460	0.19169	0.00059	56	0.977	0.1171	0.939	0.07417
12	0.20944	0.00548	0.20906	0.00076	57	0.995	0.1212	0.954	0.07808
13	0.22689	0.00643	0.22641	0.00097	58	1.012	0.1254	0.970	0.08212
14	0.24435	0.00745	0.24374	0.00121	59	1.030	0.1296	0.985	0.08629
15	0.26180	0.00856	0.26105	0.00149	60	1.047	0.1340	1.000	0.09059
16	0.27925	0.00973	0.27835	0.00181	61	1.065	0.1384	1.015	0.09502
17	0.29671	0.01098	0.29562	0.00217	62	1.082	0.1428	1.030	0.09958
18	0.31416	0.01231	0.31287	0.00257	63	1.100	0.1474	1.045	0.10428
19	0.33161	0.01371	0.33010	0.00302	64	1.117	0.1520	1.060	0.10911
20	0.34907	0.01519	0.34730	0.00352	65	1.134	0.1566	1.075	0.11408
21	0.36652	0.01675	0.36447	0.00408	66	1.152	0.1613	1.089	0.11919
22	0.38397	0.01837	0.38162	0.00468	67	1.169	0.1661	1.104	0.12443
23	0.40143	0.02008	0.39874	0.00535	68	1.187	0.1710	1.118	0.12982
24	0.41888	0.02185	0.41582	0.00607	69	1.204	0.1759	1.133	0.13535
25	0.43633	0.02370	0.43288	0.00686	70	1.222	0.1808	1.147	0.14102
26	0.45379	0.02563	0.44990	0.00771	71	1.239	0.1859	1.161	0.14683
27	0.47124	0.02763	0.46689	0.00862	72	1.257	0.1910	1.176	0.15279
28	0.48869	0.02970	0.48384	0.00961	73	1.274	0.1961	1.190	0.15889
29	0.50615	0.03185	0.50076	0.01067	74	1.292	0.2014	1.204	0.16514
30	0.52360	0.03407	0.51764	0.01180	75	1.309	0.2066	1.218	0.17154
31	0.54105	0.03637	0.53448	0.01301	76	1.326	0.2120	1.231	0.17808
32	0.55851	0.03874	0.55127	0.01429	77	1.344	0.2174	1.245	0.18477
33	0.57596	0.04118	0.56803	0.01566	78	1.361	0.2229	1.259	0.19160
34	0.59341	0.04370	0.58474	0.01711	79	1.379	0.2284	1.272	0.19859
35	0.61087	0.04628	0.60141	0.01864	80	1.396	0.2340	1.286	0.20573
36	0.62832	0.04894	0.61803	0.02027	81	1.414	0.2396	1.299	0.21301
37	0.64577	0.05168	0.63461	0.02198	82	1.431	0.2453	1.312	0.22045
38	0.66323	0.05448	0.65114	0.02378	83	1.449	0.2510	1.325	0.22804
39	0.68068	0.05736	0.66761	0.02568	84	1.466	0.2569	1.338	0.23578
40	0.69813	0.06031	0.68404	0.02767	85	1.484	0.2627	1.351	0.24367
41	0.71558	0.06333	0.70041	0.02976	86	1.501	0.2686	1.364	0.25171
42	0.73304	0.06642	0.71674	0.03195	87	1.518	0.2746	1.377	0.25990
43	0.75049	0.06958	0.73300	0.03425	88	1.536	0.2807	1.389	0.26825
44	0.76794	0.07282	0.74921	0.03664	89	1.553	0.2867	1.402	0.27675
45	0.78540	0.07612	0.76537	0.03915	90	1.571	0.2929	1.414	0.28540

Segments of Circles for Radius = 1

(English or metric units)

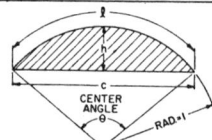

Length of arc, height of segment, length of chord, and area of segment for angles from 1 to 180 degrees and radius = 1. For other radii, multiply the values of l, h and c in the table by the given radius r, and the values for areas, by r^2, the square of the radius.
The values in the table can be used for English or metric units.

Center Angle θ, Degrees	l	h	c	Area of Segment A	Center Angle θ, Degrees	l	h	c	Area of Segment A
91	1.588	0.2991	1.427	0.2942	136	2.374	0.6254	1.854	0.8395
92	1.606	0.3053	1.439	0.3032	137	2.391	0.6335	1.861	0.8546
93	1.623	0.3116	1.451	0.3123	138	2.409	0.6416	1.867	0.8697
94	1.641	0.3180	1.463	0.3215	139	2.426	0.6498	1.873	0.8850
95	1.658	0.3244	1.475	0.3309	140	2.443	0.6580	1.879	0.9003
96	1.676	0.3309	1.486	0.3405	141	2.461	0.6662	1.885	0.9158
97	1.693	0.3374	1.498	0.3502	142	2.478	0.6744	1.891	0.9314
98	1.710	0.3439	1.509	0.3601	143	2.496	0.6827	1.897	0.9470
99	1.728	0.3506	1.521	0.3701	144	2.513	0.6910	1.902	0.9627
100	1.745	0.3572	1.532	0.3803	145	2.531	0.6993	1.907	0.9786
101	1.763	0.3639	1.543	0.3906	146	2.548	0.7076	1.913	0.9945
102	1.780	0.3707	1.554	0.4010	147	2.566	0.7160	1.918	1.0105
103	1.798	0.3775	1.565	0.4117	148	2.583	0.7244	1.923	1.0266
104	1.815	0.3843	1.576	0.4224	149	2.601	0.7328	1.927	1.0428
105	1.833	0.3912	1.587	0.4333	150	2.618	0.7412	1.932	1.0590
106	1.850	0.3982	1.597	0.4444	151	2.635	0.7496	1.936	1.0753
107	1.868	0.4052	1.608	0.4556	152	2.653	0.7581	1.941	1.0917
108	1.885	0.4122	1.618	0.4669	153	2.670	0.7666	1.945	1.1082
109	1.902	0.4193	1.628	0.4784	154	2.688	0.7750	1.949	1.1247
110	1.920	0.4264	1.638	0.4901	155	2.705	0.7836	1.953	1.1413
111	1.937	0.4336	1.648	0.5019	156	2.723	0.7921	1.956	1.1580
112	1.955	0.4408	1.658	0.5138	157	2.740	0.8006	1.960	1.1747
113	1.972	0.4481	1.668	0.5259	158	2.758	0.8092	1.963	1.1915
114	1.990	0.4554	1.677	0.5381	159	2.775	0.8178	1.967	1.2084
115	2.007	0.4627	1.687	0.5504	160	2.793	0.8264	1.970	1.2253
116	2.025	0.4701	1.696	0.5629	161	2.810	0.8350	1.973	1.2422
117	2.042	0.4775	1.705	0.5755	162	2.827	0.8436	1.975	1.2592
118	2.059	0.4850	1.714	0.5883	163	2.845	0.8522	1.978	1.2763
119	2.077	0.4925	1.723	0.6012	164	2.862	0.8608	1.981	1.2934
120	2.094	0.5000	1.732	0.6142	165	2.880	0.8695	1.983	1.3105
121	2.112	0.5076	1.741	0.6273	166	2.897	0.8781	1.985	1.3277
122	2.129	0.5152	1.749	0.6406	167	2.915	0.8868	1.987	1.3449
123	2.147	0.5228	1.758	0.6540	168	2.932	0.8955	1.989	1.3621
124	2.164	0.5305	1.766	0.6676	169	2.950	0.9042	1.991	1.3794
125	2.182	0.5383	1.774	0.6813	170	2.967	0.9128	1.992	1.3967
126	2.199	0.5460	1.782	0.6950	171	2.985	0.9215	1.994	1.4140
127	2.217	0.5538	1.790	0.7090	172	3.002	0.9302	1.995	1.4314
128	2.234	0.5616	1.798	0.7230	173	3.019	0.9390	1.996	1.4488
129	2.251	0.5695	1.805	0.7372	174	3.037	0.9477	1.997	1.4662
130	2.269	0.5774	1.813	0.7514	175	3.054	0.9564	1.998	1.4836
131	2.286	0.5853	1.820	0.7658	176	3.072	0.9651	1.999	1.5010
132	2.304	0.5933	1.827	0.7803	177	3.089	0.9738	1.999	1.5184
133	2.321	0.6013	1.834	0.7950	178	3.107	0.9825	2.000	1.5359
134	2.339	0.6093	1.841	0.8097	179	3.124	0.9913	2.000	1.5533
135	2.356	0.6173	1.848	0.8245	180	3.142	1.0000	2.000	1.5708

Mathematical Signs and Commonly Used Abbreviations

+	Plus (sign of addition)	μ	Mu (coefficient of friction)
+	Positive	π	Pi (3.1416)
−	Minus (sign of subtraction)	Σ	Sigma (sign of summation)
−	Negative	ω	⎰ Omega (angles measured
± (∓)	Plus or minus (minus or plus)		⎱ in radians)
×	⎰ Multiplied by (multiplication		⎰ Acceleration due to
	⎱ sign)	g	⎨ gravity (32.16 ft. per
.	⎰ Multiplied by (multiplication		⎩ sec. per sec.)
	⎱ sign)		⎰ Imaginary quantity
÷	Divided by (division sign)	i (or j)	⎱ ($\sqrt{-1}$)
:	Divided by (division sign)	sin	Sine
:	Is to (in proportion)	cos	Cosine
=	Equals	tan	⎫
≠	Is not equal to	(tg)	⎬ Tangent
≡	Is identical to	(tang)	⎭
::	Equals (in proportion)	cot	⎫
≅	⎱ Approximately equals	(ctg)	⎬ Cotangent
≈	⎰	sec	Secant
>	Greater than	cosec	Cosecant
<	Less than	versin	Versed sine
≧	Greater than or equal to	covers	Coversed sine
≦	Less than or equal to	$\sin^{-1}a$	⎰ Arc the sine of which
→	Approaches as a limit	arcsin a	⎱ is a
∝	Varies directly as	$(\sin a)^{-1}$	⎰ Reciprocal of sin a
∴	Therefore		⎱ $1 \div \sin a$)
$\sqrt{}$	Square root	sinh x	Hyperbolic sine of x
$\sqrt[3]{}$	Cube root	cosh x	Hyperbolic cosine of x
$\sqrt[4]{}$	4th root	Δ	Delta (increment of)
$\sqrt[n]{}$	nth root	δ	Delta (variation of)
a^2	a squared (2d power of a)	d	Differential (in calculus)
a^3	a cubed (3d power of a)	∫	Integral (in calculus)
a^4	4th power of a	$\int_b^a$	⎰ Integral between the
a^n	nth power of a		⎱ limits a and b
a^{-n}	$1 \div a^n$	!	$5! = 1 \times 2 \times 3 \times 4 \times 5$
$\dfrac{1}{n}$	Reciprocal value of n	∠	Angle
		∟	Right angle
		⊥	Perpendicular to
log	Logarithm	△	Triangle
hyp. log		⊙	Circle
nat. log	⎬ Hyperbolic, natural or	▱	Parallelogram
$\log_e$	⎨ Napierian logarithm	°	⎰ Degree (circular arc or
ln	⎩		⎱ temperature)
	⎰ Base of hyp. logarithms	′	Minutes or feet
e	⎱ (2.71828)	″	Seconds or inches
lim.	Limit value (of an expression)	a'	a prime
∞	Infinity	a''	a double prime
α	Alpha	a_1	a sub one
β	Beta ⎫	a_2	a sub two
γ	Gamma ⎬ commonly used	a_n	a sub n
θ	Theta ⎨ to denote angles	( )	Parentheses
φ	Phi ⎭	[ ]	Brackets
		{ }	Braces

Ratio and Proportion

The *ratio* between two quantities is the quotient obtained by dividing the first quantity by the second. For example, the ratio between 3 and 12 is $\frac{1}{4}$, and the ratio between 12 and 3 is 4. Ratio is generally indicated by the sign (:); thus 12 : 3 indicates the ratio of 12 to 3.

A *reciprocal* or *inverse* ratio is the reciprocal of the original ratio. Thus, the inverse ratio of 5 : 7 is 7 : 5.

In a *compound* ratio each term is the product of the corresponding terms in two or more simple ratios. Thus, when

$$8 : 2 = 4, \qquad 9 : 3 = 3, \qquad 10 : 5 = 2,$$

then the compound ratio is:

$$8 \times 9 \times 10 : 2 \times 3 \times 5 = 4 \times 3 \times 2,$$
$$720 : 30 \qquad = 24.$$

Proportion is the equality of ratios. Thus,

$$6 : 3 = 10 : 5, \quad \text{or} \quad 6 : 3 :: 10 : 5.$$

The first and last terms in a proportion are called the *extremes;* the second and third, the *means*. The product of the extremes is equal to the product of the means. Thus,

$$25 : 2 = 100 : 8 \quad \text{and} \quad 25 \times 8 = 2 \times 100.$$

If three terms in a proportion are known, the remaining term may be found by the following rules:

The first term is equal to the product of the second and third terms, divided by the fourth.

The second term is equal to the product of the first and fourth terms, divided by the third.

The third term is equal to the product of the first and fourth terms, divided by the second.

The fourth term is equal to the product of the second and third terms, divided by the first.

Examples: — Let x be the term to be found, then,

$$x : 12 = 3.5 : 21 \qquad x = \frac{12 \times 3.5}{21} = \frac{42}{21} = 2.$$

$$\tfrac{1}{4} : x = 14 : 42 \qquad x = \frac{\frac{1}{4} \times 42}{14} = \frac{1}{4} \times 3 = \frac{3}{4}$$

$$5 : 9 = x : 63 \qquad x = \frac{5 \times 63}{9} = \frac{315}{9} = 35$$

$$\tfrac{1}{4} : \tfrac{7}{8} = 4 : x \qquad x = \frac{\frac{7}{8} \times 4}{\frac{1}{4}} = \frac{3\frac{1}{2}}{\frac{1}{4}} = 14.$$

If the second and third terms are the same, either is said to be the *mean proportional* between the other two. Thus, 8 : 4 = 4 : 2, and 4 is the mean proportional between 8 and 2. The mean proportional between two numbers may be found by multiplying the numbers together, and extracting the square root of the product. Thus, the mean proportional between 3 and 12 is found as below:

$$3 \times 12 = 36, \quad \text{and} \quad \sqrt{36} = 6,$$

which is the mean proportional.

Practical Examples Involving Simple Proportion. — If it takes 18 days to assemble 4 lathes, how long would it require to assemble 14 lathes?

Let the number of days to be found be x. Then write out the proportion as below:

$$4 \ : \ 18 \ = \ 14 \ : \ x$$
$$(\text{lathes} : \text{days} = \text{lathes} : \text{days})$$

Find now the fourth term by the rule given:

$$x = \frac{18 \times 14}{4} = 63 \text{ days.}$$

Thirty-four linear feet of bar stock are required for the blanks for 100 clamping bolts. How many feet of stock would be required for 912 bolts?

Let x = total length of stock required for 912 bolts.

$$34 \ : \ 100 \ = \ x \ : \ 912$$
$$(\text{feet} : \text{bolts} = \text{feet} : \text{bolts})$$

Then, the third term $x = \dfrac{34 \times 912}{100} = 310$ feet, approximately.

Inverse Proportion. — In an inverse proportion, as one of the items involved *increases,* the corresponding item in the proportion *decreases,* or vice versa. For example, a factory employing 270 men completes a given number of typewriters weekly, the number of working hours being 44 per week. How many men would be required for the same production if the working hours were reduced to 40 per week?

The time per week is in an inverse proportion to the number of men employed; the shorter the time, the more men. The inverse proportion is written:

$$270 : x = 40 : 44$$

(men, 44-hour basis: men, 40-hour basis = time, 40-hour basis: time, 44-hour basis) Thus

$$\frac{270}{x} = \frac{40}{44} \quad \text{and} \quad x = \frac{270 \times 44}{40} = 297 \text{ men.}$$

Problems Involving Both Simple and Inverse Proportions. — If two groups of data are related both by direct (simple) and inverse proportions among the various quantities, then a simple mathematical relation that may be used in solving problems is as follows:

$$\frac{\text{Product of all directly proportional items in first group}}{\text{Product of all inversely proportional items in first group}}$$

$$= \frac{\text{Product of all directly proportional items in second group}}{\text{Product of all inversely proportional items in second group}}$$

Example: If a man capable of turning 65 studs in a day of 10 hours is paid $6.50 per hour, how much per hour ought a man be paid who turns 72 studs in a 9-hour day, if compensated in the same proportion?

The first group of data in this problem consists of the number of hours worked by the first man, his hourly wage, and the number of studs which he produces per day; the second group contains similar data for the second man except for his unknown hourly wage which may be indicated by x.

The labor cost per stud, as may be seen, is directly proportional to the number of hours worked and the hourly wage. These quantities, therefore, are used in the numerators of the fractions in the formula. The labor cost per stud is inversely proportional to the number of studs produced per day. (The greater the number of studs produced in a given time the less the cost per stud.) The numbers of studs per day, therefore, are placed in the denominators of the fractions in the formula. Thus,

$$\frac{10 \times 6.50}{65} = \frac{9 \times x}{72} \qquad x = \frac{10 \times 6.50 \times 72}{65 \times 9} = \$8.00 \text{ per hour}$$

Problems involving belt pulleys or gears also use inverse proportions in the solution of speed and diameter problems. An electric motor running at 1750 rpm has a 5-in.-diameter pulley. What size pulley is required on the driven shaft to turn it at 875 rpm? Since an increase in the driven pulley diameter will decrease its speed, these terms are in inverse proportion to each other. To solve this problem, first set up the ratio of motor speed to driven shaft speed (1750:875). Then the ratio of motor pulley diameter, 5, to driven shaft diameter, X, is set up and used inversely. Thus,

$$\frac{1750}{875} = \frac{X}{5}$$

and $X = 8750 \div 875 = 10$ in. diameter.

Powers, Roots, and Reciprocals

The *square* of a number (or quantity) is the product of that number multiplied by itself. Thus, the square of 9 is $9 \times 9 = 81$. The square of a number is indicated by the *exponent* (^{2}), thus: $9^2 = 9 \times 9 = 81$.

The *cube* or *third power* of a number is the product obtained by using that number as a factor three times. Thus, the cube of 4 is $4 \times 4 \times 4 = 64$, and is written 4^3.

If a number is used as a factor four or five times, respectively, the product is the fourth or fifth power. Thus $3^4 = 3 \times 3 \times 3 \times 3 = 81$, and $2^5 = 2 \times 2 \times 2 \times 2 \times 2 = 32$. A number can be raised to any power by using it as a factor the required number of times.

The *square root* of a given number is that number which, when multiplied by itself, will give a product equal to the given number. The square root of 16 (written $\sqrt{16}$) equals 4, because $4 \times 4 = 16$.

The *cube root* of a given number is that number which, when used as a factor three times, will give a product equal to the given number. Thus, the cube root of 64 (written $\sqrt[3]{64}$) equals 4, because $4 \times 4 \times 4 = 64$.

The fourth, fifth, etc., roots of a given number are those numbers which when used as factors four, five, etc., times, will give as a product the given number. Thus $\sqrt[4]{16} = 2$, because $2 \times 2 \times 2 \times 2 = 16$.

In some formulas there may be such expressions as $(a^2)^3$ and $a^{3/2}$. The first of these, $(a^2)^3$, means that the number a is first to be squared, a^2, and the result then cubed to give a^6. Thus, $(a^2)^3$ is equivalent to a^6 which is obtained by *multiplying* the exponents 2 and 3. Similarly, $a^{3/2}$ may be interpreted as the cube of the square root of a, $(\sqrt{a})^3$, or $(a^{1/2})^3$, so that, for example, $16^{3/2} = (\sqrt{16})^3 = 64$.

The multiplications required for raising numbers to powers and the extracting of roots are greatly facilitated by the use of logarithms. The extracting of the square root and cube root by the regular arithmetical methods is a slow and cumbersome operation, and any roots can be more rapidly found by using logarithms.

When the power to which a number is to be raised is not an integer, say, 1.62, then the use of either logarithms or a scientific calculator becomes the only practical means of solution.

The *reciprocal R* of a number N is obtained by dividing 1 by the number; $R = 1/N$. Reciprocals are useful in some calculations because they avoid the use of negative characteristics as in calculations with logarithms. The tables of Logarithms of Gear Ratios were compiled by using the reciprocals of gear ratios to simplify entering the tables. For example, if the gear ratio is 0.45, the logarithm of the reciprocal, $1/0.45 = 2.22222$, is 0.34679 and this logarithm is found next to the ratio 100:45 in the table.

Powers of Ten Notation

Powers of ten notation is used to simplify calculations and insure accuracy, particularly with respect to the position of decimal points, and also simplifies the expression of numbers which are so large or so small as to be unwieldy. For example, the metric (SI) pressure unit pascal is equivalent to 0.00000986923 atmosphere or 0.0001450377 pound/inch^2. In powers of ten notation these figures are 9.86923×10^{-6} atmosphere and 1.450377×10^{-4} pound/inch^2. The notation also facilitates adaptation of numbers for electronic data processing and computer readout.

Expressing Numbers in Powers of Ten Notation. — In this system of notation every number is expressed by two factors, one of which is some integer from 1 to 9 followed by a decimal and the other is some power of 10.

Thus, 10,000 is expressed as 1.0000×10^4 and 10,463 as 1.0463×10^4. The number 43 is expressed 4.3×10 and 568 is expressed 5.68×10^2.

In the case of decimals, the number 0.0001 which as a fraction is $1/10,000$ is expressed as 1×10^{-4} and 0.0001463 is expressed as 1.463×10^{-4}. The decimal 0.498 is expressed as 4.98×10^{-1} and 0.03146 is expressed as 3.146×10^{-2}.

Rules for Converting any Number to Powers of Ten Notation. — Any number can be converted to the powers of ten notation by means of one of two rules.

Rule 1: If the number is a whole number or a whole number and a decimal so that it has digits to the left of the decimal point, the decimal point is moved a sufficient number of places to the *left* to bring it to the immediate right of the first digit. With the decimal point shifted to this position, the number so written comprises the *first* factor when written in powers of ten notation.

The number of places that the decimal point is moved to the left to bring it immediately to the right of the first digit is the *positive* index or power of 10 that comprises the *second* factor when written in powers of ten notation.

Thus, to write 4639 in this notation, the decimal point is moved three places to the left giving the two factors: 4.639×10^3. Similarly,

$$431.412 = 4.31412 \times 10^2$$
$$986388 = 9.86388 \times 10^5$$

Rule 2: If the number is a decimal, i.e., it has digits entirely to the right of the decimal point, then the decimal point is moved a sufficient number of places to the *right* to bring it immediately to the right of the first digit. With the decimal point shifted to this position, the number so written comprises the *first* factor when written in powers of ten notation.

The number of places that the decimal point is moved to the *right* to bring it immediately to the right of the first digit is the *negative* index or power of 10 that follows the number when written in powers of ten notation.

Thus, to bring the decimal point in 0.005721 to the immediate right of the first digit which is 5, it must be moved *three* places to the right, giving the two factors: 5.721×10^{-3}. Similarly,

$$0.469 = 4.69 \times 10^{-1}$$
$$0.0000516 = 5.16 \times 10^{-5}$$

Multiplying Numbers Written in Powers of Ten Notation. — When multiplying two numbers written in the powers of ten notation together, the procedure is as follows:

1. Multiply the first factor of one number by the first factor of the other to obtain the first factor of the product.

2. Add the index of the second factor (which is some power of 10) of one number

to the index of the second factor of the other number to obtain the index of the second factor (which is some power of 10) in the product. Thus:

$$(4.31 \times 10^{-2}) \times (9.0125 \times 10) =$$

$$(4.31 \times 9.0125) \times 10^{-2+1} = 38.844 \times 10^{-1}$$

$$(5.986 \times 10^4) \times (4.375 \times 10^3) =$$

$$(5.986 \times 4.375) \times 10^{4+3} = 26.189 \times 10^7$$

In the preceding calculations neither of the results shown are in the conventional powers of ten form since the first factor in each case has two digits. In the conventional powers of ten notation the results would be:

$$38.844 \times 10^{-1} = 3.884 \times 10^0 = 3.884$$

since $10^0 = 1$, and

$$26.189 \times 10^7 = 2.619 \times 10^8$$

in each case rounding the first factor off to three decimal places.

When multiplying several numbers written in this notation together, the procedure is the same. All of the first factors are multiplied together to get the first factor of the product and all of the indices of the respective powers of ten are added together, taking into account their respective signs, to get the index of the second factor of the product. Thus $(4.02 \times 10^{-3}) \times (3.987 \times 10) \times (4.863 \times 10^5) = (4.02 \times 3.987 \times 4.863) \times (10^{-3+1+5}) = 77.94 \times 10^3 = 7.79 \times 10^4$ rounding off the first factor to two decimal places.

Dividing Numbers Written in Powers of Ten Notation. — When dividing one number by another when both are written in this notation, the procedure is as follows:

1. Divide the first factor of the dividend by the first factor of the divisor to get the first factor of the quotient.

2. Subtract the index of the second factor of the divisor from the index of the second factor of the dividend, taking into account their respective signs, to get the index of the second factor of the quotient. Thus:

$$(4.31 \times 10^{-2}) \div (9.0125 \times 10) =$$

$$(4.31 \div 9.0125) \times (10^{-2-1}) = 0.4782 \times 10^{-3} = 4.782 \times 10^{-4}$$

It can be seen, then, that where several numbers of different magnitudes are to be multiplied and divided this system of notation is helpful.

Example: Find the quotient of $\dfrac{250 \times 4698 \times 0.00039}{43678 \times 0.002 \times 0.0147}$

Solution: Changing all of these numbers to powers of ten notation and performing the operations indicated:

$$\frac{(2.5 \times 10^2) \times (4.698 \times 10^3) \times (3.9 \times 10^{-4})}{(4.3678 \times 10^4) \times (2 \times 10^{-3}) \times (1.47 \times 10^{-2})}$$

$$= \frac{(2.5 \times 4.698 \times 3.9)\,(10^{2+3-4})}{(4.3678 \times 2 \times 1.47)\,(10^{4-3-2})} = \frac{45.8055 \times 10}{12.8413 \times 10^{-1}}$$

$$= 3.5670 \times 10^{1-(-1)}$$

$$= 3.5670 \times 10^2$$

$$= 356.70$$

Properties of the Circle*

Circumference of circle of diameter $1 = \pi = 3.14159265$

Circumference of circle $= 2\pi r = \pi d$

Diameter of circle $=$ circumference $\times 0.31831$

Diameter of circle of equal periphery as square $=$ side $\times 1.27324$

Side of square of equal periphery as circle $=$ diameter $\times 0.78540$

Diameter of circle circumscribed about square $=$ side $\times 1.41421$

Side of square inscribed in circle $=$ diameter $\times 0.70711$

Arc, $\quad l = \dfrac{\pi r \theta}{180} = 0.017453 r\theta$

Angle, $\quad \theta = \dfrac{180°l}{\pi r} = 57.29578\,\dfrac{l}{r}$

Radius, $r = \dfrac{4b^2 + c^2}{8b}$, Diameter, $d = \dfrac{4b^2 + c^2}{4b}$

Chord, $c = 2\sqrt{2br - b^2} = 2r\sin\dfrac{\theta}{2} = d\sin\dfrac{\theta}{2}$

Rise, $\quad b = r - \dfrac{1}{2}\sqrt{4r^2 - c^2} = \dfrac{c}{2}\tan\dfrac{\theta}{4} = 2r\sin^2\dfrac{\theta}{4}$

Rise, $\quad b = r + y - \sqrt{r^2 - x^2}, \qquad y = b - r + \sqrt{r^2 - x^2}$

$$x = \sqrt{r^2 - (r + y - b)^2}$$

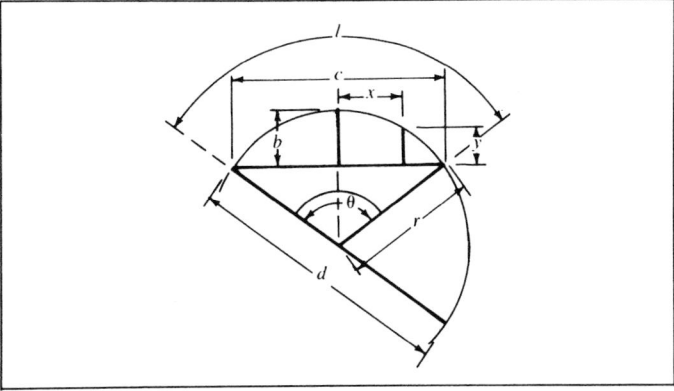

*Courtesy of the Society of Manufacturing Engineers.

Geometrical Propositions

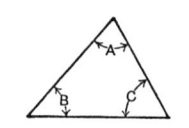

The sum of the three angles in a triangle always equals 180 degrees. Hence, if two angles are known, the third angle can always be found.

$$A + B + C = 180° \qquad A = 180° - (B + C)$$
$$B = 180° - (A + C) \qquad C = 180° - (A + B)$$

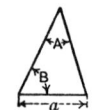 

If one side and two angles in one triangle are equal to one side and similarly located angles in another triangle, then the remaining two sides and angle are also equal.

If $a = a_1$, $A = A_1$ and $B = B_1$, then the two other sides and the remaining angle are also equal.

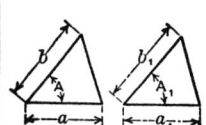

If two sides and the angle between them in one triangle are equal to two sides and a similarly located angle in another triangle, then the remaining side and angles are also equal.

If $a = a_1$, $b = b_1$ and $A = A_1$, then the remaining side and angles are also equal.

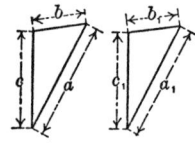

If the three sides in one triangle are equal to the three sides of another triangle, then the angles in the two triangles are also equal.

If $a = a_1$, $b = b_1$ and $c = c_1$, then the angles between the respective sides are also equal.

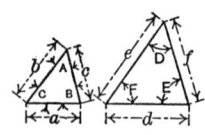

If the three sides of one triangle are proportional to corresponding sides in another triangle, then the triangles are called *similar*, and the angles in the one are equal to the angles in the other.

If $a : b : c = d : e : f$, then $A = D$, $B = E$ and $C = F$.

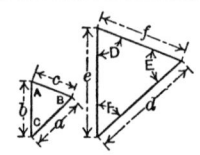

If the angles in one triangle are equal to the angles of another triangle, then the triangles are similar and their corresponding sides are proportional.

If $A = D$, $B = E$ and $C = F$, then $a : b : c = d : e : f$.

Geometrical Propositions

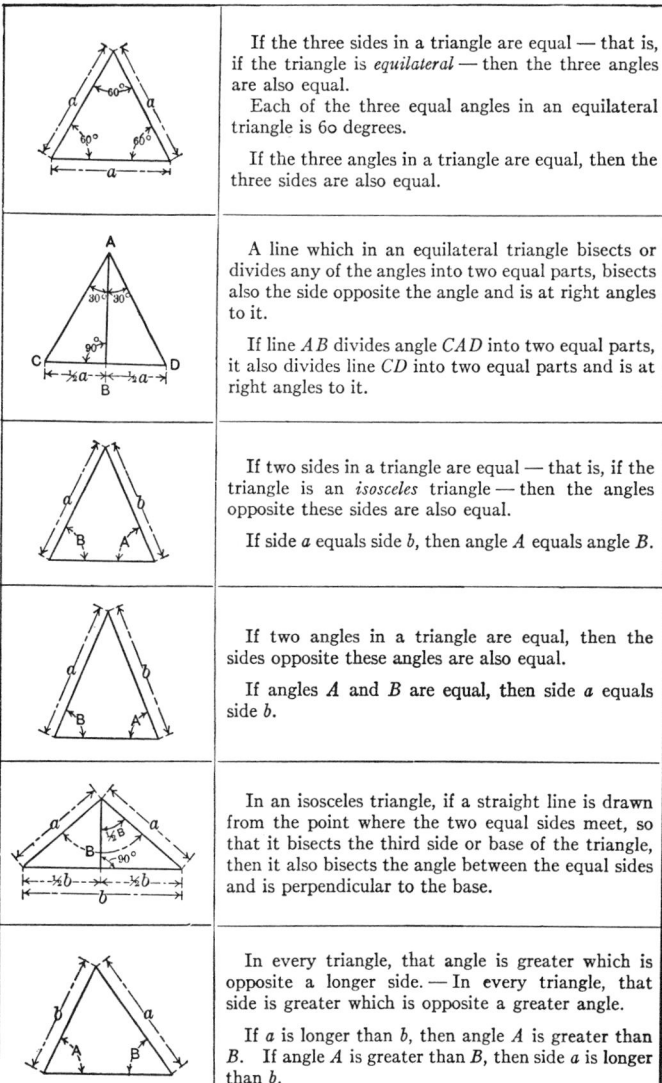

If the three sides in a triangle are equal — that is, if the triangle is *equilateral* — then the three angles are also equal.

Each of the three equal angles in an equilateral triangle is 60 degrees.

If the three angles in a triangle are equal, then the three sides are also equal.

A line which in an equilateral triangle bisects or divides any of the angles into two equal parts, bisects also the side opposite the angle and is at right angles to it.

If line AB divides angle CAD into two equal parts, it also divides line CD into two equal parts and is at right angles to it.

If two sides in a triangle are equal — that is, if the triangle is an *isosceles* triangle — then the angles opposite these sides are also equal.

If side a equals side b, then angle A equals angle B.

If two angles in a triangle are equal, then the sides opposite these angles are also equal.

If angles A and B are equal, then side a equals side b.

In an isosceles triangle, if a straight line is drawn from the point where the two equal sides meet, so that it bisects the third side or base of the triangle, then it also bisects the angle between the equal sides and is perpendicular to the base.

In every triangle, that angle is greater which is opposite a longer side. — In every triangle, that side is greater which is opposite a greater angle.

If a is longer than b, then angle A is greater than B. If angle A is greater than B, then side a is longer than b.

Geometrical Propositions

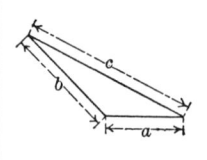

In every triangle, the sum of the lengths of two sides is always greater than the length of the third.

Side a + side b is always greater than side c.

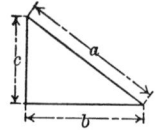

In a right-angled triangle, the square of the hypotenuse or the side opposite the right angle is equal to the sum of the squares on the two sides which form the right angle.

$$a^2 = b^2 + c^2.$$

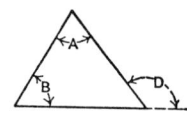

If one side of a triangle is produced, then the exterior angle is equal to the sum of the two interior opposite angles.

Angle D = angle A + angle B.

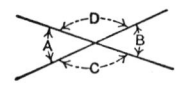

If two lines intersect, then the opposite angles formed by the intersecting lines are equal.

Angle A = angle B.
Angle C = angle D.

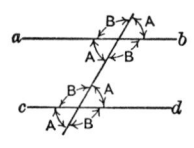

If a line intersects two parallel lines, then the corresponding angles formed by the intersecting line and the parallel lines are equal.

Lines ab and cd are parallel. Then all the angles designated A are equal, and all those designated B are equal.

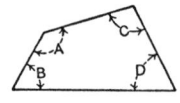

In any figure having four sides, the sum of the interior angles equals 360 degrees.

$A + B + C + D$ = 360 degrees.

Geometrical Propositions

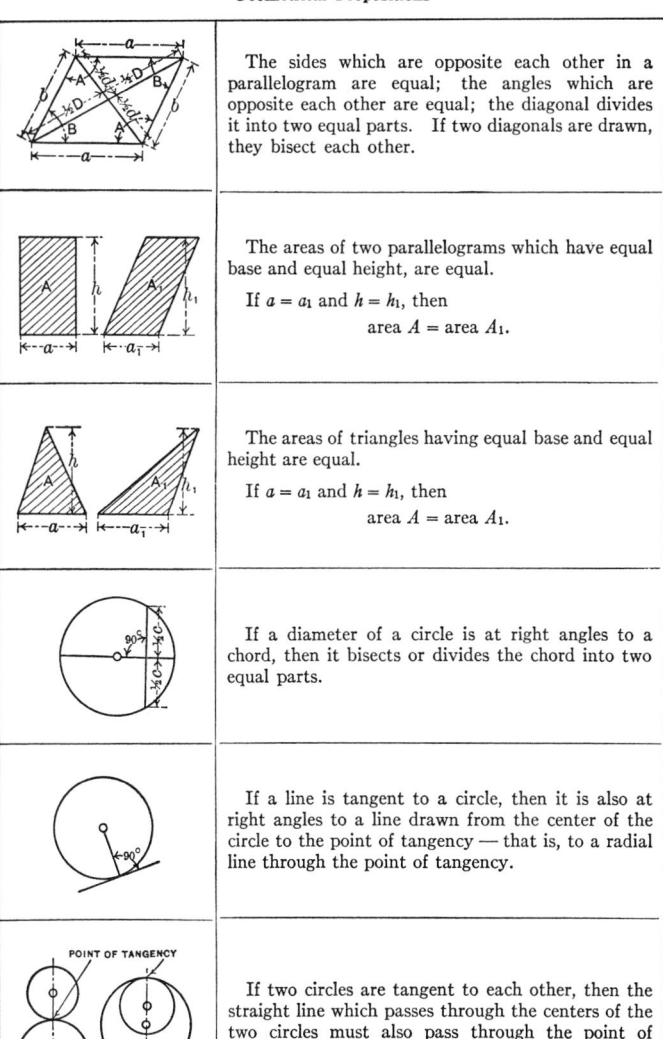

The sides which are opposite each other in a parallelogram are equal; the angles which are opposite each other are equal; the diagonal divides it into two equal parts. If two diagonals are drawn, they bisect each other.

The areas of two parallelograms which have equal base and equal height, are equal.

If $a = a_1$ and $h = h_1$, then

$$\text{area } A = \text{area } A_1.$$

The areas of triangles having equal base and equal height are equal.

If $a = a_1$ and $h = h_1$, then

$$\text{area } A = \text{area } A_1.$$

If a diameter of a circle is at right angles to a chord, then it bisects or divides the chord into two equal parts.

If a line is tangent to a circle, then it is also at right angles to a line drawn from the center of the circle to the point of tangency — that is, to a radial line through the point of tangency.

If two circles are tangent to each other, then the straight line which passes through the centers of the two circles must also pass through the point of tangency.

Geometrical Propositions

If from a point without a circle tangents are drawn to a circle, the two tangents are equal and make equal angles with the chord joining the points of tangency.

The angle between a tangent and a chord drawn from the point of tangency equals one-half the angle at the center subtended by the chord.

Angle $B = \frac{1}{2}$ angle A.

The angle between a tangent and a chord drawn from the point of tangency equals the angle at the periphery subtended by the chord.

Angle B, between tangent ab and chord cd, equals angle A subtended at the periphery by chord cd.

All angles having their vertex at the periphery of a circle and subtended by the same chord are equal.

Angles A, B and C, all subtended by chord cd, are equal.

If an angle at the circumference of a circle, between two chords, is subtended by the same arc as the angle at the center, between two radii, then the angle at the circumference is equal to one-half of the angle at the center.

Angle $A = \frac{1}{2}$ angle B.

A=LESS THAN 90° B=MORE THAN 90°

An angle subtended by a chord in a circular segment larger than one-half the circle is an acute angle — an angle less than 90 degrees. An angle subtended by a chord in a circular segment less than one-half the circle is an obtuse angle — an angle greater than 90 degrees.

Geometrical Propositions

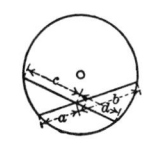

If two chords intersect each other in a circle, then the rectangle of the segments of the one equals the rectangle of the segments of the other.

$$a \times b = c \times d.$$

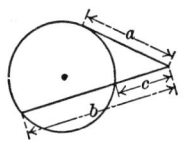

If from a point outside of a circle two lines are drawn, one of which intersects the circle while the other is tangent to it, then the rectangle contained by the total length of the intersecting line, and that part of it which is between the outside point and the periphery, equals the square of the tangent.

$$a^2 = b \times c.$$

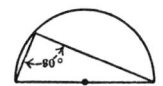

If a triangle is inscribed in a semi-circle, the angle opposite the diameter is a right (90-degree) angle.

All angles at the periphery of a circle, subtended by the diameter, are right (90-degree) angles.

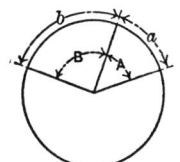

The length of circular arcs of the same circle are proportional to the corresponding angles at the center.

$$A : B = a : b.$$

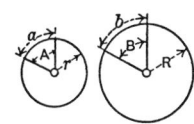

The length of circular arcs having the same center angle are proportional to the length of the radii.

If $A = B$, then $a : b = r : R$.

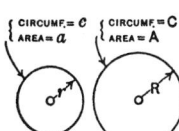

The circumferences of two circles are proportional to their radii.

The areas of two circles are proportional to the squares of their radii.

$$c : C = r : R.$$
$$a : A = r^2 : R^2.$$

Geometrical Constructions

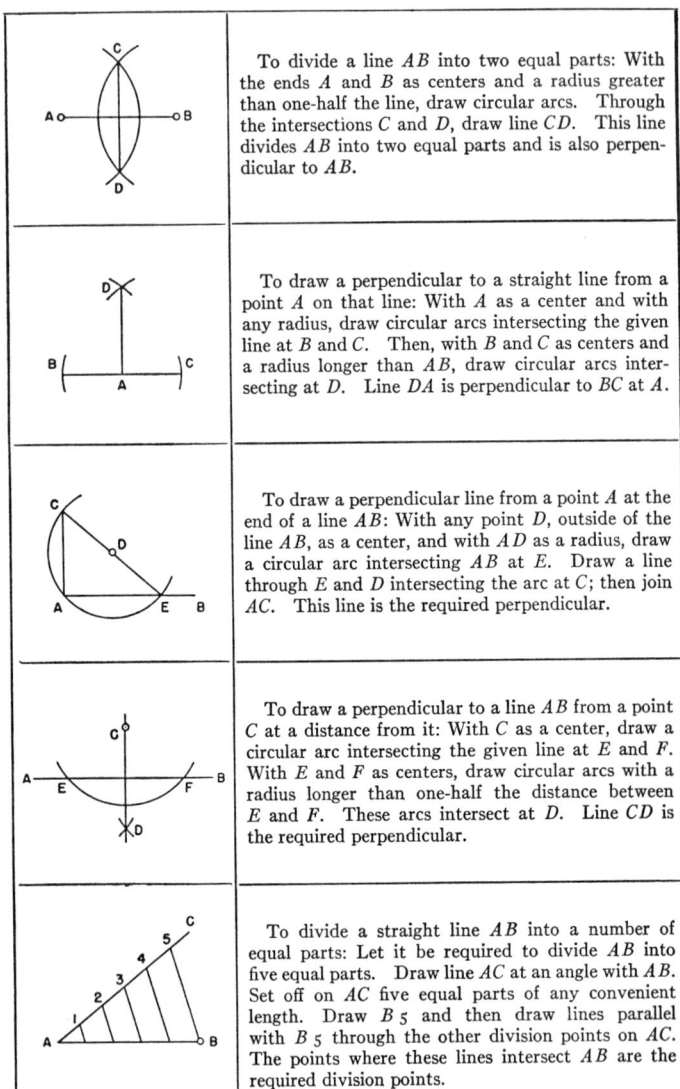

To divide a line AB into two equal parts: With the ends A and B as centers and a radius greater than one-half the line, draw circular arcs. Through the intersections C and D, draw line CD. This line divides AB into two equal parts and is also perpendicular to AB.

To draw a perpendicular to a straight line from a point A on that line: With A as a center and with any radius, draw circular arcs intersecting the given line at B and C. Then, with B and C as centers and a radius longer than AB, draw circular arcs intersecting at D. Line DA is perpendicular to BC at A.

To draw a perpendicular line from a point A at the end of a line AB: With any point D, outside of the line AB, as a center, and with AD as a radius, draw a circular arc intersecting AB at E. Draw a line through E and D intersecting the arc at C; then join AC. This line is the required perpendicular.

To draw a perpendicular to a line AB from a point C at a distance from it: With C as a center, draw a circular arc intersecting the given line at E and F. With E and F as centers, draw circular arcs with a radius longer than one-half the distance between E and F. These arcs intersect at D. Line CD is the required perpendicular.

To divide a straight line AB into a number of equal parts: Let it be required to divide AB into five equal parts. Draw line AC at an angle with AB. Set off on AC five equal parts of any convenient length. Draw $B\,5$ and then draw lines parallel with $B\,5$ through the other division points on AC. The points where these lines intersect AB are the required division points.

Geometrical Constructions

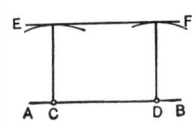	To draw a straight line parallel to a given line AB, at a given distance from it: With any points C and D on AB as centers, draw circular arcs with the given distance as radius. Line EF, drawn to touch the circular arcs, is the required parallel line.
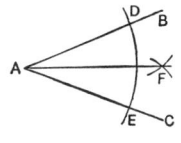	To bisect or divide an angle BAC into two equal parts: With A as a center and any radius, draw arc DE. With D and E as centers and a radius greater than one-half DE, draw circular arcs intersecting at F. Line AF divides the angle into two equal parts.
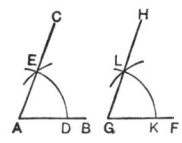	To draw an angle upon a line AB, equal to a given angle FGH: With point G as a center and with any radius, draw arc KL. With A as a center and with the same radius, draw arc DE. Make arc DE equal to KL and draw AC through E. Angle BAC then equals angle FGH.
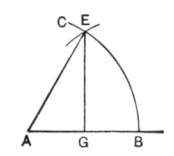	To lay out a 60-degree angle: With A as a center and any radius, draw an arc BC. With point B as a center and AB as a radius, draw an arc intersecting at E the arc just drawn. EAB is a 60-degree angle. A 30-degree angle may be obtained either by dividing a 60-degree angle into two equal parts, or by drawing a line EG perpendicular to AB. Angle AEG is then 30 degrees.
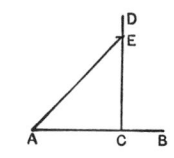	To draw a 45-degree angle: From point A on line AB set off a distance AC. Draw the perpendicular DC and set off a distance CE equal to AC. Draw AE. Angle EAC is a 45-degree angle.
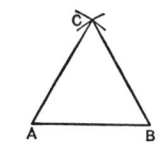	To draw an equilateral triangle, the length of the sides of which equals AB: With A and B as centers and AB as radius, draw circular arcs intersecting at C. Draw AC and BC. Then ABC is an equilateral triangle.

MATHEMATICS

Geometrical Constructions

	To draw a circular arc with a given radius through two given points A and B: With A and B as centers, and the given radius as radius, draw circular arcs intersecting at C. With C as a center, and the same radius, draw a circular arc through A and B.
	To find the center of a circle or of an arc of a circle: Select three points on the periphery of the circle, as A, B and C. With each of these points as a center and the same radius, describe arcs intersecting each other. Through the points of intersection draw lines DE and FG. Point H where these lines intersect is the center of the circle.
	To draw a tangent to a circle from a given point on the circumference: Through the point of tangency A, draw a radial line BC. At point A, draw a line EF at right angles to BC. This line is the required tangent.
	To divide a circular arc AB into two equal parts: With A and B as centers, and a radius larger than half the distance between A and B, draw circular arcs intersecting at C and D. Line CD divides arc AB into two equal parts at E.
	To describe a circle about a triangle: Divide the sides AB and AC into two equal parts, and from the division points E and F draw lines at right angles to the sides. These lines intersect at G. With G as a center and GA as a radius, draw circle ABC.
	To inscribe a circle in a triangle: Bisect two of the angles, A and B, by lines intersecting at D. From D draw a line DE perpendicular to one of the sides, and with DE as a radius, draw circle EFG.

Geometrical Problems

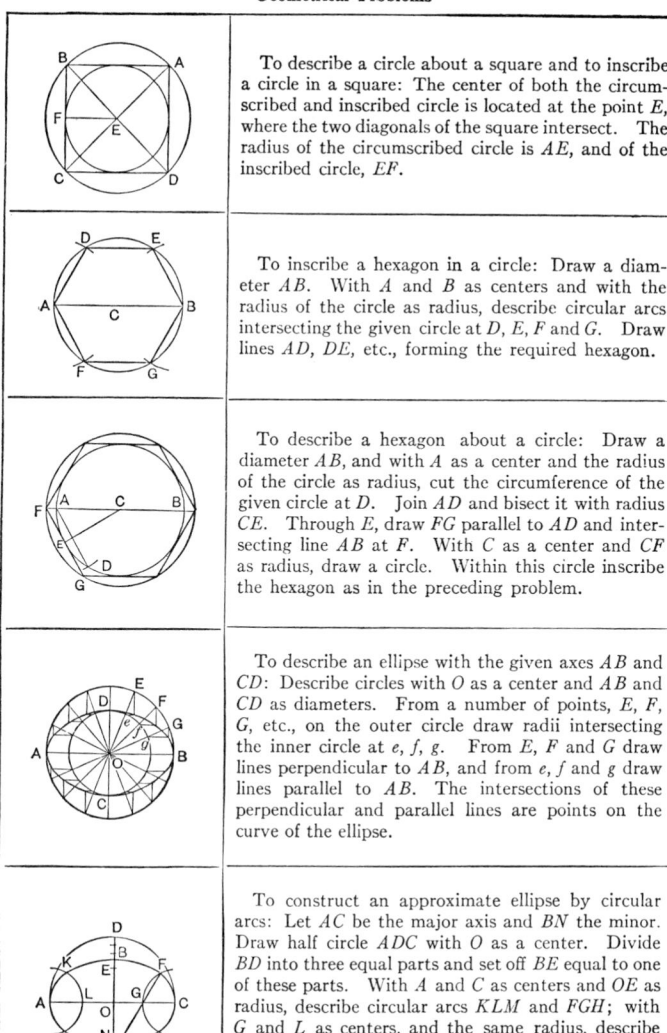

To describe a circle about a square and to inscribe a circle in a square: The center of both the circumscribed and inscribed circle is located at the point E, where the two diagonals of the square intersect. The radius of the circumscribed circle is AE, and of the inscribed circle, EF.

To inscribe a hexagon in a circle: Draw a diameter AB. With A and B as centers and with the radius of the circle as radius, describe circular arcs intersecting the given circle at D, E, F and G. Draw lines AD, DE, etc., forming the required hexagon.

To describe a hexagon about a circle: Draw a diameter AB, and with A as a center and the radius of the circle as radius, cut the circumference of the given circle at D. Join AD and bisect it with radius CE. Through E, draw FG parallel to AD and intersecting line AB at F. With C as a center and CF as radius, draw a circle. Within this circle inscribe the hexagon as in the preceding problem.

To describe an ellipse with the given axes AB and CD: Describe circles with O as a center and AB and CD as diameters. From a number of points, E, F, G, etc., on the outer circle draw radii intersecting the inner circle at e, f, g. From E, F and G draw lines perpendicular to AB, and from e, f and g draw lines parallel to AB. The intersections of these perpendicular and parallel lines are points on the curve of the ellipse.

To construct an approximate ellipse by circular arcs: Let AC be the major axis and BN the minor. Draw half circle ADC with O as a center. Divide BD into three equal parts and set off BE equal to one of these parts. With A and C as centers and OE as radius, describe circular arcs KLM and FGH; with G and L as centers, and the same radius, describe arcs FCH and KAM. Through F and G draw line FP, and with P as a center draw the arc FBK. Arc HNM is drawn in the same manner.

Geometrical Constructions

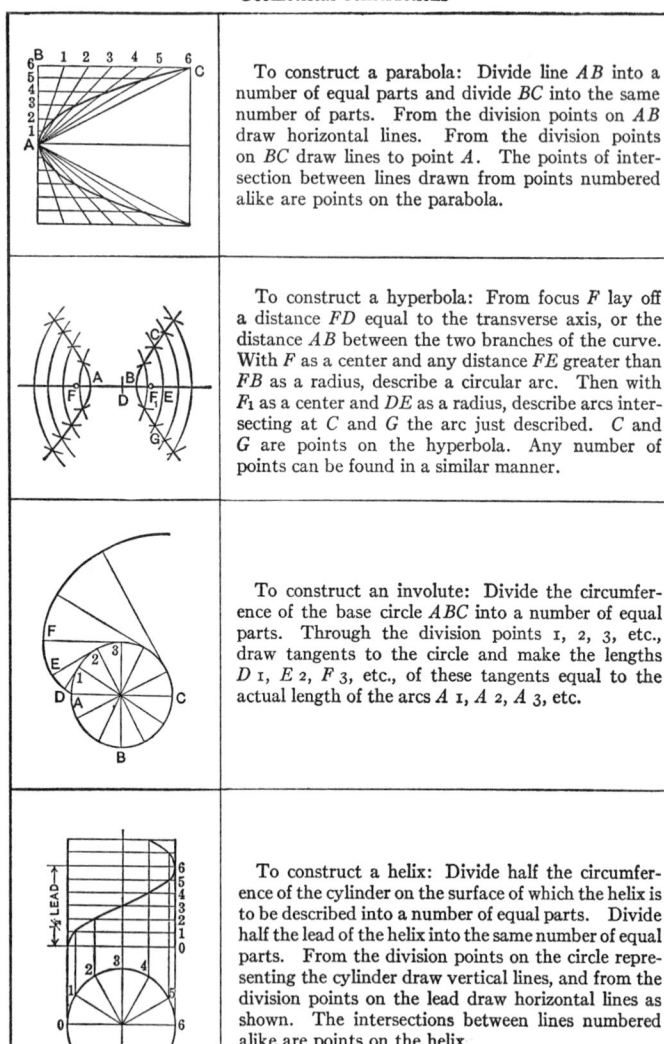

To construct a parabola: Divide line AB into a number of equal parts and divide BC into the same number of parts. From the division points on AB draw horizontal lines. From the division points on BC draw lines to point A. The points of intersection between lines drawn from points numbered alike are points on the parabola.

To construct a hyperbola: From focus F lay off a distance FD equal to the transverse axis, or the distance AB between the two branches of the curve. With F as a center and any distance FE greater than FB as a radius, describe a circular arc. Then with F_1 as a center and DE as a radius, describe arcs intersecting at C and G the arc just described. C and G are points on the hyperbola. Any number of points can be found in a similar manner.

To construct an involute: Divide the circumference of the base circle ABC into a number of equal parts. Through the division points 1, 2, 3, etc., draw tangents to the circle and make the lengths D 1, E 2, F 3, etc., of these tangents equal to the actual length of the arcs A 1, A 2, A 3, etc.

To construct a helix: Divide half the circumference of the cylinder on the surface of which the helix is to be described into a number of equal parts. Divide half the lead of the helix into the same number of equal parts. From the division points on the circle representing the cylinder draw vertical lines, and from the division points on the lead draw horizontal lines as shown. The intersections between lines numbered alike are points on the helix.

SOLUTION OF TRIANGLES

Any figure bounded by three straight lines is called a triangle. Any one of the three lines may be called the base, and the line drawn from the angle opposite the base at right angles to it is called the height or altitude of the triangle.

If all the three sides of a triangle are of equal length, the triangle is called *equilateral*. Each one of the three angles in an equilateral triangle equals 60 degrees. If two sides are of equal length, the triangle is an *isosceles* triangle. If one angle is a right or 90-degree angle, the triangle is a *right* or *right-angled* triangle. The side opposite the right angle is called the *hypotenuse*.

If all the angles are less than 90 degrees, the triangle is called an *acute* or *acute-angled* triangle. If one of the angles is larger than 90 degrees, the triangle is called an *obtuse-angled* triangle. Both acute and obtuse-angled triangles are known under the common name of *oblique-angled* triangles. The sum of the three angles in every triangle is 180 degrees.

The sides and angles of any triangle which are not known can be found when: 1. All the three sides; 2. Two sides and one angle; or 3. One side and two angles, are given. In other words, if a triangle is considered as consisting of six parts, three angles and three sides, the unknown parts can be determined when any three parts are given, provided at least one of the given parts is a side.

Functions of Angles. — The functions of angles used in solving triangles are sine, cosine, tangent, cotangent, secant, and cosecant. These expressions are usually abbreviated as follows:

sin = sine,	cot = cotangent,
cos = cosine,	sec = secant,
tan = tangent,	cosec = cosecant.

If in a right-angled triangle (see the illustration in the table below) the lengths of the three sides are represented by a, b, and c, and the angles opposite each of these sides by A, B, and C, then the side c opposite the right angle is the hypotenuse;

Trigonometrical Functions of Angles

The *sine* of an angle equals the opposite side divided by the hypotenuse. Hence, $\sin B = b \div c$, and $\sin A = a \div c$.

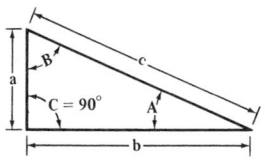

The *cosine* of an angle equals the adjacent side divided by the hypotenuse. Hence, $\cos B = a \div c$, and $\cos A = b \div c$.

The *tangent* of an angle equals the opposite side divided by the adjacent side. Hence, $\tan B = b \div a$, and $\tan A = a \div b$.

The *cotangent* of an angle equals the adjacent side divided by the opposite side. Hence, $\cot B = a \div b$, and $\cot A = b \div a$.

The *secant* of an angle equals the hypotenuse divided by the adjacent side. Hence, $\sec B = c \div a$, and $\sec A = c \div b$.

The *cosecant* of an angle equals the hypotenuse divided by the opposite side. Hence, $\operatorname{cosec} B = c \div b$, and $\operatorname{cosec} A = c \div a$.

It should be noted that the functions of the angles can be found in this manner only when the triangle is right-angled.

side *b* is called the *side adjacent* to angle *A* and is also the *side opposite* to angle *B;* side *a* is the side adjacent to angle *B* and the side opposite to angle *A*. The meanings of the various functions of angles can be explained by the aid of a right-angled triangle.

The following relation exists between the angular functions of the two acute angles in a right-angled triangle: The sine of angle *B* equals the cosine of angle *A;* the tangent of angle *B* equals the cotangent of angle *A*, and *vice versa*. The sum of the two acute angles in a right-angled triangle always equals 90 degrees; hence, when one angle is known, the other can easily be found. When any two angles together make 90 degrees, one is called the *complement* of the other, and in that case the sine of the one equals the cosine of the other, and the tangent of the one equals the cotangent of the other.

The Law of Sines. — In any triangle, any side is to the sine of the angle opposite that side as any other side is to the sine of the angle opposite that side. If *a*, *b*, and *c* are the sides, and *A*, *B*, and *C* their opposite angles, respectively, then:

$$\frac{a}{\sin A} = \frac{b}{\sin B} = \frac{c}{\sin C}, \text{ so that:}$$

$$a = \frac{b \sin A}{\sin B}, \text{ or, } a = \frac{c \sin A}{\sin C};$$

$$b = \frac{a \sin B}{\sin A}, \text{ or, } b = \frac{c \sin B}{\sin C};$$

$$c = \frac{a \sin C}{\sin A}, \text{ or, } c = \frac{b \sin C}{\sin B}$$

The Law of Cosines. — In any triangle, the square of any side is equal to the sum of the squares of the other two sides minus twice their product times the cosine of the included angle; or if *a*, *b* and *c* be the sides and *A*, *B*, and *C* are the opposite angles, respectively, then:

$$a^2 = b^2 + c^2 - 2bc \cos A$$
$$b^2 = a^2 + c^2 - 2ac \cos B$$
$$c^2 = a^2 + b^2 - 2ab \cos C$$

These two laws, together with the proposition that the sum of the three angles equals 180 degrees, are the basis of all formulas relating to the solution of triangles.

Formulas for the solution of right-angled and oblique-angled triangles, arranged in tabular form, are given on the following pages.

Signs of Trigonometric Functions. — On page 100 a diagram, "Signs of Trigonometric Functions," is given. This diagram shows the proper sign (+ or −) for the trigonometric functions of angles in each of the four quadrants, 0 to 90, 90 to 180, 180 to 270, and 270 to 360 degrees. Thus, the cosine of an angle between 90 and 180 degrees is negative; the sine of the same angle is positive.

Trigonometric Identities. — Trigonometric identities are formulas that show the relationship between different trigonometric functions. They may be used to change the form of some trigonometric expressions to simplify calculations. For example, if a formula has a term, $2 \sin A \cos A$, the equivalent but simpler term $\sin 2A$ may be substituted. The identities given below may themselves be combined or rearranged in various ways to form new identities.

1. *Basic:*
$$\tan A = \frac{\sin A}{\cos A} = \frac{1}{\cot A} \qquad \sec A = \frac{1}{\cos A} \qquad \csc A = \frac{1}{\sin A}$$

2. *Negative-Angle:*
$$\sin(-A) = -\sin A \qquad \cos(-A) = \cos A \qquad \tan(-A) = -\tan A$$

3. *Pythagorean:*
$$\sin^2 A + \cos^2 A = 1 \qquad 1 + \tan^2 A = \sec^2 A \qquad 1 + \cot^2 A = \csc^2 A$$

4. *Sum and Difference of Angles:*
$$\tan(A+B) = \frac{\tan A + \tan B}{1 - \tan A \tan B} \qquad \cot(A+B) = \frac{\cot A \cot B - 1}{\cot B + \cot A}$$
$$\tan(A-B) = \frac{\tan A - \tan B}{1 + \tan A \tan B} \qquad \cot(A-B) = \frac{\cot A \cot B + 1}{\cot B - \cot A}$$
$$\sin(A+B) = \sin A \cos B + \cos A \sin B \qquad \cos(A+B) = \cos A \cos B - \sin A \sin B$$
$$\sin(A-B) = \sin A \cos B - \cos A \sin B \qquad \cos(A-B) = \cos A \cos B + \sin A \sin B$$

5. *Double-Angle:*
$$\cos 2A = \cos^2 A - \sin^2 A = 2\cos^2 A - 1 = 1 - 2\sin^2 A$$
$$\sin 2A = 2 \sin A \cos A \qquad \tan 2A = \frac{2 \tan A}{1 - \tan^2 A} = \frac{2}{\cot A - \tan A}$$

6. *Half-Angle:*
$$\sin \tfrac{1}{2}A = \sqrt{\tfrac{1}{2}(1 - \cos A)} \qquad \cos \tfrac{1}{2}A = \sqrt{\tfrac{1}{2}(1 + \cos A)}$$
$$\tan \tfrac{1}{2}A = \sqrt{\frac{1 - \cos A}{1 + \cos A}} = \frac{1 - \cos A}{\sin A} = \frac{\sin A}{1 + \cos A}$$

7. *Product-to-Sum:*
$$\sin A \cos B = \tfrac{1}{2}[\sin(A+B) + \sin(A-B)]$$
$$\cos A \cos B = \tfrac{1}{2}[\cos(A+B) + \cos(A-B)]$$
$$\sin A \sin B = \tfrac{1}{2}[\cos(A-B) - \cos(A+B)]$$
$$\tan A \tan B = \frac{\tan A + \tan B}{\cot A + \cot B}$$

8. *Sum and Difference of Functions:*
$$\sin A + \sin B = 2[\sin \tfrac{1}{2}(A+B) \cos \tfrac{1}{2}(A-B)]$$
$$\sin A - \sin B = 2[\sin \tfrac{1}{2}(A-B) \cos \tfrac{1}{2}(A+B)]$$
$$\cos A + \cos B = 2[\cos \tfrac{1}{2}(A+B) \cos \tfrac{1}{2}(A-B)]$$
$$\cos A - \cos B = -2[\sin \tfrac{1}{2}(A+B) \sin \tfrac{1}{2}(A-B)]$$
$$\tan A + \tan B = \frac{\sin(A+B)}{\cos A \cos B} \qquad \cot A + \cot B = \frac{\sin(B+A)}{\sin A \sin B}$$
$$\tan A - \tan B = \frac{\sin(A-B)}{\cos A \cos B} \qquad \cot A - \cot B = \frac{\sin(B-A)}{\sin A \sin B}$$

Solution of Right-angled Triangles

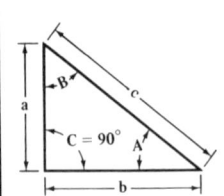

As shown in the illustration, the sides of the right-angled triangle are designated a and b and the hypotenuse, c. The angles opposite each of these sides are designated A and B, respectively.

Angle C, opposite the hypotenuse c is the right angle, and is therefore always one of the known quantities.

Sides and Angles Known	Formulas for Sides and Angles to be Found		
Side a; side b	$c = \sqrt{a^2 + b^2}$	$\tan A = \dfrac{a}{b}$	$B = 90° - A$
Side a; hypotenuse c	$b = \sqrt{c^2 - a^2}$	$\sin A = \dfrac{a}{c}$	$B = 90° - A$
Side b; hypotenuse c	$a = \sqrt{c^2 - b^2}$	$\sin B = \dfrac{b}{c}$	$A = 90° - B$
Hypotenuse c; angle B ...	$b = c \times \sin B$	$a = c \times \cos B$	$A = 90° - B$
Hypotenuse c; angle A ...	$b = c \times \cos A$	$a = c \times \sin A$	$B = 90° - A$
Side b; angle B	$c = \dfrac{b}{\sin B}$	$a = b \times \cot B$	$A = 90° - B$
Side b; angle A	$c = \dfrac{b}{\cos A}$	$a = b \times \tan A$	$B = 90° - A$
Side a; angle B	$c = \dfrac{a}{\cos B}$	$b = a \times \tan B$	$A = 90° - B$
Side a; angle A	$c = \dfrac{a}{\sin A}$	$b = a \times \cot A$	$B = 90° - A$

Examples of the Solution of Right-angled Triangles (English and metric units)

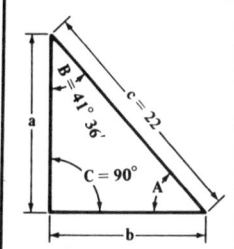

Hypotenuse and one angle known:
$$c = 22 \text{ inches}; \quad B = 41° 36'.$$
Then, by the formulas given on the preceding page:
$$a = c \times \cos B = 22 \times \cos 41° 36' = 22 \times 0.74780$$
$$= 16.4516 \text{ inches.}$$
$$b = c \times \sin B = 22 \times \sin 41° 36' = 22 \times 0.66393$$
$$= 14.6065 \text{ inches.}$$
$$A = 90° - B = 90° - 41° 36' = 48° 24'.$$

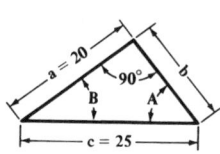

Hypotenuse and one side known:
$$c = 25 \text{ centimeters}; \quad a = 20 \text{ centimeters.}$$
From the formulas on the preceding page:
$$b = \sqrt{c^2 - a^2} = \sqrt{25^2 - 20^2} = \sqrt{625 - 400}$$
$$= \sqrt{225} = 15 \text{ centimeters.}$$
$$\sin A = \frac{a}{c} = \frac{20}{25} = 0.8$$
Hence, $A = 53° 8'.$
$$B = 90° - A = 90° - 53° 8' = 36° 52'.$$

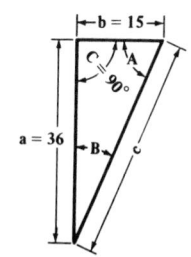

Two sides known:
$$a = 36 \text{ inches}; \quad b = 15 \text{ inches.}$$
Then, by the formulas given on the preceding page:
$$c = \sqrt{a^2 + b^2} = \sqrt{36^2 + 15^2} = \sqrt{1296 + 225}$$
$$= \sqrt{1521} = 39 \text{ inches.}$$
$$\tan A = \frac{a}{b} = \frac{36}{15} = 2.4$$
Hence, $A = 67° 23'.$
$$B = 90° - A = 90° - 67° 23' = 22° 37'.$$

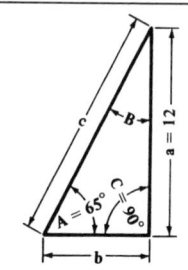

One side and one angle known:
$$a = 12 \text{ meters}; \quad A = 65°.$$
Then, by the formulas given on the preceding page:
$$c = \frac{a}{\sin A} = \frac{12}{\sin 65°} = \frac{12}{0.90631} = 13.2405 \text{ meters.}$$
$$b = a \times \cot A = 12 \times \cot 65° = 12 \times 0.46631$$
$$= 5.5957 \text{ meters.}$$
$$B = 90° - A = 90° - 65° = 25°.$$

Solution of Oblique-angled Triangles

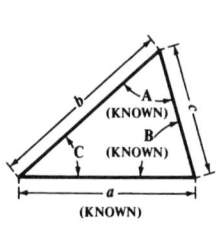	One side and two angles known: Call the known side a, the angle opposite it A, and the other known angle B. Then: $C = 180° - (A + B)$; or if angles B and C are given, but not A, then $A = 180° - (B + C)$. $$C = 180° - (A + B)$$ $$b = \frac{a \times \sin B}{\sin A} \qquad c = \frac{a \times \sin C}{\sin A}$$ $$\text{Area} = \frac{a \times b \times \sin C}{2}$$
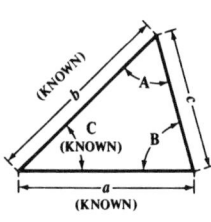	Two sides and the angle between them known: Call the known sides a and b, and the known angle between them C. Then: $$\tan A = \frac{a \times \sin C}{b - (a \times \cos C)}$$ $$B = 180° - (A + C) \qquad c = \frac{a \times \sin C}{\sin A}$$ Side c may also be found directly as below: $$c = \sqrt{a^2 + b^2 - (2ab \times \cos C)}$$ $$\text{Area} = \frac{a \times b \times \sin C}{2}$$
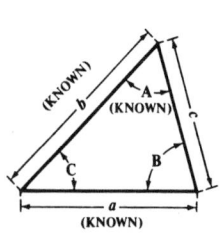	Two sides and the angle opposite one of the sides known: Call the known angle A, the side opposite it a, and the other known side b. Then: $$\sin B = \frac{b \times \sin A}{a} \qquad C = 180° - (A + B)$$ $$c = \frac{a \times \sin C}{\sin A} \qquad \text{Area} = \frac{a \times b \times \sin C}{2}$$ If, in the above, angle B > angle A but <90°, then a second solution B_2, C_2, c_2 exists for which: $B_2 = 180° - B$; $\quad C_2 = 180° - (A + B_2)$; $\quad c_2 = (a \times \sin C_2) \div \sin A$; Area = $(a \times b \times \sin C_2) \div 2$. If $a \geqq b$, then the first solution only exists. If $a < b \times \sin A$, then no solution exists.
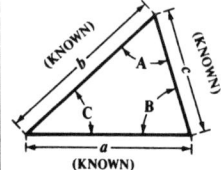	All three sides known: Call the sides a, b, and c, and the angles opposite them, A, B, and C. Then: $$\cos A = \frac{b^2 + c^2 - a^2}{2bc} \qquad \sin B = \frac{b \times \sin A}{a}$$ $$C = 180° - (A + B) \qquad \text{Area} = \frac{a \times b \times \sin C}{2}$$

Examples of the Solution of Oblique-angled Triangles (English and metric units)

	Side and angles known: $\quad a = 5$ centimeters; $\quad A = 80°$; $\quad B = 62°$ Then, by the formulas on the opposite page: $C = 180° - (80° + 62°) = 180° - 142° = 38°$. $b = \dfrac{a \times \sin B}{\sin A} = \dfrac{5 \times \sin 62°}{\sin 80°} = \dfrac{5 \times 0.88295}{0.98481} = 4.483$ centimeters. $c = \dfrac{a \times \sin C}{\sin A} = \dfrac{5 \times \sin 38°}{\sin 80°} = \dfrac{5 \times 0.61566}{0.98481} = 3.126$ centimeters.
	Sides and angle known: $\quad a = 9$ inches; $\quad b = 8$ inches; $\quad C = 35°$. $\tan A = \dfrac{a \times \sin C}{b - (a \times \cos C)} = \dfrac{9 \times \sin 35°}{8 - (9 \times \cos 35°)}$ $\qquad = \dfrac{9 \times 0.57358}{8 - (9 \times 0.81915)} = \dfrac{5.16222}{0.62765} = 8.22468$. Hence, $\qquad A = 83° 4'$. $B = 180° - (A + C) = 180° - 118° 4' = 61° 56'$. $c = \dfrac{a \times \sin C}{\sin A} = \dfrac{9 \times 0.57358}{0.99269} = 5.2$ inches.
	Sides and angle known: $a = 20$ centimeters; $b = 17$ centimeters; $A = 61°$. $\qquad \sin B = \dfrac{b \times \sin A}{a} = \dfrac{17 \times \sin 61°}{20}$ $\qquad\qquad = \dfrac{17 \times 0.87462}{20} = 0.74343$. Hence, $\qquad B = 48° 1'$. $C = 180° - (A + B) = 180° - 109° 1' = 70° 59'$. $c = \dfrac{a \times \sin C}{\sin A} = \dfrac{20 \times \sin 70° 59'}{\sin 61°} = \dfrac{20 \times 0.94542}{0.87462}$ $\qquad = 21.62$ centimeters.
	Sides known: $\quad a = 8$ inches; $\quad b = 9$ inches; $\quad c = 10$ inches. $\cos A = \dfrac{b^2 + c^2 - a^2}{2bc} = \dfrac{9^2 + 10^2 - 8^2}{2 \times 9 \times 10}$ $\qquad = \dfrac{81 + 100 - 64}{180} = \dfrac{117}{180} = 0.65000$. Hence, $\qquad A = 49° 27'$. $\sin B = \dfrac{b \times \sin A}{a} = \dfrac{9 \times 0.75984}{8} = 0.85482$. Hence, $\qquad B = 58° 44'$. $C = 180° - (A + B) = 180° - 108° 11' = 71° 49'$.

Tables of Trigonometric Functions. — The numerical values for the natural or trigonometric functions for all angles from 0 to 360 degrees are given in the tables, pages 101 to 145. The chart below shows how to entry the table.

How to Enter Table of Trigonometric Functions

For Angles from	Enter Table for Degrees and Function	Minutes	For Angles from	Enter Table for Degrees and Function	Minutes
0° to 45°	at top	at left	180° to 225°	at top	at left
45° to 90°	at bottom	at right	225° to 270°	at bottom	at right
90° to 135°	at bottom	at left	270° to 315°	at bottom	at left
135° to 180°	at top	at right	315° to 360°	at top	at right
Examples: The sine of 26° is 0.43837; of 126°, 0.80902; of 226°, −0.71934.					

Signs of Trigonometric Functions

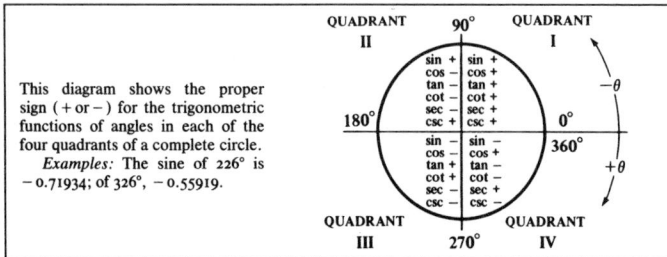

This diagram shows the proper sign (+ or −) for the trigonometric functions of angles in each of the four quadrants of a complete circle.
Examples: The sine of 226° is −0.71934; of 326°, −0.55919.

Useful Relationships Among Angles

Angle / Function	θ	$-\theta$	$90° \pm \theta$	$180° \pm \theta$	$270° \pm \theta$	$360° \pm \theta$
sin	$\sin \theta$	$-\sin \theta$	$+\cos \theta$	$\mp \sin \theta$	$-\cos \theta$	$\pm \sin \theta$
cos	$\cos \theta$	$+\cos \theta$	$\mp \sin \theta$	$-\cos \theta$	$\pm \sin \theta$	$+\cos \theta$
tan	$\tan \theta$	$-\tan \theta$	$\mp \cot \theta$	$\pm \tan \theta$	$\mp \cot \theta$	$\pm \tan \theta$
cot	$\cot \theta$	$-\cot \theta$	$\mp \tan \theta$	$\pm \cot \theta$	$\mp \tan \theta$	$\pm \cot \theta$
sec	$\sec \theta$	$+\sec \theta$	$\mp \csc \theta$	$-\sec \theta$	$\pm \csc \theta$	$+\sec \theta$
csc	$\csc \theta$	$-\csc \theta$	$+\sec \theta$	$\mp \csc \theta$	$-\sec \theta$	$\pm \csc \theta$
Examples: $\cos 270° - \theta = -\sin \theta$; $\tan 90° + \theta = -\cot \theta$.						

Involute Functions. — Involute functions are used in certain formulas relating to the design and measurement of gear teeth as well as measurement of threads over wires. Included in the trigonometric tables, pages 101 to 145, are values for the involute functions of angles of from 0 to 90 degrees. These involute functions were calculated from the formula: Involute of $\theta = \tan \theta - \pi \times \theta \div 180$ when θ is given in degrees.

Sevolute Functions. — Sevolute functions are used in certain spline calculations. They may be computed by subtracting the involute of an angle from the secant of the angle. Thus, sevolute 20° = sec 20° − inv 20° = 1.0642 − 0.014904 = 1.0493.

Versed Sine and Versed Cosine. — These functions are sometimes used in formulas for segments of a circle and may be obtained by using the trigonometric tables together with the relationships: versed sine $\theta = 1 - \cos \theta$; and versed cosine $\theta = 1 - \sin \theta$.

0° or 180° Trigonometric and Involute Functions 179° or 359°

M	Sine	Cosine	Tan.	Cotan.	Secant	Cosec.	Involute 0°–1°	READ UP	M
0	0.00000	1.0000	0.00000	Infinite	1.0000	Infinite	.0000000	Infinite	60
1	.00029	.0000	.00029	3437.7	.0000	3437.7	.0000000	3436.2	59
2	.00058	.0000	.00058	1718.9	.0000	1718.9	.0000000	1717.3	58
3	.00087	.0000	.00087	1145.9	.0000	1145.9	.0000000	1144.3	57
4	.00116	.0000	.00116	859.44	.0000	859.44	.0000000	857.87	56
5	0.00145	1.0000	0.00145	687.55	1.0000	687.55	.0000000	685.98	55
6	.00175	.0000	.00175	572.96	.0000	572.96	.0000000	571.39	54
7	.00204	.0000	.00204	491.11	.0000	491.11	.0000000	489.54	53
8	.00233	.0000	.00233	429.72	.0000	429.72	.0000000	428.15	52
9	.00262	.0000	.00262	381.97	.0000	381.97	.0000000	380.40	51
10	0.00291	1.00000	0.00291	343.77	1.0000	343.78	.0000000	342.21	50
11	.00320	.99999	.00320	312.52	.0000	312.52	.0000000	310.95	49
12	.00349	.99999	.00349	286.48	.0000	286.48	.0000000	284.91	48
13	.00378	.99999	.00378	264.44	.0000	264.44	.0000000	262.87	47
14	.00407	.99999	.00407	245.55	.0000	245.55	.0000000	243.99	46
15	0.00436	0.99999	0.00436	229.18	1.0000	229.18	.0000000	227.62	45
16	.00465	.99999	.00465	214.86	.0000	214.86	.0000000	213.29	44
17	.00495	.99999	.00495	202.22	.0000	202.22	.0000000	200.65	43
18	.00524	.99999	.00524	190.98	.0000	190.99	.0000000	189.42	42
19	.00553	.99998	.00553	180.93	.0000	180.93	.0000001	179.37	41
20	0.00582	0.99998	0.00582	171.89	1.0000	171.89	.0000001	170.32	40
21	.00611	.99998	.00611	163.70	.0000	163.70	.0000001	162.14	39
22	.00640	.99998	.00640	156.26	.0000	156.26	.0000001	154.69	38
23	.00669	.99998	.00669	149.47	.0000	149.47	.0000001	147.90	37
24	.00698	.99998	.00698	143.24	.0000	143.24	.0000001	141.67	36
25	0.00727	0.99997	0.00727	137.51	1.0000	137.51	.0000001	135.94	35
26	.00756	.99997	.00756	132.22	.0000	132.22	.0000001	130.66	34
27	.00785	.99997	.00785	127.32	.0000	127.33	.0000002	125.76	33
28	.00814	.99997	.00815	122.77	.0000	122.78	.0000002	121.21	32
29	.00844	.99996	.00844	118.54	.0000	118.54	.0000002	116.98	31
30	0.00873	0.99996	0.00873	114.59	1.0000	114.59	.0000002	113.03	30
31	.00902	.99996	.00902	110.89	.0000	110.90	.0000002	109.33	29
32	.00931	.99996	.00931	107.43	.0000	107.43	.0000003	105.86	28
33	.00960	.99995	.00960	104.17	.0000	104.18	.0000003	102.61	27
34	.00989	.99995	.00989	101.11	.0000	101.11	.0000003	99.546	26
35	0.01018	0.99995	0.01018	98.218	1.0001	98.223	.0000004	96.657	25
36	.01047	.99995	.01047	95.489	.0001	95.495	.0000004	93.929	24
37	.01076	.99994	.01076	92.908	.0001	92.914	.0000004	91.348	23
38	.01105	.99994	.01105	90.463	.0001	90.469	.0000005	88.904	22
39	.01134	.99994	.01135	88.144	.0001	88.149	.0000005	86.584	21
40	0.01164	0.99993	0.01164	85.940	1.0001	85.946	.0000005	84.381	20
41	.01193	.99993	.01193	83.844	.0001	83.849	.0000006	82.285	19
42	.01222	.99993	.01222	81.847	.0001	81.853	.0000006	80.288	18
43	.01251	.99992	.01251	79.943	.0001	79.950	.0000007	78.385	17
44	.01280	.99992	.01280	78.126	.0001	78.133	.0000007	76.568	16
45	0.01309	0.99991	0.01309	76.390	1.0001	76.397	.0000007	74.832	15
46	.01338	.99991	.01338	74.729	.0001	74.736	.0000008	73.172	14
47	.01367	.99991	.01367	73.139	.0001	73.146	.0000009	71.582	13
48	.01396	.99990	.01396	71.615	.0001	71.622	.0000009	70.058	12
49	.01425	.99990	.01425	70.153	.0001	70.160	.0000010	68.597	11
50	0.01454	0.99989	0.01455	68.750	1.0001	68.757	.0000010	67.194	10
51	.01483	.99989	.01484	67.402	.0001	67.409	.0000011	65.846	9
52	.01513	.99989	.01513	66.105	.0001	66.113	.0000012	64.550	8
53	.01542	.99988	.01542	64.858	.0001	64.866	.0000012	63.303	7
54	.01571	.99988	.01571	63.657	.0001	63.665	.0000013	62.102	6
55	0.01600	0.99987	0.01600	62.499	1.0001	62.507	.0000014	60.944	5
56	.01629	.99987	.01629	61.383	.0001	61.391	.0000014	59.828	4
57	.01658	.99986	.01658	60.306	.0001	60.314	.0000015	58.752	3
58	.01687	.99986	.01687	59.266	.0001	59.274	.0000016	57.712	2
59	.01716	.99985	.01716	58.261	.0001	58.270	.0000017	56.708	1
60	0.01745	0.99985	0.01746	57.290	1.0002	57.299	.0000018	55.737	0
M	Cosine	Sine	Cotan.	Tan.	Cosec.	Secant	READ DOWN	89°–90° Involute	M

90° or 270°   89° or 269°

1° or 181° **Trigonometric and Involute Functions** **178° or 358°**

M	Sine	Cosine	Tan.	Cotan.	Secant	Cosec.	Involute 1°−2°	READ UP	M
0	0.01745	0.99985	0.01746	57.290	1.0002	57.299	.0000018	55.737	60
1	.01774	.99984	.01775	56.351	.0002	56.359	.0000019	54.798	59
2	.01803	.99984	.01804	55.442	.0002	55.451	.0000020	53.889	58
3	.01832	.99983	.01833	54.561	.0002	54.570	.0000021	53.009	57
4	.01862	.99983	.01862	53.709	.0002	53.718	.0000022	52.156	56
5	0.01891	0.99982	0.01891	52.882	1.0002	52.892	.0000023	51.330	55
6	.01920	.99982	.01920	52.081	.0002	52.090	.0000024	50.529	54
7	.01949	.99981	.01949	51.303	.0002	51.313	.0000025	49.752	53
8	.01978	.99980	.01978	50.549	.0002	50.558	.0000026	48.997	52
9	.02007	.99980	.02007	49.816	.0002	49.826	.0000027	48.265	51
10	0.02036	0.99979	0.02036	49.104	1.0002	49.114	.0000028	47.553	50
11	.02065	.99979	.02066	48.412	.0002	48.422	.0000029	46.862	49
12	.02094	.99978	.02095	47.740	.0002	47.750	.0000031	46.190	48
13	.02123	.99977	.02124	47.085	.0002	47.096	.0000032	45.536	47
14	.02152	.99977	.02153	46.449	.0002	46.460	.0000033	44.900	46
15	0.02181	0.99976	0.02182	45.829	1.0002	45.840	.0000035	44.280	45
16	.02211	.99976	.02211	45.226	.0002	45.237	.0000036	43.677	44
17	.02240	.99975	.02240	44.639	.0003	44.650	.0000037	43.090	43
18	.02269	.99974	.02269	44.066	.0003	44.077	.0000039	42.518	42
19	.02298	.99974	.02298	43.508	.0003	43.520	.0000040	41.960	41
20	0.02327	0.99973	0.02328	42.964	1.0003	42.976	.0000042	41.417	40
21	.02356	.99972	.02357	42.433	.0003	42.445	.0000044	40.886	39
22	.02385	.99972	.02386	41.916	.0003	41.928	.0000045	40.369	38
23	.02414	.99971	.02415	41.411	.0003	41.423	.0000047	39.864	37
24	.02443	.99970	.02444	40.917	.0003	40.930	.0000049	39.371	36
25	0.02472	0.99969	0.02473	40.436	1.0003	40.448	.0000050	38.890	35
26	.02501	.99969	.02502	39.965	.0003	39.978	.0000052	38.420	34
27	.02530	.99968	.02531	39.506	.0003	39.519	.0000054	37.960	33
28	.02560	.99967	.02560	39.057	.0003	39.070	.0000055	37.512	32
29	.02589	.99966	.02589	38.618	.0003	38.631	.0000058	37.073	31
30	0.02618	0.99966	0.02619	38.188	1.0003	38.202	.0000060	36.644	30
31	.02647	.99965	.02648	37.769	.0004	37.782	.0000062	36.224	29
32	.02676	.99964	.02677	37.358	.0004	37.371	.0000064	35.814	28
33	.02705	.99963	.02706	36.956	.0004	36.970	.0000066	35.412	27
34	.02734	.99963	.02735	36.563	.0004	36.576	.0000068	35.019	26
35	0.02763	0.99962	0.02764	36.178	1.0004	36.191	.0000070	34.634	25
36	.02792	.99961	.02793	35.801	.0004	35.815	.0000073	34.258	24
37	.02821	.99960	.02822	35.431	.0004	35.445	.0000075	33.889	23
38	.02850	.99959	.02851	35.070	.0004	35.084	.0000077	33.527	22
39	.02879	.99959	.02881	34.715	.0004	34.730	.0000080	33.173	21
40	0.02908	0.99958	0.02910	34.368	1.0004	34.382	.0000082	32.826	20
41	.02938	.99957	.02939	34.027	.0004	34.042	.0000085	32.486	19
42	.02967	.99956	.02968	33.694	.0004	33.708	.0000087	32.152	18
43	.02996	.99955	.02997	33.366	.0004	33.381	.0000090	31.825	17
44	.03025	.99954	.03026	33.045	.0005	33.060	.0000092	31.505	16
45	0.03054	0.99953	0.03055	32.730	1.0005	32.746	.0000095	31.190	15
46	.03083	.99952	.03084	32.421	.0005	32.437	.0000098	30.881	14
47	.03112	.99952	.03114	32.118	.0005	32.134	.0000101	30.578	13
48	.03141	.99951	.03143	31.821	.0005	31.836	.0000103	30.281	12
49	.03170	.99950	.03172	31.528	.0005	31.544	.0000106	29.989	11
50	0.03199	0.99949	0.03201	31.242	1.0005	31.258	.0000109	29.703	10
51	.03228	.99948	.03230	30.960	.0005	30.976	.0000112	29.421	9
52	.03257	.99947	.03259	30.683	.0005	30.700	.0000115	29.145	8
53	.03286	.99946	.03288	30.412	.0005	30.428	.0000118	28.874	7
54	.03316	.99945	.03317	30.145	.0006	30.161	.0000122	28.607	6
55	0.03345	0.99944	0.03346	29.882	1.0006	29.899	.0000125	28.345	5
56	.03374	.99943	.03376	29.624	.0006	29.641	.0000128	28.087	4
57	.03403	.99942	.03405	29.371	.0006	29.388	.0000131	27.834	3
58	.03432	.99941	.03434	29.122	.0006	29.139	.0000135	27.586	2
59	.03461	.99940	.03463	28.877	.0006	28.894	.0000138	27.341	1
60	0.03490	0.99939	0.03492	28.636	1.0006	28.654	.0000142	27.100	0
M	Cosine	Sine	Cotan.	Tan.	Cosec.	Secant	READ DOWN	88°−89°	M
								Involute	

2° or 182°   Trigonometric and Involute Functions   177° or 357°

M	Sine	Cosine	Tan.	Cotan.	Secant	Cosec.	Involute 2°–3°	READ UP	M
0	0.03490	0.99939	0.03492	28.636	1.0006	28.654	.0000142	27.100	60
1	.03519	.99938	.03521	28.399	.0006	28.417	.0000145	26.864	59
2	.03548	.99937	.03550	28.166	.0006	28.184	.0000149	26.631	58
3	.03577	.99936	.03579	27.937	.0006	27.955	.0000153	26.402	57
4	.03606	.99935	.03609	27.712	.0007	27.730	.0000157	26.177	56
5	0.03635	0.99934	0.03638	27.490	1.0007	27.508	.0000160	25.955	55
6	.03664	.99933	.03667	27.271	.0007	27.290	.0000164	25.737	54
7	.03693	.99932	.03696	27.057	.0007	27.075	.0000168	25.523	53
8	.03723	.99931	.03725	26.845	.0007	26.864	.0000172	25.311	52
9	.03752	.99930	.03754	26.637	.0007	26.655	.0000176	25.103	51
10	0.03781	0.99929	0.03783	26.432	1.0007	26.451	.0000180	24.899	50
11	.03810	.99927	.03812	26.230	.0007	26.249	.0000185	24.697	49
12	.03839	.99926	.03842	26.031	.0007	26.050	.0000189	24.498	48
13	.03868	.99925	.03871	25.835	.0007	25.854	.0000193	24.303	47
14	.03897	.99924	.03900	25.642	.0008	25.661	.0000198	24.110	46
15	0.03926	0.99923	0.03929	25.452	1.0008	25.471	.0000202	23.920	45
16	.03955	.99922	.03958	25.264	.0008	25.284	.0000207	23.733	44
17	.03984	.99921	.03987	25.080	.0008	25.100	.0000211	23.549	43
18	.04013	.99919	.04016	24.898	.0008	24.918	.0000216	23.367	42
19	.04042	.99918	.04046	24.719	.0008	24.739	.0000220	23.188	41
20	0.04071	0.99917	0.04075	24.542	1.0008	24.562	.0000225	23.012	40
21	.04100	.99916	.04104	24.368	.0008	24.388	.0000230	22.838	39
22	.04129	.99915	.04133	24.196	.0009	24.216	.0000235	22.666	38
23	.04159	.99913	.04162	24.026	.0009	24.047	.0000240	22.497	37
24	.04188	.99912	.04191	23.859	.0009	23.880	.0000245	22.330	36
25	0.04217	0.99911	0.04220	23.695	1.0009	23.716	.0000250	22.166	35
26	.04246	.99910	.04250	23.532	.0009	23.553	.0000256	22.004	34
27	.04275	.99909	.04279	23.372	.0009	23.393	.0000261	21.844	33
28	.04304	.99907	.04308	23.214	.0009	23.235	.0000266	21.686	32
29	.04333	.99906	.04337	23.058	.0009	23.079	.0000272	21.530	31
30	0.04362	0.99905	0.04366	22.904	1.0010	22.926	.0000277	21.377	30
31	.04391	.99904	.04395	22.752	.0010	22.774	.0000283	21.225	29
32	.04420	.99902	.04424	22.602	.0010	22.624	.0000288	21.075	28
33	.04449	.99901	.04454	22.454	.0010	22.476	.0000294	20.928	27
34	.04478	.99900	.04483	22.308	.0010	22.330	.0000300	20.782	26
35	0.04507	0.99898	0.04512	22.164	1.0010	22.187	.0000306	20.638	25
36	.04536	.99897	.04541	22.022	.0010	22.044	.0000312	20.496	24
37	.04565	.99896	.04570	21.881	.0010	21.904	.0000318	20.356	23
38	.04594	.99894	.04599	21.743	.0011	21.766	.0000324	20.218	22
39	.04623	.99893	.04628	21.606	.0011	21.629	.0000330	20.081	21
40	0.04653	0.99892	0.04658	21.470	1.0011	21.494	.0000336	19.946	20
41	.04682	.99890	.04687	21.337	.0011	21.360	.0000343	19.813	19
42	.04711	.99889	.04716	21.205	.0011	21.229	.0000349	19.681	18
43	.04740	.99888	.04745	21.075	.0011	21.098	.0000356	19.551	17
44	.04769	.99886	.04774	20.946	.0011	20.970	.0000362	19.423	16
45	0.04798	0.99885	0.04803	20.819	1.0012	20.843	.0000369	19.296	15
46	.04827	.99883	.04833	20.693	.0012	20.717	.0000376	19.171	14
47	.04856	.99882	.04862	20.569	.0012	20.593	.0000382	19.047	13
48	.04885	.99881	.04891	20.446	.0012	20.471	.0000389	18.925	12
49	.04914	.99879	.04920	20.325	.0012	20.350	.0000396	18.804	11
50	0.04943	0.99878	0.04949	20.206	1.0012	20.230	.0000403	18.684	10
51	.04972	.99876	.04978	20.087	.0012	20.112	.0000411	18.566	9
52	.05001	.99875	.05007	19.970	.0013	19.995	.0000418	18.449	8
53	.05030	.99873	.05037	19.855	.0013	19.880	.0000425	18.334	7
54	.05059	.99872	.05066	19.740	.0013	19.766	.0000433	18.220	6
55	0.05088	0.99870	0.05095	19.627	1.0013	19.653	.0000440	18.107	5
56	.05117	.99869	.05124	19.516	.0013	19.541	.0000448	17.996	4
57	.05146	.99867	.05153	19.405	.0013	19.431	.0000455	17.886	3
58	.05175	.99866	.05182	19.296	.0013	19.322	.0000463	17.777	2
59	.05205	.99864	.05212	19.188	.0014	19.214	.0000471	17.669	1
60	0.05234	0.99863	0.05241	19.081	1.0014	19.107	.0000479	17.563	0

M	Cosine	Sine	Cotan.	Tan.	Cosec.	Secant	READ DOWN	87°–88° Involute	M

92° or 272°   87° or 267°

3° or 183° Trigonometric and Involute Functions 176° or 356°

M	Sine	Cosine	Tan.	Cotan.	Secant	Cosec.	Involute 3°–4°	READ UP	M
0	0.05234	0.99863	0.05241	19.081	1.0014	19.107	.0000479	17.563	60
1	.05263	.99861	.05270	18.976	.0014	19.002	.0000487	17.457	59
2	.05292	.99860	.05299	18.871	.0014	18.898	.0000495	17.353	58
3	.05321	.99858	.05328	18.768	.0014	18.794	.0000503	17.250	57
4	.05350	.99857	.05357	18.666	.0014	18.692	.0000512	17.148	56
5	0.05379	0.99855	0.05387	18.564	1.0014	18.591	.0000520	17.047	55
6	.05408	.99854	.05416	18.464	.0015	18.492	.0000529	16.948	54
7	.05437	.99852	.05445	18.366	.0015	18.393	.0000537	16.849	53
8	.05466	.99851	.05474	18.268	.0015	18.295	.0000546	16.752	52
9	.05495	.99849	.05503	18.171	.0015	18.198	.0000555	16.655	51
10	0.05524	0.99847	0.05533	18.075	1.0015	18.103	.0000563	16.559	50
11	.05553	.99846	.05562	17.980	.0015	18.008	.0000572	16.465	49
12	.05582	.99844	.05591	17.886	.0016	17.914	.0000581	16.371	48
13	.05611	.99842	.05620	17.793	.0016	17.822	.0000591	16.279	47
14	.05640	.99841	.05649	17.702	.0016	17.730	.0000600	16.187	46
15	0.05669	0.99839	0.05678	17.611	1.0016	17.639	.0000609	16.096	45
16	.05698	.99838	.05708	17.521	.0016	17.549	.0000619	16.007	44
17	.05727	.99836	.05737	17.431	.0016	17.460	.0000628	15.918	43
18	.05756	.99834	.05766	17.343	.0017	17.372	.0000638	15.830	42
19	.05785	.99833	.05795	17.256	.0017	17.285	.0000647	15.743	41
20	0.05814	0.99831	0.05824	17.169	1.0017	17.198	.0000657	15.657	40
21	.05844	.99829	.05854	17.084	.0017	17.113	.0000667	15.571	39
22	.05873	.99827	.05883	16.999	.0017	17.028	.0000677	15.487	38
23	.05902	.99826	.05912	16.915	.0017	16.945	.0000687	15.403	37
24	.05931	.99824	.05941	16.832	.0018	16.862	.0000698	15.320	36
25	0.05960	0.99822	0.05970	16.750	1.0018	16.779	.0000708	15.238	35
26	.05989	.99821	.05999	16.668	.0018	16.698	.0000718	15.157	34
27	.06018	.99819	.06029	16.587	.0018	16.618	.0000729	15.077	33
28	.06047	.99817	.06058	16.507	.0018	16.538	.0000739	14.997	32
29	.06076	.99815	.06087	16.428	.0019	16.459	.0000750	14.918	31
30	0.06105	0.99813	0.06116	16.350	1.0019	16.380	.0000761	14.840	30
31	.06134	.99812	.06145	16.272	.0019	16.303	.0000772	14.763	29
32	.06163	.99810	.06175	16.195	.0019	16.226	.0000783	14.686	28
33	.06192	.99808	.06204	16.119	.0019	16.150	.0000794	14.610	27
34	.06221	.99806	.06233	16.043	.0019	16.075	.0000805	14.535	26
35	0.06250	0.99804	0.06262	15.969	1.0020	16.000	.0000817	14.460	25
36	.06279	.99803	.06291	15.895	.0020	15.926	.0000828	14.387	24
37	.06308	.99801	.06321	15.821	.0020	15.853	.0000840	14.313	23
38	.06337	.99799	.06350	15.748	.0020	15.780	.0000851	14.241	22
39	.06366	.99797	.06379	15.676	.0020	15.708	.0000863	14.169	21
40	0.06395	0.99795	0.06408	15.605	1.0021	15.637	.0000875	14.098	20
41	.06424	.99793	.06438	15.534	.0021	15.566	.0000887	14.027	19
42	.06453	.99792	.06467	15.464	.0021	15.496	.0000899	13.958	18
43	.06482	.99790	.06496	15.394	.0021	15.427	.0000911	13.888	17
44	.06511	.99788	.06525	15.325	.0021	15.358	.0000924	13.820	16
45	0.06540	0.99786	0.06554	15.257	1.0021	15.290	.0000936	13.752	15
46	.06569	.99784	.06584	15.189	.0022	15.222	.0000949	13.684	14
47	.06598	.99782	.06613	15.122	.0022	15.155	.0000961	13.617	13
48	.06627	.99780	.06642	15.056	.0022	15.089	.0000974	13.551	12
49	.06656	.99778	.06671	14.990	.0022	15.023	.0000987	13.486	11
50	0.06685	0.99776	0.06700	14.924	1.0022	14.958	.0001000	13.421	10
51	.06714	.99774	.06730	14.860	.0023	14.893	.0001013	13.356	9
52	.06743	.99772	.06759	14.795	.0023	14.829	.0001026	13.292	8
53	.06773	.99770	.06788	14.732	.0023	14.766	.0001040	13.229	7
54	.06802	.99768	.06817	14.669	.0023	14.703	.0001053	13.166	6
55	0.06831	0.99766	0.06847	14.606	1.0023	14.640	.0001067	13.103	5
56	.06860	.99764	.06876	14.544	.0024	14.578	.0001080	13.042	4
57	.06889	.99762	.06905	14.482	.0024	14.517	.0001094	12.980	3
58	.06918	.99760	.06934	14.421	.0024	14.456	.0001108	12.920	2
59	.06947	.99758	.06963	14.361	.0024	14.395	.0001122	12.859	1
60	0.06976	0.99756	0.06993	14.301	1.0024	14.336	.0001136	12.800	0
M	Cosine	Sine	Cotan.	Tan.	Cosec.	Secant	READ DOWN	86°–87°	M
								Involute	

4° or 184° Trigonometric and Involute Functions 175° or 355°

M	Sine	Cosine	Tan.	Cotan.	Secant	Cosec.	Involute 4°–5°	READ UP	M
0	0.06976	0.99756	0.06993	14.301	1.0024	14.336	.0001136	12.800	60
1	.07005	.99754	.07022	14.241	.0025	14.276	.0001151	12.740	59
2	.07034	.99752	.07051	14.182	.0025	14.217	.0001165	12.682	58
3	.07063	.99750	.07080	14.124	.0025	14.159	.0001180	12.623	57
4	.07092	.99748	.07110	14.065	.0025	14.101	.0001194	12.566	56
5	0.07121	0.99746	0.07139	14.008	1.0025	14.044	.0001209	12.508	55
6	.07150	.99744	.07168	13.951	.0026	13.987	.0001224	12.451	54
7	.07179	.99742	.07197	13.894	.0026	13.930	.0001239	12.395	53
8	.07208	.99740	.07227	13.838	.0026	13.874	.0001254	12.339	52
9	.07237	.99738	.07256	13.782	.0026	13.818	.0001269	12.284	51
10	0.07266	0.99736	0.07285	13.727	1.0027	13.763	.0001285	12.229	50
11	.07295	.99734	.07314	13.672	.0027	13.708	.0001300	12.174	49
12	.07324	.99731	.07344	13.617	.0027	13.654	.0001316	12.120	48
13	.07353	.99729	.07373	13.563	.0027	13.600	.0001332	12.066	47
14	.07382	.99727	.07402	13.510	.0027	13.547	.0001347	12.013	46
15	0.07411	0.99725	0.07431	13.457	1.0028	13.494	.0001363	11.960	45
16	.07440	.99723	.07461	13.404	.0028	13.441	.0001380	11.908	44
17	.07469	.99721	.07490	13.352	.0028	13.389	.0001396	11.855	43
18	.07498	.99719	.07519	13.300	.0028	13.337	.0001412	11.804	42
19	.07527	.99716	.07548	13.248	.0028	13.286	.0001429	11.753	41
20	0.07556	0.99714	0.07578	13.197	1.0029	13.235	.0001445	11.702	40
21	.07585	.99712	.07607	13.146	.0029	13.184	.0001462	11.651	39
22	.07614	.99710	.07636	13.096	.0029	13.134	.0001479	11.601	38
23	.07643	.99707	.07665	13.046	.0029	13.084	.0001496	11.551	37
24	.07672	.99705	.07695	12.996	.0030	13.035	.0001513	11.502	36
25	0.07701	0.99703	0.07724	12.947	1.0030	12.985	.0001530	11.453	35
26	.07730	.99701	.07753	12.898	.0030	12.937	.0001548	11.405	34
27	.07759	.99699	.07782	12.850	.0030	12.888	.0001565	11.356	33
28	.07788	.99696	.07812	12.801	.0030	12.840	.0001583	11.309	32
29	.07817	.99694	.07841	12.754	.0031	12.793	.0001601	11.261	31
30	0.07846	0.99692	0.07870	12.706	1.0031	12.745	.0001619	11.214	30
31	.07875	.99689	.07899	12.659	.0031	12.699	.0001637	11.167	29
32	.07904	.99687	.07929	12.612	.0031	12.652	.0001655	11.121	28
33	.07933	.99685	.07958	12.566	.0032	12.606	.0001674	11.075	27
34	.07962	.99683	.07987	12.520	.0032	12.560	.0001692	11.029	26
35	0.07991	0.99680	0.08017	12.474	1.0032	12.514	.0001711	10.983	25
36	.08020	.99678	.08046	12.429	.0032	12.469	.0001729	10.938	24
37	.08049	.99676	.08075	12.384	.0033	12.424	.0001748	10.894	23
38	.08078	.99673	.08104	12.339	.0033	12.379	.0001767	10.849	22
39	.08107	.99671	.08134	12.295	.0033	12.335	.0001787	10.805	21
40	0.08136	0.99668	0.08163	12.251	1.0033	12.291	.0001806	10.761	20
41	.08165	.99666	.08192	12.207	.0034	12.248	.0001825	10.718	19
42	.08194	.99664	.08221	12.163	.0034	12.204	.0001845	10.674	18
43	.08223	.99661	.08251	12.120	.0034	12.161	.0001865	10.632	17
44	.08252	.99659	.08280	12.077	.0034	12.119	.0001885	10.589	16
45	0.08281	0.99657	0.08309	12.035	1.0034	12.076	.0001905	10.547	15
46	.08310	.99654	.08339	11.992	.0035	12.034	.0001925	10.505	14
47	.08339	.99652	.08368	11.950	.0035	11.992	.0001945	10.463	13
48	.08368	.99649	.08397	11.909	.0035	11.951	.0001965	10.422	12
49	.08397	.99647	.08427	11.867	.0035	11.909	.0001986	10.381	11
50	0.08426	0.99644	0.08456	11.826	1.0036	11.868	.0002007	10.340	10
51	.08455	.99642	.08485	11.785	.0036	11.828	.0002028	10.299	9
52	.08484	.99639	.08514	11.745	.0036	11.787	.0002049	10.259	8
53	.08513	.99637	.08544	11.705	.0036	11.747	.0002070	10.219	7
54	.08542	.99635	.08573	11.664	.0037	11.707	.0002091	10.179	6
55	0.08571	0.99632	0.08602	11.625	1.0037	11.668	.0002113	10.140	5
56	.08600	.99630	.08632	11.585	.0037	11.628	.0002134	10.101	4
57	.08629	.99627	.08661	11.546	.0037	11.589	.0002156	10.062	3
58	.08658	.99625	.08690	11.507	.0038	11.551	.0002178	10.023	2
59	.08687	.99622	.08720	11.468	.0038	11.512	.0002200	9.9847	1
60	0.08716	0.99619	0.08749	11.430	1.0038	11.474	.0002222	9.9465	0

| M | Cosine | Sine | Cotan. | Tan. | Cosec. | Secant | READ DOWN | 85°–86° Involute | M |

94° or 274°   85° or 265°

5° or 185° Trigonometric and Involute Functions 174° or 354°

M	Sine	Cosine	Tan.	Cotan.	Secant	Cosec.	Involute 5°–6°	READ UP	M
0	0.08716	0.99619	0.08749	11.430	1.0038	11.474	.0002222	9.9465	60
1	.08745	.99617	.08778	11.392	.0038	11.436	.0002244	9.9086	59
2	.08774	.99614	.08807	11.354	.0039	11.398	.0002267	9.8710	58
3	.08803	.99612	.08837	11.316	.0039	11.360	.0002289	9.8336	57
4	.08831	.99609	.08866	11.279	.0039	11.323	.0002312	9.7965	56
5	0.08860	0.99607	0.08895	11.242	1.0039	11.286	.0002335	9.7596	55
6	.08889	.99604	.08925	11.205	.0040	11.249	.0002358	9.7230	54
7	.08918	.99602	.08954	11.168	.0040	11.213	.0002382	9.6866	53
8	.08947	.99599	.08983	11.132	.0040	11.176	.0002405	9.6504	52
9	.08976	.99596	.09013	11.095	.0041	11.140	.0002429	9.6145	51
10	0.09005	0.99594	0.09042	11.059	1.0041	11.105	.0002452	9.5788	50
11	.09034	.99591	.09071	11.024	.0041	11.069	.0002476	9.5433	49
12	.09063	.99588	.09101	10.988	.0041	11.034	.0002500	9.5081	48
13	.09092	.99586	.09130	10.953	.0042	10.998	.0002524	9.4731	47
14	.09121	.99583	.09159	10.918	.0042	10.963	.0002549	9.4383	46
15	0.09150	0.99580	0.09189	10.883	1.0042	10.929	.0002573	9.4038	45
16	.09179	.99578	.09218	10.848	.0042	10.894	.0002598	9.3694	44
17	.09208	.99575	.09247	10.814	.0043	10.860	.0002622	9.3353	43
18	.09237	.99572	.09277	10.780	.0043	10.826	.0002647	9.3014	42
19	.09266	.99570	.09306	10.746	.0043	10.792	.0002673	9.2677	41
20	0.09295	0.99567	0.09335	10.712	1.0043	10.758	.0002698	9.2342	40
21	.09324	.99564	.09365	10.678	.0044	10.725	.0002723	9.2009	39
22	.09353	.99562	.09394	10.645	.0044	10.692	.0002749	9.1679	38
23	.09382	.99559	.09423	10.612	.0044	10.659	.0002775	9.1350	37
24	.09411	.99556	.09453	10.579	.0045	10.626	.0002801	9.1023	36
25	0.09440	0.99553	0.09482	10.546	1.0045	10.593	.0002827	9.0699	35
26	.09469	.99551	.09511	10.514	.0045	10.561	.0002853	9.0376	34
27	.09498	.99548	.09541	10.481	.0045	10.529	.0002879	9.0056	33
28	.09527	.99545	.09570	10.449	.0046	10.497	.0002906	8.9737	32
29	.09556	.99542	.09600	10.417	.0046	10.465	.0002933	8.9421	31
30	0.09585	0.99540	0.09629	10.385	1.0046	10.433	.0002959	8.9106	30
31	.09614	.99537	.09658	10.354	.0047	10.402	.0002986	8.8793	29
32	.09642	.99534	.09688	10.322	.0047	10.371	.0003014	8.8482	28
33	.09671	.99531	.09717	10.291	.0047	10.340	.0003041	8.8173	27
34	.09700	.99528	.09746	10.260	.0047	10.309	.0003069	8.7866	26
35	.09729	.99526	.09776	10.229	1.0048	10.278	.0003096	8.7561	25
36	.09758	.99523	.09805	10.199	.0048	10.248	.0003124	8.7257	24
37	.09787	.99520	.09834	10.168	.0048	10.217	.0003152	8.6956	23
38	.09816	.99517	.09864	10.138	.0049	10.187	.0003180	8.6656	22
39	.09845	.99514	.09893	10.108	.0049	10.157	.0003209	8.6358	21
40	.09874	.99511	.09923	10.078	1.0049	10.128	.0003237	8.6061	20
41	.09903	.99508	.09952	10.048	.0049	10.098	.0003266	8.5767	19
42	.09932	.99506	.09981	10.019	.0050	10.068	.0003295	8.5474	18
43	.09961	.99503	.10011	9.9893	.0050	10.039	.0003324	8.5183	17
44	.09990	.99500	.10040	9.9601	.0050	10.010	.0003353	8.4893	16
45	0.10019	0.99497	0.10069	9.9310	1.0051	9.9812	.0003383	8.4606	15
46	.10048	.99494	.10099	9.9021	.0051	9.9525	.0003412	8.4320	14
47	.10077	.99491	.10128	9.8734	.0051	9.9239	.0003442	8.4035	13
48	.10106	.99488	.10158	9.8448	.0051	9.8955	.0003472	8.3752	12
49	.10135	.99485	.10187	9.8164	.0052	9.8672	.0003502	8.3471	11
50	0.10164	0.99482	0.10216	9.7882	1.0052	9.8391	.0003532	8.3192	10
51	.10192	.99479	.10246	9.7601	.0052	9.8112	.0003563	8.2914	9
52	.10221	.99476	.10275	9.7322	.0053	9.7834	.0003593	8.2638	8
53	.10250	.99473	.10305	9.7044	.0053	9.7558	.0003624	8.2363	7
54	.10279	.99470	.10334	9.6768	.0053	9.7283	.0003655	8.2090	6
55	0.10308	0.99467	0.10363	9.6493	1.0054	9.7010	.0003686	8.1818	5
56	.10337	.99464	.10393	9.6220	.0054	9.6739	.0003718	8.1548	4
57	.10366	.99461	.10422	9.5949	.0054	9.6469	.0003749	8.1280	3
58	.10395	.99458	.10452	9.5679	.0054	9.6200	.0003781	8.1012	2
59	.10424	.99455	.10481	9.5411	.0055	9.5933	.0003813	8.0747	1
60	0.10453	0.99452	0.10510	9.5144	1.0055	9.5668	.0003845	8.0483	0
M	Cosine	Sine	Cotan.	Tan.	Cosec.	Secant	READ DOWN	84°–85° Involute	M

6° or 186°   **Trigonometric and Involute Functions**   **173° or 353°**

M	Sine	Cosine	Tan.	Cotan.	Secant	Cosec.	Involute 6°-7°	READ UP	M
0	0.10453	0.99452	0.10510	9.5144	1.0055	9.5668	.0003845	8.0483	60
1	.10482	.99449	.10540	.4878	.0055	.5404	.0003877	8.0220	59
2	.10511	.99446	.10569	.4614	.0056	.5141	.0003909	7.9959	58
3	.10540	.99443	.10599	.4352	.0056	.4880	.0003942	7.9699	57
4	.10569	.99440	.10628	.4090	.0056	.4620	.0003975	7.9441	56
5	0.10597	0.99437	0.10657	9.3831	1.0057	9.4362	.0004008	7.9184	55
6	.10626	.99434	.10687	.3572	.0057	.4105	.0004041	7.8929	54
7	.10655	.99431	.10716	.3315	.0057	.3850	.0004074	7.8675	53
8	.10684	.99428	.10746	.3060	.0058	.3596	.0004108	7.8422	52
9	.10713	.99424	.10775	.2806	.0058	.3343	.0004141	7.8171	51
10	0.10742	0.99421	0.10805	9.2553	1.0058	9.3092	.0004175	7.7921	50
11	.10771	.99418	.10834	.2302	.0059	.2842	.0004209	7.7673	49
12	.10800	.99415	.10863	.2052	.0059	.2593	.0004244	7.7426	48
13	.10829	.99412	.10893	.1803	.0059	.2346	.0004278	7.7180	47
14	.10858	.99409	.10922	.1555	.0059	.2100	.0004313	7.6935	46
15	0.10887	0.99406	0.10952	9.1309	1.0060	9.1855	.0004347	7.6692	45
16	.10916	.99402	.10981	.1065	.0060	.1612	.0004382	7.6450	44
17	.10945	.99399	.11011	.0821	.0060	.1370	.0004417	7.6210	43
18	.10973	.99396	.11040	.0579	.0061	.1129	.0004453	7.5970	42
19	.11002	.99393	.11070	.0338	.0061	.0890	.0004488	7.5732	41
20	0.11031	0.99390	0.11099	9.0098	1.0061	9.0652	.0004524	7.5496	40
21	.11060	.99386	.11128	8.9860	.0062	.0415	.0004560	7.5260	39
22	.11089	.99383	.11158	.9623	.0062	.0179	.0004596	7.5026	38
23	.11118	.99380	.11187	.9387	.0062	8.9944	.0004632	7.4793	37
24	.11147	.99377	.11217	.9152	.0063	.9711	.0004669	7.4561	36
25	0.11176	0.99374	0.11246	8.8918	1.0063	8.9479	.0004706	7.4330	35
26	.11205	.99370	.11276	.8686	.0063	.9248	.0004743	7.4101	34
27	.11234	.99367	.11305	.8455	.0064	.9019	.0004780	7.3873	33
28	.11263	.99364	.11335	.8225	.0064	.8790	.0004817	7.3646	32
29	.11291	.99360	.11364	.7996	.0064	.8563	.0004854	7.3420	31
30	0.11320	0.99357	0.11394	8.7769	1.0065	8.8337	.0004892	7.3195	30
31	.11349	.99354	.11423	.7542	.0065	.8112	.0004930	7.2972	29
32	.11378	.99351	.11452	.7317	.0065	.7888	.0004968	7.2750	28
33	.11407	.99347	.11482	.7093	.0066	.7665	.0005006	7.2528	27
34	.11436	.99344	.11511	.6870	.0066	.7444	.0005045	7.2308	26
35	0.11465	0.99341	0.11541	8.6648	1.0066	8.7223	.0005083	7.2089	25
36	.11494	.99337	.11570	.6427	.0067	.7004	.0005122	7.1871	24
37	.11523	.99334	.11600	.6208	.0067	.6786	.0005161	7.1655	23
38	.11552	.99330	.11629	.5989	.0067	.6569	.0005200	7.1439	22
39	.11580	.99327	.11659	.5772	.0068	.6353	.0005240	7.1225	21
40	0.11609	0.99324	0.11688	8.5555	1.0068	8.6138	.0005280	7.1011	20
41	.11638	.99320	.11718	.5340	.0068	.5924	.0005319	7.0799	19
42	.11667	.99317	.11747	.5126	.0069	.5711	.0005359	7.0587	18
43	.11696	.99313	.11777	.4913	.0069	.5500	.0005400	7.0377	17
44	.11725	.99310	.11806	.4701	.0069	.5289	.0005440	7.0168	16
45	0.11754	0.99307	0.11836	8.4490	1.0070	8.5079	.0005481	6.9960	15
46	.11783	.99303	.11865	.4280	.0070	.4871	.0005522	6.9753	14
47	.11812	.99300	.11895	.4071	.0070	.4663	.0005563	6.9546	13
48	.11840	.99296	.11924	.3863	.0071	.4457	.0005604	6.9341	12
49	.11869	.99293	.11954	.3656	.0071	.4251	.0005645	6.9137	11
50	0.11898	0.99290	0.11983	8.3450	1.0072	8.4047	.0005687	6.8934	10
51	.11927	.99286	.12013	.3245	.0072	.3843	.0005729	6.8732	9
52	.11956	.99283	.12042	.3041	.0072	.3641	.0005771	6.8531	8
53	.11985	.99279	.12072	.2838	.0073	.3439	.0005813	6.8331	7
54	.12014	.99276	.12101	.2636	.0073	.3238	.0005856	6.8132	6
55	0.12043	0.99272	0.12131	8.2434	1.0073	8.3039	.0005898	6.7934	5
56	.12071	.99269	.12160	.2234	.0074	.2840	.0005941	6.7737	4
57	.12100	.99265	.12190	.2035	.0074	.2642	.0005985	6.7540	3
58	.12129	.99262	.12219	.1837	.0074	.2446	.0006028	6.7345	2
59	.12158	.99258	.12249	.1640	.0075	.2250	.0006071	6.7151	1
60	0.12187	0.99255	0.12278	8.1443	1.0075	8.2055	.0006115	6.6957	0
M	Cosine	Sine	Cotan.	Tan.	Cosec.	Secant	READ DOWN	83°-84° Involute	M

96° or 276°   **83° or 263°**

M	Sine	Cosine	Tan.	Cotan.	Secant	Cosec.	Involute 7°-8°	READ UP	M
0	0.12187	0.99255	0.12278	8.1443	1.0075	8.2055	.0006115	6.6957	60
1	.12216	.99251	.12308	.1248	.0075	.1861	.0006159	6.6765	59
2	.12245	.99248	.12338	.1054	.0076	.1668	.0006203	6.6573	58
3	.12274	.99244	.12367	.0860	.0076	.1476	.0006248	6.6383	57
4	.12302	.99240	.12397	.0667	.0077	.1285	.0006292	6.6193	56
5	0.12331	0.99237	0.12426	8.0476	1.0077	8.1095	.0006337	6.6004	55
6	.12360	.99233	.12456	.0285	.0077	.0905	.0006382	6.5816	54
7	.12389	.99230	.12485	.0095	.0078	.0717	.0006427	6.5629	53
8	.12418	.99226	.12515	7.9906	.0078	.0529	.0006473	6.5443	52
9	.12447	.99222	.12544	.9718	.0078	.0342	.0006518	6.5258	51
10	0.12476	0.99219	0.12574	7.9530	1.0079	8.0156	.0006564	6.5073	50
11	.12504	.99215	.12603	.9344	.0079	7.9971	.0006610	6.4890	49
12	.12533	.99211	.12633	.9158	.0079	.9787	.0006657	6.4707	48
13	.12562	.99208	.12662	.8973	.0080	.9604	.0006703	6.4525	47
14	.12591	.99204	.12692	.8789	.0080	.9422	.0006750	6.4344	46
15	0.12620	0.99200	0.12722	7.8606	1.0081	7.9240	.0006797	6.4164	45
16	.12649	.99197	.12751	.8424	.0081	.9059	.0006844	6.3985	44
17	.12678	.99193	.12781	.8243	.0081	.8879	.0006892	6.3806	43
18	.12706	.99189	.12810	.8062	.0082	.8700	.0006939	6.3628	42
19	.12735	.99186	.12840	.7882	.0082	.8522	.0006987	6.3451	41
20	0.12764	0.99182	0.12869	7.7704	1.0082	7.8344	.0007035	6.3275	40
21	.12793	.99178	.12899	.7525	.0083	.8168	.0007083	6.3100	39
22	.12822	.99175	.12929	.7348	.0083	.7992	.0007132	6.2926	38
23	.12851	.99171	.12958	.7171	.0084	.7817	.0007181	6.2752	37
24	.12880	.99167	.12988	.6996	.0084	.7642	.0007230	6.2579	36
25	0.12908	0.99163	0.13017	7.6821	1.0084	7.7469	.0007279	6.2407	35
26	.12937	.99160	.13047	.6647	.0085	.7296	.0007328	6.2236	34
27	.12966	.99156	.13076	.6473	.0085	.7124	.0007378	6.2065	33
28	.12995	.99152	.13106	.6301	.0086	.6953	.0007428	6.1896	32
29	.13024	.99148	.13136	.6129	.0086	.6783	.0007478	6.1727	31
30	0.13053	0.99144	0.13165	7.5958	1.0086	7.6613	.0007528	6.1559	30
31	.13081	.99141	.13195	.5787	.0087	.6444	.0007579	6.1391	29
32	.13110	.99137	.13224	.5618	.0087	.6276	.0007629	6.1224	28
33	.13139	.99133	.13254	.5449	.0087	.6109	.0007680	6.1058	27
34	.13168	.99129	.13284	.5281	.0088	.5942	.0007732	6.0893	26
35	0.13197	0.99125	0.13313	7.5113	1.0088	7.5776	.0007783	6.0729	25
36	.13226	.99122	.13343	.4947	.0089	.5611	.0007835	6.0565	24
37	.13254	.99118	.13372	.4781	.0089	.5446	.0007887	6.0402	23
38	.13283	.99114	.13402	.4615	.0089	.5282	.0007939	6.0240	22
39	.13312	.99110	.13432	.4451	.0090	.5119	.0007991	6.0078	21
40	0.13341	0.99106	0.13461	7.4287	1.0090	7.4957	.0008044	5.9917	20
41	.13370	.99102	.13491	.4124	.0091	.4795	.0008096	5.9757	19
42	.13399	.99098	.13521	.3962	.0091	.4635	.0008150	5.9598	18
43	.13427	.99094	.13550	.3800	.0091	.4474	.0008203	5.9439	17
44	.13456	.99091	.13580	.3639	.0092	.4315	.0008256	5.9281	16
45	0.13485	0.99087	0.13609	7.3479	1.0092	7.4156	.0008310	5.9123	15
46	.13514	.99083	.13639	.3319	.0093	.3998	.0008364	5.8967	14
47	.13543	.99079	.13669	.3160	.0093	.3840	.0008418	5.8811	13
48	.13572	.99075	.13698	.3002	.0093	.3684	.0008473	5.8655	12
49	.13600	.99071	.13728	.2844	.0094	.3527	.0008527	5.8500	11
50	0.13629	0.99067	0.13758	7.2687	1.0094	7.3372	.0008582	5.8346	10
51	.13658	.99063	.13787	.2531	.0095	.3217	.0008638	5.8193	9
52	.13687	.99059	.13817	.2375	.0095	.3063	.0008693	5.8040	8
53	.13716	.99055	.13846	.2220	.0095	.2909	.0008749	5.7888	7
54	.13744	.99051	.13876	.2066	.0096	.2757	.0008805	5.7737	6
55	0.13773	0.99047	0.13906	7.1912	1.0096	7.2604	.0008861	5.7586	5
56	.13802	.99043	.13935	.1759	.0097	.2453	.0008917	5.7436	4
57	.13831	.99039	.13965	.1607	.0097	.2302	.0008974	5.7287	3
58	.13860	.99035	.13995	.1455	.0097	.2152	.0009031	5.7138	2
59	.13889	.99031	.14024	.1304	.0098	.2002	.0009088	5.6990	1
60	0.13917	0.99027	0.14054	7.1154	1.0098	7.1853	.0009145	5.6842	0
M	Cosine	Sine	Cotan.	Tan.	Cosec.	Secant	READ DOWN	82°-83° Involute	M

8° or 188°   Trigonometric and Involute Functions   171° or 351°

M	Sine	Cosine	Tan.	Cotan.	Secant	Cosec.	Involute 8°–9°	READ UP	M
0	0.13917	0.99027	0.14054	7.1154	1.0098	7.1853	.0009145	5.6842	60
1	.13946	.99023	.14084	.1004	.0099	.1705	.0009203	5.6695	59
2	.13975	.99019	.14113	.0855	.0099	.1557	.0009260	5.6549	58
3	.14004	.99015	.14143	.0706	.0100	.1410	.0009318	5.6403	57
4	.14033	.99011	.14173	.0558	.0100	.1263	.0009377	5.6258	56
5	0.14061	0.99006	0.14202	7.0410	1.0100	7.1117	.0009435	5.6113	55
6	.14090	.99002	.14232	.0264	.0101	.0972	.0009494	5.5969	54
7	.14119	.98998	.14262	.0117	.0101	.0827	.0009553	5.5826	53
8	.14148	.98994	.14291	6.9972	.0102	.0683	.0009612	5.5683	52
9	.14177	.98990	.14321	.9827	.0102	.0539	.0009672	5.5541	51
10	0.14205	0.98986	0.14351	6.9682	1.0102	7.0396	.0009732	5.5400	50
11	.14234	.98982	.14381	.9538	.0103	.0254	.0009792	5.5259	49
12	.14263	.98978	.14410	.9395	.0103	.0112	.0009852	5.5118	48
13	.14292	.98973	.14440	.9252	.0104	6.9971	.0009913	5.4979	47
14	.14320	.98969	.14470	.9110	.0104	.9830	.0009973	5.4839	46
15	0.14349	0.98965	0.14499	6.8969	1.0105	6.9690	.0010034	5.4701	45
16	.14378	.98961	.14529	.8828	.0105	.9550	.0010096	5.4563	44
17	.14407	.98957	.14559	.8687	.0105	.9411	.0010157	5.4425	43
18	.14436	.98953	.14588	.8547	.0106	.9273	.0010219	5.4288	42
19	.14464	.98948	.14618	.8408	.0106	.9135	.0010281	5.4152	41
20	0.14493	0.98944	0.14648	6.8269	1.0107	6.8998	.0010343	5.4016	40
21	.14522	.98940	.14678	.8131	.0107	.8861	.0010406	5.3881	39
22	.14551	.98936	.14707	.7994	.0108	.8725	.0010469	5.3746	38
23	.14580	.98931	.14737	.7856	.0108	.8589	.0010532	5.3612	37
24	.14608	.98927	.14767	.7720	.0108	.8454	.0010595	5.3478	36
25	0.14637	0.98923	0.14796	6.7584	1.0109	6.8320	.0010659	5.3345	35
26	.14666	.98919	.14826	.7448	.0109	.8186	.0010722	5.3212	34
27	.14695	.98914	.14856	.7313	.0110	.8052	.0010786	5.3080	33
28	.14723	.98910	.14886	.7179	.0110	.7919	.0010851	5.2949	32
29	.14752	.98906	.14915	.7045	.0111	.7787	.0010915	5.2818	31
30	0.14781	0.98902	0.14945	6.6912	1.0111	6.7655	.0010980	5.2687	30
31	.14810	.98897	.14975	.6779	.0112	.7523	.0011045	5.2557	29
32	.14838	.98893	.15005	.6646	.0112	.7392	.0011111	5.2428	28
33	.14867	.98889	.15034	.6514	.0112	.7262	.0011176	5.2299	27
34	.14896	.98884	.15064	.6383	.0113	.7132	.0011242	5.2170	26
35	0.14925	0.98880	0.15094	6.6252	1.0113	6.7003	.0011308	5.2042	25
36	.14954	.98876	.15124	.6122	.0114	.6874	.0011375	5.1915	24
37	.14982	.98871	.15153	.5992	.0114	.6745	.0011441	5.1788	23
38	.15011	.98867	.15183	.5863	.0115	.6618	.0011508	5.1662	22
39	.15040	.98863	.15213	.5734	.0115	.6490	.0011575	5.1536	21
40	0.15069	0.98858	0.15243	6.5606	1.0116	6.6363	.0011643	5.1410	20
41	.15097	.98854	.15272	.5478	.0116	.6237	.0011711	5.1285	19
42	.15126	.98849	.15302	.5350	.0116	.6111	.0011779	5.1161	18
43	.15155	.98845	.15332	.5223	.0117	.5986	.0011847	5.1037	17
44	.15184	.98841	.15362	.5097	.0117	.5861	.0011915	5.0913	16
45	0.15212	0.98836	0.15391	6.4971	1.0118	6.5736	.0011984	5.0790	15
46	.15241	.98832	.15421	.4846	.0118	.5612	.0012053	5.0668	14
47	.15270	.98827	.15451	.4721	.0119	.5489	.0012122	5.0546	13
48	.15299	.98823	.15481	.4596	.0119	.5366	.0012192	5.0424	12
49	.15327	.98818	.15511	.4472	.0120	.5243	.0012262	5.0303	11
50	0.15356	0.98814	0.15540	6.4348	1.0120	6.5121	.0012332	5.0182	10
51	.15385	.98809	.15570	.4225	.0120	.4999	.0012402	5.0062	9
52	.15414	.98805	.15600	.4103	.0121	.4878	.0012473	4.9942	8
53	.15442	.98800	.15630	.3980	.0121	.4757	.0012544	4.9823	7
54	.15471	.98796	.15660	.3859	.0122	.4637	.0012615	4.9704	6
55	0.15500	0.98791	0.15689	6.3737	1.0122	6.4517	.0012687	4.9586	5
56	.15529	.98787	.15719	.3617	.0123	.4398	.0012758	4.9468	4
57	.15557	.98782	.15749	.3496	.0123	.4279	.0012830	4.9350	3
58	.15586	.98778	.15779	.3376	.0124	.4160	.0012903	4.9233	2
59	.15615	.98773	.15809	.3257	.0124	.4042	.0012975	4.9117	1
60	0.15643	0.98769	0.15838	6.3138	1.0125	6.3925	.0013048	4.9000	0
M	Cosine	Sine	Cotan.	Tan.	Cosec.	Secant	READ DOWN	81°–82° Involute	M

98° or 278°   81° or 261°

9° or 189°   Trigonometric and Involute Functions   170° or 350°

M	Sine	Cosine	Tan.	Cotan.	Secant	Cosec.	Involute 9°–10°	READ UP	M
0	0.15643	0.98769	0.15838	6.3138	1.0125	6.3925	.0013048	4.9000	60
1	.15672	.98764	.15868	.3019	.0125	.3807	.0013121	4.8885	59
2	.15701	.98760	.15898	.2901	.0126	.3691	.0013195	4.8769	58
3	.15730	.98755	.15928	.2783	.0126	.3574	.0013268	4.8654	57
4	.15758	.98751	.15958	.2666	.0127	.3458	.0013342	4.8540	56
5	0.15787	0.98746	0.15988	6.2549	1.0127	6.3343	.0013416	4.8426	55
6	.15816	.98741	.16017	.2432	.0127	.3228	.0013491	4.8312	54
7	.15845	.98737	.16047	.2316	.0128	.3113	.0013566	4.8199	53
8	.15873	.98732	.16077	.2200	.0128	.2999	.0013641	4.8086	52
9	.15902	.98728	.16107	.2085	.0129	.2885	.0013716	4.7974	51
10	0.15931	0.98723	0.16137	6.1970	1.0129	6.2772	.0013792	4.7862	50
11	.15959	.98718	.16167	.1856	.0130	.2659	.0013868	4.7751	49
12	.15988	.98714	.16196	.1742	.0130	.2546	.0013944	4.7640	48
13	.16017	.98709	.16226	.1628	.0131	.2434	.0014020	4.7529	47
14	.16046	.98704	.16256	.1515	.0131	.2323	.0014097	4.7419	46
15	0.16074	0.98700	0.16286	6.1402	1.0132	6.2211	.0014174	4.7309	45
16	.16103	.98695	.16316	.1290	.0132	.2100	.0014251	4.7199	44
17	.16132	.98690	.16346	.1178	.0133	.1990	.0014329	4.7090	43
18	.16160	.98686	.16376	.1066	.0133	.1880	.0014407	4.6982	42
19	.16189	.98681	.16405	.0955	.0134	.1770	.0014485	4.6873	41
20	0.16218	0.98676	0.16435	6.0844	1.0134	6.1661	.0014563	4.6765	40
21	.16246	.98671	.16465	.0734	.0135	.1552	.0014642	4.6658	39
22	.16275	.98667	.16495	.0624	.0135	.1443	.0014721	4.6551	38
23	.16304	.98662	.16525	.0514	.0136	.1335	.0014800	4.6444	37
24	.16333	.98657	.16555	.0405	.0136	.1227	.0014880	4.6338	36
25	0.16361	0.98652	0.16585	6.0296	1.0137	6.1120	.0014960	4.6232	35
26	.16390	.98648	.16615	.0188	.0137	.1013	.0015040	4.6126	34
27	.16419	.98643	.16645	.0080	.0138	.0906	.0015120	4.6021	33
28	.16447	.98638	.16674	5.9972	.0138	.0800	.0015201	4.5916	32
29	.16476	.98633	.16704	.9865	.0139	.0694	.0015282	4.5812	31
30	0.16505	0.98629	0.16734	5.9758	1.0139	6.0589	.0015363	4.5708	30
31	.16533	.98624	.16764	.9651	.0140	.0483	.0015445	4.5604	29
32	.16562	.98619	.16794	.9545	.0140	.0379	.0015527	4.5501	28
33	.16591	.98614	.16824	.9439	.0141	.0274	.0015609	4.5398	27
34	.16620	.98609	.16854	.9333	.0141	.0170	.0015691	4.5295	26
35	0.16648	0.98604	0.16884	5.9228	1.0142	6.0067	.0015774	4.5193	25
36	.16677	.98600	.16914	.9124	.0142	5.9963	.0015857	4.5091	24
37	.16706	.98595	.16944	.9019	.0143	.9860	.0015941	4.4990	23
38	.16734	.98590	.16974	.8915	.0143	.9758	.0016024	4.4888	22
39	.16763	.98585	.17004	.8811	.0144	.9656	.0016108	4.4788	21
40	0.16792	0.98580	0.17033	5.8708	1.0144	5.9554	.0016193	4.4687	20
41	.16820	.98575	.17063	.8605	.0145	.9452	.0016277	4.4587	19
42	.16849	.98570	.17093	.8502	.0145	.9351	.0016362	4.4487	18
43	.16878	.98565	.17123	.8400	.0146	.9250	.0016447	4.4388	17
44	.16906	.98561	.17153	.8298	.0146	.9150	.0016533	4.4289	16
45	0.16935	0.98556	0.17183	5.8197	1.0147	5.9049	.0016618	4.4190	15
46	.16964	.98551	.17213	.8095	.0147	.8950	.0016704	4.4092	14
47	.16992	.98546	.17243	.7994	.0148	.8850	.0016791	4.3994	13
48	.17021	.98541	.17273	.7894	.0148	.8751	.0016877	4.3896	12
49	.17050	.98536	.17303	.7794	.0149	.8652	.0016964	4.3799	11
50	0.17078	0.98531	0.17333	5.7694	1.0149	5.8554	.0017051	4.3702	10
51	.17107	.98526	.17363	.7594	.0150	.8456	.0017139	4.3605	9
52	.17136	.98521	.17393	.7495	.0150	.8358	.0017227	4.3509	8
53	.17164	.98516	.17423	.7396	.0151	.8261	.0017315	4.3413	7
54	.17193	.98511	.17453	.7297	.0151	.8164	.0017403	4.3317	6
55	0.17222	0.98506	0.17483	5.7199	1.0152	5.8067	.0017492	4.3222	5
56	.17250	.98501	.17513	.7101	.0152	.7970	.0017581	4.3127	4
57	.17279	.98496	.17543	.7004	.0153	.7874	.0017671	4.3032	3
58	.17308	.98491	.17573	.6906	.0153	.7778	.0017760	4.2938	2
59	.17336	.98486	.17603	.6809	.0154	.7683	.0017850	4.2844	1
60	0.17365	0.98481	0.17633	5.6713	1.0154	5.7588	.0017941	4.2750	0
M	Cosine	Sine	Cotan.	Tan.	Cosec.	Secant	READ DOWN	80°–81° Involute	M

10° or 190° Trigonometric and Involute Functions 169° or 349°

M	Sine	Cosine	Tan.	Cotan.	Secant	Cosec.	Involute 10°–11°	READ UP	M
0	0.17365	0.98481	0.17633	5.6713	1.0154	5.7588	.0017941	4.2750	60
1	.17393	.98476	.17663	.6617	.0155	.7493	.0018031	4.2657	59
2	.17422	.98471	.17693	.6521	.0155	.7398	.0018122	4.2564	58
3	.17451	.98466	.17723	.6425	.0156	.7304	.0018213	4.2471	57
4	.17479	.98461	.17753	.6329	.0156	.7210	.0018305	4.2378	56
5	0.17508	0.98455	0.17783	5.6234	1.0157	5.7117	.0018397	4.2286	55
6	.17537	.98450	.17813	.6140	.0157	.7023	.0018489	4.2194	54
7	.17565	.98445	.17843	.6045	.0158	.6930	.0018581	4.2103	53
8	.17594	.98440	.17873	.5951	.0158	.6838	.0018674	4.2012	52
9	.17623	.98435	.17903	.5857	.0159	.6745	.0018767	4.1921	51
10	0.17651	0.98430	0.17933	5.5764	1.0160	5.6653	.0018860	4.1830	50
11	.17680	.98425	.17963	.5671	.0160	.6562	.0018954	4.1740	49
12	.17708	.98420	.17993	.5578	.0161	.6470	.0019048	4.1650	48
13	.17737	.98414	.18023	.5485	.0161	.6379	.0019142	4.1560	47
14	.17766	.98409	.18053	.5393	.0162	.6288	.0019237	4.1471	46
15	0.17794	0.98404	0.18083	5.5301	1.0162	5.6198	.0019332	4.1382	45
16	.17823	.98399	.18113	.5209	.0163	.6107	.0019427	4.1293	44
17	.17852	.98394	.18143	.5118	.0163	.6017	.0019523	4.1204	43
18	.17880	.98389	.18173	.5026	.0164	.5928	.0019619	4.1116	42
19	.17909	.98383	.18203	.4936	.0164	.5838	.0019715	4.1028	41
20	0.17937	0.98378	0.18233	5.4845	1.0165	5.5749	.0019812	4.0941	40
21	.17966	.98373	.18263	.4755	.0165	.5660	.0019909	4.0853	39
22	.17995	.98368	.18293	.4665	.0166	.5572	.0020006	4.0766	38
23	.18023	.98362	.18323	.4575	.0166	.5484	.0020103	4.0679	37
24	.18052	.98357	.18353	.4486	.0167	.5396	.0020201	4.0593	36
25	0.18081	0.98352	0.18384	5.4397	1.0168	5.5308	.0020299	4.0507	35
26	.18109	.98347	.18414	.4308	.0168	.5221	.0020398	4.0421	34
27	.18138	.98341	.18444	.4219	.0169	.5134	.0020496	4.0335	33
28	.18166	.98336	.18474	.4131	.0169	.5047	.0020596	4.0250	32
29	.18195	.98331	.18504	.4043	.0170	.4960	.0020695	4.0165	31
30	0.18224	0.98325	0.18534	5.3955	1.0170	5.4874	.0020795	4.0080	30
31	.18252	.98320	.18564	.3868	.0171	.4788	.0020895	3.9995	29
32	.18281	.98315	.18594	.3781	.0171	.4702	.0020995	3.9911	28
33	.18309	.98310	.18624	.3694	.0172	.4617	.0021096	3.9827	27
34	.18338	.98304	.18654	.3607	.0173	.4532	.0021197	3.9743	26
35	0.18367	0.98299	0.18684	5.3521	1.0173	5.4447	.0021298	3.9660	25
36	.18395	.98294	.18714	.3435	.0174	.4362	.0021400	3.9577	24
37	.18424	.98288	.18745	.3349	.0174	.4278	.0021502	3.9494	23
38	.18452	.98283	.18775	.3263	.0175	.4194	.0021605	3.9411	22
39	.18481	.98277	.18805	.3178	.0175	.4110	.0021707	3.9329	21
40	0.18509	0.98272	0.18835	5.3093	1.0176	5.4026	.0021810	3.9247	20
41	.18538	.98267	.18865	.3008	.0176	.3943	.0021914	3.9165	19
42	.18567	.98261	.18895	.2924	.0177	.3860	.0022017	3.9083	18
43	.18595	.98256	.18925	.2839	.0178	.3777	.0022121	3.9002	17
44	.18624	.98250	.18955	.2755	.0178	.3695	.0022226	3.8921	16
45	0.18652	0.98245	0.18986	5.2672	1.0179	5.3612	.0022330	3.8840	15
46	.18681	.98240	.19016	.2588	.0179	.3530	.0022435	3.8759	14
47	.18710	.98234	.19046	.2505	.0180	.3449	.0022541	3.8679	13
48	.18738	.98229	.19076	.2422	.0180	.3367	.0022646	3.8599	12
49	.18767	.98223	.19106	.2339	.0181	.3286	.0022752	3.8519	11
50	0.18795	0.98218	0.19136	5.2257	1.0181	5.3205	.0022859	3.8439	10
51	.18824	.98212	.19166	.2174	.0182	.3124	.0022965	3.8360	9
52	.18852	.98207	.19197	.2092	.0183	.3044	.0023073	3.8281	8
53	.18881	.98201	.19227	.2011	.0183	.2963	.0023180	3.8202	7
54	.18910	.98196	.19257	.1929	.0184	.2883	.0023288	3.8124	6
55	0.18938	0.98190	0.19287	5.1848	1.0184	5.2804	.0023396	3.8045	5
56	.18967	.98185	.19317	.1767	.0185	.2724	.0023504	3.7967	4
57	.18995	.98179	.19347	.1686	.0185	.2645	.0023613	3.7889	3
58	.19024	.98174	.19378	.1606	.0186	.2566	.0023722	3.7812	2
59	.19052	.98168	.19408	.1526	.0187	.2487	.0023831	3.7735	1
60	0.19081	0.98163	0.19438	5.1446	1.0187	5.2408	.0023941	3.7657	0
M	Cosine	Sine	Cotan.	Tan.	Cosec.	Secant	READ DOWN	79°–80° Involute	M

100° or 280° 79° or 259°

11° or 191°   **Trigonometric and Involute Functions**   **168° or 348°**

M	Sine	Cosine	Tan.	Cotan.	Secant	Cosec.	Involute 11°–12°	READ UP	M
0	0.19081	0.98163	0.19438	5.1446	1.0187	5.2408	.0023941	3.7657	60
1	.19109	.98157	.19468	.1366	.0188	.2330	.0024051	3.7581	59
2	.19138	.98152	.19498	.1286	.0188	.2252	.0024161	3.7504	58
3	.19167	.98146	.19529	.1207	.0189	.2174	.0024272	3.7428	57
4	.19195	.98140	.19559	.1128	.0189	.2097	.0024383	3.7351	56
5	0.19224	0.98135	0.19589	5.1049	1.0190	5.2019	.0024495	3.7275	55
6	.19252	.98129	.19619	.0970	.0191	.1942	.0024607	3.7200	54
7	.19281	.98124	.19649	.0892	.0191	.1865	.0024719	3.7124	53
8	.19309	.98118	.19680	.0814	.0192	.1789	.0024831	3.7049	52
9	.19338	.98112	.19710	.0736	.0192	.1712	.0024944	3.6974	51
10	0.19366	0.98107	0.19740	5.0658	1.0193	5.1636	.0025057	3.6899	50
11	.19395	.98101	.19770	.0581	.0194	.1560	.0025171	3.6825	49
12	.19423	.98096	.19801	.0504	.0194	.1484	.0025285	3.6750	48
13	.19452	.98090	.19831	.0427	.0195	.1409	.0025399	3.6676	47
14	.19481	.98084	.19861	.0350	.0195	.1333	.0025513	3.6603	46
15	0.19509	0.98079	0.19891	5.0273	1.0196	5.1258	.0025628	3.6529	45
16	.19538	.98073	.19921	.0197	.0197	.1183	.0025744	3.6456	44
17	.19566	.98067	.19952	.0121	.0197	.1109	.0025859	3.6382	43
18	.19595	.98061	.19982	.0045	.0198	.1034	.0025975	3.6309	42
19	.19623	.98056	.20012	4.9969	.0198	.0960	.0026091	3.6237	41
20	0.19652	0.98050	0.20042	4.9894	1.0199	5.0886	.0026208	3.6164	40
21	.19680	.98044	.20073	.9819	.0199	.0813	.0026325	3.6092	39
22	.19709	.98039	.20103	.9744	.0200	.0739	.0026443	3.6020	38
23	.19737	.98033	.20133	.9669	.0201	.0666	.0026560	3.5948	37
24	.19766	.98027	.20164	.9594	.0201	.0593	.0026678	3.5876	36
25	0.19794	0.98021	0.20194	4.9520	1.0202	5.0520	.0026797	3.5805	35
26	.19823	.98016	.20224	.9446	.0202	.0447	.0026916	3.5734	34
27	.19851	.98010	.20254	.9372	.0203	.0375	.0027035	3.5663	33
28	.19880	.98004	.20285	.9298	.0204	.0302	.0027154	3.5592	32
29	.19908	.97998	.20315	.9225	.0204	.0230	.0027274	3.5521	31
30	0.19937	0.97992	0.20345	4.9152	1.0205	5.0159	.0027394	3.5451	30
31	.19965	.97987	.20376	.9078	.0205	.0087	.0027515	3.5381	29
32	.19994	.97981	.20406	.9006	.0206	.0016	.0027636	3.5311	28
33	.20022	.97975	.20436	.8933	.0207	4.9944	.0027757	3.5241	27
34	.20051	.97969	.20466	.8860	.0207	.9873	.0027879	3.5171	26
35	0.20079	0.97963	0.20497	4.8788	1.0208	4.9803	.0028001	3.5102	25
36	.20108	.97958	.20527	.8716	.0209	.9732	.0028123	3.5033	24
37	.20136	.97952	.20557	.8644	.0209	.9662	.0028246	3.4964	23
38	.20165	.97946	.20588	.8573	.0210	.9591	.0028369	3.4895	22
39	.20193	.97940	.20618	.8501	.0210	.9521	.0028493	3.4827	21
40	0.20222	0.97934	0.20648	4.8430	1.0211	4.9452	.0028616	3.4758	20
41	.20250	.97928	.20679	.8359	.0212	.9382	.0028741	3.4690	19
42	.20279	.97922	.20709	.8288	.0212	.9313	.0028865	3.4622	18
43	.20307	.97916	.20739	.8218	.0213	.9244	.0028990	3.4555	17
44	.20336	.97910	.20770	.8147	.0213	.9175	.0029115	3.4487	16
45	0.20364	0.97905	0.20800	4.8077	1.0214	4.9106	.0029241	3.4420	15
46	.20393	.97899	.20830	.8007	.0215	.9037	.0029367	3.4353	14
47	.20421	.97893	.20861	.7937	.0215	.8969	.0029494	3.4286	13
48	.20450	.97887	.20891	.7867	.0216	.8901	.0029620	3.4219	12
49	.20478	.97881	.20921	.7799	.0217	.8833	.0029747	3.4152	11
50	0.20507	0.97875	0.20952	4.7729	1.0217	4.8765	.0029875	3.4086	10
51	.20535	.97869	.20982	.7659	.0218	.8697	.0030003	3.4020	9
52	.20563	.97863	.21013	.7591	.0218	.8630	.0030131	3.3954	8
53	.20592	.97857	.21043	.7522	.0219	.8563	.0030260	3.3888	7
54	.20620	.97851	.21073	.7453	.0219	.8496	.0030389	3.3822	6
55	0.20649	0.97845	0.21104	4.7385	1.0220	4.8429	.0030518	3.3757	5
56	.20677	.97839	.21134	.7317	.0221	.8362	.0030648	3.3692	4
57	.20706	.97833	.21164	.7249	.0222	.8296	.0030778	3.3627	3
58	.20734	.97827	.21195	.7181	.0222	.8229	.0030908	3.3562	2
59	.20763	.97821	.21225	.7114	.0223	.8163	.0031039	3.3497	1
60	0.20791	0.97815	0.21256	4.7046	1.0223	4.8097	.0031171	3.3433	0
M	Cosine	Sine	Cotan.	Tan.	Cosec.	Secant	READ DOWN	78°–79° Involute	M

101° or 281°   **78° or 258°**

12° or 192°   Trigonometric and Involute Functions   167° or 347°

M	Sine	Cosine	Tan.	Cotan.	Secant	Cosec.	Involute 12°-13°	READ UP	M
0	0.20791	0.97815	0.21256	4.7046	1.0223	4.8097	.0031171	3.3433	60
1	.20820	.97809	.21286	.6979	.0224	.8032	.0031302	3.3368	59
2	.20848	.97803	.21316	.6912	.0225	.7966	.0031434	3.3304	58
3	.20877	.97797	.21347	.6845	.0225	.7901	.0031566	3.3240	57
4	.20905	.97791	.21377	.6779	.0226	.7836	.0031699	3.3177	56
5	0.20933	0.97784	0.21408	4.6712	1.0227	4.7771	.0031832	3.3113	55
6	.20962	.97778	.21438	.6646	.0227	.7706	.0031966	3.3050	54
7	.20990	.97772	.21469	.6580	.0228	.7641	.0032100	3.2987	53
8	.21019	.97766	.21499	.6514	.0228	.7577	.0032234	3.2923	52
9	.21047	.97760	.21529	.6448	.0229	.7512	.0032369	3.2861	51
10	0.21076	0.97754	0.21560	4.6382	1.0230	4.7448	.0032504	3.2798	50
11	.21104	.97748	.21590	.6317	.0230	.7384	.0032639	3.2735	49
12	.21132	.97742	.21621	.6252	.0231	.7321	.0032775	3.2673	48
13	.21161	.97735	.21651	.6187	.0232	.7257	.0032911	3.2611	47
14	.21189	.97729	.21682	.6122	.0232	.7194	.0033048	3.2549	46
15	0.21218	0.97723	0.21712	4.6057	1.0233	4.7130	.0033185	3.2487	45
16	.21246	.97717	.21743	.5993	.0234	.7067	.0033322	3.2426	44
17	.21275	.97711	.21773	.5928	.0234	.7004	.0033460	3.2364	43
18	.21303	.97705	.21804	.5864	.0235	.6942	.0033598	3.2303	42
19	.21331	.97698	.21834	.5800	.0236	.6879	.0033736	3.2242	41
20	0.21360	0.97692	0.21864	4.5736	1.0236	4.6817	.0033875	3.2181	40
21	.21388	.97686	.21895	.5673	.0237	.6755	.0034014	3.2120	39
22	.21417	.97680	.21925	.5609	.0238	.6693	.0034154	3.2060	38
23	.21445	.97673	.21956	.5546	.0238	.6631	.0034294	3.1999	37
24	.21474	.97667	.21986	.5483	.0239	.6569	.0034434	3.1939	36
25	0.21502	0.97661	0.22017	4.5420	1.0240	4.6507	.0034575	3.1879	35
26	.21530	.97655	.22047	.5357	.0240	.6446	.0034716	3.1819	34
27	.21559	.97648	.22078	.5294	.0241	.6385	.0034858	3.1759	33
28	.21587	.97642	.22108	.5232	.0241	.6324	.0035000	3.1699	32
29	.21616	.97636	.22139	.5169	.0242	.6263	.0035142	3.1640	31
30	0.21644	0.97630	0.22169	4.5107	1.0243	4.6202	.0035285	3.1581	30
31	.21672	.97623	.22200	.5045	.0243	.6142	.0035428	3.1522	29
32	.21701	.97617	.22231	.4983	.0244	.6081	.0035572	3.1463	28
33	.21729	.97611	.22261	.4922	.0245	.6021	.0035716	3.1404	27
34	.21758	.97604	.22292	.4860	.0245	.5961	.0035860	3.1345	26
35	0.21786	0.97598	0.22322	4.4799	1.0246	4.5901	.0036005	3.1287	25
36	.21814	.97592	.22353	.4737	.0247	.5841	.0036150	3.1229	24
37	.21843	.97585	.22383	.4676	.0247	.5782	.0036296	3.1170	23
38	.21871	.97579	.22414	.4615	.0248	.5722	.0036441	3.1112	22
39	.21899	.97573	.22444	.4555	.0249	.5663	.0036588	3.1055	21
40	0.21928	0.97566	0.22475	4.4494	1.0249	4.5604	.0036735	3.0997	20
41	.21956	.97560	.22505	.4434	.0250	.5545	.0036882	3.0939	19
42	.21985	.97553	.22536	.4373	.0251	.5486	.0037029	3.0882	18
43	.22013	.97547	.22567	.4313	.0251	.5428	.0037177	3.0825	17
44	.22041	.97541	.22597	.4253	.0252	.5369	.0037325	3.0768	16
45	0.22070	0.97534	0.22628	4.4194	1.0253	4.5311	.0037474	3.0711	15
46	.22098	.97528	.22658	.4134	.0253	.5253	.0037623	3.0654	14
47	.22126	.97521	.22689	.4075	.0254	.5195	.0037773	3.0598	13
48	.22155	.97515	.22719	.4015	.0255	.5137	.0037923	3.0541	12
49	.22183	.97508	.22750	.3956	.0256	.5079	.0038073	3.0485	11
50	0.22212	0.97502	0.22781	4.3897	1.0256	4.5022	.0038224	3.0429	10
51	.22240	.97496	.22811	.3838	.0257	.4964	.0038375	3.0373	9
52	.22268	.97489	.22842	.3779	.0258	.4907	.0038527	3.0317	8
53	.22297	.97483	.22872	.3721	.0258	.4850	.0038679	3.0261	7
54	.22325	.97476	.22903	.3662	.0259	.4793	.0038831	3.0206	6
55	0.22353	0.97470	0.22934	4.3604	1.0260	4.4736	.0038984	3.0150	5
56	.22382	.97463	.22964	.3546	.0260	.4679	.0039137	3.0095	4
57	.22410	.97457	.22995	.3488	.0261	.4623	.0039291	3.0040	3
58	.22438	.97450	.23026	.3430	.0262	.4566	.0039445	2.9985	2
59	.22467	.97444	.23056	.3372	.0262	.4510	.0039599	2.9930	1
60	0.22495	0.97437	0.23087	4.3315	1.0263	4.4454	.0039754	2.9876	0
M	Cosine	Sine	Cotan.	Tan.	Cosec.	Secant	READ DOWN	77°-78° Involute	M

102° or 282°   77° or 257°

13° or 193°   **Trigonometric and Involute Functions**   **166° or 346°**

M	Sine	Cosine	Tan.	Cotan.	Secant	Cosec.	Involute 13°–14°	READ UP	M
0	0.22495	0.97437	0.23087	4.3315	1.0263	4.4454	.0039754	2.9876	60
1	.22523	.97430	.23117	.3257	.0264	.4398	.0039909	2.9821	59
2	.22552	.97424	.23148	.3200	.0264	.4342	.0040065	2.9767	58
3	.22580	.97417	.23179	.3143	.0265	.4287	.0040221	2.9713	57
4	.22608	.97411	.23209	.3086	.0266	.4231	.0040377	2.9659	56
5	0.22637	0.97404	0.23240	4.3029	1.0266	4.4176	.0040534	2.9605	55
6	.22665	.97398	.23271	.2972	.0267	.4121	.0040692	2.9551	54
7	.22693	.97391	.23301	.2916	.0268	.4066	.0040849	2.9497	53
8	.22722	.97384	.23332	.2859	.0269	.4011	.0041007	2.9444	52
9	.22750	.97378	.23363	.2803	.0269	.3956	.0041166	2.9390	51
10	0.22778	0.97371	0.23393	4.2747	1.0270	4.3901	.0041325	2.9337	50
11	.22807	.97365	.23424	.2691	.0271	.3847	.0041484	2.9284	49
12	.22835	.97358	.23455	.2635	.0271	.3792	.0041644	2.9231	48
13	.22863	.97351	.23485	.2580	.0272	.3738	.0041804	2.9178	47
14	.22892	.97345	.23516	.2524	.0273	.3684	.0041965	2.9126	46
15	0.22920	0.97338	0.23547	4.2468	1.0273	4.3630	.0042126	2.9073	45
16	.22948	.97331	.23578	.2413	.0274	.3576	.0042288	2.9021	44
17	.22977	.97325	.23608	.2358	.0275	.3522	.0042450	2.8968	43
18	.23005	.97318	.23639	.2303	.0276	.3469	.0042612	2.8916	42
19	.23033	.97311	.23670	.2248	.0276	.3415	.0042775	2.8864	41
20	0.23062	0.97304	0.23700	4.2193	1.0277	4.3362	.0042938	2.8812	40
21	.23090	.97298	.23731	.2139	.0278	.3309	.0043101	2.8761	39
22	.23118	.97291	.23762	.2084	.0278	.3256	.0043266	2.8709	38
23	.23146	.97284	.23793	.2030	.0279	.3203	.0043430	2.8658	37
24	.23175	.97278	.23823	.1976	.0280	.3150	.0043595	2.8606	36
25	0.23203	0.97271	0.23854	4.1922	1.0281	4.3098	.0043760	2.8555	35
26	.23231	.97264	.23885	.1868	.0281	.3045	.0043926	2.8504	34
27	.23260	.97257	.23916	.1814	.0282	.2993	.0044092	2.8453	33
28	.23288	.97251	.23946	.1760	.0283	.2941	.0044259	2.8402	32
29	.23316	.97244	.23977	.1706	.0283	.2889	.0044426	2.8352	31
30	0.23345	0.97237	0.24008	4.1653	1.0284	4.2837	.0044593	2.8301	30
31	.23373	.97230	.24039	.1600	.0285	.2785	.0044761	2.8251	29
32	.23401	.97223	.24069	.1547	.0286	.2733	.0044929	2.8201	28
33	.23429	.97217	.24100	.1493	.0286	.2681	.0045098	2.8150	27
34	.23458	.97210	.24131	.1441	.0287	.2630	.0045267	2.8100	26
35	0.23486	0.97203	0.24162	4.1388	1.0288	4.2579	.0045437	2.8050	25
36	.23514	.97196	.24193	.1335	.0288	.2527	.0045607	2.8001	24
37	.23542	.97189	.24223	.1282	.0289	.2476	.0045777	2.7951	23
38	.23571	.97182	.24254	.1230	.0290	.2425	.0045948	2.7902	22
39	.23599	.97176	.24285	.1178	.0291	.2375	.0046120	2.7852	21
40	0.23627	0.97169	0.24316	4.1126	1.0291	4.2324	.0046291	2.7803	20
41	.23656	.97162	.24347	.1074	.0292	.2273	.0046464	2.7754	19
42	.23684	.97155	.24377	.1022	.0293	.2223	.0046636	2.7705	18
43	.23712	.97148	.24408	.0970	.0294	.2173	.0046809	2.7656	17
44	.23740	.97141	.24439	.0918	.0294	.2122	.0046983	2.7607	16
45	0.23769	0.97134	0.24470	4.0867	1.0295	4.2072	.0047157	2.7558	15
46	.23797	.97127	.24501	.0815	.0296	.2022	.0047331	2.7510	14
47	.23825	.97120	.24532	.0764	.0297	.1973	.0047506	2.7462	13
48	.23853	.97113	.24562	.0713	.0297	.1923	.0047681	2.7413	12
49	.23882	.97106	.24593	.0662	.0298	.1873	.0047857	2.7365	11
50	0.23910	0.97100	0.24624	4.0611	1.0299	4.1824	.0048033	2.7317	10
51	.23938	.97093	.24655	.0560	.0299	.1774	.0048210	2.7269	9
52	.23966	.97086	.24686	.0509	.0300	.1725	.0048387	2.7221	8
53	.23995	.97079	.24717	.0459	.0301	.1676	.0048564	2.7174	7
54	.24023	.97072	.24747	.0408	.0302	.1627	.0048742	2.7126	6
55	0.24051	0.97065	0.24778	4.0358	1.0302	4.1578	.0048921	2.7079	5
56	.24079	.97058	.24809	.0308	.0303	.1529	.0049099	2.7031	4
57	.24108	.97051	.24840	.0257	.0304	.1481	.0049279	2.6984	3
58	.24136	.97044	.24871	.0207	.0305	.1432	.0049458	2.6937	2
59	.24164	.97037	.24902	.0158	.0305	.1384	.0049638	2.6890	1
60	0.24192	0.97030	0.24933	4.0108	1.0306	4.1336	.0049819	2.6843	0
M	Cosine	Sine	Cotan.	Tan.	Cosec.	Secant	READ DOWN	76°–77° Involute	M

14° or 194°   Trigonometric and Involute Functions   165° or 345°

M	Sine	Cosine	Tan.	Cotan.	Secant	Cosec.	Involute 14°-15°	READ UP	M
0	0.24192	0.97030	0.24933	4.0108	1.0306	4.1336	.0049819	2.6843	60
1	.24220	.97023	.24964	.0058	.0307	.1287	.0050000	2.6797	59
2	.24249	.97015	.24995	.0009	.0308	.1239	.0050182	2.6750	58
3	.24277	.97008	.25026	3.9959	.0308	.1191	.0050364	2.6703	57
4	.24305	.97001	.25056	.9910	.0309	.1144	.0050546	2.6657	56
5	0.24333	0.96994	0.25087	3.9861	1.0310	4.1096	.0050729	2.6611	55
6	.24362	.96987	.25118	.9812	.0311	.1048	.0050912	2.6565	54
7	.24390	.96980	.25149	.9763	.0311	.1001	.0051096	2.6519	53
8	.24418	.96973	.25180	.9714	.0312	.0954	.0051280	2.6473	52
9	.24446	.96966	.25211	.9665	.0313	.0906	.0051465	2.6427	51
10	0.24474	0.96959	0.25242	3.9617	1.0314	4.0859	.0051650	2.6381	50
11	.24503	.96952	.25273	.9568	.0314	.0812	.0051835	2.6336	49
12	.24531	.96945	.25304	.9520	.0315	.0765	.0052021	2.6290	48
13	.24559	.96937	.25335	.9471	.0316	.0718	.0052208	2.6245	47
14	.24587	.96930	.25366	.9423	.0317	.0672	.0052395	2.6199	46
15	0.24615	0.96923	0.25397	3.9375	1.0317	4.0625	.0052582	2.6154	45
16	.24644	.96916	.25428	.9327	.0318	.0579	.0052770	2.6109	44
17	.24672	.96909	.25459	.9279	.0319	.0532	.0052958	2.6064	43
18	.24700	.96902	.25490	.9232	.0320	.0486	.0053147	2.6019	42
19	.24728	.96894	.25521	.9184	.0321	.0440	.0053336	2.5975	41
20	0.24756	0.96887	0.25552	3.9136	1.0321	4.0394	.0053526	2.5930	40
21	.24784	.96880	.25583	.9089	.0322	.0348	.0053716	2.5886	39
22	.24813	.96873	.25614	.9042	.0323	.0302	.0053907	2.5841	38
23	.24841	.96866	.25645	.8995	.0324	.0256	.0054098	2.5797	37
24	.24869	.96858	.25676	.8947	.0324	.0211	.0054289	2.5753	36
25	0.24897	0.96851	0.25707	3.8900	1.0325	4.0165	.0054481	2.5709	35
26	.24925	.96844	.25738	.8854	.0326	.0120	.0054674	2.5665	34
27	.24954	.96837	.25769	.8807	.0327	.0075	.0054867	2.5621	33
28	.24982	.96829	.25800	.8760	.0327	.0029	.0055060	2.5577	32
29	.25010	.96822	.25831	.8714	.0328	3.9984	.0055254	2.5533	31
30	0.25038	0.96815	0.25862	3.8667	1.0329	3.9939	.0055448	2.5490	30
31	.25066	.96807	.25893	.8621	.0330	.9894	.0055643	2.5446	29
32	.25094	.96800	.25924	.8575	.0331	.9850	.0055838	2.5403	28
33	.25122	.96793	.25955	.8528	.0331	.9805	.0056034	2.5360	27
34	.25151	.96786	.25986	.8482	.0332	.9760	.0056230	2.5317	26
35	0.25179	0.96778	0.26017	3.8436	1.0333	3.9716	.0056427	2.5274	25
36	.25207	.96771	.26048	.8391	.0334	.9672	.0056624	2.5231	24
37	.25235	.96764	.26079	.8345	.0334	.9627	.0056822	2.5188	23
38	.25263	.96756	.26110	.8299	.0335	.9583	.0057020	2.5145	22
39	.25291	.96749	.26141	.8254	.0336	.9539	.0057218	2.5103	21
40	0.25320	0.96742	0.26172	3.8208	1.0337	3.9495	.0057417	2.5060	20
41	.25348	.96734	.26203	.8163	.0338	.9451	.0057617	2.5018	19
42	.25376	.96727	.26235	.8118	.0338	.9408	.0057817	2.4975	18
43	.25404	.96719	.26266	.8073	.0339	.9364	.0058017	2.4933	17
44	.25432	.96712	.26297	.8028	.0340	.9320	.0058218	2.4891	16
45	0.25460	0.96705	0.26328	3.7983	1.0341	3.9277	.0058420	2.4849	15
46	.25488	.96697	.26359	.7938	.0342	.9234	.0058622	2.4807	14
47	.25516	.96690	.26390	.7893	.0342	.9190	.0058824	2.4765	13
48	.25545	.96682	.26421	.7848	.0343	.9147	.0059027	2.4724	12
49	.25573	.96675	.26452	.7804	.0344	.9104	.0059230	2.4682	11
50	0.25601	0.96667	0.26483	3.7760	1.0345	3.9061	.0059434	2.4640	10
51	.25629	.96660	.26515	.7715	.0346	.9018	.0059638	2.4599	9
52	.25657	.96653	.26546	.7671	.0346	.8976	.0059843	2.4558	8
53	.25685	.96645	.26577	.7627	.0347	.8933	.0060048	2.4516	7
54	.25713	.96638	.26608	.7583	.0348	.8890	.0060254	2.4475	6
55	0.25741	0.96630	0.26639	3.7539	1.0349	3.8848	.0060460	2.4434	5
56	.25769	.96623	.26670	.7495	.0350	.8806	.0060667	2.4393	4
57	.25798	.96615	.26701	.7451	.0350	.8763	.0060874	2.4353	3
58	.25826	.96608	.26733	.7408	.0351	.8721	.0061081	2.4312	2
59	.25854	.96600	.26764	.7364	.0352	.8679	.0061289	2.4271	1
60	0.25882	0.96593	0.26795	3.7321	1.0353	3.8637	.0061498	2.4231	0
M	Cosine	Sine	Cotan.	Tan.	Cosec.	Secant	READ DOWN	75°-76° Involute	M

15° or 195°   Trigonometric and Involute Functions   164° or 344°

M	Sine	Cosine	Tan.	Cotan.	Secant	Cosec.	Involute 15°–16°	READ UP	M
0	0.25882	0.96593	0.26795	3.7321	1.0353	3.8637	.0061498	2.4231	60
1	.25910	.96585	.26826	.7277	.0354	.8595	.0061707	2.4190	59
2	.25938	.96578	.26857	.7234	.0354	.8553	.0061917	2.4150	58
3	.25966	.96570	.26888	.7191	.0355	.8512	.0062127	2.4109	57
4	.25994	.96562	.26920	.7148	.0356	.8470	.0062337	2.4069	56
5	0.26022	0.96555	0.26951	3.7105	1.0357	3.8428	.0062548	2.4029	55
6	.26050	.96547	.26982	.7062	.0358	.8387	.0062760	2.3989	54
7	.26079	.96540	.27013	.7019	.0358	.8346	.0062972	2.3949	53
8	.26107	.96532	.27044	.6976	.0359	.8304	.0063184	2.3909	52
9	.26135	.96524	.27076	.6933	.0360	.8263	.0063397	2.3870	51
10	0.26163	0.96517	0.27107	3.6891	1.0361	3.8222	.0063611	2.3830	50
11	.26191	.96509	.27138	.6848	.0362	.8181	.0063825	2.3791	49
12	.26219	.96502	.27169	.6806	.0363	.8140	.0064039	2.3751	48
13	.26247	.96494	.27201	.6764	.0363	.8100	.0064254	2.3712	47
14	.26275	.96486	.27232	.6722	.0364	.8059	.0064470	2.3672	46
15	0.26303	0.96479	0.27263	3.6680	1.0365	3.8018	.0064686	2.3633	45
16	.26331	.96471	.27294	.6638	.0366	.7978	.0064902	2.3594	44
17	.26359	.96463	.27326	.6596	.0367	.7937	.0065119	2.3555	43
18	.26387	.96456	.27357	.6554	.0367	.7897	.0065337	2.3516	42
19	.26415	.96449	.27388	.6512	.0368	.7857	.0065555	2.3477	41
20	0.26443	0.96440	0.27419	3.6470	1.0369	3.7817	.0065773	2.3439	40
21	.26471	.96433	.27451	.6429	.0370	.7777	.0065992	2.3400	39
22	.26500	.96425	.27482	.6387	.0371	.7737	.0066211	2.3361	38
23	.26528	.96417	.27513	.6346	.0372	.7697	.0066431	2.3323	37
24	.26556	.96410	.27545	.6305	.0372	.7657	.0066652	2.3285	36
25	0.26584	0.96402	0.27576	3.6264	1.0373	3.7617	.0066873	2.3246	35
26	.26612	.96394	.27607	.6222	.0374	.7577	.0067094	2.3208	34
27	.26640	.96386	.27638	.6181	.0375	.7538	.0067316	2.3170	33
28	.26668	.96379	.27670	.6140	.0376	.7498	.0067539	2.3132	32
29	.26696	.96371	.27701	.6100	.0377	.7459	.0067762	2.3094	31
30	0.26724	0.96363	0.27732	3.6059	1.0377	3.7420	.0067985	2.3056	30
31	.26752	.96355	.27764	.6018	.0378	.7381	.0068209	2.3018	29
32	.26780	.96347	.27795	.5978	.0379	.7341	.0068434	2.2981	28
33	.26808	.96340	.27826	.5937	.0380	.7302	.0068659	2.2943	27
34	.26836	.96332	.27858	.5897	.0381	.7263	.0068884	2.2906	26
35	0.26864	0.96324	0.27889	3.5856	1.0382	3.7225	.0069110	2.2868	25
36	.26892	.96316	.27921	.5816	.0382	.7186	.0069337	2.2831	24
37	.26920	.96308	.27952	.5776	.0383	.7147	.0069564	2.2793	23
38	.26948	.96301	.27983	.5736	.0384	.7108	.0069791	2.2756	22
39	.26976	.96293	.28015	.5696	.0385	.7070	.0070019	2.2719	21
40	0.27004	0.96285	0.28046	3.5656	1.0386	3.7032	.0070248	2.2682	20
41	.27032	.96277	.28077	.5616	.0387	.6993	.0070477	2.2645	19
42	.27060	.96269	.28109	.5576	.0388	.6955	.0070706	2.2608	18
43	.27088	.96261	.28140	.5536	.0388	.6917	.0070936	2.2572	17
44	.27116	.96253	.28172	.5497	.0389	.6879	.0071167	2.2535	16
45	0.27144	0.96246	0.28203	3.5457	1.0390	3.6840	.0071398	2.2498	15
46	.27172	.96238	.28234	.5418	.0391	.6803	.0071630	2.2462	14
47	.27200	.96230	.28266	.5379	.0392	.6765	.0071862	2.2425	13
48	.27228	.96222	.28297	.5339	.0393	.6727	.0072095	2.2389	12
49	.27256	.96214	.28329	.5300	.0394	.6689	.0072328	2.2353	11
50	0.27284	0.96206	0.28360	3.5261	1.0394	3.6652	.0072561	2.2316	10
51	.27312	.96198	.28391	.5222	.0395	.6614	.0072796	2.2280	9
52	.27340	.96190	.28423	.5183	.0396	.6576	.0073030	2.2244	8
53	.27368	.96182	.28454	.5144	.0397	.6539	.0073266	2.2208	7
54	.27396	.96174	.28486	.5105	.0398	.6502	.0073501	2.2172	6
55	0.27424	0.96166	0.28517	3.5067	1.0399	3.6465	.0073738	2.2137	5
56	.27452	.96158	.28549	.5028	.0400	.6427	.0073975	2.2101	4
57	.27480	.96150	.28580	.4989	.0400	.6390	.0074212	2.2065	3
58	.27508	.96142	.28612	.4951	.0401	.6353	.0074450	2.2030	2
59	.27536	.96134	.28643	.4912	.0402	.6316	.0074688	2.1994	1
60	0.27564	0.96126	0.28675	3.4874	1.0403	3.6280	.0074927	2.1959	0
M	Cosine	Sine	Cotan.	Tan.	Cosec.	Secant	READ DOWN	74°–75° Involute	M

105° or 285°   74° or 254°

16° or 196°   Trigonometric and Involute Functions   163° or 343°

M	Sine	Cosine	Tan.	Cotan.	Secant	Cosec.	Involute 16°–17°	READ UP	M
0	0.27564	0.96126	0.28675	3.4874	1.0403	3.6280	.0074927	2.1959	60
1	.27592	.96118	.28706	.4836	.0404	.6243	.0075166	2.1923	59
2	.27620	.96110	.28738	.4798	.0405	.6206	.0075406	2.1888	58
3	.27648	.96102	.28769	.4760	.0406	.6169	.0075647	2.1853	57
4	.27676	.96094	.28801	.4722	.0406	.6133	.0075888	2.1818	56
5	0.27704	0.96086	0.28832	3.4684	1.0407	3.6097	.0076130	2.1783	55
6	.27731	.96078	.28864	.4646	.0408	.6060	.0076372	2.1748	54
7	.27759	.96070	.28895	.4608	.0409	.6024	.0076614	2.1713	53
8	.27787	.96062	.28927	.4570	.0410	.5988	.0076857	2.1678	52
9	.27815	.96054	.28958	.4533	.0411	.5951	.0077101	2.1643	51
10	0.27843	0.96046	0.28990	3.4495	1.0412	3.5915	.0077345	2.1609	50
11	.27871	.96037	.29021	.4458	.0413	.5879	.0077590	2.1574	49
12	.27899	.96029	.29053	.4420	.0413	.5843	.0077835	2.1540	48
13	.27927	.96021	.29084	.4383	.0414	.5808	.0078081	2.1505	47
14	.27955	.96013	.29116	.4346	.0415	.5772	.0078327	2.1471	46
15	0.27983	0.96005	0.29147	3.4308	1.0416	3.5736	.0078574	2.1437	45
16	.28011	.95997	.29179	.4271	.0417	.5700	.0078822	2.1402	44
17	.28039	.95989	.29210	.4234	.0418	.5665	.0079069	2.1368	43
18	.28067	.95981	.29242	.4197	.0419	.5629	.0079318	2.1334	42
19	.28095	.95972	.29274	.4160	.0420	.5594	.0079567	2.1300	41
20	0.28123	0.95964	0.29305	3.4124	1.0421	3.5559	.0079817	2.1266	40
21	.28150	.95956	.29337	.4087	.0421	.5523	.0080067	2.1233	39
22	.28178	.95948	.29368	.4050	.0422	.5488	.0080317	2.1199	38
23	.28206	.95940	.29400	.4014	.0423	.5453	.0080568	2.1165	37
24	.28234	.95931	.29432	.3977	.0424	.5418	.0080820	2.1131	36
25	0.28262	0.95923	0.29463	3.3941	1.0425	3.5383	.0081072	2.1098	35
26	.28290	.95915	.29495	.3904	.0426	.5348	.0081325	2.1064	34
27	.28318	.95907	.29526	.3868	.0427	.5313	.0081578	2.1031	33
28	.28346	.95898	.29558	.3832	.0428	.5279	.0081832	2.0998	32
29	.28374	.95890	.29590	.3796	.0429	.5244	.0082087	2.0964	31
30	0.28402	0.95882	0.29621	3.3759	1.0429	3.5209	.0082342	2.0931	30
31	.28429	.95874	.29653	.3723	.0430	.5175	.0082597	2.0898	29
32	.28457	.95865	.29685	.3687	.0431	.5140	.0082853	2.0865	28
33	.28485	.95857	.29716	.3652	.0432	.5106	.0083110	2.0832	27
34	.28513	.95849	.29748	.3616	.0433	.5072	.0083367	2.0799	26
35	0.28541	0.95841	0.29780	3.3580	1.0434	3.5037	.0083625	2.0766	25
36	.28569	.95832	.29811	.3544	.0435	.5003	.0083883	2.0734	24
37	.28597	.95824	.29843	.3509	.0436	.4969	.0084142	2.0701	23
38	.28625	.95816	.29875	.3473	.0437	.4935	.0084401	2.0668	22
39	.28652	.95807	.29906	.3438	.0438	.4901	.0084661	2.0636	21
40	0.28680	0.95799	0.29938	3.3402	1.0439	3.4867	.0084921	2.0603	20
41	.28708	.95791	.29970	.3367	.0439	.4833	.0085182	2.0571	19
42	.28736	.95782	.30001	.3332	.0440	.4799	.0085444	2.0538	18
43	.28764	.95774	.30033	.3297	.0441	.4766	.0085706	2.0506	17
44	.28792	.95766	.30065	.3261	.0442	.4732	.0085969	2.0474	16
45	0.28820	0.95757	0.30097	3.3226	1.0443	3.4699	.0086232	2.0442	15
46	.28847	.95749	.30128	.3191	.0444	.4665	.0086496	2.0410	14
47	.28875	.95740	.30160	.3156	.0445	.4632	.0086760	2.0378	13
48	.28903	.95732	.30192	.3122	.0446	.4598	.0087025	2.0346	12
49	.28931	.95724	.30224	.3087	.0447	.4565	.0087290	2.0314	11
50	0.28959	0.95715	0.30255	3.3052	1.0448	3.4532	.0087556	2.0282	10
51	.28987	.95707	.30287	.3017	.0449	.4499	.0087823	2.0250	9
52	.29015	.95698	.30319	.2983	.0450	.4465	.0088090	2.0219	8
53	.29042	.95690	.30351	.2948	.0450	.4432	.0088358	2.0187	7
54	.29070	.95681	.30382	.2914	.0451	.4399	.0088626	2.0156	6
55	0.29098	0.95673	0.30414	3.2879	1.0452	3.4367	.0088895	2.0124	5
56	.29126	.95664	.30446	.2845	.0453	.4334	.0089164	2.0093	4
57	.29154	.95656	.30478	.2811	.0454	.4301	.0089434	2.0061	3
58	.29182	.95647	.30509	.2777	.0455	.4268	.0089704	2.0030	2
59	.29209	.95639	.30541	.2743	.0456	.4236	.0089975	1.9999	1
60	0.29237	0.95630	0.30573	3.2709	1.0457	3.4203	.0090247	1.9968	0
M	Cosine	Sine	Cotan.	Tan.	Cosec.	Secant	READ DOWN	73°–74° Involute	M

106° or 286°   73° or 253°

17° or 197°   Trigonometric and Involute Functions   162° or 342°

M	Sine	Cosine	Tan.	Cotan.	Secant	Cosec.	Involute 17°–18°	READ UP	M
0	0.29237	0.95630	0.30573	3.2709	1.0457	3.4203	.0090247	1.9968	60
1	.29265	.95622	.30605	.2675	.0458	.4171	.0090519	1.9937	59
2	.29293	.95613	.30637	.2641	.0459	.4138	.0090792	1.9906	58
3	.29321	.95605	.30669	.2607	.0460	.4106	.0091065	1.9875	57
4	.29348	.95596	.30700	.2573	.0461	.4073	.0091339	1.9844	56
5	0.29376	0.95588	0.30732	3.2539	1.0462	3.4041	.0091614	1.9813	55
6	.29404	.95579	.30764	.2506	.0463	.4009	.0091889	1.9782	54
7	.29432	.95571	.30796	.2472	.0463	.3977	.0092164	1.9751	53
8	.29460	.95562	.30828	.2438	.0464	.3945	.0092440	1.9721	52
9	.29487	.95554	.30860	.2405	.0465	.3913	.0092717	1.9690	51
10	0.29515	0.95545	0.30891	3.2371	1.0466	3.3881	.0092994	1.9660	50
11	.29543	.95536	.30923	.2338	.0467	.3849	.0093272	1.9629	49
12	.29571	.95528	.30955	.2305	.0468	.3817	.0093551	1.9599	48
13	.29599	.95519	.30987	.2272	.0469	.3785	.0093830	1.9568	47
14	.29626	.95511	.31019	.2238	.0470	.3754	.0094109	1.9538	46
15	0.29654	0.95502	0.31051	3.2205	1.0471	3.3722	.0094390	1.9508	45
16	.29682	.95493	.31083	.2172	.0472	.3691	.0094670	1.9478	44
17	.29710	.95485	.31115	.2139	.0473	.3659	.0094952	1.9448	43
18	.29737	.95476	.31147	.2106	.0474	.3628	.0095234	1.9418	42
19	.29765	.95467	.31178	.2073	.0475	.3596	.0095516	1.9388	41
20	0.29793	0.95459	0.31210	3.2041	1.0476	3.3565	.0095799	1.9358	40
21	.29821	.95450	.31242	.2008	.0477	.3534	.0096083	1.9328	39
22	.29849	.95441	.31274	.1975	.0478	.3502	.0096367	1.9298	38
23	.29876	.95433	.31306	.1943	.0479	.3471	.0096652	1.9269	37
24	.29904	.95424	.31338	.1910	.0480	.3440	.0096937	1.9239	36
25	0.29932	0.95415	0.31370	3.1878	1.0480	3.3409	.0097223	1.9209	35
26	.29960	.95407	.31402	.1845	.0481	.3378	.0097510	1.9180	34
27	.29987	.95398	.31434	.1813	.0482	.3347	.0097797	1.9150	33
28	.30015	.95389	.31466	.1780	.0483	.3317	.0098085	1.9121	32
29	.30043	.95380	.31498	.1748	.0484	.3286	.0098373	1.9092	31
30	0.30071	0.95372	0.31530	3.1716	1.0485	3.3255	.0098662	1.9062	30
31	.30098	.95363	.31562	.1684	.0486	.3224	.0098951	1.9033	29
32	.30126	.95354	.31594	.1652	.0487	.3194	.0099241	1.9004	28
33	.30154	.95345	.31626	.1620	.0488	.3163	.0099532	1.8975	27
34	.30182	.95337	.31658	.1588	.0489	.3133	.0099823	1.8946	26
35	0.30209	0.95328	0.31690	3.1556	1.0490	3.3102	.010012	1.8917	25
36	.30237	.95319	.31722	.1524	.0491	.3072	.010041	1.8888	24
37	.30265	.95310	.31754	.1492	.0492	.3042	.010070	1.8859	23
38	.30292	.95301	.31786	.1460	.0493	.3012	.010099	1.8830	22
39	.30320	.95293	.31818	.1429	.0494	.2981	.010129	1.8801	21
40	0.30348	0.95284	0.31850	3.1397	1.0495	3.2951	.010158	1.8773	20
41	.30376	.95275	.31882	.1366	.0496	.2921	.010188	1.8744	19
42	.30403	.95266	.31914	.1334	.0497	.2891	.010217	1.8715	18
43	.30431	.95257	.31946	.1303	.0498	.2861	.010247	1.8687	17
44	.30459	.95248	.31978	.1271	.0499	.2831	.010277	1.8658	16
45	0.30486	0.95240	0.32010	3.1240	1.0500	3.2801	.010307	1.8630	15
46	.30514	.95231	.32042	.1209	.0501	.2772	.010336	1.8602	14
47	.30542	.95222	.32074	.1178	.0502	.2742	.010366	1.8573	13
48	.30570	.95213	.32106	.1146	.0503	.2712	.010396	1.8545	12
49	.30597	.95204	.32139	.1115	.0504	.2683	.010426	1.8517	11
50	0.30625	0.95195	0.32171	3.1084	1.0505	3.2653	.010456	1.8489	10
51	.30653	.95186	.32203	.1053	.0506	.2624	.010487	1.8461	9
52	.30680	.95177	.32235	.1022	.0507	.2594	.010517	1.8433	8
53	.30708	.95168	.32267	.0991	.0508	.2565	.010547	1.8405	7
54	.30736	.95159	.32299	.0961	.0509	.2535	.010577	1.8377	6
55	0.30763	0.95150	0.32331	3.0930	1.0510	3.2506	.010608	1.8349	5
56	.30791	.95142	.32363	.0899	.0511	.2477	.010638	1.8321	4
57	.30819	.95133	.32396	.0868	.0512	.2448	.010669	1.8293	3
58	.30846	.95124	.32428	.0838	.0513	.2419	.010699	1.8266	2
59	.30874	.95115	.32460	.0807	.0514	.2390	.010730	1.8238	1
60	0.30902	0.95106	0.32492	3.0777	1.0515	3.2361	.010760	1.8210	0
M	Cosine	Sine	Cotan.	Tan.	Cosec.	Secant	READ DOWN	72°–73° Involute	M

18° or 198° Trigonometric and Involute Functions 161° or 341°

M	Sine	Cosine	Tan.	Cotan.	Secant	Cosec.	Involute 18°-19°	READ UP	M
0	0.30902	0.95106	0.32492	3.0777	1.0515	3.2361	.010760	1.8210	60
1	.30929	.95097	.32524	.0746	.0516	.2332	.010791	1.8183	59
2	.30957	.95088	.32556	.0716	.0517	.2303	.010822	1.8155	58
3	.30985	.95079	.32588	.0686	.0518	.2274	.010853	1.8128	57
4	.31012	.95070	.32621	.0655	.0519	.2245	.010884	1.8101	56
5	0.31040	0.95061	0.32653	3.0625	1.0520	3.2217	.010915	1.8073	55
6	.31068	.95052	.32685	.0595	.0521	.2188	.010946	1.8046	54
7	.31095	.95043	.32717	.0565	.0522	.2159	.010977	1.8019	53
8	.31123	.95033	.32749	.0535	.0523	.2131	.011008	1.7992	52
9	.31151	.95024	.32782	.0505	.0524	.2102	.011039	1.7965	51
10	0.31178	0.95015	0.32814	3.0475	1.0525	3.2074	.011071	1.7938	50
11	.31206	.95006	.32846	.0445	.0526	.2045	.011102	1.7911	49
12	.31233	.94997	.32878	.0415	.0527	.2017	.011133	1.7884	48
13	.31261	.94988	.32911	.0385	.0528	.1989	.011165	1.7857	47
14	.31289	.94979	.32943	.0356	.0529	.1960	.011196	1.7830	46
15	0.31316	0.94970	0.32975	3.0326	1.0530	3.1932	.011228	1.7803	45
16	.31344	.94961	.33007	.0296	.0531	.1904	.011260	1.7776	44
17	.31372	.94952	.33040	.0267	.0532	.1876	.011291	1.7750	43
18	.31399	.94943	.33072	.0237	.0533	.1848	.011323	1.7723	42
19	.31427	.94933	.33104	.0208	.0534	.1820	.011355	1.7697	41
20	0.31454	0.94924	0.33136	3.0178	1.0535	3.1792	.011387	1.7670	40
21	.31482	.94915	.33169	.0149	.0536	.1764	.011419	1.7644	39
22	.31510	.94906	.33201	.0120	.0537	.1736	.011451	1.7617	38
23	.31537	.94897	.33233	.0090	.0538	.1708	.011483	1.7591	37
24	.31565	.94888	.33266	.0061	.0539	.1681	.011515	1.7565	36
25	0.31593	0.94878	0.33298	3.0032	1.0540	3.1653	.011547	1.7538	35
26	.31620	.94869	.33330	.0003	.0541	.1625	.011580	1.7512	34
27	.31648	.94860	.33363	2.9974	.0542	.1598	.011612	1.7486	33
28	.31675	.94851	.33395	.9945	.0543	.1570	.011644	1.7460	32
29	.31703	.94842	.33427	.9916	.0544	.1543	.011677	1.7434	31
30	0.31730	0.94832	0.33460	2.9887	1.0545	3.1515	.011709	1.7408	30
31	.31758	.94823	.33492	.9858	.0546	.1488	.011742	1.7382	29
32	.31786	.94814	.33524	.9829	.0547	.1461	.011775	1.7356	28
33	.31813	.94805	.33557	.9800	.0548	.1433	.011807	1.7330	27
34	.31841	.94795	.33589	.9772	.0549	.1406	.011840	1.7304	26
35	0.31868	0.94786	0.33621	2.9743	1.0550	3.1379	.011873	1.7278	25
36	.31896	.94777	.33654	.9714	.0551	.1352	.011906	1.7253	24
37	.31923	.94768	.33686	.9686	.0552	.1325	.011939	1.7227	23
38	.31951	.94758	.33718	.9657	.0553	.1298	.011972	1.7201	22
39	.31979	.94749	.33751	.9629	.0554	.1271	.012005	1.7176	21
40	0.32006	0.94740	0.33783	2.9600	1.0555	3.1244	.012038	1.7150	20
41	.32034	.94730	.33816	.9572	.0556	.1217	.012071	1.7125	19
42	.32061	.94721	.33848	.9544	.0557	.1190	.012105	1.7100	18
43	.32089	.94712	.33881	.9515	.0558	.1163	.012138	1.7074	17
44	.32116	.94702	.33913	.9487	.0559	.1137	.012172	1.7049	16
45	0.32144	0.94693	0.33945	2.9459	1.0560	3.1110	.012205	1.7024	15
46	.32171	.94684	.33978	.9431	.0561	.1083	.012239	1.6998	14
47	.32199	.94674	.34010	.9403	.0563	.1057	.012272	1.6973	13
48	.32227	.94665	.34043	.9375	.0564	.1030	.012306	1.6948	12
49	.32254	.94656	.34075	.9347	.0565	.1004	.012340	1.6923	11
50	0.32282	0.94646	0.34108	2.9319	1.0566	3.0977	.012373	1.6898	10
51	.32309	.94637	.34140	.9291	.0567	.0951	.012407	1.6873	9
52	.32337	.94627	.34173	.9263	.0568	.0925	.012441	1.6848	8
53	.32364	.94618	.34205	.9235	.0569	.0898	.012475	1.6823	7
54	.32392	.94609	.34238	.9208	.0570	.0872	.012509	1.6798	6
55	0.32419	0.94599	0.34270	2.9180	1.0571	3.0846	.012543	1.6774	5
56	.32447	.94590	.34303	.9152	.0572	.0820	.012578	1.6749	4
57	.32474	.94580	.34335	.9125	.0573	.0794	.012612	1.6724	3
58	.32502	.94571	.34368	.9097	.0574	.0768	.012646	1.6699	2
59	.32529	.94561	.34400	.9070	.0575	.0742	.012681	1.6675	1
60	0.32557	0.94552	0.34433	2.9042	1.0576	3.0716	.012715	1.6650	0
M	Cosine	Sine	Cotan.	Tan.	Cosec.	Secant	READ DOWN	71°-72° Involute	M

108° or 288° 71° or 251°

19° or 199°   **Trigonometric and Involute Functions**   **160° or 340°**

M	Sine	Cosine	Tan.	Cotan.	Secant	Cosec.	Involute 19°–20°	READ UP	M
0	0.32557	0.94552	0.34433	2.9042	1.0576	3.0716	.012715	1.6650	60
1	.32584	.94542	.34465	.9015	.0577	.0690	.012750	1.6626	59
2	.32612	.94533	.34498	.8987	.0578	.0664	.012784	1.6601	58
3	.32639	.94523	.34530	.8960	.0579	.0638	.012819	1.6577	57
4	.32667	.94514	.34563	.8933	.0580	.0612	.012854	1.6553	56
5	0.32694	0.94504	0.34596	2.8905	1.0582	3.0586	.012888	1.6528	55
6	.32722	.94495	.34628	.8878	.0583	.0561	.012923	1.6504	54
7	.32749	.94485	.34661	.8851	.0584	.0535	.012958	1.6480	53
8	.32777	.94476	.34693	.8824	.0585	.0509	.012993	1.6455	52
9	.32804	.94466	.34726	.8797	.0586	.0484	.013028	1.6431	51
10	0.32832	0.94457	0.34758	2.8770	1.0587	3.0458	.013063	1.6407	50
11	.32859	.94447	.34791	.8743	.0588	.0433	.013098	1.6383	49
12	.32887	.94438	.34824	.8716	.0589	.0407	.013134	1.6359	48
13	.32914	.94428	.34856	.8689	.0590	.0382	.013169	1.6335	47
14	.32942	.94418	.34889	.8662	.0591	.0357	.013204	1.6311	46
15	0.32969	0.94409	0.34922	2.8636	1.0592	3.0331	.013240	1.6287	45
16	.32997	.94399	.34954	.8609	.0593	.0306	.013275	1.6264	44
17	.33024	.94390	.34987	.8582	.0594	.0281	.013311	1.6240	43
18	.33051	.94380	.35020	.8556	.0595	.0256	.013346	1.6216	42
19	.33079	.94370	.35052	.8529	.0597	.0231	.013382	1.6192	41
20	0.33106	0.94361	0.35085	2.8502	1.0598	3.0206	.013418	1.6169	40
21	.33134	.94351	.35118	.8476	.0599	.0181	.013454	1.6145	39
22	.33161	.94342	.35150	.8449	.0600	.0156	.013490	1.6122	38
23	.33189	.94332	.35183	.8423	.0601	.0131	.013526	1.6098	37
24	.33216	.94322	.35216	.8397	.0602	.0106	.013562	1.6075	36
25	0.33244	0.94313	0.35248	2.8370	1.0603	3.0081	.013598	1.6051	35
26	.33271	.94303	.35281	.8344	.0604	.0056	.013634	1.6028	34
27	.33298	.94293	.35314	.8318	.0605	.0031	.013670	1.6004	33
28	.33326	.94284	.35346	.8291	.0606	.0007	.013707	1.5981	32
29	.33353	.94274	.35379	.8265	.0607	2.9982	.013743	1.5958	31
30	0.33381	0.94264	0.35412	2.8239	1.0608	2.9957	.013779	1.5935	30
31	.33408	.94254	.35445	.8213	.0610	.9933	.013816	1.5911	29
32	.33436	.94245	.35477	.8187	.0611	.9908	.013852	1.5888	28
33	.33463	.94235	.35510	.8161	.0612	.9884	.013889	1.5865	27
34	.33490	.94225	.35543	.8135	.0613	.9859	.013926	1.5842	26
35	0.33518	0.94215	0.35576	2.8109	1.0614	2.9835	.013963	1.5819	25
36	.33545	.94206	.35608	.8083	.0615	.9811	.013999	1.5796	24
37	.33573	.94196	.35641	.8057	.0616	.9786	.014036	1.5773	23
38	.33600	.94186	.35674	.8032	.0617	.9762	.014073	1.5750	22
39	.33627	.94176	.35707	.8006	.0618	.9738	.014110	1.5728	21
40	0.33655	0.94167	0.35740	2.7980	1.0619	2.9713	.014148	1.5705	20
41	.33682	.94157	.35772	.7955	.0621	.9689	.014185	1.5682	19
42	.33710	.94147	.35805	.7929	.0622	.9665	.014222	1.5659	18
43	.33737	.94137	.35838	.7903	.0623	.9641	.014259	1.5637	17
44	.33764	.94127	.35871	.7878	.0624	.9617	.014297	1.5614	16
45	0.33792	0.94118	0.35904	2.7852	1.0625	2.9593	.014334	1.5591	15
46	.33819	.94108	.35937	.7827	.0626	.9569	.014372	1.5569	14
47	.33846	.94098	.35969	.7801	.0627	.9545	.014409	1.5546	13
48	.33874	.94088	.36002	.7776	.0628	.9521	.014447	1.5524	12
49	.33901	.94078	.36035	.7751	.0629	.9498	.014485	1.5501	11
50	0.33929	0.94068	0.36068	2.7725	1.0631	2.9474	.014523	1.5479	10
51	.33956	.94058	.36101	.7700	.0632	.9450	.014560	1.5457	9
52	.33983	.94049	.36134	.7675	.0633	.9426	.014598	1.5434	8
53	.34011	.94039	.36167	.7650	.0634	.9403	.014636	1.5412	7
54	.34038	.94029	.36199	.7625	.0635	.9379	.014674	1.5390	6
55	0.34065	0.94019	0.36232	2.7600	1.0636	2.9355	.014713	1.5368	5
56	.34093	.94009	.36265	.7575	.0637	.9332	.014751	1.5346	4
57	.34120	.93999	.36298	.7550	.0638	.9308	.014789	1.5324	3
58	.34147	.93989	.36331	.7525	.0640	.9285	.014827	1.5301	2
59	.34175	.93979	.36364	.7500	.0641	.9261	.014866	1.5279	1
60	0.34202	0.93969	0.36397	2.7475	1.0642	2.9238	.014904	1.5257	0
M	Cosine	Sine	Cotan.	Tan.	Cosec.	Secant	READ DOWN	70°–71° Involute	M

20° or 200°   Trigonometric and Involute Functions   159° or 339°

M	Sine	Cosine	Tan.	Cotan.	Secant	Cosec.	Involute 20°–21°	READ UP	M
0	0.34202	0.93969	0.36397	2.7475	1.0642	2.9238	.014904	1.5257	60
1	.34229	.93959	.36430	.7450	.0643	.9215	.014943	1.5236	59
2	.34257	.93949	.36463	.7425	.0644	.9191	.014982	1.5214	58
3	.34284	.93939	.36496	.7400	.0645	.9168	.015020	1.5192	57
4	.34311	.93929	.36529	.7376	.0646	.9145	.015059	1.5170	56
5	0.34339	0.93919	0.36562	2.7351	1.0647	2.9122	.015098	1.5148	55
6	.34366	.93909	.36595	.7326	.0649	.9099	.015137	1.5126	54
7	.34393	.93899	.36628	.7302	.0650	.9075	.015176	1.5105	53
8	.34421	.93889	.36661	.7277	.0651	.9052	.015215	1.5083	52
9	.34448	.93879	.36694	.7253	.0652	.9029	.015254	1.5061	51
10	0.34475	0.93869	0.36727	2.7228	1.0653	2.9006	.015293	1.5040	50
11	.34503	.93859	.36760	.7204	.0654	.8983	.015333	1.5018	49
12	.34530	.93849	.36793	.7179	.0655	.8960	.015372	1.4997	48
13	.34557	.93839	.36826	.7155	.0657	.8938	.015411	1.4975	47
14	.34584	.93829	.36859	.7130	.0658	.8915	.015451	1.4954	46
15	0.34612	0.93819	0.36892	2.7106	1.0659	2.8892	.015490	1.4933	45
16	.34639	.93809	.36925	.7082	.0660	.8869	.015530	1.4911	44
17	.34666	.93799	.36958	.7058	.0661	.8846	.015570	1.4890	43
18	.34694	.93789	.36991	.7034	.0662	.8824	.015609	1.4869	42
19	.34721	.93779	.37024	.7009	.0663	.8801	.015649	1.4847	41
20	0.34748	0.93769	0.37057	2.6985	1.0665	2.8779	.015689	1.4826	40
21	.34775	.93759	.37090	.6961	.0666	.8756	.015729	1.4805	39
22	.34803	.93748	.37123	.6937	.0667	.8733	.015769	1.4784	38
23	.34830	.93738	.37157	.6913	.0668	.8711	.015809	1.4763	37
24	.34857	.93728	.37190	.6889	.0669	.8688	.015849	1.4742	36
25	0.34884	0.93718	0.37223	2.6865	1.0670	2.8666	.015890	1.4721	35
26	.34912	.93708	.37256	.6841	.0671	.8644	.015930	1.4700	34
27	.34939	.93698	.37289	.6818	.0673	.8621	.015971	1.4679	33
28	.34966	.93688	.37322	.6794	.0674	.8599	.016011	1.4658	32
29	.34993	.93677	.37355	.6770	.0675	.8577	.016052	1.4637	31
30	0.35021	0.93667	0.37388	2.6746	1.0676	2.8555	.016092	1.4616	30
31	.35048	.93657	.37422	.6723	.0677	.8532	.016133	1.4595	29
32	.35075	.93647	.37455	.6699	.0678	.8510	.016174	1.4575	28
33	.35102	.93637	.37488	.6675	.0680	.8488	.016214	1.4554	27
34	.35130	.93626	.37521	.6652	.0681	.8466	.016255	1.4533	26
35	0.35157	0.93616	0.37554	2.6628	1.0682	2.8444	.016296	1.4513	25
36	.35184	.93606	.37588	.6605	.0683	.8422	.016337	1.4492	24
37	.35211	.93596	.37621	.6581	.0684	.8400	.016379	1.4471	23
38	.35239	.93585	.37654	.6558	.0685	.8378	.016420	1.4451	22
39	.35266	.93575	.37687	.6534	.0687	.8356	.016461	1.4430	21
40	0.35293	0.93565	0.37720	2.6511	1.0688	2.8334	.016502	1.4410	20
41	.35320	.93555	.37754	.6488	.0689	.8312	.016544	1.4389	19
42	.35347	.93544	.37787	.6464	.0690	.8291	.016585	1.4369	18
43	.35375	.93534	.37820	.6441	.0691	.8269	.016627	1.4349	17
44	.35402	.93524	.37853	.6418	.0692	.8247	.016669	1.4328	16
45	0.35429	0.93514	0.37887	2.6395	1.0694	2.8225	.016710	1.4308	15
46	.35456	.93503	.37920	.6371	.0695	.8204	.016752	1.4288	14
47	.35484	.93493	.37953	.6348	.0696	.8182	.016794	1.4268	13
48	.35511	.93483	.37986	.6325	.0697	.8161	.016836	1.4248	12
49	.35538	.93472	.38020	.6302	.0698	.8139	.016878	1.4227	11
50	0.35565	0.93462	0.38053	2.6279	1.0700	2.8117	.016920	1.4207	10
51	.35592	.93452	.38086	.6256	.0701	.8096	.016962	1.4187	9
52	.35619	.93441	.38120	.6233	.0702	.8075	.017004	1.4167	8
53	.35647	.93431	.38153	.6210	.0703	.8053	.017047	1.4147	7
54	.35674	.93420	.38186	.6187	.0704	.8032	.017089	1.4127	6
55	0.35701	0.93410	0.38220	2.6165	1.0705	2.8010	.017132	1.4107	5
56	.35728	.93400	.38253	.6142	.0707	.7989	.017174	1.4087	4
57	.35755	.93389	.38286	.6119	.0708	.7968	.017217	1.4067	3
58	.35782	.93379	.38320	.6096	.0709	.7947	.017259	1.4048	2
59	.35810	.93368	.38353	.6074	.0710	.7925	.017302	1.4028	1
60	0.35837	0.93358	0.38386	2.6051	1.0711	2.7904	.017345	1.4008	0
M	Cosine	Sine	Cotan.	Tan.	Cosec.	Secant	READ DOWN	69°–70° Involute	M

21° or 201°   Trigonometric and Involute Functions   **158° or 338°**

M	Sine	Cosine	Tan.	Cotan.	Secant	Cosec.	Involute 21°–22°	READ UP	M
0	0.35837	0.93358	0.38386	2.6051	1.0711	2.7904	.017345	1.4008	60
1	.35864	.93348	.38420	.6028	.0713	.7883	.017388	1.3988	59
2	.35891	.93337	.38453	.6006	.0714	.7862	.017431	1.3969	58
3	.35918	.93327	.38487	.5983	.0715	.7841	.017474	1.3949	57
4	.35945	.93316	.38520	.5961	.0716	.7820	.017517	1.3929	56
5	0.35973	0.93306	0.38553	2.5938	1.0717	2.7799	.017560	1.3910	55
6	.36000	.93295	.38587	.5916	.0719	.7778	.017603	1.3890	54
7	.36027	.93285	.38620	.5893	.0720	.7757	.017647	1.3871	53
8	.36054	.93274	.38654	.5871	.0721	.7736	.017690	1.3851	52
9	.36081	.93264	.38687	.5848	.0722	.7715	.017734	1.3832	51
10	0.36108	0.93253	0.38721	2.5826	1.0723	2.7695	.017777	1.3812	50
11	.36135	.93243	.38754	.5804	.0725	.7674	.017821	1.3793	49
12	.36162	.93232	.38787	.5782	.0726	.7653	.017865	1.3774	48
13	.36190	.93222	.38821	.5759	.0727	.7632	.017908	1.3754	47
14	.36217	.93211	.38854	.5737	.0728	.7612	.017952	1.3735	46
15	0.36244	0.93201	0.38888	2.5715	1.0730	2.7591	.017996	1.3716	45
16	.36271	.93190	.38921	.5693	.0731	.7570	.018040	1.3697	44
17	.36298	.93180	.38955	.5671	.0732	.7550	.018084	1.3677	43
18	.36325	.93169	.38988	.5649	.0733	.7529	.018129	1.3658	42
19	.36352	.93159	.39022	.5627	.0734	.7509	.018173	1.3639	41
20	0.36379	0.93148	0.39055	2.5605	1.0736	2.7488	.018217	1.3620	40
21	.36406	.93137	.39089	.5583	.0737	.7468	.018262	1.3601	39
22	.36434	.93127	.39122	.5561	.0738	.7447	.018306	1.3582	38
23	.36461	.93116	.39156	.5539	.0739	.7427	.018351	1.3563	37
24	.36488	.93106	.39190	.5517	.0740	.7407	.018395	1.3544	36
25	0.36515	0.93095	0.39223	2.5495	1.0742	2.7386	.018440	1.3525	35
26	.36542	.93084	.39257	.5473	.0743	.7366	.018485	1.3506	34
27	.36569	.93074	.39290	.5452	.0744	.7346	.018530	1.3487	33
28	.36596	.93063	.39324	.5430	.0745	.7325	.018575	1.3469	32
29	.36623	.93052	.39357	.5408	.0747	.7305	.018620	1.3450	31
30	0.36650	0.93042	0.39391	2.5386	1.0748	2.7285	.018665	1.3431	30
31	.36677	.93031	.39425	.5365	.0749	.7265	.018710	1.3412	29
32	.36704	.93020	.39458	.5343	.0750	.7245	.018755	1.3394	28
33	.36731	.93010	.39492	.5322	.0752	.7225	.018800	1.3375	27
34	.36758	.92999	.39526	.5300	.0753	.7205	.018846	1.3356	26
35	0.36785	0.92988	0.39559	2.5279	1.0754	2.7185	.018891	1.3338	25
36	.36812	.92978	.39593	.5257	.0755	.7165	.018937	1.3319	24
37	.36839	.92967	.39626	.5236	.0757	.7145	.018983	1.3301	23
38	.36867	.92956	.39660	.5214	.0758	.7125	.019028	1.3282	22
39	.36894	.92945	.39694	.5193	.0759	.7105	.019074	1.3264	21
40	0.36921	0.92935	0.39727	2.5172	1.0760	2.7085	.019120	1.3245	20
41	.36948	.92924	.39761	.5150	.0761	.7065	.019166	1.3227	19
42	.36975	.92913	.39795	.5129	.0763	.7046	.019212	1.3208	18
43	.37002	.92902	.39829	.5108	.0764	.7026	.019258	1.3190	17
44	.37029	.92892	.39862	.5086	.0765	.7006	.019304	1.3172	16
45	0.37056	0.92881	0.39896	2.5065	1.0766	2.6986	.019350	1.3153	15
46	.37083	.92870	.39930	.5044	.0768	.6967	.019397	1.3135	14
47	.37110	.92859	.39963	.5023	.0769	.6947	.019443	1.3117	13
48	.37137	.92849	.39997	.5002	.0770	.6927	.019490	1.3099	12
49	.37164	.92838	.40031	.4981	.0771	.6908	.019536	1.3080	11
50	0.37191	0.92827	0.40065	2.4960	1.0773	2.6888	.019583	1.3062	10
51	.37218	.92816	.40098	.4939	.0774	.6869	.019630	1.3044	9
52	.37245	.92805	.40132	.4918	.0775	.6849	.019676	1.3026	8
53	.37272	.92794	.40166	.4897	.0777	.6830	.019723	1.3008	7
54	.37299	.92784	.40200	.4876	.0778	.6811	.019770	1.2990	6
55	0.37326	0.92773	0.40234	2.4855	1.0779	2.6791	.019817	1.2972	5
56	.37353	.92762	.40267	.4834	.0780	.6772	.019864	1.2954	4
57	.37380	.92751	.40301	.4813	.0782	.6752	.019912	1.2936	3
58	.37407	.92740	.40335	.4792	.0783	.6733	.019959	1.2918	2
59	.37434	.92729	.40369	.4771	.0784	.6714	.020006	1.2900	1
60	0.37461	0.92718	0.40403	2.4751	1.0785	2.6695	.020054	1.2883	0
M	Cosine	Sine	Cotan.	Tan.	Cosec.	Secant	READ DOWN	68°–69° Involute	M

22° or 202° Trigonometric and Involute Functions 157° or 337°

M	Sine	Cosine	Tan.	Cotan.	Secant	Cosec.	Involute 22°–23°	READ UP	M
0	0.37461	0.92718	0.40403	2.4751	1.0785	2.6695	.020054	1.2883	60
1	.37488	.92707	.40436	.4730	.0787	.6675	.020101	1.2865	59
2	.37515	.92697	.40470	.4709	.0788	.6656	.020149	1.2847	58
3	.37542	.92686	.40504	.4689	.0789	.6637	.020197	1.2829	57
4	.37569	.92675	.40538	.4668	.0790	.6618	.020244	1.2812	56
5	0.37595	.92664	0.40572	2.4648	1.0792	2.6599	.020292	1.2794	55
6	.37622	.92653	.40606	.4627	.0793	.6580	.020340	1.2776	54
7	.37649	.92642	.40640	.4606	.0794	.6561	.020388	1.2759	53
8	.37676	.92631	.40674	.4586	.0796	.6542	.020436	1.2741	52
9	.37703	.92620	.40707	.4566	.0797	.6523	.020484	1.2723	51
10	0.37730	0.92609	0.40741	2.4545	1.0798	2.6504	.020533	1.2706	50
11	.37757	.92598	.40775	.4525	.0799	.6485	.020581	1.2688	49
12	.37784	.92587	.40809	.4504	.0801	.6466	.020629	1.2671	48
13	.37811	.92576	.40843	.4484	.0802	.6447	.020678	1.2653	47
14	.37838	.92565	.40877	.4464	.0803	.6429	.020726	1.2636	46
15	0.37865	0.92554	0.40911	2.4443	1.0804	2.6410	.020775	1.2619	45
16	.37892	.92543	.40945	.4423	.0806	.6391	.020824	1.2601	44
17	.37919	.92532	.40979	.4403	.0807	.6372	.020873	1.2584	43
18	.37946	.92521	.41013	.4383	.0808	.6354	.020921	1.2567	42
19	.37973	.92510	.41047	.4362	.0810	.6335	.020970	1.2549	41
20	0.37999	0.92499	0.41081	2.4342	1.0811	2.6316	.021019	1.2532	40
21	.38026	.92488	.41115	.4322	.0812	.6298	.021069	1.2515	39
22	.38053	.92477	.41149	.4302	.0814	.6279	.021118	1.2498	38
23	.38080	.92466	.41183	.4282	.0815	.6260	.021167	1.2481	37
24	.38107	.92455	.41217	.4262	.0816	.6242	.021217	1.2463	36
25	0.38134	0.92443	0.41251	2.4242	1.0817	2.6223	.021266	1.2446	35
26	.38161	.92432	.41285	.4222	.0819	.6205	.021316	1.2429	34
27	.38188	.92421	.41319	.4202	.0820	.6186	.021365	1.2412	33
28	.38215	.92410	.41353	.4182	.0821	.6168	.021415	1.2395	32
29	.38241	.92399	.41387	.4162	.0823	.6150	.021465	1.2378	31
30	0.38268	0.92388	0.41421	2.4142	1.0824	2.6131	.021514	1.2361	30
31	.38295	.92377	.41455	.4122	.0825	.6113	.021564	1.2344	29
32	.38322	.92366	.41490	.4102	.0827	.6095	.021614	1.2327	28
33	.38349	.92355	.41524	.4083	.0828	.6076	.021665	1.2310	27
34	.38376	.92343	.41558	.4063	.0829	.6058	.021715	1.2294	26
35	0.38403	0.92332	0.41592	2.4043	1.0830	2.6040	.021765	1.2277	25
36	.38430	.92321	.41626	.4023	.0832	.6022	.021815	1.2260	24
37	.38456	.92310	.41660	.4004	.0833	.6003	.021866	1.2243	23
38	.38483	.92299	.41694	.3984	.0834	.5985	.021916	1.2226	22
39	.38510	.92287	.41728	.3964	.0836	.5967	.021967	1.2210	21
40	0.38537	0.92276	0.41763	2.3945	1.0837	2.5949	.022018	1.2193	20
41	.38564	.92265	.41797	.3925	.0838	.5931	.022068	1.2176	19
42	.38591	.92254	.41831	.3906	.0840	.5913	.022119	1.2160	18
43	.38617	.92243	.41865	.3886	.0841	.5895	.022170	1.2143	17
44	.38644	.92231	.41899	.3867	.0842	.5877	.022221	1.2127	16
45	0.38671	0.92220	0.41933	2.3847	1.0844	2.5859	.022272	1.2110	15
46	.38698	.92209	.41968	.3828	.0845	.5841	.022324	1.2093	14
47	.38725	.92198	.42002	.3808	.0846	.5823	.022375	1.2077	13
48	.38752	.92186	.42036	.3789	.0848	.5805	.022426	1.2060	12
49	.38778	.92175	.42070	.3770	.0849	.5788	.022478	1.2044	11
50	0.38805	0.92164	0.42105	2.3750	1.0850	2.5770	.022529	1.2028	10
51	.38832	.92152	.42139	.3731	.0852	.5752	.022581	1.2011	9
52	.38859	.92141	.42173	.3712	.0853	.5734	.022632	1.1995	8
53	.38886	.92130	.42207	.3692	.0854	.5716	.022684	1.1978	7
54	.38912	.92119	.42242	.3673	.0856	.5699	.022736	1.1962	6
55	0.38939	0.92107	0.42276	2.3654	1.0857	2.5681	.022788	1.1946	5
56	.38966	.92096	.42310	.3635	.0858	.5663	.022840	1.1930	4
57	.38993	.92085	.42345	.3616	.0860	.5646	.022892	1.1913	3
58	.39020	.92073	.42379	.3597	.0861	.5628	.022944	1.1897	2
59	.39046	.92062	.42413	.3578	.0862	.5611	.022997	1.1881	1
60	0.39073	0.92050	0.42447	2.3559	1.0864	2.5593	.023049	1.1865	0
M	Cosine	Sine	Cotan.	Tan.	Cosec.	Secant	READ DOWN	67°–68° Involute	M

112° or 292°   **67° or 247°**

23° or 203°   **Trigonometric and Involute Functions**   **156° or 336°**

M	Sine	Cosine	Tan.	Cotan.	Secant	Cosec.	Involute 23°–24°	READ UP	M
0	0.39073	0.92050	0.42447	2.3559	1.0864	2.5593	.023049	1.1865	60
1	.39100	.92039	.42482	.3539	.0865	.5576	.023102	1.1849	59
2	.39127	.92028	.42516	.3520	.0866	.5558	.023154	1.1833	58
3	.39153	.92016	.42551	.3501	.0868	.5541	.023207	1.1817	57
4	.39180	.92005	.42585	.3483	.0869	.5523	.023259	1.1800	56
5	0.39207	0.91994	0.42619	2.3464	1.0870	2.5506	.023312	1.1784	55
6	.39234	.91982	.42654	.3445	.0872	.5488	.023365	1.1768	54
7	.39260	.91971	.42688	.3426	.0873	.5471	.023418	1.1752	53
8	.39287	.91959	.42722	.3407	.0874	.5454	.023471	1.1736	52
9	.39314	.91948	.42757	.3388	.0876	.5436	.023524	1.1721	51
10	0.39341	0.91936	0.42791	2.3369	1.0877	2.5419	.023577	1.1705	50
11	.39367	.91925	.42826	.3351	.0878	.5402	.023631	1.1689	49
12	.39394	.91914	.42860	.3332	.0880	.5384	.023684	1.1673	48
13	.39421	.91902	.42894	.3313	.0881	.5367	.023738	1.1657	47
14	.39448	.91891	.42929	.3294	.0883	.5350	.023791	1.1641	46
15	0.39474	0.91879	0.42963	2.3276	1.0884	2.5333	.023845	1.1626	45
16	.39501	.91868	.42998	.3257	.0885	.5316	.023899	1.1610	44
17	.39528	.91856	.43032	.3238	.0887	.5299	.023952	1.1594	43
18	.39555	.91845	.43067	.3220	.0888	.5282	.024006	1.1578	42
19	.39581	.91833	.43101	.3201	.0889	.5264	.024060	1.1563	41
20	0.39608	0.91822	0.43136	2.3183	1.0891	2.5247	.024114	1.1547	40
21	.39635	.91810	.43170	.3164	.0892	.5230	.024169	1.1531	39
22	.39661	.91799	.43205	.3146	.0893	.5213	.024223	1.1516	38
23	.39688	.91787	.43239	.3127	.0895	.5196	.024277	1.1500	37
24	.39715	.91775	.43274	.3109	.0896	.5180	.024332	1.1485	36
25	0.39741	0.91764	0.43308	2.3090	1.0898	2.5163	.024386	1.1469	35
26	.39768	.91752	.43343	.3072	.0899	.5146	.024441	1.1454	34
27	.39795	.91741	.43378	.3053	.0900	.5129	.024495	1.1438	33
28	.39822	.91729	.43412	.3035	.0902	.5112	.024550	1.1423	32
29	.39848	.91718	.43447	.3017	.0903	.5095	.024605	1.1407	31
30	0.39875	0.91706	0.43481	2.2998	1.0904	2.5078	.024660	1.1392	30
31	.39902	.91694	.43516	.2980	.0906	.5062	.024715	1.1377	29
32	.39928	.91683	.43550	.2962	.0907	.5045	.024770	1.1361	28
33	.39955	.91671	.43585	.2944	.0909	.5028	.024825	1.1346	27
34	.39982	.91660	.43620	.2925	.0910	.5012	.024881	1.1331	26
35	0.40008	0.91648	0.43654	2.2907	1.0911	2.4995	.024936	1.1315	25
36	.40035	.91636	.43689	.2889	.0913	.4978	.024992	1.1300	24
37	.40062	.91625	.43724	.2871	.0914	.4962	.025047	1.1285	23
38	.40088	.91613	.43758	.2853	.0915	.4945	.025103	1.1270	22
39	.40115	.91601	.43793	.2835	.0917	.4928	.025159	1.1254	21
40	0.40141	0.91590	0.43828	2.2817	1.0918	2.4912	.025214	1.1239	20
41	.40168	.91578	.43862	.2799	.0920	.4895	.025270	1.1224	19
42	.40195	.91566	.43897	.2781	.0921	.4879	.025326	1.1209	18
43	.40221	.91555	.43932	.2763	.0922	.4862	.025382	1.1194	17
44	.40248	.91543	.43966	.2745	.0924	.4846	.025439	1.1179	16
45	0.40275	0.91531	0.44001	2.2727	1.0925	2.4830	.025495	1.1164	15
46	.40301	.91519	.44036	.2709	.0927	.4813	.025551	1.1149	14
47	.40328	.91508	.44071	.2691	.0928	.4797	.025608	1.1134	13
48	.40355	.91496	.44105	.2673	.0929	.4780	.025664	1.1119	12
49	.40381	.91484	.44140	.2655	.0931	.4764	.025721	1.1104	11
50	0.40408	0.91472	0.44175	2.2637	1.0932	2.4748	.025778	1.1089	10
51	.40434	.91461	.44210	.2620	.0934	.4731	.025834	1.1074	9
52	.40461	.91449	.44244	.2602	.0935	.4715	.025891	1.1059	8
53	.40488	.91437	.44279	.2584	.0936	.4699	.025948	1.1044	7
54	.40514	.91425	.44314	.2566	.0938	.4683	.026005	1.1030	6
55	0.40541	0.91414	0.44349	2.2549	1.0939	2.4667	.026062	1.1015	5
56	.40567	.91402	.44384	.2531	.0941	.4650	.026120	1.1000	4
57	.40594	.91390	.44418	.2513	.0942	.4634	.026177	1.0985	3
58	.40621	.91378	.44453	.2496	.0944	.4618	.026235	1.0971	2
59	.40647	.91366	.44488	.2478	.0945	.4602	.026292	1.0956	1
60	0.40674	0.91355	0.44523	2.2460	1.0946	2.4586	.026350	1.0941	0
M	Cosine	Sine	Cotan.	Tan.	Cosec.	Secant	READ DOWN	66°–67° Involute	M

24° or 204°   Trigonometric and Involute Functions   155° or 335°

M	Sine	Cosine	Tan.	Cotan.	Secant	Cosec.	Involute 24°–25°	READ UP	M
0	0.40674	0.91355	0.44523	2.2460	1.0946	2.4586	.026350	1.0941	60
1	.40700	.91343	.44558	.2443	.0948	.4570	.026407	1.0927	59
2	.40727	.91331	.44593	.2425	.0949	.4554	.026465	1.0912	58
3	.40753	.91319	.44627	.2408	.0951	.4538	.026523	1.0897	57
4	.40780	.91307	.44662	.2390	.0952	.4522	.026581	1.0883	56
5	0.40806	0.91295	0.44697	2.2373	1.0953	2.4506	.026639	1.0868	55
6	.40833	.91283	.44732	.2355	.0955	.4490	.026697	1.0854	54
7	.40860	.91272	.44767	.2338	.0956	.4474	.026756	1.0839	53
8	.40886	.91260	.44802	.2320	.0958	.4458	.026814	1.0825	52
9	.40913	.91248	.44837	.2303	.0959	.4442	.026872	1.0810	51
10	.40939	0.91236	0.44872	2.2286	1.0961	2.4426	.026931	1.0796	50
11	.40966	.91224	.44907	.2268	.0962	.4411	.026989	1.0781	49
12	.40992	.91212	.44942	.2251	.0963	.4395	.027048	1.0767	48
13	.41019	.91200	.44977	.2234	.0965	.4379	.027107	1.0752	47
14	.41045	.91188	.45012	.2216	.0966	.4363	.027166	1.0738	46
15	0.41072	0.91176	0.45047	2.2199	1.0968	2.4348	.027225	1.0724	45
16	.41098	.91164	.45082	.2182	.0969	.4332	.027284	1.0709	44
17	.41125	.91152	.45117	.2165	.0971	.4316	.027343	1.0695	43
18	.41151	.91140	.45152	.2148	.0972	.4300	.027402	1.0681	42
19	.41178	.91128	.45187	.2130	.0974	.4285	.027462	1.0666	41
20	0.41204	0.91116	0.45222	2.2113	1.0975	2.4269	.027521	1.0652	40
21	.41231	.91104	.45257	.2096	.0976	.4254	.027581	1.0638	39
22	.41257	.91092	.45292	.2079	.0978	.4238	.027640	1.0624	38
23	.41284	.91080	.45327	.2062	.0979	.4222	.027700	1.0610	37
24	.41310	.91068	.45362	.2045	.0981	.4207	.027760	1.0596	36
25	0.41337	0.91056	0.45397	2.2028	1.0982	2.4191	.027820	1.0581	35
26	.41363	.91044	.45432	.2011	.0984	.4176	.027880	1.0567	34
27	.41390	.91032	.45467	.1994	.0985	.4160	.027940	1.0553	33
28	.41416	.91020	.45502	.1977	.0987	.4145	.028000	1.0539	32
29	.41443	.91008	.45538	.1960	.0988	.4130	.028060	1.0525	31
30	0.41469	0.90996	0.45573	2.1943	1.0989	2.4114	.028121	1.0511	30
31	.41496	.90984	.45608	.1926	.0991	.4099	.028181	1.0497	29
32	.41522	.90972	.45643	.1909	.0992	.4083	.028242	1.0483	28
33	.41549	.90960	.45678	.1892	.0994	.4068	.028302	1.0469	27
34	.41575	.90948	.45713	.1876	.0995	.4053	.028363	1.0455	26
35	0.41602	0.90936	0.45748	2.1859	1.0997	2.4038	.028424	1.0441	25
36	.41628	.90924	.45784	.1842	.0998	.4022	.028485	1.0427	24
37	.41655	.90911	.45819	.1825	.1000	.4007	.028546	1.0414	23
38	.41681	.90899	.45854	.1808	.1001	.3992	.028607	1.0400	22
39	.41707	.90887	.45889	.1792	.1003	.3977	.028668	1.0386	21
40	0.41734	0.90875	0.45924	2.1775	1.1004	2.3961	.028729	1.0372	20
41	.41760	.90863	.45960	.1758	.1006	.3946	.028791	1.0358	19
42	.41787	.90851	.45995	.1742	.1007	.3931	.028852	1.0345	18
43	.41813	.90839	.46030	.1725	.1009	.3916	.028914	1.0331	17
44	.41840	.90826	.46065	.1708	.1010	.3901	.028976	1.0317	16
45	0.41866	0.90814	0.46101	2.1692	1.1011	2.3886	.029037	1.0303	15
46	.41892	.90802	.46136	.1675	.1013	.3871	.029099	1.0290	14
47	.41919	.90790	.46171	.1659	.1014	.3856	.029161	1.0276	13
48	.41945	.90778	.46206	.1642	.1016	.3841	.029223	1.0262	12
49	.41972	.90766	.46242	.1625	.1017	.3826	.029285	1.0249	11
50	0.41998	0.90753	0.46277	2.1609	1.1019	2.3811	.029348	1.0235	10
51	.42024	.90741	.46312	.1592	.1020	.3796	.029410	1.0222	9
52	.42051	.90729	.46348	.1576	.1022	.3781	.029472	1.0208	8
53	.42077	.90717	.46383	.1560	.1023	.3766	.029535	1.0195	7
54	.42104	.90704	.46418	.1543	.1025	.3751	.029598	1.0181	6
55	0.42130	0.90692	0.46454	2.1527	1.1026	2.3736	.029660	1.0168	5
56	.42156	.90680	.46489	.1510	.1028	.3721	.029723	1.0154	4
57	.42183	.90668	.46525	.1494	.1029	.3706	.029786	1.0141	3
58	.42209	.90655	.46560	.1478	.1031	.3692	.029849	1.0127	2
59	.42235	.90643	.46595	.1461	.1032	.3677	.029912	1.0114	1
60	0.42262	0.90631	0.46631	2.1445	1.1034	2.3662	.029975	1.0100	0
M	Cosine	Sine	Cotan.	Tan.	Cosec.	Secant	READ DOWN	65°–66° Involute	M

114° or 294°   **65° or 245°**

25° or 205°   **Trigonometric and Involute Functions**   **154° or 334°**

M	Sine	Cosine	Tan.	Cotan.	Secant	Cosec.	Involute 25°-26°	READ UP	M
0	0.42262	0.90631	0.46631	2.1445	1.1034	2.3662	.029975	1.0100	60
1	.42288	.90618	.46666	.1429	.1035	.3647	.030039	1.0087	59
2	.42315	.90606	.46702	.1413	.1037	.3633	.030102	1.0074	58
3	.42341	.90594	.46737	.1396	.1038	.3618	.030166	1.0060	57
4	.42367	.90582	.46772	.1380	.1040	.3603	.030229	1.0047	56
5	0.42394	0.90569	0.46808	2.1364	1.1041	2.3588	.030293	1.0034	55
6	.42420	.90557	.46843	.1348	.1043	.3574	.030357	1.0021	54
7	.42446	.90545	.46879	.1332	.1044	.3559	.030420	1.0007	53
8	.42473	.90532	.46914	.1315	.1046	.3545	.030484	0.9994	52
9	.42499	.90520	.46950	.1299	.1047	.3530	.030549	0.9981	51
10	0.42525	0.90507	0.46985	2.1283	1.1049	2.3515	.030613	0.9968	50
11	.42552	.90495	.47021	.1267	.1050	.3501	.030677	0.9954	49
12	.42578	.90483	.47056	.1251	.1052	.3486	.030741	0.9941	48
13	.42604	.90470	.47092	.1235	.1053	.3472	.030806	0.9928	47
14	.42631	.90458	.47128	.1219	.1055	.3457	.030870	0.9915	46
15	0.42657	0.90446	0.47163	2.1203	1.1056	2.3443	.030935	0.9902	45
16	.42683	.90433	.47199	.1187	.1058	.3428	.031000	0.9889	44
17	.42709	.90421	.47234	.1171	.1059	.3414	.031065	0.9876	43
18	.42736	.90408	.47270	.1155	.1061	.3400	.031130	0.9863	42
19	.42762	.90396	.47305	.1139	.1062	.3385	.031195	0.9850	41
20	0.42788	0.90383	0.47341	2.1123	1.1064	2.3371	.031260	0.9837	40
21	.42815	.90371	.47377	.1107	.1066	.3356	.031325	0.9824	39
22	.42841	.90358	.47412	.1092	.1067	.3342	.031390	0.9811	38
23	.42867	.90346	.47448	.1076	.1069	.3328	.031456	0.9798	37
24	.42894	.90334	.47483	.1060	.1070	.3314	.031521	0.9785	36
25	0.42920	0.90321	0.47519	2.1044	1.1072	2.3299	.031587	0.9772	35
26	.42946	.90309	.47555	.1028	.1073	.3285	.031653	0.9759	34
27	.42972	.90296	.47590	.1013	.1075	.3271	.031718	0.9747	33
28	.42999	.90284	.47626	.0997	.1076	.3257	.031784	0.9734	32
29	.43025	.90271	.47662	.0981	.1078	.3242	.031850	0.9721	31
30	0.43051	0.90259	0.47698	2.0965	1.1079	2.3228	.031917	0.9708	30
31	.43077	.90246	.47733	.0950	.1081	.3214	.031983	0.9695	29
32	.43104	.90233	.47769	.0934	.1082	.3200	.032049	0.9683	28
33	.43130	.90221	.47805	.0918	.1084	.3186	.032116	0.9670	27
34	.43156	.90208	.47840	.0903	.1085	.3172	.032182	0.9657	26
35	0.43182	0.90196	0.47876	2.0887	1.1087	2.3158	.032249	0.9644	25
36	.43209	.90183	.47912	.0872	.1089	.3144	.032315	0.9632	24
37	.43235	.90171	.47948	.0856	.1090	.3130	.032382	0.9619	23
38	.43261	.90158	.47984	.0840	.1092	.3115	.032449	0.9606	22
39	.43287	.90146	.48019	.0825	.1093	.3101	.032516	0.9594	21
40	0.43313	0.90133	0.48055	2.0809	1.1095	2.3088	.032583	0.9581	20
41	.43340	.90120	.48091	.0794	.1096	.3074	.032651	0.9569	19
42	.43366	.90108	.48127	.0778	.1098	.3060	.032718	0.9556	18
43	.43392	.90095	.48163	.0763	.1099	.3046	.032785	0.9543	17
44	.43418	.90082	.48198	.0748	.1101	.3032	.032853	0.9531	16
45	0.43445	0.90070	0.48234	2.0732	1.1102	2.3018	.032920	0.9518	15
46	.43471	.90057	.48270	.0717	.1104	.3004	.032988	0.9506	14
47	.43497	.90044	.48306	.0701	.1106	.2990	.033056	0.9493	13
48	.43523	.90032	.48342	.0686	.1107	.2976	.033124	0.9481	12
49	.43549	.90019	.48378	.0671	.1109	.2962	.033192	0.9469	11
50	0.43575	0.90007	0.48414	2.0655	1.1110	2.2949	.033260	0.9456	10
51	.43602	.89994	.48450	.0640	.1112	.2935	.033328	0.9444	9
52	.43628	.89981	.48486	.0625	.1113	.2921	.033397	0.9431	8
53	.43654	.89968	.48521	.0609	.1115	.2907	.033465	0.9419	7
54	.43680	.89956	.48557	.0594	.1117	.2894	.033534	0.9407	6
55	0.43706	0.89943	0.48593	2.0579	1.1118	2.2880	.033602	0.9394	5
56	.43733	.89930	.48629	.0564	.1120	.2866	.033671	0.9382	4
57	.43759	.89918	.48665	.0549	.1121	.2853	.033740	0.9370	3
58	.43785	.89905	.48701	.0533	.1123	.2839	.033809	0.9357	2
59	.43811	.89892	.48737	.0518	.1124	.2825	.033878	0.9345	1
60	0.43837	0.89879	0.48773	2.0503	1.1126	2.2812	.033947	0.9333	0
M	Cosine	Sine	Cotan.	Tan.	Cosec.	Secant	READ DOWN	64°-65° Involute	M

26° or 206°   Trigonometric and Involute Functions   153° or 333°

M	Sine	Cosine	Tan.	Cotan.	Secant	Cosec.	Involute 26°–27°	READ UP	M
0	0.43837	0.89879	0.48773	2.0503	1.1126	2.2812	.033947	.93329	60
1	.43863	.89867	.48809	.0488	.1128	.2798	.034016	.93207	59
2	.43889	.89854	.48845	.0473	.1129	.2785	.034086	.93085	58
3	.43916	.89841	.48881	.0458	.1131	.2771	.034155	.92963	57
4	.43942	.89828	.48917	.0443	.1132	.2757	.034225	.92842	56
5	0.43968	0.89816	0.48953	2.0428	1.1134	2.2744	.034294	.92720	55
6	.43994	.89803	.48989	.0413	.1136	.2730	.034364	.92599	54
7	.44020	.89790	.49026	.0398	.1137	.2717	.034434	.92478	53
8	.44046	.89777	.49062	.0383	.1139	.2703	.034504	.92357	52
9	.44072	.89764	.49098	.0368	.1140	.2690	.034574	.92236	51
10	0.44098	0.89752	0.49134	2.0353	1.1142	2.2677	.034644	.92115	50
11	.44124	.89739	.49170	.0338	.1143	.2663	.034714	.91995	49
12	.44151	.89726	.49206	.0323	.1145	.2650	.034785	.91875	48
13	.44177	.89713	.49242	.0308	.1147	.2636	.034855	.91755	47
14	.44203	.89700	.49278	.0293	.1148	.2623	.034926	.91635	46
15	0.44229	0.89687	0.49315	2.0278	1.1150	2.2610	.034996	.91515	45
16	.44255	.89674	.49351	.0263	.1151	.2596	.035067	.91396	44
17	.44281	.89662	.49387	.0248	.1153	.2583	.035138	.91276	43
18	.44307	.89649	.49423	.0233	.1155	.2570	.035209	.91157	42
19	.44333	.89636	.49459	.0219	.1156	.2556	.035280	.91038	41
20	0.44359	0.89623	0.49495	2.0204	1.1158	2.2543	.035352	.90919	40
21	.44385	.89610	.49532	.0189	.1159	.2530	.035423	.90801	39
22	.44411	.89597	.49568	.0174	.1161	.2517	.035494	.90682	38
23	.44437	.89584	.49604	.0160	.1163	.2504	.035566	.90564	37
24	.44464	.89571	.49640	.0145	.1164	.2490	.035637	.90446	36
25	0.44490	0.89558	0.49677	2.0130	1.1166	2.2477	.035709	.90328	35
26	.44516	.89545	.49713	.0115	.1168	.2464	.035781	.90210	34
27	.44542	.89532	.49749	.0101	.1169	.2451	.035853	.90092	33
28	.44568	.89519	.49786	.0086	.1171	.2438	.035925	.89975	32
29	.44594	.89506	.49822	.0072	.1172	.2425	.035997	.89858	31
30	0.44620	0.89493	0.49858	2.0057	1.1174	2.2412	.036069	.89741	30
31	.44646	.89480	.49894	.0042	.1176	.2399	.036142	.89624	29
32	.44672	.89467	.49931	.0028	.1177	.2385	.036214	.89507	28
33	.44698	.89454	.49967	.0013	.1179	.2372	.036287	.89390	27
34	.44724	.89441	.50004	1.9999	.1180	.2359	.036359	.89274	26
35	0.44750	0.89428	0.50040	1.9984	1.1182	2.2346	.036432	.89158	25
36	.44776	.89415	.50076	.9970	.1184	.2333	.036505	.89042	24
37	.44802	.89402	.50113	.9955	.1185	.2320	.036578	.88926	23
38	.44828	.89389	.50149	.9941	.1187	.2308	.036651	.88810	22
39	.44854	.89376	.50185	.9926	.1189	.2295	.036724	.88694	21
40	0.44880	0.89363	0.50222	1.9912	1.1190	2.2282	.036798	.88579	20
41	.44906	.89350	.50258	.9897	.1192	.2269	.036871	.88464	19
42	.44932	.89337	.50295	.9883	.1194	.2256	.036945	.88349	18
43	.44958	.89324	.50331	.9868	.1195	.2243	.037018	.88234	17
44	.44984	.89311	.50368	.9854	.1197	.2230	.037092	.88119	16
45	0.45010	0.89298	0.50404	1.9840	1.1198	2.2217	.037166	.88004	15
46	.45036	.89285	.50441	.9825	.1200	.2205	.037240	.87890	14
47	.45062	.89272	.50477	.9811	.1202	.2192	.037314	.87776	13
48	.45088	.89259	.50514	.9797	.1203	.2179	.037388	.87662	12
49	.45114	.89245	.50550	.9782	.1205	.2166	.037462	.87548	11
50	0.45140	0.89232	0.50587	1.9768	1.1207	2.2153	.037537	.87434	10
51	.45166	.89219	.50623	.9754	.1208	.2141	.037611	.87320	9
52	.45192	.89206	.50660	.9740	.1210	.2128	.037686	.87207	8
53	.45218	.89193	.50696	.9725	.1212	.2115	.037761	.87094	7
54	.45243	.89180	.50733	.9711	.1213	.2103	.037835	.86980	6
55	0.45269	0.89167	0.50769	1.9697	1.1215	2.2090	.037910	.86868	5
56	.45295	.89153	.50806	.9683	.1217	.2077	.037985	.86755	4
57	.45321	.89140	.50843	.9669	.1218	.2065	.038060	.86642	3
58	.45347	.89127	.50879	.9654	.1220	.2052	.038136	.86530	2
59	.45373	.89114	.50916	.9640	.1222	.2039	.038211	.86417	1
60	0.45399	0.89101	0.50953	1.9626	1.1223	2.2027	.038287	.86305	0
M	Cosine	Sine	Cotan.	Tan.	Cosec.	Secant	READ DOWN	63°–64° Involute	M

116° or 296°   63° or 243°

27° or 207° **Trigonometric and Involute Functions** **152° or 332°**

M	Sine	Cosine	Tan.	Cotan.	Secant	Cosec.	Involute 27°–28°	READ UP	M
0	0.45399	0.89101	0.50953	1.9626	1.1223	2.2027	.038287	.86305	60
1	.45425	.89087	.50989	.9612	.1225	.2014	.038362	.86193	59
2	.45451	.89074	.51026	.9598	.1227	.2002	.038438	.86082	58
3	.45477	.89061	.51063	.9584	.1228	.1989	.038514	.85970	57
4	.45503	.89048	.51099	.9570	.1230	.1977	.038590	.85858	56
5	0.45529	0.89035	0.51136	1.9556	1.1232	2.1964	.038666	.85747	55
6	.45554	.89021	.51173	.9542	.1233	.1952	.038742	.85636	54
7	.45580	.89008	.51209	.9528	.1235	.1939	.038818	.85525	53
8	.45606	.88995	.51246	.9514	.1237	.1927	.038894	.85414	52
9	.45632	.88981	.51283	.9500	.1238	.1914	.038971	.85303	51
10	0.45658	0.88968	0.51319	1.9486	1.1240	2.1902	.039047	.85193	50
11	.45684	.88955	.51356	.9472	.1242	.1890	.039124	.85082	49
12	.45710	.88942	.51393	.9458	.1243	.1877	.039201	.84972	48
13	.45736	.88928	.51430	.9444	.1245	.1865	.039278	.84862	47
14	.45762	.88915	.51467	.9430	.1247	.1852	.039355	.84752	46
15	0.45787	0.88902	0.51503	1.9416	1.1248	2.1840	.039432	.84643	45
16	.45813	.88888	.51540	.9402	.1250	.1828	.039509	.84533	44
17	.45839	.88875	.51577	.9388	.1252	.1815	.039586	.84424	43
18	.45865	.88862	.51614	.9375	.1253	.1803	.039664	.84314	42
19	.45891	.88848	.51651	.9361	.1255	.1791	.039741	.84205	41
20	0.45917	0.88835	0.51688	1.9347	1.1257	2.1779	.039819	.84096	40
21	.45942	.88822	.51724	.9333	.1259	.1766	.039897	.83987	39
22	.45968	.88808	.51761	.9319	.1260	.1754	.039974	.83879	38
23	.45994	.88795	.51798	.9306	.1262	.1742	.040052	.83770	37
24	.46020	.88782	.51835	.9292	.1264	.1730	.040131	.83662	36
25	0.46046	0.88768	0.51872	1.9278	1.1265	2.1718	.040209	.83554	35
26	.46072	.88755	.51909	.9265	.1267	.1705	.040287	.83446	34
27	.46097	.88741	.51946	.9251	.1269	.1693	.040366	.83338	33
28	.46123	.88728	.51983	.9237	.1270	.1681	.040444	.83230	32
29	.46149	.88715	.52020	.9223	.1272	.1669	.040523	.83123	31
30	0.46175	0.88701	0.52057	1.9210	1.1274	2.1657	.040602	.83015	30
31	.46201	.88688	.52094	.9196	.1276	.1645	.040680	.82908	29
32	.46226	.88674	.52131	.9183	.1277	.1633	.040759	.82801	28
33	.46252	.88661	.52168	.9169	.1279	.1621	.040838	.82694	27
34	.46278	.88647	.52205	.9155	.1281	.1609	.040918	.82587	26
35	0.46304	0.88634	0.52242	1.9142	1.1282	2.1596	.040997	.82480	25
36	.46330	.88620	.52279	.9128	.1284	.1584	.041076	.82374	24
37	.46355	.88607	.52316	.9115	.1286	.1572	.041156	.82267	23
38	.46381	.88593	.52353	.9101	.1288	.1560	.041236	.82161	22
39	.46407	.88580	.52390	.9088	.1289	.1549	.041316	.82055	21
40	0.46433	0.88566	0.52427	1.9074	1.1291	2.1537	.041395	.81949	20
41	.46458	.88553	.52464	.9061	.1293	.1525	.041475	.81844	19
42	.46484	.88539	.52501	.9047	.1294	.1513	.041556	.81738	18
43	.46510	.88526	.52538	.9034	.1296	.1501	.041636	.81632	17
44	.46536	.88512	.52575	.9020	.1298	.1489	.041716	.81527	16
45	0.46561	0.88499	0.52613	1.9007	1.1300	2.1477	.041797	.81422	15
46	.46587	.88485	.52650	.8993	.1301	.1465	.041877	.81317	14
47	.46613	.88472	.52687	.8980	.1303	.1453	.041958	.81212	13
48	.46639	.88458	.52724	.8967	.1305	.1441	.042039	.81107	12
49	.46664	.88445	.52761	.8953	.1307	.1430	.042120	.81003	11
50	0.46690	0.88431	0.52798	1.8940	1.1308	2.1418	.042201	.80898	10
51	.46716	.88417	.52836	.8927	.1310	.1406	.042282	.80794	9
52	.46742	.88404	.52873	.8913	.1312	.1394	.042363	.80690	8
53	.46767	.88390	.52910	.8900	.1313	.1382	.042444	.80586	7
54	.46793	.88377	.52947	.8887	.1315	.1371	.042526	.80482	6
55	0.46819	0.88363	0.52985	1.8873	1.1317	2.1359	.042607	.80378	5
56	.46844	.88349	.53022	.8860	.1319	.1347	.042689	.80275	4
57	.46870	.88336	.53059	.8847	.1320	.1336	.042771	.80172	3
58	.46896	.88322	.53096	.8834	.1322	.1324	.042853	.80068	2
59	.46921	.88308	.53134	.8820	.1324	.1312	.042935	.79965	1
60	0.46947	0.88295	0.53171	1.8807	1.1326	2.1301	.043017	.79862	0
M	Cosine	Sine	Cotan.	Tan.	Cosec.	Secant	READ DOWN	62°–63° Involute	M

28° or 208° **Trigonometric and Involute Functions** **151° or 331°**

M	Sine	Cosine	Tan.	Cotan.	Secant	Cosec.	Involute 28°–29°	READ UP	M
0	0.46947	0.88295	0.53171	1.8807	1.1326	2.1301	.043017	.79862	60
1	.46973	.88281	.53208	.8794	.1327	.1289	.043100	.79759	59
2	.46999	.88267	.53246	.8781	.1329	.1277	.043182	.79657	58
3	.47024	.88254	.53283	.8768	.1331	.1266	.043264	.79554	57
4	.47050	.88240	.53320	.8755	.1333	.1254	.043347	.79452	56
5	0.47076	0.88226	0.53358	1.8741	1.1334	2.1242	.043430	.79350	55
6	.47101	.88213	.53395	.8728	.1336	.1231	.043513	.79247	54
7	.47127	.88199	.53432	.8715	.1338	.1219	.043596	.79146	53
8	.47153	.88185	.53470	.8702	.1340	.1208	.043679	.79044	52
9	.47178	.88172	.53507	.8689	.1342	.1196	.043762	.78942	51
10	0.47204	0.88158	0.53545	1.8676	1.1343	2.1185	.043845	.78841	50
11	.47229	.88144	.53582	.8663	.1345	.1173	.043929	.78739	49
12	.47255	.88130	.53620	.8650	.1347	.1162	.044012	.78638	48
13	.47281	.88117	.53657	.8637	.1349	.1150	.044096	.78537	47
14	.47306	.88103	.53694	.8624	.1350	.1139	.044180	.78436	46
15	0.47332	0.88089	0.53732	1.8611	1.1352	2.1127	.044264	.78335	45
16	.47358	.88075	.53769	.8598	.1354	.1116	.044348	.78234	44
17	.47383	.88062	.53807	.8585	.1356	.1105	.044432	.78134	43
18	.47409	.88048	.53844	.8572	.1357	.1093	.044516	.78033	42
19	.47434	.88034	.53882	.8559	.1359	.1082	.044601	.77933	41
20	0.47460	0.88020	0.53920	1.8546	1.1361	2.1070	.044685	.77833	40
21	.47486	.88006	.53957	.8533	.1363	.1059	.044770	.77733	39
22	.47511	.87993	.53995	.8520	.1365	.1048	.044855	.77633	38
23	.47537	.87979	.54032	.8507	.1366	.1036	.044939	.77533	37
24	.47562	.87965	.54070	.8495	.1368	.1025	.045024	.77434	36
25	0.47588	0.87951	0.54107	1.8482	1.1370	2.1014	.045110	.77334	35
26	.47614	.87937	.54145	.8469	.1372	.1002	.045195	.77235	34
27	.47639	.87923	.54183	.8456	.1374	.0991	.045280	.77136	33
28	.47665	.87909	.54220	.8443	.1375	.0980	.045366	.77037	32
29	.47690	.87896	.54258	.8430	.1377	.0969	.045451	.76938	31
30	0.47716	0.87882	0.54296	1.8418	1.1379	2.0957	.045537	.76839	30
31	.47741	.87868	.54333	.8405	.1381	.0946	.045623	.76741	29
32	.47767	.87854	.54371	.8392	.1383	.0935	.045709	.76642	28
33	.47793	.87840	.54409	.8379	.1384	.0924	.045795	.76544	27
34	.47818	.87826	.54446	.8367	.1386	.0913	.045881	.76446	26
35	0.47844	0.87812	0.54484	1.8354	1.1388	2.0901	.045967	.76348	25
36	.47869	.87798	.54522	.8341	.1390	.0890	.046054	.76250	24
37	.47895	.87784	.54560	.8329	.1392	.0879	.046140	.76152	23
38	.47920	.87770	.54597	.8316	.1393	.0868	.046227	.76054	22
39	.47946	.87756	.54635	.8303	.1395	.0857	.046313	.75957	21
40	0.47971	0.87743	0.54673	1.8291	1.1397	2.0846	.046400	.75859	20
41	.47997	.87729	.54711	.8278	.1399	.0835	.046487	.75762	19
42	.48022	.87715	.54748	.8265	.1401	.0824	.046575	.75665	18
43	.48048	.87701	.54786	.8253	.1402	.0813	.046662	.75568	17
44	.48073	.87687	.54824	.8240	.1404	.0802	.046749	.75471	16
45	0.48099	0.87673	0.54862	1.8228	1.1406	2.0791	.046837	.75375	15
46	.48124	.87659	.54900	.8215	.1408	.0779	.046924	.75278	14
47	.48150	.87645	.54938	.8202	.1410	.0768	.047012	.75181	13
48	.48175	.87631	.54975	.8190	.1412	.0757	.047100	.75085	12
49	.48201	.87617	.55013	.8177	.1413	.0747	.047188	.74989	11
50	0.48226	0.87603	0.55051	1.8165	1.1415	2.0736	.047276	.74893	10
51	.48252	.87589	.55089	.8152	.1417	.0725	.047364	.74797	9
52	.48277	.87575	.55127	.8140	.1419	.0714	.047452	.74701	8
53	.48303	.87561	.55165	.8127	.1421	.0703	.047541	.74606	7
54	.48328	.87546	.55203	.8115	.1423	.0692	.047630	.74510	6
55	0.48354	0.87532	0.55241	1.8103	1.1424	2.0681	.047718	.74415	5
56	.48379	.87518	.55279	.8090	.1426	.0670	.047807	.74319	4
57	.48405	.87504	.55317	.8078	.1428	.0659	.047896	.74224	3
58	.48430	.87490	.55355	.8065	.1430	.0648	.047985	.74129	2
59	.48456	.87476	.55393	.8053	.1432	.0637	.048074	.74034	1
60	0.48481	0.87462	0.55431	1.8040	1.1434	2.0627	.048164	.73940	0
M	Cosine	Sine	Cotan.	Tan.	Cosec.	Secant	READ DOWN	61°–62° Involute	M

118° or 298° **61° or 241°**

29° or 209°   **Trigonometric and Involute Functions**   **150° or 330°**

M	Sine	Cosine	Tan.	Cotan.	Secant	Cosec.	Involute 29°–30°	READ UP	M
0	0.48481	0.87462	0.55431	1.8040	1.1434	2.0627	.048164	.73940	60
1	.48506	.87448	.55469	.8028	.1435	.0616	.048253	.73845	59
2	.48532	.87434	.55507	.8016	.1437	.0605	.048343	.73751	58
3	.48557	.87420	.55545	.8003	.1439	.0594	.048432	.73656	57
4	.48583	.87406	.55583	.7991	.1441	.0583	.048522	.73562	56
5	0.48608	0.87391	0.55621	1.7979	1.1443	2.0573	.048612	.73468	55
6	.48634	.87377	.55659	.7966	.1445	.0562	.048702	.73374	54
7	.48659	.87363	.55697	.7954	.1446	.0551	.048792	.73280	53
8	.48684	.87349	.55736	.7942	.1448	.0540	.048883	.73186	52
9	.48710	.87335	.55774	.7930	.1450	.0530	.048973	.73093	51
10	0.48735	0.87321	0.55812	1.7917	1.1452	2.0519	.049063	.72999	50
11	.48761	.87306	.55850	.7905	.1454	.0508	.049154	.72906	49
12	.48786	.87292	.55888	.7893	.1456	.0498	.049245	.72813	48
13	.48811	.87278	.55926	.7881	.1458	.0487	.049336	.72720	47
14	.48837	.87264	.55964	.7868	.1460	.0476	.049427	.72627	46
15	0.48862	0.87250	0.56003	1.7856	1.1461	2.0466	.049518	.72534	45
16	.48888	.87235	.56041	.7844	.1463	.0455	.049609	.72441	44
17	.48913	.87221	.56079	.7832	.1465	.0445	.049701	.72349	43
18	.48938	.87207	.56117	.7820	.1467	.0434	.049792	.72256	42
19	.48964	.87193	.56156	.7808	.1469	.0423	.049884	.72164	41
20	0.48989	0.87178	0.56194	1.7796	1.1471	2.0413	.049976	.72072	40
21	.49014	.87164	.56232	.7783	.1473	.0402	.050068	.71980	39
22	.49040	.87150	.56270	.7771	.1474	.0392	.050160	.71888	38
23	.49065	.87136	.56309	.7759	.1476	.0381	.050252	.71796	37
24	.49090	.87121	.56347	.7747	.1478	.0371	.050344	.71704	36
25	0.49116	0.87107	0.56385	1.7735	1.1480	2.0360	.050437	.71613	35
26	.49141	.87093	.56424	.7723	.1482	.0350	.050529	.71521	34
27	.49166	.87079	.56462	.7711	.1484	.0339	.050622	.71430	33
28	.49192	.87064	.56501	.7699	.1486	.0329	.050715	.71339	32
29	.49217	.87050	.56539	.7687	.1488	.0318	.050808	.71248	31
30	0.49242	0.87036	0.56577	1.7675	1.1490	2.0308	.050901	.71157	30
31	.49268	.87021	.56616	.7663	.1491	.0297	.050994	.71066	29
32	.49293	.87007	.56654	.7651	.1493	.0287	.051087	.70975	28
33	.49318	.86993	.56693	.7639	.1495	.0276	.051181	.70885	27
34	.49344	.86978	.56731	.7627	.1497	.0266	.051274	.70794	26
35	0.49369	0.86964	0.56769	1.7615	1.1499	2.0256	.051368	.70704	25
36	.49394	.86949	.56808	.7603	.1501	.0245	.051462	.70614	24
37	.49419	.86935	.56846	.7591	.1503	.0235	.051556	.70524	23
38	.49445	.86921	.56885	.7579	.1505	.0225	.051650	.70434	22
39	.49470	.86906	.56923	.7567	.1507	.0214	.051744	.70344	21
40	0.49495	0.86892	0.56962	1.7556	1.1509	2.0204	.051838	.70254	20
41	.49521	.86878	.57000	.7544	.1510	.0194	.051933	.70165	19
42	.49546	.86863	.57039	.7532	.1512	.0183	.052027	.70075	18
43	.49571	.86849	.57078	.7520	.1514	.0173	.052122	.69986	17
44	.49596	.86834	.57116	.7508	.1516	.0163	.052217	.69897	16
45	0.49622	0.86820	0.57155	1.7496	1.1518	2.0152	.052312	.69808	15
46	.49647	.86805	.57193	.7485	.1520	.0142	.052407	.69719	14
47	.49672	.86791	.57232	.7473	.1522	.0132	.052502	.69630	13
48	.49697	.86777	.57271	.7461	.1524	.0122	.052597	.69541	12
49	.49723	.86762	.57309	.7449	.1526	.0112	.052693	.69452	11
50	0.49748	0.86748	0.57348	1.7437	1.1528	2.0101	.052788	.69364	10
51	.49773	.86733	.57386	.7426	.1530	.0091	.052884	.69275	9
52	.49798	.86719	.57425	.7414	.1532	.0081	.052980	.69187	8
53	.49824	.86704	.57464	.7402	.1533	.0071	.053076	.69099	7
54	.49849	.86690	.57503	.7391	.1535	.0061	.053172	.69011	6
55	0.49874	0.86675	0.57541	1.7379	1.1537	2.0051	.053268	.68923	5
56	.49899	.86661	.57580	.7367	.1539	.0040	.053365	.68835	4
57	.49924	.86646	.57619	.7355	.1541	.0030	.053461	.68748	3
58	.49950	.86632	.57657	.7344	.1543	.0020	.053558	.68660	2
59	.49975	.86617	.57696	.7332	.1545	.0010	.053655	.68573	1
60	0.50000	0.86603	0.57735	1.7321	1.1547	2.0000	.053751	.68485	0
M	Cosine	Sine	Cotan.	Tan.	Cosec.	Secant	READ DOWN	60°–61° Involute	M

30° or 210° **Trigonometric and Involute Functions** **149° or 329°**

M	Sine	Cosine	Tan.	Cotan.	Secant	Cosec.	Involute 30°-31°	READ UP	M
0	0.50000	0.86603	0.57735	1.7321	1.1547	2.0000	.053751	.68485	60
1	.50025	.86588	.57774	.7309	.1549	1.9990	.053849	.68398	59
2	.50050	.86573	.57813	.7297	.1551	.9980	.053946	.68311	58
3	.50076	.86559	.57851	.7286	.1553	.9970	.054043	.68224	57
4	.50101	.86544	.57890	.7274	.1555	.9960	.054140	.68137	56
5	0.50126	0.86530	0.57929	1.7262	1.1557	1.9950	.054238	.68050	55
6	.50151	.86515	.57968	.7251	.1559	.9940	.054336	.67964	54
7	.50176	.86501	.58007	.7239	.1561	.9930	.054433	.67877	53
8	.50201	.86486	.58046	.7228	.1563	.9920	.054531	.67791	52
9	.50227	.86471	.58085	.7216	.1565	.9910	.054629	.67705	51
10	0.50252	0.86457	0.58124	1.7205	1.1566	1.9900	.054728	.67618	50
11	.50277	.86442	.58162	.7193	.1568	.9890	.054826	.67532	49
12	.50302	.86427	.58201	.7182	.1570	.9880	.054924	.67447	48
13	.50327	.86413	.58240	.7170	.1572	.9870	.055023	.67361	47
14	.50352	.86398	.58279	.7159	.1574	.9860	.055122	.67275	46
15	0.50377	0.86384	0.58318	1.7147	1.1576	1.9850	.055221	.67189	45
16	.50403	.86369	.58357	.7136	.1578	.9840	.055320	.67104	44
17	.50428	.86354	.58396	.7124	.1580	.9830	.055419	.67019	43
18	.50453	.86340	.58435	.7113	.1582	.9821	.055518	.66933	42
19	.50478	.86325	.58474	.7102	.1584	.9811	.055617	.66848	41
20	0.50503	0.86310	0.58513	1.7090	1.1586	1.9801	.055717	.66763	40
21	.50528	.86295	.58552	.7079	.1588	.9791	.055817	.66678	39
22	.50553	.86281	.58591	.7067	.1590	.9781	.055916	.66593	38
23	.50578	.86266	.58631	.7056	.1592	.9771	.056016	.66509	37
24	.50603	.86251	.58670	.7045	.1594	.9762	.056116	.66424	36
25	0.50628	0.86237	0.58709	1.7033	1.1596	1.9752	.056217	.66340	35
26	.50654	.86222	.58748	.7022	.1598	.9742	.056317	.66255	34
27	.50679	.86207	.58787	.7011	.1600	.9732	.056417	.66171	33
28	.50704	.86192	.58826	.6999	.1602	.9722	.056518	.66087	32
29	.50729	.86178	.58865	.6988	.1604	.9713	.056619	.66003	31
30	0.50754	0.86163	0.58905	1.6977	1.1606	1.9703	.056720	.65919	30
31	.50779	.86148	.58944	.6965	.1608	.9693	.056821	.65835	29
32	.50804	.86133	.58983	.6954	.1610	.9684	.056922	.65752	28
33	.50829	.86119	.59022	.6943	.1612	.9674	.057023	.65668	27
34	.50854	.86104	.59061	.6932	.1614	.9664	.057124	.65585	26
35	0.50879	0.86089	0.59101	1.6920	1.1616	1.9654	.057226	.65501	25
36	.50904	.86074	.59140	.6909	.1618	.9645	.057328	.65418	24
37	.50929	.86059	.59179	.6898	.1620	.9635	.057429	.65335	23
38	.50954	.86045	.59218	.6887	.1622	.9625	.057531	.65252	22
39	.50979	.86030	.59258	.6875	.1624	.9616	.057633	.65169	21
40	0.51004	0.86015	0.59297	1.6864	1.1626	1.9606	.057736	.65086	20
41	.51029	.86000	.59336	.6853	.1628	.9597	.057838	.65004	19
42	.51054	.85985	.59376	.6842	.1630	.9587	.057940	.64921	18
43	.51079	.85970	.59415	.6831	.1632	.9577	.058043	.64839	17
44	.51104	.85956	.59454	.6820	.1634	.9568	.058146	.64756	16
45	0.51129	0.85941	0.59494	1.6808	1.1636	1.9558	.058249	.64674	15
46	.51154	.85926	.59533	.6797	.1638	.9549	.058352	.64592	14
47	.51179	.85911	.59573	.6786	.1640	.9539	.058455	.64510	13
48	.51204	.85896	.59612	.6775	.1642	.9530	.058558	.64428	12
49	.51229	.85881	.59651	.6764	.1644	.9520	.058662	.64346	11
50	0.51254	0.85866	0.59691	1.6753	1.1646	1.9511	.058765	.64265	10
51	.51279	.85851	.59730	.6742	.1648	.9501	.058869	.64183	9
52	.51304	.85836	.59770	.6731	.1650	.9492	.058973	.64102	8
53	.51329	.85821	.59809	.6720	.1652	.9482	.059077	.64020	7
54	.51354	.85806	.59849	.6709	.1654	.9473	.059181	.63939	6
55	0.51379	0.85792	0.59888	1.6698	1.1656	1.9463	.059285	.63858	5
56	.51404	.85777	.59928	.6687	.1658	.9454	.059390	.63777	4
57	.51429	.85762	.59967	.6676	.1660	.9444	.059494	.63696	3
58	.51454	.85747	.60007	.6665	.1662	.9435	.059599	.63615	2
59	.51479	.85732	.60046	.6654	.1664	.9425	.059704	.63534	1
60	0.51504	0.85717	0.60086	1.6643	1.1666	1.9416	.059809	.63454	0
M	Cosine	Sine	Cotan.	Tan.	Cosec.	Secant	READ DOWN	59°-60° Involute	M

120° or 300° **59° or 239°**

31° or 211°   Trigonometric and Involute Functions   148° or 328°

M	Sine	Cosine	Tan.	Cotan.	Secant	Cosec.	Involute 31°–32°	READ UP	M
0	0.51504	0.85717	0.60086	1.6643	1.1666	1.9416	.059809	.63454	60
1	.51529	.85702	.60126	.6632	.1668	.9407	.059914	.63373	59
2	.51554	.85687	.60165	.6621	.1670	.9397	.060019	.63293	58
3	.51579	.85672	.60205	.6610	.1672	.9388	.060124	.63212	57
4	.51604	.85657	.60245	.6599	.1675	.9379	.060230	.63132	56
5	0.51628	0.85642	0.60284	1.6588	1.1677	1.9369	.060335	.63052	55
6	.51653	.85627	.60324	.6577	.1679	.9360	.060441	.62972	54
7	.51678	.85612	.60364	.6566	.1681	.9351	.060547	.62892	53
8	.51703	.85597	.60403	.6555	.1683	.9341	.060653	.62812	52
9	.51728	.85582	.60443	.6545	.1685	.9332	.060759	.62733	51
10	0.51753	0.85567	0.60483	1.6534	1.1687	1.9323	.060866	.62653	50
11	.51778	.85551	.60522	.6523	.1689	.9313	.060972	.62574	49
12	.51803	.85536	.60562	.6512	.1691	.9304	.061079	.62494	48
13	.51828	.85521	.60602	.6501	.1693	.9295	.061186	.62415	47
14	.51852	.85506	.60642	.6490	.1695	.9285	.061292	.62336	46
15	0.51877	0.85491	0.60681	1.6479	1.1697	1.9276	.061400	.62257	45
16	.51902	.85476	.60721	.6469	.1699	.9267	.061507	.62178	44
17	.51927	.85461	.60761	.6458	.1701	.9258	.061614	.62099	43
18	.51952	.85446	.60801	.6447	.1703	.9249	.061721	.62020	42
19	.51977	.85431	.60841	.6436	.1705	.9239	.061829	.61942	41
20	0.52002	0.85416	0.60881	1.6426	1.1707	1.9230	.061937	.61863	40
21	.52026	.85401	.60921	.6415	.1710	.9221	.062045	.61785	39
22	.52051	.85385	.60960	.6404	.1712	.9212	.062153	.61706	38
23	.52076	.85370	.61000	.6393	.1714	.9203	.062261	.61628	37
24	.52101	.85355	.61040	.6383	.1716	.9194	.062369	.61550	36
25	0.52126	0.85340	0.61080	1.6372	1.1718	1.9184	.062478	.61472	35
26	.52151	.85325	.61120	.6361	.1720	.9175	.062586	.61394	34
27	.52175	.85310	.61160	.6351	.1722	.9166	.062695	.61316	33
28	.52200	.85294	.61200	.6340	.1724	.9157	.062804	.61239	32
29	.52225	.85279	.61240	.6329	.1726	.9148	.062913	.61161	31
30	0.52250	0.85264	0.61280	1.6319	1.1728	1.9139	.063022	.61083	30
31	.52275	.85249	.61320	.6308	.1730	.9130	.063131	.61006	29
32	.52299	.85234	.61360	.6297	.1732	.9121	.063241	.60929	28
33	.52324	.85218	.61400	.6287	.1735	.9112	.063350	.60851	27
34	.52349	.85203	.61440	.6276	.1737	.9103	.063460	.60774	26
35	0.52374	0.85188	0.61480	1.6265	1.1739	1.9094	.063570	.60697	25
36	.52399	.85173	.61520	.6255	.1741	.9084	.063680	.60620	24
37	.52423	.85157	.61561	.6244	.1743	.9075	.063790	.60544	23
38	.52448	.85142	.61601	.6234	.1745	.9066	.063901	.60467	22
39	.52473	.85127	.61641	.6223	.1747	.9057	.064011	.60390	21
40	0.52498	0.85112	0.61681	1.6212	1.1749	1.9048	.064122	.60314	20
41	.52522	.85096	.61721	.6202	.1751	.9039	.064232	.60237	19
42	.52547	.85081	.61761	.6191	.1753	.9031	.064343	.60161	18
43	.52572	.85066	.61801	.6181	.1756	.9022	.064454	.60085	17
44	.52597	.85051	.61842	.6170	.1758	.9013	.064565	.60009	16
45	0.52621	0.85035	0.61882	1.6160	1.1760	1.9004	.064677	.59933	15
46	.52646	.85020	.61922	.6149	.1762	.8995	.064788	.59857	14
47	.52671	.85005	.61962	.6139	.1764	.8986	.064900	.59781	13
48	.52696	.84989	.62003	.6128	.1766	.8977	.065012	.59705	12
49	.52720	.84974	.62043	.6118	.1768	.8968	.065123	.59630	11
50	0.52745	0.84959	0.62083	1.6107	1.1770	1.8959	.065236	.59554	10
51	.52770	.84943	.62124	.6097	.1773	.8950	.065348	.59479	9
52	.52794	.84928	.62164	.6087	.1775	.8941	.065460	.59403	8
53	.52819	.84913	.62204	.6076	.1777	.8933	.065573	.59328	7
54	.52844	.84897	.62245	.6066	.1779	.8924	.065685	.59253	6
55	0.52869	0.84882	0.62285	1.6055	1.1781	1.8915	.065798	.59178	5
56	.52893	.84866	.62325	.6045	.1783	.8906	.065911	.59103	4
57	.52918	.84851	.62366	.6034	.1785	.8897	.066024	.59028	3
58	.52942	.84836	.62406	.6024	.1788	.8888	.066137	.58954	2
59	.52967	.84820	.62446	.6014	.1790	.8880	.066250	.58879	1
60	0.52992	0.84805	0.62487	1.6003	1.1792	1.8871	.066364	.58804	0
M	Cosine	Sine	Cotan.	Tan.	Cosec.	Secant	READ DOWN	58°–59° Involute	M

32° or 212° Trigonometric and Involute Functions 147° or 327°

M	Sine	Cosine	Tan.	Cotan.	Secant	Cosec.	Involute 32°-33°	READ UP	M
0	0.52992	0.84805	0.62487	1.6003	1.1792	1.8871	.066364	.58804	60
1	.53017	.84789	.62527	.5993	.1794	.8862	.066478	.58730	59
2	.53041	.84774	.62568	.5983	.1796	.8853	.066591	.58656	58
3	.53066	.84758	.62608	.5972	.1798	.8844	.066705	.58581	57
4	.53091	.84743	.62649	.5962	.1800	.8836	.066819	.58507	56
5	0.53115	0.84728	0.62689	1.5952	1.1803	1.8827	.066934	.58433	55
6	.53140	.84712	.62730	.5941	.1805	.8818	.067048	.58359	54
7	.53164	.84697	.62770	.5931	.1807	.8810	.067163	.58285	53
8	.53189	.84681	.62811	.5921	.1809	.8801	.067277	.58211	52
9	.53214	.84666	.62852	.5911	.1811	.8792	.067392	.58138	51
10	0.53238	0.84650	0.62892	1.5900	1.1813	1.8783	.067507	.58064	50
11	.53263	.84635	.62933	.5890	.1815	.8775	.067622	.57991	49
12	.53288	.84619	.62973	.5880	.1818	.8766	.067738	.57917	48
13	.53312	.84604	.63014	.5869	.1820	.8757	.067853	.57844	47
14	.53337	.84588	.63055	.5859	.1822	.8749	.067969	.57771	46
15	0.53361	0.84573	0.63095	1.5849	1.1824	1.8740	.068084	.57698	45
16	.53386	.84557	.63136	.5839	.1826	.8731	.068200	.57625	44
17	.53411	.84542	.63177	.5829	.1828	.8723	.068316	.57552	43
18	.53435	.84526	.63217	.5818	.1831	.8714	.068432	.57479	42
19	.53460	.84511	.63258	.5808	.1833	.8706	.068549	.57406	41
20	0.53484	0.84495	0.63299	1.5798	1.1835	1.8697	.068665	.57333	40
21	.53509	.84480	.63340	.5788	.1837	.8688	.068782	.57261	39
22	.53534	.84464	.63380	.5778	.1839	.8680	.068899	.57188	38
23	.53558	.84448	.63421	.5768	.1842	.8671	.069016	.57116	37
24	.53583	.84433	.63462	.5757	.1844	.8663	.069133	.57044	36
25	0.53607	0.84417	0.63503	1.5747	1.1846	1.8654	.069250	.56972	35
26	.53632	.84402	.63544	.5737	.1848	.8646	.069367	.56900	34
27	.53656	.84386	.63584	.5727	.1850	.8637	.069485	.56828	33
28	.53681	.84370	.63625	.5717	.1852	.8629	.069602	.56756	32
29	.53705	.84355	.63666	.5707	.1855	.8620	.069720	.56684	31
30	0.53730	0.84339	0.63707	1.5697	1.1857	1.8612	.069838	.56612	30
31	.53754	.84324	.63748	.5687	.1859	.8603	.069956	.56540	29
32	.53779	.84308	.63789	.5677	.1861	.8595	.070075	.56469	28
33	.53804	.84292	.63830	.5667	.1863	.8586	.070193	.56398	27
34	.53828	.84277	.63871	.5657	.1866	.8578	.070312	.56326	26
35	0.53853	0.84261	0.63912	1.5647	1.1868	1.8569	.070430	.56255	25
36	.53877	.84245	.63953	.5637	.1870	.8561	.070549	.56184	24
37	.53902	.84230	.63994	.5627	.1872	.8552	.070668	.56113	23
38	.53926	.84214	.64035	.5617	.1875	.8544	.070788	.56042	22
39	.53951	.84198	.64076	.5607	.1877	.8535	.070907	.55971	21
40	0.53975	0.84182	0.64117	1.5597	1.1879	1.8527	.071026	.55900	20
41	.54000	.84167	.64158	.5587	.1881	.8519	.071146	.55829	19
42	.54024	.84151	.64199	.5577	.1883	.8510	.071266	.55759	18
43	.54049	.84135	.64240	.5567	.1886	.8502	.071386	.55688	17
44	.54073	.84120	.64281	.5557	.1888	.8494	.071506	.55618	16
45	0.54097	0.84104	0.64322	1.5547	1.1890	1.8485	.071626	.55547	15
46	.54122	.84088	.64363	.5537	.1892	.8477	.071747	.55477	14
47	.54146	.84072	.64404	.5527	.1895	.8468	.071867	.55407	13
48	.54171	.84057	.64446	.5517	.1897	.8460	.071988	.55337	12
49	.54195	.84041	.64487	.5507	.1899	.8452	.072109	.55267	11
50	0.54220	0.84025	0.64528	1.5497	1.1901	1.8443	.072230	.55197	10
51	.54244	.84009	.64569	.5487	.1903	.8435	.072351	.55127	9
52	.54269	.83994	.64610	.5477	.1906	.8427	.072473	.55057	8
53	.54293	.83978	.64652	.5468	.1908	.8419	.072594	.54988	7
54	.54317	.83962	.64693	.5458	.1910	.8410	.072716	.54918	6
55	0.54342	0.83946	0.64734	1.5448	1.1912	1.8402	.072838	.54849	5
56	.54366	.83930	.64775	.5438	.1915	.8394	.072959	.54779	4
57	.54391	.83915	.64817	.5428	.1917	.8385	.073082	.54710	3
58	.54415	.83899	.64858	.5418	.1919	.8377	.073204	.54641	2
59	.54440	.83883	.64899	.5408	.1921	.8369	.073326	.54572	1
60	0.54464	0.83867	0.64941	1.5399	1.1924	1.8361	.073449	.54503	0
M	Cosine	Sine	Cotan.	Tan.	Cosec.	Secant	READ DOWN	57°-58° Involute	M

33° or 213°   Trigonometric and Involute Functions   146° or 326°

M	Sine	Cosine	Tan.	Cotan.	Secant	Cosec.	Involute 33°–34°	READ UP	M
0	0.54464	0.83867	0.64941	1.5399	1.1924	1.8361	.073449	.54503	60
1	.54488	.83851	.64982	.5389	.1926	.8353	.073572	.54434	59
2	.54513	.83835	.65024	.5379	.1928	.8344	.073695	.54365	58
3	.54537	.83819	.65065	.5369	.1930	.8336	.073818	.54296	57
4	.54561	.83804	.65106	.5359	.1933	.8328	.073941	.54228	56
5	0.54586	0.83788	0.65148	1.5350	1.1935	1.8320	.074064	.54159	55
6	.54610	.83772	.65189	.5340	.1937	.8312	.074188	.54090	54
7	.54635	.83756	.65231	.5330	.1939	.8303	.074312	.54022	53
8	.54659	.83740	.65272	.5320	.1942	.8295	.074435	.53954	52
9	.54683	.83724	.65314	.5311	.1944	.8287	.074559	.53885	51
10	0.54708	0.83708	0.65355	1.5301	1.1946	1.8279	.074684	.53817	50
11	.54732	.83692	.65397	.5291	.1949	.8271	.074808	.53749	49
12	.54756	.83676	.65438	.5282	.1951	.8263	.074932	.53681	48
13	.54781	.83660	.65480	.5272	.1953	.8255	.075057	.53613	47
14	.54805	.83645	.65521	.5262	.1955	.8247	.075182	.53546	46
15	0.54829	0.83629	0.65563	1.5253	1.1958	1.8238	.075307	.53478	45
16	.54854	.83613	.65604	.5243	.1960	.8230	.075432	.53410	44
17	.54878	.83597	.65646	.5233	.1962	.8222	.075557	.53343	43
18	.54902	.83581	.65688	.5224	.1964	.8214	.075683	.53275	42
19	.54927	.83565	.65729	.5214	.1967	.8206	.075808	.53208	41
20	0.54951	0.83549	0.65771	1.5204	1.1969	1.8198	.075934	.53141	40
21	.54975	.83533	.65813	.5195	.1971	.8190	.076060	.53073	39
22	.54999	.83517	.65854	.5185	.1974	.8182	.076186	.53006	38
23	.55024	.83501	.65896	.5175	.1976	.8174	.076312	.52939	37
24	.55048	.83485	.65938	.5166	.1978	.8166	.076439	.52872	36
25	0.55072	0.83469	0.65980	1.5156	1.1981	1.8158	.076565	.52805	35
26	.55097	.83453	.66021	.5147	.1983	.8150	.076692	.52739	34
27	.55121	.83437	.66063	.5137	.1985	.8142	.076819	.52672	33
28	.55145	.83421	.66105	.5127	.1987	.8134	.076946	.52605	32
29	.55169	.83405	.66147	.5118	.1990	.8126	.077073	.52539	31
30	0.55194	0.83389	0.66189	1.5108	1.1992	1.8118	.077200	.52472	30
31	.55218	.83373	.66230	.5099	.1944	.8110	.077328	.52406	29
32	.55242	.83356	.66272	.5089	.1997	.8102	.077455	.52340	28
33	.55266	.83340	.66314	.5080	.1999	.8094	.077583	.52274	27
34	.55291	.83324	.66356	.5070	.2001	.8086	.077711	.52207	26
35	0.55315	0.83308	0.66398	1.5061	1.2004	1.8078	.077839	.52141	25
36	.55339	.83292	.66440	.5051	.2006	.8070	.077968	.52076	24
37	.55363	.83276	.66482	.5042	.2008	.8062	.078096	.52010	23
38	.55388	.83260	.66524	.5032	.2011	.8055	.078225	.51944	22
39	.55412	.83244	.66566	.5023	.2013	.8047	.078354	.51878	21
40	0.55436	0.83228	0.66608	1.5013	1.2015	1.8039	.078483	.51813	20
41	.55460	.83212	.66650	.5004	.2018	.8031	.078612	.51747	19
42	.55484	.83195	.66692	.4994	.2020	.8023	.078741	.51682	18
43	.55509	.83179	.66734	.4985	.2022	.8015	.078871	.51616	17
44	.55533	.83163	.66776	.4975	.2025	.8007	.079000	.51551	16
45	0.55557	0.83147	0.66818	1.4966	1.2027	1.8000	.079130	.51486	15
46	.55581	.83131	.66860	.4957	.2029	.7992	.079260	.51421	14
47	.55605	.83115	.66902	.4947	.2032	.7984	.079390	.51356	13
48	.55630	.83098	.66944	.4938	.2034	.7976	.079520	.51291	12
49	.55654	.83082	.66986	.4928	.2036	.7968	.079651	.51226	11
50	0.55678	0.83066	0.67028	1.4919	1.2039	1.7960	.079781	.51161	10
51	.55702	.83050	.67071	.4910	.2041	.7953	.079912	.51096	9
52	.55726	.83034	.67113	.4900	.2043	.7945	.080043	.51032	8
53	.55750	.83017	.67155	.4891	.2046	.7937	.080174	.50967	7
54	.55775	.83001	.67197	.4882	.2048	.7929	.080305	.50903	6
55	0.55799	0.82985	0.67239	1.4872	1.2050	1.7922	.080437	.50838	5
56	.55823	.82969	.67282	.4863	.2053	.7914	.080569	.50774	4
57	.55847	.82953	.67324	.4854	.2055	.7906	.080700	.50710	3
58	.55871	.82936	.67366	.4844	.2057	.7898	.080832	.50646	2
59	.55895	.82920	.67409	.4835	.2060	.7891	.080964	.50582	1
60	0.55919	0.82904	0.67451	1.4826	1.2062	1.7883	.081097	.50518	0
M	Cosine	Sine	Cotan.	Tan.	Cosec.	Secant	READ DOWN	56°–57° Involute	M

34° or 214°   Trigonometric and Involute Functions   145° or 325°

M	Sine	Cosine	Tan.	Cotan.	Secant	Cosec.	Involute 34°–35°	READ UP	M
0	0.55919	0.82904	0.67451	1.4826	1.2062	1.7883	.081097	.50518	60
1	.55943	.82887	.67493	.4816	.2065	.7875	.081229	.50454	59
2	.55968	.82871	.67536	.4807	.2067	.7868	.081362	.50390	58
3	.55992	.82855	.67578	.4798	.2069	.7860	.081494	.50326	57
4	.56016	.82839	.67620	.4788	.2072	.7852	.081627	.50263	56
5	0.56040	0.82822	0.67663	1.4779	1.2074	1.7844	.081760	.50199	55
6	.56064	.82806	.67705	.4770	.2076	.7837	.081894	.50135	54
7	.56088	.82790	.67748	.4761	.2079	.7829	.082027	.50072	53
8	.56112	.82773	.67790	.4751	.2081	.7821	.082161	.50009	52
9	.56136	.82757	.67832	.4742	.2084	.7814	.082294	.49945	51
10	0.56160	0.82741	0.67875	1.4733	1.2086	1.7806	.082428	.49882	50
11	.56184	.82724	.67917	.4724	.2088	.7799	.082562	.49819	49
12	.56208	.82708	.67960	.4715	.2091	.7791	.082697	.49756	48
13	.56232	.82692	.68002	.4705	.2093	.7783	.082831	.49693	47
14	.56256	.82675	.68045	.4696	.2096	.7776	.082966	.49630	46
15	0.56280	0.82659	0.68088	1.4687	1.2098	1.7768	.083100	.49568	45
16	.56305	.82643	.68130	.4678	.2100	.7761	.083235	.49505	44
17	.56329	.82626	.68173	.4669	.2103	.7753	.083371	.49442	43
18	.56353	.82610	.68215	.4659	.2105	.7745	.083506	.49380	42
19	.56377	.82593	.68258	.4650	.2108	.7738	.083641	.49317	41
20	0.56401	0.82577	0.68301	1.4641	1.2110	1.7730	.083777	.49255	40
21	.56425	.82561	.68343	.4632	.2112	.7723	.083913	.49192	39
22	.56449	.82544	.68386	.4623	.2115	.7715	.084049	.49130	38
23	.56473	.82528	.68429	.4614	.2117	.7708	.084185	.49068	37
24	.56497	.82511	.68471	.4605	.2120	.7700	.084321	.49006	36
25	0.56521	0.82495	0.68514	1.4596	1.2122	1.7693	.084457	.48944	35
26	.56545	.82478	.68557	.4586	.2124	.7685	.084594	.48882	34
27	.56569	.82462	.68600	.4577	.2127	.7678	.084731	.48820	33
28	.56593	.82446	.68642	.4568	.2129	.7670	.084868	.48758	32
29	.56617	.82429	.68685	.4559	.2132	.7663	.085005	.48697	31
30	0.56641	0.82413	0.68728	1.4550	1.2134	1.7655	.085142	.48635	30
31	.56665	.82396	.68771	.4541	.2136	.7648	.085280	.48574	29
32	.56689	.82380	.68814	.4532	.2139	.7640	.085418	.48512	28
33	.56713	.82363	.68857	.4523	.2141	.7633	.085555	.48451	27
34	.56736	.82347	.68900	.4514	.2144	.7625	.085693	.48389	26
35	0.56760	0.82330	0.68942	1.4505	1.2146	1.7618	.085832	.48328	25
36	.56784	.82314	.68985	.4496	.2149	.7610	.085970	.48267	24
37	.56808	.82297	.69028	.4487	.2151	.7603	.086108	.48206	23
38	.56832	.82281	.69071	.4478	.2154	.7596	.086247	.48145	22
39	.56856	.82264	.69114	.4469	.2156	.7588	.086386	.48084	21
40	0.56880	0.82248	0.69157	1.4460	1.2158	1.7581	.086525	.48023	20
41	.56904	.82231	.69200	.4451	.2161	.7573	.086664	.47962	19
42	.56928	.82214	.69243	.4442	.2163	.7566	.086804	.47902	18
43	.56952	.82198	.69286	.4433	.2166	.7559	.086943	.47841	17
44	.56976	.82181	.69329	.4424	.2168	.7551	.087083	.47780	16
45	0.57000	0.82165	0.69372	1.4415	1.2171	1.7544	.087223	.47720	15
46	.57024	.82148	.69416	.4406	.2173	.7537	.087363	.47660	14
47	.57047	.82132	.69459	.4397	.2176	.7529	.087503	.47599	13
48	.57071	.82115	.69502	.4388	.2178	.7522	.087644	.47539	12
49	.57095	.82098	.69545	.4379	.2181	.7515	.087784	.47479	11
50	0.57119	0.82082	0.69588	1.4370	1.2183	1.7507	.087925	.47419	10
51	.57143	.82065	.69631	.4361	.2185	.7500	.088066	.47359	9
52	.57167	.82048	.69675	.4352	.2188	.7493	.088207	.47299	8
53	.57191	.82032	.69718	.4344	.2190	.7485	.088348	.47239	7
54	.57215	.82015	.69761	.4335	.2193	.7478	.088490	.47179	6
55	0.57238	0.81999	0.69804	1.4326	1.2195	1.7471	.088631	.47119	5
56	.57262	.81982	.69847	.4317	.2198	.7463	.088773	.47060	4
57	.57286	.81965	.69891	.4308	.2200	.7456	.088915	.47000	3
58	.57310	.81949	.69934	.4299	.2203	.7449	.089057	.46940	2
59	.57334	.81932	.69977	.4290	.2205	.7442	.089200	.46881	1
60	0.57358	0.81915	0.70021	1.4281	1.2208	1.7434	.089342	.46822	0
M	Cosine	Sine	Cotan.	Tan.	Cosec.	Secant	READ DOWN	55°–56° Involute	M

124° or 304°   55° or 235°

35° or 215° Trigonometric and Involute Functions 144° or 324°

M	Sine	Cosine	Tan.	Cotan.	Secant	Cosec.	Involute 35°–36°	READ UP	M
0	0.57358	0.81915	0.70021	1.4281	1.2208	1.7434	.089342	.46822	60
1	.57381	.81899	.70064	.4273	.2210	.7427	.089485	.46762	59
2	.57405	.81882	.70107	.4264	.2213	.7420	.089628	.46703	58
3	.57429	.81865	.70151	.4255	.2215	.7413	.089771	.46644	57
4	.57453	.81848	.70194	.4246	.2218	.7406	.089914	.46585	56
5	0.57477	0.81832	0.70238	1.4237	1.2220	1.7398	.090058	.46526	55
6	.57501	.81815	.70281	.4229	.2223	.7391	.090201	.46467	54
7	.57524	.81798	.70325	.4220	.2225	.7384	.090345	.46408	53
8	.57548	.81782	.70368	.4211	.2228	.7377	.090489	.46349	52
9	.57572	.81765	.70412	.4202	.2230	.7370	.090633	.46291	51
10	0.57596	0.81748	0.70455	1.4193	1.2233	1.7362	.090777	.46232	50
11	.57619	.81731	.70499	.4185	.2235	.7355	.090922	.46173	49
12	.57643	.81714	.70542	.4176	.2238	.7348	.091067	.46115	48
13	.57667	.81698	.70586	.4167	.2240	.7341	.091211	.46057	47
14	.57691	.81681	.70629	.4158	.2243	.7334	.091356	.45998	46
15	0.57715	0.81664	0.70673	1.4150	1.2245	1.7327	.091502	.45940	45
16	.57738	.81647	.70717	.4141	.2248	.7320	.091647	.45882	44
17	.57762	.81631	.70760	.4132	.2250	.7312	.091793	.45824	43
18	.57786	.81614	.70804	.4124	.2253	.7305	.091938	.45766	42
19	.57810	.81597	.70848	.4115	.2255	.7298	.092084	.45708	41
20	0.57833	0.81580	0.70891	1.4106	1.2258	1.7291	.092230	.45650	40
21	.57857	.81563	.70935	.4097	.2260	.7284	.092377	.45592	39
22	.57881	.81546	.70979	.4089	.2263	.7277	.092523	.45534	38
23	.57904	.81530	.71023	.4080	.2265	.7270	.092670	.45476	37
24	.57928	.81513	.71066	.4071	.2268	.7263	.092816	.45419	36
25	0.57952	0.81496	0.71110	1.4063	1.2271	1.7256	.092963	.45361	35
26	.57976	.81479	.71154	.4054	.2273	.7249	.093111	.45304	34
27	.57999	.81462	.71198	.4045	.2276	.7242	.093258	.45246	33
28	.58023	.81445	.71242	.4037	.2278	.7235	.093406	.45189	32
29	.58047	.81428	.71285	.4028	.2281	.7228	.093553	.45132	31
30	0.58070	0.81412	0.71329	1.4019	1.2283	1.7221	.093701	.45074	30
31	.58094	.81395	.71373	.4011	.2286	.7213	.093849	.45017	29
32	.58118	.81378	.71417	.4002	.2288	.7206	.093998	.44960	28
33	.58141	.81361	.71461	.3994	.2291	.7199	.094146	.44903	27
34	.58165	.81344	.71505	.3985	.2293	.7192	.094295	.44846	26
35	0.58189	0.81327	0.71549	1.3976	1.2296	1.7185	.094443	.44789	25
36	.58212	.81310	.71593	.3968	.2299	.7179	.094592	.44733	24
37	.58236	.81293	.71637	.3959	.2301	.7172	.094742	.44676	23
38	.58260	.81276	.71681	.3951	.2304	.7165	.094891	.44619	22
39	.58283	.81259	.71725	.3942	.2306	.7158	.095041	.44563	21
40	0.58307	0.81242	0.71769	1.3934	1.2309	1.7151	.095190	.44506	20
41	.58330	.81225	.71813	.3925	.2311	.7144	.095340	.44450	19
42	.58354	.81208	.71857	.3916	.2314	.7137	.095490	.44393	18
43	.58378	.81191	.71901	.3908	.2317	.7130	.095641	.44337	17
44	.58401	.81174	.71946	.3899	.2319	.7123	.095791	.44281	16
45	0.58425	0.81157	0.71990	1.3891	1.2322	1.7116	.095942	.44225	15
46	.58449	.81140	.72034	.3882	.2324	.7109	.096093	.44169	14
47	.58472	.81123	.72078	.3874	.2327	.7102	.096244	.44113	13
48	.58496	.81106	.72122	.3865	.2329	.7095	.096395	.44057	12
49	.58519	.81089	.72167	.3857	.2332	.7088	.096546	.44001	11
50	0.58543	0.81072	0.72211	1.3848	1.2335	1.7081	.096698	.43945	10
51	.58567	.81055	.72255	.3840	.2337	.7075	.096850	.43889	9
52	.58590	.81038	.72299	.3831	.2340	.7068	.097002	.43833	8
53	.58614	.81021	.72344	.3823	.2342	.7061	.097154	.43778	7
54	.58637	.81004	.72388	.3814	.2345	.7054	.097306	.43722	6
55	0.58661	0.80987	0.72432	1.3806	1.2348	1.7047	.097459	.43667	5
56	.58684	.80970	.72477	.3798	.2350	.7040	.097611	.43611	4
57	.58708	.80953	.72521	.3789	.2353	.7033	.097764	.43556	3
58	.58731	.80936	.72565	.3781	.2355	.7027	.097917	.43501	2
59	.58755	.80919	.72610	.3772	.2358	.7020	.098071	.43446	1
60	0.58779	0.80902	0.72654	1.3764	1.2361	1.7013	.098224	.43390	0
M	Cosine	Sine	Cotan.	Tan.	Cosec.	Secant	READ DOWN	54°–55° Involute	M

36° or 216°　Trigonometric and Involute Functions　143° or 323°

M	Sine	Cosine	Tan.	Cotan.	Secant	Cosec.	Involute 36°–37°	READ UP	M
0	0.58779	0.80902	0.72654	1.3764	1.2361	1.7013	.098224	.43390	60
1	.58802	.80885	.72699	.3755	.2363	.7006	.098378	.43335	59
2	.58826	.80867	.72743	.3747	.2366	.6999	.098532	.43280	58
3	.58849	.80850	.72788	.3739	.2369	.6993	.098686	.43225	57
4	.58873	.80833	.72832	.3730	.2371	.6986	.098840	.43171	56
5	0.58896	0.80816	0.72877	1.3722	1.2374	1.6979	.098994	.43116	55
6	.58920	.80799	.72921	.3713	.2376	.6972	.099149	.43061	54
7	.58943	.80782	.72966	.3705	.2379	.6966	.099304	.43006	53
8	.58967	.80765	.73010	.3697	.2382	.6959	.099459	.42952	52
9	.58990	.80748	.73055	.3688	.2384	.6952	.099614	.42897	51
10	0.59014	0.80730	0.73100	1.3680	1.2387	1.6945	.099769	.42843	50
11	.59037	.80713	.73144	.3672	.2390	.6939	.099925	.42788	49
12	.59061	.80696	.73189	.3663	.2392	.6932	.10008	.42734	48
13	.59084	.80679	.73234	.3655	.2395	.6925	.10024	.42680	47
14	.59108	.80662	.73278	.3647	.2397	.6918	.10039	.42625	46
15	0.59131	0.80644	0.73323	1.3638	1.2400	1.6912	.10055	.42571	45
16	.59154	.80627	.73368	.3630	.2403	.6905	.10070	.42517	44
17	.59178	.80610	.73413	.3622	.2405	.6898	.10086	.42463	43
18	.59201	.80593	.73457	.3613	.2408	.6892	.10102	.42409	42
19	.59225	.80576	.73502	.3605	.2411	.6885	.10118	.42355	41
20	0.59248	0.80558	0.73547	1.3597	1.2413	1.6878	.10133	.42302	40
21	.59272	.80541	.73592	.3588	.2416	.6871	.10149	.42248	39
22	.59295	.80524	.73637	.3580	.2419	.6865	.10165	.42194	38
23	.59318	.80507	.73681	.3572	.2421	.6858	.10181	.42141	37
24	.59342	.80489	.73726	.3564	.2424	.6852	.10196	.42087	36
25	0.59365	0.80472	0.73771	1.3555	1.2427	1.6845	.10212	.42034	35
26	.59389	.80455	.73816	.3547	.2429	.6838	.10228	.41980	34
27	.59412	.80438	.73861	.3539	.2432	.6832	.10244	.41927	33
28	.59436	.80420	.73906	.3531	.2435	.6825	.10260	.41874	32
29	.59459	.80403	.73951	.3522	.2437	.6818	.10276	.41820	31
30	0.59482	0.80386	0.73996	1.3514	1.2440	1.6812	.10292	.41767	30
31	.59506	.80368	.74041	.3506	.2443	.6805	.10308	.41714	29
32	.59529	.80351	.74086	.3498	.2445	.6799	.10323	.41661	28
33	.59552	.80334	.74131	.3490	.2448	.6792	.10339	.41608	27
34	.59576	.80316	.74176	.3481	.2451	.6785	.10355	.41555	26
35	0.59599	0.80299	0.74221	1.3473	1.2453	1.6779	.10371	.41502	25
36	.59622	.80282	.74267	.3465	.2456	.6772	.10388	.41450	24
37	.59646	.80264	.74312	.3457	.2459	.6766	.10404	.41397	23
38	.59669	.80247	.74357	.3449	.2462	.6759	.10420	.41344	22
39	.59693	.80230	.74402	.3440	.2464	.6753	.10436	.41292	21
40	0.59716	0.80212	0.74447	1.3432	1.2467	1.6746	.10452	.41239	20
41	.59739	.80195	.74492	.3424	.2470	.6739	.10468	.41187	19
42	.59763	.80178	.74538	.3416	.2472	.6733	.10484	.41134	18
43	.59786	.80160	.74583	.3408	.2475	.6726	.10500	.41082	17
44	.59809	.80143	.74628	.3400	.2478	.6720	.10516	.41030	16
45	0.59832	0.80125	0.74674	1.3392	1.2480	1.6713	.10533	.40977	15
46	.59856	.80108	.74719	.3384	.2483	.6707	.10549	.40925	14
47	.59879	.80091	.74764	.3375	.2486	.6700	.10565	.40873	13
48	.59902	.80073	.74810	.3367	.2489	.6694	.10581	.40821	12
49	.59926	.80056	.74855	.3359	.2491	.6687	.10598	.40769	11
50	0.59949	0.80038	0.74900	1.3351	1.2494	1.6681	.10614	.40717	10
51	.59972	.80021	.74946	.3343	.2497	.6674	.10630	.40666	9
52	.59995	.80003	.74991	.3335	.2499	.6668	.10647	.40614	8
53	.60019	.79986	.75037	.3327	.2502	.6661	.10663	.40562	7
54	.60042	.79968	.75082	.3319	.2505	.6655	.10679	.40511	6
55	0.60065	0.79951	0.75128	1.3311	1.2508	1.6649	.10696	.40459	5
56	.60089	.79934	.75173	.3303	.2510	.6642	.10712	.40407	4
57	.60112	.79916	.75219	.3295	.2513	.6636	.10729	.40356	3
58	.60135	.79899	.75264	.3287	.2516	.6629	.10745	.40305	2
59	.60158	.79881	.75310	.3278	.2519	.6623	.10762	.40253	1
60	0.60182	0.79864	0.75355	1.3270	1.2521	1.6616	.10778	.40202	0
M	Cosine	Sine	Cotan.	Tan.	Cosec.	Secant	READ DOWN	53°–54° Involute	M

37° or 217°   **Trigonometric and Involute Functions**   **142° or 322°**

M	Sine	Cosine	Tan.	Cotan.	Secant	Cosec.	Involute 37°–38°	READ UP	M
0	0.60182	0.79864	0.75355	1.3270	1.2521	1.6616	.10778	.40202	60
1	.60205	.79846	.75401	.3262	.2524	.6610	.10795	.40151	59
2	.60228	.79829	.75447	.3254	.2527	.6604	.10811	.40100	58
3	.60251	.79811	.75492	.3246	.2530	.6597	.10828	.40049	57
4	.60274	.79793	.75538	.3238	.2532	.6591	.10844	.39998	56
5	0.60298	0.79776	0.75584	1.3230	1.2535	1.6584	.10861	.39947	55
6	.60321	.79758	.75629	.3222	.2538	.6578	.10878	.39896	54
7	.60344	.79741	.75675	.3214	.2541	.6572	.10894	.39845	53
8	.60367	.79723	.75721	.3206	.2543	.6565	.10911	.39794	52
9	.60390	.79706	.75767	.3198	.2546	.6559	.10928	.39743	51
10	0.60414	0.79688	0.75812	1.3190	1.2549	1.6553	.10944	.39693	50
11	.60437	.79671	.75858	.3182	.2552	.6546	.10961	.39642	49
12	.60460	.79653	.75904	.3175	.2554	.6540	.10978	.39592	48
13	.60483	.79635	.75950	.3167	.2557	.6534	.10995	.39541	47
14	.60506	.79618	.75996	.3159	.2560	.6527	.11011	.39491	46
15	0.60529	0.79600	0.76042	1.3151	1.2563	1.6521	.11028	.39441	45
16	.60553	.79583	.76088	.3143	.2566	.6515	.11045	.39390	44
17	.60576	.79565	.76134	.3135	.2568	.6508	.11062	.39340	43
18	.60599	.79547	.76180	.3127	.2571	.6502	.11079	.39290	42
19	.60622	.79530	.76226	.3119	.2574	.6496	.11096	.39240	41
20	0.60645	0.79512	0.76272	1.3111	1.2577	1.6489	.11113	.39190	40
21	.60668	.79494	.76318	.3103	.2579	.6483	.11130	.39140	39
22	.60691	.79477	.76364	.3095	.2582	.6477	.11146	.39090	38
23	.60714	.79459	.76410	.3087	.2585	.6471	.11163	.39040	37
24	.60738	.79441	.76456	.3079	.2588	.6464	.11180	.38990	36
25	0.60761	0.79424	0.76502	1.3072	1.2591	1.6458	.11197	.38941	35
26	.60784	.79406	.76548	.3064	.2593	.6452	.11215	.38891	34
27	.60807	.79388	.76594	.3056	.2596	.6446	.11232	.38841	33
28	.60830	.79371	.76640	.3048	.2599	.6439	.11249	.38792	32
29	.60853	.79353	.76686	.3040	.2602	.6433	.11266	.38742	31
30	0.60876	0.79335	0.76733	1.3032	1.2605	1.6427	.11283	.38693	30
31	.60899	.79318	.76779	.3024	.2608	.6421	.11300	.38643	29
32	.60922	.79300	.76825	.3017	.2610	.6414	.11317	.38594	28
33	.60945	.79282	.76871	.3009	.2613	.6408	.11334	.38545	27
34	.60968	.79264	.76918	.3001	.2616	.6402	.11352	.38496	26
35	0.60991	0.79247	0.76964	1.2993	1.2619	1.6396	.11369	.38446	25
36	.61015	.79229	.77010	.2985	.2622	.6390	.11386	.38397	24
37	.61038	.79211	.77057	.2977	.2624	.6383	.11403	.38348	23
38	.61061	.79193	.77103	.2970	.2627	.6377	.11421	.38299	22
39	.61084	.79176	.77149	.2962	.2630	.6371	.11438	.38251	21
40	0.61107	0.79158	0.77196	1.2954	1.2633	1.6365	.11455	.38202	20
41	.61130	.79140	.77242	.2946	.2636	.6359	.11473	.38153	19
42	.61153	.79122	.77289	.2938	.2639	.6353	.11490	.38104	18
43	.61176	.79105	.77335	.2931	.2641	.6346	.11507	.38055	17
44	.61199	.79087	.77382	.2923	.2644	.6340	.11525	.38007	16
45	0.61222	0.79069	0.77428	1.2915	1.2647	1.6334	.11542	.37958	15
46	.61245	.79051	.77475	.2907	.2650	.6328	.11560	.37910	14
47	.61268	.79033	.77521	.2900	.2653	.6322	.11577	.37861	13
48	.61291	.79016	.77568	.2892	.2656	.6316	.11595	.37813	12
49	.61314	.78998	.77615	.2884	.2659	.6310	.11612	.37765	11
50	0.61337	0.78980	0.77661	1.2876	1.2661	1.6303	.11630	.37716	10
51	.61360	.78962	.77708	.2869	.2664	.6297	.11647	.37668	9
52	.61383	.78944	.77754	.2861	.2667	.6291	.11665	.37620	8
53	.61406	.78926	.77801	.2853	.2670	.6285	.11682	.37572	7
54	.61429	.78908	.77848	.2846	.2673	.6279	.11700	.37524	6
55	0.61451	0.78891	0.77895	1.2838	1.2676	1.6273	.11718	.37476	5
56	.61474	.78873	.77941	.2830	.2679	.6267	.11735	.37428	4
57	.61497	.78855	.77988	.2822	.2682	.6261	.11753	.37380	3
58	.61520	.78837	.78035	.2815	.2684	.6255	.11771	.37332	2
59	.61543	.78819	.78082	.2807	.2687	.6249	.11788	.37285	1
60	0.61566	0.78801	0.78129	1.2799	1.2690	1.6243	.11806	.37237	0
M	Cosine	Sine	Cotan.	Tan.	Cosec.	Secant	READ DOWN	52°–53° Involute	M

38° or 218° Trigonometric and Involute Functions 141° or 321°

M	Sine	Cosine	Tan.	Cotan.	Secant	Cosec.	Involute 38°–39°	READ UP	M
0	0.61566	0.78801	0.78129	1.2799	1.2690	1.6243	.11806	.37237	60
1	.61589	.78783	.78175	.2792	.2693	.6237	.11824	.37189	59
2	.61612	.78765	.78222	.2784	.2696	.6231	.11842	.37142	58
3	.61635	.78747	.78269	.2776	.2699	.6225	.11859	.37094	57
4	.61658	.78729	.78316	.2769	.2702	.6219	.11877	.37047	56
5	0.61681	0.78711	0.78363	1.2761	1.2705	1.6213	.11895	.36999	55
6	.61704	.78694	.78410	.2753	.2708	.6207	.11913	.36952	54
7	.61726	.78676	.78457	.2746	.2710	.6201	.11931	.36905	53
8	.61749	.78658	.78504	.2738	.2713	.6195	.11949	.36858	52
9	.61772	.78640	.78551	.2731	.2716	.6189	.11967	.36810	51
10	0.61795	0.78622	0.78598	1.2723	1.2719	1.6183	.11985	.36763	50
11	.61818	.78604	.78645	.2715	.2722	.6177	.12003	.36716	49
12	.61841	.78586	.78692	.2708	.2725	.6171	.12021	.36669	48
13	.61864	.78568	.78739	.2700	.2728	.6165	.12039	.36622	47
14	.61887	.78550	.78786	.2693	.2731	.6159	.12057	.36575	46
15	0.61909	0.78532	0.78834	1.2685	1.2734	1.6153	.12075	.36529	45
16	.61932	.78514	.78881	.2677	.2737	.6147	.12093	.36482	44
17	.61955	.78496	.78928	.2670	.2740	.6141	.12111	.36435	43
18	.61978	.78478	.78975	.2662	.2742	.6135	.12129	.36388	42
19	.62001	.78460	.79022	.2655	.2745	.6129	.12147	.36342	41
20	0.62024	0.78442	0.79070	1.2647	1.2748	1.6123	.12165	.36295	40
21	.62046	.78424	.79117	.2640	.2751	.6117	.12184	.36249	39
22	.62069	.78405	.79164	.2632	.2754	.6111	.12202	.36202	38
23	.62092	.78387	.79212	.2624	.2757	.6105	.12220	.36156	37
24	.62115	.78369	.79259	.2617	.2760	.6099	.12238	.36110	36
25	0.62138	0.78351	0.79306	1.2609	1.2763	1.6093	.12257	.36063	35
26	.62160	.78333	.79354	.2602	.2766	.6087	.12275	.36017	34
27	.62183	.78315	.79401	.2594	.2769	.6082	.12293	.35971	33
28	.62206	.78297	.79449	.2587	.2772	.6076	.12312	.35925	32
29	.62229	.78279	.79496	.2579	.2775	.6070	.12330	.35879	31
30	0.62251	0.78261	0.79544	1.2572	1.2778	1.6064	.12348	.35833	30
31	.62274	.78243	.79591	.2564	.2781	.6058	.12367	.35787	29
32	.62297	.78225	.79639	.2557	.2784	.6052	.12385	.35741	28
33	.62320	.78206	.79686	.2549	.2787	.6046	.12404	.35695	27
34	.62342	.78188	.79734	.2542	.2790	.6040	.12422	.35649	26
35	.62365	0.78170	0.79781	1.2534	1.2793	1.6035	.12441	.35604	25
36	.62388	.78152	.79829	.2527	.2796	.6029	.12459	.35558	24
37	.62411	.78134	.79877	.2519	.2799	.6023	.12478	.35512	23
38	.62433	.78116	.79924	.2512	.2802	.6017	.12496	.35467	22
39	.62456	.78098	.79972	.2504	.2804	.6011	.12515	.35421	21
40	0.62479	0.78079	0.80020	1.2497	1.2807	1.6005	.12534	.35376	20
41	.62502	.78061	.80067	.2489	.2810	.6000	.12552	.35330	19
42	.62524	.78043	.80115	.2482	.2813	.5994	.12571	.35285	18
43	.62547	.78025	.80163	.2475	.2816	.5988	.12590	.35240	17
44	.62570	.78007	.80211	.2467	.2819	.5982	.12608	.35194	16
45	0.62592	0.77988	0.80258	1.2460	1.2822	1.5976	.12627	.35149	15
46	.62615	.77970	.80306	.2452	.2825	.5971	.12646	.35104	14
47	.62638	.77952	.80354	.2445	.2828	.5965	.12664	.35059	13
48	.62660	.77934	.80402	.2437	.2831	.5959	.12683	.35014	12
49	.62683	.77916	.80450	.2430	.2834	.5953	.12702	.34969	11
50	0.62706	0.77897	0.80498	1.2423	1.2837	1.5948	.12721	.34924	10
51	.62728	.77879	.80546	.2415	.2840	.5942	.12740	.34879	9
52	.62751	.77861	.80594	.2408	.2843	.5936	.12759	.34834	8
53	.62774	.77843	.80642	.2401	.2846	.5930	.12778	.34790	7
54	.62796	.77824	.80690	.2393	.2849	.5925	.12797	.34745	6
55	0.62819	0.77806	0.80738	1.2386	1.2852	1.5919	.12815	.34700	5
56	.62842	.77788	.80786	.2378	.2855	.5913	.12834	.34656	4
57	.62864	.77769	.80834	.2371	.2859	.5907	.12853	.34611	3
58	.62887	.77751	.80882	.2364	.2862	.5902	.12872	.34567	2
59	.62909	.77733	.80930	.2356	.2865	.5896	.12891	.34522	1
60	0.62932	0.77715	0.80978	1.2349	1.2868	1.5890	.12911	.34478	0
M	Cosine	Sine	Cotan.	Tan.	Cosec.	Secant	READ DOWN	51°–52° Involute	M

128° or 308°   **51° or 231°**

39° or 219° Trigonometric and Involute Functions 140° or 320°

M	Sine	Cosine	Tan.	Cotan.	Secant	Cosec.	Involute 39°–40°	READ UP	M
0	0.62932	0.77715	0.80978	1.2349	1.2868	1.5890	.12911	.34478	60
1	.62955	.77696	.81027	.2342	.2871	.5884	.12930	.34434	59
2	.62977	.77678	.81075	.2334	.2874	.5879	.12949	.34389	58
3	.63000	.77660	.81123	.2327	.2877	.5873	.12968	.34345	57
4	.63022	.77641	.81171	.2320	.2880	.5867	.12987	.34301	56
5	0.63045	0.77623	0.81220	1.2312	1.2883	1.5862	.13006	.34257	55
6	.63068	.77605	.81268	.2305	.2886	.5856	.13025	.34213	54
7	.63090	.77586	.81316	.2298	.2889	.5850	.13045	.34169	53
8	.63113	.77568	.81364	.2290	.2892	.5845	.13064	.34125	52
9	.63135	.77550	.81413	.2283	.2895	.5839	.13083	.34081	51
10	0.63158	0.77531	0.81461	1.2276	1.2898	1.5833	.13102	.34037	50
11	.63180	.77513	.81510	.2268	.2901	.5828	.13122	.33993	49
12	.63203	.77494	.81558	.2261	.2904	.5822	.13141	.33949	48
13	.63225	.77476	.81606	.2254	.2907	.5816	.13160	.33906	47
14	.63248	.77458	.81655	.2247	.2910	.5811	.13180	.33862	46
15	0.63271	0.77439	0.81703	1.2239	1.2913	1.5805	.13199	.33818	45
16	.63293	.77421	.81752	.2232	.2916	.5799	.13219	.33775	44
17	.63316	.77402	.81800	.2225	.2919	.5794	.13238	.33731	43
18	.63338	.77384	.81849	.2218	.2923	.5788	.13258	.33688	42
19	.63361	.77366	.81898	.2210	.2926	.5783	.13277	.33645	41
20	0.63383	0.77347	0.81946	1.2203	1.2929	1.5777	.13297	.33601	40
21	.63406	.77329	.81995	.2196	.2932	.5771	.13316	.33558	39
22	.63428	.77310	.82044	.2189	.2935	.5766	.13336	.33515	38
23	.63451	.77292	.82092	.2181	.2938	.5760	.13355	.33471	37
24	.63473	.77273	.82141	.2174	.2941	.5755	.13375	.33428	36
25	0.63496	0.77255	0.82190	1.2167	1.2944	1.5749	.13395	.33385	35
26	.63518	.77236	.82238	.2160	.2947	.5744	.13414	.33342	34
27	.63541	.77218	.82287	.2153	.2950	.5738	.13434	.33299	33
28	.63563	.77199	.82336	.2145	.2953	.5732	.13454	.33256	32
29	.63585	.77181	.82385	.2138	.2957	.5727	.13473	.33213	31
30	0.63608	0.77162	0.82434	1.2131	1.2960	1.5721	.13493	.33171	30
31	.63630	.77144	.82483	.2124	.2963	.5716	.13513	.33128	29
32	.63653	.77125	.82531	.2117	.2966	.5710	.13533	.33085	28
33	.63675	.77107	.82580	.2109	.2969	.5705	.13553	.33042	27
34	.63698	.77088	.82629	.2102	.2972	.5699	.13572	.33000	26
35	0.63720	0.77070	0.82678	1.2095	1.2975	1.5694	.13592	.32957	25
36	.63742	.77051	.82727	.2088	.2978	.5688	.13612	.32915	24
37	.63765	.77033	.82776	.2081	.2981	.5683	.13632	.32872	23
38	.63787	.77014	.82825	.2074	.2985	.5677	.13652	.32830	22
39	.63810	.76996	.82874	.2066	.2988	.5672	.13672	.32787	21
40	0.63832	0.76977	0.82923	1.2059	1.2991	1.5666	.13692	.32745	20
41	.63854	.76959	.82972	.2052	.2994	.5661	.13712	.32703	19
42	.63877	.76940	.83022	.2045	.2997	.5655	.13732	.32661	18
43	.63899	.76921	.83071	.2038	.3000	.5650	.13752	.32618	17
44	.63922	.76903	.83120	.2031	.3003	.5644	.13772	.32576	16
45	0.63944	0.76884	0.83169	1.2024	1.3007	1.5639	.13792	.32534	15
46	.63966	.76866	.83218	.2017	.3010	.5633	.13812	.32492	14
47	.63989	.76847	.83268	.2009	.3013	.5628	.13833	.32450	13
48	.64011	.76828	.83317	.2002	.3016	.5622	.13853	.32408	12
49	.64033	.76810	.83366	.1995	.3019	.5617	.13873	.32366	11
50	0.64056	0.76791	0.83415	1.1988	1.3022	1.5611	.13893	.32324	10
51	.64078	.76772	.83465	.1981	.3026	.5606	.13913	.32283	9
52	.64100	.76754	.83514	.1974	.3029	.5601	.13934	.32241	8
53	.64123	.76735	.83564	.1967	.3032	.5595	.13954	.32199	7
54	.64145	.76717	.83613	.1960	.3035	.5590	.13974	.32158	6
55	0.64167	0.76698	0.83662	1.1953	1.3038	1.5584	.13995	.32116	5
56	.64190	.76679	.83712	.1946	.3041	.5579	.14015	.32075	4
57	.64212	.76661	.83761	.1939	.3045	.5573	.14035	.32033	3
58	.64234	.76642	.83811	.1932	.3048	.5568	.14056	.31992	2
59	.64256	.76623	.83860	.1925	.3051	.5563	.14076	.31950	1
60	0.64279	0.76604	0.83910	1.1918	1.3054	1.5557	.14097	.31909	0
M	Cosine	Sine	Cotan.	Tan.	Cosec.	Secant	READ DOWN	50°–51°	M
								Involute	

40° or 220° Trigonometric and Involute Functions **139° or 319°**

M	Sine	Cosine	Tan.	Cotan.	Secant	Cosec.	Involute 40°–41°	READ UP	M
0	0.64279	0.76604	0.83910	1.1918	1.3054	1.5557	.14097	.31909	60
1	.64301	.76586	.83960	.1910	.3057	.5552	.14117	.31868	59
2	.64323	.76567	.84009	.1903	.3060	.5546	.14138	.31826	58
3	.64346	.76548	.84059	.1896	.3064	.5541	.14158	.31785	57
4	.64368	.76530	.84108	.1889	.3067	.5536	.14179	.31744	56
5	0.64390	0.76511	0.84158	1.1882	1.3070	1.5530	.14200	.31703	55
6	.64412	.76492	.84208	.1875	.3073	.5525	.14220	.31662	54
7	.64435	.76473	.84258	.1868	.3076	.5520	.14241	.31621	53
8	.64457	.76455	.84307	.1861	.3080	.5514	.14261	.31580	52
9	.64479	.76436	.84357	.1854	.3083	.5509	.14282	.31539	51
10	0.64501	0.76417	0.84407	1.1847	1.3086	1.5504	.14303	.31498	50
11	.64524	.76398	.84457	.1840	.3089	.5498	.14324	.31457	49
12	.64546	.76380	.84507	.1833	.3093	.5493	.14344	.31417	48
13	.64568	.76361	.84556	.1826	.3096	.5488	.14365	.31376	47
14	.64590	.76342	.84606	.1819	.3099	.5482	.14386	.31335	46
15	0.64612	0.76323	0.84656	1.1812	1.3102	1.5477	.14407	.31295	45
16	.64635	.76304	.84706	.1806	.3105	.5472	.14428	.31254	44
17	.64657	.76286	.84756	.1799	.3109	.5466	.14448	.31214	43
18	.64679	.76267	.84806	.1792	.3112	.5461	.14469	.31173	42
19	.64701	.76248	.84856	.1785	.3115	.5456	.14490	.31133	41
20	0.64723	0.76229	0.84906	1.1778	1.3118	1.5450	.14511	.31092	40
21	.64746	.76210	.84956	.1771	.3122	.5445	.14532	.31052	39
22	.64768	.76192	.85006	.1764	.3125	.5440	.14553	.31012	38
23	.64790	.76173	.85057	.1757	.3128	.5435	.14574	.30971	37
24	.64812	.76154	.85107	.1750	.3131	.5429	.14595	.30931	36
25	0.64834	0.76135	0.85157	1.1743	1.3135	1.5424	.14616	.30891	35
26	.64856	.76116	.85207	.1736	.3138	.5419	.14638	.30851	34
27	.64878	.76097	.85257	.1729	.3141	.5413	.14659	.30811	33
28	.64901	.76078	.85308	.1722	.3144	.5408	.14680	.30771	32
29	.64923	.76059	.85358	.1715	.3148	.5403	.14701	.30731	31
30	0.64945	0.76041	0.85408	1.1708	1.3151	1.5398	.14722	.30691	30
31	.64967	.76022	.85458	.1702	.3154	.5392	.14743	.30651	29
32	.64989	.76003	.85509	.1695	.3157	.5387	.14765	.30611	28
33	.65011	.75984	.85559	.1688	.3161	.5382	.14786	.30572	27
34	.65033	.75965	.85609	.1681	.3164	.5377	.14807	.30532	26
35	0.65055	0.75946	0.85660	1.1674	1.3167	1.5372	.14829	.30492	25
36	.65077	.75927	.85710	.1667	.3171	.5366	.14850	.30453	24
37	.65100	.75908	.85761	.1660	.3174	.5361	.14871	.30413	23
38	.65122	.75889	.85811	.1653	.3177	.5356	.14893	.30374	22
39	.65144	.75870	.85862	.1647	.3180	.5351	.14914	.30334	21
40	0.65166	0.75851	0.85912	1.1640	1.3184	1.5345	.14936	.30295	20
41	.65188	.75832	.85963	.1633	.3187	.5340	.14957	.30255	19
42	.65210	.75813	.86014	.1626	.3190	.5335	.14979	.30216	18
43	.65232	.75794	.86064	.1619	.3194	.5330	.15000	.30177	17
44	.65254	.75775	.86115	.1612	.3197	.5325	.15022	.30137	16
45	0.65276	0.75756	0.86166	1.1606	1.3200	1.5320	.15043	.30098	15
46	.65298	.75738	.86216	.1599	.3203	.5314	.15065	.30059	14
47	.65320	.75719	.86267	.1592	.3207	.5309	.15087	.30020	13
48	.65342	.75700	.86318	.1585	.3210	.5304	.15108	.29981	12
49	.65364	.75680	.86368	.1578	.3213	.5299	.15130	.29942	11
50	0.65386	0.75661	0.86419	1.1571	1.3217	1.5294	.15152	.29903	10
51	.65408	.75642	.86470	.1565	.3220	.5289	.15173	.29864	9
52	.65430	.75623	.86521	.1558	.3223	.5283	.15195	.29825	8
53	.65452	.75604	.86572	.1551	.3227	.5278	.15217	.29786	7
54	.65474	.75585	.86623	.1544	.3230	.5273	.15239	.29747	6
55	0.65496	0.75566	0.86674	1.1538	1.3233	1.5268	.15261	.29709	5
56	.65518	.75547	.86725	.1531	.3237	.5263	.15282	.29670	4
57	.65540	.75528	.86776	.1524	.3240	.5258	.15304	.29631	3
58	.65562	.75509	.86827	.1517	.3243	.5253	.15326	.29593	2
59	.65584	.75490	.86878	.1510	.3247	.5248	.15348	.29554	1
60	0.65606	0.75471	0.86929	1.1504	1.3250	1.5243	.15370	.29516	0
M	Cosine	Sine	Cotan.	Tan.	Cosec.	Secant	READ DOWN	49°–50° Involute	M

130° or 310° **49° or 229°**

41° or 221°   **Trigonometric and Involute Functions**   **138° or 318°**

M	Sine	Cosine	Tan.	Cotan.	Secant	Cosec.	Involute 41°–42°	READ UP	M
0	0.65606	0.75471	0.86929	1.1504	1.3250	1.5243	.15370	.29516	60
1	.65628	.75452	.86980	.1497	.3253	.5237	.15392	.29477	59
2	.65650	.75433	.87031	.1490	.3257	.5232	.15414	.29439	58
3	.65672	.75414	.87082	.1483	.3260	.5227	.15436	.29400	57
4	.65694	.75395	.87133	.1477	.3264	.5222	.15458	.29362	56
5	0.65716	0.75375	0.87184	1.1470	1.3267	1.5217	.15480	.29324	55
6	.65738	.75356	.87236	.1463	.3270	.5212	.15503	.29286	54
7	.65759	.75337	.87287	.1456	.3274	.5207	.15525	.29247	53
8	.65781	.75318	.87338	.1450	.3277	.5202	.15547	.29209	52
9	.65803	.75299	.87389	.1443	.3280	.5197	.15569	.29171	51
10	0.65825	0.75280	0.87441	1.1436	1.3284	1.5192	.15591	.29133	50
11	.65847	.75261	.87492	.1430	.3287	.5187	.15614	.29095	49
12	.65869	.75241	.87543	.1423	.3291	.5182	.15636	.29057	48
13	.65891	.75222	.87595	.1416	.3294	.5177	.15658	.29019	47
14	.65913	.75203	.87646	.1410	.3297	.5172	.15680	.28981	46
15	0.65935	0.75184	0.87698	1.1403	1.3301	1.5167	.15703	.28943	45
16	.65956	.75165	.87749	.1396	.3304	.5162	.15725	.28906	44
17	.65978	.75146	.87801	.1389	.3307	.5156	.15748	.28868	43
18	.66000	.75126	.87852	.1383	.3311	.5151	.15770	.28830	42
19	.66022	.75107	.87904	.1376	.3314	.5146	.15793	.28792	41
20	0.66044	0.75088	0.87955	1.1369	1.3318	1.5141	.15815	.28755	40
21	.66066	.75069	.88007	.1363	.3321	.5136	.15838	.28717	39
22	.66088	.75050	.88059	.1356	.3325	.5131	.15860	.28680	38
23	.66109	.75030	.88110	.1349	.3328	.5126	.15883	.28642	37
24	.66131	.75011	.88162	.1343	.3331	.5121	.15905	.28605	36
25	0.66153	0.74992	0.88214	1.1336	1.3335	1.5116	.15928	.28567	35
26	.66175	.74973	.88265	.1329	.3338	.5111	.15950	.28530	34
27	.66197	.74953	.88317	.1323	.3342	.5107	.15973	.28493	33
28	.66218	.74934	.88369	.1316	.3345	.5102	.15996	.28455	32
29	.66240	.74915	.88421	.1310	.3348	.5097	.16019	.28418	31
30	0.66262	0.74896	0.88473	1.1303	1.3352	1.5092	.16041	.28381	30
31	.66284	.74876	.88524	.1296	.3355	.5087	.16064	.28344	29
32	.66306	.74857	.88576	.1290	.3359	.5082	.16087	.28307	28
33	.66327	.74838	.88628	.1283	.3362	.5077	.16110	.28270	27
34	.66349	.74818	.88680	.1276	.3366	.5072	.16133	.28233	26
35	0.66371	0.74799	0.88732	1.1270	1.3369	1.5067	.16156	.28196	25
36	.66393	.74780	.88784	.1263	.3373	.5062	.16178	.28159	24
37	.66414	.74760	.88836	.1257	.3376	.5057	.16201	.28122	23
38	.66436	.74741	.88888	.1250	.3380	.5052	.16224	.28085	22
39	.66458	.74722	.88940	.1243	.3383	.5047	.16247	.28048	21
40	0.66480	0.74703	0.88992	1.1237	1.3386	1.5042	.16270	.28012	20
41	.66501	.74683	.89045	.1230	.3390	.5037	.16293	.27975	19
42	.66523	.74664	.89097	.1224	.3393	.5032	.16317	.27938	18
43	.66545	.74644	.89149	.1217	.3397	.5027	.16340	.27902	17
44	.66566	.74625	.89201	.1211	.3400	.5023	.16363	.27865	16
45	0.66588	0.74606	0.89253	1.1204	1.3404	1.5018	.16386	.27828	15
46	.66610	.74586	.89306	.1197	.3407	.5013	.16409	.27792	14
47	.66632	.74567	.89358	.1191	.3411	.5008	.16432	.27755	13
48	.66653	.74548	.89410	.1184	.3414	.5003	.16456	.27719	12
49	.66675	.74528	.89463	.1178	.3418	.4998	.16479	.27683	11
50	0.66697	0.74509	0.89515	1.1171	1.3421	1.4993	.16502	.27646	10
51	.66718	.74489	.89567	.1165	.3425	.4988	.16525	.27610	9
52	.66740	.74470	.89620	.1158	.3428	.4984	.16549	.27574	8
53	.66762	.74451	.89672	.1152	.3432	.4979	.16572	.27538	7
54	.66783	.74431	.89725	.1145	.3435	.4974	.16596	.27501	6
55	0.66805	0.74412	0.89777	1.1139	1.3439	1.4969	.16619	.27465	5
56	.66827	.74392	.89830	.1132	.3442	.4964	.16642	.27429	4
57	.66848	.74373	.89883	.1126	.3446	.4959	.16666	.27393	3
58	.66870	.74353	.89935	.1119	.3449	.4954	.16689	.27357	2
59	.66891	.74334	.89988	.1113	.3453	.4950	.16713	.27321	1
60	0.66913	0.74314	0.90040	1.1106	1.3456	1.4945	.16737	.27285	0
M	Cosine	Sine	Cotan.	Tan.	Cosec.	Secant	READ DOWN	48°–49° Involute	M

42° or 222°   Trigonometric and Involute Functions   137° or 317°

M	Sine	Cosine	Tan.	Cotan.	Secant	Cosec.	Involute 42°–43°	READ UP	M
0	0.66913	0.74314	0.90040	1.1106	1.3456	1.4945	.16737	.27285	60
1	.66935	.74295	.90093	.1100	.3460	.4940	.16760	.27250	59
2	.66956	.74276	.90146	.1093	.3463	.4935	.16784	.27214	58
3	.66978	.74256	.90199	.1087	.3467	.4930	.16807	.27178	57
4	.66999	.74237	.90251	.1080	.3470	.4925	.16831	.27142	56
5	0.67021	0.74217	0.90304	1.1074	1.3474	1.4921	.16855	.27107	55
6	.67043	.74198	.90357	.1067	.3478	.4916	.16879	.27071	54
7	.67064	.74178	.90410	.1061	.3481	.4911	.16902	.27035	53
8	.67086	.74159	.90463	.1054	.3485	.4906	.16926	.27000	52
9	.67107	.74139	.90516	.1048	.3488	.4901	.16950	.26964	51
10	0.67129	0.74120	0.90569	1.1041	1.3492	1.4897	.16974	.26929	50
11	.67151	.74100	.90621	.1035	.3495	.4892	.16998	.26893	49
12	.67172	.74080	.90674	.1028	.3499	.4887	.17022	.26858	48
13	.67194	.74061	.90727	.1022	.3502	.4882	.17045	.26823	47
14	.67215	.74041	.90781	.1016	.3506	.4878	.17069	.26787	46
15	0.67237	0.74022	0.90834	1.1009	1.3510	1.4873	.17093	.26752	45
16	.67258	.74002	.90887	.1003	.3513	.4868	.17117	.26717	44
17	.67280	.73983	.90940	.0996	.3517	.4863	.17142	.26682	43
18	.67301	.73963	.90993	.0990	.3520	.4859	.17166	.26646	42
19	.67323	.73944	.91046	.0983	.3524	.4854	.17190	.26611	41
20	0.67344	0.73924	0.91099	1.0977	1.3527	1.4849	.17214	.26576	40
21	.67366	.73904	.91153	.0971	.3531	.4844	.17238	.26541	39
22	.67387	.73885	.91206	.0964	.3535	.4840	.17262	.26506	38
23	.67409	.73865	.91259	.0958	.3538	.4835	.17286	.26471	37
24	.67430	.73846	.91313	.0951	.3542	.4830	.17311	.26436	36
25	0.67452	0.73826	0.91366	1.0945	1.3545	1.4825	.17335	.26401	35
26	.67473	.73806	.91419	.0939	.3549	.4821	.17359	.26367	34
27	.67495	.73787	.91473	.0932	.3553	.4816	.17383	.26332	33
28	.67516	.73767	.91526	.0926	.3556	.4811	.17408	.26297	32
29	.67538	.73747	.91580	.0919	.3560	.4807	.17432	.26262	31
30	0.67559	0.73728	0.91633	1.0913	1.3563	1.4802	.17457	.26228	30
31	.67580	.73708	.91687	.0907	.3567	.4797	.17481	.26193	29
32	.67602	.73688	.91740	.0900	.3571	.4792	.17506	.26159	28
33	.67623	.73669	.91794	.0894	.3574	.4788	.17530	.26124	27
34	.67645	.73649	.91847	.0888	.3578	.4783	.17555	.26089	26
35	0.67666	0.73629	0.91901	1.0881	1.3582	1.4778	.17579	.26055	25
36	.67688	.73610	.91955	.0875	.3585	.4774	.17604	.26021	24
37	.67709	.73590	.92008	.0869	.3589	.4769	.17628	.25986	23
38	.67730	.73570	.92062	.0862	.3592	.4764	.17653	.25952	22
39	.67752	.73551	.92116	.0856	.3596	.4760	.17678	.25918	21
40	0.67773	0.73531	0.92170	1.0850	1.3600	1.4755	.17702	.25883	20
41	.67795	.73511	.92224	.0843	.3603	.4750	.17727	.25849	19
42	.67816	.73491	.92277	.0837	.3607	.4746	.17752	.25815	18
43	.67837	.73472	.92331	.0831	.3611	.4741	.17777	.25781	17
44	.67859	.73452	.92385	.0824	.3614	.4737	.17801	.25747	16
45	0.67880	0.73432	0.92439	1.0818	1.3618	1.4732	.17826	.25713	15
46	.67901	.73413	.92493	.0812	.3622	.4727	.17851	.25679	14
47	.67923	.73393	.92547	.0805	.3625	.4723	.17876	.25645	13
48	.67944	.73373	.92601	.0799	.3629	.4718	.17901	.25611	12
49	.67965	.73353	.92655	.0793	.3632	.4713	.17926	.25577	11
50	0.67987	0.73333	0.92709	1.0786	1.3636	1.4709	.17951	.25543	10
51	.68008	.73314	.92763	.0780	.3640	.4704	.17976	.25509	9
52	.68029	.73294	.92817	.0774	.3644	.4700	.18001	.25475	8
53	.68051	.73274	.92872	.0768	.3647	.4695	.18026	.25442	7
54	.68072	.73254	.92926	.0761	.3651	.4690	.18051	.25408	6
55	0.68093	0.73234	0.92980	1.0755	1.3655	1.4686	.18076	.25374	5
56	.68115	.73215	.93034	.0749	.3658	.4681	.18101	.25341	4
57	.68136	.73195	.93088	.0742	.3662	.4677	.18127	.25307	3
58	.68157	.73175	.93143	.0736	.3666	.4672	.18152	.25273	2
59	.68179	.73155	.93197	.0730	.3670	.4667	.18177	.25240	1
60	0.68200	0.73135	0.93252	1.0724	1.3673	1.4663	.18202	.25206	0
M	Cosine	Sine	Cotan.	Tan.	Cosec.	Secant	READ DOWN	47°–48° Involute	M

132° or 312°   **47° or 227°**

Trigonometric and Involute Functions

M	Sine	Cosine	Tan.	Cotan.	Secant	Cosec.	Involute 43°-44°	READ UP	M
0	0.68200	0.73135	0.93252	1.0724	1.3673	1.4663	.18202	.25206	60
1	.68221	.73116	.93306	.0717	.3677	.4658	.18228	.25173	59
2	.68242	.73096	.93360	.0711	.3681	.4654	.18253	.25140	58
3	.68264	.73076	.93415	.0705	.3684	.4649	.18278	.25106	57
4	.68285	.73056	.93469	.0699	.3688	.4645	.18304	.25073	56
5	0.68306	0.73036	0.93524	1.0692	1.3692	1.4640	.18329	.25040	55
6	.68327	.73016	.93578	.0686	.3696	.4635	.18355	.25006	54
7	.68349	.72996	.93633	.0680	.3699	.4631	.18380	.24973	53
8	.68370	.72976	.93688	.0674	.3703	.4626	.18406	.24940	52
9	.68391	.72957	.93742	.0668	.3707	.4622	.18431	.24907	51
10	0.68412	0.72937	0.93797	1.0661	1.3711	1.4617	.18457	.24874	50
11	.68434	.72917	.93852	.0655	.3714	.4613	.18482	.24841	49
12	.68455	.72897	.93906	.0649	.3718	.4608	.18508	.24808	48
13	.68476	.72877	.93961	.0643	.3722	.4604	.18534	.24775	47
14	.68497	.72857	.94016	.0637	.3726	.4599	.18559	.24742	46
15	0.68518	0.72837	0.94071	1.0630	1.3729	1.4595	.18585	.24709	45
16	.68539	.72817	.94125	.0624	.3733	.4590	.18611	.24676	44
17	.68561	.72797	.94180	.0618	.3737	.4586	.18637	.24643	43
18	.68582	.72777	.94235	.0612	.3741	.4581	.18662	.24611	42
19	.68603	.72757	.94290	.0606	.3744	.4577	.18688	.24578	41
20	0.68624	0.72737	0.94345	1.0599	1.3748	1.4572	.18714	.24545	40
21	.68645	.72717	.94400	.0593	.3752	.4568	.18740	.24512	39
22	.68666	.72697	.94455	.0587	.3756	.4563	.18766	.24480	38
23	.68688	.72677	.94510	.0851	.3759	.4559	.18792	.24447	37
24	.68709	.72657	.94565	.0575	.3763	.4554	.18818	.24415	36
25	0.68730	0.72637	0.94620	1.0569	1.3767	1.4550	.18844	.24382	35
26	.68751	.72617	.94676	.0562	.3771	.4545	.18870	.24350	34
27	.68772	.72597	.94731	.0556	.3775	.4541	.18896	.24317	33
28	.68793	.72577	.94786	.0550	.3778	.4536	.18922	.24285	32
29	.68814	.72557	.94841	.0544	.3782	.4532	.18948	.24253	31
30	0.68835	0.72537	0.94896	1.0538	1.3786	1.4527	.18975	.24220	30
31	.68857	.72517	.94952	.0532	.3790	.4523	.19001	.24188	29
32	.68878	.72497	.95007	.0526	.3794	.4518	.19027	.24156	28
33	.68899	.72477	.95062	.0519	.3797	.4514	.19053	.24123	27
34	.68920	.72457	.95118	.0513	.3801	.4510	.19080	.24091	26
35	0.68941	0.72437	0.95173	1.0507	1.3805	1.4505	.19106	.24059	25
36	.68962	.72417	.95229	.0501	.3809	.4501	.19132	.24027	24
37	.68983	.72397	.95284	.0495	.3813	.4496	.19159	.23995	23
38	.69004	.72377	.95340	.0489	.3817	.4492	.19185	.23963	22
39	.69025	.72357	.95395	.0483	.3820	.4487	.19212	.23931	21
40	0.69046	0.72337	0.95451	1.0477	1.3824	1.4483	.19238	.23899	20
41	.69067	.72317	.95506	.0470	.3828	.4479	.19265	.23867	19
42	.69088	.72297	.95562	.0464	.3832	.4474	.19291	.23835	18
43	.69109	.72277	.95618	.0458	.3836	.4470	.19318	.23803	17
44	.69130	.72257	.95673	.0452	.3840	.4465	.19344	.23772	16
45	0.69151	0.72236	0.95729	1.0446	1.3843	1.4461	.19371	.23740	15
46	.69172	.72216	.95785	.0440	.3847	.4457	.19398	.23708	14
47	.69193	.72196	.95841	.0434	.3851	.4452	.19424	.23676	13
48	.69214	.72176	.95897	.0428	.3855	.4448	.19451	.23645	12
49	.69235	.72156	.95952	.0422	.3859	.4443	.19478	.23613	11
50	0.69256	0.72136	0.96008	1.0416	1.3863	1.4439	.19505	.23582	10
51	.69277	.72116	.96064	.0410	.3867	.4435	.19532	.23550	9
52	.69298	.72095	.96120	.0404	.3871	.4430	.19558	.23519	8
53	.69319	.72075	.96176	.0398	.3874	.4426	.19585	.23487	7
54	.69340	.72055	.96232	.0392	.3878	.4422	.19612	.23456	6
55	0.69361	0.72035	0.96288	1.0385	1.3882	1.4417	.19639	.23424	5
56	.69382	.72015	.96344	.0379	.3886	.4413	.19666	.23393	4
57	.69403	.71995	.96400	.0373	.3890	.4409	.19693	.23362	3
58	.69424	.71974	.96457	.0367	.3894	.4404	.19720	.23330	2
59	.69445	.71954	.96513	.0361	.3898	.4400	.19747	.23299	1
60	0.69466	0.71934	0.96569	1.0355	1.3902	1.4396	.19774	.23268	0
M	Cosine	Sine	Cotan.	Tan.	Cosec.	Secant	READ DOWN	46°-47° Involute	M

44° or 224° Trigonometric and Involute Functions **135° or 315°**

M	Sine	Cosine	Tan.	Cotan.	Secant	Cosec.	Involute 44°-45°	READ UP	M
0	0.69466	0.71934	0.96569	1.0335	1.3902	1.4396	.19774	.23268	60
1	.69487	.71914	.96625	.0349	.3906	.4391	.19802	.23237	59
2	.69508	.71894	.96681	.0343	.3909	.4387	.19829	.23206	58
3	.69529	.71873	.96738	.0337	.3913	.4383	.19856	.23174	57
4	.69549	.71853	.96794	.0331	.3917	.4378	.19883	.23143	56
5	0.69570	0.71833	0.96850	1.0325	1.3921	1.4374	.19910	.23112	55
6	.69591	.71813	.96907	.0319	.3925	.4370	.19938	.23081	54
7	.69612	.71792	.96963	.0313	.3929	.4365	.19965	.23050	53
8	.69633	.71772	.97020	.0307	.3933	.4361	.19992	.23020	52
9	.69654	.71752	.97076	.0301	.3937	.4357	.20020	.22989	51
10	0.69675	0.71732	0.97133	1.0295	1.3941	1.4352	.20047	.22958	50
11	.69696	.71711	.97189	.0289	.3945	.4348	.20075	.22927	49
12	.69717	.71691	.97246	.0283	.3949	.4344	.20102	.22896	48
13	.69737	.71671	.97302	.0277	.3953	.4340	.20130	.22865	47
14	.69758	.71650	.97359	.0271	.3957	.4335	.20157	.22835	46
15	0.69779	0.71630	0.97416	1.0265	1.3961	1.4331	.20185	.22804	45
16	.69800	.71610	.97472	.0259	.3965	.4327	.20212	.22773	44
17	.69821	.71590	.97529	.0253	.3969	.4322	.20240	.22743	43
18	.69842	.71569	.97586	.0247	.3972	.4318	.20268	.22712	42
19	.69862	.71549	.97643	.0241	.3976	.4314	.20296	.22682	41
20	0.69883	0.71529	0.97700	1.0235	1.3980	1.4310	.20323	.22651	40
21	.69904	.71508	.97756	.0230	.3984	.4305	.20351	.22621	39
22	.69925	.71488	.97813	.0224	.3988	.4301	.20379	.22590	38
23	.69946	.71468	.97870	.0218	.3992	.4297	.20407	.22560	37
24	.69966	.71447	.97927	.0212	.3996	.4293	.20435	.22530	36
25	0.69987	0.71427	0.97984	1.0206	1.4000	1.4288	.20463	.22499	35
26	.70008	.71407	.98041	.0200	.4004	.4284	.20490	.22469	34
27	.70029	.71386	.98098	.0194	.4008	.4280	.20518	.22439	33
28	.70049	.71366	.98155	.0188	.4012	.4276	.20546	.22409	32
29	.70070	.71345	.98213	.0182	.4016	.4271	.20575	.22378	31
30	0.70091	0.71325	0.98270	1.0176	1.4020	1.4267	.20603	.22348	30
31	.70112	.71305	.98327	.0170	.4024	.4263	.20631	.22318	29
32	.70132	.71284	.98384	.0164	.4028	.4259	.20659	.22288	28
33	.70153	.71264	.98441	.0158	.4032	.4255	.20687	.22258	27
34	.70174	.71243	.98499	.0152	.4036	.4250	.20715	.22228	26
35	0.70195	0.71223	0.98556	1.0147	1.4040	1.4246	.20743	.22198	25
36	.70215	.71203	.98613	.0141	.4044	.4242	.20772	.22168	24
37	.70236	.71182	.98671	.0135	.4048	.4238	.20800	.22138	23
38	.70257	.71162	.98728	.0129	.4052	.4234	.20828	.22108	22
39	.70277	.71141	.98786	.0123	.4057	.4229	.20857	.22079	21
40	0.70298	0.71121	0.98843	1.0117	1.4061	1.4225	.20885	.22049	20
41	.70319	.71100	.98901	.0111	.4065	.4221	.20914	.22019	19
42	.70339	.71080	.98958	.0105	.4069	.4217	.20942	.21989	18
43	.70360	.71059	.99016	.0099	.4073	.4213	.20971	.21960	17
44	.70381	.71039	.99073	.0093	.4077	.4208	.20999	.21930	16
45	0.70401	0.71019	0.99131	1.0088	1.4081	1.4204	.21028	.21900	15
46	.70422	.70998	.99189	.0082	.4085	.4200	.21056	.21871	14
47	.70443	.70978	.99247	.0076	.4089	.4196	.21085	.21841	13
48	.70463	.70957	.99304	.0070	.4093	.4192	.21114	.21812	12
49	.70484	.70937	.99362	.0064	.4097	.4188	.21142	.21782	11
50	0.70505	0.70916	0.99420	1.0058	1.4101	1.4183	.21171	.21753	10
51	.70525	.70896	.99478	.0052	.4105	.4179	.21200	.21723	9
52	.70546	.70875	.99536	.0047	.4109	.4175	.21229	.21694	8
53	.70567	.70855	.99594	.0041	.4113	.4171	.21257	.21665	7
54	.70587	.70834	.99652	.0035	.4118	.4167	.21286	.21635	6
55	0.70608	0.70813	0.99710	1.0029	1.4122	1.4163	.21315	.21606	5
56	.70628	.70793	.99768	.0023	.4126	.4159	.21344	.21577	4
57	.70649	.70772	.99826	.0017	.4130	.4154	.21373	.21548	3
58	.70670	.70752	.99884	.0012	.4134	.4150	.21402	.21518	2
59	.70690	.70731	.99942	.0006	.4138	.4146	.21431	.21489	1
60	0.70711	0.70711	1.00000	1.0000	1.4142	1.4142	.21460	.21460	0
M	Cosine	Sine	Cotan.	Tan.	Cosec.	Secant	READ DOWN	45°-46° Involute	M

134° or 314° **45° or 225°**

Conversion Tables of Angular Measure. — The accompanying tables of degrees, minutes, and seconds into radians; radians into degrees, minutes, and seconds; radians into degrees and decimals of a degree; and minutes and seconds into decimals of a degree and vice versa facilitate the conversion of measurements.

Example: The Degrees, Minutes, and Seconds into Radians Table is used to find the number of radians in 324 degrees, 25 minutes, 13 seconds as follows:

$$
\begin{array}{rl}
300 \text{ degrees} & = 5.235988 \text{ radians} \\
20 \text{ degrees} & = 0.349066 \text{ radian} \\
4 \text{ degrees} & = 0.069813 \text{ radian} \\
25 \text{ minutes} & = 0.007272 \text{ radian} \\
13 \text{ seconds} & = 0.000063 \text{ radian} \\
\hline
324°25'13'' & = 5.662202 \text{ radians}
\end{array}
$$

Example: The Radians into Degrees and Decimals of a Degree, and Radians into Degrees, Minutes and Seconds Tables are used to find the number of decimal degrees or degrees, minutes and seconds in 0.734 radian as follows:

$$
\begin{array}{rl}
0.7 \text{ radian} & = 40.1070 \text{ degrees} \\
0.03 \text{ radian} & = 1.7189 \text{ degrees} \\
0.004 \text{ radian} & = 0.2292 \text{ degree} \\
\hline
0.734 \text{ radian} & = 42.0551 \text{ degrees}
\end{array}
\qquad
\begin{array}{rl}
0.7 \text{ radian} & = 40° \ 6'25'' \\
0.03 \text{ radian} & = 1°43' \ 8'' \\
0.004 \text{ radian} & = 0°13'45'' \\
\hline
0.734 \text{ radian} & = 41°62'78'' \text{ or } 42°3'18''
\end{array}
$$

Degrees, Minutes, and Seconds into Radians
(Based on 180 degrees = π radians)

Degrees into Radians

Deg.	Rad.	Deg.	Rad.	Deg.	Rad.	Deg.	Rad.	Deg.	Rad.	Deg.	Rad.
1000	17.453293	100	1.745329	10	0.174533	1	0.017453	0.1	0.001745	0.01	0.000175
2000	34.906585	200	3.490659	20	0.349066	2	0.034907	0.2	0.003491	0.02	0.000349
3000	52.359878	300	5.235988	30	0.523599	3	0.052360	0.3	0.005236	0.03	0.000524
4000	69.813170	400	6.981317	40	0.698132	4	0.069813	0.4	0.006981	0.04	0.000698
5000	87.266463	500	8.726646	50	0.872665	5	0.087266	0.5	0.008727	0.05	0.000873
6000	104.719755	600	10.471976	60	1.047198	6	0.104720	0.6	0.010472	0.06	0.001047
7000	122.173048	700	12.217305	70	1.221730	7	0.122173	0.7	0.012217	0.07	0.001222
8000	139.626340	800	13.962634	80	1.396263	8	0.139626	0.8	0.013963	0.08	0.001396
9000	157.079633	900	15.707963	90	1.570796	9	0.157080	0.9	0.015708	0.09	0.001571
10000	174.532925	1000	17.453293	100	1.745329	10	0.174533	1.0	0.017453	0.10	0.001745

Minutes into Radians

Min.	Rad.	Min.	Rad.	Min.	Rad.	Min.	Rad.	Min.	Rad.	Min.	Rad.
1	0.000291	11	0.003200	21	0.006109	31	0.009018	41	0.011926	51	0.014835
2	0.000582	12	0.003491	22	0.006400	32	0.009308	42	0.012217	52	0.015126
3	0.000873	13	0.003782	23	0.006690	33	0.009599	43	0.012508	53	0.015417
4	0.001164	14	0.004072	24	0.006981	34	0.009890	44	0.012799	54	0.015708
5	0.001454	15	0.004363	25	0.007272	35	0.010181	45	0.013090	55	0.015999
6	0.001745	16	0.004654	26	0.007563	36	0.010472	46	0.013381	56	0.016290
7	0.002036	17	0.004945	27	0.007854	37	0.010763	47	0.013672	57	0.016581
8	0.002327	18	0.005236	28	0.008145	38	0.011054	48	0.013963	58	0.016872
9	0.002618	19	0.005527	29	0.008436	39	0.011345	49	0.014254	59	0.017162
10	0.002909	20	0.005818	30	0.008727	40	0.011636	50	0.014544	60	0.017453

Seconds into Radians

Sec.	Rad.	Sec.	Rad.	Sec.	Rad.	Sec.	Rad.	Sec.	Rad.	Sec.	Rad.
1	0.000005	11	0.000053	21	0.000102	31	0.000150	41	0.000199	51	0.000247
2	0.000010	12	0.000058	22	0.000107	32	0.000155	42	0.000204	52	0.000252
3	0.000015	13	0.000063	23	0.000112	33	0.000160	43	0.000208	53	0.000257
4	0.000019	14	0.000068	24	0.000116	34	0.000165	44	0.000213	54	0.000262
5	0.000024	15	0.000073	25	0.000121	35	0.000170	45	0.000218	55	0.000267
6	0.000029	16	0.000078	26	0.000126	36	0.000175	46	0.000223	56	0.000271
7	0.000034	17	0.000082	27	0.000131	37	0.000179	47	0.000228	57	0.000276
8	0.000039	18	0.000087	28	0.000136	38	0.000184	48	0.000233	58	0.000281
9	0.000044	19	0.000092	29	0.000141	39	0.000189	49	0.000238	59	0.000286
10	0.000048	20	0.000097	30	0.000145	40	0.000194	50	0.000242	60	0.000291

Radians into Degrees and Decimals of a Degree
(Based on π radians = 180 degrees)

Rad.	Deg.	Rad.	Deg.	Rad.	Deg.	Rad.	Deg.	Rad.	Deg.	Rad.	Deg.
10	572.9578	1	57.2958	0.1	5.7296	0.01	0.5730	0.001	0.0573	0.0001	0.0057
20	1145.9156	2	114.5916	0.2	11.4592	0.02	1.1459	0.002	0.1146	0.0002	0.0115
30	1718.8734	3	171.8873	0.3	17.1887	0.03	1.7189	0.003	0.1719	0.0003	0.0172
40	2291.8312	4	229.1831	0.4	22.9183	0.04	2.2918	0.004	0.2292	0.0004	0.0229
50	2864.7890	5	286.4789	0.5	28.6479	0.05	2.8648	0.005	0.2865	0.0005	0.0286
60	3437.7468	6	343.7747	0.6	34.3775	0.06	3.4377	0.006	0.3438	0.0006	0.0344
70	4010.7046	7	401.0705	0.7	40.1070	0.07	4.0107	0.007	0.4011	0.0007	0.0401
80	4583.6624	8	458.3662	0.8	45.8366	0.08	4.5837	0.008	0.4584	0.0008	0.0458
90	5156.6202	9	515.6620	0.9	51.5662	0.09	5.1566	0.009	0.5157	0.0009	0.0516
100	5729.5780	10	572.9578	1.0	57.2958	0.10	5.7296	0.010	0.5730	0.0010	0.0573

Radians into Degrees, Minutes and Seconds
(Based on π radians = 180 degrees)

Rad.	Angle	Rad.	Angle	Rad.	Angle	Rad.	Angle	Rad.	Angle	Rad.	Angle
10	572°57'28"	1	57°17'45"	0.1	5°43'46"	0.01	0°34'23"	0.001	0°3'26"	0.0001	0°0'21"
20	1145°54'56"	2	114°35'30"	0.2	11°27'33"	0.02	1°8'45"	0.002	0°6'53"	0.0002	0°0'41"
30	1718°52'24"	3	171°53'14"	0.3	17°11'19"	0.03	1°43'8"	0.003	0°10'19"	0.0003	0°1'2"
40	2291°49'52"	4	229°10'59"	0.4	22°55'6"	0.04	2°17'31"	0.004	0°13'45"	0.0004	0°1'23"
50	2864°47'20"	5	286°28'44"	0.5	28°38'52"	0.05	2°51'53"	0.005	0°17'11"	0.0005	0°1'43"
60	3437°44'48"	6	343°46'29"	0.6	34°22'39"	0.06	3°26'16"	0.006	0°20'38"	0.0006	0°2'4"
70	4010°42'16"	7	401°4'14"	0.7	40°6'25"	0.07	4°0'39"	0.007	0°24'4"	0.0007	0°2'24"
80	4583°39'44"	8	458°21'58"	0.8	45°50'12"	0.08	4°35'1"	0.008	0°27'30"	0.0008	0°2'45"
90	5156°37'13"	9	515°39'43"	0.9	51°33'58"	0.09	5°9'24"	0.009	0°30'56"	0.0009	0°3'6"
100	5729°34'41"	10	572°57'28"	1.0	57°17'45"	0.10	5°43'46"	0.010	0°34'23"	0.0010	0°3'26"

Minutes and Seconds into Decimals of a Degree and Vice Versa
(Based on 1 second = 0.00027778 degrees)

Minutes into Decimals of a Degree						Seconds into Decimals of a Degree					
Min.	Deg.	Min.	Deg.	Min.	Deg.	Sec.	Deg.	Sec.	Deg.	Sec.	Deg.
1	0.0167	21	0.3500	41	0.6833	1	0.0003	21	0.0058	41	0.0114
2	0.0333	22	0.3667	42	0.7000	2	0.0006	22	0.0061	42	0.0117
3	0.0500	23	0.3833	43	0.7167	3	0.0008	23	0.0064	43	0.0119
4	0.0667	24	0.4000	44	0.7333	4	0.0011	24	0.0067	44	0.0122
5	0.0833	25	0.4167	45	0.7500	5	0.0014	25	0.0069	45	0.0125
6	0.1000	26	0.4333	46	0.7667	6	0.0017	26	0.0072	46	0.0128
7	0.1167	27	0.4500	47	0.7833	7	0.0019	27	0.0075	47	0.0131
8	0.1333	28	0.4667	48	0.8000	8	0.0022	28	0.0078	48	0.0133
9	0.1500	29	0.4833	49	0.8167	9	0.0025	29	0.0081	49	0.0136
10	0.1667	30	0.5000	50	0.8333	10	0.0028	30	0.0083	50	0.0139
11	0.1833	31	0.5167	51	0.8500	11	0.0031	31	0.0086	51	0.0142
12	0.2000	32	0.5333	52	0.8667	12	0.0033	32	0.0089	52	0.0144
13	0.2167	33	0.5500	53	0.8833	13	0.0036	33	0.0092	53	0.0147
14	0.2333	34	0.5667	54	0.9000	14	0.0039	34	0.0094	54	0.0150
15	0.2500	35	0.5833	55	0.9167	15	0.0042	35	0.0097	55	0.0153
16	0.2667	36	0.6000	56	0.9333	16	0.0044	36	0.0100	56	0.0156
17	0.2833	37	0.6167	57	0.9500	17	0.0047	37	0.0103	57	0.0158
18	0.3000	38	0.6333	58	0.9667	18	0.0050	38	0.0106	58	0.0161
19	0.3167	39	0.6500	59	0.9833	19	0.0053	39	0.0108	59	0.0164
20	0.3333	40	0.6667	60	1	20	0.0056	40	0.0111	60	0.0167

Example 1: Convert 11'37" to decimals of a degree. From the left table, 11' = 0.1833 degree. From the right table, 37" = 0.0103 degree. Adding, 11'37" = 0.1833 + 0.0103 = 0.1936 degree.

Example 2: Convert 0.1234 degree to minutes and seconds. From the left table, 0.1167 degree = 7'. Subtracting 0.1167 from 0.1234 gives 0.0067. From the right table, 0.0067 = 24" so that 0.1234 = 7'24".

Compound Angles. — Three types of compound angles are illustrated by Figs. 1–6. The first type is shown in Figs. 1, 2 and 3; the second type in Fig. 4 and the third type in Figs. 5 and 6.

In Fig. 1 is shown what might be considered as a thread-cutting tool without front clearance. A is a known angle in plane y–y of the top surface. C is the corresponding angle in plane x–x which is at some given angle B with plane y–y. Thus, angles A and B are components of the compound angle C.

Problem Referring to Fig. 1: The angle 2 A in plane y–y is known, as is also the angle B between planes x–x and y–y. It is required to find the compound angle 2 C in plane x–x.

Solution:	Let 2 A = 60° and B = 15°
Then	$\tan C = \tan A \cos B$
	$\tan C = \tan 30° \cos 15°$
	$\tan C = 0.57735 \times 0.96592$
	$\tan C = 0.55767$
	$C = 29°8.8'$ 2 $C = 58°$ 17.6$'$

Fig. 2 shows a thread-cutting tool with front clearance angle B. Angle A equals one-half the angle between the cutting edges in plane y–y of the top surface and compound angle C is one-half the angle between the cutting edges in a plane x–x at right angles to the inclined front edge of the tool. The angle between planes y–y and x–x is, therefore, equal to clearance angle B.

Problem Referring to Fig. 2: Find the angle 2 C between the front faces of a thread-cutting tool having a known clearance angle B, which will permit the grinding of these faces so that their top edges will form the desired angle 2 A for cutting the thread.

Solution:	Let 2 A = 60° and B = 15°
Then	$\tan C = \dfrac{\tan A}{\cos B} = \dfrac{\tan 30°}{\cos 15°} = \dfrac{0.57735}{0.96592}$
	$\tan C = 0.59772$
	$C = 30°\ 52'$ 2 $C = 61°\ 44'$

In Fig. 3 is shown a form-cutting tool in which A is one-half the angle between the cutting edges in plane y–y of the top surface; B is the front clearance angle; and C is one-half the angle between the cutting edges in plane x–x at right angles to the front edges of the tool. The formula for finding angle C when angles A and B are known is the same as that for Fig. 2.

Problem Referring to Fig. 3: Find the angle 2 C between the front faces of a form-cutting tool having a known clearance angle B which will permit the grinding of these faces so that their top edges will form the desired angle 2 A for form cutting.

Solution:	Let 2 A = 46° and B = 12°
Then	$\tan C = \dfrac{\tan A}{\cos B} = \dfrac{\tan 23°}{\cos 12°} = \dfrac{0.42447}{0.97815}$
	$\tan C = 0.43395$
	$= 23°\ 27.5'$ 2 $C = 46°\ 55'$

In Fig. 4 is shown a wedge-shaped block, the top surface of which is inclined at compound angle C with the base in a plane at right angles with the base and at angle R with the front edge. Angle A in the vertical plane of the front of the plate

Formulas for Compound Angles

C = compound angle in plane x-x and is the resultant of angles A and B

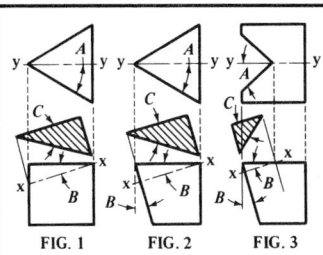

FIG. 1   FIG. 2   FIG. 3

For given angles A and B, find the resultant angle C in plane x-x. Angle B is measured in vertical plane y-y of midsection.

(Fig. 1) $\tan C = \tan A \times \cos B$

(Fig. 2) $\tan C = \dfrac{\tan A}{\cos B}$

(Fig. 3) (Same formula as for Fig. 2)

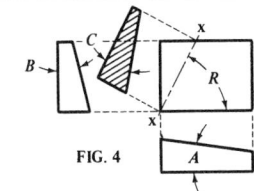

FIG. 4

Fig. 4. In machining plate to angles A and B, it is held at angle C in plane x-x. Angle of rotation R in plane parallel to base (or complement of R) is for locating plate so that plane x-x is perpendicular to axis of pivot on angle-plate or work-holding vise.

$$\tan R = \frac{\tan B}{\tan A}\ ;\ \tan C = \frac{\tan A}{\cos R}$$

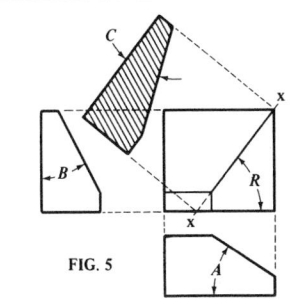

FIG. 5

Fig. 5. Angle R in horizontal plane parallel to base is angle from plane x-x to side having angle A.

$$\tan R = \frac{\tan A}{\tan B}$$

$$\tan C = \tan A \cos R = \tan B \sin R$$

Compound angle C is angle in plane x-x from base to corner formed by intersection of planes inclined to angles A and B. This formula for C may be used to find cot of complement of C_1, Fig. 6.

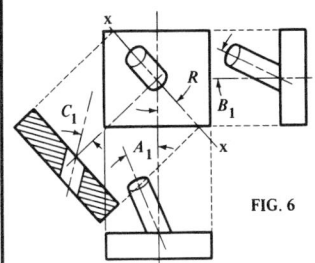

FIG. 6

Fig. 6. Angles A_1 and B_1 are measured in vertical planes of front and side elevations. Plane x-x is located by angle R from centerline or from plane of angle B_1.

$$\tan R = \frac{\tan A_1}{\tan B_1}$$

$$\tan C_1 = \frac{\tan A_1}{\sin R} = \frac{\tan B_1}{\cos R}$$

The resultant angle C_1 would be required in drilling hole for pin.

and angle B in the vertical plane of one side which is at right angles to the front are components of angle C.

Problem Referring to Fig. 4: Find the compound angle C of a wedge-shaped block having known component angles A and B in sides at right angles to each other.

Solution:

$$\text{Let } A = 47^\circ\ 14' \text{ and } B = 38^\circ\ 10'$$

$$\tan R = \frac{\tan B}{\tan A} = \frac{\tan 38^\circ\ 10'}{\tan 47^\circ\ 14'}$$

$$\tan R = \frac{0.78598}{1.0812} = 0.72695$$

$$R = 36^\circ\ 0.9'$$

$$\tan C = \frac{\tan A}{\cos R} = \frac{\tan 47^\circ\ 14'}{\cos 36^\circ\ 0.9'}$$

$$\tan C = \frac{1.0812}{0.80887} = 1.3367$$

$$C = 53^\circ\ 12'$$

In Fig. 5 is shown a four-sided block, two sides of which are at right angles to each other and to the base of the block. The other two sides are inclined at an oblique angle with the base. Angle C is a compound angle formed by the intersection of these two inclined sides and the intersection of a vertical plane passing through $x-x$, and the base of the block. The components of angle C are angles A and B and angle R is the angle in the base plane of the block between the plane of angle C and the plane of angle A.

Problem Referring to Fig. 5: Find the angles C and R in the block shown in Fig. 5 when angles A and B are known.

Solution: Let angle $A = 27^\circ$ and $B = 36^\circ$

$$\tan R = \frac{\cot B}{\cot A} = \frac{\cot 36^\circ}{\cot 27^\circ} = \frac{1.3764}{1.9626}$$

$$\tan R = 0.70131 \qquad R = 35^\circ\ 2.5'$$

$$\cot C = \sqrt{\cot^2 A + \cot^2 B} = \sqrt{(1.9626)^2 + (1.3764)^2}$$

$$\cot C = \sqrt{5.74627572} = 2.3971$$

$$C = 22^\circ\ 38.6'$$

Problem Referring to Fig. 6: A rod or pipe is inserted into a rectangular block at an angle. Angle C_1 is the compound angle of inclination (measured from the vertical) in a plane passing through the center line of the rod or pipe and at right angles to the top surface of the block. Angles A_1 and B_1 are the angles of inclination of the rod or pipe when viewed respectively in the front and side planes of the block. Angle R is the angle between the plane of angle C_1 and the plane of angle B_1. Find angles C_1 and R when a rod or pipe is inclined at known angles A_1 and B_1.

Solution: Let $A_1 = 39^\circ$ and $B_1 = 34^\circ$

Then

$$\tan C_1 = \sqrt{\tan^2 A_1 + \tan^2 B_1} = \sqrt{(0.80978)^2 + (0.67451)^2}$$

$$\tan C_1 = \sqrt{1.1107074} = 1.0539$$

$$C_1 = 46^\circ\ 30.2'$$

$$\tan R = \frac{\tan A_1}{\tan B_1} = \frac{0.80978}{0.67451}$$

$$\tan R = 1.2005 \qquad R = 50^\circ\ 12.4'$$

INSPECTION*

Micrometers.—Micrometers are linear measuring instruments used to make measurements to accuracies of .001 and .0001 of an inch and .01 and .002 of a millimeter. The most common type of micrometer is the micrometer caliper, Fig. 1. This instrument consists of several parts: The *frame* is the main unit of the micrometer caliper, which maintains the alignment of the contact points on the ends of the *spindle* and *anvil.* The *sleeve* is also attached to the frame and, on inch-type micrometers, has graduations that divide the travel of the spindle into 40 equal parts, Fig. 2a. The *thimble,* which rotates around the sleeve, is also graduated. These graduations divide the rotation of the thimble into 25 equal parts. So together, the sleeve and thimble graduations of the micrometer caliper divide an inch into 1000 (40 × 25 = 1000) parts. The *rachet stop,* located at the end of the thimble, is used to limit the pressure between the contact points. Using the ratchet stop when making measurements with a micrometer will ensure the correct measuring pressure. Some micrometer calipers also have a graduated scale on the top side of the sleeve. This scale is called the vernier scale and is used to make measurements to an accuracy of .0001 inch.

Millimeter micrometers, Fig. 2b, have the same basic construction as the inch-type micrometer, the only difference being in the graduations. The sleeve on a millimeter micrometer is graduated in half millimeter units. The thimble is divided into 50 equal parts each equal to .01 mm. So together, the sleeve and thimble graduations divide a millimeter into 100 parts. The millimeter micrometer may also have a vernier scale on the top side of the sleeve. This vernier scale further divides the micrometer reading and permits readings to an accuracy of .002 mm.

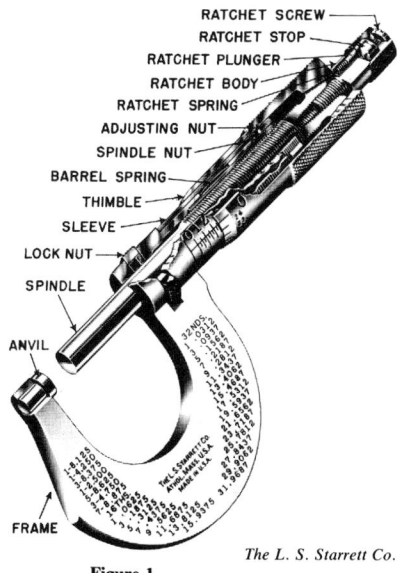

The L. S. Starrett Co.

Figure 1

*Portions of the chapter have been taken from *Print Reading for Industry,* by permission of the South-Western Publishing Co., Cincinnati, OH.

Reading An Inch Micrometer: Each graduation on the sleeve of a micrometer is equal to ¼₀″. Since micrometers are read in decimal inches, this value is expressed as .025″ (1 ÷ 40). As an aid to reading the micrometer, every fourth line is numbered. Each of these numbered lines is equal to .100″. To determine the value of the sleeve graduations, simply count the number of graduations visible to the left of the thimble and multiply by 25. Do not count any lines that are partially covered by the thimble. Next, note the value of the thimble graduation aligned with the reading line on the sleeve and add this amount to the value of the sleeve graduations. The micrometer reading shown in Fig. 3a has a sleeve value of .350″ (14 lines). The thimble graduation aligned with the reading line is 15, that is, .015″. So adding these together the total reading is expressed as .365″.

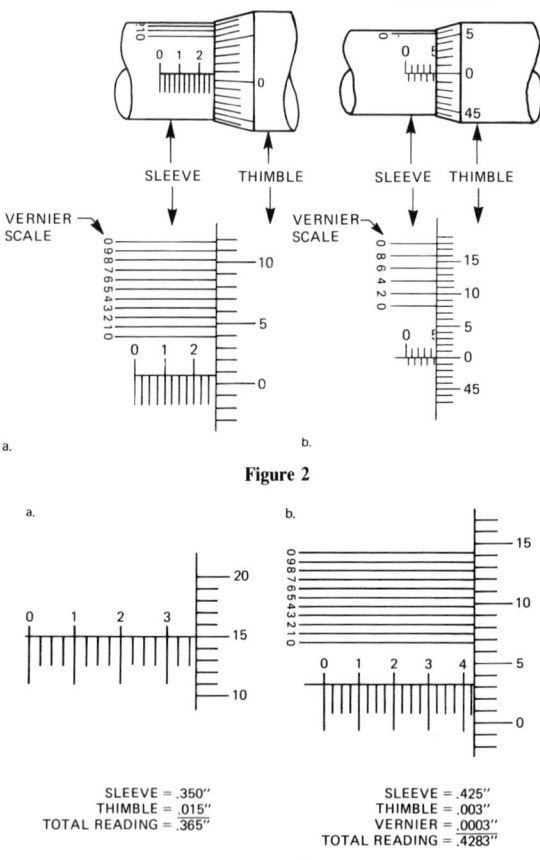

Figure 2

SLEEVE = .350″
THIMBLE = .015″
TOTAL READING = .365″

SLEEVE = .425″
THIMBLE = .003″
VERNIER = .0003″
TOTAL READING = .4283″

Figure 3

To read a vernier micrometer, simply read the micrometer as outlined above and add the value of the vernier scale. To read the vernier scale, first turn the micrometer so the complete vernier scale is visible. Now, find the one line on the vernier scale that is aligned with one of the thimble lines. There will only be one line perfectly aligned at any one time. Remember, use the value of the vernier scale, not the thimble graduation. The purpose of a vernier scale on a micrometer is to divide the space between the thimble graduations into 10 equal parts.So, when reading this micrometer, the thimble graduations may not be exactly aligned with the reading line on the sleeve. In these cases, always use the lesser value. The micrometer reading shown in Fig 3b shows how this type of micrometer should be read. The sleeve graduations visible to the left of the thimble equal .425″ (17 lines). The thimble location is between the thimble numbers 3 and 4 and the vernier is aligned on the vernier number 3. Adding all this together, the reading is expressed as a .4283″. This value should be read as 4 thousand 2 hundred and 83 ten-thousandths of an inch.

Reading A Millimeter Micrometer: When reading a millimeter micrometer, the process is identical to that used for the inch-type micrometer. First, find the value of the sleeve graduations visible to the left of the thimble. On millimeter micrometers the graduations are each equal to .5 mm. Most millimeter micrometers only number every 10 mm on the sleeve so you will have to pay close attention when counting these lines. Once you know the value of the sleeve graduations, note the value of the thimble graduation aligned with the reading line and add the two values. In the example shown in Fig 4a, the sleeve graduations equal 14.5 mm and the thimble is aligned on the number 12. Together, these values equal 14.62 mm.

To read a millimeter vernier micrometer, first determine the value of the measurement to an accuracy of .01 mm. Then find the one line on the vernier scale that is aligned with a thimble graduation and add this amount to the first reading. As shown in Fig. 4b, the sleeve graduations equal 6.5 mm. The thimble is between the numbers 5 and 6 and the vernier is aligned on the number 8. Added together this value is 6.558 mm. This value should be read as 6 and 5 hundred and 58 thousandths millimeter.

Verniers.—Verniers are another form of linear measuring tool frequently used for making measurements in the shop. The most popular type of vernier instrument is the *vernier caliper,* Fig. 5. The vernier caliper consists of a *beam,* which contains the *solid jaw* and the *main scale* graduations. The *movable jaw* is mounted on the beam

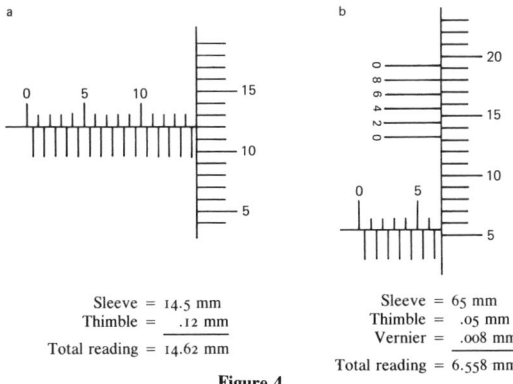

a			
	Sleeve =	14.5 mm	
	Thimble =	.12 mm	
	Total reading =	14.62 mm	

b			
	Sleeve =	65 mm	
	Thimble =	.05 mm	
	Vernier =	.008 mm	
	Total reading =	6.558 mm	

Figure 4

and can be moved and clamped in any desired position along the length of the beam with the *clamp screws*. The *vernier scale* is contained within the movable jaw and permits the vernier caliper to measure to within .001″ or .02 mm. Some types of vernier calipers also have a *fine adjustment slide* to aid in making accurate measurements. The major variations of the vernier caliper are the 25-division vernier, the 50-division vernier, the millimeter vernier, and the dial caliper.

Reading A 25-division Vernier: The beam of the 25-division vernier has a main scale that is graduated in increments of ¼₀″. Each graduation on the main scale is equal to .025″, just like the scale on the sleeve of a micrometer. The vernier plate on this vernier, like the micrometer thimble, has 25 divisions and is used to further divide the main scale graduations to .001″. To read this vernier, first locate the left zero on the vernier plate with reference to the main scale. Include all graduations to the left of the zero in your reading. Like the micrometer sleeve, the main scale graduations are also numbered at each .100″ value. The larger numbers on the main scale represent full inch graduations. You should always remember to include any full inch values in your measurements with a vernier. Once the zero is located and the value noted, find the one line on the vernier scale that is aligned with a line on the main scale and add this value to the total reading. Here again, remember to use the value of the vernier graduation and not the main scale number when determining this value. To read the vernier shown in Fig. 6a, first find the main scale value. In this case this value is 4.625″. Now find the vernier scale line that is aligned with a line on the main scale. In this example the number 16 is aligned. So, adding these values the total reading becomes 4.641″.

Reading A 50-division Vernier: The 50-division vernier is read in much the same way as the 25-division vernier. The only difference is in the main scale and vernier scale graduations. The graduations on the main scale of this vernier are equal to .050″ and the vernier scale has 50, rather than 25, divisions. To read this vernier, first find the value of the main scale graduations to the left of the zero. Then, find the one vernier scale graduation that is aligned with one of the lines on the main scale and add the values together. To read the vernier shown in Fig. 6b, first find the main scale value. In this case the value is 1.100″. Now find the vernier scale value. On this vernier the number 25 is the only number that is aligned. So, adding the values together, the total reading is 1.125″.

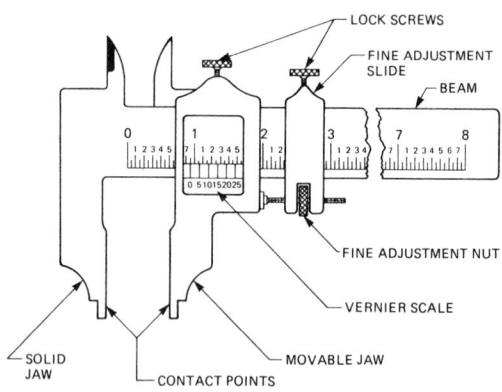

Figure 5

Reading A Millimeter Vernier: Millimeter verniers are again read in much the same way as the inch-type verniers. The main scale graduations on the millimeter vernier is graduated in full millimeter units, and the vernier scale is divided into 50 parts, each equal to .02 mm. To read this vernier, first locate the position of the zero and note the value of the reading to the left of the zero. Now, find the one line on the vernier scale that is aligned with a main scale graduation and add the two values. In the example shown in Fig. 6c, the main scale reading is 32 mm and the vernier scale is aligned at the number 58. So, adding these values the total reading is expressed as 32.58 mm.

Dial Calipers: Another variation of the vernier caliper is the *dial caliper*. These calipers are read by finding the value of the main scale graduations visible to the left of the reference edge and adding the dial reading, Fig. 7. With these calipers the dial replaces the vernier scale, generally make reading the instrument easier and faster

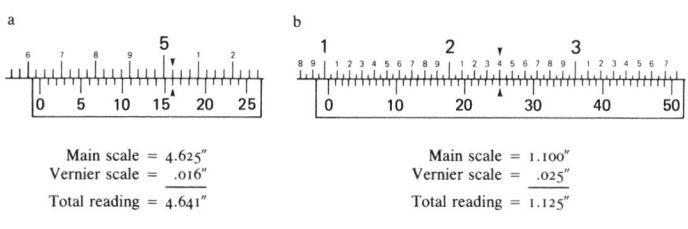

a

Main scale = 4.625″
Vernier scale = .016″

Total reading = 4.641″

b

Main scale = 1.100″
Vernier scale = .025″

Total reading = 1.125″

c

Main scale = 32.00 mm
Vernier scale = 58 mm

Total reading = 32.58 mm

Figure 6

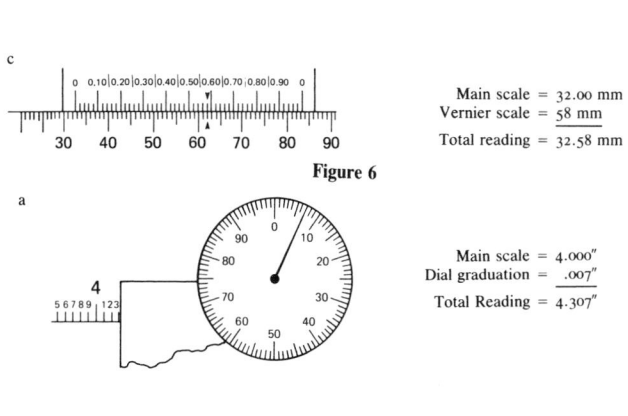

a

Main scale = 4.000″
Dial graduation = .007″

Total Reading = 4.307″

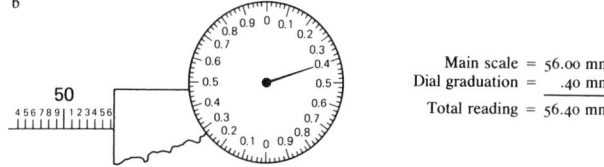

b

Main scale = 56.00 mm
Dial graduation = .40 mm

Total reading = 56.40 mm

Figure 7

than the standard vernier caliper. Inch-type dial calipers are accurate to .001″ and millimeter dial calipers are accurate to .02 mm. As shown, to read the inch-type dial caliper, Fig. 7a, the main scale value is first determined by reading the value shown to the left of the reference edge. In this reading it is 4.300″. The dial value is then read and added to the main scale value. In this case the measurement value is 4.307″. The millimeter dial vernier is read in the same way. As shown in Fig. 7b, the main scale value is 56.00 mm and the dial reading is .40 mm. Together these show a measurement value of 56.40 mm.

Protractors. — Protractors are measuring tools used to measure angles. The two basic types of protractors used in the shop are the *steel protractor* and the *vernier bevel protractor*, Fig. 8. The basic difference between these two forms of protractors is accuracy. The steel protractor, and its variations, such as the bevel protractor head on a combination square, are accurate to 1°, while the vernier bevel protractor is accurate to 5′.

The first step in measuring angles is understanding how angles are expressed.

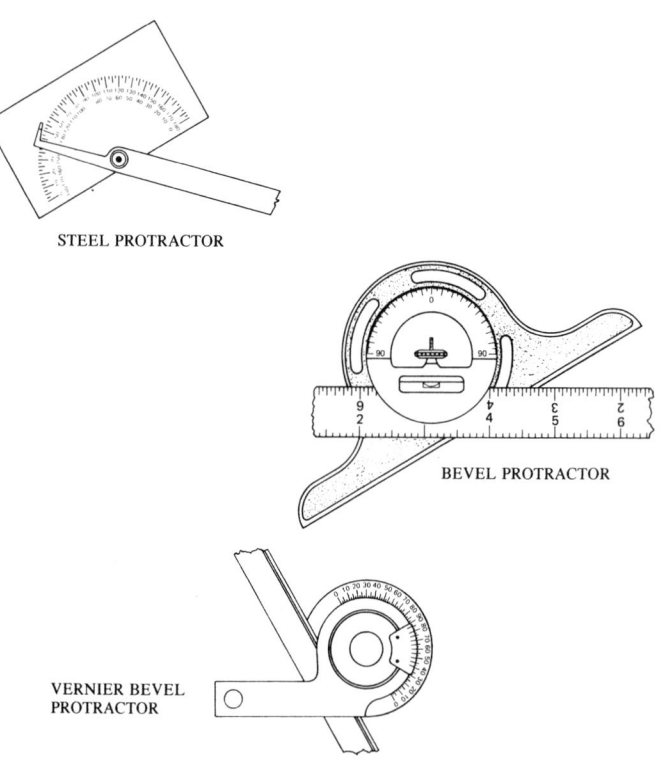

STEEL PROTRACTOR

BEVEL PROTRACTOR

VERNIER BEVEL
PROTRACTOR

Figure 8

When an angle is specified on a print, it is normally noted in units of *degrees* and *minutes*. Every circle contains 360°. Each degree contains 60'. So, if an angle were specified as 15° 30', it would be the same as 15 and ½ degrees. Minutes are also divided into *seconds* ("), but for all practical purposes, minutes are as close as you will be able to measure in the shop.

When reading a steel protractor, align the two sides of the protractor with the sides of the angle to be measured. Then, remove the protractor and note the position of the reference line against the protractor scale, Fig. 9a. Steel protractors are read directly, and the value shown on the protractor scale is the total value of the angle within 1°. Since this type of protractor is designed to be used for angles to 180°, the main scale on the protractor can be read from 0° to 180° in both directions. As shown, any measured value can be expressed in terms relative to 90° or 180°, depending the way the protractor is set. With this protractor setting, one side of the bar is set at 40° while the other side is at 140°.

When angles must be measured to an accuracy closer than 1°, a vernier bevel protractor may be used. To read a vernier bevel protractor, note the position of the zero reference line on the vernier scale. The vernier scale on this tool may be read in either direction, depending on the direction of rotation of the main scale. If the main scale is rotated to the right, or clockwise, the vernier scale to the right of the zero is used for the measurement. If, however, the main scale is rotated to the left, or counterclockwise, then the scale to the left of the zero is read. In the example shown in Fig. 9b, the main scale is rotated to the left and the zero is positioned between the 20° and 21° graduations. This means the main scale graduation is 20°. The one vernier scale graduation that is aligned with a line on the main scale is the number 15. So, adding these values together, the total reading is expressed as 20° 15".

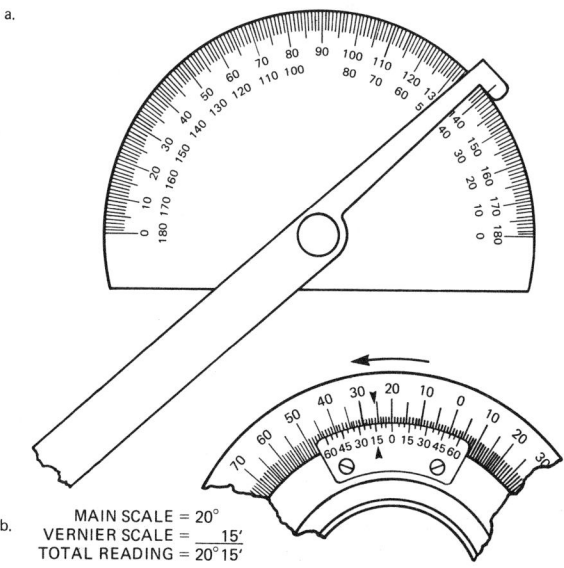

a.

b. MAIN SCALE = 20°
    VERNIER SCALE = ___15'
    TOTAL READING = 20°15'

Sine-bar. — The sine-bar is used either for very accurate angular measurements or for locating work at a given angle as, for example, in surface grinding templets, gages, etc. The sine-bar is especially useful in measuring or checking angles when the limit of accuracy is 5 minutes or less. Some bevel protractors are equipped with verniers which read to 5 minutes but the setting depends upon the alignment of graduations whereas a sine-bar usually is located by positive contact with precision gage-blocks selected for whatever dimension is required for obtaining a given angle.

Types of Sine-bars. — A sine-bar consists of a hardened, ground and lapped steel bar which has very accurate cylindrical plugs of equal diameter attached to or near each end. The form illustrated by Fig. 1 has notched ends for receiving the cylindrical plugs which are held firmly against both faces of the notch. The standard center-to-center distance C between the plugs is either 5 or 10 inches. The upper and lower sides of sine-bars are parallel to the center line of the plugs within very close limits. The body of the sine-bar ordinarily has several holes through it to reduce the weight. In the making of the sine-bar shown in Fig. 2, if too much material is removed from one locating notch, regrinding the shoulder at the opposite end would make it possible to obtain the correct center distance. That is the reason for this change in form. The type of sine-bar illustrated by Fig. 3 has the cylindrical disks or plugs attached to one side. These differences in form or arrangement do not, of course, affect the principle governing the use of the sine-bar. An accurate surface plate or master flat is always used in conjunction with a sine-bar in order to form the base from which the vertical measurements are made.

Setting 5-inch Sine-bar to Given Angle. — Since many sine-bars have a length of 5 inches, the accompanying table of constants is based upon that length. These constants represent the vertical distances H for setting a 5-inch sine-bar to the required angle. Assume that the angle is 31° 20'; the table shows that height H

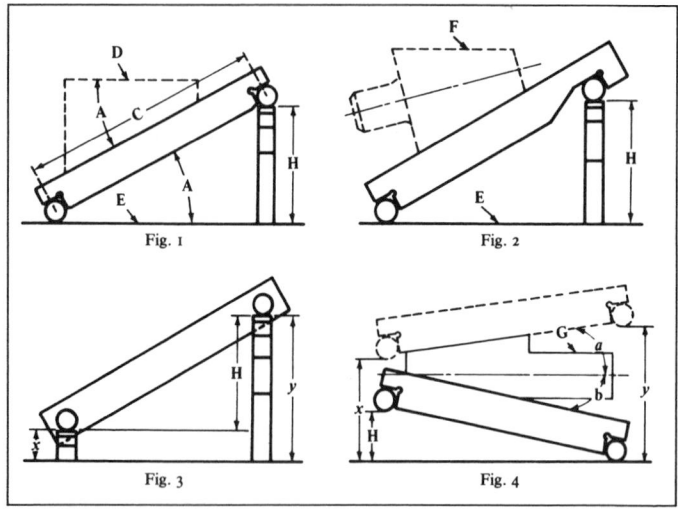

| Fig. 1 | Fig. 2 |
| Fig. 3 | Fig. 4 |

(Figs. 1, 2 and 3) should equal 2.6001 inches. Note: The constants in the table equal five times sine of the angle; thus the sine of 31° 20' (see table of trigonometric functions) is 0.52002, and 0.52002 × 5 = 2.6001 inches.

Finding Angle when Height *H* of Sine-bar is Known. — In finding the angle equivalent to a given height *H*, the table of constants is used in reverse order. To illustrate, if the height *H* is 1.4061 inches, the angle to which the sine-bar is set is 16° 20'. (Note: In using the regular table of sines, divide height *H* by length of sine-bar, find sine equal to quotient and its angle; thus 1.4061 ÷ 5 = 0.2812. Table of sines shows that this is the sine of 16° 20'.)

Checking Angle of Templet or Gage by Using Sine-bar. — Place templet or gage on sine-bar as indicated by dotted lines, Fig. 1. Clamps may be used to hold work in place. Place upper end of sine-bar on gage-blocks having total height *H* corresponding to the required angle. If upper edge *D* of work is parallel with surface plate *E*, then angle *A* of work equals angle *A* to which sine-bar is set. Parallelism between edge *D* and surface plate may be tested by checking the height at each end with a dial gage or some indicating type of comparator.

Measuring Angle of Templet or Gage with Sine-bar. — Adjust height of gage-blocks and sine-bar until edge *D*, Fig. 1, of gage or templet is parallel with surface plate *E*; then find angle corresponding to height *H* of gage-blocks. For example, if height *H* is 2.5939 inches when *D* and *E* are parallel, the table of sine-bar constants shows that angle *A* of work is 31° 15'.

Checking Taper per Foot with Sine-bar. — As an example, assume that plug gage, Fig. 2, is supposed to have a taper of 6⅛ inches per foot and taper is to be checked by using a 5-inch sine-bar. The table on page 201, Tapers per Foot and Corresponding Angles, shows that the included angle for a taper of 6⅛ inches per foot is 28° 38' 1" or practically 28° 38'. The table of sine-bar constants shows that height *H* of sine-bar should be 2.3960 inches; hence, if the upper surface *F* of the gage is parallel to surface *E* when sine-bar is at the height given, the angle corresponds to a taper of 6⅛ inches per foot.

Setting Sine-bar which has Plugs Attached to Side. — If lower plug does not rest directly upon the surface plate (see Fig. 3), note that height *H* for setting the sine-bar is the difference between heights *x* and *y* or the difference between the heights of the plugs; otherwise the procedure in setting the sine-bar and checking angles is the same as previously described.

Checking Templet Having Two Angles. — Assume that angle *a* of templet, Fig. 4, is 9 degrees, angle *b* 12 degrees, and that edge *G* is parallel to surface plate. Table shows that height *H* equals 1.03956 inches for an angle *b* of 12 degrees. For an angle *a* of 9 degrees, the table shows that the difference between measurements *x* and *y* when sine-bar is in contact with the upper edge of the templet should equal 0.78217 inch.

Setting 10-inch Sine-bar to Given Angle. — A 10-inch sine-bar may be preferable in some cases because of its longer working surface or because the longer center distance is conducive to greater precision. To obtain the vertical distances *H* for setting a 10-inch sine bar, first find the setting for a 5-inch sine bar in the following table and then multiply this setting by 2.

Example: As shown in the table, the setting for 39 degrees is 3.14660, hence the vertical height *H* for setting the 10-inch sine bar is 6.2932 inches.

Constants for Setting a 5-inch Sine-Bar — 1

Min.	0°	1°	2°	3°	4°	5°	6°	7°
0	0.00000	0.08726	0.17450	0.26168	0.34878	0.43578	0.52264	0.60935
1	0.00145	0.08872	0.17595	0.26313	0.35023	0.43723	0.52409	0.61079
2	0.00291	0.09017	0.17740	0.26458	0.35168	0.43868	0.52554	0.61223
3	0.00436	0.09162	0.17886	0.26604	0.35313	0.44013	0.52698	0.61368
4	0.00582	0.09308	0.18031	0.26749	0.35459	0.44157	0.52843	0.61512
5	0.00727	0.09453	0.18177	0.26894	0.35604	0.44302	0.52987	0.61656
6	0.00873	0.09599	0.18322	0.27039	0.35749	0.44447	0.53132	0.61801
7	0.01018	0.09744	0.18467	0.27185	0.35894	0.44592	0.53277	0.61945
8	0.01164	0.09890	0.18613	0.27330	0.36039	0.44737	0.53421	0.62089
9	0.01309	0.10035	0.18758	0.27475	0.36184	0.44882	0.53566	0.62234
10	0.01454	0.10180	0.18903	0.27620	0.36329	0.45027	0.53710	0.62378
11	0.01600	0.10326	0.19049	0.27766	0.36474	0.45171	0.53855	0.62522
12	0.01745	0.10471	0.19194	0.27911	0.36619	0.45316	0.54000	0.62667
13	0.01891	0.10617	0.19339	0.28056	0.36764	0.45461	0.54144	0.62811
14	0.02036	0.10762	0.19485	0.28201	0.36909	0.45606	0.54289	0.62955
15	0.02182	0.10907	0.19630	0.28346	0.37054	0.45751	0.54433	0.63099
16	0.02327	0.11053	0.19775	0.28492	0.37199	0.45896	0.54578	0.63244
17	0.02473	0.11198	0.19921	0.28637	0.37344	0.46040	0.54723	0.63388
18	0.02618	0.11344	0.20066	0.28782	0.37489	0.46185	0.54867	0.63532
19	0.02763	0.11489	0.20211	0.28927	0.37634	0.46330	0.55012	0.63677
20	0.02909	0.11634	0.20357	0.29072	0.37779	0.46475	0.55156	0.63821
21	0.03054	0.11780	0.20502	0.29218	0.37924	0.46620	0.55301	0.63965
22	0.03200	0.11925	0.20647	0.29363	0.38069	0.46765	0.55445	0.64109
23	0.03345	0.12071	0.20793	0.29508	0.38214	0.46909	0.55590	0.64254
24	0.03491	0.12216	0.20938	0.29653	0.38360	0.47054	0.55734	0.64398
25	0.03636	0.12361	0.21083	0.29798	0.38505	0.47199	0.55879	0.64542
26	0.03782	0.12507	0.21228	0.29944	0.38650	0.47344	0.56024	0.64686
27	0.03927	0.12652	0.21374	0.30089	0.38795	0.47489	0.56168	0.64830
28	0.04072	0.12798	0.21519	0.30234	0.38940	0.47633	0.56313	0.64975
29	0.04218	0.12943	0.21664	0.30379	0.39085	0.47778	0.56457	0.65119
30	0.04363	0.13088	0.21810	0.30524	0.39230	0.47923	0.56602	0.65263
31	0.04509	0.13234	0.21955	0.30669	0.39375	0.48068	0.56746	0.65407
32	0.04654	0.13379	0.22100	0.30815	0.39520	0.48212	0.56891	0.65551
33	0.04800	0.13525	0.22246	0.30960	0.39665	0.48357	0.57035	0.65696
34	0.04945	0.13670	0.22391	0.31105	0.39810	0.48502	0.57180	0.65840
35	0.05090	0.13815	0.22536	0.31250	0.39954	0.48647	0.57324	0.65984
36	0.05236	0.13961	0.22681	0.31395	0.40099	0.48791	0.57469	0.66128
37	0.05381	0.14106	0.22827	0.31540	0.40244	0.48936	0.57613	0.66272
38	0.05527	0.14252	0.22972	0.31686	0.40389	0.49081	0.57758	0.66417
39	0.05672	0.14397	0.23117	0.31831	0.40534	0.49226	0.57902	0.66561
40	0.05818	0.14542	0.23263	0.31976	0.40679	0.49370	0.58046	0.66705
41	0.05963	0.14688	0.23408	0.32121	0.40824	0.49515	0.58191	0.66849
42	0.06109	0.14833	0.23553	0.32266	0.40969	0.49660	0.58335	0.66993
43	0.06254	0.14979	0.23699	0.32411	0.41114	0.49805	0.58480	0.67137
44	0.06399	0.15124	0.23844	0.32556	0.41259	0.49949	0.58624	0.67281
45	0.06545	0.15269	0.23989	0.32702	0.41404	0.50094	0.58769	0.67425
46	0.06690	0.15415	0.24134	0.32847	0.41549	0.50239	0.58913	0.67570
47	0.06836	0.15560	0.24280	0.32992	0.41694	0.50383	0.59058	0.67714
48	0.06981	0.15705	0.24425	0.33137	0.41839	0.50528	0.59202	0.67858
49	0.07127	0.15851	0.24570	0.33282	0.41984	0.50673	0.59346	0.68002
50	0.07272	0.15996	0.24715	0.33427	0.42129	0.50818	0.59491	0.68146
51	0.07417	0.16141	0.24861	0.33572	0.42274	0.50962	0.59635	0.68290
52	0.07563	0.16287	0.25006	0.33717	0.42419	0.51107	0.59780	0.68434
53	0.07708	0.16432	0.25151	0.33863	0.42564	0.51252	0.59924	0.68578
54	0.07854	0.16578	0.25296	0.34008	0.42708	0.51396	0.60068	0.68722
55	0.07999	0.16723	0.25442	0.34153	0.42853	0.51541	0.60213	0.68866
56	0.08145	0.16868	0.25587	0.34298	0.42998	0.51686	0.60357	0.69010
57	0.08290	0.17014	0.25732	0.34443	0.43143	0.51830	0.60502	0.69154
58	0.08435	0.17159	0.25877	0.34588	0.43288	0.51975	0.60646	0.69298
59	0.08581	0.17304	0.26023	0.34733	0.43433	0.52120	0.60790	0.69443
60	0.08726	0.17450	0.26168	0.34878	0.43578	0.52264	0.60935	0.69587

Constants for Setting a 5-inch Sine-Bar — 2

Min.	8°	9°	10°	11°	12°	13°	14°	15°
0	0.69587	0.78217	0.86824	0.95404	1.03956	1.12476	1.20961	1.29410
1	0.69731	0.78361	0.86967	0.95547	1.04098	1.12617	1.21102	1.29550
2	0.69875	0.78505	0.87111	0.95690	1.04240	1.12759	1.21243	1.29690
3	0.70019	0.78648	0.87254	0.95833	1.04383	1.12901	1.21384	1.29831
4	0.70163	0.78792	0.87397	0.95976	1.04525	1.13042	1.21525	1.29971
5	0.70307	0.78935	0.87540	0.96118	1.04667	1.13184	1.21666	1.30112
6	0.70451	0.79079	0.87683	0.96261	1.04809	1.13326	1.21808	1.30252
7	0.70595	0.79223	0.87827	0.96404	1.04951	1.13467	1.21949	1.30393
8	0.70739	0.79366	0.87970	0.96546	1.05094	1.13609	1.22090	1.30533
9	0.70883	0.79510	0.88113	0.96689	1.05236	1.13751	1.22231	1.30673
10	0.71027	0.79653	0.88256	0.96832	1.05378	1.13892	1.22372	1.30814
11	0.71171	0.79797	0.88399	0.96974	1.05520	1.14034	1.22513	1.30954
12	0.71314	0.79941	0.88542	0.97117	1.05662	1.14175	1.22654	1.31095
13	0.71458	0.80084	0.88686	0.97260	1.05805	1.14317	1.22795	1.31235
14	0.71602	0.80228	0.88829	0.97403	1.05947	1.14459	1.22936	1.31375
15	0.71746	0.80371	0.88972	0.97545	1.06089	1.14600	1.23077	1.31516
16	0.71890	0.80515	0.89115	0.97688	1.06231	1.14742	1.23218	1.31656
17	0.72034	0.80658	0.89258	0.97830	1.06373	1.14883	1.23359	1.31796
18	0.72178	0.80802	0.89401	0.97973	1.06515	1.15025	1.23500	1.31937
19	0.72322	0.80945	0.89544	0.98116	1.06657	1.15166	1.23640	1.32077
20	0.72466	0.81089	0.89687	0.98258	1.06799	1.15308	1.23781	1.32217
21	0.72610	0.81232	0.89830	0.98401	1.06941	1.15449	1.23922	1.32357
22	0.72754	0.81376	0.89973	0.98544	1.07084	1.15591	1.24063	1.32498
23	0.72898	0.81519	0.90117	0.98686	1.07226	1.15732	1.24204	1.32638
24	0.73042	0.81663	0.90260	0.98829	1.07368	1.15874	1.24345	1.32778
25	0.73185	0.81806	0.90403	0.98971	1.07510	1.16015	1.24486	1.32918
26	0.73329	0.81950	0.90546	0.99114	1.07652	1.16157	1.24627	1.33058
27	0.73473	0.82093	0.90689	0.99256	1.07794	1.16298	1.24768	1.33199
28	0.73617	0.82237	0.90832	0.99399	1.07936	1.16440	1.24908	1.33339
29	0.73761	0.82380	0.90975	0.99541	1.08078	1.16581	1.25049	1.33479
30	0.73905	0.82524	0.91118	0.99684	1.08220	1.16723	1.25190	1.33619
31	0.74049	0.82667	0.91261	0.99826	1.08362	1.16864	1.25331	1.33759
32	0.74192	0.82811	0.91404	0.99969	1.08504	1.17006	1.25472	1.33899
33	0.74336	0.82954	0.91547	1.00112	1.08646	1.17147	1.25612	1.34040
34	0.74480	0.83098	0.91690	1.00254	1.08788	1.17288	1.25753	1.34180
35	0.74624	0.83241	0.91833	1.00396	1.08930	1.17430	1.25894	1.34320
36	0.74768	0.83384	0.91976	1.00539	1.09072	1.17571	1.26035	1.34460
37	0.74911	0.83528	0.92119	1.00681	1.09214	1.17712	1.26175	1.34600
38	0.75055	0.83671	0.92262	1.00824	1.09355	1.17854	1.26316	1.34740
39	0.75199	0.83815	0.92405	1.00966	1.09497	1.17995	1.26457	1.34880
40	0.75343	0.83958	0.92547	1.01109	1.09639	1.18136	1.26598	1.35020
41	0.75487	0.84101	0.92690	1.01251	1.09781	1.18278	1.26738	1.35160
42	0.75630	0.84245	0.92833	1.01394	1.09923	1.18419	1.26879	1.35300
43	0.75774	0.84388	0.92976	1.01536	1.10065	1.18560	1.27020	1.35440
44	0.75918	0.84531	0.93119	1.01678	1.10207	1.18702	1.27160	1.35580
45	0.76062	0.84675	0.93262	1.01821	1.10349	1.18843	1.27301	1.35720
46	0.76205	0.84818	0.93405	1.01963	1.10491	1.18984	1.27442	1.35860
47	0.76349	0.84961	0.93548	1.02106	1.10632	1.19125	1.27582	1.36000
48	0.76493	0.85105	0.93691	1.02248	1.10774	1.19267	1.27723	1.36140
49	0.76637	0.85248	0.93834	1.02390	1.10916	1.19408	1.27863	1.36280
50	0.76780	0.85391	0.93976	1.02533	1.11058	1.19549	1.28004	1.36420
51	0.76924	0.85535	0.94119	1.02675	1.11200	1.19690	1.28145	1.36560
52	0.77068	0.85678	0.94262	1.02817	1.11341	1.19832	1.28285	1.36700
53	0.77211	0.85821	0.94405	1.02960	1.11483	1.19973	1.28426	1.36840
54	0.77355	0.85965	0.94548	1.03102	1.11625	1.20114	1.28566	1.36980
55	0.77499	0.86108	0.94691	1.03244	1.11767	1.20255	1.28707	1.37119
56	0.77643	0.86251	0.94833	1.03387	1.11909	1.20396	1.28847	1.37259
57	0.77786	0.86394	0.94976	1.03529	1.12050	1.20538	1.28988	1.37399
58	0.77930	0.86538	0.95119	1.03671	1.12192	1.20679	1.29129	1.37539
59	0.78074	0.86681	0.95262	1.03814	1.12334	1.20820	1.29269	1.37679
60	0.78217	0.86824	0.95404	1.03956	1.12476	1.20961	1.29410	1.37819

Constants for Setting a 5-inch Sine-Bar — 3

Min.	16°	17°	18°	19°	20°	21°	22°	23°
0	1.37819	1.46186	1.54508	1.62784	1.71010	1.79184	1.87303	1.95366
1	1.37958	1.46325	1.54647	1.62922	1.71147	1.79320	1.87438	1.95499
2	1.38098	1.46464	1.54785	1.63059	1.71283	1.79456	1.87573	1.95633
3	1.38238	1.46603	1.54923	1.63197	1.71420	1.79591	1.87708	1.95767
4	1.38378	1.46742	1.55062	1.63334	1.71557	1.79727	1.87843	1.95901
5	1.38518	1.46881	1.55200	1.63472	1.71693	1.79863	1.87977	1.96035
6	1.38657	1.47020	1.55338	1.63609	1.71830	1.79998	1.88112	1.96169
7	1.38797	1.47159	1.55476	1.63746	1.71966	1.80134	1.88247	1.96302
8	1.38937	1.47298	1.55615	1.63884	1.72103	1.80270	1.88382	1.96436
9	1.39076	1.47437	1.55753	1.64021	1.72240	1.80405	1.88516	1.96570
10	1.39216	1.47576	1.55891	1.64159	1.72376	1.80541	1.88651	1.96704
11	1.39356	1.47715	1.56029	1.64296	1.72513	1.80677	1.88786	1.96837
12	1.39496	1.47854	1.56167	1.64433	1.72649	1.80812	1.88920	1.96971
13	1.39635	1.47993	1.56306	1.64571	1.72786	1.80948	1.89055	1.97105
14	1.39775	1.48132	1.56444	1.64708	1.72922	1.81083	1.89190	1.97238
15	1.39915	1.48271	1.56582	1.64845	1.73059	1.81219	1.89324	1.97372
16	1.40054	1.48410	1.56720	1.64983	1.73195	1.81355	1.89459	1.97506
17	1.40194	1.48549	1.56858	1.65120	1.73331	1.81490	1.89594	1.97639
18	1.40333	1.48687	1.56996	1.65257	1.73468	1.81626	1.89728	1.97773
19	1.40473	1.48826	1.57134	1.65394	1.73604	1.81761	1.89863	1.97906
20	1.40613	1.48965	1.57272	1.65532	1.73741	1.81897	1.89997	1.98040
21	1.40752	1.49104	1.57410	1.65669	1.73877	1.82032	1.90132	1.98173
22	1.40892	1.49243	1.57548	1.65806	1.74013	1.82168	1.90266	1.98307 ·
23	1.41031	1.49382	1.57687	1.65943	1.74150	1.82303	1.90401	1.98440
24	1.41171	1.49520	1.57825	1.66081	1.74286	1.82438	1.90535	1.98574
25	1.41310	1.49659	1.57963	1.66218	1.74422	1.82574	1.90670	1.98707
26	1.41450	1.49798	1.58101	1.66355	1.74559	1.82709	1.90804	1.98841
27	1.41589	1.49937	1.58238	1.66492	1.74695	1.82845	1.90939	1.98974
28	1.41729	1.50075	1.58376	1.66629	1.74831	1.82980	1.91073	1.99108
29	1.41868	1.50214	1.58514	1.66766	1.74967	1.83115	1.91207	1.99241
30	1.42008	1.50353	1.58652	1.66903	1.75104	1.83251	1.91342	1.99375
31	1.42147	1.50492	1.58790	1.67041	1.75240	1.83386	1.91476	1.99508
32	1.42287	1.50630	1.58928	1.67178	1.75376	1.83521	1.91610	1.99641
33	1.42426	1.50769	1.59066	1.67315	1.75512	1.83657	1.91745	1.99775
34	1.42565	1.50908	1.59204	1.67452	1.75649	1.83792	1.91879	1.99908
35	1.42705	1.51046	1.59342	1.67589	1.75785	1.83927	1.92013	2.00041
36	1.42844	1.51185	1.59480	1.67726	1.75921	1.84062	1.92148	2.00175
37	1.42984	1.51324	1.59617	1.67863	1.76057	1.84197	1.92282	2.00308
38	1.43123	1.51462	1.59755	1.68000	1.76193	1.84333	1.92416	2.00441
39	1.43262	1.51601	1.59893	1.68137	1.76329	1.84468	1.92550	2.00574
40	1.43402	1.51739	1.60031	1.68274	1.76465	1.84603	1.92685	2.00707
41	1.43541	1.51878	1.60169	1.68411	1.76601	1.84738	1.92819	2.00841
42	1.43680	1.52017	1.60306	1.68548	1.76737	1.84873	1.92953	2.00974
43	1.43820	1.52155	1.60444	1.68685	1.76873	1.85009	1.93087	2.01107
44	1.43959	1.52294	1.60582	1.68821	1.77010	1.85144	1.93221	2.01240
45	1.44098	1.52432	1.60720	1.68958	1.77146	1.85279	1.93355	2.01373
46	1.44237	1.52571	1.60857	1.69095	1.77282	1.85414	1.93490	2.01506
47	1.44377	1.52709	1.60995	1.69232	1.77418	1.85549	1.93624	2.01640
48	1.44516	1.52848	1.61133	1.69369	1.77553	1.85684	1.93758	2.01773
49	1.44655	1.52986	1.61271	1.69506	1.77689	1.85819	1.93892	2.01906
50	1.44794	1.53125	1.61408	1.69643	1.77825	1.85954	1.94026	2.02039
51	1.44934	1.53263	1.61546	1.69779	1.77961	1.86089	1.94160	2.02172
52	1.45073	1.53401	1.61683	1.69916	1.78097	1.86224	1.94294	2.02305
53	1.45212	1.53540	1.61821	1.70053	1.78233	1.86359	1.94428	2.02438
54	1.45351	1.53678	1.61959	1.70190	1.78369	1.86494	1.94562	2.02571
55	1.45490	1.53817	1.62096	1.70327	1.78505	1.86629	1.94696	2.02704
56	1.45629	1.53955	1.62234	1.70463	1.78641	1.86764	1.94830	2.02837
57	1.45769	1.54093	1.62371	1.70600	1.78777	1.86899	1.94964	2.02970
58	1.45908	1.54232	1.62509	1.70737	1.78912	1.87034	1.95098	2.03103
59	1.46047	1.54370	1.62647	1.70873	1.79048	1.87168	1.95232	2.03235
60	1.46186	1.54508	1.62784	1.71010	1.79184	1.87303	1.95366	2.03368

Constants for Setting a 5-inch Sine-Bar — 4

Min.	24°	25°	26°	27°	28°	29°	30°	31°
0	2.03368	2.11309	2.19186	2.26995	2.34736	2.42405	2.50000	2.57519
1	2.03501	2.11441	2.19316	2.27125	2.34864	2.42532	2.50126	2.57644
2	2.03634	2.11573	2.19447	2.27254	2.34993	2.42659	2.50252	2.57768
3	2.03767	2.11705	2.19578	2.27384	2.35121	2.42786	2.50378	2.57893
4	2.03900	2.11836	2.19708	2.27513	2.35249	2.42913	2.50504	2.58018
5	2.04032	2.11968	2.19839	2.27643	2.35378	2.43041	2.50630	2.58142
6	2.04165	2.12100	2.19970	2.27772	2.35506	2.43168	2.50755	2.58267
7	2.04298	2.12231	2.20100	2.27902	2.35634	2.43295	2.50881	2.58391
8	2.04431	2.12363	2.20231	2.28031	2.35763	2.43422	2.51007	2.58516
9	2.04563	2.12495	2.20361	2.28161	2.35891	2.43549	2.51133	2.58640
10	2.04696	2.12626	2.20492	2.28290	2.36019	2.43676	2.51259	2.58765
11	2.04829	2.12758	2.20622	2.28420	2.36147	2.43803	2.51384	2.58889
12	2.04962	2.12890	2.20753	2.28549	2.36275	2.43930	2.51510	2.59014
13	2.05094	2.13021	2.20883	2.28678	2.36404	2.44057	2.51636	2.59138
14	2.05227	2.13153	2.21014	2.28808	2.36532	2.44184	2.51761	2.59262
15	2.05359	2.13284	2.21144	2.28937	2.36660	2.44311	2.51887	2.59387
16	2.05492	2.13416	2.21275	2.29066	2.36788	2.44438	2.52013	2.59511
17	2.05625	2.13547	2.21405	2.29196	2.36916	2.44564	2.52138	2.59635
18	2.05757	2.13679	2.21536	2.29325	2.37044	2.44691	2.52264	2.59760
19	2.05890	2.13810	2.21666	2.29454	2.37172	2.44818	2.52389	2.59884
20	2.06022	2.13942	2.21796	2.29583	2.37300	2.44945	2.52515	2.60008
21	2.06155	2.14073	2.21927	2.29712	2.37428	2.45072	2.52640	2.60132
22	2.06287	2.14205	2.22057	2.29842	2.37556	2.45198	2.52766	2.60256
23	2.06420	2.14336	2.22187	2.29971	2.37684	2.45325	2.52891	2.60381
24	2.06552	2.14468	2.22318	2.30100	2.37812	2.45452	2.53017	2.60505
25	2.06685	2.14599	2.22448	2.30229	2.37940	2.45579	2.53142	2.60629
26	2.06817	2.14730	2.22578	2.30358	2.38068	2.45705	2.53268	2.60753
27	2.06949	2.14862	2.22708	2.30487	2.38196	2.45832	2.53393	2.60877
28	2.07082	2.14993	2.22839	2.30616	2.38324	2.45959	2.53519	2.61001
29	2.07214	2.15124	2.22969	2.30745	2.38452	2.46085	2.53644	2.61125
30	2.07347	2.15256	2.23099	2.30874	2.38579	2.46212	2.53769	2.61249
31	2.07479	2.15387	2.23229	2.31003	2.38707	2.46338	2.53894	2.61373
32	2.07611	2.15518	2.23359	2.31132	2.38835	2.46465	2.54020	2.61497
33	2.07744	2.15649	2.23489	2.31261	2.38963	2.46591	2.54145	2.61621
34	2.07876	2.15781	2.23619	2.31390	2.39090	2.46718	2.54270	2.61745
35	2.08008	2.15912	2.23749	2.31519	2.39218	2.46844	2.54396	2.61869
36	2.08140	2.16043	2.23880	2.31648	2.39346	2.46971	2.54521	2.61993
37	2.08273	2.16174	2.24010	2.31777	2.39474	2.47097	2.54646	2.62117
38	2.08405	2.16305	2.24140	2.31906	2.39601	2.47224	2.54771	2.62241
39	2.08537	2.16436	2.24270	2.32035	2.39729	2.47350	2.54896	2.62364
40	2.08669	2.16567	2.24400	2.32163	2.39857	2.47477	2.55021	2.62488
41	2.08801	2.16698	2.24530	2.32292	2.39984	2.47603	2.55146	2.62612
42	2.08934	2.16830	2.24659	2.32421	2.40112	2.47729	2.55271	2.62736
43	2.09066	2.16961	2.24789	2.32550	2.40239	2.47856	2.55397	2.62860
44	2.09198	2.17092	2.24919	2.32679	2.40367	2.47982	2.55522	2.62983
45	2.09330	2.17223	2.25049	2.32807	2.40494	2.48108	2.55647	2.63107
46	2.09462	2.17354	2.25179	2.32936	2.40622	2.48235	2.55772	2.63231
47	2.09594	2.17485	2.25309	2.33065	2.40749	2.48361	2.55896	2.63354
48	2.09726	2.17616	2.25439	2.33193	2.40877	2.48487	2.56021	2.63478
49	2.09858	2.17746	2.25569	2.33322	2.41004	2.48613	2.56146	2.63601
50	2.09990	2.17877	2.25698	2.33451	2.41132	2.48739	2.56271	2.63725
51	2.10122	2.18008	2.25828	2.33579	2.41259	2.48866	2.56396	2.63849
52	2.10254	2.18139	2.25958	2.33708	2.41386	2.48992	2.56521	2.63972
53	2.10386	2.18270	2.26088	2.33836	2.41514	2.49118	2.56646	2.64096
54	2.10518	2.18401	2.26217	2.33965	2.41641	2.49244	2.56771	2.64219
55	2.10650	2.18532	2.26347	2.34093	2.41769	2.49370	2.56895	2.64343
56	2.10782	2.18663	2.26477	2.34222	2.41896	2.49496	2.57020	2.64466
57	2.10914	2.18793	2.26606	2.34350	2.42023	2.49622	2.57145	2.64590
58	2.11045	2.18924	2.26736	2.34479	2.42150	2.49748	2.57270	2.64713
59	2.11177	2.19055	2.26866	2.34607	2.42278	2.49874	2.57394	2.64836
60	2.11309	2.19186	2.26995	2.34736	2.42405	2.50000	2.57519	2.64960

Constants for Setting a 5-inch Sine-Bar — 5

Min.	32°	33°	34°	35°	36°	37°	38°	39°
0	2.64960	2.72320	2.79596	2.86788	2.93893	3.00908	3.07831	3.14660
1	2.65083	2.72441	2.79717	2.86907	2.94010	3.01024	3.07945	3.14773
2	2.65206	2.72563	2.79838	2.87026	2.94128	3.01140	3.08060	3.14886
3	2.65330	2.72685	2.79958	2.87146	2.94246	3.01256	3.08174	3.14999
4	2.65453	2.72807	2.80079	2.87265	2.94363	3.01372	3.08289	3.15112
5	2.65576	2.72929	2.80199	2.87384	2.94481	3.01488	3.08403	3.15225
6	2.65699	2.73051	2.80319	2.87503	2.94598	3.01604	3.08518	3.15338
7	2.65822	2.73173	2.80440	2.87622	2.94716	3.01720	3.08632	3.15451
8	2.65946	2.73295	2.80560	2.87741	2.94833	3.01836	3.08747	3.15564
9	2.66069	2.73416	2.80681	2.87860	2.94951	3.01952	3.08861	3.15676
10	2.66192	2.73538	2.80801	2.87978	2.95068	3.02068	3.08976	3.15789
11	2.66315	2.73660	2.80921	2.88097	2.95185	3.02184	3.09090	3.15902
12	2.66438	2.73782	2.81042	2.88216	2.95303	3.02300	3.09204	3.16015
13	2.66561	2.73903	2.81162	2.88335	2.95420	3.02415	3.09318	3.16127
14	2.66684	2.74025	2.81282	2.88454	2.95538	3.02531	3.09433	3.16240
15	2.66807	2.74147	2.81402	2.88573	2.95655	3.02647	3.09547	3.16353
16	2.66930	2.74268	2.81523	2.88691	2.95772	3.02763	3.09661	3.16465
17	2.67053	2.74390	2.81643	2.88810	2.95889	3.02878	3.09775	3.16578
18	2.67176	2.74511	2.81763	2.88929	2.96007	3.02994	3.09890	3.16690
19	2.67299	2.74633	2.81883	2.89048	2.96124	3.03110	3.10004	3.16803
20	2.67422	2.74754	2.82003	2.89166	2.96241	3.03226	3.10118	3.16915
21	2.67545	2.74876	2.82123	2.89285	2.96358	3.03341	3.10232	3.17028
22	2.67668	2.74997	2.82243	2.89403	2.96475	3.03457	3.10346	3.17140
.23	2.67791	2.75119	2.82363	2.89522	2.96592	3.03572	3.10460	3.17253
24	2.67913	2.75240	2.82484	2.89641	2.96709	3.03688	3.10574	3.17365
25	2.68036	2.75362	2.82603	2.89759	2.96826	3.03803	3.10688	3.17478
26	2.68159	2.75483	2.82723	2.89878	2.96944	3.03919	3.10802	3.17590
27	2.68282	2.75605	2.82843	2.89996	2.97061	3.04034	3.10916	3.17702
28	2.68404	2.75726	2.82963	2.90115	2.97178	3.04150	3.11030	3.17815
29	2.68527	2.75847	2.83083	2.90233	2.97294	3.04265	3.11143	3.17927
30	2.68650	2.75968	2.83203	2.90351	2.97411	3.04381	3.11257	3.18039
31	2.68772	2.76090	2.83323	2.90470	2.97528	3.04496	3.11371	3.18151
32	2.68895	2.76211	2.83443	2.90588	2.97645	3.04611	3.11485	3.18264
33	2.69018	2.76332	2.83563	2.90707	2.97762	3.04727	3.11599	3.18376
34	2.69140	2.76453	2.83682	2.90825	2.97879	3.04842	3.11712	3.18488
35	2.69263	2.76575	2.83802	2.90943	2.97996	3.04957	3.11826	3.18600
36	2.69385	2.76696	2.83922	2.91061	2.98112	3.05073	3.11940	3.18712
37	2.69508	2.76817	2.84042	2.91180	2.98229	3.05188	3.12053	3.18824
38	2.69630	2.76938	2.84161	2.91298	2.98346	3.05303	3.12167	3.18936
39	2.69753	2.77059	2.84281	2.91416	2.98463	3.05418	3.12281	3.19048
40	2.69875	2.77180	2.84401	2.91534	2.98579	3.05533	3.12394	3.19160
41	2.69998	2.77301	2.84520	2.91652	2.98696	3.05648	3.12508	3.19272
42	2.70120	2.77422	2.84640	2.91771	2.98813	3.05764	3.12621	3.19384
43	2.70243	2.77543	2.84759	2.91889	2.98929	3.05879	3.12735	3.19496
44	2.70365	2.77664	2.84879	2.92007	2.99046	3.05994	3.12848	3.19608
45	2.70487	2.77785	2.84998	2.92125	2.99162	3.06109	3.12962	3.19720
46	2.70610	2.77906	2.85118	2.92243	2.99279	3.06224	3.13075	3.19831
47	2.70732	2.78027	2.85237	2.92361	2.99395	3.06339	3.13189	3.19943
48	2.70854	2.78148	2.85357	2.92479	2.99512	3.06454	3.13302	3.20055
49	2.70976	2.78269	2.85476	2.92597	2.99628	3.06568	3.13415	3.20167
50	2.71099	2.78389	2.85596	2.92715	2.99745	3.06683	3.13529	3.20278
51	2.71221	2.78510	2.85715	2.92833	2.99861	3.06798	3.13642	3.20390
52	2.71343	2.78631	2.85834	2.92950	2.99977	3.06913	3.13755	3.20502
53	2.71465	2.78752	2.85954	2.93068	3.00094	3.07028	3.13868	3.20613
54	2.71587	2.78873	2.86073	2.93186	3.00210	3.07143	3.13982	3.20725
55	2.71709	2.78993	2.86192	2.93304	3.00326	3.07257	3.14095	3.20836
56	2.71831	2.79114	2.86311	2.93422	3.00443	3.07372	3.14208	3.20948
57	2.71953	2.79235	2.86431	2.93540	3.00559	3.07487	3.14321	3.21059
58	2.72076	2.79355	2.86550	2.93657	3.00675	3.07601	3.14434	3.21171
59	2.72198	2.79476	2.86669	2.93775	3.00791	3.07716	3.14547	3.21282
60	2.72320	2.79596	2.86788	2.93893	3.00908	3.07831	3.14660	3.21394

Constants for Setting a 5-inch Sine-Bar — 6

Min.	40°	41°	42°	43°	44°	45°	46°	47°
0	3.21394	3.28030	3.34565	3.40999	3.47329	3.53553	3.59670	3.65677
1	3.21505	3.28139	3.34673	3.41106	3.47434	3.53656	3.59771	3.65776
2	3.21617	3.28249	3.34781	3.41212	3.47538	3.53759	3.59872	3.65875
3	3.21728	3.28359	3.34889	3.41318	3.47643	3.53862	3.59973	3.65974
4	3.21839	3.28468	3.34997	3.41424	3.47747	3.53965	3.60074	3.66073
5	3.21951	3.28578	3.35105	3.41531	3.47852	3.54067	3.60175	3.66172
6	3.22062	3.28688	3.35213	3.41637	3.47956	3.54170	3.60276	3.66271
7	3.22173	3.28797	3.35321	3.41743	3.48061	3.54273	3.60376	3.66370
8	3.22284	3.28907	3.35429	3.41849	3.48165	3.54375	3.60477	3.66469
9	3.22395	3.29016	3.35537	3.41955	3.48270	3.54478	3.60578	3.66568
10	3.22507	3.29126	3.35645	3.42061	3.48374	3.54580	3.60679	3.66667
11	3.22618	3.29235	3.35753	3.42168	3.48478	3.54683	3.60779	3.66766
12	3.22729	3.29345	3.35860	3.42274	3.48583	3.54785	3.60880	3.66865
13	3.22840	3.29454	3.35968	3.42380	3.48687	3.54888	3.60981	3.66964
14	3.22951	3.29564	3.36076	3.42486	3.48791	3.54990	3.61081	3.67063
15	3.23062	3.29673	3.36183	3.42591	3.48895	3.55093	3.61182	3.67161
16	3.23173	3.29782	3.36291	3.42697	3.48999	3.55195	3.61283	3.67260
17	3.23284	3.29892	3.36399	3.42803	3.49104	3.55297	3.61383	3.67359
18	3.23395	3.30001	3.36506	3.42909	3.49208	3.55400	3.61484	3.67457
19	3.23506	3.30110	3.36614	3.43015	3.49312	3.55502	3.61584	3.67556
20	3.23617	3.30219	3.36721	3.43121	3.49416	3.55604	3.61684	3.67655
21	3.23728	3.30329	3.36829	3.43227	3.49520	3.55707	3.61785	3.67753
22	3.23838	3.30438	3.36936	3.43332	3.49624	3.55809	3.61885	3.67852
23	3.23949	3.30547	3.37044	3.43438	3.49728	3.55911	3.61986	3.67950
24	3.24060	3.30656	3.37151	3.43544	3.49832	3.56013	3.62086	3.68049
25	3.24171	3.30765	3.37259	3.43649	3.49936	3.56115	3.62186	3.68147
26	3.24281	3.30874	3.37366	3.43755	3.50039	3.56217	3.62286	3.68245
27	3.24392	3.30983	3.37473	3.43861	3.50143	3.56319	3.62387	3.68344
28	3.24503	3.31092	3.37581	3.43966	3.50247	3.56421	3.62487	3.68442
29	3.24613	3.31201	3.37688	3.44072	3.50351	3.56523	3.62587	3.68540
30	3.24724	3.31310	3.37795	3.44177	3.50455	3.56625	3.62687	3.68639
31	3.24835	3.31419	3.37902	3.44283	3.50558	3.56727	3.62787	3.68737
32	3.24945	3.31528	3.38010	3.44388	3.50662	3.56829	3.62887	3.68835
33	3.25056	3.31637	3.38117	3.44494	3.50766	3.56931	3.62987	3.68933
34	3.25166	3.31746	3.38224	3.44599	3.50869	3.57033	3.63087	3.69031
35	3.25277	3.31854	3.38331	3.44704	3.50973	3.57135	3.63187	3.69130
36	3.25387	3.31963	3.38438	3.44810	3.51077	3.57236	3.63287	3.69228
37	3.25498	3.32072	3.38545	3.44915	3.51180	3.57338	3.63387	3.69326
38	3.25608	3.32181	3.38652	3.45020	3.51284	3.57440	3.63487	3.69424
39	3.25718	3.32289	3.38759	3.45126	3.51387	3.57541	3.63587	3.69522
40	3.25829	3.32398	3.38866	3.45231	3.51491	3.57643	3.63687	3.69620
41	3.25939	3.32507	3.38973	3.45336	3.51594	3.57745	3.63787	3.69718
42	3.26049	3.32615	3.39080	3.45441	3.51697	3.57846	3.63886	3.69816
43	3.26159	3.32724	3.39187	3.45546	3.51801	3.57948	3.63986	3.69913
44	3.26270	3.32832	3.39294	3.45651	3.51904	3.58049	3.64086	3.70011
45	3.26380	3.32941	3.39400	3.45757	3.52007	3.58151	3.64185	3.70109
46	3.26490	3.33049	3.39507	3.45862	3.52111	3.58252	3.64285	3.70207
47	3.26600	3.33158	3.39614	3.45967	3.52214	3.58354	3.64385	3.70305
48	3.26710	3.33266	3.39721	3.46072	3.52317	3.58455	3.64484	3.70402
49	3.26820	3.33375	3.39827	3.46177	3.52420	3.58557	3.64584	3.70500
50	3.26930	3.33483	3.39934	3.46281	3.52523	3.58658	3.64683	3.70598
51	3.27040	3.33591	3.40041	3.46386	3.52627	3.58759	3.64783	3.70695
52	3.27150	3.33700	3.40147	3.46491	3.52730	3.58861	3.64882	3.70793
53	3.27260	3.33808	3.40254	3.46596	3.52833	3.58962	3.64982	3.70890
54	3.27370	3.33916	3.40360	3.46701	3.52936	3.59063	3.65081	3.70988
55	3.27480	3.34025	3.40467	3.46806	3.53039	3.59164	3.65181	3.71085
56	3.27590	3.34133	3.40573	3.46910	3.53142	3.59266	3.65280	3.71183
57	3.27700	3.34241	3.40680	3.47015	3.53245	3.59367	3.65379	3.71280
58	3.27810	3.34349	3.40786	3.47120	3.53348	3.59468	3.65478	3.71378
59	3.27920	3.34457	3.40893	3.47225	3.53451	3.59569	3.65578	3.71475
60	3.28030	3.34565	3.40999	3.47329	3.53553	3.59670	3.65677	3.71572

Constants for Setting a 5-inch Sine-Bar — 7

Min.	48°	49°	50°	51°	52°	53°	54°	55°
0	3.71572	3.77355	3.83022	3.88573	3.94005	3.99318	4.04508	4.09576
1	3.71670	3.77450	3.83116	3.88664	3.94095	3.99405	4.04594	4.09659
2	3.71767	3.77546	3.83209	3.88756	3.94184	3.99493	4.04679	4.09743
3	3.71864	3.77641	3.83303	3.88847	3.94274	3.99580	4.04765	4.09826
4	3.71961	3.77736	3.83396	3.88939	3.94363	3.99668	4.04850	4.09909
5	3.72059	3.77831	3.83489	3.89030	3.94453	3.99755	4.04936	4.09993
6	3.72156	3.77927	3.83583	3.89122	3.94542	3.99842	4.05021	4.10076
7	3.72253	3.78022	3.83676	3.89213	3.94631	3.99930	4.05106	4.10159
8	3.72350	3.78117	3.83769	3.89304	3.94721	4.00017	4.05191	4.10242
9	3.72447	3.78212	3.83862	3.89395	3.94810	4.00104	4.05277	4.10325
10	3.72544	3.78307	3.83955	3.89487	3.94899	4.00191	4.05362	4.10409
11	3.72641	3.78402	3.84049	3.89578	3.94988	4.00279	4.05447	4.10492
12	3.72738	3.78498	3.84142	3.89669	3.95078	4.00366	4.05532	4.10575
13	3.72835	3.78593	3.84235	3.89760	3.95167	4.00453	4.05617	4.10658
14	3.72932	3.78688	3.84328	3.89851	3.95256	4.00540	4.05702	4.10741
15	3.73029	3.78782	3.84421	3.89942	3.95345	4.00627	4.05787	4.10823
16	3.73126	3.78877	3.84514	3.90033	3.95434	4.00714	4.05872	4.10906
17	3.73222	3.78972	3.84607	3.90124	3.95523	4.00801	4.05957	4.10989
18	3.73319	3.79067	3.84700	3.90215	3.95612	4.00888	4.06042	4.11072
19	3.73416	3.79162	3.84793	3.90306	3.95701	4.00975	4.06127	4.11155
20	3.73513	3.79257	3.84886	3.90397	3.95790	4.01062	4.06211	4.11238
21	3.73609	3.79352	3.84978	3.90488	3.95878	4.01148	4.06296	4.11320
22	3.73706	3.79446	3.85071	3.90579	3.95967	4.01235	4.06381	4.11403
23	3.73802	3.79541	3.85164	3.90669	3.96056	4.01322	4.06466	4.11486
24	3.73899	3.79636	3.85257	3.90760	3.96145	4.01409	4.06550	4.11568
25	3.73996	3.79730	3.85349	3.90851	3.96234	4.01495	4.06635	4.11651
26	3.74092	3.79825	3.85442	3.90942	3.96323	4.01582	4.06720	4.11733
27	3.74189	3.79919	3.85535	3.91032	3.96411	4.01669	4.06804	4.11816
28	3.74285	3.80014	3.85627	3.91123	3.96500	4.01755	4.06889	4.11898
29	3.74381	3.80109	3.85720	3.91214	3.96588	4.01842	4.06973	4.11981
30	3.74478	3.80203	3.85812	3.91304	3.96677	4.01928	4.07058	4.12063
31	3.74574	3.80297	3.85905	3.91395	3.96765	4.02015	4.07142	4.12145
32	3.74671	3.80392	3.85997	3.91485	3.96854	4.02101	4.07227	4.12228
33	3.74767	3.80486	3.86090	3.91576	3.96942	4.02188	4.07311	4.12310
34	3.74863	3.80581	3.86182	3.91666	3.97031	4.02274	4.07395	4.12392
35	3.74959	3.80675	3.86274	3.91756	3.97119	4.02361	4.07480	4.12475
36	3.75056	3.80769	3.86367	3.91847	3.97207	4.02447	4.07564	4.12557
37	3.75152	3.80863	3.86459	3.91937	3.97296	4.02533	4.07648	4.12639
38	3.75248	3.80958	3.86551	3.92027	3.97384	4.02619	4.07732	4.12721
39	3.75344	3.81052	3.86644	3.92118	3.97472	4.02706	4.07817	4.12803
40	3.75440	3.81146	3.86736	3.92208	3.97560	4.02792	4.07901	4.12885
41	3.75536	3.81240	3.86828	3.92298	3.97649	4.02878	4.07985	4.12967
42	3.75632	3.81334	3.86920	3.92388	3.97737	4.02964	4.08069	4.13049
43	3.75728	3.81428	3.87012	3.92478	3.97825	4.03050	4.08153	4.13131
44	3.75824	3.81522	3.87104	3.92568	3.97913	4.03136	4.08237	4.13213
45	3.75920	3.81616	3.87196	3.92658	3.98001	4.03222	4.08321	4.13295
46	3.76016	3.81710	3.87288	3.92748	3.98089	4.03308	4.08405	4.13377
47	3.76112	3.81804	3.87380	3.92838	3.98177	4.03394	4.08489	4.13459
48	3.76207	3.81898	3.87472	3.92928	3.98265	4.03480	4.08572	4.13540
49	3.76303	3.81992	3.87564	3.93018	3.98353	4.03566	4.08656	4.13622
50	3.76399	3.82086	3.87656	3.93108	3.98441	4.03652	4.08740	4.13704
51	3.76495	3.82180	3.87748	3.93198	3.98529	4.03738	4.08824	4.13785
52	3.76590	3.82273	3.87840	3.93288	3.98616	4.03823	4.08908	4.13867
53	3.76686	3.82367	3.87931	3.93378	3.98704	4.03909	4.08991	4.13949
54	3.76782	3.82461	3.88023	3.93468	3.98792	4.03995	4.09075	4.14030
55	3.76877	3.82554	3.88115	3.93557	3.98880	4.04081	4.09158	4.14112
56	3.76973	3.82648	3.88207	3.93647	3.98967	4.04166	4.09242	4.14193
57	3.77068	3.82742	3.88298	3.93737	3.99055	4.04252	4.09326	4.14275
58	3.77164	3.82835	3.88390	3.93826	3.99143	4.04337	4.09409	4.14356
59	3.77259	3.82929	3.88481	3.93916	3.99230	4.04423	4.09493	4.14437
60	3.77355	3.83022	3.88573	3.94005	3.99318	4.04508	4.09576	4.14519

ALLOWANCES AND TOLERANCES FOR FITS

Limits and Fits.—Fits between cylindrical parts or, briefly cylindrical fits, govern the proper assembly and performance of countless mechanisms. Clearance fits permit relative freedom of motion between a shaft and a hole—axially, radially, or both. Interference fits secure a certain amount of tightness between parts, whether these are meant to remain permanently assembled or to be taken apart from time to time. Or again, two parts may be required to fit together snugly—without apparent tightness or looseness. The designer's problem is to specify these different types of fits in such a way that the shop can produce them. This involves the adoption of two manufacturing limits for the hole and two for the shaft, and, hence, the adoption of a manufacturing tolerance on each part.

In selecting and specifying limits and fits for various applications, it is essential in the interests of interchangeable manufacturing that (1) standard definitions of terms relating to limits and fits be used; (2) preferred basic sizes be selected wherever possible to reduce material and tooling costs; (3) limits be based upon a series of preferred tolerances and allowances; and (4) a uniform system of applying tolerances (preferably unilateral) be used. These principles have been incorporated in both the American and British standards for limits and fits. Information about these standards is given beginning on page 169.

Basic Dimensions.—The basic size of a screw thread or machine part is the theoretical or nominal standard size from which variations are made. For example, a shaft may have a *basic* diameter of 2 inches, but a maximum variation of minus 0.010 inch may be permitted. The minimum hole should be of basic size in all cases where the use of standard tools represents the greatest economy. The maximum shaft should be of basic size in all cases where the use of standard purchased material, without further machining, represents the greatest economy, even though special tools are required to machine the mating part.

Tolerances.—Tolerance is the amount of variation permitted on dimensions or surfaces of machine parts. The tolerance is equal to the difference between the maximum and minimum limits of any specified dimension. For example, if the maximum limit for the diameter of a shaft is 2.000 inches and its minimum limit 1.990 inches, the tolerance for this diameter is 0.010 inch. By determining the maximum and minimum clearances required on operating surfaces, the extent of these tolerances is established. As applied to the fitting of machine parts, the word tolerance means the amount that duplicate parts are allowed to vary in size in connection with manufacturing operations, owing to unavoidable imperfections of workmanship. Tolerance may also be defined as the amount that duplicate parts are permitted to vary in size in order to secure sufficient accuracy without unnecessary refinement. The terms "tolerance" and "allowance" are often used interchangeably, but, according to common usage, *allowance* is a difference in dimensions prescribed in order to secure various classes of fits between different parts.

Unilateral and Bilateral Tolerances. — The term "unilateral tolerance" means that the total tolerance, as related to a basic dimension, is in *one* direction only. For example, if the basic dimension were 1 inch and the tolerance were expressed as 1.00 − 0.002, or as 1.00 + 0.002, these would be unilateral tolerances, since the total tolerance in each case is in one direction. On the contrary, if the tolerance were divided, so as to be partly plus and partly minus, it would be classed as "bilateral." Thus, $1.00 \begin{smallmatrix} +0.001 \\ -0.001 \end{smallmatrix}$ is an example of bilateral tolerance, because the total tolerance of 0.002 is given in two directions—plus and minus.

When unilateral tolerances are used, one of the three following methods should be used to express them:

(1) Specify limiting dimensions only as

Diameter of hole: 2.250, 2.252

Diameter of shaft: 2.249, 2.247

(2) One limiting size may be specified with its tolerances as

Diameter of hole: 2.250 + 0.002, − 0.000

Diameter of shaft: 2.249 + 0.000, − 0.002

(3) The nominal size may be specified for both parts, with a notation showing both allowance and tolerance, as

Diameter of hole: $2\frac{1}{4}$ + 0.002, − 0.000

Diameter of shaft: $2\frac{1}{4}$ − 0.001, − 0.003

Bilateral tolerances should be specified as such, usually with plus and minus tolerances of equal amount. Example of the expression of bilateral tolerances follow:

$$2 \pm 0.001 \text{ or } 2 \genfrac{}{}{0pt}{}{+\ 0.001}{-\ 0.0001}$$

Application of Tolerances. — According to common practice, tolerances are applied in such a way as to show the permissible amount of dimensional variation in the direction that is less dangerous. When a variation in either direction is equally dangerous, a bilateral tolerance should be given. When a variation in one direction is more dangerous than a variation in another, a unilateral tolerance should be given in the less dangerous direction.

For non-mating surfaces, or atmospheric fits, the tolerances may be bilateral, or unilateral, depending only on the nature of the variations that develop in manufacture. On mating surfaces, with but few exceptions, the tolerances should be unilateral.

Where tolerances are required on the distances between holes, usually they should be bilateral, as variation in either direction is usually equally dangerous. The variation in the distance between shafts carrying gears, however, should always be unilateral and plus; otherwise the gears might run too tight. A slight increase in the backlash between gears is seldom of much importance.

One exception to the use of unilateral tolerances on mating surfaces occurs when tapers are involved. In such cases either bilateral or unilateral tolerances may prove advisable, depending upon conditions. These should be determined in the same manner as the tolerances on the distances between holes. When a variation either in or out of the position of the mating taper surfaces is equally dangerous, the tolerances should be bilateral. When a variation in one direction is of less danger than a variation in the opposite direction, the tolerance should be unilateral and in the less dangerous direction.

Locating Tolerance Dimensions. — Only one dimension in the same straight line can be controlled within fixed limits. That is the distance between the cutting surface of the tool and the locating or registering surface of the part being machined. Therefore, it is incorrect to locate any point or surface with tolerances from more than one point in the same straight line.

Every part of a mechanism must be located in each plane. Every operating part must be located with proper operating allowances. After such requirements of location are met, all other surfaces should have liberal clearances. Dimensions should be given between those points or surfaces that it is essential to hold in a specific relation to each other. This applies particularly to those surfaces in each plane which control the location of other component parts. Many dimensions are relatively unimportant in this respect. It is good practice in such cases to establish a common locating-point in each plane and give, so far as possible all such dimensions from these common locating-points. The locating points on the drawing, the locating or registering points used for machining the surfaces and the locating points for measuring should all be identical.

ANSI Standard Limits and Fits (ANSI B4.1-1967, R1979). — This American National Standard for Preferred Limits and Fits for Cylindrical Parts presents definitions of terms applying to fits between plain (non-threaded) cylindrical parts and makes recommendations on preferred sizes, allowances, tolerances, and fits for use wherever they are applicable. This standard is in accord with the recommendations of American-British-Canadian (ABC) conferences up to a diameter of 20 inches. Experimental work is being carried on with the objective of reaching agreement in the range above 20 inches. The recommendations in the Standard are presented for guidance and for use where they might serve to improve and simplify products, practices, and facilites. They should have application for a wide range of products.

As revised in 1967, and reaffirmed in 1979, the definitions in ANSI B4.1 have been expanded and some of the limits in certain classes have been changed.

Factors Affecting Selection of Fits. — Many factors, such as length of engagement, bearing load, speed, lubrication, temperature, humidity, and materials, must be taken into consideration in the selection of fits for a particular application, and modifications in the ANSI recommendations may be required to satisfy extreme conditions. Subsequent adjustments may also be found desirable as a result of experience in a particular application to suit critical functional requirements or to permit optimum manufacturing economy.

Definitions. — The following terms are defined in this standard:

Nominal Size: The nominal size is the designation which is used for the purpose of general identification.

Dimension: A dimension is a geometrical characteristic such as diameter, length, angle, or center distance.

Size: Size is a designation of magnitude. When a value is assigned to a dimension, it is referred to as the size of that dimension. (It is recognized that the words "dimension" and "size" are both used at times to convey the meaning of magnitude.)

Allowance: An allowance is a prescribed difference between the maximum material limits of mating parts. (See definition of *Fit*). It is a minimum clearance (positive allowance) or maximum interference (negative allowance) between such parts.

Tolerance: A tolerance is the total permissible variation of a size. The tolerance is the difference between the limits of size.

Basic Size: The basic size is that size from which the limits of size are derived by the application of allowances and tolerances.

Design Size: The design size is the basic size with allowance applied, from which the limits of size are derived by the application of tolerances. Where there is no allowance, the design size is the same as the basic size.

Actual Size: An actual size is a measured size.

Limits of Size: The limits of size are the applicable maximum and minimum sizes.

Maximum Material Limit: A maximum material limit is that limit of size that provides the maximum amount of material for the part. Normally it is the maximum limit of size of an external dimension or the minimum limit of size of an internal dimension.*

Minimum Material Limit: A minimum material limit is that limit of size that provides the minimum amount of material for the part. Normally it is the minimum limit of size of an external dimension or the maximum limit of size of an internal dimension.*

* An example of exceptions: an exterior corner radius where the maximum radius is the minimum material limit and the minimum radius is the maximum material limit.

Tolerance Limit: A tolerance limit is the variation, positive or negative, by which a size is permitted to depart from the design size.

Unilateral Tolerance: A unilateral tolerance is a tolerance in which variation is permitted only in one direction from the design size.

Bilateral Tolerance: A bilateral tolerance is a tolerance in which variation is permitted in both directions from the design size.

Unilateral Tolerance System: A design plan which uses only unilateral tolerances is known as a Unilateral Tolerance System.

Bilateral Tolerance System: A design plan which uses only bilateral tolerances is known as a Bilateral Tolerance System.

Fit: Fit is the general term used to signify the range of tightness which may result from the application of a specific combination of allowances and tolerances in the design of mating parts.

Actual Fit: The actual fit between two mating parts is the relation existing between them with respect to the amount of clearance or interference which is present when they are assembled. (Fits are of three general types: clearance, transition, and interference.)

Clearance Fit: A clearance fit is one having limits of size so specified that a clearance always results when mating parts are assembled.

Interference Fit: An interference fit is one having limits of size so specified that an interference always results when mating parts are assembled.

Transition Fit: A transition fit is one having limits of size so specified that either a clearance or an interference may result when mating parts are assembled.

Basic Hole System: A basic hole system is a system of fits in which the design size of the hole is the basic size and the allowance, if any, is applied to the shaft.

Basic Shaft System: A basic shaft system is a system of fits in which the design size of the shaft is the basic size and the allowance, if any, is applied to the hole.

Preferred Basic Sizes. — In specifying fits, the basic size of mating parts shall be chosen from the decimal series or the fractional series in the following table. All dimensions are given in inches.

Preferred Basic Sizes

Decimal			Fractional						
0.010	2.00	8.50	$\frac{1}{64}$	0.015625	$2\frac{1}{4}$	2.2500	$9\frac{1}{2}$	9.5000
0.012	2.20	9.00	$\frac{1}{32}$	0.03125	$2\frac{1}{2}$	2.5000	10	10.0000
0.016	2.40	9.50	$\frac{1}{16}$	0.0625	$2\frac{3}{4}$	2.7500	$10\frac{1}{2}$	10.5000
0.020	2.60	10.00	$\frac{3}{32}$	0.09375	3	3.0000	11	11.0000
0.025	2.80	10.50	$\frac{1}{8}$	0.1250	$3\frac{1}{4}$	3.2500	$11\frac{1}{2}$	11.5000
0.032	3.00	11.00	$\frac{5}{32}$	0.15625	$3\frac{1}{2}$	3.5000	12	12.0000
0.040	3.20	11.50	$\frac{3}{16}$	0.1875	$3\frac{3}{4}$	3.7500	$12\frac{1}{2}$	12.5000
0.05	3.40	12.00	$\frac{1}{4}$	0.2500	4	4.0000	13	13.0000
0.06	3.60	12.50	$\frac{5}{16}$	0.3125	$4\frac{1}{4}$	4.2500	$13\frac{1}{2}$	13.5000
0.08	3.80	13.00	$\frac{3}{8}$	0.3750	$4\frac{1}{2}$	4.5000	14	14.0000
0.10	4.00	13.50	$\frac{7}{16}$	0.4375	$4\frac{3}{4}$	4.7500	$14\frac{1}{2}$	14.5000
0.12	4.20	14.00	$\frac{1}{2}$	0.5000	5	5.0000	15	15.0000
0.16	4.40	14.50	$\frac{9}{16}$	0.5625	$5\frac{1}{4}$	5.2500	$15\frac{1}{2}$	15.5000
0.20	4.60	15.00	$\frac{5}{8}$	0.6250	$5\frac{1}{2}$	5.5000	16	16.0000
0.24	4.80	15.50	$11\frac{1}{16}$	0.6875	$5\frac{3}{4}$	5.7500	$16\frac{1}{2}$	16.5000
0.30	5.00	16.00	$\frac{3}{4}$	0.7500	6	6.0000	17	17.0000
0.40	5.20	16.50	$\frac{7}{8}$	0.8750	$6\frac{1}{2}$	6.5000	$17\frac{1}{2}$	17.5000
0.50	5.40	17.00	1	1.0000	7	7.0000	18	18.0000
0.60	5.60	17.50	$1\frac{1}{4}$	1.2500	$7\frac{1}{2}$	7.5000	$18\frac{1}{2}$	18.5000
0.80	5.80	18.00	$1\frac{1}{2}$	1.5000	8	8.0000	19	19.0000
1.00	6.00	18.50	$1\frac{3}{4}$	1.7500	$8\frac{1}{2}$	8.5000	$19\frac{1}{2}$	19.5000
1.20	6.50	19.00	2	2.0000	9	9.0000	20	20.0000
1.40	7.00	19.50	.....	....	...	....	....	....
1.60	7.50	20.00						
1.80	8.00	.....	All dimensions are given in inches.					

Preferred Series of Tolerances and Allowances (In thousandths of an inch)

0.1	1	10	100	0.3	3	30	...
...	1.2	12	125	...	3.5	35	...
0.15	1.4	14	...	0.4	4	40	...
...	1.6	16	160	...	4.5	45	...
...	1.8	18	...	0.5	5	50	...
0.2	2	20	200	0.6	6	60	...
...	2.2	22	...	0.7	7	70	...
0.25	2.5	25	250	0.8	8	80	...
...	2.8	28	...	0.9	9	...	...

Standard Tolerances. — The series of standard tolerances shown in Table 1 are so arranged that for any one grade they represent approximately similar production difficulties throughout the range of sizes. This table provides a suitable range from which appropriate tolerances for holes and shafts can be selected. This enables the use of standard gages. The tolerances shown in Table 1 have been used in the succeeding tables for different classes of fits.

Table 1. ANSI Standard Tolerances (ANSI B4.1-1967, R1979)

Nominal Size, Inches		Grade									
Over	To	4	5	6	7	8	9	10	11	12	13
		Tolerances in thousandths of an inch*									
0–	0.12	0.12	0.15	0.25	0.4	0.6	1.0	1.6	2.5	4	6
0.12–	0.24	0.15	0.20	0.3	0.5	0.7	1.2	1.8	3.0	5	7
0.24–	0.40	0.15	0.25	0.4	0.6	0.9	1.4	2.2	3.5	6	9
0.40–	0.71	0.2	0.3	0.4	0.7	1.0	1.6	2.8	4.0	7	10
0.71–	1.19	0.25	0.4	0.5	0.8	1.2	2.0	3.5	5.0	8	12
1.19–	1.97	0.3	0.4	0.6	1.0	1.6	2.5	4.0	6	10	16
1.97–	3.15	0.3	0.5	0.7	1.2	1.8	3.0	4.5	7	12	18
3.15–	4.73	0.4	0.6	0.9	1.4	2.2	3.5	5	9	14	22
4.73–	7.09	0.5	0.7	1.0	1.6	2.5	4.0	6	10	16	25
7.09–	9.85	0.6	0.8	1.2	1.8	2.8	4.5	7	12	18	28
9.85–	12.41	0.6	0.9	1.2	2.0	3.0	5.0	8	12	20	30
12.41–	15.75	0.7	1.0	1.4	2.2	3.5	6	9	14	22	35
15.75–	19.69	0.8	1.0	1.6	2.5	4	6	10	16	25	40
19.69–	30.09	0.9	1.2	2.0	3	5	8	12	20	30	50
30.09–	41.49	1.0	1.6	2.5	4	6	10	16	25	40	60
41.49–	56.19	1.2	2.0	3	5	8	12	20	30	50	80
56.19–	76.39	1.6	2.5	4	6	10	16	25	40	60	100
76.39–	100.9	2.0	3	5	8	12	20	30	50	80	125
100.9–	131.9	2.5	4	6	10	16	25	40	60	100	160
131.9–	171.9	3	5	8	12	20	30	50	80	125	200
171.9–	200	4	6	10	16	25	40	60	100	160	250

* All tolerances above heavy line are in accordance with American-British-Canadian (ABC) agreements.

Relation of Machining Processes to Tolerance Grades

	MACHINING OPERATION	TOLERANCE GRADES									
		4	5	6	7	8	9	10	11	12	13
This chart may be used as a general guide to determine the machining processes that will, under normal conditions, produce work within the tolerance grades indicated.	LAPPING & HONING										
	CYLINDRICAL GRINDING										
	SURFACE GRINDING										
	DIAMOND TURNING										
	DIAMOND BORING										
	BROACHING										
	REAMING										
	TURNING										
	BORING										
	MILLING										
	PLANING & SHAPING										
	DRILLING										

ANSI Standard Fits. — Tables 2 to 6 inclusive show a series of standard types and classes of fits on a unilateral hole basis, such that the fit produced by mating parts in any one class will produce approximately similar performance throughout the range of sizes. These tables prescribe the fit for any given size, or type of fit; they also prescribe the standard limits for the mating parts which will produce the fit. The fits listed in these tables contain all of those which appear in the approved American-British-Canadian proposal.

Selection of Fits: In selecting limits of size for any application, the type of fit is determined first, based on the use or service required from the equipment being designed; then the limits of size of the mating parts are established, to insure that the desired fit will be produced.

Theoretically, an infinite number of fits could be chosen, but the number of standard fits shown in the accompanying tables should cover most applications.

Designation of Standard Fits: Standard fits are designated by means of the following symbols which facilitate reference to classes of fit for educational purposes. The symbols are not intended to be shown on manufacturing drawings; instead, sizes should be specified on drawings.

The letter symbols used are as follows:

RC   Running or Sliding Clearance Fit
LC   Locational Clearance Fit
LT · Transition Clearance or Interference Fit
LN   Locational Interference Fit
FN   Force or Shrink Fit

These letter symbols are used in conjunction with numbers representing the class of fit; thus FN 4 represents a Class 4, force fit.

Each of these symbols (two letters and a number) represents a complete fit for which the minimum and maximum clearance or interference and the limits of size for the mating parts are given directly in the tables.

Description of Fits. — The classes of fits are arranged in three general groups: running and sliding fits, locational fits, and force fits.

Running and Sliding Fits (RC): Running and sliding fits, for which limits of clearance are given in Table 2, are intended to provide a similar running performance, with suitable lubrication allowance, throughout the range of sizes. The clearances for the first two classes, used chiefly as slide fits, increase more slowly with the diameter than for the other classes, so that accurate location is maintained even at the expense of free relative motion.

These fits may be described as follows:

RC 1 *Close sliding fits* are intended for the accurate location of parts which must assemble without perceptible play.

RC 2 *Sliding fits* are intended for accurate location, but with greater maximum clearance than class RC 1. Parts made to this fit move and turn easily but are not intended to run freely, and in the larger sizes may seize with small temperature changes.

RC 3 *Precision running fits* are about the closest fits which can be expected to run freely, and are intended for precision work at slow speeds and light journal pressures, but are not suitable where appreciable temperature differences are likely to be encountered.

RC 4 *Close running fits* are intended chiefly for running fits on accurate machinery with moderate surface speeds and journal pressures, where accurate location and minimum play is desired.

RC 5 and RC 6 *Medium running fits* are intended for higher running speeds, or heavy journal pressures, or both.

RC 7 *Free running fits* are intended for use where accuracy is not essential, or where large temperature variations are likely to be encountered, or under both these conditions.

RC 8 and RC 9 *Loose running fits* are intended for use where wide commercial tolerances may be necessary, together with an allowance, on the external member.

Locational Fits (LC, LT, and LN): Locational fits are fits intended to determine only the location of the mating parts; they may provide rigid or accurate location, as with interference fits, or provide some freedom of location, as with clearance fits. Accordingly, they are divided into three groups: clearance fits (LC), transition fits (LT), and interference fits (LN).

These are described as follows:

LC *Locational clearance fits* are intended for parts which are normally stationary, but which can be freely assembled or disassembled. They range from snug fits for parts requiring accuracy of location, through the medium clearance fits for parts such as spigots, to the looser fastener fits where freedom of assembly is of prime importance.

LT *Locational transition fits* are a compromise between clearance and interference fits, for application where accuracy of location is important, but either a small amount of clearance or interference is permissible.

LN *Locational interference fits* are used where accuracy of location is of prime importance, and for parts requiring rigidity and alignment with no special requirements for bore pressure. Such fits are not intended for parts designed to transmit frictional loads from one part to another by virtue of the tightness of fit, as these conditions are covered by force fits.

Force Fits (FN): Force or shrink fits constitute a special type of interference fit, normally characterized by maintenance of constant bore pressures throughout the range of sizes. The interference therefore varies almost directly with diameter, and the difference between its minimum and maximum value is small, to maintain the resulting pressures within reasonable limits.

These fits are described as follows:

FN 1 *Light drive fits* are those requiring light assembly pressures, and produce more or less permanent assemblies. They are suitable for thin sections or long fits, or in cast-iron external members.

FN 2 *Medium drive fits* are suitable for ordinary steel parts, or for shrink fits on light sections. They are about the tightest fits that can be used with high-grade cast-iron external members.

FN 3 *Heavy drive fits* are suitable for heavier steel parts or for shrink fits in medium sections.

FN 4 and FN 5 *Force fits* are suitable for parts which can be highly stressed, or for shrink fits where the heavy pressing forces required are impractical.

Graphical Representation of Limits and Fits. — A visual comparison of the hole and shaft tolerances and the clearances or interferences provided by the various types and classes of fits can be obtained from the diagrams on page 174. These diagrams have been drawn to scale for a nominal diameter of 1 inch.

Use of Standard Fit Tables. — *Example 1:* A Class RC 1 fit is to be used in assembling a mating hole and shaft of 2-inch nominal diameter. This class of fit was selected because the application required accurate location of the parts with no perceptible play (see description of RC 1 close sliding fits). From the data in Table 2, establish the limts of size and clearance of the hole and shaft.

Maximum hole = 2 + 0.0005 = 2.0005; minimum hole = 2 inches

(*Continued on page* 183)

Graphical Representation of ANSI Standard Limits and Fits*

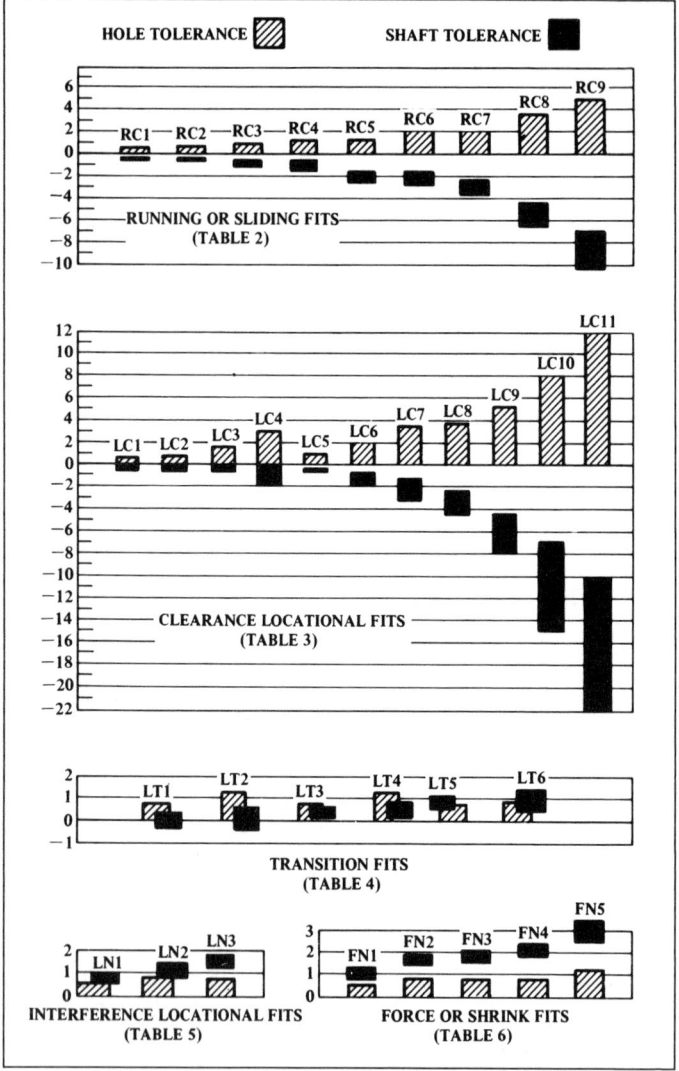

* Diagrams show disposition of hole and shaft tolerances (in thousandths of an inch) with respect to basic size (o) for a diameter of 1 inch.

Table 2. American National Standard Running and Sliding Fits (ANSI B4.1-1967, R1979)

Tolerance limits given in body of table are added or subtracted to basic size (as indicated by + or − sign) to obtain maximum and minimum sizes of mating parts.

Values shown below are in thousandths of an inch

Nominal Size Range, Inches Over — To	Class RC 1			Class RC 2			Class RC 3			Class RC 4		
	Clear-ance*	Hole H5	Shaft g4	Clear-ance*	Hole H6	Shaft g5	Clear-ance*	Hole H7	Shaft f6	Clear-ance*	Hole H8	Shaft f7
0– 0.12	0.1 / 0.45	+0.2 / 0	−0.1 / −0.25	0.1 / 0.55	+0.25 / 0	−0.1 / −0.3	0.3 / 0.95	+0.4 / 0	−0.3 / −0.55	0.3 / 1.3	+0.6 / 0	−0.3 / −0.7
0.12– 0.24	0.15 / 0.5	+0.2 / 0	−0.15 / −0.3	0.15 / 0.65	+0.3 / 0	−0.15 / −0.35	0.4 / 1.12	+0.5 / 0	−0.4 / −0.7	0.4 / 1.6	+0.7 / 0	−0.4 / −0.9
0.24– 0.40	0.2 / 0.6	+0.25 / 0	−0.2 / −0.35	0.2 / 0.85	+0.4 / 0	−0.2 / −0.45	0.5 / 1.5	+0.6 / 0	−0.5 / −0.9	0.5 / 2.0	+0.9 / 0	−0.5 / −1.1
0.40– 0.71	0.25 / 0.75	+0.3 / 0	−0.25 / −0.45	0.25 / 0.95	+0.4 / 0	−0.25 / −0.55	0.6 / 1.7	+0.7 / 0	−0.6 / −1.0	0.6 / 2.3	+1.0 / 0	−0.6 / −1.3
0.71– 1.19	0.3 / 0.95	+0.4 / 0	−0.3 / −0.55	0.3 / 1.2	+0.5 / 0	−0.3 / −0.7	0.8 / 2.1	+0.8 / 0	−0.8 / −1.3	0.8 / 2.8	+1.2 / 0	−0.8 / −1.6
1.19– 1.97	0.4 / 1.1	+0.4 / 0	−0.4 / −0.7	0.4 / 1.4	+0.6 / 0	−0.4 / −0.8	1.0 / 2.6	+1.0 / 0	−1.0 / −1.6	1.0 / 3.6	+1.6 / 0	−1.0 / −2.0
1.97– 3.15	0.4 / 1.2	+0.5 / 0	−0.4 / −0.7	0.4 / 1.6	+0.7 / 0	−0.4 / −0.9	1.2 / 3.1	+1.2 / 0	−1.2 / −1.9	1.2 / 4.2	+1.8 / 0	−1.2 / −2.4
3.15– 4.73	0.5 / 1.5	+0.6 / 0	−0.5 / −0.9	0.5 / 2.0	+0.9 / 0	−0.5 / −1.1	1.4 / 3.7	+1.4 / 0	−1.4 / −2.3	1.4 / 5.0	+2.2 / 0	−1.4 / −2.8
4.73– 7.09	0.6 / 1.8	+0.7 / 0	−0.6 / −1.1	0.6 / 2.3	+1.0 / 0	−0.6 / −1.3	1.6 / 4.2	+1.6 / 0	−1.6 / −2.6	1.6 / 5.7	+2.5 / 0	−1.6 / −3.2
7.09– 9.85	0.6 / 2.0	+0.8 / 0	−0.6 / −1.2	0.6 / 2.6	+1.2 / 0	−0.6 / −1.4	2.0 / 5.0	+1.8 / 0	−2.0 / −3.2	2.0 / 6.6	+2.8 / 0	−2.0 / −3.8
9.85–12.41	0.8 / 2.3	+0.9 / 0	−0.8 / −1.4	0.8 / 2.9	+1.2 / 0	−0.8 / −1.7	2.5 / 5.7	+2.0 / 0	−2.5 / −3.7	2.5 / 7.5	+3.0 / 0	−2.5 / −4.5
12.41–15.75	1.0 / 2.7	+1.0 / 0	−1.0 / −1.7	1.0 / 3.4	+1.4 / 0	−1.0 / −2.0	3.0 / 6.6	+2.2 / 0	−3.0 / −4.4	3.0 / 8.7	+3.5 / 0	−3.0 / −5.2
15.75–19.69	1.2 / 3.0	+1.0 / 0	−1.2 / −2.0	1.2 / 3.8	+1.6 / 0	−1.2 / −2.2	4.0 / 8.1	+2.5 / 0	−4.0 / −5.6	4.0 / 10.5	+4.0 / 0	−4.0 / −6.5

See footnotes at end of table.

Table 2 (Concluded). American National Standard Running and Sliding Fits (ANSI B4.1-1967, R1979)

Values shown below are in thousandths of an inch

Nominal Size Range, Inches (Over To)	Class RC 5 Clear*	Class RC 5 Hole H8	Class RC 5 Shaft e7	Class RC 6 Clear*	Class RC 6 Hole H9	Class RC 6 Shaft e8	Class RC 7 Clear*	Class RC 7 Hole H9	Class RC 7 Shaft d8	Class RC 8 Clear*	Class RC 8 Hole H10	Class RC 8 Shaft c9	Class RC 9 Clear*	Class RC 9 Hole H11	Class RC 9 Shaft
0- 0.12	0.6 / 1.6	+0.6 / 0	−0.6 / −1.0	0.6 / 2.2	+1.0 / 0	−0.6 / −1.2	1.0 / 2.6	+1.0 / 0	−1.0 / −1.6	2.5 / 5.1	+1.6 / 0	−2.5 / −3.5	4.0 / 8.1	+2.5 / 0	−4.0 / −5.6
0.12- 0.24	0.8 / 2.0	+0.7 / 0	−0.8 / −1.3	0.8 / 2.7	+1.2 / 0	−0.8 / −1.5	1.2 / 3.1	+1.2 / 0	−1.2 / −1.9	2.8 / 5.8	+1.8 / 0	−2.8 / −4.0	4.5 / 9.0	+3.0 / 0	−4.5 / −6.0
0.24- 0.40	1.0 / 2.5	+0.9 / 0	−1.0 / −1.6	1.0 / 3.3	+1.4 / 0	−1.0 / −1.9	1.6 / 3.9	+1.4 / 0	−1.6 / −2.5	3.0 / 6.6	+2.2 / 0	−3.0 / −4.4	5.0 / 10.7	+3.5 / 0	−5.0 / −7.2
0.40- 0.71	1.2 / 2.9	+1.0 / 0	−1.2 / −1.9	1.2 / 3.8	+1.6 / 0	−1.2 / −2.2	2.0 / 4.6	+1.6 / 0	−2.0 / −3.0	3.5 / 7.9	+2.8 / 0	−3.5 / −5.1	6.0 / 12.8	+4.0 / 0	−6.0 / −8.8
0.71- 1.19	1.6 / 3.6	+1.2 / 0	−1.6 / −2.4	1.6 / 4.8	+2.0 / 0	−1.6 / −2.8	2.5 / 5.7	+2.0 / 0	−2.5 / −3.7	4.5 / 10.0	+3.5 / 0	−4.5 / −6.5	7.0 / 15.5	+5.0 / 0	−7.0 / −10.5
1.19- 1.97	2.0 / 4.6	+1.6 / 0	−2.0 / −3.0	2.0 / 6.1	+2.5 / 0	−2.0 / −3.6	3.0 / 7.1	+2.5 / 0	−3.0 / −4.6	5.0 / 11.5	+4.0 / 0	−5.0 / −7.5	8.0 / 18.0	+6.0 / 0	−8.0 / −12.0
1.97- 3.15	2.5 / 5.5	+1.8 / 0	−2.5 / −3.7	2.5 / 7.3	+3.0 / 0	−2.5 / −4.3	4.0 / 8.8	+3.0 / 0	−4.0 / −5.8	6.0 / 13.5	+4.5 / 0	−6.0 / −9.0	9.0 / 20.5	+7.0 / 0	−9.0 / −13.5
3.15- 4.73	3.0 / 6.6	+2.2 / 0	−3.0 / −4.4	3.0 / 8.7	+3.5 / 0	−3.0 / −5.2	5.0 / 10.7	+3.5 / 0	−5.0 / −7.2	7.0 / 15.5	+5.0 / 0	−7.0 / −10.5	10.0 / 24.0	+9.0 / 0	−10.0 / −15.0
4.73- 7.09	3.5 / 7.6	+2.5 / 0	−3.5 / −5.1	3.5 / 10.0	+4.0 / 0	−3.5 / −6.0	6.0 / 12.5	+4.0 / 0	−6.0 / −8.5	8.0 / 18.0	+6.0 / 0	−8.0 / −12.0	12.0 / 28.0	+10.0 / 0	−12.0 / −18.0
7.09- 9.85	4.0 / 8.6	+2.8 / 0	−4.0 / −5.8	4.0 / 11.3	+4.5 / 0	−4.0 / −6.8	7.0 / 14.3	+4.5 / 0	−7.0 / −9.8	10.0 / 21.5	+7.0 / 0	−10.0 / −14.5	15.0 / 34.0	+12.0 / 0	−15.0 / −22.0
9.85-12.41	5.0 / 10.0	+3.0 / 0	−5.0 / −7.0	5.0 / 13.0	+5.0 / 0	−5.0 / −8.0	8.0 / 16.0	+5.0 / 0	−8.0 / −11.0	12.0 / 25.0	+8.0 / 0	−12.0 / −17.0	18.0 / 38.0	+12.0 / 0	−18.0 / −26.0
12.41-15.75	6.0 / 11.7	+3.5 / 0	−6.0 / −8.2	6.0 / 15.5	+6.0 / 0	−6.0 / −9.5	10.0 / 19.5	+6.0 / 0	−10.0 / −13.5	14.0 / 29.0	+9.0 / 0	−14.0 / −20.0	22.0 / 45.0	+14.0 / 0	−22.0 / −31.0
15.75-19.69	8.0 / 14.5	+4.0 / 0	−8.0 / −10.5	8.0 / 18.0	+6.0 / 0	−8.0 / −12.0	12.0 / 22.0	+6.0 / 0	−12.0 / −16.0	16.0 / 32.0	+10.0 / 0	−16.0 / −22.0	25.0 / 51.0	+16.0 / 0	−25.0 / −35.0

All data above heavy lines are in accord with ABC agreements. Symbols H5, g4, etc. are hole and shaft designations in ABC system. Limits for sizes above 19.69 inches are also given in the ANSI Standard.

* Pairs of values shown represent minimum and maximum amounts of clearance resulting from application of standard tolerance limits.

Table 3. American National Standard Clearance Locational Fits (ANSI B4.1-1967, R1979)

Tolerance limits given in body of table are added or subtracted to basic size (as indicated by + or − sign) to obtain maximum and minimum sizes of mating parts.

Values shown below are in thousandths of an inch

Nominal Size Range, Inches Over — To	Class LC 1 Clear-ance*	LC 1 Hole H6	LC 1 Shaft h5	Class LC 2 Clear-ance*	LC 2 Hole H7	LC 2 Shaft h6	Class LC 3 Clear-ance*	LC 3 Hole H8	LC 3 Shaft h7	Class LC 4 Clear-ance*	LC 4 Hole H10	LC 4 Shaft h9	Class LC 5 Clear-ance*	LC 5 Hole H7	LC 5 Shaft g6
0– 0.12	0 / 0.45	+0.25 / 0	0 / −0.2	0 / 0.65	+0.4 / 0	0 / −0.25	0 / 1	+0.6 / 0	0 / −0.4	0 / 2.6	+1.6 / 0	0 / −1.0	0.1 / 0.75	+0.4 / 0	−0.1 / −0.35
0.12– 0.24	0 / 0.5	+0.3 / 0	0 / −0.2	0 / 0.8	+0.5 / 0	0 / −0.3	0 / 1.2	+0.7 / 0	0 / −0.5	0 / 3.0	+1.8 / 0	0 / −1.2	0.15 / 0.95	+0.5 / 0	−0.15 / −0.45
0.24– 0.40	0 / 0.65	+0.4 / 0	0 / −0.25	0 / 1.0	+0.6 / 0	0 / −0.4	0 / 1.5	+0.9 / 0	0 / −0.6	0 / 3.6	+2.2 / 0	0 / −1.4	0.2 / 1.2	+0.6 / 0	−0.2 / −0.6
0.40– 0.71	0 / 0.7	+0.4 / 0	0 / −0.3	0 / 1.1	+0.7 / 0	0 / −0.4	0 / 1.7	+1.0 / 0	0 / −0.7	0 / 4.4	+2.8 / 0	0 / −1.6	0.25 / 1.35	+0.7 / 0	−0.25 / −0.65
0.71– 1.19	0 / 0.9	+0.5 / 0	0 / −0.4	0 / 1.3	+0.8 / 0	0 / −0.5	0 / 2	+1.2 / 0	0 / −0.8	0 / 5.5	+3.5 / 0	0 / −2.0	0.3 / 1.6	+0.8 / 0	−0.3 / −0.8
1.19– 1.97	0 / 1.0	+0.6 / 0	0 / −0.4	0 / 1.6	+1.0 / 0	0 / −0.6	0 / 2.6	+1.6 / 0	0 / −1	0 / 6.5	+4.0 / 0	0 / −2.5	0.4 / 2.0	+1.0 / 0	−0.4 / −1.0
1.97– 3.15	0 / 1.2	+0.7 / 0	0 / −0.5	0 / 1.9	+1.2 / 0	0 / −0.7	0 / 3	+1.8 / 0	0 / −1.2	0 / 7.5	+4.5 / 0	0 / −3	0.4 / 2.3	+1.2 / 0	−0.4 / −1.1
3.15– 4.73	0 / 1.5	+0.9 / 0	0 / −0.6	0 / 2.3	+1.4 / 0	0 / −0.9	0 / 3.6	+2.2 / 0	0 / −1.4	0 / 8.5	+5.0 / 0	0 / −3.5	0.5 / 2.8	+1.4 / 0	−0.5 / −1.4
4.73– 7.09	0 / 1.7	+1.0 / 0	0 / −0.7	0 / 2.6	+1.6 / 0	0 / −1.0	0 / 4.1	+2.5 / 0	0 / −1.6	0 / 10.0	+6.0 / 0	0 / −4	0.6 / 3.2	+1.6 / 0	−0.6 / −1.6
7.09– 9.85	0 / 2.0	+1.2 / 0	0 / −0.8	0 / 3.0	+1.8 / 0	0 / −1.2	0 / 4.6	+2.8 / 0	0 / −1.8	0 / 11.5	+7.0 / 0	0 / −4.5	0.6 / 3.6	+1.8 / 0	−0.6 / −1.8
9.85–12.41	0 / 2.1	+1.2 / 0	0 / −0.9	0 / 3.2	+2.0 / 0	0 / −1.2	0 / 5	+3.0 / 0	0 / −2.0	0 / 13.0	+8.0 / 0	0 / −5	0.7 / 3.9	+2.0 / 0	−0.7 / −1.9
12.41–15.75	0 / 2.4	+1.4 / 0	0 / −1.0	0 / 3.6	+2.2 / 0	0 / −1.4	0 / 5.7	+3.5 / 0	0 / −2.2	0 / 15.0	+9.0 / 0	0 / −6	0.7 / 4.3	+2.2 / 0	−0.7 / −2.1
15.75–19.69	0 / 2.6	+1.6 / 0	0 / −1.0	0 / 4.1	+2.5 / 0	0 / −1.6	0 / 6.5	+4 / 0	0 / −2.5	0 / 16.0	+10.0 / 0	0 / −6	0.8 / 4.9	+2.5 / 0	−0.8 / −2.4

See footnotes at end of table.

Table 3 (*Concluded*). **ANSI Standard Clearance Locational Fits** (ANSI B4.1-1967, R1979)

Values shown below are in thousandths of an inch

Nominal Size Range, Inches Over–To	Class LC 6 Clearance*	Class LC 6 Hole H9	Class LC 6 Shaft f8	Class LC 7 Clearance*	Class LC 7 Hole H10	Class LC 7 Shaft e9	Class LC 8 Clearance*	Class LC 8 Hole H10	Class LC 8 Shaft d9	Class LC 9 Clearance*	Class LC 9 Hole H11	Class LC 9 Shaft c10	Class LC 10 Clearance*	Class LC 10 Hole H12	Class LC 10 Shaft	Class LC 11 Clearance*	Class LC 11 Hole H13	Class LC 11 Shaft
0– 0.12	0.3 / 1.9	+1.0 / 0	−0.3 / −0.9	0.6 / 3.2	+1.6 / 0	−0.6 / −1.6	1.0 / 2.0	+1.6 / 0	−1.0 / −2.0	2.5 / 6.6	+2.5 / 0	−2.5 / −4.1	4 / 12	+4 / 0	−4 / −8	5 / 17	+6 / 0	−5 / −11
0.12– 0.24	0.4 / 2.3	+1.2 / 0	−0.4 / −1.1	0.8 / 3.8	+1.8 / 0	−0.8 / −2.0	1.2 / 4.2	+1.8 / 0	−1.2 / −2.4	2.8 / 7.6	+3.0 / 0	−2.8 / −4.6	4.5 / 14.5	+5 / 0	−4.5 / −9.5	6 / 20	+7 / 0	−6 / −13
0.24– 0.40	0.5 / 2.8	+1.4 / 0	−0.5 / −1.4	1.0 / 4.6	+2.2 / 0	−1.0 / −2.4	1.6 / 5.2	+2.2 / 0	−1.6 / −3.0	3.0 / 8.7	+3.5 / 0	−3.0 / −5.2	5 / 17	+6 / 0	−5 / −11	7 / 25	+9 / 0	−7 / −16
0.40– 0.71	0.6 / 3.2	+1.6 / 0	−0.6 / −1.6	1.2 / 5.6	+2.8 / 0	−1.2 / −2.8	2.0 / 6.4	+2.8 / 0	−2.0 / −3.6	3.5 / 10.3	+4.0 / 0	−3.5 / −6.3	6 / 20	+7 / 0	−6 / −13	8 / 28	+10 / 0	−8 / −18
0.71– 1.19	0.8 / 4.0	+2.0 / 0	−0.8 / −2.0	1.6 / 7.1	+3.5 / 0	−1.6 / −3.6	2.5 / 8.0	+3.5 / 0	−2.5 / −4.5	4.5 / 13.0	+5.0 / 0	−4.5 / −8.0	7 / 23	+8 / 0	−7 / −15	10 / 34	+12 / 0	−10 / −22
1.19– 1.97	1.0 / 5.1	+2.5 / 0	−1.0 / −2.6	2.0 / 8.5	+4.0 / 0	−2.0 / −4.5	3.6 / 9.5	+4.0 / 0	−3.0 / −5.5	5.0 / 15.0	+6 / 0	−5.0 / −9.0	8 / 28	+10 / 0	−8 / −18	12 / 44	+16 / 0	−12 / −28
1.97– 3.15	1.2 / 6.0	+3.0 / 0	−1.0 / −3.0	2.5 / 10.0	+4.5 / 0	−2.5 / −5.5	4.0 / 11.5	+4.5 / 0	−4.0 / −7.0	6.0 / 17.5	+7 / 0	−6.0 / −10.5	10 / 34	+12 / 0	−10 / −22	14 / 50	+18 / 0	−14 / −32
3.15– 4.73	1.4 / 7.1	+3.5 / 0	−1.4 / −3.6	3.0 / 11.5	+5.0 / 0	−3.0 / −6.5	5.0 / 13.5	+5.0 / 0	−5.0 / −8.5	7 / 21	+9 / 0	−7 / −12	11 / 39	+14 / 0	−11 / −25	16 / 60	+22 / 0	−16 / −38
4.73– 7.09	1.6 / 8.1	+4.0 / 0	−1.6 / −4.1	3.5 / 13.5	+6.0 / 0	−3.5 / −7.5	6 / 16	+6 / 0	−6 / −10	8 / 24	+10 / 0	−8 / −14	12 / 44	+16 / 0	−12 / −28	18 / 68	+25 / 0	−18 / −43
7.09– 9.85	2.0 / 9.3	+4.5 / 0	−2.0 / −4.8	4.0 / 15.5	+7.0 / 0	−4.0 / −8.5	7 / 18.5	+7 / 0	−7 / −11.5	10 / 29	+12 / 0	−10 / −17	16 / 52	+18 / 0	−16 / −34	22 / 78	+28 / 0	−22 / −50
9.85–12.41	2.2 / 10.2	+5.0 / 0	−2.2 / −5.2	4.5 / 17.5	+8.0 / 0	−4.5 / −9.5	7 / 20	+8 / 0	−7 / −12	12 / 32	+12 / 0	−12 / −20	20 / 60	+20 / 0	−20 / −40	28 / 88	+30 / 0	−28 / −58
12.41–15.75	2.5 / 12.0	+6.0 / 0	−2.5 / −6.0	5.0 / 20.0	+9.0 / 0	−5 / −11	8 / 23	+9 / 0	−8 / −14	14 / 37	+14 / 0	−14 / −23	22 / 66	+22 / 0	−22 / −44	30 / 100	+35 / 0	−30 / −65
15.75–19.69	2.8 / 12.8	+6.0 / 0	−2.8 / −6.8	5.0 / 21.0	+10.0 / 0	−5 / −11	9 / 25	+10 / 0	−9 / −15	16 / 42	+16 / 0	−16 / −26	25 / 75	+25 / 0	−25 / −50	35 / 115	+40 / 0	−35 / −75

All data above heavy lines are in accordance with American-British-Canadian (ABC) agreements. Symbols H6, H7, s6, etc. are hole and shaft designations in ABC system. Limits for sizes above 19.69 inches are not covered by ABC agreements but are given in the ANSI Standard.
* Pairs of values shown represent minimum and maximum amounts of interference resulting from application of standard tolerance limits.

Table 4. ANSI Standard Transition Locational Fits (ANSI B4.1-1967, R1979)

Values shown below are in thousandths of an inch

Nominal Size Range, Inches Over	To	Class LT 1 Fit*	Hole H7	Shaft js6	Class LT 2 Fit*	Hole H8	Shaft js7	Class LT 3 Fit*	Hole H7	Shaft k6	Class LT 4 Fit*	Hole H8	Shaft k7	Class LT 5 Fit*	Hole H7	Shaft n6	Class LT 6 Fit*	Hole H7	Shaft n7
0	0.12	−0.12 / +0.52	+0.4 / 0	+0.12 / −0.12	−0.2 / +0.8	+0.6 / 0	+0.2 / −0.2							−0.5 / +0.15	+0.4 / 0	+0.5 / +0.25	−0.65 / +0.15	+0.4 / 0	+0.65 / +0.25
0.12	0.24	−0.15 / +0.65	+0.5 / 0	+0.15 / −0.15	−0.25 / +0.95	+0.7 / 0	+0.25 / −0.25							−0.6 / +0.2	+0.5 / 0	+0.6 / +0.3	−0.8 / +0.2	+0.5 / 0	+0.8 / +0.3
0.24	0.40	−0.2 / +0.8	+0.6 / 0	+0.2 / −0.2	−0.3 / +1.2	+0.9 / 0	+0.3 / −0.3	−0.5 / +0.5	+0.6 / 0	+0.5 / +0.1	−0.7 / +0.8	+0.9 / 0	+0.7 / +0.1	−0.8 / +0.2	+0.6 / 0	+0.8 / +0.4	−1.0 / +0.2	+0.6 / 0	+1.0 / +0.4
0.40	0.71	−0.2 / +0.9	+0.7 / 0	+0.2 / −0.2	−0.35 / +1.35	+1.0 / 0	+0.35 / −0.35	−0.5 / +0.6	+0.7 / 0	+0.5 / +0.1	−0.8 / +0.9	+1.0 / 0	+0.8 / +0.1	−0.9 / +0.2	+0.7 / 0	+0.9 / +0.5	−1.2 / +0.2	+0.7 / 0	+1.2 / +0.5
0.71	1.19	−0.25 / +1.05	+0.8 / 0	+0.25 / −0.25	−0.4 / +1.6	+1.2 / 0	+0.4 / −0.4	−0.6 / +0.7	+0.8 / 0	+0.6 / +0.1	−0.9 / +1.1	+1.2 / 0	+0.9 / +0.1	−1.1 / +0.2	+0.8 / 0	+1.1 / +0.6	−1.4 / +0.2	+0.8 / 0	+1.4 / +0.6
1.19	1.97	−0.3 / +1.3	+1.0 / 0	+0.3 / −0.3	−0.5 / +2.1	+1.6 / 0	+0.5 / −0.5	−0.7 / +0.9	+1.0 / 0	+0.7 / +0.1	−1.1 / +1.5	+1.6 / 0	+1.1 / +0.1	−1.3 / +0.3	+1.0 / 0	+1.3 / +0.7	−1.7 / +0.3	+1.0 / 0	+1.7 / +0.7
1.97	3.15	−0.3 / +1.5	+1.2 / 0	+0.3 / −0.3	−0.6 / +2.4	+1.8 / 0	+0.6 / −0.6	−0.8 / +1.1	+1.2 / 0	+0.8 / +0.1	−1.3 / +1.7	+1.8 / 0	+1.3 / +0.1	−1.5 / +0.4	+1.2 / 0	+1.5 / +0.8	−2.0 / +0.4	+1.2 / 0	+2.0 / +0.8
3.15	4.73	−0.4 / +1.8	+1.4 / 0	+0.4 / −0.4	−0.7 / +2.9	+2.2 / 0	+0.7 / −0.7	−1.0 / +1.3	+1.4 / 0	+1.0 / +0.1	−1.5 / +2.1	+2.2 / 0	+1.5 / +0.1	−1.9 / +0.4	+1.4 / 0	+1.9 / +1.0	−2.4 / +0.4	+1.4 / 0	+2.4 / +1.0
4.73	7.09	−0.5 / +2.1	+1.6 / 0	+0.5 / −0.5	−0.8 / +3.3	+2.5 / 0	+0.8 / −0.8	−1.1 / +1.5	+1.6 / 0	+1.1 / +0.1	−1.7 / +2.4	+2.5 / 0	+1.7 / +0.1	−2.2 / +0.4	+1.6 / 0	+2.2 / +1.2	−2.8 / +0.4	+1.6 / 0	+2.8 / +1.2
7.09	9.85	−0.6 / +2.4	+1.8 / 0	+0.6 / −0.6	−0.9 / +3.7	+2.8 / 0	+0.9 / −0.9	−1.4 / +1.6	+1.8 / 0	+1.4 / +0.2	−2.0 / +2.6	+2.8 / 0	+2.0 / +0.2	−2.6 / +0.4	+1.8 / 0	+2.6 / +1.4	−3.2 / +0.4	+1.8 / 0	+3.2 / +1.4
9.85	12.41	−0.6 / +2.6	+2.0 / 0	+0.6 / −0.6	−1.0 / +4.0	+3.0 / 0	+1.0 / −1.0	−1.4 / +1.8	+2.0 / 0	+1.4 / +0.2	−2.2 / +2.8	+3.0 / 0	+2.2 / +0.2	−2.6 / +0.6	+2.0 / 0	+2.6 / +1.4	−3.4 / +0.6	+2.0 / 0	+3.4 / +1.4
12.41	15.75	−0.7 / +2.9	+2.2 / 0	+0.7 / −0.7	−1.0 / +4.5	+3.5 / 0	+1.0 / −1.0	−1.6 / +2.0	+2.2 / 0	+1.6 / +0.2	−2.4 / +3.3	+3.5 / 0	+2.4 / +0.2	−3.0 / +0.6	+2.2 / 0	+3.0 / +1.6	−3.8 / +0.6	+2.2 / 0	+3.8 / +1.6
15.75	19.69	−0.8 / +3.3	+2.5 / 0	+0.8 / −0.8	−1.2 / +5.2	+4.0 / 0	+1.2 / −1.2	−1.8 / +2.3	+2.5 / 0	+1.8 / +0.2	−2.7 / +3.8	+4.0 / 0	+2.7 / +0.2	−3.4 / +0.7	+2.5 / 0	+3.4 / +1.8	−4.3 / +0.7	+2.5 / 0	+4.3 / +1.8

All data above heavy lines are in accord with ABC agreements. Symbols H7, js6, etc. are hole and shaft designations in ABC system.
* Pairs of values shown represent maximum amount of interference (−) and maximum amount of clearance (+) resulting from application of standard tolerance limits.

Table 5. ANSI Standard Interference Locational Fits (ANSI B4.1-1967, R1979)

Tolerance limits given in body of table are added or subtracted to basic size (as indicated by + or − sign) to obtain maximum and minimum sizes of mating parts.

Nominal Size Range, Inches		Class LN 1			Class LN 2			Class LN 3		
		Limits of Inter-ference	Standard Limits		Limits of Inter-ference	Standard Limits		Limits of Inter-ference	Standard Limits	
Over	To		Hole H6	Shaft n5		Hole H7	Shaft p6		Hole H7	Shaft r6
		Values shown below are given in thousandths of an inch								
0– 0.12		0	+ 0.25	+ 0.45	0	+ 0.4	+ 0.65	0.1	+ 0.4	+ 0.75
		0.45	0	+ 0.25	0.65	0	+ 0.4	0.75	0	+ 0.5
0.12– 0.24		0	+ 0.3	+ 0.5	0	+ 0.5	+ 0.8	0.1	+ 0.5	+ 0.9
		0.5	0	+ 0.3	0.8	0	+ 0.5	0.9	0	+ 0.6
0.24– 0.40		0	+ 0.4	+ 0.65	0	+ 0.6	+ 1.0	0.2	+ 0.6	+ 1.2
		0.65	0	+ 0.4	1.0	0	+ 0.6	1.2	0	+ 0.8
0.40– 0.71		0	+ 0.4	+ 0.8	0	+ 0.7	+ 1.1	0.3	+ 0.7	+ 1.4
		0.8	0	+ 0.4	1.1	0	+ 0.7	1.4	0	+ 1.0
0.71– 1.19		0	+ 0.5	+ 1.0	0	+ 0.8	+ 1.3	0.4	+ 0.8	+ 1.7
		1.0	0	+ 0.5	1.3	0	+ 0.8	1.7	0	+ 1.2
1.19– 1.97		0	+ 0.6	+ 1.1	0	+ 1.0	+ 1.6	0.4	+ 1.0	+ 2.0
		1.1	0	+ 0.6	1.6	0	+ 1.0	2.0	0	+ 1.4
1.97– 3.15		0.1	+ 0.7	+ 1.3	0.2	+ 1.2	+ 2.1	0.4	+ 1.2	+ 2.3
		1.3	0	+ 0.8	2.1	0	+ 1.4	2.3	0	+ 1.6
3.15– 4.73		0.1	+ 0.9	+ 1.6	0.2	+ 1.4	+ 2.5	0.6	+ 1.4	+ 2.9
		1.6	0	+ 1.0	2.5	0	+ 1.6	2.9	0	+ 2.0
4.73– 7.09		0.2	+ 1.0	+ 1.9	0.2	+ 1.6	+ 2.8	0.9	+ 1.6	+ 3.5
		1.9	0	+ 1.2	2.8	0	+ 1.8	3.5	0	+ 2.5
7.09– 9.85		0.2	+ 1.2	+ 2.2	0.2	+ 1.8	+ 3.2	1.2	+ 1.8	+ 4.2
		2.2	0	+ 1.4	3.2	0	+ 2.0	4.2	0	+ 3.0
9.85–12.41		0.2	+ 1.2	+ 2.3	0.2	+ 2.0	+ 3.4	1.5	+ 2.0	+ 4.7
		2.3	0	+ 1.4	3.4	0	+ 2.2	4.7	0	+ 3.5
12.41–15.75		0.2	+ 1.4	+ 2.6	0.3	+ 2.2	+ 3.9	2.3	+ 2.2	+ 5.9
		2.6	0	+ 1.6	3.9	0	+ 2.5	5.9	0	+ 4.5
15.75–19.69		0.2	+ 1.6	+ 2.8	0.3	+ 2.5	+ 4.4	2.5	+ 2.5	+ 6.6
		2.8	0	+ 1.8	4.4	0	+ 2.8	6.6	0	+ 5.0

All data in this table are in accordance with American-British-Canadian (ABC) agreements.

Limits for sizes above 19.69 inches are not covered by ABC agreements but are given in the ANSI Standard.

Symbols H7, p6, etc. are hole and shaft designations in ABC system.

* Pairs of values shown represent minimum and maximum amounts of interference resulting from application of standard tolerance limits.

Table 6. ANSI Standard Force and Shrink Fits (ANSI B4.1-1967, R1979)

Values shown below are in thousandths of an inch

Nominal Size Range, Inches Over To	Class FN 1 Inter-ference*	Hole H6	Shaft	Class FN 2 Inter-ference*	Hole H7	Shaft s6	Class FN 3 Inter-ference*	Hole H7	Shaft t6	Class FN 4 Inter-ference*	Hole H7	Shaft u6	Class FN 5 Inter-ference*	Hole H8	Shaft x7
0–0.12	0.05 / 0.5	+0.25 / 0	+0.5 / +0.3	0.2 / 0.85	+0.4 / 0	+0.85 / +0.6				0.3 / 0.95	+0.4 / 0	+0.95 / +0.7	0.3 / 1.3	+0.6 / 0	+1.3 / +0.9
0.12–0.24	0.1 / 0.6	+0.3 / 0	+0.6 / +0.4	0.2 / 1.0	+0.5 / 0	+1.0 / +0.7				0.4 / 1.2	+0.5 / 0	+1.2 / +0.9	0.5 / 1.7	+0.7 / 0	+1.7 / +1.2
0.24–0.40	0.1 / 0.75	+0.4 / 0	+0.75 / +0.5	0.4 / 1.4	+0.6 / 0	+1.4 / +1.0				0.6 / 1.6	+0.6 / 0	+1.6 / +1.2	0.5 / 2.0	+0.9 / 0	+2.0 / +1.4
0.40–0.56	0.1 / 0.8	+0.4 / 0	+0.8 / +0.5	0.5 / 1.6	+0.7 / 0	+1.6 / +1.2				0.7 / 1.8	+0.7 / 0	+1.8 / +1.4	0.6 / 2.3	+1.0 / 0	+2.3 / +1.6
0.56–0.71	0.2 / 0.9	+0.4 / 0	+0.9 / +0.6	0.5 / 1.6	+0.7 / 0	+1.6 / +1.2				0.7 / 1.8	+0.7 / 0	+1.8 / +1.4	0.8 / 2.5	+1.0 / 0	+2.5 / +1.8
0.71–0.95	0.2 / 1.1	+0.5 / 0	+1.1 / +0.7	0.6 / 1.9	+0.8 / 0	+1.9 / +1.4	0.8 / 2.1	+0.8 / 0	+2.1 / +1.6	0.8 / 2.1	+0.8 / 0	+2.1 / +1.6	1.0 / 3.0	+1.2 / 0	+3.0 / +2.2
0.95–1.19	0.3 / 1.2	+0.5 / 0	+1.2 / +0.8	0.6 / 1.9	+0.8 / 0	+1.9 / +1.4	1.0 / 2.6	+1.0 / 0	+2.6 / +2.0	1.0 / 2.3	+0.8 / 0	+2.3 / +1.8	1.3 / 3.3	+1.2 / 0	+3.3 / +2.5
1.19–1.58	0.3 / 1.3	+0.6 / 0	+1.3 / +0.9	0.8 / 2.4	+1.0 / 0	+2.4 / +1.8	1.2 / 2.8	+1.0 / 0	+2.8 / +2.2	1.5 / 3.1	+1.0 / 0	+3.1 / +2.5	1.4 / 4.0	+1.6 / 0	+4.0 / +3.0
1.58–1.97	0.4 / 1.4	+0.6 / 0	+1.4 / +1.0	0.8 / 2.4	+1.0 / 0	+2.4 / +1.8	1.3 / 3.2	+1.2 / 0	+3.2 / +2.5	1.8 / 3.4	+1.0 / 0	+3.4 / +2.8	2.4 / 5.0	+1.6 / 0	+5.0 / +4.0
1.97–2.56	0.6 / 1.8	+0.7 / 0	+1.8 / +1.3	0.8 / 2.7	+1.2 / 0	+2.7 / +2.0	1.8 / 3.7	+1.2 / 0	+3.7 / +3.0	2.3 / 4.2	+1.2 / 0	+4.2 / +3.5	3.2 / 6.2	+1.8 / 0	+6.2 / +5.0
2.56–3.15	0.7 / 1.9	+0.7 / 0	+1.9 / +1.4	1.0 / 2.9	+1.2 / 0	+2.9 / +2.2	1.8 / 3.7	+1.2 / 0	+3.7 / +3.0	2.8 / 4.7	+1.2 / 0	+4.7 / +4.0	4.2 / 7.2	+1.8 / 0	+7.2 / +6.0
3.15–3.94	0.9 / 2.4	+0.9 / 0	+2.4 / +1.8	1.4 / 3.7	+1.4 / 0	+3.7 / +2.8	2.1 / 4.4	+1.4 / 0	+4.4 / +3.5	3.6 / 5.9	+1.4 / 0	+5.9 / +5.0	4.8 / 8.4	+2.2 / 0	+8.4 / +7.0
3.94–4.73	1.1 / 2.6	+0.9 / 0	+2.6 / +2.0	1.6 / 3.9	+1.4 / 0	+3.9 / +3.0	2.6 / 4.9	+1.4 / 0	+4.9 / +4.0	4.6 / 6.9	+1.4 / 0	+6.9 / +6.0	5.8 / 9.4	+2.2 / 0	+9.4 / +8.0

See footnotes at end of table.

Table 6 (*Concluded*). **ANSI Standard Force and Shrink Fits (ANSI B4.1-1967, R1979)**

Values shown below are in thousandths of an inch

Nominal Size Range, Inches (Over – To)	FN 1 Inter*	FN 1 Hole H6	FN 1 Shaft	FN 2 Inter*	FN 2 Hole H7	FN 2 Shaft s6	FN 3 Inter*	FN 3 Hole H7	FN 3 Shaft t6	FN 4 Inter*	FN 4 Hole H7	FN 4 Shaft u6	FN 5 Inter*	FN 5 Hole H8	FN 5 Shaft x7
4.73– 5.52	1.2 / 2.9	+1.0 / 0	+2.9 / +2.2	1.9 / 4.5	+1.6 / 0	+4.5 / +3.5	3.4 / 6.0	+1.6 / 0	+6.0 / +5.0	5.4 / 8.0	+1.6 / 0	+8.0 / +7.0	7.5 / 11.6	+2.5 / 0	+11.6 / +10.0
5.52– 6.30	1.5 / 3.2	+1.0 / 0	+3.2 / +2.5	2.4 / 5.0	+1.6 / 0	+5.0 / +4.0	3.4 / 6.0	+1.6 / 0	+6.0 / +5.0	5.4 / 8.0	+1.6 / 0	+8.0 / +7.0	9.5 / 13.6	+2.5 / 0	+13.6 / +12.0
6.30– 7.09	1.8 / 3.5	+1.0 / 0	+3.5 / +2.8	2.9 / 5.5	+1.6 / 0	+5.5 / +4.5	4.4 / 7.0	+1.6 / 0	+7.0 / +6.0	6.4 / 9.0	+1.6 / 0	+9.0 / +8.0	9.5 / 13.6	+2.5 / 0	+13.6 / +12.0
7.09– 7.88	1.8 / 3.8	+1.2 / 0	+3.8 / +3.0	3.2 / 6.2	+1.8 / 0	+6.2 / +5.0	5.2 / 8.2	+1.8 / 0	+8.2 / +7.0	7.2 / 10.2	+1.8 / 0	+10.2 / +9.0	11.2 / 15.8	+2.8 / 0	+15.8 / +14.0
7.88– 8.86	2.3 / 4.3	+1.2 / 0	+4.3 / +3.5	3.2 / 6.2	+1.8 / 0	+6.2 / +5.0	5.2 / 8.2	+1.8 / 0	+8.2 / +7.0	8.2 / 11.2	+1.8 / 0	+11.2 / +10.0	13.2 / 17.8	+2.8 / 0	+17.8 / +16.0
8.86– 9.85	2.3 / 4.3	+1.2 / 0	+4.3 / +3.5	4.2 / 7.2	+1.8 / 0	+7.2 / +6.0	6.2 / 9.2	+1.8 / 0	+9.2 / +8.0	10.2 / 13.2	+1.8 / 0	+13.2 / +12.0	13.2 / 17.8	+2.8 / 0	+17.8 / +16.0
9.85–11.03	2.8 / 4.9	+1.2 / 0	+4.9 / +4.0	4.0 / 7.2	+2.0 / 0	+7.2 / +6.0	7.0 / 10.2	+2.0 / 0	+10.2 / +9.0	10.0 / 13.2	+2.0 / 0	+13.2 / +12.0	15.0 / 20.0	+3.0 / 0	+20.0 / +18.0
11.03–12.41	2.8 / 4.9	+1.2 / 0	+4.9 / +4.0	5.0 / 8.2	+2.0 / 0	+8.2 / +7.0	7.0 / 10.2	+2.0 / 0	+10.2 / +9.0	12.0 / 15.2	+2.0 / 0	+15.2 / +14.0	17.0 / 22.0	+3.0 / 0	+22.0 / +20.0
12.41–13.98	3.1 / 5.5	+1.4 / 0	+5.5 / +4.5	5.8 / 9.4	+2.2 / 0	+9.4 / +8.0	7.8 / 11.4	+2.2 / 0	+11.4 / +10.0	13.8 / 17.4	+2.2 / 0	+17.4 / +16.0	18.5 / 24.2	+3.5 / 0	+24.2 / +22.0
13.98–15.75	3.6 / 6.1	+1.4 / 0	+6.1 / +5.0	5.8 / 9.4	+2.2 / 0	+9.4 / +8.0	9.8 / 13.4	+2.2 / 0	+13.4 / +12.0	15.8 / 19.4	+2.2 / 0	+19.4 / +18.0	21.5 / 27.2	+3.5 / 0	+27.2 / +25.0
15.75–17.72	4.4 / 7.0	+1.6 / 0	+7.0 / +6.0	6.5 / 10.6	+2.5 / 0	+10.6 / +9.0	9.5 / 13.6	+2.5 / 0	+13.6 / +12.0	17.5 / 21.6	+2.5 / 0	+21.6 / +20.0	24.0 / 30.5	+4.0 / 0	+30.5 / +28.0
17.72–19.69	4.4 / 7.0	+1.6 / 0	+7.0 / +6.0	7.5 / 11.6	+2.5 / 0	+11.6 / +10.0	11.5 / 15.6	+2.5 / 0	+15.6 / +14.0	19.5 / 23.6	+2.5 / 0	+23.6 / +22.0	26.0 / 32.5	+4.0 / 0	+32.5 / +30.0

All data above heavy lines are in accordance with American-British-Canadian (ABC) agreements. Symbols H6, H7, s6, etc. are hole and shaft designations in ABC system. Limits for sizes above 19.69 inches are not covered by ABC agreements but are given in the ANSI standard.
* Pairs of values shown represent minimum and maximum amounts of interference resulting from application of standard tolerance limits.

Maximum shaft = 2 − 0.0004 = 1.9996; minimum shaft = 2 − 0.0007 = 1.9993 inches
Minimum clearance = 0.0004; maximum clearance = 0.0012 inch

Example 2: Establish the limits for a Class LT 1 fit for a 2-inch diameter.
Maximum hole = 2 + 0.0012 = 2.0012; minimum hole = 2 inches
Maximum shaft = 2 + 0.0003 = 2.0003; minimum shaft = 2 − 0.0003 = 1.9997 inches
Maximum resulting *interference* = 0.0003; maximum resulting *clearance* = 0.0015 inch

Modified Standard Fits. — Fits having the same limits of clearance or interference as those shown in Tables 2 to 6 may sometimes have to be produced by using holes or shafts having limits of size other than those shown in these tables. This may be accomplished by using either a *Bilateral Hole (System B)* or a *Basic Shaft System (Symbol S)*. Both methods will result in non-standard holes and shafts.

Bilateral Hole Fits (Symbol B): The common case is where holes are produced with fixed tools, such as drills or reamers; to provide a longer wear life for such tools a bilateral tolerance is desired.

The symbols used for these fits are identical with those used for standard fits except that they are followed by the letter B. Thus, LC 4B is a clearance locational fit, Class 4, except that it is produced with a bilateral hole.

The limits of clearance or interference are identical with those shown in Tables 2 to 6 for the corresponding fits.

The hole tolerance, however, is changed so that the plus limit is that for one grade finer than the value shown in the tables and the minus limit equals the amount by which the plus limit was lowered. The shaft limits are both lowered by the same amount as the lower limit of size of the hole. The finer grade of tolerance required to make these modifications may be obtained from Table 1. For example, an LC 4B fit for a 6-inch diameter hole would have tolerance limits of + 4.0, − 2.0 (+ .0040 inch, − .0020 inch); the shaft would have tolerance limits of − 2.0, − 6.0 (− .0020 inch, − .0060 inch).

Basic Shaft Fits (Symbol S): For these fits the maximum size of the shaft is basic. The limits of clearance or interference are identical with those shown in Tables 2 to 6 for the corresponding fits and the symbols used for these fits are identical with those used for standard fits except that they are followed by the letter S. Thus, LC 4S is a clearance locational fit, Class 4, except that it is produced on a basic shaft basis.

The limits for hole and shaft as given in Tables 2 to 6 are increased for clearance fits (*decreased* for transition or interference fits) by the value of the upper shaft limit; that is, by the amount required to change the maximum shaft to the basic size.

American National Standard Preferred Metric Limits and Fits. — This standard (ANSI B4.2-1978) describes the ISO system of metric limits and fits for mating parts as approved for general engineering usage in the United States. It establishes: (1) the designation symbols used to define dimensional limits on drawings, material stock, related tools, gages, etc.; (2) the preferred basic sizes (first and second choices); (3) the preferred tolerance zones (first, second and third choices); (4) the preferred limits and fits for sizes (first choice only) up to and including 500 millimeters; and (5) the definitions of related terms.

The general terms "hole" and "shaft" can also be taken to refer to the space containing or contained by two parallel faces of any part, such as the width of a slot, the thickness of a key, etc.

Definitions. — The most important terms relating to limits and fits are shown in Fig. 1 and are defined as follows:

Basic Size: The size to which limits of deviation are assigned. The basic size is the same for both members of a fit. For example, it is designated by the numbers 40 in 40H7.

Deviation: The algebraic difference between a size and the corresponding basic size.

Upper Deviation: The algebraic difference between the maximum limit of size and the corresponding basic size.

Lower Deviation: The algebraic difference between the minimum limit of size and the corresponding basic size.

Fundamental Deviation: That one of the two deviations closest to the basic size. For example, it is designated by the letter H in 40H7.

Tolerance: The difference between the maximum and minimum size limits on a part.

Tolerance Zone: A zone representing the tolerance and its position in relation to the basic size.

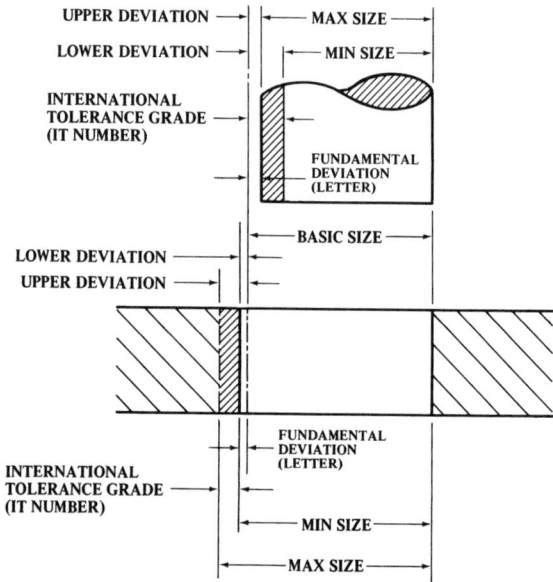

Fig. 1. Illustration of Definitions

International Tolerance Grade (IT): A group of tolerances which vary depending upon the basic size, but which provide the same relative level of accuracy within a given grade. For example, it is designated by the number 7 in 40H7 or as IT7.

Hole Basis: The system of fits where the minimum hole size is basic. The fundamental deviation for a hole basis system is H.

Shaft Basis: The system of fits where the maximum shaft size is basic. The fundamental deviation for a shaft basis system is h.

Clearance Fit: The relationship between assembled parts when clearance occurs under all tolerance conditions.

Interference Fit: The relationship between assembled parts when interference occurs under all tolerance conditions.

Transition Fit: The relationship between assembled parts when either a clearance or an interference fit can result, depending on the tolerance conditions of the mating parts.

Tolerances Designation. — An "International Tolerance grade" establishes the magnitude of the tolerance zone or the amount of part size variation allowed for external and internal dimensions alike (see Fig. 11). Tolerances are expressed in grade numbers which are consistent with International Tolerance grades identified by the prefix IT, such as IT6, IT11, etc. A smaller grade number provides a smaller tolerance zone.

A fundamental deviation establishes the position of the tolerance zone with respect to the basic size (see Fig. 1). Fundamental deviations are expressed by tolerance position letters. Capital letters are used for internal dimensions and lower case or small letters for external dimensions.

Symbols. — By combining the IT grade number and the tolerance position letter, the tolerance symbol is established which identifies the actual maximum and minimum limits of the part. The toleranced size is thus defined by the basic size of the part followed by a symbol composed of a letter and a number, such as 40H7, 40f7, etc.

A fit is indicated by the basic size common to both components, followed by a symbol corresponding to each component, the internal part symbol preceding the external part symbol, such as 40H8/f7.

Some methods of designating tolerances on drawings are:

$$\text{a. } 40H8 \qquad \text{b. } 40H8 \left(\frac{40.039}{40.000} \right) \qquad \text{c. } \left(\frac{40.039}{40.000} \right) 40H8$$

The values in parentheses indicate reference only.

Table 1. American National Standard Preferred Metric Sizes (ANSI B4.2-1978)

Basic Size, mm		Basic Size, mm		Basic Size, mm		Basic Size, mm	
1st Choice	2nd Choice	1st Choice	2nd Choice	1st Choice	2nd Choice	1st Choice	2nd Choice
1.0	. . .	. . .	7.0	25	. . .	. . .	90
. . .	1.1	8.0	. . .	. . .	26	100	. . .
1.2	. . .	. . .	9.0	. . .	28	. . .	110
. . .	1.4	10	. . .	30	. . .	120	. . .
1.6	. . .	. . .	11	. . .	32	. . .	130
. . .	1.8	12	. . .	35	. . .	140	. . .
2.0	. . .	. . .	13	. . .	38	. . .	150
. . .	2.2	14	. . .	40	. . .	160	. . .
2.5	. . .	. . .	15	. . .	42	. . .	170
. . .	2.8	16	. . .	45	. . .	180	. . .
3.0	. . .	. . .	17	. . .	48	. . .	190
. . .	3.5	18	. . .	50	. . .	200	. . .
4.0	. . .	. . .	19	. . .	55	. . .	220
. . .	4.5	20	. . .	60	. . .	250	. . .
5.0	. . .	. . .	21	. . .	65	. . .	280
. . .	5.5	22	. . .	. . .	70	300	. . .
6.0	. . .	. . .	23	. . .	75	. . .	320
. . .	6.5	. . .	24	80	. . .	. . .	. . .

Preferred Metric Sizes. — American National Standard ANSI B32.4M presents series of preferred metric sizes for round, square, rectangle, and hexagon metal products. Table 1 on the previous page gives preferred metric diameters from 1 to 320 millimeters for round metal products. Wherever possible, sizes should be selected from the Preferred Series shown in the table. A Second Preference series is also shown. A Third Preference Series not shown in the table is: 1.3, 2.1, 2.4, 2.6, 3.2, 3.8, 4.2, 4.8, 7.5, 8.5, 9.5, 36, 85, and 95.

Most of the Preferred Series of sizes are derived from the American National Standard "10 series" of preferred numbers (See Preferred Numbers in Index). Most of the Second Preference Series are derived from the "20 series" of preferred numbers. Third Preference sizes are generally from the "40 series" of preferred numbers.

For preferred metric diameters less than 1 millimeter, for preferred across flat metric sizes of square and hexagon metal products, preferred across flat metric sizes of rectangle metal products, and preferred metric lengths of metal products, reference should be made to the Standard.

Preferred Fits. — First choice tolerance zones are used to establish preferred fits in the Standard for Preferred Metric Limits and Fits ANSI B4.2, as shown in Figs. 2 and 3. A complete listing of first, second and third choice tolerance zones is given in the Standard.

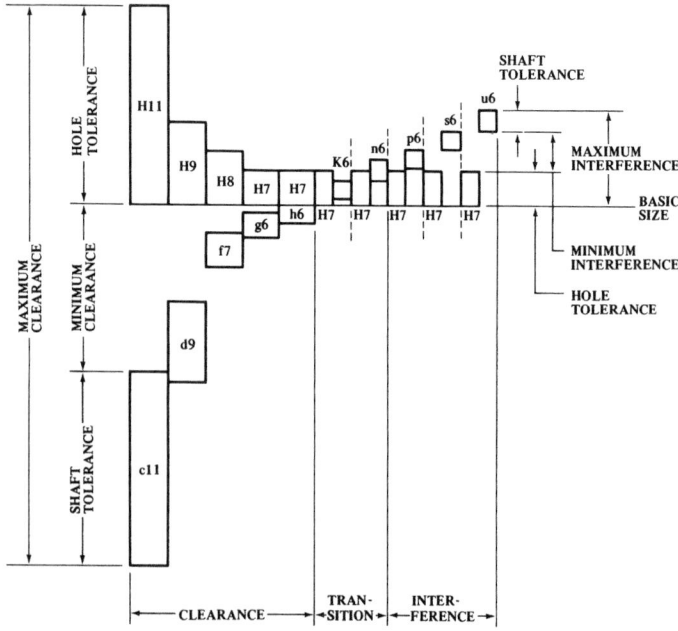

Fig. 2. Preferred Hole Basis Fits

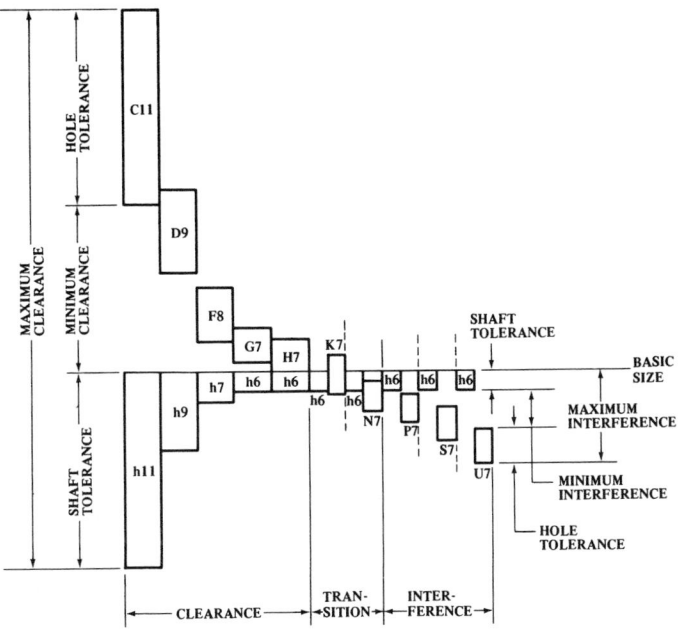

Fig. 3. Preferred Shaft Basis Fits

Hole basis fits have a fundamental deviation of H on the hole, and shaft basis fits have a fundamental deviation of h on the shaft and are shown in Fig. 2 for hole basis and Fig. 3 for shaft basis fits. A description of both types of fits which have the same relative fit condition is given in Fig. 4. Normally the hole basis system is preferred; however, when a common shaft mates with several holes, the shaft basis system should be used.

The hole basis and shaft basis fits shown in Fig. 4 are combined with the first choice sizes shown in Table 1 to form Tables 2, 3, 4, and 5 where specific limits as well as the resultant fits are tabulated.

If the required size is not tabulated in Tables 2 to 5, incl., then the preferred fit can be calculated from numerical values given in an appendix of ANSI B4.2-1978. It is anticipated that other fit conditions may be necessary to meet special requirements, and a preferred fit can be loosened or tightened simply by selecting a standard tolerance zone as given in the Standard. Information on how to calculate limit dimensions, clearances, and interferences, for non-preferred fits and sizes can also be found in an appendix of this Standard.

Applications. — Many factors such as length of engagement, bearing load, speed, lubrication, operating temperatures, humidity, surface texture, and materials must be taken into account in fit selections for a particular application. Choice of other than

(Concluded on page 190)

Fit Category	Hole Basis	Shaft Basis	DESCRIPTION	
Clearance Fits	H11/c11	C11/h11	Loose running fit for wide commercial tolerances or allowances on external members.	More Clearance ↑
	H9/d9	D9/h9	Free running fit not for use where accuracy is essential, but good for large temperature variations, high running speeds, or heavy journal pressures.	
	H8/f7	F8/h7	Close running fit for running on accurate machines and for accurate location at moderate speeds and journal pressures.	
	H7/g6	G7/h6	Sliding fit not intended to run freely, but to move and turn freely and locate accurately.	
Transition Fits	H7/h6	H7/h6	Locational clearance fit provides snug fit for locating stationary parts; but can be freely assembled and disassembled.	
	H7/k6	K7/h6	Locational transition fit for accurate location, a compromise between clearance and interference.	
	H7/n6	N7/h6	Locational transition fit for more accurate location where greater interference is permissible.	
Interference Fits	H7/p6[1]	P7/h6	Locational interference fit for parts requiring rigidity and alignment with, prime accuracy of location but without special bore pressure requirements.	
	H7/s6	S7/h6	Medium drive fit for ordinary steel parts or shrink fits on light sections, the tightest fit usable with cast iron.	
	H7/u6	U7/h6	Force fit suitable for parts which can be highly stressed or for shrink fits where the heavy pressing forces required are impractical.	More Interference ↓

[1] Transition fit for basic sizes in range from 0 through 3 mm.

Fig. 4. Description of preferred fits.

Table 2. American National Standard Preferred Hole Basis Metric Clearance Fits (ANSI B4.2-1978)

Basic Size*		Loose Running			Free Running			Close Running			Sliding			Locational Clearance		
		Hole H11	Shaft c11	Fit†	Hole H9	Shaft d9	Fit†	Hole H8	Shaft f7	Fit†	Hole H7	Shaft g6	Fit†	Hole H7	Shaft h6	Fit†
1	Max	1.060	0.940	0.180	1.025	0.980	0.070	1.014	0.994	0.030	1.010	0.998	0.018	1.010	1.000	0.016
	Min	1.000	0.880	0.060	1.000	0.955	0.020	1.000	0.984	0.006·	1.000	0.992	0.002	1.000	0.994	0.000
1.2	Max	1.260	1.140	0.180	1.225	1.180	0.070	1.214	1.194	0.030	1.210	1.198	0.018	1.210	1.200	0.016
	Min	1.200	1.080	0.060	1.200	1.155	0.020	1.200	1.184	0.006	1.200	1.192	0.002	1.200	1.194	0.000
1.6	Max	1.660	1.540	0.180	1.625	1.580	0.070	1.614	1.594	0.030	1.610	1.598	0.018	1.610	1.600	0.016
	Min	1.600	1.480	0.060	1.600	1.555	0.020	1.600	1.584	0.006	1.600	1.592	0.002	1.600	1.594	0.000
2	Max	2.060	1.940	0.180	2.025	1.980	0.070	2.014	1.994	0.030	2.010	1.998	0.018	2.010	2.000	0.016
	Min	2.000	1.880	0.060	2.000	1.955	0.020	2.000	1.984	0.006	2.000	1.992	0.002	2.000	1.994	0.000
2.5	Max	2.560	2.440	0.180	2.525	2.480	0.070	2.514	2.494	0.030	2.510	2.498	0.018	2.510	2.500	0.016
	Min	2.500	2.380	0.060	2.500	2.455	0.020	2.500	2.484	0.006	2.500	2.492	0.002	2.500	2.494	0.000
3	Max	3.060	2.940	0.180	3.025	2.980	0.070	3.014	2.994	0.030	3.010	2.998	0.018	3.010	3.000	0.016
	Min	3.000	2.880	0.060	3.000	2.955	0.020	3.000	2.984	0.006	3.000	2.992	0.002	3.000	2.994	0.000
4	Max	4.075	3.930	0.220	4.030	3.970	0.090	4.018	3.990	0.040	4.012	3.996	0.024	4.012	4.000	0.020
	Min	4.000	3.855	0.070	4.000	3.940	0.030	4.000	3.978	0.010	4.000	3.988	0.004	4.000	3.992	0.000
5	Max	5.075	4.930	0.220	5.030	4.970	0.090	5.018	4.990	0.040	5.012	4.996	0.024	5.012	5.000	0.020
	Min	5.000	4.855	0.070	5.000	4.940	0.030	5.000	4.978	0.010	5.000	4.988	0.004	5.000	4.992	0.000
6	Max	6.075	5.930	0.220	6.030	5.970	0.090	6.018	5.990	0.040	6.012	5.996	0.024	6.012	6.000	0.020
	Min	6.000	5.855	0.070	6.000	5.940	0.030	6.000	5.978	0.010	6.000	5.988	0.004	6.000	5.992	0.000
8	Max	8.090	7.920	0.260	8.036	7.960	0.112	8.022	7.987	0.050	8.015	7.995	0.029	8.015	8.000	0.024
	Min	8.000	7.830	0.080	8.000	7.924	0.040	8.000	7.972	0.013	8.000	7.986	0.005	8.000	7.991	0.000
10	Max	10.090	9.920	0.260	10.036	9.960	0.112	10.022	9.987	0.050	10.015	9.995	0.029	10.015	10.000	0.024
	Min	10.000	9.830	0.080	10.000	9.924	0.040	10.000	9.972	0.013	10.000	9.986	0.005	10.000	9.991	0.000
12	Max	12.110	11.905	0.315	12.043	11.956	0.136	12.027	11.984	0.061	12.018	11.994	0.035	12.018	12.000	0.029
	Min	12.000	11.795	0.095	12.000	11.907	0.050	12.000	11.966	0.016	12.000	11.983	0.006	12.000	11.989	0.000
16	Max	16.110	15.905	0.315	16.043	15.950	0.136	16.027	15.984	0.061	16.018	15.994	0.035	16.018	16.000	0.029
	Min	16.000	15.795	0.095	16.000	15.907	0.050	16.000	15.966	0.016	16.000	15.983	0.006	16.000	15.989	0.000
20	Max	20.130	19.890	0.370	20.052	19.935	0.169	20.033	19.980	0.074	20.021	19.993	0.041	20.021	20.000	0.034
	Min	20.000	19.760	0.110	20.000	19.883	0.065	20.000	19.959	0.020	20.000	19.980	0.007	20.000	19.987	0.000

All dimensions are in millimeters.

*The sizes shown are first choice basic sizes (see Table 1). Preferred fits for other sizes can be calculated from data given in ANSI B4.2-1978.
†All fits shown in this table have clearance.

Table 2 (*Concluded*). American National Standard Preferred Hole Basis Metric Clearance Fits (ANSI B4.2-1978)

Basic Size*		Loose Running			Free Running			Close Running			Sliding			Locational Clearance		
		Hole H11	Shaft c11	Fit†	Hole H9	Shaft d9	Fit†	Hole H8	Shaft f7	Fit†	Hole H7	Shaft g6	Fit†	Hole H7	Shaft h6	Fit†
25	Max	25.130	24.890	0.370	25.052	24.935	0.169	25.033	24.980	0.074	25.021	24.993	0.041	25.021	25.000	0.034
	Min	25.000	24.760	0.110	25.000	24.883	0.065	25.000	24.959	0.020	25.000	24.980	0.007	25.000	24.987	0.000
30	Max	30.130	29.890	0.370	30.052	29.935	0.169	30.033	29.980	0.074	30.021	29.993	0.041	30.021	30.000	0.034
	Min	30.000	29.760	0.110	30.000	29.883	0.065	30.000	29.959	0.020	30.000	29.980	0.007	30.000	29.987	0.000
40	Max	40.160	39.880	0.440	40.062	39.920	0.204	40.039	39.975	0.089	40.025	39.991	0.050	40.025	40.000	0.041
	Min	40.000	39.720	0.120	40.000	39.858	0.080	40.000	39.950	0.025	40.000	39.975	0.009	40.000	39.984	0.000
50	Max	50.160	49.870	0.450	50.062	49.920	0.204	50.039	49.975	0.089	50.025	49.991	0.050	50.025	50.000	0.041
	Min	50.000	49.710	0.130	50.000	49.858	0.080	50.000	49.950	0.025	50.000	49.975	0.009	50.000	49.984	0.000
60	Max	60.190	59.860	0.520	60.074	59.900	0.248	60.046	59.970	0.106	60.030	59.990	0.059	60.030	60.000	0.049
	Min	60.000	59.670	0.140	60.000	59.826	0.100	60.000	59.940	0.030	60.000	59.971	0.010	60.000	59.981	0.000
80	Max	80.190	79.850	0.530	80.074	79.900	0.248	80.046	79.970	0.106	80.030	79.990	0.059	80.030	80.000	0.049
	Min	80.000	79.660	0.150	80.000	79.826	0.100	80.000	79.940	0.030	80.000	79.971	0.010	80.000	79.981	0.000
100	Max	100.220	99.830	0.610	100.087	99.880	0.294	100.054	99.964	0.125	100.035	99.988	0.069	100.035	100.000	0.057
	Min	100.000	99.610	0.170	100.000	99.793	0.120	100.000	99.929	0.036	100.000	99.966	0.012	100.000	99.978	0.000
120	Max	120.220	119.820	0.620	120.087	119.880	0.294	120.054	119.964	0.125	120.035	119.988	0.069	120.035	120.000	0.057
	Min	120.000	119.600	0.180	120.000	119.793	0.120	120.000	119.929	0.036	120.000	119.966	0.012	120.000	119.978	0.000
160	Max	160.250	159.790	0.710	160.100	159.855	0.345	160.063	159.957	0.146	160.040	159.986	0.079	160.040	160.000	0.065
	Min	160.000	159.540	0.210	160.000	159.755	0.145	160.000	159.917	0.043	160.000	159.961	0.014	160.000	159.975	0.000
200	Max	200.290	199.760	0.820	200.115	199.830	0.400	200.072	199.950	0.168	200.046	199.985	0.090	200.046	200.000	0.075
	Min	200.000	199.470	0.240	200.000	199.715	0.170	200.000	199.904	0.050	200.000	199.956	0.015	200.000	199.971	0.000
250	Max	250.290	249.720	0.860	250.115	249.830	0.400	250.072	249.950	0.168	250.046	249.985	0.090	250.046	250.000	0.075
	Min	250.000	249.430	0.280	250.000	249.715	0.170	250.000	249.904	0.050	250.000	249.956	0.015	250.000	249.971	0.000
300	Max	300.320	299.670	0.970	300.130	299.810	0.450	300.081	299.944	0.189	300.052	299.983	0.101	300.052	300.000	0.084
	Min	300.000	299.350	0.330	300.000	299.680	0.190	300.000	299.892	0.056	300.000	299.951	0.017	300.000	299.968	0.000
400	Max	400.360	399.600	1.120	400.140	399.790	0.490	400.089	399.938	0.208	400.057	399.982	0.111	400.057	400.000	0.093
	Min	400.000	399.240	0.400	400.000	399.650	0.210	400.000	399.881	0.062	400.000	399.946	0.018	400.000	399.964	0.000
500	Max	500.400	499.520	1.280	500.155	499.770	0.540	500.097	499.932	0.228	500.063	499.980	0.123	500.063	500.000	0.103
	Min	500.000	499.120	0.480	500.000	499.615	0.230	500.000	499.869	0.068	500.000	499.940	0.020	500.000	499.960	0.000

All dimensions are in millimeters.
*The sizes shown are first choice basic sizes (see Table 1). Preferred fits for other sizes can be calculated from data given in ANSI B4.2-1978.
†All fits shown in this table have clearance.

Table 3. American National Standard Preferred Hole Basis Metric Transition and Interference Fits (ANSI B4.2-1978)

Basic Size*		Locational Transition Hole H7	Shaft k6	Fit†	Locational Transition Hole H7	Shaft n6	Fit†	Locational Interference Hole H7	Shaft p6	Fit†	Medium Drive Hole H7	Shaft s6	Fit†	Force Hole H7	Shaft u6	Fit†
1	Max	1.010	1.006	+0.010	1.010	1.010	+0.006	1.010	1.012	+0.004	1.010	1.020	−0.004	1.010	1.024	−0.008
	Min	1.000	1.000	−0.006	1.000	1.004	−0.010	1.000	1.006	−0.012	1.000	1.014	−0.020	1.000	1.018	−0.024
1.2	Max	1.210	1.206	+0.010	1.210	1.210	+0.006	1.210	1.212	+0.004	1.210	1.220	−0.004	1.210	1.224	−0.008
	Min	1.200	1.200	−0.006	1.200	1.204	−0.010	1.200	1.206	−0.012	1.200	1.214	−0.020	1.200	1.218	−0.024
1.6	Max	1.610	1.606	+0.010	1.610	1.610	+0.006	1.610	1.612	+0.004	1.610	1.620	−0.004	1.610	1.624	−0.008
	Min	1.600	1.600	−0.006	1.600	1.604	−0.010	1.600	1.606	−0.012	1.600	1.614	−0.020	1.600	1.618	−0.024
2	Max	2.010	2.006	+0.010	2.010	2.010	+0.006	2.010	2.012	+0.004	2.010	2.020	−0.004	2.010	2.024	−0.008
	Min	2.000	2.000	−0.006	2.000	2.004	−0.010	2.000	2.006	−0.012	2.000	2.014	−0.020	2.000	2.018	−0.024
2.5	Max	2.510	2.506	+0.010	2.510	2.510	+0.006	2.510	2.512	+0.004	2.510	2.520	−0.004	2.510	2.524	−0.008
	Min	2.500	2.500	−0.006	2.500	2.504	−0.010	2.500	2.506	−0.012	2.500	2.514	−0.020	2.500	2.518	−0.024
3	Max	3.010	3.006	+0.010	3.010	3.010	+0.006	3.010	3.012	+0.004	3.010	3.020	−0.004	3.010	3.024	−0.008
	Min	3.000	3.000	−0.006	3.000	3.004	−0.010	3.000	3.006	−0.012	3.000	3.014	−0.020	3.000	3.018	−0.024
4	Max	4.012	4.009	+0.011	4.012	4.016	+0.004	4.012	4.020	0.000	4.012	4.027	−0.007	4.012	4.031	−0.011
	Min	4.000	4.001	−0.009	4.000	4.008	−0.016	4.000	4.012	−0.020	4.000	4.019	−0.027	4.000	4.023	−0.031
5	Max	5.012	5.009	+0.011	5.012	5.016	+0.004	5.012	5.020	0.000	5.012	5.027	−0.007	5.012	5.031	−0.011
	Min	5.000	5.001	−0.009	5.000	5.008	−0.016	5.000	5.012	−0.020	5.000	5.019	−0.027	5.000	5.023	−0.031
6	Max	6.012	6.009	+0.011	6.012	6.016	+0.004	6.012	6.020	0.000	6.012	6.027	−0.007	6.012	6.031	−0.011
	Min	6.000	6.001	−0.009	6.000	6.008	−0.016	6.000	6.012	−0.020	6.000	6.019	−0.027	6.000	6.023	−0.031
8	Max	8.015	8.010	+0.014	8.015	8.019	+0.005	8.015	8.024	0.000	8.015	8.032	−0.008	8.015	8.037	−0.013
	Min	8.000	8.001	−0.010	8.000	8.010	−0.019	8.000	8.015	−0.024	8.000	8.023	−0.032	8.000	8.028	−0.037
10	Max	10.015	10.010	+0.014	10.015	10.019	+0.005	10.015	10.024	0.000	10.015	10.032	−0.008	10.015	10.037	−0.013
	Min	10.000	10.001	−0.010	10.000	10.010	−0.019	10.000	10.015	−0.024	10.000	10.023	−0.032	10.000	10.028	−0.037
12	Max	12.018	12.012	+0.017	12.018	12.023	+0.006	12.018	12.029	0.000	12.018	12.039	−0.010	12.018	12.044	−0.015
	Min	12.000	12.001	−0.012	12.000	12.012	−0.023	12.000	12.018	−0.029	12.000	12.028	−0.039	12.000	12.033	−0.044
16	Max	16.018	16.012	+0.017	16.018	16.023	+0.006	16.018	16.029	0.000	16.018	16.039	−0.010	16.018	16.044	−0.015
	Min	16.000	16.001	−0.012	16.000	16.012	−0.023	16.000	16.018	−0.029	16.000	16.028	−0.039	16.000	16.033	−0.044
20	Max	20.021	20.015	+0.019	20.021	20.028	+0.006	20.021	20.035	−0.001	20.021	20.048	−0.014	20.021	20.054	−0.020
	Min	20.000	20.002	−0.015	20.000	20.015	−0.028	20.000	20.022	−0.035	20.000	20.035	−0.048	20.000	20.041	−0.054

All dimensions are in millimeters.
*The sizes shown are first choice basic sizes (see Table 1). Preferred fits for other sizes can be calculated from data given in ANSI B4.2-1978.
†A plus sign indicates clearance; a minus sign indicates interference.

Table 3 (*Concluded*). American National Standard Preferred Hole Basis Metric Transition and Interference Fits (ANSI B4.2-1978)

Basic Size*		Locational Transition			Locational Transition			Locational Interference			Medium Drive			Force		
		Hole H7	Shaft k6	Fit†	Hole H7	Shaft n6	Fit†	Hole H7	Shaft p6	Fit†	Hole H7	Shaft s6	Fit†	Hole H7	Shaft u6	Fit†
25	Max	25.021	25.015	+0.019	25.021	25.028	+0.006	25.021	25.035	-0.001	25.021	25.048	-0.014	25.021	25.061	-0.027
	Min	25.000	25.002	-0.015	25.000	25.015	-0.028	25.000	25.022	-0.035	25.000	25.035	-0.048	25.000	25.048	-0.061
30	Max	30.021	30.015	+0.019	30.021	30.028	+0.006	30.021	30.035	-0.001	30.021	30.048	-0.014	30.021	30.061	-0.027
	Min	30.000	30.002	-0.015	30.000	30.015	-0.028	30.000	30.022	-0.035	30.000	30.035	-0.048	30.000	30.048	-0.061
40	Max	40.025	40.018	+0.023	40.025	40.033	+0.008	40.025	40.042	-0.001	40.025	40.059	-0.018	40.025	40.076	-0.035
	Min	40.000	40.002	-0.018	40.000	40.017	-0.033	40.000	40.026	-0.042	40.000	40.043	-0.059	40.000	40.060	-0.076
50	Max	50.025	50.018	+0.023	50.025	50.033	+0.008	50.025	50.042	-0.001	50.025	50.059	-0.018	50.025	50.086	-0.045
	Min	50.000	50.002	-0.018	50.000	50.017	-0.033	50.000	50.026	-0.042	50.000	50.043	-0.059	50.000	50.070	-0.086
60	Max	60.030	60.021	+0.028	60.030	60.039	+0.010	60.030	60.051	-0.002	60.030	60.072	-0.023	60.030	60.106	-0.057
	Min	60.000	60.002	-0.021	60.000	60.020	-0.039	60.000	60.032	-0.051	60.000	60.053	-0.072	60.000	60.087	-0.106
80	Max	80.030	80.021	+0.028	80.030	80.039	+0.010	80.030	80.051	-0.002	80.030	80.078	-0.029	80.030	80.121	-0.072
	Min	80.000	80.002	-0.021	80.000	80.020	-0.039	80.000	80.032	-0.051	80.000	80.059	-0.078	80.000	80.102	-0.121
100	Max	100.035	100.025	+0.032	100.035	100.045	+0.012	100.035	100.059	-0.002	100.035	100.093	-0.036	100.035	100.146	-0.089
	Min	100.000	100.003	-0.025	100.000	100.023	-0.045	100.000	100.037	-0.059	100.000	100.071	-0.093	100.000	100.124	-0.146
120	Max	120.035	120.025	+0.032	120.035	120.045	+0.012	120.035	120.059	-0.002	120.035	120.101	-0.044	120.035	120.166	-0.109
	Min	120.000	120.003	-0.025	120.000	120.023	-0.045	120.000	120.037	-0.059	120.000	120.079	-0.101	120.000	120.144	-0.166
160	Max	160.040	160.028	+0.037	160.040	160.052	+0.013	160.040	160.068	-0.003	160.040	160.125	-0.060	160.040	160.215	-0.150
	Min	160.000	160.003	-0.028	160.000	160.027	-0.052	160.000	160.043	-0.068	160.000	160.100	-0.125	160.000	160.190	-0.215
200	Max	200.046	200.033	+0.042	200.046	200.060	+0.015	200.046	200.079	-0.004	200.046	200.151	-0.076	200.046	200.265	-0.190
	Min	200.000	200.004	-0.033	200.000	200.031	-0.060	200.000	200.050	-0.079	200.000	200.122	-0.151	200.000	200.236	-0.265
250	Max	250.046	250.033	+0.042	250.046	250.060	+0.015	250.046	250.079	-0.004	250.046	250.169	-0.094	250.046	250.313	-0.238
	Min	250.000	250.004	-0.033	250.000	250.031	-0.060	250.000	250.050	-0.079	250.000	250.140	-0.169	250.000	250.284	-0.313
300	Max	300.052	300.036	+0.048	300.052	300.066	+0.018	300.052	300.088	-0.004	300.052	300.202	-0.118	300.052	300.382	-0.298
	Min	300.000	300.004	-0.036	300.000	300.034	-0.066	300.000	300.056	-0.088	300.000	300.170	-0.202	300.000	300.350	-0.382
400	Max	400.057	400.040	+0.053	400.057	400.073	+0.020	400.057	400.098	-0.005	400.057	400.244	-0.151	400.057	400.471	-0.378
	Min	400.000	400.004	-0.040	400.000	400.037	-0.073	400.000	400.062	-0.098	400.000	400.208	-0.244	400.000	400.435	-0.471
500	Max	500.063	500.045	+0.058	500.063	500.080	+0.023	500.063	500.108	-0.005	500.063	500.292	-0.189	500.063	500.580	-0.477
	Min	500.000	500.005	-0.045	500.000	500.040	-0.080	500.000	500.068	-0.108	500.000	500.252	-0.292	500.000	500.540	-0.580

All dimensions are in millimeters.

*The sizes shown are first choice basic sizes (see Table 1). Preferred fits for other sizes can be calculated from data given in ANSI B4.2-1978.

†A plus sign indicates clearance; a minus sign indicates interference.

Table 4. American National Standard Preferred Shaft Basis Metric Clearance Fits (ANSI B4.2-1978)

Basic Size*		Loose Running			Free Running			Close Running			Sliding			Locational Clearance		
		Hole C11	Shaft h11	Fit†	Hole D9	Shaft h9	Fit†	Hole F8	Shaft h7	Fit†	Hole G7	Shaft h6	Fit†	Hole H7	Shaft h6	Fit†
1	Max	1.120	1.000	0.180	1.045	1.000	0.070	1.020	1.000	0.030	1.012	1.000	0.018	1.010	1.000	0.016
	Min	1.060	0.940	0.060	1.020	0.975	0.020	1.006	0.990	0.006	1.002	0.994	0.002	1.000	0.994	0.000
1.2	Max	1.320	1.200	0.180	1.245	1.200	0.070	1.220	1.200	0.030	1.212	1.200	0.018	1.210	1.200	0.016
	Min	1.260	1.140	0.060	1.220	1.175	0.020	1.206	1.190	0.006	1.202	1.194	0.002	1.200	1.194	0.000
1.6	Max	1.720	1.600	0.180	1.645	1.600	0.070	1.620	1.600	0.030	1.612	1.600	0.018	1.610	1.600	0.016
	Min	1.660	1.540	0.060	1.620	1.575	0.020	1.606	1.590	0.006	1.602	1.594	0.002	1.600	1.594	0.000
2	Max	2.120	2.000	0.180	2.045	2.000	0.070	2.020	2.000	0.030	2.012	2.000	0.018	2.010	2.000	0.016
	Min	2.060	1.940	0.060	2.020	1.975	0.020	2.006	1.990	0.006	2.002	1.994	0.002	2.000	1.994	0.000
2.5	Max	2.620	2.500	0.180	2.545	2.500	0.070	2.520	2.500	0.030	2.512	2.500	0.018	2.510	2.500	0.016
	Min	2.560	2.440	0.060	2.520	2.475	0.020	2.506	2.490	0.006	2.502	2.494	0.002	2.500	2.494	0.000
3	Max	3.120	3.000	0.180	3.045	3.000	0.070	3.020	3.000	0.030	3.012	3.000	0.018	3.010	3.000	0.016
	Min	3.060	2.940	0.060	3.020	2.975	0.020	3.006	2.990	0.006	3.002	2.994	0.002	3.000	2.994	0.000
4	Max	4.145	4.000	0.220	4.060	4.000	0.090	4.028	4.000	0.040	4.016	4.000	0.024	4.012	4.000	0.020
	Min	4.070	3.925	0.070	4.030	3.970	0.030	4.010	3.988	0.010	4.004	3.992	0.004	4.000	3.992	0.000
5	Max	5.145	5.000	0.220	5.060	5.000	0.090	5.028	5.000	0.040	5.016	5.000	0.024	5.012	5.000	0.020
	Min	5.070	4.925	0.070	5.030	4.970	0.030	5.010	4.988	0.010	5.004	4.992	0.004	5.000	4.992	0.000
6	Max	6.145	6.000	0.220	6.060	6.000	0.090	6.028	6.000	0.040	6.016	6.000	0.024	6.012	6.000	0.020
	Min	6.070	5.925	0.070	6.030	5.970	0.030	6.010	5.988	0.010	6.004	5.992	0.004	6.000	5.992	0.000
8	Max	8.170	8.000	0.260	8.076	8.000	0.112	8.035	8.000	0.050	8.020	8.000	0.029	8.015	8.000	0.024
	Min	8.080	7.910	0.080	8.040	7.964	0.040	8.013	7.985	0.013	8.005	7.991	0.005	8.000	7.991	0.000
10	Max	10.170	10.000	0.260	10.076	10.000	0.112	10.035	10.000	0.050	10.020	10.000	0.029	10.015	10.000	0.024
	Min	10.080	9.910	0.080	10.040	9.964	0.040	10.013	9.985	0.013	10.005	9.991	0.005	10.000	9.991	0.000
12	Max	12.205	12.000	0.315	12.093	12.000	0.136	12.043	12.000	0.061	12.024	12.000	0.035	12.018	12.000	0.029
	Min	12.095	11.890	0.095	12.050	11.957	0.050	12.016	11.982	0.016	12.006	11.989	0.006	12.000	11.989	0.000
16	Max	16.205	16.000	0.315	16.093	16.000	0.136	16.043	16.000	0.061	16.024	16.000	0.035	16.018	16.000	0.029
	Min	16.095	15.890	0.095	16.050	15.957	0.050	16.016	15.982	0.016	16.006	15.989	0.006	16.000	15.989	0.000
20	Max	20.240	20.000	0.370	20.117	20.000	0.169	20.053	20.000	0.074	20.028	20.000	0.041	20.021	20.000	0.034
	Min	20.110	19.870	0.110	20.065	19.948	0.065	20.020	19.979	0.020	20.007	19.987	0.007	20.000	19.987	0.000

All dimensions are in millimeters.
*The sizes shown are first choice basic sizes (see Table 1). Preferred fits for other sizes can be calculated from data given in ANSI B4.2-1978.
†All fits shown in this table have clearance.

Table 4 (*Concluded*). American National Standard Preferred Shaft Basis Metric Clearance Fits (ANSI B4.2-1978)

Basic Size*		Loose Running			Free Running			Close Running			Sliding			Locational Clearance		
		Hole C11	Shaft h11	Fit†	Hole D9	Shaft h9	Fit†	Hole F8	Shaft h7	Fit†	Hole G7	Shaft h6	Fit†	Hole H7	Shaft h6	Fit†
25	Max	25.240	25.000	0.370	25.117	25.000	0.169	25.053	25.000	0.074	25.028	25.000	0.041	25.021	25.000	0.034
	Min	25.110	24.870	0.110	25.065	24.948	0.065	25.020	24.979	0.020	25.007	24.987	0.007	25.000	24.987	0.000
30	Max	30.240	30.000	0.370	30.117	30.000	0.169	30.053	30.000	0.074	30.028	30.000	0.041	30.021	30.000	0.034
	Min	30.110	29.870	0.110	30.065	29.948	0.065	30.020	29.979	0.020	30.007	29.987	0.007	30.000	29.987	0.000
40	Max	40.280	40.000	0.440	40.142	40.000	0.204	40.064	40.000	0.089	40.034	40.000	0.050	40.025	40.000	0.041
	Min	40.120	39.840	0.120	40.080	39.938	0.080	40.025	39.975	0.025	40.009	39.984	0.009	40.000	39.984	0.000
50	Max	50.290	50.000	0.450	50.142	50.000	0.204	50.064	50.000	0.089	50.034	50.000	0.050	50.025	50.000	0.041
	Min	50.130	49.840	0.130	50.080	49.938	0.080	50.025	49.975	0.025	50.009	49.984	0.009	50.000	49.984	0.000
60	Max	60.330	60.000	0.520	60.174	60.000	0.248	60.076	60.000	0.106	60.040	60.000	0.059	60.030	60.000	0.049
	Min	60.140	59.810	0.140	60.100	59.926	0.100	60.030	59.970	0.030	60.010	59.981	0.010	60.000	59.981	0.000
80	Max	80.340	80.000	0.530	80.174	80.000	0.248	80.076	80.000	0.106	80.040	80.000	0.059	80.030	80.000	0.049
	Min	80.150	79.810	0.150	80.100	79.926	0.100	80.030	79.970	0.030	80.010	79.981	0.010	80.000	79.981	0.000
100	Max	100.390	100.000	0.610	100.207	100.000	0.294	100.090	100.000	0.125	100.047	100.000	0.069	100.035	100.000	0.057
	Min	100.170	99.780	0.170	100.120	99.913	0.120	100.036	99.965	0.036	100.012	99.978	0.012	100.000	99.978	0.000
120	Max	120.400	120.000	0.620	120.207	120.000	0.294	120.090	120.000	0.125	120.047	120.000	0.069	120.035	120.000	0.057
	Min	120.180	119.780	0.180	120.120	119.913	0.120	120.036	119.965	0.036	120.012	119.978	0.012	120.000	119.978	0.000
160	Max	160.460	160.000	0.710	160.245	160.000	0.345	160.106	160.000	0.146	160.054	160.000	0.079	160.040	160.000	0.065
	Min	160.210	159.750	0.210	160.145	159.900	0.145	160.043	159.960	0.043	160.014	159.975	0.014	160.000	159.975	0.000
200	Max	200.530	200.000	0.820	200.285	200.000	0.400	200.122	200.000	0.168	200.061	200.000	0.090	200.046	200.000	0.075
	Min	200.240	199.710	0.240	200.170	199.885	0.170	200.050	199.954	0.050	200.015	199.971	0.015	200.000	199.971	0.000
250	Max	250.570	250.000	0.860	250.285	250.000	0.400	250.122	250.000	0.168	250.061	250.000	0.090	250.046	250.000	0.075
	Min	250.280	249.710	0.280	250.170	249.885	0.170	250.050	249.954	0.050	250.015	249.971	0.015	250.000	249.971	0.000
300	Max	300.650	300.000	0.970	300.320	300.000	0.450	300.137	300.000	0.189	300.069	300.000	0.101	300.052	300.000	0.084
	Min	300.330	299.680	0.330	300.190	299.870	0.190	300.056	299.948	0.056	300.017	299.968	0.017	300.000	299.968	0.000
400	Max	400.760	400.000	1.120	400.350	400.000	0.490	400.151	400.000	0.208	400.075	400.000	0.111	400.057	400.000	0.093
	Min	400.400	399.640	0.400	400.210	399.860	0.210	400.062	399.943	0.062	400.018	399.964	0.018	400.000	399.964	0.000
500	Max	500.880	500.000	1.280	500.385	500.000	0.540	500.165	500.000	0.228	500.083	500.000	0.123	500.063	500.000	0.103
	Min	500.480	499.600	0.480	500.230	499.845	0.230	500.068	499.937	0.068	500.020	499.960	0.020	500.000	499.960	0.000

All dimensions are in millimeters.
*The sizes shown are first choice basic sizes (see Table 1). Preferred fits for other sizes can be calculated from data given in ANSI B4.2-1978.
†All fits shown in this table have clearance.

Table 5. American National Standard Preferred Shaft Basis Metric Transition and Interference Fits (ANSI B4.2-1978)

Basic Size*		Locational Transition			Locational Transition			Locational Interference			Medium Drive			Force		
		Hole K7	Shaft h6	Fit†	Hole N7	Shaft h6	Fit†	Hole P7	Shaft h6	Fit†	Hole S7	Shaft h6	Fit†	Hole U7	Shaft h6	Fit†
1	Max	1.000	1.000	+0.006	0.996	1.000	+0.002	0.994	1.000	0.000	0.986	1.000	-0.008	0.982	1.000	-0.012
	Min	0.990	0.994	-0.010	0.986	0.994	-0.014	0.984	0.994	-0.016	0.976	0.994	-0.024	0.972	0.994	-0.028
1.2	Max	1.200	1.200	+0.006	1.196	1.200	+0.002	1.194	1.200	0.000	1.186	1.200	-0.008	1.182	1.200	-0.012
	Min	1.190	1.194	-0.010	1.186	1.194	-0.014	1.184	1.194	-0.016	1.176	1.194	-0.024	1.172	1.194	-0.028
1.6	Max	1.600	1.600	+0.006	1.596	1.600	+0.002	1.594	1.600	0.000	1.586	1.600	-0.008	1.582	1.600	-0.012
	Min	1.590	1.594	-0.010	1.586	1.594	-0.014	1.584	1.594	-0.016	1.576	1.594	-0.024	1.572	1.594	-0.028
2	Max	2.000	2.000	+0.006	1.996	2.000	+0.002	1.994	2.000	0.000	1.986	2.000	-0.008	1.982	2.000	-0.012
	Min	1.990	1.994	-0.010	1.986	1.994	-0.014	1.984	1.994	-0.016	1.976	1.994	-0.024	1.972	1.994	-0.028
2.5	Max	2.500	2.500	+0.006	2.496	2.500	+0.002	2.494	2.500	0.000	2.486	2.500	-0.008	2.482	2.500	-0.012
	Min	2.490	2.494	-0.010	2.486	2.494	-0.014	2.484	2.494	-0.016	2.476	2.494	-0.024	2.472	2.494	-0.028
3	Max	3.000	3.000	+0.006	2.996	3.000	+0.002	2.994	3.000	0.000	2.986	3.000	-0.008	2.982	3.000	-0.012
	Min	2.990	2.994	-0.010	2.986	2.994	-0.014	2.984	2.994	-0.016	2.976	2.994	-0.024	2.972	2.994	-0.028
4	Max	4.003	4.000	+0.011	3.996	4.000	+0.004	3.992	4.000	0.000	3.985	4.000	-0.007	3.981	4.000	-0.011
	Min	3.991	3.992	-0.009	3.984	3.992	-0.016	3.980	3.992	-0.020	3.973	3.992	-0.027	3.969	3.992	-0.031
5	Max	5.003	5.000	+0.011	4.996	5.000	+0.004	4.992	5.000	0.000	4.985	5.000	-0.007	4.981	5.000	-0.011
	Min	4.991	4.992	-0.009	4.984	4.992	-0.016	4.980	4.992	-0.020	4.973	4.992	-0.027	4.969	4.992	-0.031
6	Max	6.003	6.000	+0.011	5.996	6.000	+0.004	5.992	6.000	0.000	5.985	6.000	-0.007	5.981	6.000	-0.011
	Min	5.991	5.992	-0.009	5.984	5.992	-0.016	5.980	5.992	-0.020	5.973	5.992	-0.027	5.969	5.992	-0.031
8	Max	8.005	8.000	+0.014	7.996	8.000	+0.005	7.991	8.000	0.000	7.983	8.000	-0.008	7.978	8.000	-0.013
	Min	7.990	7.991	-0.010	7.981	7.991	-0.019	7.976	7.991	-0.024	7.968	7.991	-0.032	7.963	7.991	-0.037
10	Max	10.005	10.000	+0.014	9.996	10.000	+0.005	9.991	10.000	0.000	9.983	10.000	-0.008	9.978	10.000	-0.013
	Min	9.990	9.991	-0.010	9.981	9.991	-0.019	9.976	9.991	-0.024	9.968	9.991	-0.032	9.963	9.991	-0.037
12	Max	12.006	12.000	+0.017	11.995	12.000	+0.006	11.989	12.000	0.000	11.979	12.000	-0.010	11.974	12.000	-0.015
	Min	11.988	11.989	-0.012	11.977	11.989	-0.023	11.971	11.989	-0.029	11.961	11.989	-0.039	11.956	11.989	-0.044
16	Max	16.006	16.000	+0.017	15.995	16.000	+0.006	15.989	16.000	0.000	15.979	16.000	-0.010	15.974	16.000	-0.015
	Min	15.988	15.989	-0.012	15.977	15.989	-0.023	15.971	15.989	-0.029	15.961	15.989	-0.039	15.956	15.989	-0.044
20	Max	20.006	20.000	+0.019	19.993	20.000	+0.006	19.986	20.000	-0.001	19.973	20.000	-0.014	19.967	20.000	-0.020
	Min	19.985	19.987	-0.015	19.972	19.987	-0.028	19.965	19.987	-0.035	19.952	19.987	-0.048	19.946	19.987	-0.054

All dimensions are in millimeters.
*The sizes shown are first choice basic sizes (see Table 1). Preferred fits for other sizes can be calculated from data given in ANSI B4.2-1978.
†A plus sign indicates clearance and a minus sign indicates interference.

Table 5 (*Concluded*), American National Standard Preferred Shaft Basis Metric Transition and Interference Fits (ANSI B4.2-1978)

Basic Size*		Locational Transition			Locational Transition			Locational Interference			Medium Drive			Force		
		Hole K7	Shaft h6	Fit†	Hole N7	Shaft h6	Fit†	Hole P7	Shaft h6	Fit†	Hole S7	Shaft h6	Fit†	Hole U7	Shaft h6	Fit†
25	Max	25.006	25.000	+0.019	24.993	25.000	+0.006	24.986	25.000	-0.001	24.973	25.000	-0.014	24.960	25.000	-0.027
	Min	24.985	24.987	-0.015	24.972	24.987	-0.028	24.965	24.987	-0.035	24.952	24.987	-0.048	24.939	24.987	-0.061
30	Max	30.006	30.000	+0.019	29.993	30.000	+0.006	29.986	30.000	-0.001	29.973	30.000	-0.014	29.960	30.000	-0.027
	Min	29.985	29.987	-0.015	29.972	29.987	-0.028	29.965	29.987	-0.035	29.952	29.987	-0.048	29.939	29.987	-0.061
40	Max	40.007	40.000	+0.023	39.992	40.000	+0.008	39.983	40.000	-0.001	39.966	40.000	-0.018	39.949	40.000	-0.035
	Min	39.982	39.984	-0.018	39.967	39.984	-0.033	39.958	39.984	-0.042	39.941	39.984	-0.059	39.924	39.984	-0.076
50	Max	50.007	50.000	+0.023	49.992	50.000	+0.008	49.983	50.000	-0.001	49.966	50.000	-0.018	49.939	50.000	-0.045
	Min	49.982	49.984	-0.018	49.967	49.984	-0.033	49.958	49.984	-0.042	49.941	49.984	-0.059	49.914	49.984	-0.086
60	Max	60.009	60.000	+0.028	59.991	60.000	+0.010	59.979	60.000	-0.002	59.958	60.000	-0.023	59.924	60.000	-0.057
	Min	59.979	59.981	-0.021	59.961	59.981	-0.039	59.949	59.981	-0.051	59.928	59.981	-0.072	59.894	59.981	-0.106
80	Max	80.009	80.000	+0.028	79.991	80.000	+0.010	79.979	80.000	-0.002	79.952	80.000	-0.029	79.909	80.000	-0.072
	Min	79.979	79.981	-0.021	79.961	79.981	-0.039	79.949	79.981	-0.051	79.922	79.981	-0.078	79.879	79.981	-0.121
100	Max	100.010	100.000	+0.032	99.990	100.000	+0.012	99.976	100.000	-0.002	99.942	100.000	-0.036	99.889	100.000	-0.089
	Min	99.975	99.978	-0.025	99.955	99.978	-0.045	99.941	99.978	-0.059	99.907	99.978	-0.093	99.854	99.978	-0.146
120	Max	120.010	120.000	+0.032	119.990	120.000	+0.012	119.976	120.000	-0.002	119.934	120.000	-0.044	119.869	120.000	-0.109
	Min	119.975	119.978	-0.025	119.955	119.978	-0.045	119.941	119.978	-0.059	119.899	119.978	-0.101	119.834	119.978	-0.166
160	Max	160.012	160.000	+0.037	159.988	160.000	+0.013	159.972	160.000	-0.003	159.915	160.000	-0.060	159.825	160.000	-0.150
	Min	159.972	159.975	-0.028	159.948	159.975	-0.052	159.932	159.975	-0.068	159.875	159.975	-0.125	159.785	159.975	-0.215
200	Max	200.013	200.000	+0.042	199.986	200.000	+0.015	199.967	200.000	-0.004	199.895	200.000	-0.076	199.781	200.000	-0.190
	Min	199.967	199.971	-0.033	199.940	199.971	-0.060	199.921	199.971	-0.079	199.849	199.971	-0.151	199.735	199.971	-0.265
250	Max	250.013	250.000	+0.042	249.986	250.000	+0.015	249.967	250.000	-0.004	249.877	250.000	-0.094	249.733	250.000	-0.238
	Min	249.967	249.971	-0.033	249.940	249.971	-0.060	249.921	249.971	-0.079	249.831	249.971	-0.169	249.687	249.971	-0.313
300	Max	300.016	300.000	+0.048	299.986	300.000	+0.018	299.964	300.000	-0.004	299.850	300.000	-0.118	299.670	300.000	-0.298
	Min	299.964	299.968	-0.036	299.934	299.968	-0.066	299.912	299.968	-0.088	299.798	299.968	-0.202	299.618	299.968	-0.382
400	Max	400.017	400.000	+0.053	399.984	400.000	+0.020	399.959	400.000	-0.005	399.813	400.000	-0.151	399.586	400.000	-0.378
	Min	399.960	399.964	-0.040	399.927	399.964	-0.073	399.902	399.964	-0.098	399.756	399.964	-0.244	399.529	399.964	-0.477
500	Max	500.018	500.000	+0.058	499.983	500.000	+0.023	499.955	500.000	-0.005	499.771	500.000	-0.189	499.483	500.000	-0.477
	Min	499.955	499.960	-0.045	499.920	499.960	-0.080	499.892	499.960	-0.108	499.708	499.960	-0.292	499.420	499.960	-0.580

All dimensions are in millimeters.

*The sizes shown are first choice basic sizes (see Table 1). Preferred fits for other sizes can be calculated from data given in ANSI B4.2-1978.

†A plus sign indicates clearance and a minus sign indicates interference.

American National Standard Recommended Gage Usage (ANSI B4.4M-1981)

Gagemakers Tolerance		Workpiece Tolerance		
	Class	ISO Symbol*	IT Grade	Recommended Gage Usage
Rejection of Good Parts Increase ⋀	ZM	0.05 IT11	IT11	Low precision gages recommended used to inspect workpieces held to internal (hole) tolerances C11 and H11 and to external (shaft) tolerances c11 and h11.
	YM	0.05 IT9	IT9	Gages recommended used to inspect workpieces held to internal (hole) tolerances D9 and H9 and to external (shaft) tolerances d9 and h9.
	XM	0.05 IT8	IT8	Precision gages recommended used to inspect workpieces held to internal (hole) tolerances F8 and H8.
Gage Cost Increase	XXM	0.05 IT7	IT7	Recommended used for gages to inspect workpieces held to internal (hole) tolerances G7, H7, K7, N7, P7, S7 and U7 and to external (shaft) tolerances f7 and h7.
⋁	XXXM	0.05 IT6	IT6	High precision gages recommended used to inspect workpieces held to external (shaft) tolerances g6, h6, k6, n6, p6, s6 and u6.

* Gagemakers tolerance is equal to 5 per cent of workpiece tolerance or 5 per cent of applicable IT grade value (see table below).
For workpiece tolerance class values, see previous tables 2 to 5, incl.

American National Standard Gagemakers Tolerances (ANSI B4.4M-1981)

Basic Size Over	To	Class ZM (0.05 IT11)	Class YM (0.05 IT9)	Class XM (0.05 IT8)	Class XXM (0.05 IT7)	Class XXXM (0.05 IT6)
0	3	0.0030	0.0012	0.0007	0.0005	0.0003
3	6	0.0037	0.0015	0.0009	0.0006	0.0004
6	10	0.0045	0.0018	0.0011	0.0007	0.0005
10	18	0.0055	0.0021	0.0013	0.0009	0.0006
18	30	0.0065	0.0026	0.0016	0.0010	0.0007
30	50	0.0080	0.0031	0.0019	0.0012	0.0008
50	80	0.0095	0.0037	0.0023	0.0015	0.0010
80	120	0.0110	0.0043	0.0027	0.0017	0.0011
120	180	0.0125	0.0050	0.0031	0.0020	0.0013
180	250	0.0145	0.0057	0.0036	0.0023	0.0015
250	315	0.0160	0.0065	0.0040	0.0026	0.0016
315	400	0.0180	0.0070	0.0044	0.0028	0.0018
400	500	0.0200	0.0077	0.0048	0.0031	0.0020

All dimensions are in millimeters. For closer gagemakers tolerance classes than Class XXXM, specify 5 per cent of IT5, IT4, or IT3 and use the designation 0.05 IT5, 0.05 IT4, etc.

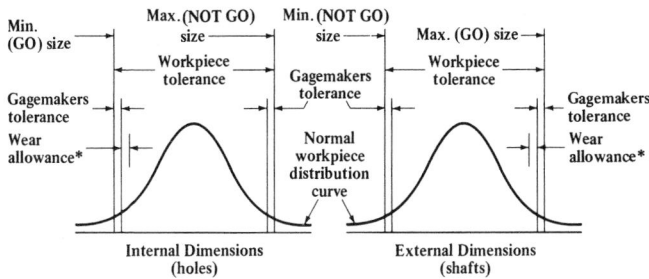

Fig. 5. Relation of Gagemakers Tolerance and Wear Allowance to Workpiece Tolerance

(*Continued from page* 187)
the preferred fits might be considered necessary to satisfy extreme conditions. Subsequent adjustments might also be desired as the result of experience in a particular application to suit critical functional requirements or to permit optimum manufacturing economy. Selection of a departure from these recommendations will depend upon consideration of the engineering and economic factors that might be involved; however, the benefits to be derived from the use of preferred fits should not be overlooked.

A general guide to machining processes which may normally be expected to produce work within the tolerances indicated by the IT grades given in ANSI B4.2-1978 is shown in the chart in Fig. 6.

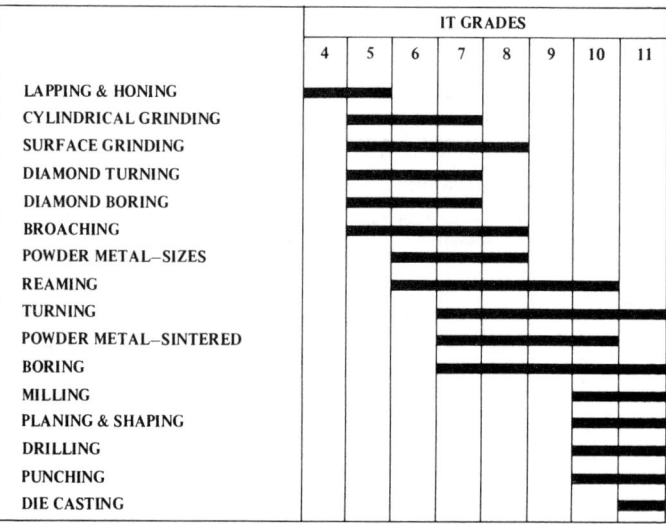

Fig. 6. Relation of Machining Processes to IT Tolerance Grades

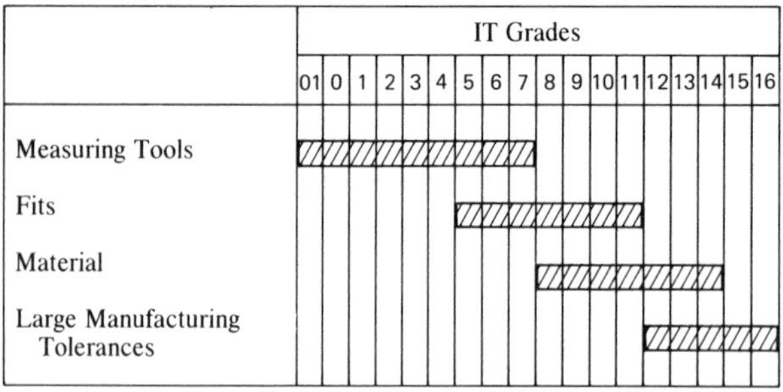

Fig. 7. Practical Usage of IT Tolerance Grades

TAPERS AND KEYS

Measuring Tapers with Vee-Block and Sine-Bar. — The taper on a conical part may be checked or found by placing the part in a vee-block which rests on the surface of a sine-plate or sine-bar as shown in the accompanying diagram. The advantage of this method is that the axis of the vee-block may be aligned with the sides of the sine-bar. Thus when the tapered part is placed in the vee-block it will be aligned perpendicular to the transverse axis of the sine-bar.

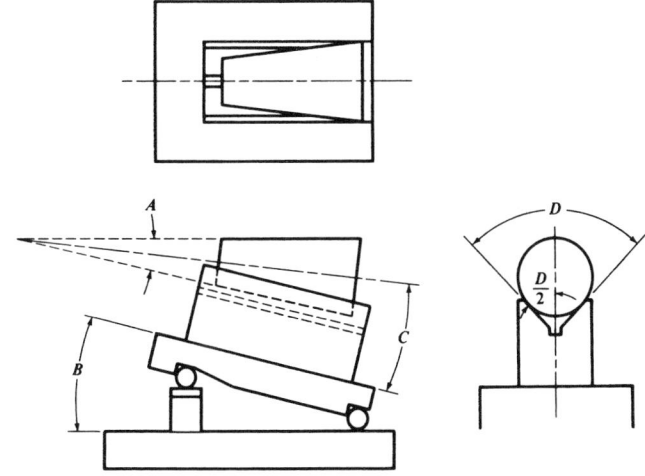

The sine-bar is set to angle $B = (C + A/2)$ where $A/2$ is one-half the included angle of the tapered part. If D is the included angle of the precision vee-block, the angle C is calculated from the formula:

$$\sin C = \frac{\sin A/2}{\sin D/2}$$

If dial indicator readings show no change across all points along the top of the taper surface, then this checks that the angle A of the taper is correct.

If the indicator readings vary, proceed as follows to find the actual angle of taper: (1) Adjust the angle of the sine-bar until the indicator reading is constant. Then find the new angle B' as explained on page 1574 in the paragraph "Measuring Angle of Templet or Gage with Sine-bar." (2) Using the angle B' calculate the actual half-angle $A'/2$ of the taper from the formula:

$$\tan \frac{A'}{2} = \frac{\sin B'}{\csc \frac{D}{2} + \cos B'}$$

The taper per foot corresponding to certain half-angles of taper may be found in the table on page 201.

Measuring Dovetail Slides. — Dovetail slides which must be machined accurately to a given width are commonly gaged by using pieces of cylindrical rod or wire and measuring as indicated by the dimensions x and y of the accompanying

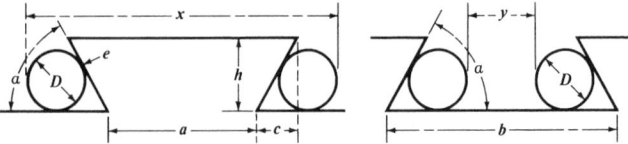

illustrations. To obtain dimension x for measuring male dovetails, add 1 to the cotangent of one-half the dovetail angle α, multiply by diameter D of the rods used, and add the product to dimension a. To obtain dimension y for measuring a female dovetail, add 1 to the cotangent of one-half the dovetail angle α, multiply by diameter D of the rod used, and subtract the result from dimension b. Expressing these rules as formulas:

$$x = D\ (1 + \cot \tfrac{1}{2}\ \alpha) + a.$$
$$y = b - D\ (1 + \cot \tfrac{1}{2}\ \alpha).$$

Dimension c equals $h \times \cot \alpha$.

The rod or wire used should be small enough so that the point of contact e is somewhat below the corner or edge of the dovetail.

Rules for Figuring Tapers

Given	To Find	Rule
The taper per foot.	The taper per inch.	Divide the taper per foot by 12.
The taper per inch.	The taper per foot.	Multiply the taper per inch by 12.
End diameters and length of taper in inches.	The taper per foot.	Subtract small diameter from large; divide by length of taper, and multiply quotient by 12.
Large diameter and length of taper in inches, and taper per foot.	Diameter at small end in inches.	Divide taper per foot by 12; multiply by length of taper, and subtract result from large diameter.
Small diameter and length of taper in inches, and taper per foot.	Diameter at large end in inches.	Divide taper per foot by 12; multiply by length of taper, and add result to small diameter.
The taper per foot and two diameters in inches.	Distance between two given diameters in inches.	Subtract small diameter from large; divide remainder by taper per foot, and multiply quotient by 12.
The taper per foot.	Amount of taper in a certain length given in inches.	Divide taper per foot by 12; multiply by given length of tapered part.

Tapers per Foot and Corresponding Angles *

Taper per Foot	Included Angle	Angle with Center Line	Taper per Foot	Included Angle	Angle with Center Line
1/64	0° 4' 29"	0° 2' 14"	1 7/8	8° 56' 4"	4° 28' 2"
1/32	0 8 57	0 4 29	1 15/16	9 13 51	4 36 56
1/16	0 17 54	0 8 57	2	9 31 38	4 45 49
3/32	0 26 51	0 13 26	2 1/8	10 7 11	5 3 36
1/8	0 35 49	0 17 54	2 1/4	10 42 42	5 21 21
5/32	0 44 46	0 22 23	2 3/8	11 18 11	5 39 5
3/16	0 53 43	0 26 51	2 1/2	11 53 37	5 56 49
7/32	1 2 40	0 31 20	2 5/8	12 29 2	6 14 31
1/4	1 11 37	0 35 49	2 3/4	13 4 24	6 32 12
9/32	1 20 34	0 40 17	2 7/8	13 39 43	6 49 52
5/16	1 29 31	0 44 46	3	14 15 0	7 7 30
11/32	1 38 28	0 49 14	3 1/8	14 50 14	7 25 7
3/8	1 47 25	0 53 43	3 1/4	15 25 26	7 42 43
13/32	1 56 22	0 58 11	3 3/8	16 0 34	8 0 17
7/16	2 5 19	1 2 40	3 1/2	16 35 39	8 17 50
15/32	2 14 16	1 7 8	3 5/8	17 10 42	8 35 21
1/2	2 23 13	1 11 37	3 3/4	17 45 41	8 52 50
17/32	2 32 10	1 16 5	3 7/8	18 20 36	9 10 18
9/16	2 41 7	1 20 33	4	18 55 29	9 27 44
19/32	2 50 4	1 25 2	4 1/8	19 30 17	9 45 9
5/8	2 59 1	1 29 30	4 1/4	20 5 3	10 2 31
21/32	3 7 57	1 33 59	4 3/8	20 39 44	10 19 52
11/16	3 16 54	1 38 27	4 1/2	21 14 22	10 37 11
23/32	3 25 51	1 42 55	4 5/8	21 48 55	10 54 28
3/4	3 34 47	1 47 24	4 3/4	22 23 25	11 11 42
25/32	3 43 44	1 51 52	4 7/8	22 57 50	11 28 55
13/16	3 52 41	1 56 20	5	23 32 12	11 46 6
27/32	4 1 37	2 0 49	5 1/8	24 6 29	12 3 14
7/8	4 10 33	2 5 17	5 1/4	24 40 41	12 20 21
29/32	4 19 30	2 9 45	5 3/8	25 14 50	12 37 25
15/16	4 28 26	2 14 13	5 1/2	25 48 53	12 54 27
31/32	4 37 23	2 18 41	5 5/8	26 22 52	13 11 26
1	4 46 19	2 23 9	5 3/4	26 56 47	13 28 23
1 1/16	5 4 11	2 32 6	5 7/8	27 30 36	13 45 18
1 1/8	5 22 3	2 41 2	6	28 4 21	14 2 10
1 3/16	5 39 55	2 49 57	6 1/8	28 38 1	14 19 0
1 1/4	5 57 47	2 58 53	6 1/4	29 11 35	14 35 48
1 5/16	6 15 38	3 7 49	6 3/8	29 45 5	14 52 32
1 3/8	6 33 29	3 16 44	6 1/2	30 18 29	15 9 15
1 7/16	6 51 19	3 25 40	6 5/8	30 51 48	15 25 54
1 1/2	7 9 10	3 34 35	6 3/4	31 25 2	15 42 31
1 9/16	7 27 0	3 43 30	6 7/8	31 58 11	15 59 5
1 5/8	7 44 49	3 52 25	7	32 31 13	16 15 37
1 11/16	8 2 38	4 1 19	7 1/8	33 4 11	16 32 5
1 3/4	8 20 27	4 10 14	7 1/4	33 37 3	16 48 31
1 13/16	8 38 16	4 19 8	7 3/8	34 9 49	17 4 54

* For conversions into decimal degrees and radians see pages 146 and 147.

Accurate Measurement of Angles and Tapers. — When great accuracy is required in the measurement of angles, or when originating tapers, disks are commonly used. The principle of the disk method of taper measurement is that if two disks of unequal diameters are placed either in contact or a certain distance apart, lines tangent to their peripheries will represent an angle or taper, the degree of which depends upon the diameters of the two disks and the distance between them. The gage shown in the accompany-

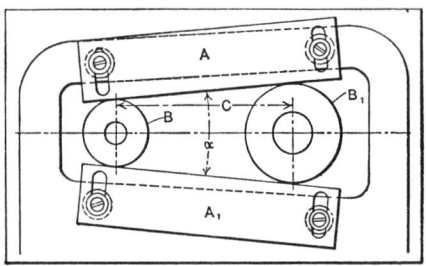

ing illustration, which is a form commonly used for originating tapers or measuring angles accurately, is set by means of disks. This gage consists of two adjustable straight-edges A and A_1, which are in contact with disks B and B_1. The angle α or the taper between the straight-edges depends, of course, upon the diameters of the disks and the center distance C, and as these three dimensions can be measured accurately, it is possible to set the gage to a given angle within very close limits. Moreover, if a record of the three dimensions is kept, the exact setting of the gage can be reproduced quickly at any time. The following rules may be used for adjusting a gage of this type, and cover all problems likely to arise in practice. Disks are also occasionally used for the setting of parts in angular positions for accurately machining them to a given angle; the rules will be found applicable to these conditions also.

To Find Angle for Given Taper per Foot. — When the taper in inches per foot is known, and the corresponding angle α is required. *Rule:* Divide the

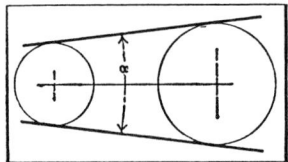

taper in inches per foot by 24; find the angle corresponding to the quotient, in a table of tangents, and double this angle.

Example: What angle α is equivalent to a taper of 1½ inch per foot?

$\dfrac{1.5}{24} = 0.0625$. The angle whose tangent is 0.0625 equals 3 degrees 35 minutes, nearly; then, 3 deg. 35 min. × 2 = 7 deg. 10 min.

To Find Angle for Given Disk Dimensions. — When the diameters D and d of the large and small disks and the center distance are given, to determine the

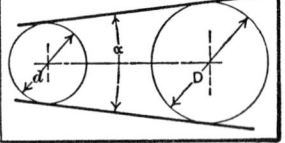

angle α. *Rule:* Divide the difference between the disk diameters by twice the center distance; find the angle corresponding to the quotient, in a table of sines, and double the angle.

Example: If the disk diameters are 1 and 1.5 inch, respectively, and the center distance is 5 inches, find the included angle α.

$\dfrac{1.5 - 1}{2 \times 5} = 0.05$. The angle whose sine is 0.05 equals 2 degrees 52 minutes; then, 2 deg. 52 min. × 2 = 5 deg. 44 min. = angle α.

To Find the Taper per Foot. — When the diameters D and d of the large and small disks and the center distance C are given, to determine the taper per foot (measured at right angles to line through disk centers).

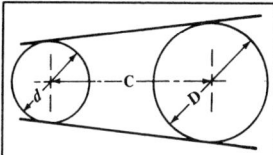

Rule: Divide the difference between the disk diameters by twice the center distance; find the angle corresponding to the quotient, in a table of sines; then find the tangent corresponding to this angle, and multiply the tangent by 24.

Example: If disk diameters are 1 and 1.5 inch, respectively, and center distance is 5 inches, find the taper per foot.

$\dfrac{1.5 - 1}{2 \times 5} = 0.05$. The angle whose sine is 0.05 equals 2 degrees 52 minutes;

$\tan 2° 52' = 0.05007$; $0.05007 \times 24 = 1.2017$ inch taper per foot.

Taper Measured at Right Angles to One Side. — When one side is taken as a base line, and the taper is measured at right angles to that side, use the following rule for determining the taper per foot. *Rule:* Divide the difference between the disk diameters D and d by twice the center distance C; find the angle corresponding to the quotient, in a table of sines; double this angle and find the corresponding tangent; then multiply the tangent by 12.

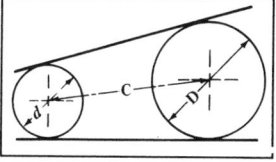

Example: If the disk diameters are 2 and 3 inches, respectively, and the center distance is 5 inches, what is the taper per foot measured at right angles to one side?

$\dfrac{3 - 2}{2 \times 5}$

$= 0.1$. The angle whose sine is 0.1 equals 5 degrees 44 minutes 21 seconds;

then, $2 \times 5° 44' 21'' = 11° 28' 42''$; $\tan 11° 28' 42'' = 0.20306$;

$0.20306 \times 12 = 2.4367$ inches taper per foot.

To Find Center Distance for a Given Taper. — When the taper, in inches per foot, is given, to determine center distance x. *Rule:* Divide the taper by 24 and find the angle corresponding to the quotient in a table of tangents; then find the sine corresponding to this angle and divide the difference between the disk diameters by twice the sine.

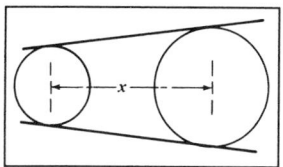

Example: Gage is to be set to ¾ inch per foot, and disk diameters are 1.25 and 1.5 inch, respectively. Find the required center distance for the disks.

$\dfrac{0.75}{24} = 0.03125$. The angle whose tangent is 0.03125 equals 1 degree 47.4 minutes;

$\sin 1° 47.4' = 0.03123$; $1.50 - 1.25 = 0.25$ inch;

$\dfrac{0.25}{2 \times 0.03123} = 4.002$ inches $=$ center distance x.

To Find Center Distance for a Given Angle. — When straight-edges must be set to a given angle α, to determine center distance x between disks of known diameter. *Rule:* Find the sine of half the angle α in a table of sines; divide the difference between the disk diameters by double this sine.

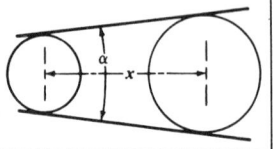

Example: If an angle α of 20 degrees is required, and the disks are 1 and 3 inches in diameter, respectively, find the required center distance x.

$$\frac{20}{2} = 10 \text{ degrees}; \qquad \sin 10° = 0.17365;$$

$$\frac{3-1}{2 \times 0.17365} = 5.759 \text{ inches} = \text{center distance } x.$$

Center Distance when Taper is Measured from One Side. — When taper is measured at right angles to one side, use the following rule for determining the center

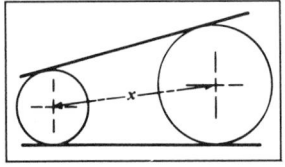

distance x. *Rule:* Divide the taper in inches per foot by 12; find the angle corresponding to the quotient, in a table of tangents; find the sine of one-half this angle, and then divide the difference between the disk diameters by double the sine.

Example: If taper measured at right angles to one side is 6.9 inches per foot, and the disks are 2 and 5 inches in diameter, respectively, what is

center distance x?

$$\frac{6.9}{12} = 0.575. \text{ The angle whose tangent is } 0.575 \text{ equals } 29 \text{ degrees } 54 \text{ minutes};$$

$$\text{then,} \qquad \frac{29 \text{ deg. } 54 \text{ min.}}{2} = 14 \text{ deg. } 57 \text{ min}; \qquad \sin 14°57' = 0.25798;$$

$$\frac{5-2}{2 \times 0.25798} = 5.814 \text{ inches, center distance.}$$

Angular Measurements with Disks in Contact. — When the two disks are to be in contact and the diameter of the small disk is known, the diameter D of the large disk for a given angle α can be obtained as follows.

Rule: Multiply twice the diameter of the small disk by the sine of one-half the required angle; divide this product by 1 minus the sine of one-half the required angle; add the quotient to the diameter of the small disk to obtain the diameter of the large disk.

Example: The required angle α is 15 degrees. Find diameter of large disk, to be in contact with a standard 1-inch reference disk.

$$\sin 7° 30' = 0.13053. \qquad 2 \times 1 \times 0.13053 = 0.26106;$$

$$\frac{0.26106}{1 - 0.13053} = 0.3002. \qquad 1 + 0.3002 = 1.3002 = \text{diameter of large disk.}$$

Table 4. American National Standard Chamfered and Square End Straight Pins
(ANSI B18.8.2-1978)

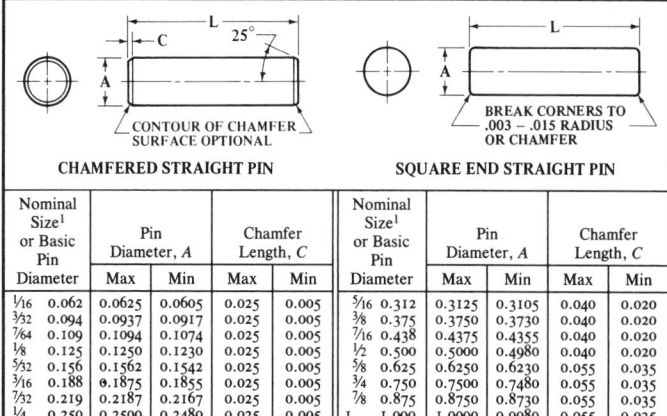

CHAMFERED STRAIGHT PIN   SQUARE END STRAIGHT PIN

Nominal Size[1] or Basic Pin Diameter	Pin Diameter, A		Chamfer Length, C		Nominal Size[1] or Basic Pin Diameter	Pin Diameter, A		Chamfer Length, C	
	Max	Min	Max	Min		Max	Min	Max	Min
1/16 0.062	0.0625	0.0605	0.025	0.005	5/16 0.312	0.3125	0.3105	0.040	0.020
3/32 0.094	0.0937	0.0917	0.025	0.005	3/8 0.375	0.3750	0.3730	0.040	0.020
7/64 0.109	0.1094	0.1074	0.025	0.005	7/16 0.438	0.4375	0.4355	0.040	0.020
1/8 0.125	0.1250	0.1230	0.025	0.005	1/2 0.500	0.5000	0.4980	0.040	0.020
5/32 0.156	0.1562	0.1542	0.025	0.005	5/8 0.625	0.6250	0.6230	0.055	0.035
3/16 0.188	0.1875	0.1855	0.025	0.005	3/4 0.750	0.7500	0.7480	0.055	0.035
7/32 0.219	0.2187	0.2167	0.025	0.005	7/8 0.875	0.8750	0.8730	0.055	0.035
1/4 0.250	0.2500	0.2480	0.025	0.005	1 1.000	1.0000	0.9980	0.055	0.035

All dimensions are in inches.
[1] Where specifying nominal size in decimals, zeros preceding decimal point are omitted.

American National Standard Taper Pins. — Taper pins have a uniform taper over the pin length with both ends crowned. Most sizes are supplied in commercial and precision classes, the latter having generally tighter tolerances and being more closely controlled in manufacture.

Diameters: The major diameter of both commercial and precision classes of pins is the diameter of the large end and is the basis for pin size. The diameter at the small end is computed by multiplying the nominal length of the pin by the factor 0.02083 and subtracting the result from the basic pin diameter. See also Table 6.

Taper: The taper on commercial class pins is 0.250 ± 0.006 inch per foot and on the precision class pins is 0.250 ± 0.004 inch per foot of length.

Materials: Unless otherwise specified, taper pins are made from AISI 1211 steel or cold drawn AISI 1212 or 1213 steel or equivalents, and no mechanical properties apply.

Hole Sizes: Under most circumstances, holes for taper pins require taper reaming. Sizes and lengths of taper pins for which standard reamers are available are given in Table 5. Drilling specifications for taper pins are given below.

Designation: Taper pins are designated by the following data in the sequence shown: Product name (noun first), class, size number (or decimal equivalent), length (fraction or three-place decimal equivalent), material, and protective finish, if required.

Examples: Pin, Taper (Commercial Class) No. 0 × ¾, Steel

Pin, Taper (Precision Class) .219 × 1.750, Steel, Zinc Plated

Drilling Specifications for Taper Pins. — When helically fluted taper pin reamers are used, the diameter of the through hole drilled prior to reaming is equal to the diameter at the small end of the taper pin. (See Table 6.) However, when straight fluted taper reamers are to be used, it may be necessary, in the case of long pins, to step drill the hole before reaming, the number and sizes of the drills to be used depending on the depth of the hole (pin length).

Table 5. American National Standard Taper Pins (ANSI B18.8.2-1978)

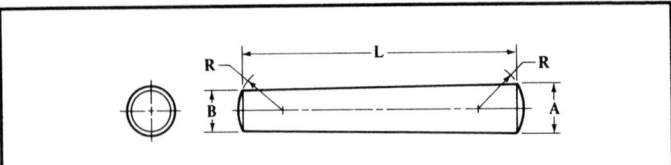

Pin Size Number and Basic Pin Diameter[1]	Major Diameter (Large End), A				End Crown Radius, R		Range of Lengths,[2] L	
	Commercial Class		Precision Class					
	Max	Min	Max	Min	Max	Min	Stand. Reamer Avail.[3]	Other
7/0 0.0625	0.0638	0.0618	0.0635	0.0625	0.072	0.052	. . .	¼–1
6/0 0.0780	0.0793	0.0773	0.0790	0.0780	0.088	0.068	. . .	¼–1½
5/0 0.0940	0.0953	0.0933	0.0950	0.0940	0.104	0.084	¼–1	1¼, 1½
4/0 0.1090	0.1103	0.1083	0.1100	0.1090	0.119	0.099	¼–1	1¼–2
3/0 0.1250	0.1263	0.1243	0.1260	0.1250	0.135	0.115	¼–1	1¼–2
2/0 0.1410	0.1423	0.1403	0.1420	0.1410	0.151	0.131	½–1¼	1½–2½
0 0.1560	0.1573	0.1553	0.1570	0.1560	0.166	0.146	½–1¼	1½–3
1 0.1720	0.1733	0.1713	0.1730	0.1720	0.182	0.162	¾–1¼	1½–3
2 0.1930	0.1943	0.1923	0.1940	0.1930	0.203	0.183	¾–1½	1¾–3
3 0.2190	0.2203	0.2183	0.2200	0.2190	0.229	0.209	¾–1¾	2–4
4 0.2500	0.2513	0.2493	0.2510	0.2500	0.260	0.240	¾–2	2¼–4
5 0.2890	0.2903	0.2883	0.2900	0.2890	0.299	0.279	1–2½	2¾–6
6 0.3410	0.3423	0.3403	0.3420	0.3410	0.351	0.331	1¼–3	3¼–6
7 0.4090	0.4103	0.4083	0.4100	0.4090	0.419	0.399	1¼–3¾	4–8
8 0.4920	0.4933	0.4913	0.4930	0.4920	0.502	0.482	1¼–4½	4¾–8
9 0.5910	0.5923	0.5903	0.5920	0.5910	0.601	0.581	1¼–5¼	5½–8
10 0.7060	0.7073	0.7053	0.7070	0.7060	0.716	0.696	1½–6	6¼–8
11 0.8600	0.8613	0.8593	. . .	. . .	0.870	0.850	. . .	2–8
12 1.0320	1.0333	1.0313	. . .	. . .	1.042	1.022	. . .	2–9
13 1.2410	1.2423	1.2403	. . .	. . .	1.251	1.231	. . .	3–11
14 1.5210	1.5223	1.5203	. . .	. . .	1.531	1.511	. . .	3–13

All dimensions are in inches.

For nominal diameters, B, see Table 6.

[1] When specifying nominal pin size in decimals, zeros preceding the decimal and in the fourth decimal place are omitted.

[2] Lengths increase in ⅛-inch steps up to 1 inch and in ¼-inch steps above 1 inch.

[3] Standard reamers are available for pin lengths in this column.

To determine the number and sizes of step drills required: (1) find the length of pin to be used at the top of the chart on page 207 and follow this length down to the intersection with that heavy line which represents the size of taper pin (see taper pin numbers at the right-hand end of each heavy line). (2) If the length of pin falls between the first and second dots, counting from the left, only one drill is required. Its size is indicated by following the nearest horizontal line from the point of intersection (of the pin length) on the heavy line over to the drill diameter values at the left. (3) If the intersection of pin length comes between the second and third dots, then two drills are required. In this case the smaller has a size corresponding to the intersection of the pin length and heavy line and the larger is the corresponding drill diameter for the intersection of one-half this length with the heavy line. (4) Should the pin length fall between the third and fourth dots, then three drills are required. The smallest will have a diameter corresponding to the intersection of the total pin length with the heavy line, the next in size will have a diameter corresponding to the intersection of two-thirds of this length with the heavy line and the largest will have a diameter corresponding to the intersection of one-third of this length with the heavy line. Where the intersection falls between two drill sizes, use the smaller.

Chart to Facilitate Selection of Number and Sizes of Drills
for Step-Drilling Prior to Taper Reaming

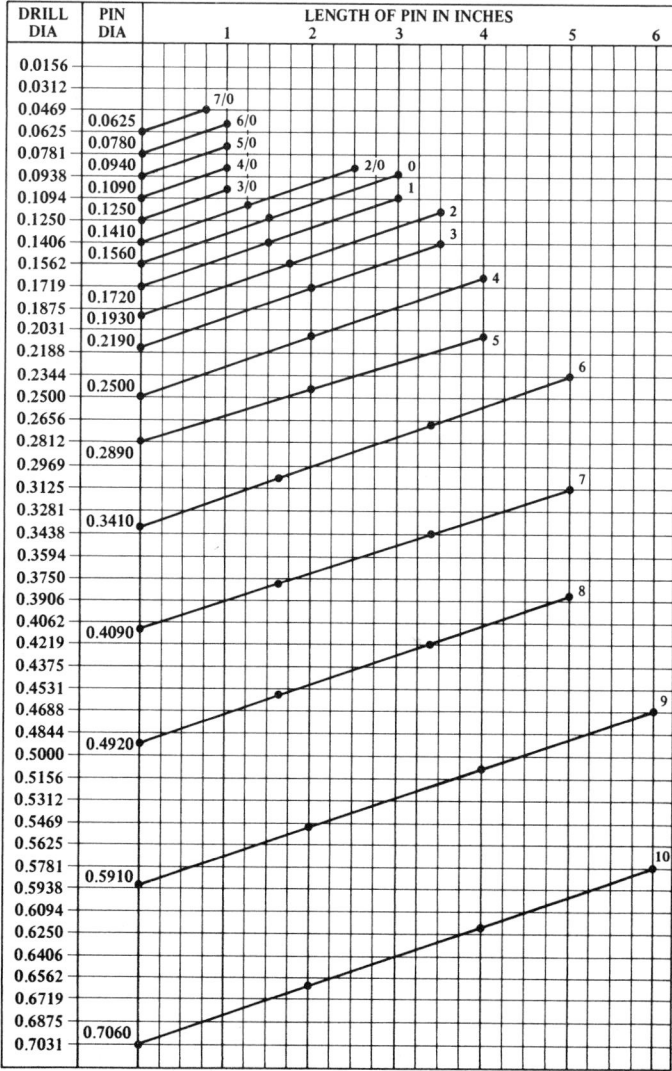

Table 6. Nominal Diameter at Small Ends of Standard Taper Pins

Pin Length in Inches	Pin Number and Small End Diameter for Given Length										
	0	1	2	3	4	5	6	7	8	9	10
¾	0.140	0.156	0.177	0.203	0.235	0.273	0.325	0.393	0.476	0.575	0.690
1	0.135	0.151	0.172	0.198	0.230	0.268	0.320	0.388	0.471	0.570	0.685
1¼	0.130	0.146	0.167	0.192	0.224	0.263	0.315	0.382	0.466	0.565	0.680
1½	0.125	0.141	0.162	0.187	0.219	0.258	0.310	0.377	0.460	0.560	0.675
1¾	0.120	0.136	0.157	0.182	0.214	0.252	0.305	0.372	0.455	0.554	0.669
2	0.114	0.130	0.151	0.177	0.209	0.247	0.299	0.367	0.450	0.549	0.664
2¼	0.109	0.125	0.146	0.172	0.204	0.242	0.294	0.362	0.445	0.544	0.659
2½	0.104	0.120	0.141	0.166	0.198	0.237	0.289	0.356	0.440	0.539	0.654
2¾	0.099	0.115	0.136	0.161	0.193	0.232	0.284	0.351	0.434	0.534	0.649
3	0.094	0.110	0.131	0.156	0.188	0.227	0.279	0.346	0.429	0.528	0.643
3¼	. . .	. . .	. . .	0.151	0.182	0.221	0.273	0.340	0.424	0.523	0.638
3½	. . .	. . .	. . .	0.146	0.177	0.216	0.268	0.335	0.419	0.518	0.633
3¾	. . .	. . .	. . .	0.141	0.172	0.211	0.263	0.330	0.414	0.513	0.628
4	. . .	. . .	. . .	0.136	0.167	0.206	0.258	0.326	0.409	0.508	0.623
4¼	. . .	. . .	. . .	0.131	0.162	0.201	0.253	0.321	0.403	0.502	0.617
4½	. . .	. . .	. . .	0.125	0.156	0.195	0.247	0.315	0.398	0.497	0.612
5	. . .	. . .	. . .	. . .	0.146	0.185	0.237	0.305	0.389	0.487	0.602
5½	. . .	. . .	. . .	. . .	. . .	. . .	. . .	0.294	0.377	0.476	0.591
6	. . .	. . .	. . .	. . .	. . .	. . .	. . .	0.284	0.367	0.466	0.581

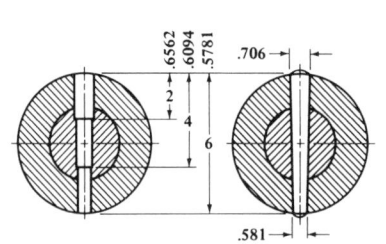

Examples: For a No. 10 taper pin 6-inches long, three drills would be used, of the sizes and for the depths shown in the accompanying diagram.

For a No. 10 taper pin 3-inches long, two drills would be used since the 3-inch length falls between the second and third dots. The first or through drill will be 0.6406 inch and the second drill, 0.6719 inch for a depth of 1½ inches.

American National Standard Grooved Pins. — These pins have three equally spaced longitudinal grooves and an expanded diameter over the crests of the ridges formed by the material displaced when the grooves are produced. The grooves are aligned with the axis of the pins.

Materials: Grooved pins are normally made from cold drawn low carbon steel wire or rod. Where additional performance is required, carbon steel pins may be supplied surface hardened and heat treated to a hardness consistent with the performance requirements. Pins may also be made from alloy steel, corrosion resistant steel, brass, Monel and other non-ferrous metals having chemical properties as agreed upon between manufacturer and purchaser.

STANDARD TAPERS

Certain types of small tools and machine parts, such as twist drills, end mills, arbors, lathe centers, etc., are provided with taper shanks which fit into spindles or sockets of corresponding taper, thus providing not only accurate alignment between the tool or other part and its supporting member, but also more or less frictional resistance for driving the tool. There are several standards for "self-holding" tapers, but the American National, Morse, and the Brown & Sharpe are the standards most widely used by American manufacturers.

The name *self-holding* has been applied to the smaller tapers — like the Morse and the Brown & Sharpe — because, where the angle of the taper is only 2 or 3 degrees, the shank of a tool is so firmly seated in its socket that there is considerable frictional resistance to any force tending to turn or rotate the tool relative to the socket. The term "self-holding" is used to distinguish relatively small tapers from the larger or *self-releasing* type. A milling machine spindle having a taper of 3½ inches per foot is an example of a self-releasing taper. The included angle in this case is over 16 degrees and the tool or arbor requires a positive locking device to prevent slipping, but the shank may be released or removed more readily than one having a smaller taper of the self-holding type.

Morse Taper. — Dimensions relating to Morse standard taper shanks and sockets may be found in an accompanying table. The taper for different numbers of Morse tapers is slightly different, but it is approximately ⅝ inch per foot in most cases. The table gives the actual tapers, accurate to five decimal places. Morse taper shanks are used on a variety of tools, and exclusively on the shanks of twist drills. Dimensions for Morse Stub Taper Shanks are given in Table 1B.

Brown & Sharpe Taper. — This standard taper is used for taper shanks on tools such as end mills and reamers, the taper being approximately ½ inch per foot for all sizes except for taper No. 10, where the taper is 0.5161 inch per foot. Brown & Sharpe taper sockets are used for many arbors, collets, and machine tool spindles, especially milling machines and grinding machines. In many cases there are a number of different lengths of sockets corresponding to the same number of taper; all these tapers, however, are of the same diameter at the small end.

Jarno Taper. — The Jarno taper was originally proposed by Oscar J. Beale of the Brown & Sharpe Mfg. Co. This taper is based on such simple formulas that practically no calculations are required when the number of taper is known. The taper per foot of all Jarno taper sizes is 0.600 inch on the diameter. The diameter at the large end is as many eighths, the diameter at the small end is as many tenths, and the length as many half inches as are indicated by the number of the taper. For example, a No. 7 Janor taper is ⅞ inch in diameter at the large end; ⁷⁄₁₀, or 0.700 inch at the small end; and ⁷⁄₂, or 3½ inches long; hence, diameter at large end = No. of taper ÷ 8; diameter at small end = No. of taper ÷ 10; length of taper = No. of taper ÷ 2. The Jarno taper is used on various machine tools, especially profiling machines and die-sinking machines. It has also been used for the headstock and tailstock spindles of some lathes.

American National Standard Machine Tapers. — This standard includes a self-holding series (Table 5, 7, 8, 9 and 10) and a steep taper series, Table 6. The self-holding taper series consists of 22 sizes which are listed in Table 5. The reference gage for the self-holding tapers is a plug gage. Table 11 gives the dimensions and tolerances for both plug and ring gages applying to this series. Tables 7 to 10 inclusive give the dimensions for self-holding taper shanks and sockets which are classified as to (1) means of transmitting torque from spindle to the tool shank, and (2) means of retaining the shank in the socket. The steep machine tapers consist of a preferred

Table 1A. Morse Standard Taper Shanks

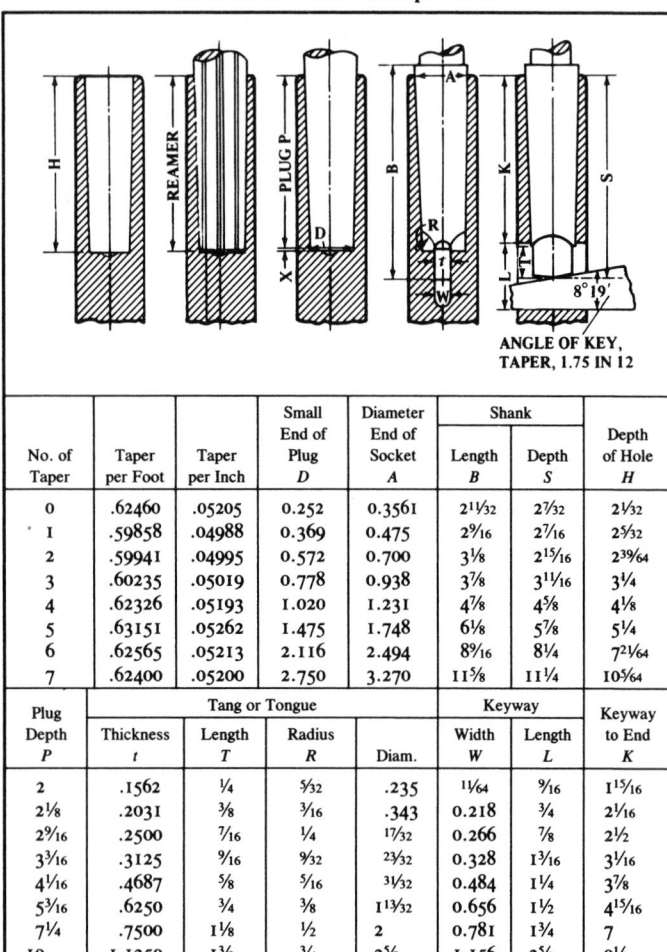

ANGLE OF KEY,
TAPER, 1.75 IN 12

No. of Taper	Taper per Foot	Taper per Inch	Small End of Plug D	Diameter End of Socket A	Shank Length B	Shank Depth S	Depth of Hole H
0	.62460	.05205	0.252	0.3561	2¹¹⁄₃₂	2⁷⁄₃₂	2¹⁄₃₂
1	.59858	.04988	0.369	0.475	2⁹⁄₁₆	2⁷⁄₁₆	2⁵⁄₃₂
2	.59941	.04995	0.572	0.700	3⅛	2¹⁵⁄₁₆	2³⁹⁄₆₄
3	.60235	.05019	0.778	0.938	3⅞	3¹¹⁄₁₆	3¼
4	.62326	.05193	1.020	1.231	4⅞	4⅝	4⅛
5	.63151	.05262	1.475	1.748	6⅛	5⅞	5¼
6	.62565	.05213	2.116	2.494	8⁹⁄₁₆	8¼	7²¹⁄₆₄
7	.62400	.05200	2.750	3.270	11⅝	11¼	10⁵⁄₆₄

Plug Depth P	Tang or Tongue Thickness t	Length T	Radius R	Diam.	Keyway Width W	Length L	Keyway to End K
2	.1562	¼	⁵⁄₃₂	.235	¹¹⁄₆₄	⁹⁄₁₆	1¹⁵⁄₁₆
2⅛	.2031	⅜	³⁄₁₆	.343	0.218	¾	2¹⁄₁₆
2⁹⁄₁₆	.2500	⁷⁄₁₆	¼	¹⁷⁄₃₂	0.266	⅞	2½
3³⁄₁₆	.3125	⁹⁄₁₆	⁹⁄₃₂	²³⁄₃₂	0.328	1³⁄₁₆	3¹⁄₁₆
4¹⁄₁₆	.4687	⅝	⁵⁄₁₆	³¹⁄₃₂	0.484	1¼	3⅞
5³⁄₁₆	.6250	¾	⅜	1¹³⁄₃₂	0.656	1½	4¹⁵⁄₁₆
7¼	.7500	1⅛	½	2	0.781	1¾	7
10	1.1250	1⅜	¾	2⅝	1.156	2⅝	9½

series (bold-face type, Table 6) and an intermediate series (light-face type). A self-holding taper is defined as "a taper with an angle small enough to hold a shank in place ordinarily by friction without holding means. (Sometimes referred to as slow taper.)" A steep taper is defined as "a taper having an angle sufficiently large to insure the easy or self-releasing feature." The term "gage line" indicates the basic diameter at or near the large end of the taper.

Table 1B. Morse Stub Taper Shanks

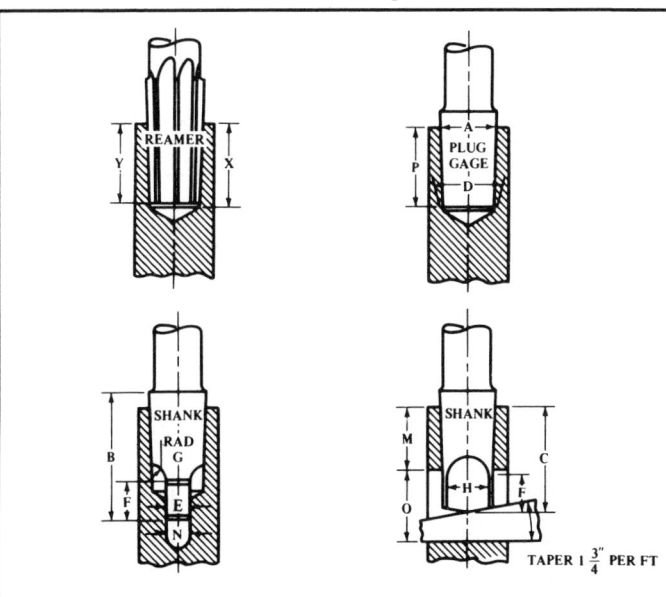

TAPER $1\frac{3}{4}''$ PER FT

No. of Taper	Taper per Foot*	Taper per Inch†	Small End of Plug,† D	Diam. End of Socket,* A	Shank Total Length, B	Shank Depth, C	Tang Thickness, E	Tang Length, F
1	.59858	.049882	.4314	.475	$1\frac{5}{16}$	$1\frac{1}{8}$	$\frac{13}{64}$	$\frac{5}{16}$
2	.59941	.049951	.6469	.700	$1\frac{11}{16}$	$1\frac{7}{16}$	$\frac{19}{64}$	$\frac{7}{16}$
3	.60235	.050196	.8753	.938	2	$1\frac{3}{4}$	$\frac{25}{64}$	$\frac{9}{16}$
4	.62326	.051938	1.1563	1.231	$2\frac{3}{8}$	$2\frac{1}{16}$	$\frac{33}{64}$	$1\frac{1}{16}$
5	.63151	.052626	1.6526	1.748	3	$2\frac{11}{16}$	$\frac{3}{4}$	$1\frac{5}{16}$

No. of Taper	Tang Radius of Mill, G	Tang Diameter, H	Plug Depth, P	Socket Min. Depth of Tapered Hole Drilled, X	Socket Min. Depth of Tapered Hole Reamed, Y	Socket End to Tang Slot, M	Tang Slot Width, N	Tang Slot Length, O
1	$\frac{3}{16}$	$\frac{13}{32}$	$\frac{7}{8}$	$\frac{5}{16}$	$\frac{29}{32}$	$\frac{25}{32}$	$\frac{7}{32}$	$\frac{23}{32}$
2	$\frac{7}{32}$	$\frac{39}{64}$	$1\frac{1}{16}$	$1\frac{5}{32}$	$1\frac{7}{64}$	$\frac{15}{16}$	$\frac{5}{16}$	$\frac{15}{16}$
3	$\frac{9}{32}$	$\frac{13}{16}$	$1\frac{1}{4}$	$1\frac{3}{8}$	$1\frac{5}{16}$	$1\frac{1}{16}$	$\frac{13}{32}$	$1\frac{1}{8}$
4	$\frac{3}{8}$	$1\frac{3}{32}$	$1\frac{7}{16}$	$1\frac{9}{16}$	$1\frac{1}{2}$	$1\frac{3}{16}$	$\frac{17}{32}$	$1\frac{3}{8}$
5	$\frac{9}{16}$	$1\frac{19}{32}$	$1\frac{13}{16}$	$1\frac{15}{16}$	$1\frac{7}{8}$	$1\frac{7}{16}$	$\frac{25}{32}$	$1\frac{3}{4}$

All dimensions in inches.
* These are basic dimensions.
† These dimensions are calculated for reference only.
Radius J is $\frac{3}{64}$, $\frac{1}{16}$, $\frac{5}{64}$, $\frac{3}{32}$, and $\frac{1}{8}$ inch respectively for Nos. 1, 2, 3, 4, and 5 tapers.

Table 2. Dimensions of Morse Taper Sleeves

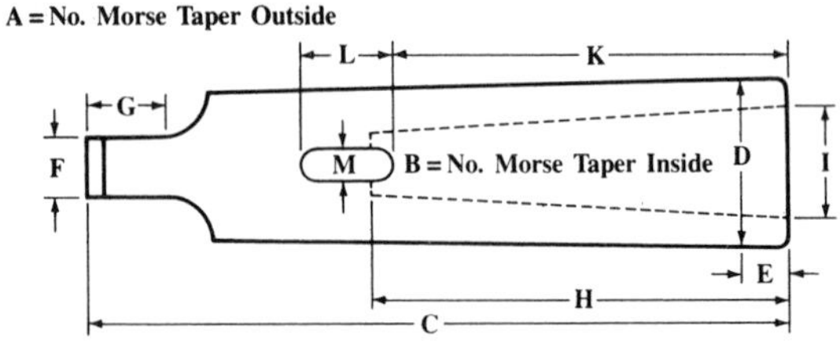

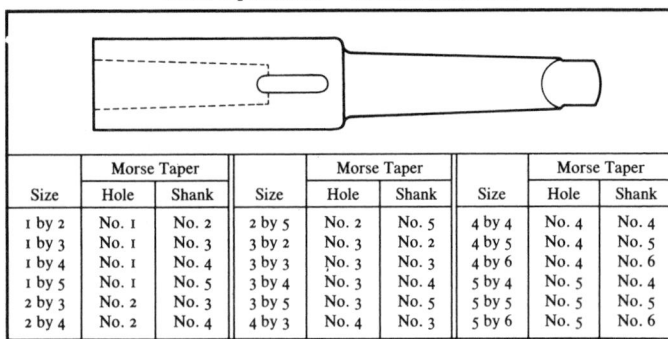

A = No. Morse Taper Outside
B = No. Morse Taper Inside

A	B	C	D	E	F	G	H	I	K	L	M
2	1	3 9/16	0.700	5/8	1/4	7/16	2 3/16	0.475	2 1/16	3/4	0.213
3	1	3 15/16	0.938	1/4	5/16	9/16	2 3/16	0.475	2 1/16	3/4	0.213
3	2	4 7/16	0.938	3/4	5/16	9/16	2 5/8	0.700	2 1/2	7/8	0.260
4	1	4 7/8	1.231	1/4	15/32	5/8	2 3/16	0.475	2 1/16	3/4	0.213
4	2	4 7/8	1.231	1/4	15/32	5/8	2 5/8	0.700	2 1/2	7/8	0.260
4	3	5 3/8	1.231	3/4	15/32	5/8	3 1/4	0.938	3 1/16	1 3/16	0.322
5	1	6 1/8	1.748	1/4	5/8	3/4	2 3/16	0.475	2 1/16	3/4	0.213
5	2	6 1/8	1.748	1/4	5/8	3/4	2 5/8	0.700	2 1/2	7/8	0.260
5	3	6 1/8	1.748	1/4	5/8	3/4	3 1/4	0.938	3 1/16	1 3/16	0.322
5	4	6 5/8	1.748	3/4	5/8	3/4	4 1/8	1.231	3 7/8	1 1/4	0.478
6	1	8 5/8	2.494	3/8	3/4	1 1/8	2 3/16	0.475	2 1/16	3/4	0.213
6	2	8 5/8	2.494	3/8	3/4	1 1/8	2 5/8	0.700	2 1/2	7/8	0.260
6	3	8 5/8	2.494	3/8	3/4	1 1/8	3 1/4	0.938	3 1/16	1 3/16	0.322
6	4	8 5/8	2.494	3/8	3/4	1 1/8	4 1/8	1.231	3 7/8	1 1/4	0.478
6	5	8 5/8	2.494	3/8	3/4	1 1/8	5 1/4	1.748	4 15/16	1 1/2	0.635
7	3	11 5/8	3.270	3/8	1 1/8	1 3/8	3 1/4	0.938	3 1/16	1 3/16	0.322
7	4	11 5/8	3.270	3/8	1 1/8	1 3/8	4 1/8	1.231	3 7/8	1 1/4	0.478
7	5	11 5/8	3.270	3/8	1 1/8	1 3/8	5 1/4	1.748	4 15/16	1 1/2	0.635
7	6	12 1/2	3.270	1 1/4	1 1/8	1 3/8	7 3/8	2.494	7	1 3/4	0.760

Morse Taper Sockets — Hole and Shank Sizes

Size	Morse Taper Hole	Morse Taper Shank	Size	Morse Taper Hole	Morse Taper Shank	Size	Morse Taper Hole	Morse Taper Shank
1 by 2	No. 1	No. 2	2 by 5	No. 2	No. 5	4 by 4	No. 4	No. 4
1 by 3	No. 1	No. 3	3 by 2	No. 3	No. 2	4 by 5	No. 4	No. 5
1 by 4	No. 1	No. 4	3 by 3	No. 3	No. 3	4 by 6	No. 4	No. 6
1 by 5	No. 1	No. 5	3 by 4	No. 3	No. 4	5 by 4	No. 5	No. 4
2 by 3	No. 2	No. 3	3 by 5	No. 3	No. 5	5 by 5	No. 5	No. 5
2 by 4	No. 2	No. 4	4 by 3	No. 4	No. 3	5 by 6	No. 5	No. 6

Table 3. Brown & Sharpe Taper Shanks

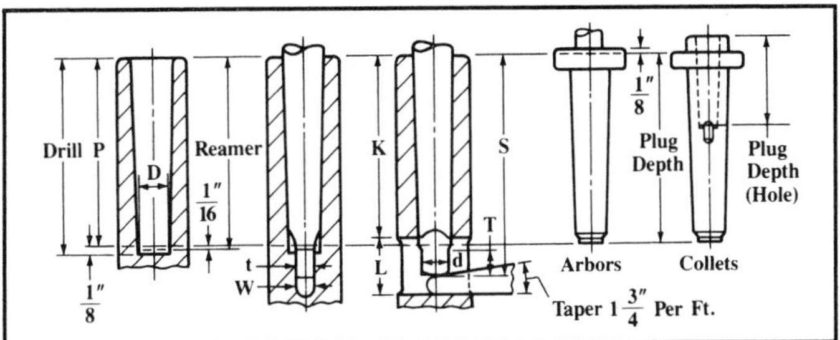
Drill P
D
Reamer
1″/16
t
W
1″/8
K
S
T
L
d
1″/8
Plug Depth
Plug Depth (Hole)
Arbors
Collets
Taper 1 3/4″ Per Ft.

Number of Taper	Taper per Foot (inch)	Diam. of Plug at Small End *D*	Plug Depth, *P*: B & S** Standard	Plug Depth, *P*: Mill. Mach. Standard	Plug Depth, *P*: Miscell.	Keyway from End of Spindle *K*	Shank Depth *S*	Length of Keyway† *L*	Width of Keyway *W*	Length of Arbor Tongue *T*	Diameter of Arbor Tongue *d*	Thickness of Arbor Tongue *t*
*1	.50200	.20000	15/16	...	...	15/16	1 3/16	3/8	.135	3/16	.170	1/8
*2	.50200	.25000	1 3/16	...	...	1 11/64	1 1/2	1/2	.166	1/4	.220	5/32
*3	.50200	.31250	1 1/2	...	...	1 15/32	1 7/8	5/8	.197	5/16	.282	3/16
			...	...	1 3/4	1 23/32	2 1/8	5/8	.197	5/16	.282	3/16
			...	...	2	1 31/32	2 3/8	5/8	.197	5/16	.282	3/16
4	.50240	.35000	...	1 1/4	...	1 13/64	1 21/32	11/16	.228	11/32	.320	7/32
			1 11/16	...	...	1 41/64	2 3/32	11/16	.228	11/32	.320	7/32
5	.50160	.45000	...	1 3/4	...	1 11/16	2 3/16	3/4	.260	3/8	.420	1/4
			...	...	2	1 15/16	2 7/16	3/4	.260	3/8	.420	1/4
			2 1/8	...	...	2 1/16	2 9/16	3/4	.260	3/8	.420	1/4
6	.50329	.50000	2 3/8	...	...	2 19/64	2 7/8	7/8	.291	7/16	.460	9/32
7	.50147	.60000	...	...	2 1/2	2 13/32	3 1/32	15/16	.322	15/32	.560	5/16
			2 7/8	...	...	2 25/32	3 13/32	15/16	.322	15/32	.560	5/16
			...	3	...	2 29/32	3 17/32	15/16	.322	15/32	.560	5/16
8	.50100	.75000	3 9/16	...	...	3 29/64	4 1/8	1	.353	1/2	.710	11/32
9	.50085	.90010	...	4	...	3 7/8	4 5/8	1 1/8	.385	9/16	.860	3/8
			4 1/4	...	...	4 1/8	4 7/8	1 1/8	.385	9/16	.860	3/8
10	.51612	1.04465	5	...	...	4 27/32	5 23/32	1 5/16	.447	21/32	1.010	7/16
			...	5 11/16	...	5 17/32	6 13/32	1 5/16	.447	21/32	1.010	7/16
			...	...	6 7/32	6 1/16	6 15/16	1 5/16	.447	21/32	1.010	7/16
11	.50100	1.24995	5 15/16	...	...	5 25/32	6 21/32	1 5/16	.447	21/32	1.210	7/16
			...	6 3/4	...	6 19/32	7 15/32	1 5/16	.447	21/32	1.210	7/16
12	.49973	1.50010	7 1/8	7 1/8	...	6 15/16	7 15/16	1 1/2	.510	3/4	1.460	1/2
			...	...	6 1/4	...	...	...	...	...	...	...
13	.50020	1.75005	7 3/4	...	...	7 9/16	8 9/16	1 1/2	.510	3/4	1.710	1/2
14	.50000	2.00000	8 1/4	8 1/4	...	8 1/32	9 5/32	1 11/16	.572	27/32	1.960	9/16
15	.50000	2.25000	8 3/4	...	...	8 17/32	9 21/32	1 11/16	.572	27/32	2.210	9/16
16	.50000	2.50000	9 1/4	...	...	9	10 1/4	1 7/8	.635	15/16	2.450	5/8
17	.50000	2.75000	9 3/4	...	...	...	...	...	...	...	...	...
18	.50000	3.00000	10 1/4	...	...	...	...	...	...	...	...	...

* Adopted by American Standards Association. ** "B & S Standard" Plug Depths are not used in all cases. † Special lengths of keyway are used instead of standard lengths in some places. Standard lengths need not be used when keyway is for driving only and not for admitting key to force out tool.

Table 4. Jarno Taper Shanks

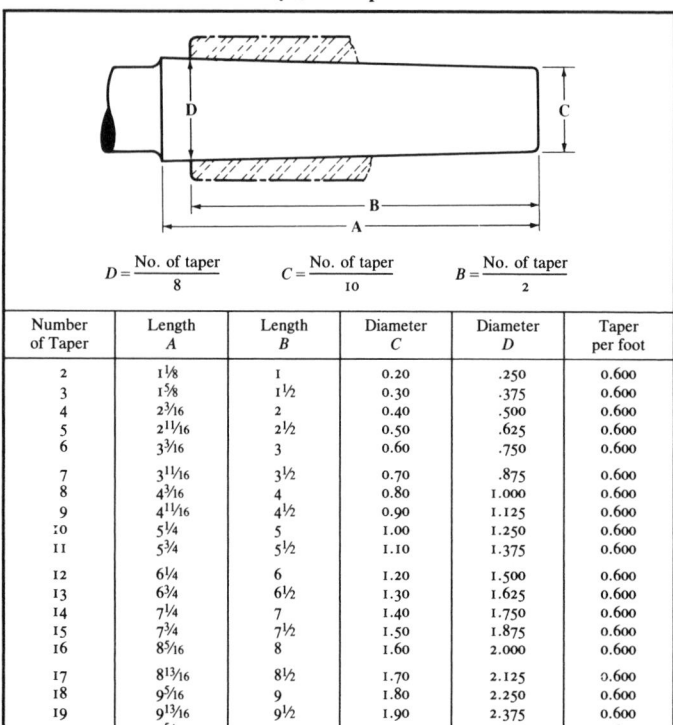

$$D = \frac{\text{No. of taper}}{8} \qquad C = \frac{\text{No. of taper}}{10} \qquad B = \frac{\text{No. of taper}}{2}$$

Number of Taper	Length A	Length B	Diameter C	Diameter D	Taper per foot
2	1⅛	1	0.20	.250	0.600
3	1⅝	1½	0.30	.375	0.600
4	2³⁄₁₆	2	0.40	.500	0.600
5	2¹¹⁄₁₆	2½	0.50	.625	0.600
6	3³⁄₁₆	3	0.60	.750	0.600
7	3¹¹⁄₁₆	3½	0.70	.875	0.600
8	4³⁄₁₆	4	0.80	1.000	0.600
9	4¹¹⁄₁₆	4½	0.90	1.125	0.600
10	5¼	5	1.00	1.250	0.600
11	5¾	5½	1.10	1.375	0.600
12	6¼	6	1.20	1.500	0.600
13	6¾	6½	1.30	1.625	0.600
14	7¼	7	1.40	1.750	0.600
15	7¾	7½	1.50	1.875	0.600
16	8⁵⁄₁₆	8	1.60	2.000	0.600
17	8¹³⁄₁₆	8½	1.70	2.125	0.600
18	9⁵⁄₁₆	9	1.80	2.250	0.600
19	9¹³⁄₁₆	9½	1.90	2.375	0.600
20	10⁵⁄₁₆	10	2.00	2.500	0.600

Tapers for Machine Tool Spindles. — Various standard tapers have been used for the taper holes in the spindles of machine tools requiring a taper hole for receiving either the shank of a cutter, an arbor, a center, or any tool or accessory requiring a tapering seat. The spindles of drilling machines and the taper shanks of twist drills are made to fit the Morse taper. For lathes, the Morse taper is generally used, but some lathes have either the Jarno, Brown & Sharpe, or a special taper. The practice of 33 lathe manufacturers is as follows: 20 use the Morse taper; 5, the Jarno; 3 use special tapers of their own; 2 use modified Morse (longer than the standard but the same taper); 2 use Reed (which is a short Jarno); 1 uses the Brown & Sharpe standard. For grinding machine centers Jarno, Morse and Brown & Sharpe tapers are used. Ten grinding machine manufacturers were divided as follows: 3 use Brown & Sharpe; 3 use Morse, and 4 use Jarno. The Brown & Sharpe taper has been extensively used for milling machine and dividing head spindles. The standard milling machine spindle adopted in 1927 by the milling machine manufacturers. of the National Machine Tool Builders' Association, has a taper of 3½ inches per foot. This comparatively steep taper was adopted to insure easy release of arbors.

Table 5. American National Standard Self-holding Tapers — Basic Dimensions (ANSI B5.10-1981)

No. of Taper	Taper per Foot	Diam. at Gage Line* A	Means of Driving and Holding				Origin of Series
.239	0.50200	0.23922					Brown & Sharpe Taper Series
.299	0.50200	0.29968					
.375	0.50200	0.37525					
1	0.59858	0.47500					Morse Taper Series
2	0.59941	0.70000					
3	0.60235	0.93800					
4	0.62326	1.23100					
4½	0.62400	1.50000	Tang Drive With Shank Held in by Friction	Tang Drive With Shank Held in by Key	Key Drive With Shank Held in by Key	Key Drive With Shank Held in by Draw-bolt	
5	0.63151	1.74800					
6	0.62565	2.49400					
7	0.62400	3.27000					
200	0.750	2.000					
250	0.750	2.500					
300	0.750	3.000					
350	0.750	3.500					¾ Inch per Foot Taper Series
400	0.750	4.000					
450	0.750	4.500					
500	0.750	5.000					
600	0.750	6.000					
800	0.750	8.000					
1000	0.750	10.000					
1200	0.750	12.000					
All dimensions given in inches.							

Table 6. ANSI Standard Steep Machine Tapers (ANSI B5.10-1981)

No. of Taper	Taper per Foot (1)	Diam. at Gage Line (2)	Length Along Axis	No. of Taper	Taper per Foot (1)	Diam. at Gage Line (2)	Length Along Axis
5	3.500	0.500	0.6875	35	3.500	1.500	2.2500
10	3.500	0.625	0.8750	40	3.500	1.750	2.5625
15	3.500	0.750	1.0625	45	3.500	2.250	3.3125
20	3.500	0.875	1.3125	50	3.500	2.750	4.0000
25	3.500	1.000	1.5625	55	3.500	3.500	5.1875
30	3.500	1.250	1.8750	60	3.500	4.250	6.3750

All dimensions given in inches.

(1) This taper corresponds to an included angle of 16°, 35′, 39.4″.

The tapers numbered 10, 20, 30, 40, 50, and 60 that are printed in heavy-faced type are designated as the "Preferred Series." The tapers numbered 5, 15, 25, 35, 45, and 55 that are printed in light-faced type are designated as the "Intermediate Series."

(2) The basic diameter at gage line is at large end of taper.

Table 7. American National Standard Taper Drive with Tang, Self-Holding Tapers (ANSI B5.10-1981)

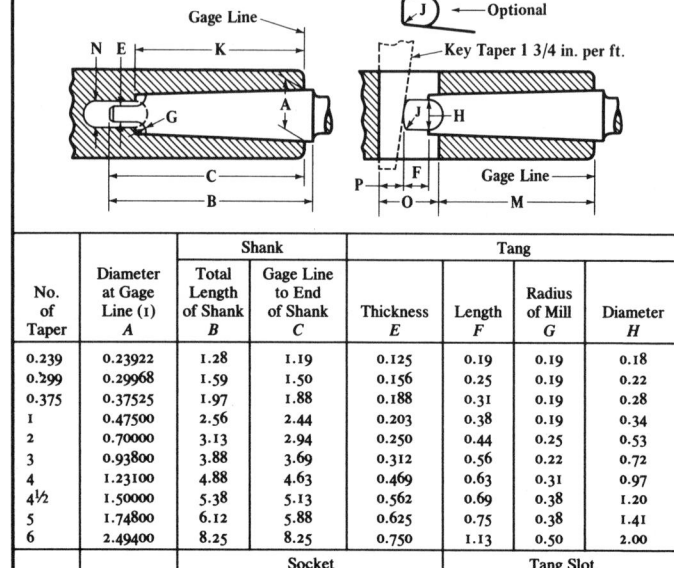

No. of Taper	Diameter at Gage Line (1) A	Shank		Tang			
		Total Length of Shank B	Gage Line to End of Shank C	Thickness E	Length F	Radius of Mill G	Diameter H
0.239	0.23922	1.28	1.19	0.125	0.19	0.19	0.18
0.299	0.29968	1.59	1.50	0.156	0.25	0.19	0.22
0.375	0.37525	1.97	1.88	0.188	0.31	0.19	0.28
1	0.47500	2.56	2.44	0.203	0.38	0.19	0.34
2	0.70000	3.13	2.94	0.250	0.44	0.25	0.53
3	0.93800	3.88	3.69	0.312	0.56	0.22	0.72
4	1.23100	4.88	4.63	0.469	0.63	0.31	0.97
4½	1.50000	5.38	5.13	0.562	0.69	0.38	1.20
5	1.74800	6.12	5.88	0.625	0.75	0.38	1.41
6	2.49400	8.25	8.25	0.750	1.13	0.50	2.00

No. of Taper	Radius J	Socket		Tang Slot			
		Min. Depth of Hole K		Gage Line to Tang Slot M	Width N	Length O	Shank End to Back of Tang Slot P
		Drilled	Reamed				
0.239	0.03	1.06	1.00	0.94	0.141	0.38	0.13
0.299	0.03	1.31	1.25	1.17	0.172	0.50	0.17
0.375	0.05	1.63	1.56	1.47	0.203	0.63	0.22
1	0.05	2.19	2.16	2.06	0.218	0.75	0.38
2	0.06	2.66	2.61	2.50	0.266	0.88	0.44
3	0.08	3.31	3.25	3.06	0.328	1.19	0.56
4	0.09	4.19	4.13	3.88	0.484	1.25	0.50
4½	0.13	4.62	4.56	4.31	0.578	1.38	0.56
5	0.13	5.31	5.25	4.94	0.656	1.50	0.56
6	0.16	7.41	7.33	7.00	0.781	1.75	0.50

All dimensions are in inches. (1) See Table 11 for plug and ring gage dimensions.

Tolerances: For shank diameter A at gage line, $+0.002 - 0.000$; for hole diameter A, $+0.000 - 0.002$. For tang thickness E up to No. 5 inclusive, $+0.000 - 0.006$; No. 6, $+0.000 - 0.008$. For width N of tang slot up to No. 5 inclusive, $+0.006 - 0.000$; No. 6, $+0.008 - 0.000$. For centrality of tang E with center line of taper, .0025 (.005 total indicator variation). These centrality tolerances also apply to the tang slot N. On rate of taper, all sizes 0.002 per foot. This tolerance may be applied on *shanks* only in the direction which *increases* the rate of taper and on *sockets* only in the direction which *decreases* the rate of taper. Tolerances for two-decimal dimensions are plus or minus 0.010, unless otherwise specified.

KEYS AND KEYSEATS

ANSI Standard Keys and Keyseats. — American National Standard, B17.1 Keys and Keyseats, based on current industry practice, was approved in 1967, and reaffirmed in 1973. This standard establishes a uniform relationship between shaft sizes and key sizes for parallel and taper keys as shown in Table 2. Other data in this standard are given in Tables 1 and 3–7. The sizes and tolerances shown are for single key applications only.

The following definitions are given in the standard:

Key: A demountable machinery part which, when assembled into keyseats, provides a positive means for transmitting torque between the shaft and hub.

Keyseat: An axially located rectangular groove in a shaft or hub.

This standard recognizes that there are two classes of stock for parallel keys used by industry. One is a close, plus toleranced key stock and the other is a broad, negative toleranced bar stock. Based on the use of two types of stock, two classes of fit are shown:

Class 1: A clearance of metal-to-metal side fit obtained by using bar stock keys and keyseat tolerances as given in Table 4. This is a relatively free fit and applies only to parallel keys.

Class 2: A side fit, with possible interference or clearance, obtained by using key stock and keyseat tolerances as given in Table 4. This is a relatively tight fit.

Class 3: This is an interference side fit and is not tabulated in Table 4 since the degree of interference has not been standardized. However, it is suggested that the top and bottom fit range given under Class 2 in Table 4, for parallel keys be used.

Key Size vs. Shaft Diameter: Shaft diameters are listed in Table 2 for identification of various key sizes and are not intended to establish shaft dimensions, tolerances or selections. For a stepped shaft, the size of a key is determined by the diameter of the shaft at the point of location of the key. Up through 6½-inch diameter shafts square keys are preferred; rectangular keys for larger shafts.

If special considerations dictate the use of a keyseat in the hub shallower than the preferred nominal depth shown, it is recommended that the tabulated preferred nominal standard keyseat be used in the shaft in all cases.

Keyseat Alignment Tolerances: A tolerance of 0.010 inch, max is provided for offset (due to parallel displacement of keyseat centerline from centerline of shaft or bore) of keyseats in shaft and bore. The following tolerances for maximum lead (due to angular displacement of keyseat centerline from centerline of shaft or bore and measured at right angles to the shaft or bore centerline) of keyseats in shaft and bore are specified: 0.002 inch for keyseat length up to and including 4 inches; 0.0005 inch per inch of length for keyseat lengths above 4 inches to and including 10 inches; and 0.005 inch for keyseat lengths above 10 inches.

ANSI Standard Woodruff Keys and Keyseats. — American National Standard B17.2 was approved in 1967, and reaffirmed in 1978. Data from this standard are shown in Tables 9, 10, and 11.

The following definitions are given in this standard:

Woodruff Key: A demountable machinery part which, when assembled into keyseats, provides a positive means for transmitting torque between the shaft and hub.

Woodruff Key Number: An identification number by which the size of key may be readily determined.

Woodruff Keyseat — Shaft: The circular pocket in which the key is retained.

Woodruff Keyseat — Hub: An axially located rectangular groove in a hub. (This has been referred to as a keyway.)

Woodruff Keyseat Milling Cutter: An arbor type or shank type milling cutter normally used for milling Woodruff keyseats in shafts.

Table 1. ANSI Standard Plain and Gib Head Keys (ANSI B17.1-1967, R1973)

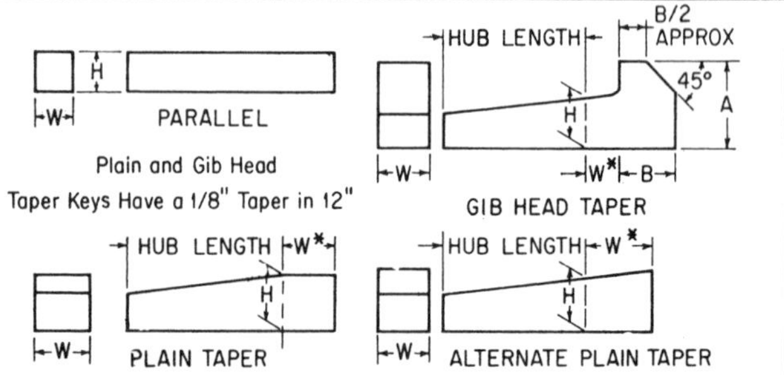

Key			Nominal Key Size Width, W: Over	Nominal Key Size Width, W: To (Incl.)	Tolerance Width, W: +	Tolerance Width, W: −	Tolerance Height, H: +	Tolerance Height, H: −
Parallel	Square	Keystock	...	1 1/4	+0.001	−0.000	+0.001	−0.000
			1 1/4	3	+0.002	−0.000	+0.002	−0.000
			3	3 1/2	+0.003	−0.000	+0.003	−0.000
		Bar Stock	...	3/4	+0.000	−0.002	+0.000	−0.002
			3/4	1 1/2	+0.000	−0.003	+0.000	−0.003
			1 1/2	2 1/2	+0.000	−0.004	+0.000	−0.004
			2 1/2	3 1/2	+0.000	−0.006	+0.000	−0.006
	Rectangular	Keystock	...	1 1/4	+0.001	−0.000	+0.005	−0.005
			1 1/4	3	+0.002	−0.000	+0.005	−0.005
			3	7	+0.003	−0.000	+0.005	−0.005
		Bar Stock	...	3/4	+0.000	−0.003	+0.000	−0.003
			3/4	1 1/2	+0.000	−0.004	+0.000	−0.004
			1 1/2	3	+0.000	−0.005	+0.000	−0.005
			3	4	+0.000	−0.006	+0.000	−0.006
			4	6	+0.000	−0.008	+0.000	−0.008
			6	7	+0.000	−0.013	+0.000	−0.013
Taper	Plain or Gib Head Square or Rectangular		...	1 1/4	+0.001	−0.000	+0.005	−0.000
			1 1/4	3	+0.002	−0.000	+0.005	−0.000
			3	7	+0.003	−0.000	+0.005	−0.000

GIB HEAD NOMINAL DIMENSIONS

Nominal Key Size Width, W	Square H	Square A	Square B	Rectangular H	Rectangular A	Rectangular B	Nominal Key Size Width, W	Square H	Square A	Square B	Rectangular H	Rectangular A	Rectangular B
1/8	1/8	1/4	1/4	3/32	3/16	1/8	1	1	1 5/8	1 1/8	3/4	1 1/4	7/8
3/16	3/16	5/16	5/16	1/8	1/4	1/4	1 1/4	1 1/4	2	1 7/16	7/8	1 3/8	1
1/4	1/4	7/16	3/8	3/16	5/16	5/16	1 1/2	1 1/2	2 3/8	1 3/4	1	1 5/8	1 1/8
5/16	5/16	1/2	7/16	1/4	7/16	3/8	1 3/4	1 3/4	2 3/4	2	1 1/2	2 3/8	1 3/4
3/8	3/8	5/8	1/2	1/4	7/16	3/8	2	2	3 1/2	2 1/4	1 1/2	2 3/8	1 3/4
1/2	1/2	7/8	5/8	3/8	5/8	1/2	2 1/2	2 1/2	4	3	1 3/4	2 3/4	2
5/8	5/8	1	3/4	7/16	3/4	9/16	3	3	5	3 1/2	2	3 1/2	2 1/4
3/4	3/4	1 1/4	7/8	1/2	7/8	5/8	3 1/2	3 1/2	6	4	2 1/2	4	3
7/8	7/8	1 3/8	1	5/8	1	3/4	...	...	...	...	...	...	...

All dimensions are given in inches.

* For locating position of dimension *H*. Tolerance does not apply.

For larger sizes the following relationships are suggested as guides for establishing *A* and *B*: $A = 1.8H$ and $B = 1.2H$.

Table 2.   Key Size Versus Shaft Diameter (ANSI B17.1-1967, R1973)

Nominal Shaft Diameter		Nominal Key Size			Nominal Keyseat Depth	
			Height, H		$H/2$	
Over	To (Incl.)	Width, W	Square	Rectangular	Square	Rectangular
5/16	7/16	3/32	3/32	....	3/64	....
7/16	9/16	1/8	1/8	3/32	1/16	3/64
9/16	7/8	3/16	3/16	1/8	3/32	1/16
7/8	1 1/4	1/4	1/4	3/16	1/8	3/32
1 1/4	1 3/8	5/16	5/16	1/4	5/32	1/8
1 3/8	1 3/4	3/8	3/8	1/4	3/16	1/8
1 3/4	2 1/4	1/2	1/2	3/8	1/4	3/16
2 1/4	2 3/4	5/8	5/8	7/16	5/16	7/32
2 3/4	3 1/4	3/4	3/4	1/2	3/8	1/4
3 1/4	3 3/4	7/8	7/8	5/8	7/16	5/16
3 3/4	4 1/2	1	1	3/4	1/2	3/8
4 1/2	5 1/2	1 1/4	1 1/4	7/8	5/8	7/16
5 1/2	6 1/2	1 1/2	1 1/2	1	3/4	1/2
6 1/2	7 1/2	1 3/4	1 3/4	1 1/2*	7/8	3/4
7 1/2	9	2	2	1 1/2	1	3/4
9	11	2 1/2	2 1/2	1 3/4	1 1/4	7/8

All dimensions are given in inches.   For larger shaft sizes, see ANSI Standard.
Square keys preferred for shaft diameters above heavy line; rectangular keys, below.
* Some key standards show 1 1/4 inches; preferred height is 1 1/2 inches.

Table 3.   Depth Control Values S and T for Shaft and Hub (ANSI B17.1-1967, R1973)

Nominal Shaft Diameter	Parallel and Taper		Parallel		Taper	
	Square	Rectangular	Square	Rectangular	Square	Rectangular
	S	S	T	T	T	T
1/2	0.430	0.445	0.560	0.544	0.535	0.519
9/16	0.493	0.509	0.623	0.607	0.598	0.582
5/8	0.517	0.548	0.709	0.678	0.684	0.653
11/16	0.581	0.612	0.773	0.742	0.748	0.717
3/4	0.644	0.676	0.837	0.806	0.812	0.781
13/16	0.708	0.739	0.900	0.869	0.875	0.844
7/8	0.771	0.802	0.964	0.932	0.939	0.907
15/16	0.796	0.827	1.051	1.019	1.026	0.994
1	0.859	0.890	1.114	1.083	1.089	1.058
1 1/16	0.923	0.954	1.178	1.146	1.153	1.121
1 1/8	0.986	1.017	1.241	1.210	1.216	1.185
1 3/16	1.049	1.080	1.304	1.273	1.279	1.248
1 1/4	1.112	1.144	1.367	1.336	1.342	1.311
1 5/16	1.137	1.169	1.455	1.424	1.430	1.399
1 3/8	1.201	1.232	1.518	1.487	1.493	1.462
1 7/16	1.225	1.288	1.605	1.543	1.580	1.518
1 1/2	1.289	1.351	1.669	1.606	1.644	1.581
1 9/16	1.352	1.415	1.732	1.670	1.707	1.645
1 5/8	1.416	1.478	1.796	1.733	1.771	1.708
1 11/16	1.479	1.541	1.859	1.796	1.834	1.771
1 3/4	1.542	1.605	1.922	1.860	1.897	1.835
1 13/16	1.527	1.590	2.032	1.970	2.007	1.945
1 7/8	1.591	1.654	2.096	2.034	2.071	2.009
1 15/16	1.655	1.717	2.160	2.097	2.135	2.072
2	1.718	1.781	2.223	2.161	2.198	2.136

All dimensions are given in inches.   See Table 4 for tolerances.

Table 3.   (*Concluded*).   **Depth Control Values S and T for Shaft and Hub** (ANSI B17.1-1967, R1973)

Nominal Shaft Diameter	Parallel and Taper		Parallel		Taper	
	Square	Rectangular	Square	Rectangular	Square	Rectangular
	S	S	T	T	T	T
2 1/16	1.782	1.844	2.287	2.224	2.262	2.199
2 1/8	1.845	1.908	2.350	2.288	2.325	2.263
2 3/16	1.909	1.971	2.414	2.351	2.389	2.326
2 1/4	1.972	2.034	2.477	2.414	2.452	2.389
2 5/16	1.957	2.051	2.587	2.493	2.562	2.468
2 3/8	2.021	2.114	2.651	2.557	2.626	2.532
2 7/16	2.084	2.178	2.714	2.621	2.689	2.596
2 1/2	2.148	2.242	2.778	2.684	2.753	2.659
2 9/16	2.211	2.305	2.841	2.748	2.816	2.723
2 5/8	2.275	2.369	2.905	2.811	2.880	.2.786
2 11/16	2.338	2.432	2.968	2.874	2.943	2.849
2 3/4	2.402	2.495	3.032	2.938	3.007	2.913
2 13/16	2.387	2.512	3.142	3.017	3.117	2.992
2 7/8	2.450	2.575	3.205	3.080	3.180	3.055
2 15/16	2.514	2.639	3.269	3.144	3.244	3.119
3	2.577	2.702	3.332	3.207	3.307	3.182
3 1/16	2.641	2.766	3.396	3.271	3.371	3.246
3 1/8	2.704	2.829	3.459	3.334	3.434	3.309
3 3/16	2.768	2.893	3.523	3.398	3.498	3.373
3 1/4	2.831	2.956	3.586	3.461	3.561	3.436
3 5/16	2.816	2.941	3.696	3.571	3.671	3.546
3 3/8	2.880	3.005	3.760	3.635	3.735	3.610
3 7/16	2.943	3.068	3.823	3.698	3.798	3.673
3 1/2	3.007	3.132	3.887	3.762	3.862	3.737
3 9/16	3.070	3.195	3.950	3.825	3.925	3.800
3 5/8	3.134	3.259	4.014	3.889	3.989	3.864
3 11/16	3.197	3.322	4.077	3.952	4.052	3.927
3 3/4	3.261	3.386	4.141	4.016	4.116	3.991
3 13/16	3.246	3.371	4.251	4.126	4.226	4.101
3 7/8	3.309	3.434	4.314	4.189	4.289	4.164
3 15/16	3.373	3.498	4.378	4.253	4.353	4.228
4	3.436	3.561	4.441	4.316	4.416	4.291
4 3/16	3.627	3.752	4.632	4.507	4.607	4.482
4 1/4	3.690	3.815	4.695	4.570	4.670	4.545
4 3/8	3.817	3.942	4.822	4.697	4.797	4.672
4 7/16	3.880	4.005	4.885	4.760	4.860	4.735
4 1/2	3.944	4.069	4.949	4.824	4.924	4.799
4 3/4	4.041	4.229	5.296	5.109	5.271	5.084
4 7/8	4.169	4.356	5.424	5.236	5.399	5.211
4 15/16	4.232	4.422	5.487	5.300	5.462	5.275
5	4.296	4.483	5.551	5.363	5.526	5.338
5 3/16	4.486	4.674	5.741	5.554	5.716	5.529
5 1/4	4.550	4.737	5.805	5.617	5.780	5.592
5 7/16	4.740	4.927	5.995	5.807	5.970	5.782
5 1/2	4.803	4.991	6.058	5.871	6.033	5.846
5 3/4	4.900	5.150	6.405	6.155	6.380	6.130
5 15/16	5.091	5.341	6.596	6.346	6.571	6.321
6	5.155	5.405	6.660	6.410	6.635	6.385
6 1/4	5.409	5.659	6.914	6.664	6.889	6.639
6 1/2	5.662	5.912	7.167	6.917	7.142	6.892
6 3/4	5.760	*5.885	7.515	*7.390	7.490	*7.365
7	6.014	*6.139	7.769	*7.644	7.744	*7.619
7 1/4	6.268	*6.393	8.023	*7.898	7.998	*7.873
7 1/2	6.521	*6.646	8.276	*8.151	8.251	*8.126
7 3/4	6.619	6.869	8.624	8.374	8.599	8.349
8	6.873	7.123	8.878	8.628	8.853	8.603
9	7.887	8.137	9.892	9.642	9.867	9.617
10	8.591	8.966	11.096	10.721	11.071	10.696
11	9.606	9.981	12.111	11.736	12.086	11.711
12	10.309	10.809	13.314	12.814	13.289	12.789
13	11.325	11.825	14.330	13.830	14.305	13.805
14	12.028	12.528	15.533	15.033	15.508	15.008
15	13.043	13.543	16.548	16.048	16.523	16.023

All dimensions given in inches.   See Table 4 for tolerances.
* 1¾ × 1½ inch key.

Table 4. ANSI Standard Fits for Parallel and Taper Keys (ANSI B17.1-1967, R1973)

Type of Key	Key Width Over	Key Width To (Incl.)	Side Fit Width Tolerance Key	Side Fit Width Tolerance Key-Seat	Side Fit Fit Range*	Top and Bottom Fit Depth Tolerance Key	Top and Bottom Fit Depth Tolerance Shaft Key-Seat	Top and Bottom Fit Depth Tolerance Hub Key-Seat	Top and Bottom Fit Fit Range*
					Class 1 Fit for Parallel Keys				
Square	...	½	+0.000 −0.002	+0.002 −0.000	0.004 CL 0.000	+0.000 −0.002	+0.000 −0.015	+0.010 −0.000	0.032 CL 0.005 CL
	½	¾	+0.000 −0.002	+0.003 −0.000	0.005 CL 0.000	+0.000 −0.002	+0.000 −0.015	+0.010 −0.000	0.032 CL 0.005 CL
	¾	1	+0.000 −0.003	+0.003 −0.000	0.006 CL 0.000	+0.000 −0.003	+0.000 −0.015	+0.010 −0.000	0.033 CL 0.005 CL
	1	1½	+0.000 −0.003	+0.004 −0.000	0.007 CL 0.000	+0.000 −0.003	+0.000 −0.015	+0.010 −0.000	0.033 CL 0.005 CL
	1½	2½	+0.000 −0.004	+0.004 −0.000	0.008 CL 0.000	+0.000 −0.004	+0.000 −0.015	+0.010 −0.000	0.034 CL 0.005 CL
	2½	3½	+0.000 −0.006	+0.004 −0.000	0.010 CL 0.000	+0.000 −0.006	+0.000 −0.015	+0.010 −0.000	0.036 CL 0.005 CL
Rectangular	...	½	+0.000 −0.003	+0.002 −0.000	0.005 CL 0.000	+0.000 −0.003	+0.000 −0.015	+0.010 −0.000	0.033 CL 0.005 CL
	½	¾	+0.000 −0.003	+0.003 −0.000	0.006 CL 0.000	+0.000 −0.003	+0.000 −0.015	+0.010 −0.000	0.033 CL 0.005 CL
	¾	1	+0.000 −0.004	+0.003 −0.000	0.007 CL 0.000	+0.000 −0.004	+0.000 −0.015	+0.010 −0.000	0.034 CL 0.005 CL
	1	1½	+0.000 −0.004	+0.004 −0.000	0.008 CL 0.000	+0.000 −0.004	+0.000 −0.015	+0.010 −0.000	0.034 CL 0.005 CL
	1½	3	+0.000 −0.005	+0.004 −0.000	0.009 CL 0.000	+0.000 −0.005	+0.000 −0.015	+0.010 −0.000	0.035 CL 0.005 CL
	3	4	+0.000 −0.006	+0.004 −0.000	0.010 CL 0.000	+0.000 −0.006	+0.000 −0.015	+0.010 −0.000	0.036 CL 0.005 CL
	4	6	+0.000 −0.008	+0.004 −0.000	0.012 CL 0.000	+0.000 −0.008	+0.000 −0.015	+0.010 −0.000	0.038 CL 0.005 CL
	6	7	+0.000 −0.013	+0.004 −0.000	0.017 CL 0.000	+0.000 −0.013	+0.000 −0.015	+0.010 −0.000	0.043 CL 0.005 CL
					Class 2 Fit for Parallel and Taper Keys				
Parallel Square	...	1¼	+0.001 −0.000	+0.002 −0.000	0.002 CL 0.001 INT	+0.001 −0.000	+0.000 −0.015	+0.010 −0.000	0.030 CL 0.004 CL
	1¼	3	+0.002 −0.000	+0.002 −0.000	0.002 CL 0.002 INT	+0.002 −0.000	+0.000 −0.015	+0.010 −0.000	0.030 CL 0.003 CL
	3	3½	+0.003 −0.000	+0.002 −0.000	0.002 CL 0.003 INT	+0.003 −0.000	+0.000 −0.015	+0.010 −0.000	0.030 CL 0.002 CL
Parallel Rectangular	...	1¼	+0.001 −0.000	+0.002 −0.000	0.002 CL 0.001 INT	+0.005 −0.000	+0.000 −0.015	+0.010 −0.000	0.035 CL 0.000 CL
	1¼	3	+0.002 −0.000	+0.002 −0.000	0.002 CL 0.002 INT	+0.005 −0.000	+0.000 −0.015	+0.010 −0.000	0.035 CL 0.000 CL
	3	7	+0.003 −0.000	+0.002 −0.000	0.002 CL 0.003 INT	+0.005 −0.000	+0.000 −0.015	+0.010 −0.000	0.035 CL 0.000 CL
Taper	...	1¼	+0.001 −0.000	+0.002 −0.000	0.002 CL 0.001 INT	+0.005 −0.000	+0.000 −0.015	+0.010 −0.000	0.005 CL 0.025 INT
	1¼	3	+0.002 −0.000	+0.002 −0.000	0.002 CL 0.002 INT	+0.005 −0.000	+0.000 −0.015	+0.010 −0.000	0.005 CL 0.025 INT
	3	§	+0.003 −0.000	+0.002 −0.000	0.002 CL 0.003 INT	+0.005 −0.000	+0.000 −0.015	+0.010 −0.000	0.005 CL 0.025 INT

All dimensions are given in inches.   See also text on page 217.
* Limits of variation.   CL = Clearance;   INT = Interference.
§ To (Incl.) 3½-inch Square and 7-inch Rectangular key widths.

222 TAPERS AND KEYS

Chamfered Keys and Filleted Keyseats. — In general practice, chamfered keys and filleted keyseats are not used. However, it is recognized that fillets in keyseats decrease stress concentrations at corners. When used, fillet radii should be as large as possible without causing excessive bearing stresses due to reduced contact area between the key and its mating parts. Keys must be chamfered or rounded to clear fillet radii. Values in Table 5 assume general conditions and should be used only as a guide when critical stresses are encountered.

Table 5. Suggested Keyseat Fillet Radius and Key Chamfer (ANSI B17.1-1967, R1973)

Keyseat Depth, $H/2$		Fillet Radius	45 deg. Chamfer	Keyseat Depth, $H/2$		Fillet Radius	45 deg. Chamfer
Over	To (Incl.)			Over	To (Incl.)		
1/16	1/4	1/32	3/64	7/8	1 1/4	3/16	7/32
1/4	1/2	1/16	5/64	1 1/4	1 3/4	1/4	9/32
1/2	7/8	1/8	5/32	1 3/4	2 1/2	3/8	13/32

All dimensions are given in inches.

Table 6. ANSI Standard Keyseat Tolerances for Electric Motor and Generator Shaft Extensions (ANSI B17.1-1967, R1973)

Keyseat Width		Width Tolerance	Depth Tolerance
Over	To (Incl.)		
. . .	1/4	+0.001 −0.001	+0.000 −0.015
1/4	3/4	+0.000 −0.002	+0.000 −0.015
3/4	1 1/4	+0.000 −0.003	+0.000 −0.015

All dimensions are given in inches.

Table 7. Set Screws for Use Over Keys* (ANSI B17.1-1967, R1973)

Nom. Shaft Diam.		Nom. Key Width	Set Screw Diam.	Nom. Shaft Diam.		Nom. Key Width	Set Screw Diam.
Over	To (Incl.)			Over	To (Incl.)		
5/16	7/16	3/32	No. 10	2 1/4	2 3/4	5/8	1/2
7/16	9/16	1/8	No. 10	2 3/4	3 1/4	3/4	5/8
9/16	7/8	3/16	1/4	3 1/4	3 3/4	7/8	3/4
7/8	1 1/4	1/4	5/16	3 3/4	4 1/2	1	3/4
1 1/4	1 3/8	5/16	3/8	4 1/2	5 1/2	1 1/4	7/8
1 3/8	1 3/4	3/8	3/8	5 1/2	6 1/2	1 1/2	1
1 3/4	2 1/4	1/2	1/2	. . .	. . .	. . .	. . .

All dimensions are given in inches.
* These set screw diameter selections are offered as a guide but their use should be dependent upon design considerations.

Table 8. Finding Depth of Keyseat and Distance from Top of Key to Bottom of Shaft

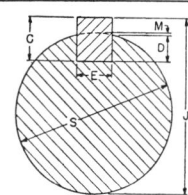

For milling keyseats, the total depth to feed cutter in from outside of shaft to bottom of keyseat is $M + D$, where D is depth of keyseat.

For checking an assembled key and shaft, caliper measurement J between top of key and bottom of shaft is used. $J = S - (M + D) + C$, where C is depth of key. For Woodruff keys, dimensions C and D can be found in Tables 9 to 11. Assuming shaft diameter S is nominal size, the tolerances on dimension J for Woodruff keys in keyslots are +0.000, −0.010 inch.

Diam. of Shaft S, Inches	Width of Keyseat, E														
	1/16	3/32	1/8	5/32	3/16	7/32	1/4	5/16	3/8	7/16	1/2	9/16	5/8	11/16	3/4
	Dimension M, Inch														
0.3125	.0032	...	...	...	...	...	...	...	...	...	...	...	...	...	...
0.3437	.0029	.0065	...	...	...	...	...	...	...	...	...	...	...	...	...
0.3750	.0026	.0060	.0107	...	...	...	...	...	...	...	...	...	...	...	...
0.4060	.0024	.0055	.0099	...	...	...	...	...	...	...	...	...	...	...	...
0.4375	.0022	.0051	.0091	...	...	...	...	...	...	...	...	...	...	...	...
0.4687	.0021	.0047	.0085	.0134	...	...	...	...	...	...	...	...	...	...	...
0.5000	.0020	.0044	.0079	.0125	...	...	...	...	...	...	...	...	...	...	...
0.5625	...	.0039	.0070	.0111	.0161	...	...	...	...	...	...	...	...	...	...
0.6250	...	.0035	.0063	.0099	.0144	.0198	...	...	...	...	...	...	...	...	...
0.6875	...	.0032	.0057	.0090	.0130	.0179	.0235	...	...	...	...	...	...	...	...
0.7500	...	.0029	.0052	.0082	.0119	.0163	.0214	.0341	...	...	...	...	...	...	...
0.8125	...	.0027	.0048	.0076	.0110	.0150	.0197	.0312	...	...	...	...	...	...	...
0.8750	...	.0025	.0045	.0070	.0102	.0139	.0182	.0288	...	...	...	...	...	...	...
0.9375	...	...	.0042	.0066	.0095	.0129	.0170	.0263	.0391	...	...	...	...	...	...
1.0000	...	...	.0039	.0061	.0089	.0121	.0159	.0250	.0365	...	...	...	...	...	...
1.0625	...	...	.0037	.0058	.0083	.0114	.0149	.0235	.0342	...	...	...	...	...	...
1.1250	...	...	.0035	.0055	.0079	.0107	.0141	.0221	.0322	.0443	...	...	...	...	...
1.1875	...	...	.0033	.0052	.0074	.0102	.0133	.0209	.0304	.0418	...	...	...	...	...
1.2500	...	...	.0031	.0049	.0071	.0097	.0126	.0198	.0288	.0395	...	...	...	...	...
1.3750	...	...	...	.0045	.0064	.0088	.0115	.0180	.0261	.0357	.0471	...	...	...	...
1.5000	...	...	...	.0041	.0059	.0080	.0105	.0165	.0238	.0326	.0429	...	...	...	...
1.6250	...	...	...	.0038	.0054	.0074	.0097	.0152	.0219	.0300	.0394	.0502	...	...	...
1.7500	...	...	...	...	.0050	.0069	.0090	.0141	.0203	.0278	.0365	.0464	...	...	...
1.8750	...	...	...	...	.0047	.0064	.0084	.0131	.0189	.0259	.0340	.0432	.0536	...	...
2.0000	...	...	...	...	.0044	.0060	.0078	.0123	.0177	.0242	.0318	.0404	.0501	...	...
2.1250	...	...	...	...	...	.0056	.0074	.0116	.0167	.0228	.0298	.0379	.0470	.0572	.0684
2.2500	...	...	...	...	...	...	.0070	.0109	.0157	.0215	.0281	.0357	.0443	.0538	.0643
2.3750	...	...	...	...	...	...	...	.0103	.0149	.0203	.0266	.0338	.0419	.0509	.0608
2.5000	...	...	...	...	...	...	...	...	.0141	.0193	.0253	.0321	.0397	.0482	.0576
2.6250	...	...	...	...	...	...	...	...	.0135	.0184	.0240	.0305	.0377	.0457	.0547
2.7500	...	...	...	...	...	...	...	...	...	.0175	.0229	.0291	.0360	.0437	.0521
2.8750	...	...	...	...	...	...	...	...	...	.0168	.0219	.0278	.0344	.0417	.0498
3.0000	...	...	...	...	...	...	...	...	...	...	.0210	.0266	.0329	.0399	.0476

Depths for Milling Keyseats. — The above table has been compiled to facilitate the accurate milling of keyseats. This table gives the distance M (see illustration accompanying table) between the top of the shaft and a line passing through the upper corners or edges of the keyseat. Dimension M is calculated by the formula: $M = \frac{1}{2}(S - \sqrt{S^2 - E^2})$ where S is diameter of shaft and E is width of keyseat. A simple approximate formula that gives M to within 0.001 inch is: $M = E^2 \div 4S$.

Table 9. ANSI Standard Woodruff Keys (ANSI B17.2-1967, R1978)

Key No.	Nominal Key Size $W \times B$	Actual Length F +0.000 −0.010	Height of Key				Distance Below Center E
			C		D		
			Max.	Min.	Max.	Min.	
202	1/16 × 1/4	0.248	0.109	0.104	0.109	0.104	1/64
202.5	1/16 × 5/16	0.311	0.140	0.135	0.140	0.135	1/64
302.5	3/32 × 5/16	0.311	0.140	0.135	0.140	0.135	1/64
203	1/16 × 3/8	0.374	0.172	0.167	0.172	0.167	1/64
303	3/32 × 3/8	0.374	0.172	0.167	0.172	0.167	1/64
403	1/8 × 3/8	0.374	0.172	0.167	0.172	0.167	1/64
204	1/16 × 1/2	0.491	0.203	0.198	0.194	0.188	3/64
304	3/32 × 1/2	0.491	0.203	0.198	0.194	0.188	3/64
404	1/8 × 1/2	0.491	0.203	0.198	0.194	0.188	3/64
305	3/32 × 5/8	0.612	0.250	0.245	0.240	0.234	1/16
405	1/8 × 5/8	0.612	0.250	0.245	0.240	0.234	1/16
505	5/32 × 5/8	0.612	0.250	0.245	0.240	0.234	1/16
605	3/16 × 5/8	0.612	0.250	0.245	0.240	0.234	1/16
406	1/8 × 3/4	0.740	0.313	0.308	0.303	0.297	1/16
506	5/32 × 3/4	0.740	0.313	0.308	0.303	0.297	1/16
606	3/16 × 3/4	0.740	0.313	0.308	0.303	0.297	1/16
806	1/4 × 3/4	0.740	0.313	0.308	0.303	0.297	1/16
507	5/32 × 7/8	0.866	0.375	0.370	0.365	0.359	1/16
607	3/16 × 7/8	0.866	0.375	0.370	0.365	0.359	1/16
707	7/32 × 7/8	0.866	0.375	0.370	0.365	0.359	1/16
807	1/4 × 7/8	0.866	0.375	0.370	0.365	0.359	1/16
608	3/16 × 1	0.992	0.438	0.433	0.428	0.422	1/16
708	7/32 × 1	0.992	0.438	0.433	0.428	0.422	1/16
808	1/4 × 1	0.992	0.438	0.433	0.428	0.422	1/16
1008	5/16 × 1	0.992	0.438	0.433	0.428	0.422	1/16
1208	3/8 × 1	0.992	0.438	0.433	0.428	0.422	1/16
609	3/16 × 1 1/8	1.114	0.484	0.479	0.475	0.469	5/64
709	7/32 × 1 1/8	1.114	0.484	0.479	0.475	0.469	5/64
809	1/4 × 1 1/8	1.114	0.484	0.479	0.475	0.469	5/64
1009	5/16 × 1 1/8	1.114	0.484	0.479	0.475	0.469	5/64
610	3/16 × 1 1/4	1.240	0.547	0.542	0.537	0.531	5/64
710	7/32 × 1 1/4	1.240	0.547	0.542	0.537	0.531	5/64
810	1/4 × 1 1/4	1.240	0.547	0.542	0.537	0.531	5/64
1010	5/16 × 1 1/4	1.240	0.547	0.542	0.537	0.531	5/64
1210	3/8 × 1 1/4	1.240	0.547	0.542	0.537	0.531	5/64
811	1/4 × 1 3/8	1.362	0.594	0.589	0.584	0.578	3/32
1011	5/16 × 1 3/8	1.362	0.594	0.589	0.584	0.578	3/32
1211	3/8 × 1 3/8	1.362	0.594	0.589	0.584	0.578	3/32
812	1/4 × 1 1/2	1.484	0.641	0.636	0.631	0.625	7/64
1012	5/16 × 1 1/2	1.484	0.641	0.636	0.631	0.625	7/64
1212	3/8 × 1 1/2	1.484	0.641	0.636	0.631	0.625	7/64

All dimensions are given in inches.

The key numbers indicate nominal key dimensions. The last two digits give the nominal diameter B in eighths of an inch and the digits preceding the last two give the nominal width W in thirty-seconds of an inch.

Table 10. ANSI Standard Woodruff Keys (ANSI B17.2-1967, R1978)

Key No.	Nominal Key Size $W \times B$	Actual Length F +0.000 −0.010	Height of Key				Distance Below Center E
			C		D		
			Max.	Min.	Max.	Min.	
617-1	$\frac{3}{16} \times 2\frac{1}{8}$	1.380	0.406	0.401	0.396	0.390	$2\frac{1}{32}$
817-1	$\frac{1}{4} \times 2\frac{1}{8}$	1.380	0.406	0.401	0.396	0.390	$2\frac{1}{32}$
1017-1	$\frac{5}{16} \times 2\frac{1}{8}$	1.380	0.406	0.401	0.396	0.390	$2\frac{1}{32}$
1217-1	$\frac{3}{8} \times 2\frac{1}{8}$	1.380	0.406	0.401	0.396	0.390	$2\frac{1}{32}$
617	$\frac{3}{16} \times 2\frac{1}{8}$	1.723	0.531	0.526	0.521	0.515	$1\frac{7}{32}$
817	$\frac{1}{4} \times 2\frac{1}{8}$	1.723	0.531	0.526	0.521	0.515	$1\frac{7}{32}$
1017	$\frac{5}{16} \times 2\frac{1}{8}$	1.723	0.531	0.526	0.521	0.515	$1\frac{7}{32}$
1217	$\frac{3}{8} \times 2\frac{1}{8}$	1.723	0.531	0.526	0.521	0.515	$1\frac{7}{32}$
822-1	$\frac{1}{4} \times 2\frac{3}{4}$	2.000	0.594	0.589	0.584	0.578	$2\frac{5}{32}$
1022-1	$\frac{5}{16} \times 2\frac{3}{4}$	2.000	0.594	0.589	0.584	0.578	$2\frac{5}{32}$
1222-1	$\frac{3}{8} \times 2\frac{3}{4}$	2.000	0.594	0.589	0.584	0.578	$2\frac{5}{32}$
1422-1	$\frac{7}{16} \times 2\frac{3}{4}$	2.000	0.594	0.589	0.584	0.578	$2\frac{5}{32}$
1622-1	$\frac{1}{2} \times 2\frac{3}{4}$	2.000	0.594	0.589	0.584	0.578	$2\frac{5}{32}$
822	$\frac{1}{4} \times 2\frac{3}{4}$	2.317	0.750	0.745	0.740	0.734	$\frac{5}{8}$
1022	$\frac{5}{16} \times 2\frac{3}{4}$	2.317	0.750	0.745	0.740	0.734	$\frac{5}{8}$
1222	$\frac{3}{8} \times 2\frac{3}{4}$	2.317	0.750	0.745	0.740	0.734	$\frac{5}{8}$
1422	$\frac{7}{16} \times 2\frac{3}{4}$	2.317	0.750	0.745	0.740	0.734	$\frac{5}{8}$
1622	$\frac{1}{2} \times 2\frac{3}{4}$	2.317	0.750	0.745	0.740	0.734	$\frac{5}{8}$
1228	$\frac{3}{8} \times 3\frac{1}{2}$	2.880	0.938	0.933	0.928	0.922	$1\frac{3}{16}$
1428	$\frac{7}{16} \times 3\frac{1}{2}$	2.880	0.938	0.933	0.928	0.922	$1\frac{3}{16}$
1628	$\frac{1}{2} \times 3\frac{1}{2}$	2.880	0.938	0.933	0.928	0.922	$1\frac{3}{16}$
1828	$\frac{9}{16} \times 3\frac{1}{2}$	2.880	0.938	0.933	0.928	0.922	$1\frac{3}{16}$
2028	$\frac{5}{8} \times 3\frac{1}{2}$	2.880	0.938	0.933	0.928	0.922	$1\frac{3}{16}$
2228	$1\frac{1}{16} \times 3\frac{1}{2}$	2.880	0.938	0.933	0.928	0.922	$1\frac{3}{16}$
2428	$\frac{3}{4} \times 3\frac{1}{2}$	2.880	0.938	0.933	0.928	0.922	$1\frac{3}{16}$

All dimensions are given in inches.

The key numbers indicate nominal key dimensions. The last two digits give the nominal diameter B in eighths of an inch and the digits preceding the last two give the nominal width W in thirty-seconds of an inch.

The key numbers with the -1 designation, while representing the nominal key size have a shorter length F and due to a greater distance below center E are less in height than the keys of the same number without the -1 designation.

Table 11. ANSI Keyseat Dimensions for Woodruff Keys (ANSI B17.2-1967, R1978)

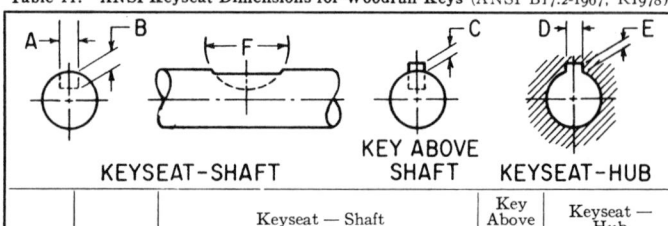

KEYSEAT–SHAFT   KEY ABOVE SHAFT   KEYSEAT–HUB

| Key No. | Nominal Size Key | Keyseat — Shaft | | | | | Key Above Shaft | Keyseat — Hub | |
| | | Width A* | | Depth B | Diameter F | | Height C | Width D | Depth E |
		Min.	Max.	+0.005 −0.000	Min.	Max.	+0.005 −0.005	+0.002 −0.000	+0.005 −0.000
202	1/16 × 1/4	0.0615	0.0630	0.0728	0.250	0.268	0.0312	0.0635	0.0372
202.5	1/16 × 5/16	0.0615	0.0630	0.1038	0.312	0.330	0.0312	0.0635	0.0372
302.5	3/32 × 5/16	0.0928	0.0943	0.0882	0.312	0.330	0.0469	0.0948	0.0529
203	1/16 × 3/8	0.0615	0.0630	0.1358	0.375	0.393	0.0312	0.0635	0.0372
303	3/32 × 3/8	0.0928	0.0943	0.1202	0.375	0.393	0.0469	0.0948	0.0529
403	1/8 × 3/8	0.1240	0.1255	0.1045	0.375	0.393	0.0625	0.1260	0.0685
204	1/16 × 1/2	0.0615	0.0630	0.1668	0.500	0.518	0.0312	0.0635	0.0372
304	3/32 × 1/2	0.0928	0.0943	0.1511	0.500	0.518	0.0469	0.0948	0.0529
404	1/8 × 1/2	0.1240	0.1255	0.1355	0.500	0.518	0.0625	0.1260	0.0685
305	3/32 × 5/8	0.0928	0.0943	0.1981	0.625	0.643	0.0469	0.0948	0.0529
405	1/8 × 5/8	0.1240	0.1255	0.1825	0.625	0.643	0.0625	0.1260	0.0685
505	5/32 × 5/8	0.1553	0.1568	0.1669	0.625	0.643	0.0781	0.1573	0.0841
605	3/16 × 5/8	0.1863	0.1880	0.1513	0.625	0.643	0.0937	0.1885	0.0997
406	1/8 × 3/4	0.1240	0.1255	0.2455	0.750	0.768	0.0625	0.1260	0.0685
506	5/32 × 3/4	0.1553	0.1568	0.2299	0.750	0.768	0.0781	0.1573	0.0841
606	3/16 × 3/4	0.1863	0.1880	0.2143	0.750	0.768	0.0937	0.1885	0.0997
806	1/4 × 3/4	0.2487	0.2505	0.1830	0.750	0.768	0.1250	0.2510	0.1310
507	5/32 × 7/8	0.1553	0.1568	0.2919	0.875	0.895	0.0781	0.1573	0.0841
607	3/16 × 7/8	0.1863	0.1880	0.2763	0.875	0.895	0.0937	0.1885	0.0997
707	7/32 × 7/8	0.2175	0.2193	0.2607	0.875	0.895	0.1093	0.2198	0.1153
807	1/4 × 7/8	0.2487	0.2505	0.2450	0.875	0.895	0.1250	0.2510	0.1310
608	3/16 × 1	0.1863	0.1880	0.3393	1.000	1.020	0.0937	0.1885	0.0997
708	7/32 × 1	0.2175	0.2193	0.3237	1.000	1.020	0.1093	0.2198	0.1153
808	1/4 × 1	0.2487	0.2505	0.3080	1.000	1.020	0.1250	0.2510	0.1310
1008	5/16 × 1	0.3111	0.3130	0.2768	1.000	1.020	0.1562	0.3135	0.1622
1208	3/8 × 1	0.3735	0.3755	0.2455	1.000	1.020	0.1875	0.3760	0.1935
609	3/16 × 1 1/8	0.1863	0.1880	0.3853	1.125	1.145	0.0937	0.1885	0.0997
709	7/32 × 1 1/8	0.2175	0.2193	0.3697	1.125	1.145	0.1093	0.2198	0.1153
809	1/4 × 1 1/8	0.2487	0.2505	0.3540	1.125	1.145	0.1250	0.2510	0.1310
1009	5/16 × 1 1/8	0.3111	0.3130	0.3228	1.125	1.145	0.1562	0.3135	0.1622
610	3/16 × 1 1/4	0.1863	0.1880	0.4483	1.250	1.273	0.0937	0.1885	0.0997
710	7/32 × 1 1/4	0.2175	0.2193	0.4327	1.250	1.273	0.1093	0.2198	0.1153
810	1/4 × 1 1/4	0.2487	0.2505	0.4170	1.250	1.273	0.1250	0.2510	0.1310
1010	5/16 × 1 1/4	0.3111	0.3130	0.3858	1.250	1.273	0.1562	0.3135	0.1622
1210	3/8 × 1 1/4	0.3735	0.3755	0.3545	1.250	1.273	0.1875	0.3760	0.1935
811	1/4 × 1 3/8	0.2487	0.2505	0.4640	1.375	1.398	0.1250	0.2510	0.1310
1011	5/16 × 1 3/8	0.3111	0.3130	0.4328	1.375	1.398	0.1562	0.3135	0.1622
1211	3/8 × 1 3/8	0.3735	0.3755	0.4015	1.375	1.398	0.1875	0.3760	0.1935

All dimensions are given in inches.
* See footnote at end of table.

Table 11 *(Concluded)*. **ANSI Standard Keyseat Dimensions for Woodruff Keys**
(ANSI B17.2-1967, R1978)

Key No.	Nominal Size Key	Keyseat — Shaft					Key Above Shaft	Keyseat — Hub	
		Width A *		Depth B	Diameter F		Height C	Width D	Depth E
		Min.	Max.	+0.005 −0.000	Min.	Max.	+0.005 −0.005	+0.002 −0.000	+0.005 −0.000
812	¼ × 1½	0.2487	0.2505	0.5110	1.500	1.523	0.1250	0.2510	0.1310
1012	⁵⁄₁₆ × 1½	0.3111	0.3130	0.4798	1.500	1.523	0.1562	0.3135	0.1622
1212	⅜ × 1½	0.3735	0.3755	0.4485	1.500	1.523	0.1875	0.3760	0.1935
617-1	³⁄₁₆ × 2⅛	0.1863	0.1880	0.3073	2.125	2.160	0.0937	0.1885	0.0997
817-1	¼ × 2⅛	0.2487	0.2505	0.2760	2.125	2.160	0.1250	0.2510	0.1310
1017-1	⁵⁄₁₆ × 2⅛	0.3111	0.3130	0.2448	2.125	2.160	0.1562	0.3135	0.1622
1217-1	⅜ × 2⅛	0.3735	0.3755	0.2135	2.125	2.160	0.1875	0.3760	0.1935
617	³⁄₁₆ × 2⅛	0.1863	0.1880	0.4323	2.125	2.160	0.0937	0.1885	0.0997
817	¼ × 2⅛	0.2487	0.2505	0.4010	2.125	2.160	0.1250	0.2510	0.1310
1017	⁵⁄₁₆ × 2⅛	0.3111	0.3130	0.3698	2.125	2.160	0.1562	0.3135	0.1622
1217	⅜ × 2⅛	0.3735	0.3755	0.3385	2.125	2.160	0.1875	0.3760	0.1935
822-1	¼ × 2¾	0.2487	0.2505	0.4640	2.750	2.785	0.1250	0.2510	0.1310
1022-1	⁵⁄₁₆ × 2¾	0.3111	0.3130	0.4328	2.750	2.785	0.1562	0.3135	0.1622
1222-1	⅜ × 2¾	0.3735	0.3755	0.4015	2.750	2.785	0.1875	0.3760	0.1935
1422-1	⁷⁄₁₆ × 2¾	0.4360	0.4380	0.3703	2.750	2.785	0.2187	0.4385	0.2247
1622-1	½ × 2¾	0.4985	0.5005	0.3390	2.750	2.785	0.2500	0.5010	0.2560
822	¼ × 2¾	0.2487	0.2505	0.6200	2.750	2.785	0.1250	0.2510	0.1310
1022	⁵⁄₁₆ × 2¾	0.3111	0.3130	0.5888	2.750	2.785	0.1562	0.3135	0.1622
1222	⅜ × 2¾	0.3735	0.3755	0.5575	2.750	2.785	0.1875	0.3760	0.1935
1422	⁷⁄₁₆ × 2¾	0.4360	0.4380	0.5263	2.750	2.785	0.2187	0.4385	0.2247
1622	½ × 2¾	0.4985	0.5005	0.4950	2.750	2.785	0.2500	0.5010	0.2560
1228	⅜ × 3½	0.3735	0.3755	0.7455	3.500	3.535	0.1875	0.3760	0.1935
1428	⁷⁄₁₆ × 3½	0.4360	0.4380	0.7143	3.500	3.535	0.2187	0.4385	0.2247
1628	½ × 3½	0.4985	0.5005	0.6830	3.500	3.535	0.2500	0.5010	0.2560
1828	⁹⁄₁₆ × 3½	0.5610	0.5630	0.6518	3.500	3.535	0.2812	0.5635	0.2872
2028	⅝ × 3½	0.6235	0.6255	0.6205	3.500	3.535	0.3125	0.6260	0.3185
2228	¹¹⁄₁₆ × 3½	0.6860	0.6880	0.5893	3.500	3.535	0.3437	0.6885	0.3497
2428	¾ × 3½	0.7485	0.7505	0.5580	3.500	3.535	0.3750	0.7510	0.3810

All dimensions are given in inches.

* These Width *A* values were set with the maximum keyseat (shaft) width as that figure which will receive a key with the greatest amount of looseness consistent with assuring the key's sticking in the keyseat (shaft). Minimum keyseat width is that figure permitting the largest shaft distortion acceptable when assembling maximum key in minimum keyseat.

Dimensions *A*, *B*, *C*, *D* are taken at side intersection.

Cotters. — A cotter is a form of key that is used to connect rods, etc., that are subjected either to tension or compression or both, the cotter being subjected to shearing stresses at two transverse cross-sections. When taper cotters are used for drawing and holding parts together, if the cotter is held in place by the friction between the bearing surfaces, the taper should not be too great. Ordinarily a taper varying from ¼ to ½ inch per foot is used for plain cotters. When a set-screw or other device is used to prevent the cotter from backing out of its slot, the taper may vary from 1½ to 2 inches per foot.

British Preferred Lengths of Plain (Parallel or Taper) and Gib-head Keys, Rectangular and Square Section (B.S. 46: Part 1: 1958 Appendix)

All dimensions are in inches

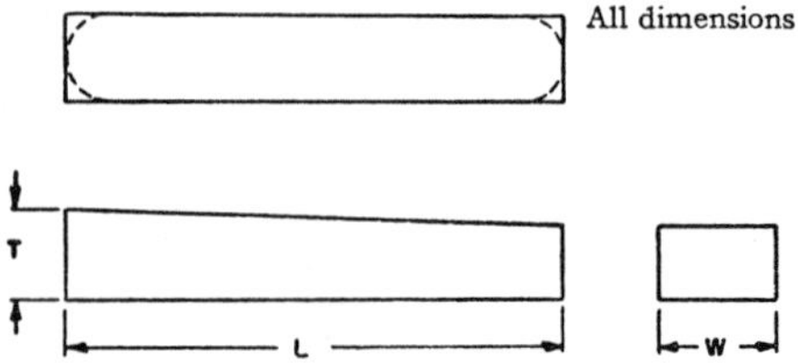

Plain Key Size $W \times T$	¾	1	1¼	1½	1¾	2	2¼	2½	2¾	3	3½	4	4½	5	6
⅛ × ⅛	X	X													
³⁄₁₆ × ³⁄₁₆	X	X	X	X	X	X									
¼ × ¼	X	X	X	X	X	X	X	X	X	X					
⁵⁄₁₆ × ¼	X	X	X	X	X	X	X	X	X	X	X				
⁵⁄₁₆ × ⁵⁄₁₆	X	X	X	X	X	X	X	X	X	X					
⅜ × ¼		X	X	X	X	X	X	X	X	X	X				
⅜ × ⅜		X	X	X	X	X	X	X	X	X	X	X			
⁷⁄₁₆ × ⁵⁄₁₆			X	X	X	X	X	X	X	X	X	X			
⁷⁄₁₆ × ⁷⁄₁₆				X	X	X	X	X	X	X	X	X			
½ × ⁵⁄₁₆					X	X	X	X	X	X	X	X			
½ × ½						X	X	X	X	X	X	X			
⅝ × ⁷⁄₁₆							X	X	X	X	X	X	X		
⅝ × ⅝								X	X	X	X	X	X	X	
¾ × ½										X	X	X	X	X	
¾ × ¾											X	X	X	X	X
⅞ × ⅝											X	X	X	X	X

All dimensions are in inches

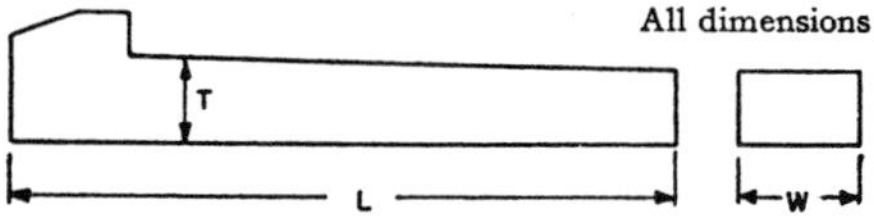

Gib-head Key Size, $W \times T$	1½	1¾	2	2¼	2½	2¾	3	3½	4	4½	5	5½	6	6½	7	7½	8
³⁄₁₆ × ³⁄₁₆	X	X	X	X	X												
¼ × ¼	X	X	X	X	X	X	X										
⁵⁄₁₆ × ¼			X	X	X	X	X	X	X								
⁵⁄₁₆ × ⁵⁄₁₆			X	X	X	X	X	X	X	X							
⅜ × ¼			X	X	X	X	X	X	X	X							
⅜ × ⅜			X	X	X	X	X	X	X	X							
⁷⁄₁₆ × ⁵⁄₁₆					X	X	X	X	X	X	X						
⁷⁄₁₆ × ⁷⁄₁₆						X	X	X	X	X	X						
½ × ⁷⁄₁₆							X	X	X	X	X	X					
½ × ½							X	X	X	X	X	X					
⅝ × ⁷⁄₁₆								X	X	X	X	X	X				
⅝ × ⅝									X	X	X	X	X	X			
¾ × ½										X	X	X	X	X			
¾ × ¾										X	X	X	X	X	X		
⅞ × ⅝											X	X	X	X	X		
⅞ × ⅞											X	X	X	X	X	X	
1 × ¾												X	X	X	X	X	X
1 × 1													X	X	X	X	X

SCREW THREAD SYSTEMS

Screw Thread Forms. — Of the various screw thread forms which have been developed, the most used are those having symmetrical sides inclined at equal angles with a vertical center line through the thread apex. Present-day examples of such threads would include the Unified, the Whitworth and the Acme forms. One of the early forms was the Sharp V which is now used only occasionally. Symmetrical threads are relatively easy to manufacture and inspect and hence are widely used on mass-produced general-purpose threaded fasteners of all types. In addition to general-purpose fastener applications, certain threads are used to repeatedly move or translate machine parts against heavy loads. For these so-called translation threads a stronger form is required. The most widely used translation thread forms are the square, the Acme, and the buttress. Of these, the square thread is the most efficient, but it is also the most difficult to cut owing to its parallel sides and it cannot be adjusted to compensate for wear. Although less efficient, the Acme form of thread has none of the disadvantages of the square form and has the advantage of being somewhat stronger. The buttress form is used for translation of loads in one direction only because of its non-symmetrical form and combines the high efficiency and strength of the square thread with the case of cutting and adjustment of the Acme thread.

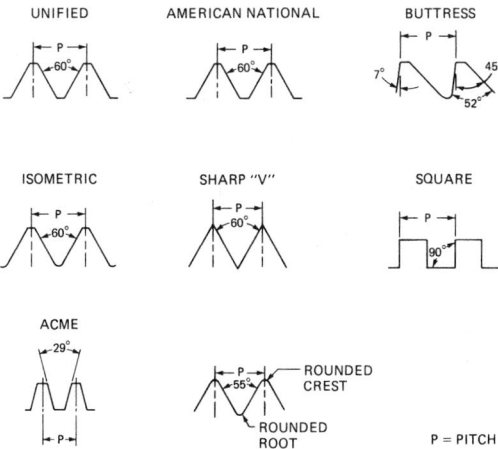

UNIFIED AMERICAN NATIONAL BUTTRESS

ISOMETRIC SHARP "V" SQUARE

ACME

ROUNDED CREST

ROUNDED ROOT

P = PITCH

Parts of a Screw Thread

Major Diameter. The largest diameter of the thread. This is the distance between the crests of the thread measured perpendicular to the thread axis.

Pitch Diameter. The diameter of the thread used to establish the relationship, or fit, between an internal and external thread. The pitch diameter is the distance between the pitch points measured perpendicular to the thread axis. The pitch points are the points on the thread where the thread ridge and the space between the threads are the same width.

Minor Diameter. The smallest diameter of the thread. This is the distance between the roots of the thread measured perpendicular to the thread axis.

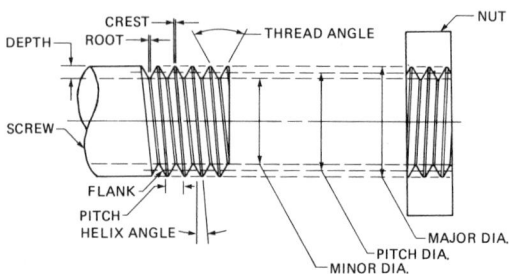

Thread Angle. The included angle of the thread form.

Helix Angle. The angle of the thread ridge measured from a plane perpendicular to the thread axis.

Pitch. The distance between the same points on adjacent threads. This is also the linear distance the thread will travel in one revolution.

Depth. The distance between the crest and root of a thread measured perpendicular to the axis of the thread on one side.

Flank. The sides of the thread that connect the root with the crest of each thread.

Root. The surface of the thread that joins the flanks of adjacent threads. The distance between the roots on opposite sides of the thread is called the root, or minor, diameter.

Crest. The surface of a thread that joins the flanks of the same thread. The distance between the crests on opposite sides of the thread is called the outside, or major, diameter.

Thread Designations.—Screw threads must conform to specific dimensions to work properly. To interpret thread dimensions correctly you must know how threads are specified. Since the Unified, Metric, Acme, and pipe are the most common thread forms, this discussion will cover just these four.

Unified: The size of a Unified thread is shown with a series of numbers and letters. This series follows the pattern shown in Fig. 1a, and must be read from left to right in the sequence indicated. The first number in the series (¼) shows the nominal diameter of the thread. This number represents the fractional diameter, the decimal diameter, or the number of the screw size. The second number (20) indicates the number of

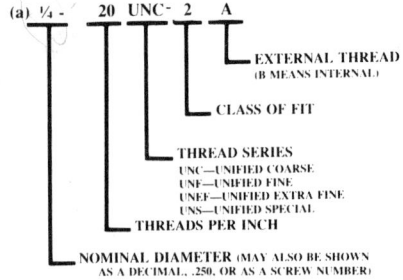

Figure 1a

threads per inch. The next group of letters (UNC) shows the thread series. In this example, Unified coarse is indicated. The last number and letter set (2A) indicates the thread class, or class of fit. The standard classes are 1, 2, and 3. Number 1 is the loosest fit and number 3 is the tightest fit. The letters used here are A and B. The A means the thread is external and the B is used to designate internal threads.

In some cases other information is added to the standard designation, Fig. 1b. Here the LH means left hand. All threads are considered to be right hand unless this LH designation is used. The last entry in this thread designation indicates the length of thread. This length may be shown in the thread designation or, as is more often the case, as a separate dimension on the print.

Metric: Metric threads are also designated by a series of numbers and letters. As shown in Fig. 2, the first letter and number (M10) indicate the nominal diameter of the thread. The letter M is used to identify all ISO (International Standards Organization) or American National Standard M Profile metric threads. The second number (1.5) specifies the pitch, or distance between adjacent threads, in millimeters. The last number and letter combination (4g) indicate the class of fit. In this designation lower case letters indicate external threads and upper case letters are used with internal threads. As with Unified threads, the abbreviation LH after the thread designation also means the thread is left handed.

Acme: Acme threads are also designated by their nominal diameter, threads per inch, thread series, and fit, Fig. 3. The principal differences are the word ACME and the different thread classes. The thread classes commonly used for Acme threads are 2, 3, 4, and 5, with 2 the loosest and 5 the tightest. The letter G indicates a general purpose thread. If the letter C is used, in place of the letter G, it indicates a centralizing

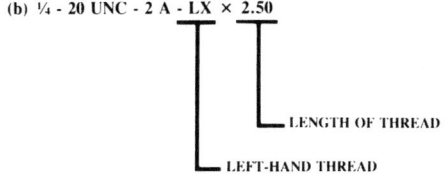

Figure 1b

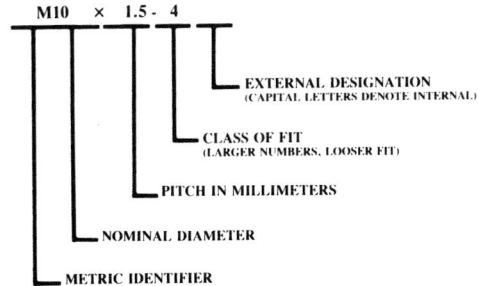

Figure 2

Acme thread. The LH, when used, means the thread is left handed. There is also a Stub Acme thread which is designated, for example, as ½-20 Stub Acme without a letter symbol.

Pipe: Pipe threads are specified on a print by the nominal size and threads per inch, Fig. 4. The letters NPT are used to specify taper pipe and NPS is used for straight pipe.

In addition to the thread designations discussed here there are also several other thread designations in common use today. To find specific information about these thread types, or more detailed information about those designations mentioned here, you should consult MACHINERY'S HANDBOOK for all relevant data.

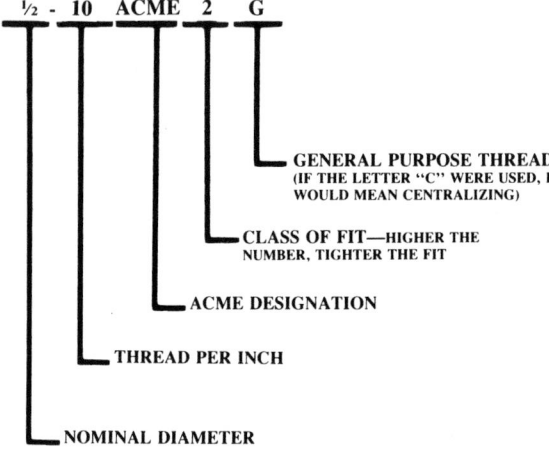

Figure 3

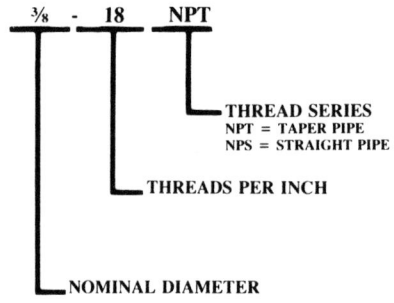

Figure 4

Tap Drill Sizes and Percentage of Thread (Unified Threads)

Taps Size—Threads per Inch	Drills				Percentage of Full Thread	Nut Minor Diameter, min* max
	No.	in.	mm	Decimal		
0–80	•••	•••	1.30	0.0512	54	0.0465
	•••	•••	1.25	0.0492	68	0.0514
	•••	•••	1.20	0.0472	79	
1–64	•••	1/16	•••	0.0625	51	0.0561
	•••	•••	1.55	0.0610	59	0.0623
	53	•••	•••	0.0595	66	
	•••	•••	1.50	0.0590	69	
	•••	•••	1.45	0.0571	77	
1–72	52	•••	•••	0.0635	53	0.0580
	•••	1/16	•••	0.0625	58	0.0635
	•••	•••	1.55	0.0610	66	
	53	•••	•••	0.0595	75	
	•••	•••	1.50	0.0590	77	
2–56	49	•••	•••	0.0730	56	0.0667
	•••	•••	1.80	0.0709	65	0.0737
	50	•••	•••	0.0700	69	
	•••	•••	1.75	0.0689	74	
2–64	•••	•••	1.90	0.0748	55	0.0691
	49	•••	•••	0.0730	64	0.0753
	•••	•••	1.85	0.0728	65	
	•••	•••	1.80	0.0709	74	
	50	•••	•••	0.0700	79	
3–48	•••	•••	2.15	0.0846	53	0.0764
	•••	•••	2.10	0.0827	60	0.0845
	45	•••	•••	0.0820	63	
	46	•••	•••	0.0810	66	
	47	•••	•••	0.0785	76	
	•••	5/64	•••	0.0781	77	
3–56	•••	•••	2.20	0.0866	53	0.0797
	44	•••	•••	0.0860	56	0.0865
	•••	•••	2.15	0.0846	62	
	•••	•••	2.10	0.0827	70	
	45	•••	•••	0.0820	73	
	46	•••	•••	0.0801	77	
	•••	•••	2.05	0.0807	79	
4–36	43	•••	•••	0.0890	64	0.0821**
	44	•••	•••	0.0860	72	0.0882**
4–40	•••	•••	2.40	0.0945	54	0.0849
	42	•••	•••	0.0935	57	0.0939
	•••	•••	2.35	0.0925	60	
	•••	•••	2.30	0.0905	66	
	43	•••	•••	0.0890	71	
	•••	•••	2.20	0.0866	78	

Courtesy of SME

Tap Drill Sizes and Percentage of Thread (Unified Threads) *(Continued)*

Taps Size— Threads per Inch	Drills				Percentage of Full Thread	Nut Minor Diameter, min* max
	No.	in.	mm	Deci-mal		
4–48	···	···	2.45	0.0964	57	0.0894
	41	···	···	0.0960	59	0.0968
	···	···	2.40	0.0945	64	
	···	3/32	···	0.0938	68	
	42	···	···	0.0935	68	
	···	···	2.35	0.0925	72	
	···	···	2.30	0.0905	79	
5–40	···	···	2.70	0.1063	57	0.0979
	37	···	···	0.1040	65	0.1062
	···	···	2.60	0.1024	70	
	38	···	···	0.1015	72	
	39	···	···	0.0995	78	
5–44	···	···	2.75	0.1083	57	0.1004
	36	···	···	0.1065	62	0.1079
	37	···	···	0.1040	71	
	···	···	2.60	0.1024	76	
	38	···	···	0.1015	79	
6–32	33	···	···	0.1130	62	0.1042
	34	···	···	0.1110	67	0.1130
	···	···	2.80	0.1102	68	
	35	···	···	0.1100	69	
	···	7/64	···	0.1094	70	
	···	···	2.75	0.1083	73	
	36	···	···	0.1065	77	
6–40	31	···	···	0.1200	55	0.1109
	···	···	3.00	0.1181	61	0.1186
	32	···	···	0.1160	68	
	···	···	2.90	0.1142	73	
	33	···	···	0.1130	77	
8–32	···	9/64	···	0.1406	57	0.1302
	28	···	···	0.1405	58	0.1389
	···	···	3.50	0.1378	65	
	29	···	···	0.1360	69	
	···	···	3.40	0.1339	74	
8–36	27	···	···	0.1440	55	0.1339
	···	···	3.60	0.1417	62	0.1416
	···	9/64	···	0.1406	68	
	28	···	···	0.1405	65	
	···	···	3.50	0.1378	73	
	29	···	···	0.1360	78	

Courtesy of SME

Tap Drill Sizes and Percentage of Thread (Unified Threads) *(Continued)*

Taps Size—Threads per Inch	Drills				Percentage of Full Thread	Nut Minor Diameter, min* max
	No.	in.	mm	Deci-mal		
10–24	21	···	···	0.1590	57	0.1449
	22	···	···	0.1570	61	0.1555
	···	⁵⁄₃₂	···	0.1563	62	
	23	···	···	0.1540	66	
	24	···	···	0.1520	67	
	25	···	···	0.1496	75	
	26	···	···	0.1470	79	
10–32	19	···	···	0.1660	59	0.1562
	···	···	4.1	0.1614	70	0.1641
	20	···	···	0.1610	71	
	21	···	···	0.1590	76	
12–24	14	···	···	0.1820	63	0.1709
	15	···	···	0.1800	67	0.1708
	16	···	···	0.1770	72	
	17	···	···	0.1730	79	
12–28	···	³⁄₁₆	···	0.1875	61	0.1773
	13	···	···	0.1850	67	0.1857
	14	···	···	0.1820	73	
	15	···	···	0.1800	78	
¼–20	4	···	···	0.2090	63	0.1959
	5	···	···	0.2055	69	0.2067
	6	···	···	0.2040	71	
	···	¹³⁄₆₄	···	0.2031	72	
	7	···	···	0.2010	75	
	8	···	···	0.1990	79	
¼–27	2	···	···	0.2210	61	0.212
	···	⁷⁄₃₂	···	0.2187	65	0.218**
	3	···	···	0.2130	77	
¼–28	2	···	···	0.2210	63	0.2113
	···	⁷⁄₃₂	···	0.2187	67	0.2190
	···	···	5.5	0.2165	72	
	3	···	···	0.2130	80	
¼–32	···	···	5.7	0.2244	63	0.2162
	2	···	···	0.2210	71	0.2208**
	···	⁷⁄₂₈	···	0.2187	77	
⁵⁄₁₆–18	···	N	···	0.2660	64	0.2524
	···	¹⁷⁄₆₄	···	0.2656	65	0.2630
	···	G	···	0.2610	71	
	···	F	···	0.2570	77	

Courtesy of SME

Tap Drill Sizes and Percentage of Thread (Unified Threads) *(Continued)*

Taps Size— Threads per Inch	Drills				Percentage of Full Thread	Nut Minor Diameter, min* max
	No.	in.	mm	Deci-mal		
5/16–24	···	J	···	0.2770	66	0.2674
	···	···	7.0	0.2756	68	0.2754
	···	I	···	0.2720	75	
5/16–27	···	9/32	···	0.2812	65	0.2718
	···	J	···	0.2770	74	0.2792**
5/16–32	···	···	7.3	0.2841	62	0.2787
	···	···	7.2	0.2835	71	0.2833**
	···	9/32	···	0.2812	77	
3/8–16	···	P	···	0.3230	64	0.3073
	···	···	8.1	0.3189	69	0.3182
	···	O	···	0.3160	72	
	···	5/16	···	0.3125	77	
3/8–24	···	R	···	0.3390	67	0.3299
	···	···	8.5	0.3345	74	0.3372
	···	Q	···	0.3320	79	
3/8–27	···	11/32	···	0.3437	65	0.3347
	···	R	···	0.3390	75	0.3416**
7/16–14	···	···	9.7	0.3818	60	0.3602
	···	V	···	0.3770	65	0.3717
	···	3/8	···	0.3750	67	
	···	U	···	0.3680	75	
7/16–20	···	X	···	0.3970	62	0.3834
	···	···	10.0	0.3937	67	0.3916
	···	25/64	···	0.3906	72	
	···	W	···	0.3860	79	
7/16–24	···	Y	···	0.4040	62	0.3925
	···	X	···	0.3970	74	0.3985**
7/16–27	···	13/32	···	0.4062	65	0.3982
	···	Y	···	0.4040	70	0.4043**
1/2–12	···	27/64	···	0.4219	72	0.4098 0.4223
1/2–13	···	7/16	···	0.4375	62	0.4167
	···	···	11.0	0.4331	68	0.4284
	···	27/64	···	0.4219	78	
1/2–20	···	29/64	···	0.4531	72	0.4459
	†	···	···	0.4492	78	0.4537
1/2–27	···	15/32	···	0.4687	65	0.4618
	†	···	···	0.4640	75	0.4672**

Courtesy of SME

Tap Drill Sizes and Percentage of Thread (Unified Threads) *(Concluded)*

Taps Size—Threads per Inch	Drills				Percentage of Full Thread	Nut Minor Diameter, min* max
	No.	in.	mm	Decimal		
9/16–12	...	1/2	...	0.5000	58	0.4723
	...	...	12.5	0.4921	65	0.4843
	...	31/64	...	0.4844	72	
9/16–18	...	33/64	...	0.5156	65	0.5024
	...	...	13.0	0.5118	70	0.5106
9/16–27	...	17/32	...	0.5312	65	0.5234
	†	...	...	0.5265	75	0.5273**
5/8–11	...	...	14.0	0.5512	62	0.5266
	...	35/64	...	0.5469	66	0.5391
	...	17/32	...	0.5312	79	
5/8–18	...	37/64	...	0.5781	65	0.5649
	...	...	14.5	0.5709	75	0.5730
5/8–27	...	19/32	...	0.5937	65	0.5860
	†	...	...	0.5890	75	0.5912**
3/4–10	...	...	17.0	0.6693	62	0.6417
	...	21/32	...	0.6563	72	0.6545
	...	...	16.5	0.6496	77	
3/4–16	...	45/64	...	0.7031	58	0.6823
	...	...	17.5	0.6890	75	0.6908
	...	11/16	...	0.6875	77	
3/4–27	...	23/32	...	0.7187	65	0.7102
	†	...	...	0.7140	75	0.7164†
7/8–9	...	25/32	...	0.7813	65	0.7547
	...	49/64	...	0.7656	76	0.7679
7/8–14	...	13/16	...	0.8125	67	0.7977
	...	...	20.5	0.8071	73	0.8068
	...					
7/8–18	...	53/64	...	0.8281	65	0.8149
	†	...	...	0.8210	75	0.8223**
7/8–27	...	27/32	...	0.8437	65	0.8340
	†	...	...	0.8390	75	0.8406†
1–8	...	...	23.0	0.9055	58	0.8647
	...	57/64	...	0.8906	67	0.8797
	...	7/8	...	0.8750	77	
1–12	...	59/64	...	0.9219	72	0.9098
						0.9198

* Unified or American Standard Threads.
** Not based on Unified or American Thread Standards.
† Special drill required for this size as the next size smaller gives too great a percentage of thread.

Courtesy of SME

Tap Drill Sizes and Percentage of Thread (Meteric Threads)

Metric Tap Size	Recommended Metric Drill				Closest Recommended Inch Drill				
	Tape Drill Size, mm	Theoretical Percentage of Thread	Probable Hole Size, mm	Percentage of Thread	Tap Drill Size	Tap Drill Equiv., in.	Theoretical Percentage of Thread	Probable Hole Size, in.	Percentage of Thread
M1.6x0.35	1.25	77	1.288	69	#55	0.0520	61	0.0535	53
	1.3	66	1.339	59					
M1.8x0.35	1.45	77	1.488	69	#53	0.0575	64	0.0610	55
	1.5	66	1.539	57					
M2x0.4	1.6	77	1.643	69	#52	0.0635	74	0.0652	66
M2.2x0.45	1.75	77	1.793	70	#50	0.0700	72	0.0717	65
M2.5x0.45	2.05	77	2.098	69	#45	0.0820	71	0.0839	63
M3x0.5	2.5	77	2.558	68	#39	0.0995	73	0.1018	64
M3.5x0.6	2.9	77	2.967	68	#32	0.1160	71	0.1186	63
M4x0.7	3.3	77	3.373	69	#29	0.1360	60	0.1389	52
	3.4	66	3.475	58					
M4.5x0.75	3.75	77	3.830	69	#25	0.1495	72	0.1527	64
M5x0.8	4.2	77	4.282	69	#18	0.1695	67	0.1730	58
M6x1	5	77	5.095	70	#8	0.1990	73	0.2028	65
M6.3x1	5.3	77	5.396	70	#4	0.2090	76	0.2128	69
M7x1	6	77	6.096	70	B	0.2380	74	0.2418	66
M8x1.25	6.75	77	6.853	71	1	0.2720	67	0.2761	61
	6.8	74	6.904	68					
M8x1	7	77	7.104	69	J	0.2770	74	0.2811	66
M10x1.5	8.5	77	8.611	71	K	0.3390	71	0.3434	66
M10x1.25	8.75	77	8.867	70	S	0.3480	71	0.3526	64
M12x1.75	10.2	79	10.319	74	13/32	0.4062	74	0.4109	69
	10.3	75	10.419	70	Z	0.4130	66	0.4177	61

Thread	Drill (mm)	%	Drill (mm)	%	Inch (frac)	Inch (dec)	%	Inch (dec)	%
M12X1.25	11	62	11.120	54	27/64	0.4219	79	0.4266	72
M14X2	12	77	12.121	72	31/64	0.4844	65	0.4892	61
M14X1.5	12.5	77	12.621	71	1/2	0.5000	67	0.5048	60
M16X2	14	77	14.125	72	9/16	0.5625	66	0.5674	61
M16X1.5	14.5	77	14.625	71	37/64	0.5781	68	0.5830	61
M18X2.5	15.5	77	15.626	73	5/8	0.6250	65	0.6300	62
M18X1.5	16.5	77	16.627	70	21/32	0.6562	68	0.6612	62
M20X2.5	17.5	77	17.633	73	45/64	0.7031	66	0.7083	62
M20X1.5	18.5	77	18.631	70	47/64	0.7344	69	0.7396	62
M22X2.5	19.5	77	19.632	73	25/32	0.7812	66	0.7864	62
M22X1.5	20.5	77	20.632	70	13/16	0.8125	70	0.8177	63
M24X3	21	77	21.151	73	27/32	0.8437	66	0.8496	62
M24X2	22	77	22.149	71	7/8	0.8750	68	0.8809	63
M27X3	24	77	24.158	73	61/64	0.9531	72	0.9593	68
M27X2	25	77	25.179	70	63/64	0.9844	77	0.9914	70
M30X3.5	26.5	77			1 1/16	1.0625	66		
M30X2	28	77			1 7/64	1.1094	70		
M33X3.5	29.5	77	Reaming Recommended		1 11/64	1.1719	71	Reaming Recommended	
M33X2	31	77			1 15/64	1.2344	63		
M36X4	32	77			1 17/64	1.2656	74		
M36X3	32	77			1 19/64	1.2969	78		
					1 5/16	1.3125	68		
M39X4	35	77			1 3/8	1.3750	78		
					1 25/64	1.3906	71		
M39X3	36	77			1 27/64	1.4219	74		

Courtesy of Cleveland Twist Drill Co.

Tap Drill Sizes (Pipe Threads)

Taper Pipe		Straight Pipe	
Thread	Tap Drill	Thread	Tap Drill
$\frac{1}{8}$–27	R	$\frac{1}{8}$–27	S
$\frac{1}{4}$–18	$\frac{7}{16}$	$\frac{1}{4}$–18	$\frac{29}{64}$
$\frac{3}{8}$–18	$\frac{37}{64}$	$\frac{3}{8}$–18	$\frac{19}{32}$
$\frac{1}{2}$–14	$\frac{23}{32}$	$\frac{1}{2}$–14	$\frac{47}{64}$
$\frac{3}{4}$–14	$\frac{59}{64}$	$\frac{3}{4}$–14	$\frac{15}{16}$
1–11$\frac{1}{2}$	1$\frac{5}{32}$	1–11$\frac{1}{2}$	1$\frac{3}{16}$
1$\frac{1}{4}$–11$\frac{1}{2}$	1$\frac{1}{2}$	1$\frac{1}{4}$–11$\frac{1}{2}$	1$\frac{33}{64}$
1$\frac{1}{2}$–11$\frac{1}{2}$	1$\frac{47}{64}$	1$\frac{1}{2}$–11$\frac{1}{2}$	1$\frac{3}{4}$
2–11$\frac{1}{2}$	2$\frac{7}{32}$	2–11$\frac{1}{2}$	2$\frac{7}{32}$
2$\frac{1}{2}$–8	2$\frac{5}{8}$	2$\frac{1}{2}$–8	2$\frac{21}{32}$
3–8	3$\frac{1}{4}$	3–8	3$\frac{9}{32}$
3$\frac{1}{2}$–8	3$\frac{3}{4}$	3$\frac{1}{2}$–8	3$\frac{25}{32}$
4–8	4$\frac{1}{4}$	4–8	4$\frac{9}{32}$

Courtesy of SME

General Threading Formulas.* — *Tap Drill Sizes*

Unified threads = $D_m - 1.0825 \times P \times \%$ American Standard Threads = $D_m - 1.2990 \times P \times \%$ ISO Metric Threads = $D_m - 1.0825 \times P \times \%$ (All values in mm)
Determining Percentage of Thread:
Unified Threads = $\dfrac{D_m - S}{1.0825 - P}$ American Standard Threads = $\dfrac{D_m - S}{1.2990 - P}$ ISO Metric Threads = $\dfrac{D_m - S}{1.0825 - P}$ (All values in mm)
Determining Machine Screw Sizes:
$$N = \frac{D_m - 0.060}{0.013}$$ $D_m = N \times 0.013 + 0.060$

In the preceding formulas:
    D_m = Major diameter
    P = Pitch
    $\%$ = Percentage of full thread
    S = Size of selected tap drill
    N = Number of mahine screw

Courtesy of the Society of Manufacturing Engineers.

Table 1. Internal Metric Thread — M Profile Limiting Dimensions
(ANSI B1.13M-1979)

Basic Thread Designation	Tol Class	Minor Diameter D_1		Pitch Diameter D_2			Major Diameter D	
		Min	Max	Min	Max	Tol	Min	Max†
M1.6 × 0.35	6H	1.221	1.321	1.373	1.458	0.085	1.600	1.736
M2 × 0.4	6H	1.567	1.679	1.740	1.830	0.090	2.000	2.148
M2.5 × 0.45	6H	2.013	2.138	2.208	2.303	0.095	2.500	2.660
M3 × 0.5	6H	2.459	2.599	2.675	2.775	0.100	3.000	3.172
M3.5 × 0.6	6H	2.850	3.010	3.110	3.222	0.112	3.500	3.699
M4 × 0.7	6H	3.242	3.422	3.545	3.663	0.118	4.000	4.219
M5 × 0.8	6H	4.134	4.334	4.480	4.605	0.125	5.000	5.240
M6 × 1	6H	4.917	5.153	5.350	5.500	0.150	6.000	6.294
M8 × 1.25	6H	6.647	6.912	7.188	7.348	0.160	8.000	8.340
M8 × 1	6H	6.917	7.153	7.350	7.500	0.150	8.000	8.294
M10 × 1.5	6H	8.376	8.676	9.026	9.206	0.180	10.000	10.396
M10 × 1.25	6H	8.647	8.912	9.188	9.348	0.160	10.000	10.340
M10 × 0.75	6H	9.188	9.378	9.513	9.645	0.132	10.000	10.240
M12 × 1.75	6H	10.106	10.441	10.863	11.063	0.200	12.000	12.453
M12 × 1.5	6H	10.376	10.676	11.026	11.216	0.190	12.000	12.406
M12 × 1.25	6H	10.647	10.912	11.188	11.368	0.180	12.000	12.360
M12 × 1	6H	10.917	11.153	11.350	11.510	0.160	12.000	12.304
M14 × 2	6H	11.835	12.210	12.701	12.913	0.212	14.000	14.501
M14 × 1.5 M14 × 1.25*	6H	12.376	12.676	13.026	13.216	0.190	14.000	14.406
M15 × 1	6H	13.917	14.153	14.350	14.510	0.160	15.000	15.304
M16 × 2	6H	13.835	14.210	14.701	14.913	0.212	16.000	16.501
M16 × 1.5	6H	14.376	14.676	15.026	15.216	0.190	16.000	16.406
M17 × 1	6H	15.917	16.153	16.350	16.510	0.160	17.000	17.304
M18 × 1.5	6H	16.376	16.676	17.026	17.216	0.190	18.000	18.406
M20 × 2.5	6H	17.294	17.744	18.376	18.600	0.224	20.000	20.585
M20 × 1.5	6H	18.376	18.676	19.026	19.216	0.190	20.000	20.406
M20 × 1	6H	18.917	19.153	19.350	19.510	0.160	20.000	20.304
M22 × 2.5	6H	19.294	19.744	20.376	20.600	0.224	22.000	22.585
M22 × 1.5	6H	20.376	20.676	21.026	21.216	0.190	22.000	22.406
M24 × 3	6H	20.752	21.252	22.051	22.316	0.265	24.000	24.698
M24 × 2	6H	21.835	22.210	22.701	22.925	0.224	24.000	24.513
M25 × 1.5	6H	23.376	23.676	24.026	24.226	0.200	25.000	25.416
M27 × 3	6H	23.752	24.252	25.051	25.316	0.265	27.000	27.698
M27 × 2	6H	24.835	25.210	25.701	25.925	0.224	27.000	27.513
M30 × 3.5	6H	26.211	26.771	27.727	28.007	0.280	30.000	30.785
M30 × 2	6H	27.835	28.210	28.701	28.925	0.224	30.000	30.513
M30 × 1.5	6H	28.376	28.676	29.026	29.226	0.200	30.000	30.416
M33 × 2	6H	30.835	31.210	31.701	31.925	0.224	33.000	33.513
M35 × 1.5	6H	33.376	33.676	34.026	34.226	0.200	35.000	35.416

All dimensions are in millimeters.   * Special thread for spark plugs only.
† This is a reference dimension used in design of tools, etc. In dimensioning internal threads it is not normally specified. Generally, major diameter acceptance is based upon maximum material condition gaging.

Table 1. *(Concluded).* **Internal Metric Thread — M Profile Limiting Dimensions**
(ANSI B1.13M-1979)

Basic Thread Designation	Tol Class	Minor Diameter D_1		Pitch Diameter D_2			Major Diameter D	
		Min	Max	Min	Max	Tol	Min	Max†
M36 × 4	6H	31.670	32.270	33.402	33.702	0.300	36.000	36.877
M36 × 2	6H	33.835	34.210	34.701	34.925	0.224	36.000	36.513
M39 × 2	6H	36.835	37.210	37.701	37.925	0.224	39.000	39.513
M40 × 1.5	6H	38.376	38.676	39.026	39.226	0.200	40.000	40.416
M42 × 4.5	6H	37.129	37.799	39.077	39.392	0.315	42.000	42.965
M42 × 2	6H	39.835	40.210	40.701	40.925	0.224	42.000	42.513
M45 × 1.5	6H	43.376	43.676	44.026	44.226	0.200	45.000	45.416
M48 × 5	6H	42.587	43.297	44.752	45.087	0.335	48.000	49.057
M48 × 2	6H	45.835	46.210	46.701	46.937	0.236	48.000	48.525
M50 × 1.5	6H	48.376	48.676	49.026	49.238	0.212	50.000	50.428
M55 × 1.5	6H	53.376	53.676	54.026	54.238	0.212	55.000	55.428
M56 × 5.5	6H	50.046	50.796	52.428	52.783	0.355	56.000	57.149
M56 × 2	6H	53.835	54.210	54.701	54.937	0.236	56.000	56.525
M60 × 1.5	6H	58.376	58.676	59.026	59.238	0.212	60.000	60.428
M64 × 6	6H	57.505	58.305	60.103	60.478	0.375	64.000	65.241
M64 × 2	6H	61.835	62.210	62.701	62.937	0.236	64.000	64.525
M65 × 1.5	6H	63.376	63.676	64.026	64.238	0.212	65.000	65.428
M70 × 1.5	6H	68.376	68.676	69.026	69.238	0.212	70.000	70.428
M72 × 6	6H	65.505	66.305	68.103	68.478	0.375	72.000	73.241
M72 × 2	6H	69.835	70.210	70.701	70.937	0.236	72.000	72.525
M75 × 1.5	6H	73.376	73.676	74.026	74.238	0.212	75.000	75.428
M80 × 6	6H	73.505	74.305	76.103	76.478	0.375	80.000	81.241
M80 × 2	6H	77.835	78.210	78.701	78.937	0.236	80.000	80.525
M80 × 1.5	6H	78.376	78.676	79.026	79.238	0.212	80.000	80.428
M85 × 2	6H	82.835	83.210	83.701	83.937	0.236	85.000	85.525
M90 × 6	6H	83.505	84.305	86.103	86.478	0.375	90.000	91.241
M90 × 2	6H	87.835	88.210	88.701	88.937	0.236	90.000	90.525
M95 × 2	6H	92.835	93.210	93.701	93.951	0.250	95.000	95.539
M100 × 6	6H	93.505	94.305	96.103	96.503	0.400	100.000	101.266
M100 × 2	6H	97.835	98.210	96.701	98.951	0.250	100.000	100.539
M105 × 2	6H	102.835	103.210	103.701	103.951	0.250	105.000	105.539
M110 × 2	6H	107.835	108.210	108.701	108.951	0.250	110.000	110.539
M120 × 2	6H	117.835	118.210	118.701	118.951	0.250	120.000	120.539
M130 × 2	6H	127.835	128.210	128.701	128.951	0.250	130.000	130.539
M140 × 2	6H	137.835	138.210	138.701	138.951	0.250	140.000	140.539
M150 × 2	6H	147.835	148.210	148.701	148.951	0.250	150.000	150.539
M160 × 3	6H	156.752	157.252	158.051	158.351	0.300	160.000	160.733
M170 × 3	6H	166.752	167.252	168.051	168.351	0.300	170.000	170.733
M180 × 3	6H	176.752	177.252	178.051	178.351	0.300	180.000	180.733
M190 × 3	6H	186.752	187.252	188.051	188.386	0.335	190.000	190.768
M200 × 3	6H	196.752	197.252	198.051	198.386	0.335	200.000	200.768

All dimensions are in millimeters. † This is a reference dimension used in design of tools, etc. In dimensioning internal threads it is not normally specified. Generally, major diameter acceptance is based upon maximum material condition gaging.

Table 2. External Metric Thread — M Profile Limiting Dimensions
(ANSI B1.13M-1979)

Basic Thd. Desig.	Toler. Class	Allow. es^{**}	Major Diam.* d		Pitch Diam.* d_2			Minor Diam., d_1*	Minor Diam., d_3†
			Max	Min	Max	Min	Tol.	Max	Min
M1.6 × 0.35	6g	0.019	1.581	1.496	1.354	1.291	0.063	1.202	1.075
M1.6 × 0.35	4g6g	0.019	1.581	1.496	1.354	1.314	0.040	1.202	1.098
M2 × 0.4	6g	0.019	1.981	1.886	1.721	1.654	0.067	1.548	1.408
M2 × 0.4	4g6g	0.019	1.981	1.886	1.721	1.679	0.042	1.548	1.433
M2.5 × 0.45	6g	0.020	2.480	2.380	2.188	2.117	0.071	1.993	1.840
M2.5 × 0.45	4g6g	0.020	2.480	2.380	2.188	2.143	0.045	1.993	1.866
M3 × 0.5	6g	0.020	2.980	2.874	2.655	2.580	0.075	2.439	2.272
M3 × 0.5	4g6g	0.020	2.980	2.874	2.655	2.607	0.048	2.439	2.299
M3.5 × 0.6	6g	0.021	3.479	3.354	3.089	3.004	0.085	2.829	2.635
M3.5 × 0.6	4g6g	0.021	3.479	3.354	3.089	3.036	0.053	2.829	2.667
M4 × 0.7	6g	0.022	3.978	3.838	3.523	3.433	0.090	3.220	3.002
M4 × 0.7	4g6g	0.022	3.978	3.838	3.523	3.467	0.056	3.220	3.036
M5 × 0.8	6g	0.024	4.976	4.826	4.456	4.361	0.095	4.110	3.869
M5 × 0.8	4g6g	0.024	4.976	4.826	4.456	4.396	0.060	4.110	3.904
M6 × 1	6g	0.026	5.974	5.794	5.324	5.212	0.112	4.891	4.596
M6 × 1	4g6g	0.026	5.974	5.794	5.324	5.253	0.071	4.891	4.637
M8 × 1.25	6g	0.028	7.972	7.760	7.160	7.042	0.118	6.619	6.272
M8 × 1.25	4g6g	0.028	7.972	7.760	7.160	7.085	0.075	6.619	6.315
M8 × 1	6g	0.026	7.974	7.794	7.324	7.212	0.112	6.891	6.596
M8 × 1	4g6g	0.026	7.974	7.794	7.324	7.253	0.071	6.891	6.637
M10 × 1.5	6g	0.032	9.968	9.732	8.994	8.862	0.132	8.344	7.938
M10 × 1.5	4g6g	0.032	9.968	9.732	8.994	8.909	0.085	8.344	7.985
M10 × 1.25	6g	0.028	9.972	9.760	9.160	9.042	0.118	8.619	8.272
M10 × 1.25	4g6g	0.028	9.972	9.760	9.160	9.085	0.075	8.619	8.315
M10 × 0.75	6g	0.022	9.978	9.838	9.491	9.391	0.100	9.166	8.929
M10 × 0.75	4g6g	0.022	9.978	9.838	9.491	9.428	0.063	9.166	8.966
M12 × 1.75	6g	0.034	11.966	11.701	10.829	10.679	0.150	10.072	9.601
M12 × 1.75	4g6g	0.034	11.966	11.701	10.829	10.734	0.095	10.072	9.656
M12 × 1.5	6g	0.032	11.968	11.732	10.994	10.854	0.140	10.344	9.930
M12 × 1.25	6g	0.028	11.972	11.760	11.160	11.028	0.132	10.619	10.258
M12 × 1.25	4g6g	0.028	11.972	11.760	11.160	11.075	0.085	10.619	10.305
M12 × 1	6g	0.026	11.974	11.794	11.324	11.206	0.118	10.891	10.590
M12 × 1	4g6g	0.026	11.974	11.794	11.324	11.249	0.075	10.891	10.633
M14 × 2	6g	0.038	13.962	13.682	12.663	12.503	0.160	11.797	11.271
M14 × 2	4g6g	0.038	13.962	13.682	12.663	12.563	0.100	11.797	11.331
M14 × 1.5	6g	0.032	13.968	13.732	12.994	12.854	0.140	12.344	11.930
M14 × 1.5	4g6g	0.032	13.968	13.732	12.994	12.904	0.090	12.344	11.980
M15 × 1	6g	0.026	14.974	14.794	14.324	14.206	0.118	13.891	13.590
M15 × 1	4g6g	0.026	14.974	14.794	14.324	14.249	0.075	13.891	13.633
M16 × 2	6g	0.038	15.962	15.682	14.663	14.503	0.160	13.797	13.271
M16 × 2	4g6g	0.038	15.962	15.682	14.663	14.563	0.100	13.797	13.331

All dimensions are in millimeters. For footnotes see end of table.

Table 2. *(Continued).* **External Metric Thread — M Profile Limiting Dimensions**
(ANSI B1.13M-1979)

Basic Thd. Desig.	Toler. Class	Allow. es**	Major Diam.* d		Pitch Diam.* d_2			Minor Diam., d_1*	Minor Diam., d_3†
			Max	Min	Max	Min	Tol.	Max	Min
M16 × 1.5	6g	0.032	15.968	15.732	14.994	14.854	0.140	14.344	13.930
M16 × 1.5	4g6g	0.032	15.968	15.732	14.994	14.904	0.090	14.344	13.980
M17 × 1	6g	0.026	16.974	16.794	16.324	16.206	0.118	15.891	15.590
M17 × 1	4g6g	0.026	16.974	16.794	16.324	16.249	0.075	15.891	15.633
M18 × 1.5	6g	0.032	17.968	17.732	16.994	16.854	0.140	16.344	15.930
M18 × 1.5	4g6g	0.032	17.968	17.732	16.994	16.904	0.090	16.344	15.980
M20 × 2.5	6g	0.042	19.958	19.623	18.334	18.164	0.170	17.252	16.624
M20 × 2.5	4g6g	0.042	19.958	19.623	18.334	18.228	0.106	17.252	16.688
M20 × 1.5	6g	0.032	19.968	19.732	18.994	18.854	0.140	18.344	17.930
M20 × 1.5	4g6g	0.032	19.968	19.732	18.994	18.904	0.090	18.344	17.980
M20 × 1	6g	0.026	19.974	19.794	19.324	19.206	0.118	18.891	18.590
M20 × 1	4g6g	0.026	19.974	19.794	19.324	19.249	0.075	18.891	18.633
M22 × 2.5	6g	0.042	21.958	21.623	20.334	20.164	0.170	19.252	18.624
M22 × 1.5	6g	0.032	21.968	21.732	20.994	20.854	0.140	20.344	19.930
M22 × 1.5	4g6g	0.032	21.968	21.732	20.994	20.904	0.090	20.344	19.980
M24 × 3	6g	0.048	23.952	23.577	22.003	21.803	0.200	20.704	19.955
M24 × 3	4g6g	0.048	23.952	23.577	22.003	21.878	0.125	20.704	20.030
M24 × 2	6g	0.038	23.962	23.682	22.663	22.493	0.170	21.797	21.261
M24 × 2	4g6g	0.038	23.962	23.682	22.663	22.557	0.106	21.797	21.325
M25 × 1.5	6g	0.032	24.968	24.732	23.994	23.844	0.150	23.344	22.920
M25 × 1.5	4g6g	0.032	24.968	24.732	23.994	23.899	0.095	23.344	22.975
M27 × 3	6g	0.048	26.952	26.577	25.003	24.803	0.200	23.744	22.955
M27 × 2	6g	0.038	26.962	26.682	25.663	25.493	0.170	24.797	24.261
M27 × 2	4g6g	0.038	29.962	26.682	25.663	25.557	0.106	24.797	24.325
M30 × 3.5	6g	0.053	29.947	29.522	27.674	27.462	0.212	26.158	25.306
M30 × 3.5	4g6g	0.053	29.947	29.522	27.674	27.542	0.132	26.158	25.386
M30 × 2	6g	0.038	29.962	29.682	28.663	28.493	0.170	27.797	27.261
M30 × 2	4g6g	0.038	29.962	29.682	28.663	28.557	0.106	27.797	27.325
M30 × 1.5	6g	0.032	29.968	29.732	28.994	28.844	0.150	28.344	27.920
M30 × 1.5	4g6g	0.032	29.968	29.732	28.994	28.899	0.095	28.344	27.975
M33 × 2	6g	0.038	32.962	32.682	31.663	31.493	0.170	30.797	30.261
M33 × 2	4g6g	0.038	32.962	32.682	31.663	31.557	0.106	30.797	30.325
M35 × 1.5	6g	0.032	34.968	34.732	33.994	33.844	0.150	33.344	33.920
M36 × 4	6g	0.060	35.940	35.465	33.342	33.118	0.224	31.610	30.654
M36 × 4	4g6g	0.060	35.940	35.465	33.342	33.202	0.140	31.610	30.738
M36 × 2	6g	0.038	35.962	35.682	34.663	34.493	0.170	33.797	33.261
M36 × 2	4g6g	0.038	35.962	35.682	34.663	34.557	0.106	33.797	33.325
M39 × 2	6g	0.038	38.962	38.682	37.663	37.493	0.170	36.797	36.261
M39 × 2	4g6g	0.038	38.962	38.682	37.663	37.557	0.106	36.797	36.325
M40 × 1.5	6g	0.032	39.968	39.732	38.994	38.844	0.150	38.344	37.920
M40 × 1.5	4g6g	0.032	39.968	39.732	38.994	38.899	0.095	38.344	37.975

All dimensions are in millimeters.
For footnotes see end of table.

Table 2. *(Continued).* **External Metric Thread — M Profile Limiting Dimensions**
(ANSI B1.13M-1979)

Basic Thd. Desig.	Toler. Class	Allow. es**	Major Diam.* d		Pitch Diam.* d2			Minor Diam., d1*	Minor Diam., d3†
			Max	Min	Max	Min	Tol.	Max	Min
M42 × 4.5	6g	0.063	41.937	41.437	39.014	38.778	0.236	37.066	36.006
M42 × 4.5	4g6g	0.063	41.937	41.437	39.014	38.864	0.150	37.066	36.092
M42 × 2	6g	0.038	41.962	41.682	40.663	40.493	0.170	39.797	39.261
M42 × 2	4g6g	0.038	41.962	41.682	40.663	40.557	0.106	39.797	39.325
M45 × 1.5	6g	0.032	44.968	44.732	43.994	43.844	0.150	43.344	42.920
M45 × 1.5	4g6g	0.032	44.968	44.732	43.994	43.899	0.095	43.344	42.975
M48 × 5	6g	0.071	47.929	47.399	44.681	44.431	0.250	42.516	41.351
M48 × 5	4g6g	0.071	47.929	47.399	44.681	44.521	0.160	42.516	41.441
M48 × 2	6g	0.038	47.962	47.682	46.663	46.483	0.180	45.797	45.251
M48 × 2	4g6g	0.038	47.962	47.682	46.663	46.551	0.112	45.797	45.319
M50 × 1.5	6g	0.032	49.968	49.732	48.994	48.834	0.160	48.344	47.910
M50 × 1.5	4g6g	0.032	49.968	49.732	48.994	48.894	0.100	48.344	47.970
M55 × 1.5	6g	0.032	54.968	54.732	53.994	53.834	0.160	53.344	52.910
M55 × 1.5	4g6g	0.032	54.968	54.732	53.994	53.894	0.100	53.344	52.970
M56 × 5.5	6g	0.075	55.925	55.365	52.353	52.088	0.265	49.971	48.700
M56 × 5.5	4g6g	0.075	55.925	55.365	52.353	52.183	0.170	49.971	48.795
M56 × 2	6g	0.038	55.962	55.682	54.663	54.483	0.180	53.797	53.251
M56 × 2	4g6g	0.038	55.962	55.682	54.663	54.551	0.112	53.797	53.319
M60 × 1.5	6g	0.032	59.968	59.732	58.994	58.834	0.160	58.344	57.910
M60 × 1.5	4g6g	0.032	59.968	59.732	58.994	58.894	0.100	58.344	57.970
M64 × 6	6g	0.080	63.920	63.320	60.023	59.743	0.280	57.425	56.047
M64 × 6	4g6g	0.080	63.920	63.320	60.023	59.843	0.180	57.425	56.147
M64 × 2	6g	0.038	63.962	63.682	62.663	62.483	0.180	61.797	61.251
M64 × 2	4g6g	0.038	63.962	63.682	62.663	62.551	0.112	61.797	61.319
M65 × 1.5	6g	0.032	64.968	64.732	63.994	63.834	0.160	63.344	62.910
M65 × 1.5	4g6g	0.032	64.968	64.732	63.994	63.894	0.100	63.344	62.970
M70 × 1.5	6g	0.032	69.968	69.732	68.994	68.834	0.160	68.344	67.910
M70 × 1.5	4g6g	0.032	69.968	69.732	68.994	68.894	0.100	68.344	67.970
M72 × 6	6g	0.080	71.920	71.320	68.023	67.743	0.280	65.425	64.047
M72 × 6	4g6g	0.080	71.920	71.320	68.023	67.843	0.180	65.425	64.147
M72 × 2	6g	0.038	71.962	71.682	70.663	70.483	0.180	69.797	69.251
M72 × 2	4g6g	0.038	71.962	71.682	70.663	70.551	0.112	69.797	69.319
M75 × 1.5	6g	0.032	74.968	74.732	73.994	73.834	0.160	73.344	72.910
M75 × 1.5	4g6g	0.032	74.968	74.732	73.994	73.894	0.100	73.344	72.970
M80 × 6	6g	0.080	79.920	79.320	76.023	75.743	0.280	73.425	72.047
M80 × 6	4g6g	0.080	79.920	79.320	76.023	75.843	0.180	73.425	72.147
M80 × 2	6g	0.038	79.962	79.682	78.663	78.483	0.180	77.797	77.251
M80 × 2	4g6g	0.038	79.962	79.682	78.663	78.551	0.112	77.797	77.319
M80 × 1.5	6g	0.032	79.968	79.732	78.994	78.834	0.160	78.344	77.910
M80 × 1.5	4g6g	0.032	79.968	79.732	78.994	78.894	0.100	78.344	77.970
M85 × 2	6g	0.038	84.962	84.682	83.663	83.483	0.180	82.797	82.251
M85 × 2	4g6g	0.038	84.962	84.682	83.663	83.551	0.112	82.797	82.319

All dimensions are in millimeters.
For footnotes see end of table.

Table 2. *(Concluded).* **External Metric Thread — M Profile Limiting Dimensions**
(ANSI B1.13M-1979)

Basic Thd. Desig.	Toler. Class	Allow. es^{**}	Major Diam.* d		Pitch Diam.* d_2			Minor Diam., d_1*	Minor Diam., d_3†
			Max	Min	Max	Min	Tol.	Max	Min
M90 × 6	6g	0.080	89.920	89.320	86.023	85.743	0.280	83.425	82.047
M90 × 6	4g6g	0.080	89.920	89.320	86.023	85.843	0.180	83.425	82.147
M90 × 2	6g	0.038	89.962	89.682	88.663	88.483	0.180	87.797	87.251
M90 × 2	4g6g	0.038	89.962	89.682	88.663	88.551	0.112	87.797	87.319
M95 × 2	6g	0.038	94.962	94.682	93.663	93.473	0.190	92.797	92.241
M95 × 2	4g6g	0.038	94.962	94.682	93.663	93.545	0.118	92.797	92.313
M100 × 6	6g	0.080	99.920	99.320	96.023	95.723	0.300	93.425	92.027
M100 × 6	4g6g	0.080	99.920	99.320	96.023	95.833	0.190	93.425	92.137
M100 × 2	6g	0.038	99.962	99.682	98.663	98.473	0.190	97.797	97.241
M100 × 2	4g6g	0.038	99.962	99.682	98.663	98.545	0.118	97.797	97.313
M105 × 2	6g	0.038	104.962	104.682	103.663	103.473	0.190	102.797	102.241*
M105 × 2	4g6g	0.038	104.962	104.682	103.663	103.545	0.118	102.797	102.313
M110 × 2	6g	0.038	109.962	109.682	108.663	108.473	0.190	107.797	107.241
M110 × 2	4g6g	0.038	109.962	109.682	108.663	108.545	0.118	107.797	107.313
M120 × 2	6g	0.038	119.962	119.682	118.663	118.473	0.190	117.797	117.241
M120 × 2	4g6g	0.038	119.962	119.682	118.663	118.545	0.118	117.797	117.313
M130 × 2	6g	0.038	129.962	129.682	128.663	128.473	0.190	127.797	127.241
M130 × 2	4g6g	0.038	129.962	129.682	128.663	128.545	0.118	127.797	127.313
M140 × 2	6g	0.038	139.962	139.682	138.663	138.473	0.190	137.797	137.241
M140 × 2	4g6g	0.038	139.962	139.682	138.663	138.545	0.118	137.797	137.313
M150 × 2	6g	0.038	149.962	149.682	148.663	148.473	0.190	147.797	147.241
M150 × 2	4g6g	0.038	149.962	149.682	148.663	148.545	0.118	147.797	147.313
M160 × 3	6g	0.048	159.952	159.577	158.003	157.779	0.224	156.704	155.931
M160 × 3	4g6g	0.048	159.952	159.577	158.003	157.863	0.140	156.704	156.015
M170 × 3	6g	0.048	169.952	169.577	168.003	167.779	0.224	166.704	165.931
M170 × 3	4g6g	0.048	169.952	169.577	168.003	167.863	0.140	166.704	166.015
M180 × 3	6g	0.048	179.952	179.577	178.003	177.779	0.224	176.704	175.931
M180 × 3	4g6g	0.048	179.952	179.577	178.003	177.863	0.140	176.704	176.015
M190 × 3	6g	0.048	189.952	189.577	188.003	187.753	0.250	186.704	185.905
M190 × 3	4g6g	0.048	189.952	189.577	188.003	187.843	0.160	186.704	185.995
M200 × 3	6g	0.048	199.952	199.577	198.003	197.753	0.250	196.704	195.905
M200 × 3	4g6g	0.048	199.952	199.577	198.003	197.843	0.160	196.704	195.995

All dimensions are in millimeters.

* (Flat root) For screw threads at maximum limits of tolerance position *h*, add the absolute value *es* to the maximum diameters required. For maximum major diameter this is the basic thread size listed in Table 1 as Minimum Major Diameter (D_{min}); for maximum pitch diameter this is the same as listed in Table 1 as Minimum Pitch Diameter ($D_{2\ min}$); and for maximum minor diameter this is the same as listed in Table 1 as Minimum Minor Diameter ($D_{1\ min}$).

** *es* is an absolute value.

† (Rounded root) This reference dimension is used in the design of tools, etc. In dimensioning external threads it is not normally specified. Generally minor diameter acceptance is based upon maximum material condition gaging.

Table 3. **Basic Dimensions, American National Standard Taper Pipe Threads,[1] NPT**
(ANSI B2.1-1968)

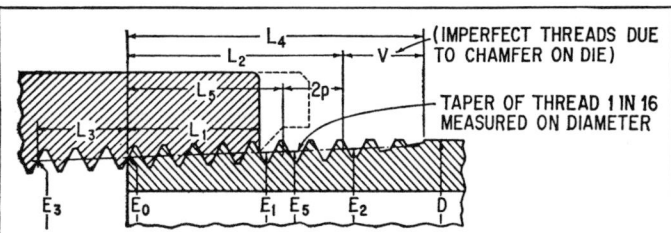

For all dimensions see corresponding reference letters in table.

Angle between sides of thread is 60 degrees. Taper of thread, on diameter, is ¾ inch per foot. Angle of taper with centerline is 1°47'.

The basic maximum thread height, h, of the truncated thread is 0.8 × pitch of thread. The crest and root are truncated a minimum of 0.033 × pitch for all pitches.

Nominal Pipe Size	Outside Diam. of Pipe, D	Threads per Inch, n	Pitch of Thread, p	Pitch Diameter at Beginning of External Thread, E_0	Handtight Engagement		Effective Thread, External	
					Length,[2] L_1 In.	Diam.,[3] E_1	Length,[4] L_2 In.	Diam., E_2
¹⁄₁₆	0.3125	27	0.03704	0.27118	0.160	0.28118	0.2611	0.28750
⅛	0.405	27	0.03704	0.36351	0.1615	0.37360	0.2639	0.38000
¼	0.540	18	0.05556	0.47739	0.2278	0.49163	0.4018	0.50250
⅜	0.675	18	0.05556	0.61201	0.240	0.62701	0.4078	0.63750
½	0.840	14	0.07143	0.75843	0.320	0.77843	0.5337	0.79179
¾	1.050	14	0.07143	0.96768	0.339	0.98887	0.5457	1.00179
1	1.315	11½	0.08696	1.21363	0.400	1.23863	0.6828	1.25630
1¼	1.660	11½	0.08696	1.55713	0.420	1.58338	0.7068	1.60130
1½	1.900	11½	0.08696	1.79609	0.420	1.82234	0.7235	1.84130
2	2.375	11½	0.08696	2.26902	0.436	2.29627	0.7565	2.31630
2½	2.875	8	0.12500	2.71953	0.682	2.76216	1.1375	2.79062
3	3.500	8	0.12500	3.34062	0.766	3.38850	1.2000	3.41562
3½	4.000	8	0.12500	3.83750	0.821	3.88881	1.2500	3.91562
4	4.500	8	0.12500	4.33438	0.844	4.38712	1.3000	4.41562
5	5.563	8	0.12500	5.39073	0.937	5.44929	1.4063	5.47862
6	6.625	8	0.12500	6.44609	0.958	6.50597	1.5125	6.54062
8	8.625	8	0.12500	8.43359	1.063	8.50003	1.7125	8.54062
10	10.750	8	0.12500	10.54531	1.210	10.62094	1.9250	10.66562
12	12.750	8	0.12500	12.53281	1.360	12.61781	2.1250	12.66562
14 OD	14.000	8	0.12500	13.77500	1.562	13.87262	2.2500	13.91562
16 OD	16.000	8	0.12500	15.76250	1.812	15.87575	2.4500	15.91562
18 OD	18.000	8	0.12500	17.75000	2.000	17.87500	2.6500	17.91562
20 OD	20.000	8	0.12500	19.73750	2.125	19.87031	2.8500	19.91562
24 OD	24.000	8	0.12500	23.71250	2.375	23.86094	3.2500	23.91562

All dimensions given in inches.

[1] The basic dimensions of the ANSI Standard Taper Pipe Thread are given in inches to four or five decimal places. While this implies a greater degree of precision than is ordinarily attained, these dimensions are the basis of gage dimensions and are so expressed for the purpose of eliminating errors in computations.

[2] Also length of thin ring gage and length from gaging notch to small end of plug gage.

[3] Also pitch diameter at gaging notch (handtight plane).

[4] Also length of plug gage.

Table 3. (*Concluded*). **Basic Dimensions, American National Standard Taper Pipe Threads, NPT** (ANSI B2.1-1968)

Nominal Pipe Size	Wrench Makeup Length for Internal Thread		Vanish Thread, (3.47 thds.), V	Overall Length External Thread, L_4	Nominal Perfect External Threads[5]		Height of Thread, h	Basic Minor Diam. at Small End of Pipe,[6] K_0
	Length,[7] L_3	Diam., E_3			Length, L_5	Diam., E_5		
1/16	0.1111	0.26424	0.1285	0.3896	0.1870	0.28287	0.02963	0.2416
1/8	0.1111	0.35656	0.1285	0.3924	0.1898	0.37537	0.02963	0.3339
1/4	0.1667	0.46697	0.1928	0.5946	0.2907	0.49556	0.04444	0.4329
3/8	0.1667	0.60160	0.1928	0.6006	0.2967	0.63056	0.04444	0.5676
1/2	0.2143	0.74504	0.2478	0.7815	0.3909	0.78286	0.05714	0.7013
3/4	0.2143	0.95429	0.2478	0.7935	0.4029	0.99286	0.05714	0.9105
1	0.2609	1.19733	0.3017	0.9845	0.5089	1.24543	0.06957	1.1441
1 1/4	0.2609	1.54083	0.3017	1.0085	0.5329	1.59043	0.06957	1.4876
1 1/2	0.2609	1.77978	0.3017	1.0252	0.5496	1.83043	0.06957	1.7265
2	0.2609	2.25272	0.3017	1.0582	0.5826	2.30543	0.06957	2.1995
2 1/2	0.25008[8]	2.70391	0.4337	1.5712	0.8875	2.77500	0.100000	2.6195
3	0.25008[8]	3.32500	0.4337	1.6337	0.9500	3.40000	0.100000	3.2406
3 1/2	0.2500	3.82188	0.4337	1.6837	1.0000	3.90000	0.100000	3.7375
4	0.2500	4.31875	0.4337	1.7337	1.0500	4.40000	0.100000	4.2344
5	0.2500	5.37511	0.4337	1.8400	1.1563	5.46300	0.100000	5.2907
6	0.2500	6.43047	0.4337	1.9462	1.2625	6.52500	0.100000	6.3461
8	0.2500	8.41797	0.4337	2.1462	1.4625	8.52500	0.100000	8.3336
10	0.2500	10.52969	0.4337	2.3587	1.6750	10.65000	0.100000	10.4453
12	0.2500	12.51719	0.4337	2.5587	1.8750	12.65000	0.100000	12.4328
14 OD	0.2500	13.75938	0.4337	2.6837	2.0000	13.90000	0.100000	13.6750
16 OD	0.2500	15.74688	0.4337	2.8837	2.2000	15.90000	0.100000	15.6625
18 OD	0.2500	17.73438	0.4337	3.0837	2.4000	17.90000	0.100000	17.6500
20 OD	0.2500	19.72188	0.4337	3.2837	2.6000	19.90000	0.100000	19.6375
24 OD	0.2500	23.69688	0.4337	3.6837	3.0000	23.90000	0.100000	23.6125

[5] The length L_5 from the end of the pipe determines the plane beyond which the thread form is imperfect at the crest. The next two threads are perfect at the root. At this plane the cone formed by the crests of the thread intersects the cylinder forming the external surface of the pipe. $L_5 = L_2 - 2p$.

[6] Given as information for use in selecting tap drills.

[7] Three threads for 2-inch size and smaller; two threads for larger sizes.

[8] Military Specification MIL — P — 7105 gives the wrench makeup as three threads for 3 in. and smaller. The E_3 dimensions are then as follows: Size 2 1/2 in., 2.69609 and size 3 in., 3.31719.

Increase in diameter per thread is equal to 0.0625/n.

Engagement between External and Internal Taper Threads. — The normal length of engagement between external and internal taper threads when screwed together handtight is shown as L_1 in Table 3. This length is controlled by the construction and use of the pipe thread gages. It is recognized that in special applications, such as flanges for high-pressure work, longer thread engagement is used, in which case the pitch diameter E_1 (Table 3) is maintained and the pitch diameter E_0 at the end of the pipe is proportionately smaller.

Railing Joint Taper Pipe Threads, NPTR. — Railing joints require a rigid mechanical thread joint with external and internal taper threads. The external thread is basically the same as the ANSI Standard Taper Pipe Thread, except that sizes 1/2 through 2 inches are shortened by 3 threads and sizes 2 1/2 through 4 inches are shortened by 4 threads to permit the use of the larger end of the pipe thread. A recess in the fitting covers the last scratch or imperfect threads on the pipe.

Table 4. American National Standard Straight Pipe Threads for Mechanical Joints, NPSM and NPSI. (ANSI B2.1-1968)

Nominal Pipe Size	Threads per Inch	Allowance	External Thread				Internal Thread			
			Major Diameter		Pitch Diameter		Minor Diameter		Pitch Diameter	
			Max.[2]	Min.	Max.	Min.	Min.[2]	Max.	Min.[1]	Max.
Free-fitting Mechanical Joints for Fixtures — NPSM										
⅛	27	0.0011	0.397	0.390	0.3725	0.3689	0.358	0.364	0.3736	0.3783
¼	18	0.0013	0.526	0.517	0.4903	0.4859	0.468	0.481	0.4916	0.4974
⅜	18	0.0014	0.662	0.653	0.6256	0.6211	0.603	0.612	0.6270	0.6329
½	14	0.0015	0.823	0.813	0.7769	0.7718	0.747	0.759	0.7784	0.7851
¾	14	0.0016	1.034	1.024	0.9873	0.9820	0.958	0.970	0.9889	0.9958
1	11½	0.0017	1.293	1.281	1.2369	1.2311	1.201	1.211	1.2386	1.2462
1¼	11½	0.0018	1.638	1.626	1.5816	1.5756	1.546	1.555	1.5834	1.5912
1½	11½	0.0018	1.877	1.865	1.8205	1.8144	1.785	1.794	1.8223	1.8302
2	11½	0.0019	2.351	2.339	2.2944	2.2882	2.259	2.268	2.2963	2.3044
2½	8	0.0022	2.841	2.826	2.7600	2.7526	2.708	2.727	2.7622	2.7720
3	8	0.0023	3.467	3.452	3.3862	3.3786	3.334	3.353	3.3885	3.3984
3½	8	0.0023	3.968	3.953	3.8865	3.8788	3.835	3.848	3.8888	3.8988
4	8	0.0023	4.466	4.451	4.3848	4.3771	4.333	4.346	4.3871	4.3971
5	8	0.0024	5.528	5.513	5.4469	5.4390	5.395	5.408	5.4493	5.4598
6	8	0.0024	6.585	6.570	6.5036	6.4955	6.452	6.464	6.5060	6.5165
Loose-fitting Mechanical Joints for Locknut Connections — NPSL										
⅛	27	...	0.409	...	0.3840	0.3805	0.362	...	0.3863	0.3898
¼	18	...	0.541	...	0.5038	0.4986	0.470	...	0.5073	0.5125
⅜	18	...	0.678	...	0.6409	0.6357	0.607	...	0.6444	0.6496
½	14	...	0.844	...	0.7963	0.7896	0.753	...	0.8008	0.8075
¾	14	...	1.054	...	1.0067	1.0000	0.964	...	1.0112	1.0179
1	11½	...	1.318	...	1.2604	1.2523	1.208	...	1.2658	1.2739
1¼	11½	...	1.663	...	1.6051	1.5970	1.553	...	1.6106	1.6187
1½	11½	...	1.902	...	1.8441	1.8360	1.792	...	1.8495	1.8576
2	11½	...	2.376	...	2.3180	2.3099	2.265	...	2.3234	2.3315
2½	8	...	2.877	...	2.7934	2.7817	2.718	...	2.8012	2.8129
3	8	...	3.503	...	3.4198	3.4081	3.344	...	3.4276	3.4393
3½	8	...	4.003	...	3.9201	3.9084	3.845	...	3.9279	3.9396
4	8	...	4.502	...	4.4184	4.4067	4.343	...	4.4262	4.4379
5	8	...	5.564	...	5.4805	5.4688	5.405	...	5.4884	5.5001
6	8	...	6.620	...	6.5372	6.5255	6.462	...	6.5450	6.5567
8	8	...	8.615	...	8.5313	8.5196	8.456	...	8.5391	8.5508
10	8	...	10.735	...	10.6522	10.6405	10.577	...	10.6600	10.6717
12	8	...	12.732	...	12.6491	12.6374	12.574	...	12.6569	12.6686

All dimensions are given in inches.

Notes for Free-fitting Fixture Threads:

[1] This is the same as the pitch diameter at end of internal thread, E_1 *Basic.* (See Table 3.) The minor diameters of external threads and major diameters of internal threads are those as produced by commercial straight pipe dies and commercial ground straight pipe taps.

The major diameter of the external thread has been calculated on the basis of a truncation of $0.10825p$, and the minor diameter of the internal thread has been calculated on the basis of a truncation of $0.21651p$, to provide no interference at crest and root when product is gaged with gages made in accordance with the Standard.

Notes for Loose-fitting Locknut Threads:

[2] As the ANSI Standard Straight Pipe Thread form of thread is maintained, the major and the minor diameters of the internal thread and the minor diameter of the external thread vary with the pitch diameter. The major diameter of the external thread is usually determined by the diameter of the pipe. These theoretical diameters result from adding the depth of the truncated thread $(0.666025 \times p)$ to the maximum pitch diameters, and it should be understood that commercial pipe will not always have these maximum major diameters.

The locknut thread is established on the basis of retaining the greatest possible amount of metal thickness between the bottom of the thread and the inside of the pipe.

In order that a locknut may fit loosely on the externally threaded part, an allowance equal to the "increase in pitch diameter per turn" is provided, with a tolerance of 1½ turns for both external and internal threads.

Table 5. Standard Series and Selected Combinations† — Unified Screw Threads

Nominal Size, Threads per Inch, and Series Designation^f	Class	Allowance	Major Diameter Max^b	Major Diameter Min	Major Diameter Min^d	Pitch Diameter Max^b	Pitch Diameter Min	Minor Diameter^a	Class	Minor Diameter^a Min	Minor Diameter^a Max	Pitch Diameter Min	Pitch Diameter Max	Major Diameter Min
0–80 UNF	2A	0.0005	0.0595	0.0563	—	0.0514	0.0496	0.0442	2B	0.0465	0.0514	0.0519	0.0542	0.0600
	3A	0.0000	0.0600	0.0568	—	0.0519	0.0506	0.0447	3B	0.0465	0.0514	0.0519	0.0536	0.0600
1–64 UNC	2A	0.0006	0.0724	0.0686	—	0.0623	0.0603	0.0532	2B	0.0561	0.0623	0.0629	0.0655	0.0730
	3A	0.0000	0.0730	0.0692	—	0.0629	0.0614	0.0538	3B	0.0561	0.0623	0.0629	0.0648	0.0730
1–72 UNF	2A	0.0006	0.0724	0.0689	—	0.0634	0.0615	0.0554	2B	0.0580	0.0635	0.0640	0.0665	0.0730
	3A	0.0000	0.0730	0.0695	—	0.0640	0.0626	0.0560	3B	0.0580	0.0635	0.0640	0.0659	0.0730
2–56 UNC	2A	0.0006	0.0854	0.0813	—	0.0738	0.0717	0.0635	2B	0.0667	0.0737	0.0744	0.0772	0.0860
	3A	0.0000	0.0860	0.0819	—	0.0744	0.0728	0.0641	3B	0.0667	0.0737	0.0744	0.0765	0.0860
2–64 UNF	2A	0.0006	0.0854	0.0816	—	0.0753	0.0733	0.0662	2B	0.0691	0.0753	0.0759	0.0786	0.0860
	3A	0.0000	0.0860	0.0822	—	0.0759	0.0744	0.0668	3B	0.0691	0.0753	0.0759	0.0779	0.0860
3–48 UNC	2A	0.0007	0.0983	0.0938	—	0.0848	0.0825	0.0727	2B	0.0764	0.0845	0.0855	0.0885	0.0990
	3A	0.0000	0.0990	0.0945	—	0.0855	0.0838	0.0734	3B	0.0764	0.0845	0.0855	0.0877	0.0990
3–56 UNF	2A	0.0007	0.0983	0.0942	—	0.0867	0.0845	0.0764	2B	0.0797	0.0865	0.0874	0.0902	0.0990
	3A	0.0000	0.0990	0.0949	—	0.0874	0.0858	0.0771	3B	0.0797	0.0865	0.0874	0.0895	0.0990
4–40 UNC	2A	0.0008	0.1112	0.1061	—	0.0950	0.0925	0.0805	2B	0.0849	0.0939	0.0958	0.0991	0.1120
	3A	0.0000	0.1120	0.1069	—	0.0958	0.0939	0.0813	3B	0.0849	0.0939	0.0958	0.0982	0.1120
4–48 UNF	2A	0.0007	0.1113	0.1068	—	0.0978	0.0954	0.0857	2B	0.0894	0.0968	0.0985	0.1016	0.1120
	3A	0.0000	0.1120	0.1075	—	0.0985	0.0967	0.0864	3B	0.0894	0.0968	0.0985	0.1008	0.1120
5–40 UNC	2A	0.0008	0.1242	0.1191	—	0.1080	0.1054	0.0935	2B	0.0979	0.1062	0.1088	0.1121	0.1250
	3A	0.0000	0.1250	0.1199	—	0.1088	0.1069	0.0943	3B	0.0979	0.1062	0.1088	0.1113	0.1250
5–44 UNF	2A	0.0007	0.1243	0.1195	—	0.1095	0.1070	0.0964	2B	0.1004	0.1079	0.1102	0.1134	0.1250
	3A	0.0000	0.1250	0.1202	—	0.1102	0.1083	0.0971	3B	0.1004	0.1079	0.1102	0.1126	0.1250
6–32 UNC	2A	0.0008	0.1372	0.1312	—	0.1169	0.1141	0.0989	2B	0.104	0.114	0.1177	0.1214	0.1380
	3A	0.0000	0.1380	0.1320	—	0.1177	0.1156	0.0997	3B	0.1040	0.1140	0.1177	0.1204	0.1380
6–40 UNF	2A	0.0008	0.1372	0.1321	—	0.1210	0.1184	0.1065	2B	0.111	0.119	0.1218	0.1252	0.1380
	3A	0.0000	0.1380	0.1339	—	0.1218	0.1198	0.1073	3B	0.1110	0.1186	0.1218	0.1243	0.1380
8–32 UNC	2A	0.0009	0.1631	0.1571	—	0.1428	0.1399	0.1248	2B	0.130	0.139	0.1437	0.1475	0.1640
	3A	0.0000	0.1640	0.1580	—	0.1437	0.1415	0.1257	3B	0.1300	0.1389	0.1437	0.1465	0.1640

Designation	Class								Class					
8-36 UNF	2A	0.0008	0.1633	0.1577	—	0.1453	0.1424	0.1291	2B	0.134	0.142	0.1460	0.1496	0.1640
	3A	0.0000	0.1640	0.1585	—	0.1460	0.1439	0.1299	3B	0.1340	0.1416	0.1460	0.1487	0.1640
10-24 UNC	2A	0.0010	0.1890	0.1818	—	0.1619	0.1586	0.1379	2B	0.145	0.156	0.1629	0.1672	0.1900
	3A	0.0000	0.1900	0.1828	—	0.1629	0.1604	0.1389	3B	0.1450	0.1555	0.1629	0.1661	0.1900
10-28 UNS	2A	0.0010	0.1890	0.1825	—	0.1658	0.1625	0.1452	2B	0.151	0.160	0.1668	0.1711	0.1900
10-32 UNF	2A	0.0009	0.1891	0.1831	—	0.1688	0.1658	0.1508	2B	0.156	0.164	0.1697	0.1736	0.1900
	3A	0.0000	0.1900	0.1840	—	0.1697	0.1674	0.1517	3B	0.1560	0.1641	0.1697	0.1726	0.1900
10-36 UNS	2A	0.0009	0.1891	0.1836	—	0.1711	0.1681	0.1550	2B	0.160	0.166	0.1720	0.1759	0.1900
10-40 UNS	2A	0.0009	0.1891	0.1840	—	0.1729	0.1700	0.1584	2B	0.163	0.169	0.1738	0.1775	0.1900
10-48 UNS	2A	0.0008	0.1892	0.1847	—	0.1757	0.1731	0.1636	2B	0.167	0.172	0.1765	0.1799	0.1900
10-56 UNS	2A	0.0007	0.1893	0.1852	—	0.1777	0.1752	0.1674	2B	0.171	0.175	0.1784	0.1816	0.1900
12-24 UNC	2A	0.0010	0.2150	0.2078	—	0.1879	0.1845	0.1639	2B	0.171	0.181	0.1889	0.1933	0.2160
	3A	0.0000	0.2160	0.2088	—	0.1889	0.1863	0.1649	3B	0.1710	0.1807	0.1889	0.1922	0.2160
12-28 UNF	2A	0.0010	0.2150	0.2085	—	0.1918	0.1886	0.1712	2B	0.177	0.186	0.1928	0.1970	0.2160
	3A	0.0000	0.2160	0.2095	—	0.1928	0.1904	0.1722	3B	0.1770	0.1857	0.1928	0.1959	0.2160
12-32 UNEF	2A	0.0009	0.2151	0.2091	—	0.1948	0.1917	0.1768	2B	0.182	0.190	0.1957	0.1998	0.2160
	3A	0.0000	0.2160	0.2100	—	0.1957	0.1933	0.1777	3B	0.1820	0.1895	0.1957	0.1988	0.2160
12-36 UNS	2A	0.0009	0.2151	0.2096	—	0.1971	0.1941	0.1810	2B	0.186	0.192	0.1980	0.2019	0.2160
12-40 UNS	2A	0.0009	0.2151	0.2100	—	0.1989	0.1960	0.1844	2B	0.189	0.195	0.1998	0.2035	0.2160
12-48 UNS	2A	0.0008	0.2152	0.2107	—	0.2017	0.1991	0.1896	2B	0.193	0.198	0.2025	0.2059	0.2160
12-56 UNS	2A	0.0007	0.2153	0.2112	—	0.2037	0.2012	0.1934	2B	0.197	0.201	0.2044	0.2076	0.2160
¼-20 UNC	1A	0.0011	0.2489	0.2367	—	0.2164	0.2108	0.1876	1B	0.196	0.207	0.2175	0.2248	0.2500
	2A	0.0011	0.2489	0.2408	0.2367	0.2164	0.2127	0.1876	2B	0.196	0.207	0.2175	0.2223	0.2500
	3A	0.0000	0.2500	0.2419	—	0.2175	0.2147	0.1887	3B	0.1960	0.2067	0.2175	0.2211	0.2500
¼-24 UNS	2A	0.0011	0.2489	0.2417	—	0.2218	0.2181	0.1978	2B	0.205	0.215	0.2229	0.2277	0.2500
¼-27 UNS	2A	0.0010	0.2490	0.2423	—	0.2249	0.2214	0.2036	2B	0.210	0.219	0.2259	0.2304	0.2500
¼-28 UNF	1A	0.0010	0.2490	0.2392	—	0.2258	0.2208	0.2052	1B	0.211	0.220	0.2268	0.2333	0.2500
	2A	0.0010	0.2490	0.2425	—	0.2258	0.2225	0.2052	2B	0.211	0.220	0.2268	0.2311	0.2500
	3A	0.0000	0.2500	0.2435	—	0.2268	0.2243	0.2062	3B	0.2110	0.2190	0.2268	0.2300	0.2500
¼-32 UNEF	2A	0.0010	0.2490	0.2430	—	0.2287	0.2255	0.2107	2B	0.216	0.224	0.2297	0.2339	0.2500
	3A	0.0000	0.2500	0.2440	—	0.2297	0.2273	0.2117	3B	0.2160	0.2229	0.2297	0.2338	0.2500

† Use UNS threads only if Standard Series do not meet requirements. See footnotes b, d, e, and f at end of table.

Table 5. (Continued). Standard Series and Selected Combinations† — Unified Screw Threads

Nominal Size, Threads per Inch, and Series Designation^f		External^c							Internal^t					
				Major Diameter			Pitch Diameter		Minor Diam-eter		Minor Diameter^e		Pitch Diameter		Major Diameter
	Class	Allow-ance	Max^b	Min	Min^d	Max^b	Min		Class	Min	Max	Min	Max	Min
1/4-36 UNS	2A	0.0009	0.2491	0.2436	—	0.2311	0.2280	0.2150	2B	0.220	0.226	0.2330	0.2360	0.2500
1/4-40 UNS	2A	0.0009	0.2491	0.2440	—	0.2329	0.2290	0.2184	2B	0.223	0.229	0.2338	0.2376	0.2500
1/4-48 UNS	2A	0.0008	0.2492	0.2447	—	0.2357	0.2330	0.2236	2B	0.227	0.232	0.2365	0.2401	0.2500
1/4-56 UNS	2A	0.0008	0.2492	0.2451	—	0.2376	0.2350	0.2273	2B	0.231	0.235	0.2384	0.2417	0.2500
5/16-18 UNC	1A	0.0012	0.3113	0.2982	—	0.2752	0.2691	0.2431	1B	0.252	0.265	0.2764	0.2843	0.3125
	2A	0.0012	0.3113	0.3026	0.2982	0.2752	0.2712	0.2431	2B	0.252	0.265	0.2764	0.2817	0.3125
	3A	0.0000	0.3125	0.3038	—	0.2764	0.2734	0.2443	3B	0.2520	0.2630	0.2764	0.2803	0.3125
5/16-20 UN	2A	0.0012	0.3113	0.3032	—	0.2788	0.2748	0.2500	2B	0.258	0.270	0.2800	0.2852	0.3125
	3A	0.0000	0.3125	0.3044	—	0.2800	0.2770	0.2512	3B	0.2580	0.2662	0.2800	0.2839	0.3125
5/16-24 UNF	1A	0.0011	0.3114	0.3006	—	0.2843	0.2788	0.2603	1B	0.267	0.277	0.2854	0.2925	0.3125
	2A	0.0011	0.3114	0.3042	—	0.2843	0.2806	0.2603	2B	0.267	0.277	0.2854	0.2902	0.3125
	3A	0.0000	0.3125	0.3053	—	0.2854	0.2827	0.2614	3B	0.2670	0.2754	0.2854	0.2890	0.3125
5/16-27 UNS	2A	0.0010	0.3115	0.3048	—	0.2874	0.2839	0.2661	2B	0.272	0.281	0.2884	0.2929	0.3125
5/16-28 UN	2A	0.0010	0.3115	0.3050	—	0.2883	0.2849	0.2677	2B	0.274	0.282	0.2893	0.2937	0.3125
	3A	0.0000	0.3125	0.3060	—	0.2893	0.2867	0.2687	3B	0.2740	0.2801	0.2893	0.2926	0.3125
5/16-32 UNEF	2A	0.0010	0.3115	0.3055	—	0.2912	0.2880	0.2732	2B	0.279	0.286	0.2922	0.2964	0.3125
	3A	0.0000	0.3125	0.3065	—	0.2922	0.2898	0.2742	3B	0.2790	0.2847	0.2922	0.2953	0.3125
5/16-36 UNS	2A	0.0009	0.3116	0.3061	—	0.2936	0.2905	0.2775	2B	0.282	0.289	0.2945	0.2985	0.3125
5/16-40 UNS	2A	0.0009	0.3116	0.3065	—	0.2954	0.2925	0.2809	2B	0.285	0.291	0.2963	0.3001	0.3125
5/16-48 UNS	2A	0.0008	0.3117	0.3072	—	0.2982	0.2955	0.2861	2B	0.290	0.295	0.2990	0.3026	0.3125
3/8-16 UNC	1A	0.0013	0.3737	0.3595	—	0.3331	0.3266	0.2970	1B	0.307	0.321	0.3344	0.3429	0.3750
	2A	0.0013	0.3737	0.3643	0.3595	0.3331	0.3287	0.2970	2B	0.307	0.321	0.3344	0.3401	0.3750
	3A	0.0000	0.3750	0.3656	—	0.3344	0.3311	0.2983	3B	0.3070	0.3182	0.3344	0.3387	0.3750
3/8-18 UNS	2A	0.0013	0.3737	0.3650	—	0.3376	0.3333	0.3055	2B	0.315	0.328	0.3389	0.3445	0.3750
3/8-20 UN	2A	0.0012	0.3738	0.3657	—	0.3413	0.3372	0.3125	2B	0.321	0.332	0.3425	0.3479	0.3750
	3A	0.0000	0.3750	0.3669	—	0.3425	0.3394	0.3137	3B	0.3210	0.3287	0.3425	0.3465	0.3750
3/8-24 UNF	1A	0.0011	0.3739	0.3631	—	0.3468	0.3411	0.3228	1B	0.330	0.340	0.3479	0.3553	0.3750
	2A	0.0011	0.3739	0.3667	—	0.3468	0.3430	0.3228	2B	0.330	0.340	0.3479	0.3538	0.3750

3/8–24 UNF	3A	0.0000	0.3750	0.3678	—	0.3479	0.3450	0.3239	3B	0.3300	0.3372	0.3479	0.3516	0.3750
3/8–27 UNS	2A	0.0011	0.3739	0.3672	—	0.3498	0.3462	0.3285	2B	0.335	0.344	0.3509	0.3556	0.3750
3/8–28 UN	2A	0.0011	0.3739	0.3674	—	0.3507	0.3471	0.3301	2B	0.336	0.345	0.3518	0.3564	0.3750
	3A	0.0000	0.3750	0.3685	—	0.3518	0.3491	0.3312	3B	0.3360	0.3426	0.3518	0.3553	0.3750
3/8–32 UNEF	2A	0.0010	0.3740	0.3680	—	0.3537	0.3503	0.3357	2B	0.341	0.349	0.3547	0.3591	0.3750
	3A	0.0000	0.3750	0.3690	—	0.3547	0.3522	0.3367	3B	0.3410	0.3469	0.3547	0.3580	0.3750
3/8–36 UNS	2A	0.0010	0.3740	0.3685	—	0.3560	0.3528	0.3399	2B	0.345	0.352	0.3570	0.3612	0.3750
3/8–40 UNS	2A	0.0009	0.3741	0.3690	—	0.3579	0.3548	0.3434	2B	0.348	0.354	0.3588	0.3628	0.3750
0.390–27 UNS	2A	0.0011	0.3889	0.3822	—	0.3648	0.3612	0.3435	2B	0.350	0.359	0.3659	0.3706	0.3900
7/16–14 UNC	1A	0.0014	0.4361	0.4206	—	0.3897	0.3826	0.3485	1B	0.360	0.376	0.3911	0.4003	0.4375
	2A	0.0014	0.4361	0.4258	0.4206	0.3897	0.3850	0.3485	2B	0.360	0.376	0.3911	0.3972	0.4375
	3A	0.0000	0.4375	0.4272	—	0.3911	0.3876	0.3499	3B	0.3600	0.3717	0.3911	0.3957	0.4375
7/16–16 UN	2A	0.0014	0.4361	0.4267	—	0.3955	0.3909	0.3594	2B	0.370	0.384	0.3969	0.4028	0.4375
	3A	0.0000	0.4375	0.4281	—	0.3969	0.3935	0.3608	3B	0.3700	0.3783	0.3969	0.4009	0.4375
7/16–18 UNS	2A	0.0013	0.4362	0.4275	—	0.4001	0.3958	0.3680	2B	0.377	0.390	0.4014	0.4070	0.4375
7/16–20 UNF	1A	0.0013	0.4362	0.4240	—	0.4037	0.3975	0.3749	1B	0.383	0.395	0.4050	0.4131	0.4375
	2A	0.0013	0.4362	0.4281	—	0.4037	0.3995	0.3749	2B	0.383	0.395	0.4050	0.4104	0.4375
	3A	0.0000	0.4375	0.4294	—	0.4050	0.4019	0.3762	3B	0.3830	0.3916	0.4050	0.4091	0.4375
7/16–24 UNS	2A	0.0011	0.4364	0.4292	—	0.4093	0.4055	0.3853	2B	0.392	0.402	0.4104	0.4153	0.4375
7/16–27 UNS	2A	0.0011	0.4364	0.4297	—	0.4123	0.4087	0.3910	2B	0.397	0.406	0.4134	0.4181	0.4375
7/16–28 UNEF	2A	0.0011	0.4364	0.4299	—	0.4132	0.4096	0.3926	2B	0.399	0.407	0.4143	0.4189	0.4375
	3A	0.0000	0.4375	0.4310	—	0.4143	0.4116	0.3937	3B	0.3990	0.4051	0.4143	0.4178	0.4375
7/16–32 UN	2A	0.0010	0.4365	0.4305	—	0.4162	0.4128	0.3982	2B	0.404	0.411	0.4172	0.4216	0.4375
	3A	0.0000	0.4375	0.4315	—	0.4172	0.4147	0.3992	3B	0.4040	0.4094	0.4172	0.4205	0.4375
1/2–12 UNS	2A	0.0016	0.4984	0.4870	—	0.4443	0.4389	0.3962	2B	0.410	0.428	0.4459	0.4529	0.5000
	3A	0.0000	0.5000	0.4886	—	0.4459	0.4419	0.3978	3B	0.4100	0.4223	0.4459	0.4511	0.5000
1/2–13 UNC	1A	0.0015	0.4985	0.4822	—	0.4485	0.4411	0.4041	1B	0.417	0.434	0.4500	0.4597	0.5000
	2A	0.0015	0.4985	0.4876	0.4822	0.4485	0.4435	0.4041	2B	0.417	0.434	0.4500	0.4565	0.5000
	3A	0.0000	0.5000	0.4891	—	0.4500	0.4463	0.4056	3B	0.4170	0.4284	0.4500	0.4548	0.5000
1/2–14 UNS	2A	0.0015	0.4985	0.4882	—	0.4521	0.4471	0.4109	2B	0.423	0.438	0.4536	0.4601	0.5000

† Use UNS threads only if Standard Series do not meet requirements. See footnotes *b*, *d*, *e*, and *f* at end of table.

Table 5. (Continued). Standard Series and Selected Combinations† — Unified Screw Threads

Nominal Size, Threads per Inch, and Series Designation^f	Class	Allowance	Major Diameter Max^b	Major Diameter Min	Major Diameter Min^d	Pitch Diameter Max^b	Pitch Diameter Min	Minor Diameter	Class	Minor Diameter^e Min	Minor Diameter^e Max	Pitch Diameter Min	Pitch Diameter Max	Major Diameter Min
½–16 UN	2A	0.0014	0.4986	0.4892	—	0.4580	0.4533	0.4219	2B	0.432	0.446	0.4594	0.4655	0.5000
	3A	0.0000	0.5000	0.4906	—	0.4594	0.4559	0.4233	3B	0.4320	0.4408	0.4594	0.4640	0.5000
½–18 UNS	2A	0.0013	0.4987	0.4900	—	0.4626	0.4582	0.4305	2B	0.440	0.453	0.4639	0.4697	0.5000
½–20 UNF	1A	0.0013	0.4987	0.4865	—	0.4662	0.4598	0.4374	1B	0.446	0.457	0.4675	0.4759	0.5000
	2A	0.0013	0.4987	0.4906	—	0.4662	0.4619	0.4374	2B	0.446	0.457	0.4675	0.4731	0.5000
	3A	0.0000	0.5000	0.4919	—	0.4675	0.4643	0.4387	3B	0.4460	0.4537	0.4675	0.4717	0.5000
½–24 UNS	2A	0.0012	0.4988	0.4916	—	0.4717	0.4678	0.4477	2B	0.455	0.465	0.4729	0.4780	0.5000
½–27 UNS	2A	0.0011	0.4989	0.4922	—	0.4748	0.4711	0.4535	2B	0.460	0.469	0.4759	0.4807	0.5000
½–28 UNEF	2A	0.0011	0.4989	0.4924	—	0.4757	0.4720	0.4551	2B	0.461	0.470	0.4768	0.4816	0.5000
	3A	0.0000	0.5000	0.4935	—	0.4768	0.4740	0.4562	3B	0.4610	0.4676	0.4768	0.4804	0.5000
½–32 UN	2A	0.0010	0.4990	0.4930	—	0.4787	0.4752	0.4607	2B	0.466	0.474	0.4797	0.4842	0.5000
	3A	0.0000	0.5000	0.4940	—	0.4797	0.4771	0.4617	3B	0.4660	0.4719	0.4797	0.4831	0.5000
⁹⁄₁₆–12 UNC	1A	0.0016	0.5609	0.5437	—	0.5068	0.4990	0.4587	1B	0.472	0.490	0.5084	0.5186	0.5625
	2A	0.0016	0.5609	0.5495	0.5437	0.5068	0.5016	0.4587	2B	0.472	0.490	0.5084	0.5152	0.5625
	3A	0.0000	0.5625	0.5511	—	0.5084	0.5045	0.4603	3B	0.4720	0.4843	0.5084	0.5135	0.5625
⁹⁄₁₆–14 UNS	2A	0.0015	0.5610	0.5507	—	0.5146	0.5096	0.4734	2B	0.485	0.501	0.5161	0.5226	0.5625
⁹⁄₁₆–16 UN	2A	0.0014	0.5611	0.5517	—	0.5205	0.5158	0.4844	2B	0.495	0.509	0.5219	0.5280	0.5625
	3A	0.0000	0.5625	0.5531	—	0.5219	0.5184	0.4858	3B	0.4950	0.5033	0.5219	0.5265	0.5625
⁹⁄₁₆–18 UNF	1A	0.0014	0.5611	0.5480	—	0.5250	0.5182	0.4929	1B	0.502	0.515	0.5264	0.5353	0.5625
	2A	0.0014	0.5611	0.5524	—	0.5250	0.5205	0.4929	2B	0.502	0.515	0.5264	0.5323	0.5625
	3A	0.0000	0.5625	0.5538	—	0.5264	0.5230	0.4943	3B	0.5020	0.5106	0.5264	0.5308	0.5625
⁹⁄₁₆–20 UN	2A	0.0013	0.5612	0.5531	—	0.5287	0.5245	0.4999	2B	0.508	0.520	0.5300	0.5355	0.5625
	3A	0.0000	0.5625	0.5544	—	0.5300	0.5268	0.5012	3B	0.5080	0.5162	0.5300	0.5341	0.5625
⁹⁄₁₆–24 UNEF	2A	0.0012	0.5613	0.5541	—	0.5342	0.5303	0.5102	2B	0.517	0.527	0.5354	0.5405	0.5625
	3A	0.0000	0.5625	0.5553	—	0.5354	0.5325	0.5114	3B	0.5170	0.5244	0.5354	0.5392	0.5625
⁹⁄₁₆–27 UNS	2A	0.0011	0.5614	0.5547	—	0.5373	0.5336	0.5160	2B	0.522	0.531	0.5384	0.5432	0.5625
⁹⁄₁₆–28 UN	2A	0.0011	0.5614	0.5549	—	0.5382	0.5345	0.5176	2B	0.524	0.532	0.5393	0.5441	0.5625
	3A	0.0000	0.5625	0.5560	—	0.5393	0.5365	0.5187	3B	0.5240	0.5301	0.5393	0.5429	0.5625
⁹⁄₁₆–32 UN	2A	0.0010	0.5615	0.5555	—	0.5412	0.5377	0.5232	2B	0.529	0.536	0.5422	0.5467	0.5625
	3A	0.0000	0.5625	0.5565	—	0.5422	0.5396	0.5242	3B	0.5290	0.5344	0.5422	0.5456	0.5625

Designation	Class												
5⁄8–11 UNC	1A	0.0016	0.6234	0.6052	—	0.5644	0.5561	0.5119	0.527	0.546	0.5660	0.5767	0.6250
	2A	0.0016	0.6234	0.6113	0.6052	0.5644	0.5589	0.5119	0.527	0.546	0.5660	0.5732	0.6250
	3A	0.0000	0.6250	0.6129	—	0.5660	0.5619	0.5135	0.5270	0.5391	0.5660	0.5714	0.6250
5⁄8–12 UN	2A	0.0016	0.6234	0.6120	—	0.5693	0.5639	0.5212	0.535	0.553	0.5709	0.5780	0.6250
	3A	0.0000	0.6250	0.6136	—	0.5709	0.5668	0.5228	0.5350	0.5463	0.5709	0.5762	0.6250
5⁄8–14 UNS	2A	0.0015	0.6235	0.6132	—	0.5771	0.5720	0.5359	0.548	0.564	0.5786	0.5852	0.6250
5⁄8–16 UN	2A	0.0014	0.6236	0.6142	—	0.5830	0.5782	0.5469	0.557	0.571	0.5844	0.5906	0.6250
	3A	0.0000	0.6250	0.6156	—	0.5844	0.5808	0.5483	0.5570	0.5658	0.5844	0.5890	0.6250
5⁄8–18 UNF	1A	0.0014	0.6236	0.6105	—	0.5875	0.5805	0.5554	0.565	0.578	0.5889	0.5980	0.6250
	2A	0.0014	0.6236	0.6149	—	0.5875	0.5828	0.5554	0.565	0.578	0.5889	0.5949	0.6250
	3A	0.0000	0.6250	0.6163	—	0.5889	0.5854	0.5568	0.5650	0.5730	0.5889	0.5934	0.6250
5⁄8–20 UN	2A	0.0013	0.6237	0.6156	—	0.5912	0.5869	0.5624	0.571	0.582	0.5925	0.5981	0.6250
	3A	0.0000	0.6250	0.6169	—	0.5925	0.5893	0.5637	0.5710	0.5787	0.5925	0.5997	0.6250
5⁄8–24 UNEF	2A	0.0012	0.6238	0.6166	—	0.5967	0.5927	0.5727	0.580	0.590	0.5979	0.6031	0.6250
	3A	0.0000	0.6250	0.6178	—	0.5979	0.5949	0.5739	0.5800	0.5869	0.5979	0.6018	0.6250
5⁄8–27 UNS	2A	0.0011	0.6239	0.6172	—	0.5998	0.5960	0.5785	0.585	0.594	0.6009	0.6059	0.6250
5⁄8–28 UN	2A	0.0011	0.6239	0.6174	—	0.6007	0.5969	0.5801	0.586	0.595	0.6018	0.6067	0.6250
	3A	0.0000	0.6250	0.6185	—	0.6018	0.5990	0.5812	0.5860	0.5936	0.6018	0.6055	0.6250
5⁄8–32 UN	2A	0.0011	0.6239	0.6179	—	0.6036	0.6000	0.5856	0.591	0.599	0.6047	0.6093	0.6250
	3A	0.0000	0.6250	0.6190	—	0.6047	0.6020	0.5867	0.5910	0.5969	0.6047	0.6082	0.6250
11⁄16–12 UN	2A	0.0016	0.6859	0.6745	—	0.6318	0.6264	0.5837	0.597	0.615	0.6334	0.6405	0.6875
	3A	0.0000	0.6875	0.6761	—	0.6334	0.6293	0.5853	0.5970	0.6085	0.6334	0.6387	0.6875
11⁄16–16 UN	2A	0.0014	0.6861	0.6767	—	0.6455	0.6407	0.6094	0.620	0.634	0.6469	0.6531	0.6875
	3A	0.0000	0.6875	0.6781	—	0.6469	0.6433	0.6108	0.6200	0.6283	0.6469	0.6515	0.6875
11⁄16–20 UN	2A	0.0013	0.6862	0.6781	—	0.6537	0.6494	0.6249	0.633	0.645	0.6550	0.6606	0.6875
	3A	0.0000	0.6875	0.6794	—	0.6550	0.6518	0.6262	0.6330	0.6412	0.6550	0.6592	0.6875
11⁄16–24 UNEF	2A	0.0012	0.6863	0.6791	—	0.6592	0.6552	0.6352	0.642	0.652	0.6604	0.6656	0.6875
	3A	0.0000	0.6875	0.6803	—	0.6604	0.6574	0.6364	0.6420	0.6494	0.6604	0.6643	0.6875
11⁄16–28 UN	2A	0.0011	0.6864	0.6799	—	0.6632	0.6594	0.6426	0.649	0.657	0.6643	0.6692	0.6875
	3A	0.0000	0.6875	0.6810	—	0.6643	0.6615	0.6437	0.6490	0.6551	0.6643	0.6680	0.6875
11⁄16–32 UN	2A	0.0011	0.6864	0.6804	—	0.6661	0.6625	0.6481	0.654	0.661	0.6672	0.6718	0.6875
	3A	0.0000	0.6875	0.6815	—	0.6672	0.6645	0.6492	0.6540	0.6594	0.6672	0.6707	0.6875

† Use UNS threads only if Standard Series do not meet requirements. See footnotes *b*, *d*, *e*, and *f* at end of table.

Table 5. *(Continued).* **Standard Series and Selected Combinations† — Unified Screw Threads**

Nominal Size, Threads per Inch, and Series Designation^f	Class	Allowance	Major Diameter Max^b	Major Diameter Min	Major Diameter Min^d	Pitch Diameter Max^b	Pitch Diameter Min	Minor Diameter^d	Class	Minor Diameter^c Min	Minor Diameter^c Max	Pitch Diameter Min	Pitch Diameter Max	Major Diameter Min
3/4–10 UNC	1A	0.0018	0.7482	0.7288	—	0.6832	0.6744	0.6255	1B	0.642	0.663	0.6850	0.6965	0.7500
	2A	0.0018	0.7482	0.7353	0.7288	0.6832	0.6773	0.6255	2B	0.642	0.663	0.6850	0.6927	0.7500
	3A	0.0000	0.7500	0.7371	—	0.6850	0.6806	0.6273	3B	0.6420	0.6545	0.6850	0.6907	0.7500
3/4–12 UN	2A	0.0017	0.7483	0.7369	—	0.6942	0.6887	0.6461	2B	0.660	0.678	0.6959	0.7031	0.7500
	3A	0.0000	0.7500	0.7386	—	0.6959	0.6918	0.6478	3B	0.6600	0.6707	0.6959	0.7013	0.7500
3/4–14 UNS	2A	0.0015	0.7485	0.7382	—	0.7021	0.6970	0.6609	2B	0.673	0.688	0.7036	0.7103	0.7500
3/4–16 UNF	1A	0.0015	0.7485	0.7343	—	0.7079	0.7004	0.6718	1B	0.682	0.696	0.7094	0.7192	0.7500
	2A	0.0015	0.7485	0.7391	—	0.7079	0.7029	0.6718	2B	0.682	0.696	0.7094	0.7159	0.7500
	3A	0.0000	0.7500	0.7406	—	0.7094	0.7056	0.6733	3B	0.6820	0.6908	0.7094	0.7143	0.7500
3/4–18 UNS	2A	0.0014	0.7486	0.7399	—	0.7125	0.7079	0.6804	2B	0.690	0.703	0.7139	0.7199	0.7500
3/4–20 UNEF	2A	0.0013	0.7487	0.7406	—	0.7162	0.7118	0.6874	2B	0.696	0.707	0.7175	0.7232	0.7500
	3A	0.0000	0.7500	0.7419	—	0.7175	0.7142	0.6887	3B	0.6960	0.7037	0.7175	0.7218	0.7500
3/4–24 UNS	2A	0.0012	0.7488	0.7416	—	0.7217	0.7176	0.6977	2B	0.705	0.715	0.7229	0.7282	0.7500
3/4–27 UNS	2A	0.0012	0.7488	0.7421	—	0.7247	0.7208	0.7034	2B	0.710	0.719	0.7259	0.7310	0.7500
3/4–28 UN	2A	0.0012	0.7488	0.7423	—	0.7256	0.7218	0.7050	2B	0.711	0.720	0.7268	0.7318	0.7500
	3A	0.0000	0.7500	0.7435	—	0.7268	0.7239	0.7062	3B	0.7110	0.7176	0.7268	0.7305	0.7500
3/4–32 UN	2A	0.0011	0.7489	0.7429	—	0.7286	0.7250	0.7106	2B	0.716	0.724	0.7297	0.7344	0.7500
	3A	0.0000	0.7500	0.7440	—	0.7297	0.7270	0.7117	3B	0.7160	0.7219	0.7297	0.7333	0.7500
13/16–12 UN	2A	0.0017	0.8108	0.7994	—	0.7567	0.7512	0.7086	2B	0.722	0.740	0.7584	0.7656	0.8125
	3A	0.0000	0.8125	0.8011	—	0.7584	0.7543	0.7103	3B	0.7220	0.7329	0.7584	0.7638	0.8125
13/16–16 UN	2A	0.0015	0.8110	0.8016	—	0.7704	0.7655	0.7343	2B	0.745	0.759	0.7719	0.7782	0.8125
	3A	0.0000	0.8125	0.8031	—	0.7719	0.7683	0.7358	3B	0.7450	0.7533	0.7719	0.7766	0.8125
13/16–20 UNEF	2A	0.0013	0.8112	0.8031	—	0.7787	0.7743	0.7498	2B	0.758	0.770	0.7800	0.7857	0.8125
	3A	0.0000	0.8125	0.8044	—	0.7800	0.7767	0.7512	3B	0.7580	0.7662	0.7800	0.7843	0.8125
13/16–28 UN	2A	0.0012	0.8113	0.8048	—	0.7881	0.7843	0.7675	2B	0.774	0.782	0.7893	0.7943	0.8125
	3A	0.0000	0.8125	0.8060	—	0.7893	0.7864	0.7687	3B	0.7740	0.7801	0.7893	0.7930	0.8125
13/16–32 UN	2A	0.0011	0.8114	0.8054	—	0.7911	0.7875	0.7731	2B	0.779	0.786	0.7922	0.7969	0.8125
	3A	0.0000	0.8125	0.8065	—	0.7922	0.7895	0.7742	3B	0.7790	0.7844	0.7922	0.7958	0.8125

Size	Class	Allowance							Class					
7/8–9 UNC	1A	0.0019	0.8731	0.8523	—	0.8009	0.7914	0.7368	1B	0.755	0.778	0.8028	0.8151	0.8750
	2A	0.0019	0.8731	0.8592	0.8523	0.8009	0.7946	0.7368	2B	0.755	0.778	0.8028	0.8110	0.8750
	3A	0.0000	0.8750	0.8611	—	0.8028	0.7981	0.7387	3B	0.7550	0.7681	0.8028	0.8089	0.8750
7/8–10 UNS	2A	0.0018	0.8732	0.8603	—	0.8082	0.8022	0.7505	2B	0.767	0.788	0.8100	0.8178	0.8750
7/8–12 UN	2A	0.0017	0.8733	0.8619	—	0.8192	0.8137	0.7711	2B	0.785	0.8032	0.8209	0.8281	0.8750
	3A	0.0000	0.8750	0.8636	—	0.8209	0.8168	0.7728	3B	0.7850	0.795	0.8209	0.8263	0.8750
7/8–14 UNF	1A	0.0016	0.8734	0.8579	—	0.8270	0.8189	0.7858	1B	0.798	0.814	0.8286	0.8392	0.8750
	2A	0.0016	0.8734	0.8631	—	0.8270	0.8216	0.7858	2B	0.798	0.814	0.8286	0.8356	0.8750
	3A	0.0000	0.8750	0.8647	—	0.8286	0.8245	0.7874	3B	0.7980	0.8068	0.8286	0.8339	0.8750
7/8–16 UN	2A	0.0015	0.8735	0.8641	—	0.8329	0.8280	0.7968	2B	0.807	0.821	0.8344	0.8407	0.8750
	3A	0.0000	0.8750	0.8656	—	0.8344	0.8308	0.7983	3B	0.8070	0.8158	0.8344	0.8391	0.8750
7/8–18 UNS	2A	0.0014	0.8736	0.8649	—	0.8375	0.8329	0.8054	2B	0.815	0.828	0.8389	0.8449	0.8750
7/8–20 UNEF	2A	0.0013	0.8737	0.8656	—	0.8412	0.8368	0.8124	2B	0.821	0.832	0.8425	0.8482	0.8750
	3A	0.0000	0.8750	0.8669	—	0.8425	0.8392	0.8137	3B	0.8210	0.8287	0.8425	0.8468	0.8750
7/8–24 UNS	2A	0.0012	0.8738	0.8666	—	0.8467	0.8426	0.8227	2B	0.830	0.840	0.8479	0.8532	0.8750
7/8–27 UNS	2A	0.0012	0.8738	0.8671	—	0.8497	0.8458	0.8284	2B	0.835	0.844	0.8509	0.8560	0.8750
7/8–28 UN	2A	0.0012	0.8738	0.8673	—	0.8506	0.8468	0.8300	2B	0.836	0.845	0.8518	0.8568	0.8750
	3A	0.0000	0.8750	0.8685	—	0.8518	0.8489	0.8312	3B	0.8360	0.8426	0.8518	0.8555	0.8750
7/8–32 UN	2A	0.0011	0.8739	0.8679	—	0.8536	0.8500	0.8356	2B	0.841	0.849	0.8547	0.8594	0.8750
	3A	0.0000	0.8750	0.8690	—	0.8547	0.8520	0.8367	3B	0.8410	0.8469	0.8547	0.8583	0.8750
15/16–12 UN	2A	0.0017	0.9358	0.9244	—	0.8817	0.8760	0.8336	2B	0.847	0.865	0.8834	0.8908	0.9375
	3A	0.0000	0.9375	0.9261	—	0.8834	0.8793	0.8353	3B	0.8470	0.8575	0.8834	0.8889	0.9375
15/16–16 UN	2A	0.0015	0.9360	0.9266	—	0.8954	0.8904	0.8593	2B	0.870	0.884	0.8969	0.9034	0.9375
	3A	0.0000	0.9375	0.9281	—	0.8969	0.8932	0.8608	3B	0.8700	0.8783	0.8969	0.9018	0.9375
15/16–20 UNEF	2A	0.0014	0.9361	0.9280	—	0.9036	0.8991	0.8748	2B	0.883	0.895	0.9050	0.9109	0.9375
	3A	0.0000	0.9375	0.9294	—	0.9050	0.9016	0.8762	3B	0.8830	0.8912	0.9050	0.9094	0.9375
15/16–28 UN	2A	0.0012	0.9363	0.9298	—	0.9131	0.9091	0.8925	2B	0.899	0.907	0.9143	0.9195	0.9375
	3A	0.0000	0.9375	0.9310	—	0.9143	0.9113	0.8937	3B	0.8990	0.9061	0.9143	0.9182	0.9375
15/16–32 UN	2A	0.0011	0.9364	0.9304	—	0.9161	0.9123	0.8981	2B	0.904	0.911	0.9172	0.9221	0.9375
	3A	0.0000	0.9375	0.9315	—	0.9172	0.9144	0.8992	3B	0.9040	0.9994	0.9172	0.9209	0.9375
1–8 UNC	1A	0.0020	0.9980	0.9755	—	0.9168	0.9067	0.8446	1B	0.865	0.890	0.9188	0.9320	1.0000
	2A	0.0020	0.9980	0.9830	0.9755	0.9168	0.9100	0.8446	2B	0.865	0.890	0.9188	0.9276	1.0000
	3A	0.0000	1.0000	0.9850	—	0.9188	0.9137	0.8466	3B	0.8650	0.8797	0.9188	0.9254	1.0000

† Use UNS threads only if Standard Series do not meet requirements. See footnotes *b*, *d*, *e*, and *f* at end of table.

Table 5. (Continued). Standard Series and Selected Combinations† — Unified Screw Threads

Nominal Size, Threads per Inch, and Series Designation[f]	Class	Allow-ance	Major Diameter Max[b]	Major Diameter Min	Major Diameter Min[d]	Pitch Diameter Max[b]	Pitch Diameter Min	Minor Diameter	Class	Minor Diameter Min	Minor Diameter Max	Pitch Diameter Min	Pitch Diameter Max	Major Diameter Min
1-10 UNS	2A	0.0018	0.9982	0.9853	—	0.9332	0.9270	0.8755	2B	0.892	0.913	0.9350	0.9430	1.0000
1-12 UNF	1A	0.0018	0.9982	0.9810	—	0.9441	0.9353	0.8960	1B	0.910	0.928	0.9459	0.9573	1.0000
	2A	0.0018	0.9982	0.9868	—	0.9441	0.9382	0.8960	2B	0.910	0.928	0.9459	0.9535	1.0000
	3A	0.0000	1.0000	0.9886	—	0.9459	0.9415	0.8978	3B	0.9100	0.9198	0.9459	0.9516	1.0000
1-14 UNS	1A	0.0017	0.9983	0.9828	—	0.9519	0.9435	0.9107	1B	0.923	0.938	0.9536	0.9645	1.0000
	2A	0.0017	0.9983	0.9880	—	0.9519	0.9463	0.9107	2B	0.923	0.938	0.9536	0.9609	1.0000
	3A	0.0000	1.0000	0.9897	—	0.9536	0.9494	0.9124	3B	0.9230	0.9315	0.9536	0.9590	1.0000
1-16 UN	2A	0.0015	0.9985	0.9891	—	0.9579	0.9529	0.9218	2B	0.932	0.946	0.9594	0.9659	1.0000
	3A	0.0000	1.0000	0.9906	—	0.9594	0.9557	0.9233	3B	0.9320	0.9408	0.9594	0.9643	1.0000
1-18 UNS	2A	0.0014	0.9986	0.9899	—	0.9625	0.9578	0.9304	2B	0.940	0.953	0.9639	0.9701	1.0000
1-20 UNEF	2A	0.0014	0.9986	0.9905	—	0.9661	0.9616	0.9373	2B	0.946	0.957	0.9675	0.9734	1.0000
	3A	0.0000	1.0000	0.9919	—	0.9675	0.9641	0.9387	3B	0.9460	0.9537	0.9675	0.9719	1.0000
1-24 UNS	2A	0.0013	0.9987	0.9915	—	0.9716	0.9674	0.9476	2B	0.955	0.965	0.9729	0.9784	1.0000
1-27 UNS	2A	0.0012	0.9988	0.9921	—	0.9747	0.9707	0.9534	2B	0.960	0.969	0.9759	0.9811	1.0000
1-28 UN	2A	0.0012	0.9988	0.9923	—	0.9756	0.9716	0.9550	2B	0.961	0.970	0.9768	0.9820	1.0000
	3A	0.0000	1.0000	0.9935	—	0.9768	0.9738	0.9562	3B	0.9610	0.9676	0.9768	0.9807	1.0000
1-32 UN	2A	0.0011	0.9989	0.9929	—	0.9786	0.9748	0.9606	2B	0.966	0.974	0.9797	0.9846	1.0000
	3A	0.0000	1.0000	0.9940	—	0.9797	0.9769	0.9617	3B	0.9660	0.9719	0.9797	0.9834	1.0000
1 1/16-8 UN	2A	0.0020	1.0605	1.0455	—	0.9793	0.9725	0.9071	2B	0.927	0.952	0.9813	0.9902	1.0625
	3A	0.0000	1.0625	1.0475	—	0.9813	0.9762	0.9091	3B	0.9270	0.9422	0.9813	0.9880	1.0625
1 1/16-12 UN	2A	0.0017	1.0608	1.0494	—	1.0067	1.0010	0.9586	2B	0.972	0.990	1.0084	1.0158	1.0625
	3A	0.0000	1.0625	1.0511	—	1.0084	1.0042	0.9603	3B	0.9720	0.9823	1.0084	1.0139	1.0625
1 1/16-16 UN	2A	0.0015	1.0610	1.0516	—	1.0204	1.0154	0.9843	2B	0.995	1.009	1.0219	1.0284	1.0625
	3A	0.0000	1.0625	1.0531	—	1.0219	1.0182	0.9858	3B	0.9950	1.0033	1.0219	1.0268	1.0625
1 1/16-18 UNEF	2A	0.0014	1.0611	1.0524	—	1.0250	1.0203	0.9929	2B	1.002	1.015	1.0264	1.0326	1.0625
	3A	0.0000	1.0625	1.0538	—	1.0264	1.0228	0.9943	3B	1.0020	1.0105	1.0264	1.0310	1.0625
1 1/16-20 UN	2A	0.0014	1.0611	1.0530	—	1.0286	1.0241	0.9998	2B	1.008	1.020	1.0300	1.0359	1.0625
	3A	0.0000	1.0625	1.0544	—	1.0300	1.0266	1.0012	3B	1.0080	1.0162	1.0300	1.0344	1.0625

Size	Class	Allow.	Maj Dia Max	Maj Dia Min	UNR Min	PD Max	PD Min	Minor Dia	Class	Minor Min	Minor Max	PD Min	PD Max	Maj Min
1 1/16-28 UN	2A	0.0012	1.0613	1.0548	—	1.0381	1.0341	1.0175	2B	1.024	1.032	1.0393	1.0445	1.0625
	3A	0.0000	1.0625	1.0560	—	1.0393	1.0363	1.0187	3B	1.0240	1.0301	1.0393	1.0432	1.0625
1 1/8-7 UNC	1A	0.0022	1.1228	1.0982	—	1.0300	1.0191	0.9475	1B	0.970	0.998	1.0322	1.0463	1.1250
	2A	0.0022	1.1228	1.1064	—	1.0300	1.0228	0.9475	2B	0.970	0.998	1.0322	1.0416	1.1250
	3A	0.0000	1.1250	1.1086	1.0982	1.0322	1.0268	0.9497	3B	0.9700	0.9875	1.0322	1.0393	1.1250
1 1/8-8 UN	2A	0.0021	1.1229	1.1079	—	1.0417	1.0348	0.9695	2B	0.990	1.015	1.0438	1.0528	1.1250
	3A	0.0000	1.1250	1.1100	1.0004	1.0438	1.0386	0.9716	3B	0.9900	1.0047	1.0438	1.0505	1.1250
1 1/8-10 UNS	2A	0.0018	1.1232	1.1103	—	1.0582	1.0520	1.0005	2B	1.017	1.038	1.0600	1.0680	1.1250
1 1/8-12 UNF	1A	0.0018	1.1232	1.1060	—	1.0691	1.0601	1.0210	1B	1.035	1.053	1.0709	1.0826	1.1250
	2A	0.0018	1.1232	1.1118	—	1.0691	1.0631	1.0210	2B	1.035	1.053	1.0709	1.0787	1.1250
	3A	0.0000	1.1250	1.1136	—	1.0709	1.0664	1.0228	3B	1.0350	1.0448	1.0709	1.0768	1.1250
1 1/8-14 UNS	2A	0.0016	1.1234	1.1131	—	1.0770	1.0717	1.0358	2B	1.048	1.064	1.0786	1.0855	1.1250
1 1/8-16 UN	2A	0.0015	1.1235	1.1141	—	1.0829	1.0779	1.0468	2B	1.057	1.071	1.0844	1.0909	1.1250
	3A	0.0000	1.1250	1.1156	—	1.0844	1.0807	1.0483	3B	1.0570	1.0658	1.0844	1.0893	1.1250
1 1/8-18 UNEF	2A	0.0014	1.1236	1.1149	—	1.0875	1.0828	1.0554	2B	1.065	1.078	1.0889	1.0951	1.1250
	3A	0.0000	1.1250	1.1163	—	1.0889	1.0853	1.0568	3B	1.0650	1.0730	1.0889	1.0935	1.1250
1 1/8-20 UN	2A	0.0014	1.1236	1.1155	—	1.0911	1.0866	1.0623	2B	1.071	1.082	1.0925	1.0984	1.1250
	3A	0.0000	1.1250	1.1169	—	1.0925	1.0891	1.0637	3B	1.0710	1.0787	1.0925	1.0969	1.1250
1 1/8-24 UNS	2A	0.0013	1.1237	1.1165	—	1.0966	1.0924	1.0726	2B	1.080	1.090	1.0979	1.1034	1.1250
1 1/8-28 UN	2A	0.0012	1.1238	1.1173	—	1.1006	1.0966	1.0800	2B	1.086	1.095	1.1018	1.1070	1.1250
	3A	0.0000	1.1250	1.1185	—	1.1018	1.0988	1.0812	3B	1.0860	1.0926	1.1018	1.1057	1.1250
1 3/16-8 UN	2A	0.0021	1.1854	1.1704	—	1.1042	1.0972	1.0320	2B	1.052	1.077	1.1063	1.1154	1.1875
	3A	0.0000	1.1875	1.1725	—	1.1063	1.1011	1.0341	3B	1.0520	1.0672	1.1063	1.1131	1.1875
1 3/16-12 UN	2A	0.0017	1.1858	1.1744	—	1.1317	1.1259	1.0836	2B	1.097	1.115	1.1334	1.1409	1.1875
	3A	0.0000	1.1875	1.1761	—	1.1334	1.1291	1.0853	3B	1.0970	1.1073	1.1334	1.1390	1.1875
1 3/16-16 UN	2A	0.0015	1.1860	1.1766	—	1.1454	1.1403	1.1093	2B	1.120	1.134	1.1469	1.1535	1.1875
	3A	0.0000	1.1875	1.1781	—	1.1469	1.1431	1.1108	3B	1.1200	1.1283	1.1469	1.1519	1.1875
1 3/16-18 UNEF	2A	0.0015	1.1860	1.1773	—	1.1499	1.1450	1.1178	2B	1.127	1.140	1.1514	1.1577	1.1875
	3A	0.0000	1.1875	1.1788	—	1.1514	1.1478	1.1193	3B	1.1270	1.1355	1.1514	1.1561	1.1875
1 3/16-20 UN	2A	0.0014	1.1861	1.1780	—	1.1536	1.1489	1.1248	2B	1.133	1.145	1.1550	1.1611	1.1875
	3A	0.0000	1.1875	1.1794	—	1.1550	1.1515	1.1262	3B	1.1330	1.1412	1.1550	1.1595	1.1875
1 3/16-28 UN	2A	0.0012	1.1863	1.1798	—	1.1631	1.1590	1.1425	2B	1.149	1.157	1.1643	1.1696	1.1875
	3A	0.0000	1.1875	1.1810	—	1.1643	1.1612	1.1437	3B	1.1490	1.1551	1.1643	1.1683	1.1875

† Use UNS threads only if Standard Series do not meet requirements. See footnotes b, d, e, and f at end of table.

Table 5. (Continued). Standard Series and Selected Combinations† — Unified Screw Threads

Nominal Size, Threads per Inch, and Series Designation^f	Class	Allowance	External Major Diameter Max^b	Min	Min^d	Pitch Diameter Max^b	Min	Minor Diameter	Internal Class	Minor Diameter^e Min	Max	Pitch Diameter Min	Max	Major Diameter Min
1¼-7 UNC	1A	0.0022	1.2478	1.2233	—	1.1550	1.1439	1.0725	1B	1.095	1.123	1.1572	1.1716	1.2500
	2A	0.0022	1.2478	1.2314	1.2232	1.1550	1.1476	1.0725	2B	1.095	1.123	1.1572	1.1668	1.2500
	3A	0.0000	1.2500	1.2336	1.2254	1.1572	1.1517	1.0747	3B	1.0950	1.1125	1.1572	1.1644	1.2500
1¼-8 UN	2A	0.0021	1.2479	1.2329	1.2254	1.1667	1.1597	1.0945	2B	1.115	1.140	1.1688	1.1780	1.2500
	3A	0.0000	1.2500	1.2350	—	1.1688	1.1635	1.0966	3B	1.1150	1.1297	1.1688	1.1757	1.2500
1¼-10 UNS	2A	0.0019	1.2481	1.2352		1.1831	1.1768	1.1254	2B	1.142	1.163	1.1850	1.1932	1.2500
1¼-12 UNF	1A	0.0018	1.2481	1.2310		1.1941	1.1849	1.1460	1B	1.160	1.178	1.1959	1.2079	1.2500
	2A	0.0018	1.2482	1.2368		1.1941	1.1879	1.1460	2B	1.160	1.178	1.1959	1.2039	1.2500
	3A	0.0000	1.2500	1.2386		1.1959	1.1913	1.1478	3B	1.1600	1.1698	1.1959	1.2019	1.2500
1¼-14 UNS	2A	0.0016	1.2484	1.2381		1.2020	1.1966	1.1608	2B	1.173	1.188	1.2036	1.2106	1.2500
1¼-16 UN	2A	0.0015	1.2485	1.2391		1.2079	1.2028	1.1718	2B	1.182	1.196	1.2094	1.2160	1.2500
	3A	0.0000	1.2500	1.2406		1.2094	1.2056	1.1733	3B	1.1820	1.1908	1.2094	1.2144	1.2500
1¼-18 UNEF	2A	0.0015	1.2485	1.2398		1.2124	1.2075	1.1803	2B	1.190	1.203	1.2139	1.2202	1.2500
	3A	0.0000	1.2500	1.2413		1.2139	1.2103	1.1818	3B	1.1900	1.1980	1.2139	1.2186	1.2500
1¼-20 UN	2A	0.0014	1.2486	1.2405		1.2161	1.2114	1.1873	2B	1.196	1.207	1.2175	1.2236	1.2500
	3A	0.0000	1.2500	1.2419		1.2175	1.2140	1.1887	3B	1.1960	1.2037	1.2175	1.2220	1.2500
1¼-24 UNS	2A	0.0013	1.2487	1.2415		1.2216	1.2173	1.1976	2B	1.205	1.215	1.2229	1.2285	1.2500
1¼-28 UN	2A	0.0012	1.2488	1.2423		1.2256	1.2215	1.2050	2B	1.211	1.220	1.2268	1.2321	1.2500
	3A	0.0000	1.2500	1.2435		1.2268	1.2237	1.2062	3B	1.2110	1.2176	1.2268	1.2308	1.2500
1⁵⁄₁₆-8 UN	2A	0.0021	1.3104	1.2954		1.2292	1.2221	1.1570	2B	1.177	1.202	1.2313	1.2405	1.3125
	3A	0.0000	1.3125	1.2975		1.2313	1.2260	1.1591	3B	1.1770	1.1922	1.2313	1.2382	1.3125
1⁵⁄₁₆-12 UN	2A	0.0017	1.3108	1.2994		1.2567	1.2509	1.2086	2B	1.222	1.240	1.2584	1.2659	1.3125
	3A	0.0000	1.3125	1.3011		1.2584	1.2541	1.2103	3B	1.2220	1.2323	1.2584	1.2640	1.3125
1⁵⁄₁₆-16 UN	2A	0.0015	1.3110	1.3016		1.2704	1.2653	1.2343	2B	1.245	1.259	1.2719	1.2785	1.3125
	3A	0.0000	1.3125	1.3031		1.2719	1.2681	1.2358	3B	1.2450	1.2533	1.2719	1.2769	1.3125
1⁵⁄₁₆-18 UNEF	2A	0.0015	1.3110	1.3023		1.2749	1.2700	1.2428	2B	1.252	1.265	1.2764	1.2827	1.3125
	3A	0.0000	1.3125	1.3038		1.2764	1.2728	1.2443	3B	1.2520	1.2605	1.2764	1.2811	1.3125
1⁵⁄₁₆-20 UN	2A	0.0014	1.3111	1.3030		1.2786	1.2739	1.2498	2B	1.258	1.270	1.2800	1.2861	1.3125
	3A	0.0000	1.3125	1.3044		1.2800	1.2765	1.2512	3B	1.2580	1.2662	1.2800	1.2845	1.3125

Identification	Class	Allow.							Class					
1⁵/₁₆-28 UN	2A	0.0012	1.3113	1.3048	—	1.2881	1.2840	1.2675	2B	1.274	1.282	1.2893	1.2946	1.3125
	3A	0.0000	1.3125	1.3060	—	1.2893	1.2862	1.2687	3B	1.2740	1.2861	1.2893	1.2933	1.3125
1³/₈-6 UNC	1A	0.0024	1.3726	1.3453	—	1.2643	1.2523	1.1681	1B	1.195	1.225	1.2667	1.2822	1.3750
	2A	0.0024	1.3726	1.3544	—	1.2643	1.2563	1.1681	2B	1.195	1.225	1.2667	1.2771	1.3750
	3A	0.0000	1.3750	1.3568	1.3453	1.2667	1.2607	1.1705	3B	1.1950	1.2146	1.2667	1.2745	1.3750
1³/₈-8 UN	2A	0.0022	1.3728	1.3578	—	1.2916	1.2844	1.2194	2B	1.240	1.265	1.2938	1.3031	1.3750
	3A	0.0000	1.3750	1.3600	1.3503	1.2938	1.2884	1.2216	3B	1.2400	1.2547	1.2938	1.3008	1.3750
1³/₈-10 UNS	2A	0.0019	1.3731	1.3602	—	1.3081	1.3018	1.2504	2B	1.267	1.288	1.3100	1.3182	1.3750
1³/₈-12 UNF	1A	0.0019	1.3731	1.3559	—	1.3190	1.3096	1.2709	1B	1.285	1.303	1.3209	1.3332	1.3750
	2A	0.0019	1.3731	1.3617	—	1.3190	1.3127	1.2709	2B	1.285	1.303	1.3209	1.3291	1.3750
	3A	0.0000	1.3750	1.3636	—	1.3209	1.3162	1.2728	3B	1.2850	1.2948	1.3209	1.3270	1.3750
1³/₈-14 UNS	2A	0.0016	1.3734	1.3631	—	1.3270	1.3216	1.2858	2B	1.298	1.314	1.3286	1.3356	1.3750
1³/₈-16 UN	2A	0.0015	1.3735	1.3641	—	1.3329	1.3278	1.2968	2B	1.307	1.321	1.3344	1.3410	1.3750
	3A	0.0000	1.3750	1.3656	—	1.3344	1.3306	1.2983	3B	1.3070	1.3158	1.3344	1.3394	1.3750
1³/₈-18 UNEF	2A	0.0015	1.3735	1.3648	—	1.3374	1.3325	1.3053	2B	1.315	1.328	1.3389	1.3452	1.3750
	3A	0.0000	1.3750	1.3663	—	1.3389	1.3353	1.3068	3B	1.3150	1.3230	1.3389	1.3436	1.3750
1³/₈-20 UN	2A	0.0014	1.3736	1.3655	—	1.3411	1.3364	1.3123	2B	1.321	1.332	1.3425	1.3486	1.3750
	3A	0.0000	1.3750	1.3669	—	1.3425	1.3390	1.3137	3B	1.3210	1.3287	1.3425	1.3470	1.3750
1³/₈-28 UN	2A	0.0012	1.3738	1.3673	—	1.3506	1.3465	1.3300	2B	1.336	1.345	1.3518	1.3571	1.3750
	3A	0.0000	1.3750	1.3685	—	1.3518	1.3487	1.3312	3B	1.3360	1.3426	1.3518	1.3558	1.3750
1⁷/₁₆-6 UN	2A	0.0024	1.4351	1.4169	—	1.3268	1.3188	1.2306	2B	1.257	1.288	1.3292	1.3396	1.4375
	3A	0.0000	1.4375	1.4193	—	1.3292	1.3233	1.2330	3B	1.2570	1.2771	1.3292	1.3370	1.4375
1⁷/₁₆-8 UN	2A	0.0022	1.4353	1.4203	—	1.3541	1.3469	1.2819	2B	1.302	1.327	1.3563	1.3657	1.4375
	3A	0.0000	1.4375	1.4225	—	1.3563	1.3509	1.2841	3B	1.3020	1.3172	1.3563	1.3634	1.4375
1⁷/₁₆-12 UN	2A	0.0018	1.4357	1.4243	—	1.3816	1.3757	1.3335	2B	1.347	1.365	1.3834	1.3910	1.4375
	3A	0.0000	1.4375	1.4261	—	1.3834	1.3790	1.3353	3B	1.3470	1.3573	1.3834	1.3891	1.4375
1⁷/₁₆-16 UN	2A	0.0016	1.4359	1.4265	—	1.3953	1.3901	1.3592	2B	1.370	1.384	1.3969	1.4037	1.4375
	3A	0.0000	1.4375	1.4281	—	1.3969	1.3930	1.3608	3B	1.3700	1.3783	1.3969	1.4020	1.4375
1⁷/₁₆-18 UNEF	2A	0.0015	1.4360	1.4273	—	1.3999	1.3949	1.3678	2B	1.377	1.390	1.4014	1.4079	1.4375
	3A	0.0000	1.4375	1.4288	—	1.4014	1.3977	1.3693	3B	1.3770	1.3855	1.4014	1.4062	1.4375
1⁷/₁₆-20 UN	2A	0.0014	1.4361	1.4280	—	1.4036	1.3988	1.3748	2B	1.383	1.395	1.4050	1.4112	1.4375
	3A	0.0000	1.4375	1.4294	—	1.4050	1.4014	1.3762	3B	1.3830	1.3912	1.4050	1.4096	1.4375
1⁷/₁₆-28 UN	2A	0.0013	1.4362	1.4297	—	1.4130	1.4088	1.3924	2B	1.399	1.407	1.4143	1.4198	1.4375
	3A	0.0000	1.4375	1.4310	—	1.4143	1.4112	1.3937	3B	1.3990	1.4051	1.4143	1.4184	1.4375

† Use UNS threads only if Standard Series do not meet requirements. See footnotes b, d, e, and f at end of table.

Table 5. (Continued). Standard Series and Selected Combinations† — Unified Screw Threads

Nominal Size, Threads per Inch, and Series Designation^f	Class	Allowance	External Major Diameter Max^b	Min	Min^d	External Pitch Diameter Max^b	Min	Minor Diameter	Internal Class	Minor Diameter^e Min	Max	Pitch Diameter Min	Max	Major Diame'er Min
1½–6 UNC	1A	0.0024	1.4976	1.4703	—	1.3893	1.3772	1.2931	1B	1.320	1.350	1.3917	1.4075	1.5000
	2A	0.0024	1.4976	1.4794	1.4703	1.3893	1.3812	1.2931	2B	1.320	1.350	1.3917	1.4022	1.5000
	3A	0.0000	1.5000	1.4818	1.4753	1.3917	1.3856	1.2955	3B	1.3200	1.3396	1.3917	1.3996	1.5000
1½–8 UN	2A	0.0022	1.4978	1.4828	1.4753	1.4166	1.4093	1.3444	2B	1.365	1.390	1.4188	1.4283	1.5000
	3A	0.0000	1.5000	1.4850	—	1.4188	1.4133	1.3466	3B	1.3650	1.3797	1.4188	1.4259	1.5000
1½–10 UNS	2A	0.0019	1.4981	1.4852	—	1.4331	1.4267	1.3754	2B	1.392	1.413	1.4350	1.4433	1.5000
1½–12 UNF	1A	0.0019	1.4981	1.4809	—	1.4440	1.4344	1.3959	1B	1.410	1.428	1.4459	1.4584	1.5000
	2A	0.0019	1.4981	1.4867	—	1.4440	1.4376	1.3959	2B	1.410	1.428	1.4459	1.4542	1.5000
	3A	0.0000	1.5000	1.4886	—	1.4459	1.4411	1.3978	3B	1.4100	1.4198	1.4459	1.4522	1.5000
1½–14 UNS	2A	0.0017	1.4983	1.4880	—	1.4519	1.4464	1.4107	2B	1.423	1.438	1.4536	1.4608	1.5000
1½–16 UN	2A	0.0016	1.4984	1.4890	—	1.4578	1.4526	1.4217	2B	1.432	1.446	1.4594	1.4662	1.5000
	3A	0.0000	1.5000	1.4906	—	1.4594	1.4555	1.4233	3B	1.4320	1.4408	1.4594	1.4645	1.5000
1½–18 UNEF	2A	0.0015	1.4985	1.4898	—	1.4624	1.4574	1.4303	2B	1.440	1.452	1.4639	1.4704	1.5000
	3A	0.0000	1.5000	1.4913	—	1.4639	1.4602	1.4318	3B	1.4400	1.4480	1.4639	1.4687	1.5000
1½–20 UN	2A	0.0014	1.4986	1.4905	—	1.4661	1.4613	1.4373	2B	1.446	1.457	1.4675	1.4737	1.5000
	3A	0.0000	1.5000	1.4919	—	1.4675	1.4639	1.4387	3B	1.4460	1.4537	1.4675	1.4721	1.5000
1½–24 UNS	2A	0.0013	1.4987	1.4915	—	1.4716	1.4672	1.4476	2B	1.455	1.465	1.4729	1.4787	1.5000
1½–28 UN	2A	0.0013	1.4987	1.4922	—	1.4755	1.4713	1.4549	2B	1.461	1.470	1.4768	1.4823	1.5000
	3A	0.0000	1.5000	1.4935	—	1.4768	1.4737	1.4562	3B	1.4610	1.4676	1.4768	1.4809	1.5000
1⁹⁄₁₆–6 UN	2A	0.0024	1.5601	1.5419	—	1.4518	1.4436	1.3556	2B	1.382	1.413	1.4542	1.4648	1.5625
	3A	0.0000	1.5625	1.5443	—	1.4542	1.4481	1.3580	3B	1.3820	1.4021	1.4542	1.4622	1.5625
1⁹⁄₁₆–8 UN	2A	0.0022	1.5603	1.5453	—	1.4791	1.4717	1.4069	2B	1.427	1.452	1.4813	1.4909	1.5625
	3A	0.0000	1.5625	1.5475	—	1.4813	1.4758	1.4091	3B	1.4270	1.4422	1.4813	1.4885	1.5625
1⁹⁄₁₆–12 UN	2A	0.0018	1.5607	1.5493	—	1.5066	1.5007	1.4585	2B	1.472	1.490	1.5084	1.5160	1.5625
	3A	0.0000	1.5625	1.5511	—	1.5084	1.5040	1.4603	3B	1.4720	1.4823	1.5084	1.5141	1.5625
1⁹⁄₁₆–16 UN	2A	0.0016	1.5609	1.5515	—	1.5203	1.5151	1.4842	2B	1.495	1.509	1.5219	1.5287	1.5625
	3A	0.0000	1.5625	1.5531	—	1.5219	1.5186	1.4858	3B	1.4950	1.5033	1.5219	1.5270	1.5625
1⁹⁄₁₆–18 UNEF	2A	0.0015	1.5610	1.5523	—	1.5249	1.5199	1.4928	2B	1.502	1.515	1.5264	1.5329	1.5625
	3A	0.0000	1.5625	1.5538	—	1.5264	1.5227	1.4943	3B	1.5020	1.5105	1.5264	1.5312	1.5625

1	2	3	4	5	6	7	8	9	10	11	12	13	14	Designation
1.5625	1.5362	1.5300	1.520	1.508	2B	1.4998	1.5238	1.5286	—	1.5530	1.5611	0.0014	2A	1⁹⁄₁₆-20 UN
1.5625	1.5346	1.5300	1.5162	1.5080	3B	1.5012	1.5264	1.5300	—	1.5544	1.5625	0.0000	3A	
1.6250	1.5274	1.5167	1.475	1.445	2B	1.4180	1.5060	1.5142	—	1.6043	1.6225	0.0025	2A	1⅝-6 UN
1.6250	1.5247	1.5167	1.4646	1.4450	3B	1.4205	1.5105	1.5167	—	1.6068	1.6250	0.0000	3A	
1.6250	1.5535	1.5438	1.515	1.490	2B	1.4694	1.5342	1.5416	1.6003	1.6078	1.6228	0.0022	2A	1⅝-8 UN
1.6250	1.5510	1.5438	1.5047	1.4900	3B	1.4716	1.5382	1.5438	—	1.6100	1.6250	0.0000	3A	
1.6250	1.5683	1.5600	1.538	1.517	2B	1.5004	1.5517	1.5581	—	1.6102	1.6231	0.0019	2A	1⅝-10 UNS
1.6250	1.5785	1.5709	1.553	1.535	2B	1.5210	1.5633	1.5691	—	1.6118	1.6232	0.0018	2A	1⅝-12 UN
1.6250	1.5766	1.5709	1.5448	1.5350	3B	1.5228	1.5665	1.5709	—	1.6136	1.6250	0.0000	3A	
1.6250	1.5858	1.5786	1.564	1.548	2B	1.5357	1.5714	1.5769	—	1.6130	1.6233	0.0017	2A	1⅝-14 UNS
1.6250	1.5911	1.5844	1.571	1.557	2B	1.5467	1.5776	1.5828	—	1.6140	1.6234	0.0016	2A	1⅝-16 UN
1.6250	1.5895	1.5844	1.5658	1.5570	3B	1.5483	1.5805	1.5844	—	1.6156	1.6250	0.0000	3A	
1.6250	1.5954	1.5889	1.578	1.565	2B	1.5553	1.5824	1.5874	—	1.6148	1.6235	0.0015	2A	1⅝-18 UNEF
1.6250	1.5937	1.5889	1.5730	1.5650	3B	1.5568	1.5852	1.5889	—	1.6163	1.6250	0.0000	3A	
1.6250	1.5987	1.5925	1.582	1.571	2B	1.5623	1.5863	1.5911	—	1.6155	1.6236	0.0014	2A	1⅝-20 UN
1.6250	1.5971	1.5925	1.5787	1.5710	3B	1.5637	1.5889	1.5925	—	1.6169	1.6250	0.0000	3A	
1.6250	1.6037	1.5979	1.590	1.580	2B	1.5726	1.5922	1.5966	—	1.6165	1.6237	0.0013	2A	1⅝-24 UNS
1.6875	1.5900	1.5792	1.538	1.507	2B	1.4805	1.5684	1.5767	—	1.6668	1.6850	0.0025	2A	1¹¹⁄₁₆-6 UN
1.6875	1.5873	1.5792	1.5271	1.5070	3B	1.4830	1.5730	1.5792	—	1.6693	1.6875	0.0000	3A	
1.6875	1.6160	1.6063	1.577	1.553	2B	1.5319	1.5966	1.6041	—	1.6703	1.6853	0.0022	2A	1¹¹⁄₁₆-8 UN
1.6875	1.6136	1.6063	1.5672	1.5520	3B	1.5341	1.6007	1.6063	—	1.6725	1.6875	0.0000	3A	
1.6875	1.6412	1.6334	1.615	1.597	2B	1.5835	1.6256	1.6316	—	1.6743	1.6857	0.0018	2A	1¹¹⁄₁₆-12 UN
1.6875	1.6392	1.6334	1.6070	1.5970	3B	1.5853	1.6289	1.6334	—	1.6761	1.6875	0.0000	3A	
1.6875	1.6538	1.6469	1.634	1.620	2B	1.6092	1.6400	1.6453	—	1.6765	1.6859	0.0016	2A	1¹¹⁄₁₆-16 UN
1.6875	1.6521	1.6469	1.6283	1.6200	3B	1.6108	1.6429	1.6469	—	1.6781	1.6875	0.0000	3A	
1.6875	1.6580	1.6514	1.640	1.627	2B	1.6178	1.6448	1.6499	—	1.6773	1.6860	0.0015	2A	1¹¹⁄₁₆-18 UNEF
1.6875	1.6563	1.6514	1.6355	1.6270	3B	1.6193	1.6476	1.6514	—	1.6788	1.6875	0.0000	3A	
1.6875	1.6613	1.6550	1.645	1.633	2B	1.6247	1.6487	1.6535	—	1.6779	1.6860	0.0015	2A	1¹¹⁄₁₆-20 UN
1.6875	1.6597	1.6550	1.6412	1.6330	3B	1.6262	1.6514	1.6550	—	1.6794	1.6875	0.0000	3A	
1.7500	1.6375	1.6201	1.568	1.534	1B	1.5019	1.6040	1.6174	1.7165	1.7165	1.7473	0.0027	1A	1¾-5 UNC
1.7500	1.6317	1.6201	1.568	1.534	2B	1.5019	1.6085	1.6174	—	1.7268	1.7473	0.0027	2A	
1.7500	1.6288	1.6201	1.5575	1.5340	3B	1.5046	1.6134	1.6201		1.7295	1.7500	0.0000	3A	

† Use UNS threads only if Standard Series do not meet requirements. See footnotes b, d, e, and f at end of table.

Table 5. *(Continued).* Standard Series and Selected Combinations† — Unified Screw Threads

Nominal Size, Threads per Inch, and Series Designation^f	Class	Allow-ance	Major Diameter Max^b	Major Diameter Min	Major Diameter Min^d	Pitch Diameter Max^b	Pitch Diameter Min	Minor Diam-eter	Class	Minor Diameter^e Min	Minor Diameter^e Max	Pitch Diameter Min	Pitch Diameter Max	Major Diameter Min
1³/₄–6 UN	2A	0.0025	1.7475	1.7293	—	1.6392	1.6309	1.5430	2B	1.570	1.600	1.6417	1.6525	1.7500
	3A	0.0000	1.7500	1.7318	—	1.6417	1.6354	1.5455	3B	1.5700	1.5896	1.6417	1.6498	1.7500
1³/₄–8 UN	2A	0.0023	1.7477	1.7327	1.7252	1.6665	1.6590	1.5943	2B	1.615	1.640	1.6688	1.6786	1.7500
	3A	0.0000	1.7500	1.7350	—	1.6688	1.6632	1.5966	3B	1.6150	1.6297	1.6688	1.6762	1.7500
1³/₄–10 UNS	2A	0.0019	1.7481	1.7352	—	1.6831	1.6766	1.6254	2B	1.642	1.663	1.6850	1.6934	1.7500
1³/₄–12 UN	2A	0.0018	1.7482	1.7368	—	1.6941	1.6881	1.6460	2B	1.660	1.678	1.6959	1.7037	1.7500
	3A	0.0000	1.7500	1.7386	—	1.6959	1.6914	1.6478	3B	1.6600	1.6698	1.6959	1.7017	1.7500
1³/₄–14 UNS	2A	0.0017	1.7483	1.7380	—	1.7019	1.6963	1.6607	2B	1.673	1.688	1.7036	1.7109	1.7500
1³/₄–16 UN	2A	0.0016	1.7484	1.7390	—	1.7078	1.7025	1.6717	2B	1.682	1.696	1.7094	1.7163	1.7500
	3A	0.0000	1.7500	1.7406	—	1.7094	1.7054	1.6733	3B	1.6820	1.6908	1.7094	1.7146	1.7500
1³/₄–18 UNS	2A	0.0015	1.7485	1.7398	—	1.7124	1.7073	1.6803	2B	1.690	1.703	1.7139	1.7205	1.7500
1³/₄–20 UN	2A	0.0015	1.7485	1.7404	—	1.7160	1.7112	1.6872	2B	1.696	1.707	1.7175	1.7238	1.7500
	3A	0.0000	1.7500	1.7419	—	1.7175	1.7139	1.6887	3B	1.6960	1.7037	1.7175	1.7222	1.7500
1¹³/₁₆–8 UN	2A	0.0025	1.8100	1.7918	—	1.7017	1.6933	1.6555	2B	1.632	1.663	1.7042	1.7151	1.8125
	3A	0.0000	1.8125	1.7943	—	1.7042	1.6979	1.6591	3B	1.6320	1.6521	1.7042	1.7124	1.8125
1¹³/₁₆–12 UN	2A	0.0018	1.8107	1.7993	—	1.7566	1.7506	1.7085	2B	1.722	1.740	1.7584	1.7662	1.8125
	3A	0.0000	1.8125	1.8011	—	1.7584	1.7539	1.7103	3B	1.7220	1.7320	1.7584	1.7642	1.8125
1¹³/₁₆–16 UN	2A	0.0016	1.8109	1.8015	—	1.7703	1.7650	1.7342	2B	1.745	1.759	1.7719	1.7788	1.8125
	3A	0.0000	1.8125	1.8031	—	1.7719	1.7679	1.7358	3B	1.7450	1.7533	1.7719	1.7771	1.8125
1¹³/₁₆–20 UN	2A	0.0015	1.8110	1.8029	—	1.7785	1.7737	1.7497	2B	1.758	1.770	1.7800	1.7863	1.8125
	3A	0.0000	1.8125	1.8044	—	1.7800	1.7764	1.7512	3B	1.7580	1.7662	1.7800	1.7847	1.8125
1⅞–6 UN	2A	0.0025	1.8725	1.8543	—	1.7642	1.7558	1.6680	2B	1.695	1.725	1.7667	1.7777	1.8750
	3A	0.0000	1.8750	1.8568	—	1.7667	1.7604	1.6705	3B	1.6950	1.7146	1.7667	1.7749	1.8750
1⅞–8 UN	2A	0.0023	1.8727	1.8577	1.8502	1.7915	1.7838	1.7193	2B	1.740	1.765	1.7938	1.8038	1.8750
	3A	0.0000	1.8750	1.8600	—	1.7938	1.7881	1.7216	3B	1.7400	1.7547	1.7938	1.8013	1.8750
1⅞–10 UNS	2A	0.0019	1.8731	1.8602	—	1.8081	1.8016	1.7504	2B	1.767	1.788	1.8100	1.8184	1.8750

Identification						Class							Class	
1⅞–12 UN	1.8750 / 1.8750	1.8287 / 1.8267	1.8209 / 1.8209	1.803 / 1.7948	1.785 / 1.7850	2B / 3B	1.7710 / 1.7728	1.8131 / 1.8164	1.8191 / 1.8209	— / —	1.8618 / 1.8636	1.8732 / 1.8750	0.0018 / 0.0000	2A / 3A
1⅞–14 UNS	1.8750	1.8359	1.8286	1.814	1.798	2B	1.7857	1.8213	1.8269	—	1.8630	1.8733	0.0017	2A
1⅞–16 UN	1.8750 / 1.8750	1.8413 / 1.8396	1.8344 / 1.8344	1.821 / 1.8158	1.807 / 1.8070	2B / 3B	1.7967 / 1.7983	1.8275 / 1.8304	1.8338 / 1.8344	— / —	1.8640 / 1.8656	1.8734 / 1.8750	0.0016 / 0.0000	2A / 3A
1⅞–18 UNS	1.8750	1.8455	1.8389	1.828	1.815	2B	1.8053	1.8323	1.8374	—	1.8648	1.8735	0.0015	2A
1⅞–20 UN	1.8750 / 1.8750	1.8488 / 1.8472	1.8425 / 1.8425	1.832 / 1.8287	1.821 / 1.8210	2B / 3B	1.8122 / 1.8137	1.8362 / 1.8389	1.8410 / 1.8425	— / —	1.8654 / 1.8669	1.8735 / 1.8750	0.0015 / 0.0000	2A / 3A
1 15/16–6 UN	1.9375 / 1.9375	1.8403 / 1.8375	1.8292 / 1.8292	1.788 / 1.7771	1.757 / 1.7570	2B / 3B	1.7304 / 1.7330	1.8181 / 1.8228	1.8266 / 1.8292	— / —	1.9167 / 1.9193	1.9349 / 1.9375	0.0026 / 0.0000	2A / 3A
1 15/16–8 UN	1.9375 / 1.9375	1.8663 / 1.8638	1.8563 / 1.8563	1.827 / 1.8172	1.802 / 1.8020	2B / 3B	1.7818 / 1.7841	1.8463 / 1.8505	1.8540 / 1.8563	— / —	1.9202 / 1.9225	1.9352 / 1.9375	0.0023 / 0.0000	2A / 3A
1 15/16–12 UN	1.9375 / 1.9375	1.8913 / 1.8893	1.8834 / 1.8834	1.865 / 1.8570	1.847 / 1.8470	2B / 3B	1.8335 / 1.8353	1.8755 / 1.8789	1.8816 / 1.8834	— / —	1.9243 / 1.9261	1.9357 / 1.9375	0.0018 / 0.0000	2A / 3A
1 15/16–16 UN	1.9375 / 1.9375	1.9039 / 1.9021	1.8969 / 1.8969	1.884 / 1.8783	1.870 / 1.8700	2B / 3B	1.8592 / 1.8608	1.8899 / 1.8929	1.8953 / 1.8969	— / —	1.9265 / 1.9281	1.9359 / 1.9375	0.0016 / 0.0000	2A / 3A
1 15/16–20 UN	1.9375 / 1.9375	1.9114 / 1.9098	1.9050 / 1.9050	1.895 / 1.8912	1.883 / 1.8830	2B / 3B	1.8747 / 1.8762	1.8986 / 1.9013	1.9035 / 1.9050	— / —	1.9279 / 1.9294	1.9360 / 1.9375	0.0015 / 0.0000	2A / 3A
2–4½ UNC	2.0000 / 2.0000 / 2.0000	1.8743 / 1.8681 / 1.8650	1.8557 / 1.8557 / 1.8557	1.795 / 1.795 / 1.7861	1.759 / 1.759 / 1.7590	1B / 2B / 3B	1.7245 / 1.7245 / 1.7274	1.8385 / 1.8433 / 1.8486	1.8528 / 1.8538 / 1.8557	1.9641 / — / —	1.9641 / 1.9751 / 1.9780	1.9971 / 1.9971 / 2.0000	0.0029 / 0.0029 / 0.0000	1A / 2A / 3A
2–6 UN	2.0000 / 2.0000	1.9028 / 1.9000	1.8917 / 1.8917	1.850 / 1.8396	1.820 / 1.8200	2B / 3B	1.7929 / 1.7955	1.8805 / 1.8853	1.8891 / 1.8917	— / —	1.9792 / 1.9818	1.9974 / 2.0000	0.0026 / 0.0000	2A / 3A
2–8 UN	2.0000 / 2.0000	1.9288 / 1.9264	1.9188 / 1.9188	1.890 / 1.8797	1.865 / 1.8650	2B / 3B	1.8443 / 1.8466	1.9087 / 1.9130	1.9165 / 1.9188	1.9752 / —	1.9827 / 1.9850	1.9977 / 2.0000	0.0023 / 0.0000	2A / 3A
2–10 UNS	2.0000	1.9435	1.9350	1.913	1.892	2B	1.8753	1.9265	1.9330	—	1.9851	1.9980	0.0020	2A
2–12 UN	2.0000 / 2.0000	1.9538 / 1.9518	1.9459 / 1.9459	1.928 / 1.9198	1.910 / 1.9100	2B / 3B	1.8960 / 1.8978	1.9380 / 1.9414	1.9441 / 1.9459	— / —	1.9868 / 1.9886	1.9982 / 2.0000	0.0018 / 0.0000	2A / 3A
2–14 UNS	2.0000	1.9610	1.9536	1.938	1.923	2B	1.9107	1.9462	1.9519	—	1.9880	1.9983	0.0017	2A
2–16 UN	2.0000 / 2.0000	1.9664 / 1.9646	1.9594 / 1.9594	1.946 / 1.9408	1.932 / 1.9320	2B / 3B	1.9217 / 1.9233	1.9524 / 1.9554	1.9578 / 1.9594	— / —	1.9890 / 1.9906	1.9984 / 2.0000	0.0016 / 0.0000	2A / 3A
2–18 UNS	2.0000	1.9706	1.9639	1.953	1.940	2B	1.9303	1.9573	1.9624	—	1.9898	1.9985	0.0015	2A

† Use UNS threads only if Standard Series do not meet requirements. See footnotes b, d, e, and f at end of table.

Table 5. (Continued). Standard Series and Selected Combinations† — Unified Screw Threads

Nominal Size, Threads per Inch, and Series Designation^f	Class	Allowance	Major Dia. Max^b	Major Dia. Min	Major Dia. Min^d	Pitch Dia. Max^b	Pitch Dia. Min	Minor Diameter	Class	Minor Dia. Min	Minor Dia. Max	Pitch Dia. Min	Pitch Dia. Max	Major Dia. Min
2–20 UN	2A	0.0015	1.9985	1.9904		1.9660	1.9611	1.9372	2B	1.946	1.957	1.9675	1.9739	2.0000
	3A	0.0000	2.0000	1.9919		1.9675	1.9638	1.9387	3B	1.9460	1.9537	1.9675	1.9723	2.0000
2 1/16–16 UNS	2A	0.0016	2.0609	2.0515		2.0203	2.0149	1.9842	2B	1.995	2.009	2.0219	2.0289	2.0625
	3A	0.0000	2.0625	2.0531		2.0219	2.0179	1.9858	3B	1.9950	2.0033	2.0219	2.0271	2.0625
2 1/8–6 UN	2A	0.0026	2.1224	2.1042		2.0141	2.0054	1.9179	2B	1.945	1.975	2.0167	2.0280	2.1250
	3A	0.0000	2.1250	2.1068	2.1001	2.0167	2.0102	1.9205	3B	1.9450	1.9646	2.0167	2.0251	2.1250
2 1/8–8 UN	2A	0.0024	2.1226	2.1076		2.0414	2.0335	1.9692	2B	1.990	2.015	2.0438	2.0540	2.1250
	3A	0.0000	2.1250	2.1100		2.0438	2.0379	1.9716	3B	1.9900	2.0047	2.0438	2.0515	2.1250
2 1/8–12 UN	2A	0.0018	2.1232	2.1118		2.0691	2.0630	2.0210	2B	2.035	2.053	2.0709	2.0788	2.1250
	3A	0.0000	2.1250	2.1136		2.0709	2.0664	2.0228	3B	2.0350	2.0448	2.0709	2.0768	2.1250
2 1/8–16 UN	2A	0.0016	2.1234	2.1140		2.0828	2.0774	2.0467	2B	2.057	2.071	2.0844	2.0914	2.1250
	3A	0.0000	2.1250	2.1156		2.0844	2.0803	2.0483	3B	2.0570	2.0658	2.0844	2.0896	2.1250
2 1/8–20 UN	2A	0.0015	2.1235	2.1154		2.0910	2.0861	2.0622	2B	2.071	2.082	2.0925	2.0989	2.1250
	3A	0.0000	2.1250	2.1169		2.0925	2.0888	2.0637	3B	2.0710	2.0787	2.0925	2.0973	2.1250
2 3/16–16 UNS	2A	0.0016	2.1859	2.1765		2.1453	2.1399	2.1092	2B	2.120	2.134	2.1469	2.1539	2.1875
	3A	0.0000	2.1875	2.1781		2.1469	2.1428	2.1108	3B	2.1200	2.1283	2.1469	2.1521	2.1875
2 1/4–4 1/2 UNC	1A	0.0029	2.2471	2.2141		2.1028	2.0882	1.9745	1B	2.009	2.045	2.1057	2.1247	2.2500
	2A	0.0029	2.2471	2.2251		2.1028	2.0931	1.9745	2B	2.009	2.045	2.1057	2.1183	2.2500
	3A	0.0000	2.2500	2.2280	2.2141	2.1057	2.0984	1.9774	3B	2.0090	2.0361	2.1057	2.1152	2.2500
2 1/4–6 UN	2A	0.0026	2.2474	2.2292		2.1391	2.1304	2.0429	2B	2.070	2.100	2.1417	2.1531	2.2500
	3A	0.0000	2.2500	2.2318	2.2251	2.1417	2.1351	2.0455	3B	2.0700	2.0896	2.1417	2.1502	2.2500
2 1/4–8 UN	2A	0.0024	2.2476	2.2326		2.1664	2.1584	2.0942	2B	2.115	2.140	2.1688	2.1792	2.2500
	3A	0.0000	2.2500	2.2350		2.1688	2.1628	2.0966	3B	2.1150	2.1297	2.1688	2.1766	2.2500
2 1/4–10 UNS	2A	0.0020	2.2480	2.2351		2.1830	2.1765	2.1253	2B	2.142	2.163	2.1850	2.1935	2.2500
2 1/4–12 UN	2A	0.0018	2.2482	2.2368		2.1941	2.1880	2.1460	2B	2.160	2.178	2.1959	2.2038	2.2500
	3A	0.0000	2.2500	2.2386		2.1959	2.1914	2.1478	3B	2.1600	2.1698	2.1959	2.2018	2.2500
2 1/4–14 UNS	2A	0.0017	2.2483	2.2380		2.2019	2.1962	2.1607	2B	2.173	2.188	2.2036	2.2110	2.2500
2 1/4–16 UN	2A	0.0016	2.2484	2.2390		2.2078	2.2024	2.1717	2B	2.182	2.196	2.2094	2.2164	2.2500
	3A	0.0000	2.2500	2.2406		2.2094	2.2053	2.1733	3B	2.1820	2.1908	2.2094	2.2146	2.2500

Sizes and Threads Per Inch	Class	Allowance	Major Dia, Max	Major Dia, Min	—	Pitch Dia, Max	Pitch Dia, Min	Minor Dia, Max	Class	Minor Dia, Min	Minor Dia, Max	Pitch Dia, Min	Pitch Dia, Max	Major Dia, Min
2¼–18 UNS	2A	0.0015	2.2485	2.2398	—	2.2124	2.2073	2.1803	2B	2.190	2.207	2.2139	2.2206	2.2500
	3A	0.0000	2.2500	2.2404	—	2.2139	2.2101	2.1822	3B	2.1900	2.203	2.2139	2.2189	2.2500
2¼–20 UN	2A	0.0015	2.2485	2.2404	—	2.2160	2.2111	2.1872	2B	2.196	2.207	2.2175	2.2239	2.2500
	3A	0.0000	2.2500	2.2419	—	2.2175	2.2137	2.1887	3B	2.1960	2.2037	2.2175	2.2223	2.2500
2⁵⁄₁₆–16 UNS	2A	0.0017	2.3108	2.3014	—	2.2702	2.2647	2.2341	2B	2.245	2.259	2.2719	2.2791	2.3125
	3A	0.0000	2.3125	2.3031	—	2.2719	2.2678	2.2358	3B	2.2450	2.2533	2.2719	2.2773	2.3125
2⅜–6 UN	2A	0.0027	2.3723	2.3541	—	2.2640	2.2551	2.1678	2B	2.195	2.226	2.2667	2.2782	2.3750
	3A	0.0000	2.3750	2.3568	—	2.2667	2.2601	2.1705	3B	2.1950	2.2146	2.2667	2.2753	2.3750
2⅜–8 UN	2A	0.0024	2.3726	2.3576	—	2.2914	2.2833	2.2192	2B	2.240	2.265	2.2938	2.3043	2.3750
	3A	0.0000	2.3750	2.3600	—	2.2938	2.2878	2.2216	3B	2.2400	2.2552	2.2938	2.3017	2.3750
2⅜–12 UN	2A	0.0019	2.3731	2.3617	—	2.3190	2.3128	2.2709	2B	2.285	2.303	2.3209	2.3390	2.3750
	3A	0.0000	2.3750	2.3636	—	2.3209	2.3163	2.2728	3B	2.2850	2.2948	2.3209	2.3369	2.3750
2⅜–16 UN	2A	0.0017	2.3733	2.3639	—	2.3327	2.3272	2.2966	2B	2.307	2.321	2.3344	2.3416	2.3750
	3A	0.0000	2.3750	2.3656	—	2.3344	2.3303	2.2983	3B	2.3070	2.3158	2.3344	2.3398	2.3750
2⅜–20 UN	2A	0.0015	2.3735	2.3654	—	2.3410	2.3359	2.3122	2B	2.321	2.332	2.3425	2.3491	2.3750
	3A	0.0000	2.3750	2.3669	—	2.3425	2.3387	2.3137	3B	2.3210	2.3287	2.3425	2.3475	2.3750
2⁷⁄₁₆–16 UNS	2A	0.0017	2.4358	2.4264	—	2.3952	2.3897	2.3591	2B	2.370	2.384	2.3969	2.4041	2.4375
	3A	0.0000	2.4375	2.4281	—	2.3969	2.3928	2.3608	3B	2.3700	2.3783	2.3969	2.4023	2.4375
2½–4 UNC	1A	0.0031	2.4969	2.4612	—	2.3345	2.3190	2.1902	1B	2.229	2.267	2.3376	2.3578	2.5000
	2A	0.0031	2.4969	2.4731	2.4612	2.3345	2.3241	2.1902	2B	2.229	2.267	2.3376	2.3511	2.5000
	3A	0.0000	2.5000	2.4762	—	2.3376	2.3398	2.1933	3B	2.2290	2.2594	2.3376	2.3477	2.5000
2½–6 UN	2A	0.0027	2.4973	2.4791	—	2.3890	2.3800	2.2928	2B	2.320	2.350	2.3917	2.4033	2.5000
	3A	0.0000	2.5000	2.4818	—	2.3917	2.3850	2.2955	3B	2.3200	2.3396	2.3917	2.4004	2.5000
2½–8 UN	2A	0.0024	2.4976	2.4826	2.4751	2.4164	2.4082	2.3442	2B	2.365	2.390	2.4188	2.4294	2.5000
	3A	0.0000	2.5000	2.4850	—	2.4188	2.4127	2.3466	3B	2.3650	2.3797	2.4188	2.4268	2.5000
2½–10 UNS	2A	0.0020	2.4980	2.4851	—	2.4330	2.4263	2.3753	2B	2.392	2.413	2.4350	2.4437	2.5000
2½–12 UN	2A	0.0019	2.4981	2.4867	—	2.4440	2.4378	2.3959	2B	2.410	2.428	2.4459	2.4540	2.5000
	3A	0.0000	2.5000	2.4886	—	2.4459	2.4413	2.3978	3B	2.4100	2.4198	2.4459	2.4519	2.5000
2½–14 UNS	2A	0.0017	2.4983	2.4880	—	2.4519	2.4461	2.4107	2B	2.423	2.438	2.4536	2.4612	2.5000
2½–16 UN	2A	0.0017	2.4983	2.4889	—	2.4577	2.4522	2.4216	2B	2.432	2.466	2.4594	2.4666	2.5000
	3A	0.0000	2.5000	2.4906	—	2.4594	2.4553	2.4233	3B	2.4320	2.4408	2.4594	2.4648	2.5000
2½–18 UNS	2A	0.0016	2.4984	2.4897	—	2.4623	2.4570	2.4302	2B	2.440	2.453	2.4639	2.4708	2.5000
2½–20 UN	2A	0.0015	2.4985	2.4904	—	2.4660	2.4609	2.4372	2B	2.446	2.457	2.4675	2.4741	2.5000
	3A	0.0000	2.5000	2.4919	—	2.4675	2.4637	2.4387	3B	2.4460	2.4537	2.4675	2.4725	2.5000

† Use UNS threads only if Standard Series do not meet requirements. See footnotes *b, d, e,* and *f* at end of table.

Table 5. (Continued). Standard Series and Selected Combinations† — Unified Screw Threads

Nominal Size, Threads per Inch, and Series Designation^f	Class	Allow-ance	External^c Major Diameter Max^b	Min	Min^d	Pitch Diameter Max^b	Min	Minor Diam-eter	Class	Internal Minor Diameter Min	Max^e	Pitch Diameter Min	Max	Major Diameter Min
2⅝–6 UN	2A	0.0027	2.6223	2.6041	—	2.5140	2.5050	2.4178	2B	2.445	2.475	2.5167	2.5285	2.6250
	3A	0.0000	2.6250	2.6068	—	2.5167	2.5099	2.4205	3B	2.4450	2.4646	2.5167	2.5255	2.6250
2⅝–8 UN	2A	0.0025	2.6225	2.6075	—	2.5413	2.5331	2.4691	2B	2.490	2.515	2.5438	2.5545	2.6250
	3A	0.0000	2.6250	2.6100	—	2.5438	2.5376	2.4716	3B	2.4900	2.5052	2.5438	2.5518	2.6250
2⅝–12 UN	2A	0.0019	2.6231	2.6117	—	2.5690	2.5628	2.5209	2B	2.535	2.553	2.5709	2.5790	2.6250
	3A	0.0000	2.6250	2.6136	—	2.5709	2.5663	2.5228	3B	2.5350	2.5448	2.5709	2.5769	2.6250
2⅝–16 UN	2A	0.0017	2.6233	2.6139	—	2.5827	2.5772	2.5466	2B	2.557	2.571	2.5844	2.5916	2.6250
	3A	0.0000	2.6250	2.6156	—	2.5844	2.5803	2.5483	3B	2.5570	2.5558	2.5844	2.5898	2.6250
2⅝–20 UN	2A	0.0015	2.6235	2.6154	—	2.5910	2.5859	2.5622	2B	2.571	2.582	2.5925	2.5991	2.6250
	3A	0.0000	2.6250	2.6169	—	2.5925	2.5887	2.5637	3B	2.5710	2.5787	2.5925	2.5975	2.6250
2¾–4 UNC	1A	0.0032	2.7468	2.7111	—	2.5844	2.5686	2.4401	1B	2.479	2.517	2.5876	2.6082	2.7500
	2A	0.0032	2.7468	2.7130	2.7111	2.5844	2.5739	2.4401	2B	2.479	2.517	2.5876	2.6013	2.7500
	3A	0.0000	2.7500	2.7262	—	2.5876	2.5797	2.4433	3B	2.4790	2.5094	2.5876	2.5979	2.7500
2¾–6 UN	2A	0.0027	2.7473	2.7291	—	2.6390	2.6299	2.5428	2B	2.570	2.600	2.6417	2.6536	2.7500
	3A	0.0000	2.7500	2.7318	—	2.6417	2.6349	2.5455	3B	2.5700	2.5896	2.6417	2.6506	2.7500
2¾–8 UN	2A	0.0025	2.7475	2.7325	2.7250	2.6663	2.6580	2.5941	2B	2.615	2.640	2.6688	2.6796	2.7500
	3A	0.0000	2.7500	2.7350	—	2.6688	2.6625	2.5966	3B	2.6150	2.6297	2.6688	2.6769	2.7500
2¾–10 UNS	2A	0.0020	2.7480	2.7351	—	2.6830	2.6763	2.6253	2B	2.642	2.663	2.6850	2.6937	2.7500
2¾–12 UN	2A	0.0019	2.7481	2.7367	—	2.6940	2.6878	2.6459	2B	2.660	2.678	2.6959	2.7040	2.7500
	3A	0.0000	2.7500	2.7386	—	2.6959	2.6913	2.6478	3B	2.6600	2.6698	2.6959	2.7019	2.7500
2¾–14 UNS	2A	0.0017	2.7483	2.7380	—	2.7019	2.6961	2.6607	2B	2.673	2.688	2.7036	2.7112	2.7500
2¾–16 UN	2A	0.0017	2.7483	2.7389	—	2.7077	2.7022	2.6716	2B	2.682	2.696	2.7094	2.7166	2.7500
	3A	0.0000	2.7500	2.7406	—	2.7094	2.7053	2.6733	3B	2.6820	2.6908	2.7094	2.7148	2.7500
2¾–18 UNS	2A	0.0016	2.7484	2.7397	—	2.7123	2.7070	2.6802	2B	2.690	2.703	2.7139	2.7208	2.7500
2¾–20 UN	2A	0.0015	2.7485	2.7404	—	2.7160	2.7109	2.6872	2B	2.696	2.707	2.7175	2.7241	2.7500
	3A	0.0000	2.7500	2.7419	—	2.7175	2.7137	2.6887	3B	2.6960	2.7037	2.7175	2.7225	2.7500
2⅞–6 UN	2A	0.0028	2.8722	2.8540	—	2.7639	2.7547	2.6677	2B	2.695	2.725	2.7667	2.7787	2.8750
	3A	0.0000	2.8750	2.8568	—	2.7667	2.7598	2.6705	3B	2.6950	2.7146	2.7667	2.7757	2.8750

Size	Class	Allow.	Ext. Major Max	Ext. Major Min	(UNR)	Ext. Pitch Max	Ext. Pitch Min	Ext. Minor	Class	Int. Minor Min	Int. Minor Max	Int. Pitch Min	Int. Pitch Max	Int. Major Min
2⅞–8 UN	2A	0.0025	2.8725	2.8575	—	2.7913	2.7829	2.7191	2B	2.740	2.765	2.7938	2.8048	2.8750
	3A	0.0000	2.8750	2.8600	—	2.7938	2.7875	2.7216	3B	2.7400	2.7552	2.7938	2.8020	2.8750
2⅞–12 UN	2A	0.0019	2.8731	2.8617	—	2.8190	2.8127	2.7709	2B	2.785	2.803	2.8209	2.8291	2.8750
	3A	0.0000	2.8750	2.8636	—	2.8209	2.8162	2.7728	3B	2.7850	2.7948	2.8209	2.8271	2.8750
2⅞–16 UN	2A	0.0017	2.8733	2.8639	—	2.8327	2.8271	2.7966	2B	2.807	2.821	2.8344	2.8417	2.8750
	3A	0.0000	2.8750	2.8656	—	2.8344	2.8302	2.7983	3B	2.8070	2.8158	2.8344	2.8399	2.8750
2⅞–20 UN	2A	0.0016	2.8734	2.8653	—	2.8409	2.8357	2.8121	2B	2.821	2.832	2.8425	2.8493	2.8750
	3A	0.0000	2.8750	2.8669	—	2.8425	2.8386	2.8137	3B	2.8210	2.8287	2.8425	2.8476	2.8750
3–4 UNC	1A	0.0032	2.9968	2.9611	—	2.8344	2.8183	2.6901	1B	2.729	2.767	2.8376	2.8585	3.0000
	2A	0.0032	2.9968	2.9730	2.9611	2.8344	2.8237	2.6901	2B	2.729	2.767	2.8376	2.8515	3.0000
	3A	0.0000	3.0000	2.9762	—	2.8376	2.8296	2.6933	3B	2.7290	2.7594	2.8376	2.8486	3.0000
3–6 UN	2A	0.0028	2.9972	2.9790	—	2.8889	2.8796	2.7927	2B	2.820	2.850	2.8917	2.9038	3.0000
	3A	0.0000	3.0000	2.9818	—	2.8917	2.8847	2.7955	3B	2.8200	2.8396	2.8917	2.9008	3.0000
3–8 UN	2A	0.0026	2.9974	2.9824	2.9749	2.9162	2.9077	2.8440	2B	2.865	2.890	2.9188	2.9299	3.0000
	3A	0.0000	3.0000	2.9850	—	2.9188	2.9124	2.8466	3B	2.8650	2.8797	2.9188	2.9271	3.0000
3–10 UNS	2A	0.0020	2.9980	2.9851	—	2.9330	2.9262	2.8753	2B	2.892	2.913	2.9350	2.9439	3.0000
3–12 UN	2A	0.0019	2.9981	2.9867	—	2.9440	2.9377	2.8959	2B	2.910	2.928	2.9459	2.9541	3.0000
	3A	0.0000	3.0000	2.9886	—	2.9459	2.9412	2.8978	3B	2.9100	2.9198	2.9459	2.9521	3.0000
3–14 UNS	2A	0.0018	2.9982	2.9879	—	2.9518	2.9459	2.9106	2B	2.923	2.938	2.9536	2.9613	3.0000
3–16 UN	2A	0.0017	2.9983	2.9889	—	2.9577	2.9521	2.9216	2B	2.932	2.946	2.9594	2.9667	3.0000
	3A	0.0000	3.0000	2.9906	—	2.9594	2.9552	2.9233	3B	2.9320	2.9408	2.9594	2.9649	3.0000
3–18 UNS	2A	0.0016	2.9984	2.9897	—	2.9623	2.9569	2.9302	2B	2.940	2.953	2.9639	2.9709	3.0000
3–20 UN	2A	0.0016	2.9984	2.9903	—	2.9659	2.9607	2.9371	2B	2.946	2.957	2.9675	2.9743	3.0000
	3A	0.0000	3.0000	2.9919	—	2.9675	2.9636	2.9387	3B	2.9460	2.9537	2.9675	2.9726	3.0000
3⅛–6 UN	2A	0.0028	3.1222	3.1040	—	3.0139	3.0045	2.9177	2B	2.945	2.975	3.0167	3.0289	3.1250
	3A	0.0000	3.1250	3.1068	—	3.0167	3.0097	2.9205	3B	2.9450	2.9646	3.0167	3.0259	3.1250
3⅛–8 UN	2A	0.0026	3.1224	3.1074	—	3.0412	3.0326	2.9690	2B	2.990	3.015	3.0438	3.0550	3.1250
	3A	0.0000	3.1250	3.1100	—	3.0438	3.0374	2.9716	3B	2.9900	3.0052	3.0438	3.0522	3.1250
3⅛–12 UN	2A	0.0019	3.1231	3.1117	—	3.0690	3.0627	3.0209	2B	3.035	3.053	3.0709	3.0791	3.1250
	3A	0.0000	3.1250	3.1136	—	3.0709	3.0662	3.0228	3B	3.0350	3.0448	3.0709	3.0771	3.1250
3⅛–16 UN	2A	0.0017	3.1233	3.1139	—	3.0827	3.0771	3.0466	2B	3.057	3.071	3.0844	3.0917	3.1250
	3A	0.0000	3.1250	3.1156	—	3.0844	3.0802	3.0483	3B	3.0570	3.0658	3.0844	3.0899	3.1250

† Use UNS threads only if Standard Series do not meet requirements. See footnotes b, d, e, and f at end of table.

Table 5. (Continued). Standard Series and Selected Combinations† — Unified Screw Threads

Nominal Size, Threads per Inch, and Series Designation^f	Class	Allowance	External Major Diameter Max^b	Min	Min^d	External Pitch Diameter Max^b	Min	Minor Diameter	Class	Internal Minor Diameter^e Min	Max	Internal Pitch Diameter Min	Max	Major Diameter Min
3¼-4 UNC	1A	0.0033	3.2467	3.2110	—	3.0843	3.0680	2.9400	1B	2.979	3.017	3.0876	3.1088	3.2500
	2A	0.0033	3.2467	3.2229	3.2110	3.0843	3.0734	2.9400	2B	2.979	3.017	3.0876	3.1017	3.2500
	3A	0.0000	3.2500	3.2262	—	3.0876	3.0794	2.9433	3B	2.9790	3.0094	3.0876	3.0982	3.2500
3¼-6 UN	2A	0.0028	3.2472	3.2290	—	3.1389	3.1294	3.0427	2B	3.070	3.100	3.1417	3.1540	3.2500
	3A	0.0000	3.2500	3.2318	—	3.1417	3.1346	3.0455	3B	3.0700	3.0896	3.1417	3.1509	3.2500
3¼-8 UN	2A	0.0026	3.2474	3.2324	3.2249	3.1662	3.1575	3.0940	2B	3.115	3.140	3.1688	3.1801	3.2500
	3A	0.0000	3.2500	3.2350	—	3.1688	3.1623	3.0966	3B	3.1150	3.1297	3.1688	3.1772	3.2500
3¼-10 UNS	2A	0.0020	3.2480	3.2351	—	3.1830	3.1762	3.1253	2B	3.142	3.163	3.1850	3.1939	3.2500
3¼-12 UN	2A	0.0019	3.2481	3.2367	—	3.1940	3.1877	3.1459	2B	3.160	3.178	3.1959	3.2041	3.2500
	3A	0.0000	3.2500	3.2386	—	3.1959	3.1912	3.1478	3B	3.1600	3.1698	3.1959	3.2021	3.2500
3¼-14 UNS	2A	0.0018	3.2482	3.2379	—	3.2018	3.1959	3.1606	2B	3.173	3.188	3.2036	3.2113	3.2500
3¼-16 UN	2A	0.0017	3.2483	3.2389	—	3.2077	3.2021	3.1716	2B	3.182	3.196	3.2094	3.2167	3.2500
	3A	0.0000	3.2500	3.2406	—	3.2094	3.2052	3.1733	3B	3.1820	3.1908	3.2094	3.2149	3.2500
3¼-18 UNS	2A	0.0016	3.2484	3.2397	—	3.2123	3.2069	3.1802	2B	3.190	3.203	3.2139	3.2209	3.2500
3⅜-6 UN	2A	0.0029	3.3721	3.3539	—	3.2638	3.2543	3.1676	2B	3.195	3.225	3.2667	3.2791	3.3750
	3A	0.0000	3.3750	3.3568	—	3.2667	3.2595	3.1705	3B	3.1950	3.2146	3.2667	3.2760	3.3750
3⅜-8 UN	2A	0.0026	3.3724	3.3574	—	3.2912	3.2824	3.2190	2B	3.240	3.265	3.2938	3.3052	3.3750
	3A	0.0000	3.3750	3.3600	—	3.2938	3.2876	3.2216	3B	3.2400	3.2552	3.2938	3.3023	3.3750
3⅜-12 UN	2A	0.0019	3.3731	3.3617	—	3.3190	3.3126	3.2709	2B	3.285	3.303	3.3209	3.3293	3.3750
	3A	0.0000	3.3750	3.3636	—	3.3209	3.3161	3.2728	3B	3.2850	3.2948	3.3209	3.3272	3.3750
3⅜-16 UN	2A	0.0017	3.3733	3.3639	—	3.3327	3.3269	3.2966	2B	3.307	3.321	3.3344	3.3419	3.3750
	3A	0.0000	3.3750	3.3656	—	3.3344	3.3301	3.2983	3B	3.3070	3.3158	3.3344	3.3400	3.3750
3½-4 UNC	1A	0.0033	3.4967	3.4610	—	3.3343	3.3177	3.1900	1B	3.229	3.267	3.3376	3.3591	3.5000
	2A	0.0033	3.4967	3.4729	3.4610	3.3343	3.3233	3.1900	2B	3.229	3.267	3.3376	3.3519	3.5000
	3A	0.0000	3.5000	3.4762	—	3.3376	3.3293	3.1933	3B	3.2290	3.2594	3.3376	3.3484	3.5000
3½-6 UN	2A	0.0029	3.4971	3.4789	—	3.3888	3.3792	3.2926	2B	3.320	3.350	3.3917	3.4042	3.5000
	3A	0.0000	3.5000	3.4818	—	3.3917	3.3845	3.2955	3B	3.3200	3.3396	3.3917	3.4011	3.5000
3½-8 UN	2A	0.0026	3.4974	3.4824	3.4749	3.4162	3.4074	3.3440	2B	3.365	3.390	3.4188	3.4303	3.5000
	3A	0.0000	3.5000	3.4850	—	3.4188	3.4122	3.3466	3B	3.3650	3.3797	3.4188	3.4274	3.5000

Sizes and Threads per Inch	Class	Allowance	Major Dia Max	Major Dia Min	(UNR)	Pitch Dia Max	Pitch Dia Min	Minor Dia Min (Ext)	Class	Minor Dia Min (Int)	Minor Dia Max (Int)	Pitch Dia Min	Pitch Dia Max	Major Dia Min (Int)
3½–10 UNS	2A	0.0021	3.4979	3.4850	—	3.4329	3.4260	3.3752	2B	3.392	3.413	3.4350	3.4440	3.5000
3½–12 UN	2A	0.0019	3.4981	3.4867	—	3.4440	3.4376	3.3959	2B	3.410	3.428	3.4459	3.4543	3.5000
	3A	0.0000	3.5000	3.4886	—	3.4459	3.4411	3.3978	3B	3.4100	3.4198	3.4459	3.4522	3.5000
3½–14 UNS	2A	0.0018	3.4982	3.4879	—	3.4518	3.4457	3.4106	2B	3.423	3.438	3.4536	3.4615	3.5000
3½–16 UN	2A	0.0017	3.4983	3.4889	—	3.4577	3.4519	3.4216	2B	3.432	3.446	3.4594	3.4669	3.5000
	3A	0.0000	3.5000	3.4906	—	3.4594	3.4551	3.4333	3B	3.4320	3.4408	3.4594	3.4650	3.5000
3½–18 UNS	2A	0.0017	3.4983	3.4896	—	3.4622	3.4567	3.4301	2B	3.440	3.453	3.4639	3.4711	3.5000
3⅝–6 UN	2A	0.0029	3.6221	3.6039	—	3.5138	3.5041	3.4176	2B	3.445	3.475	3.5167	3.5293	3.6250
	3A	0.0000	3.6250	3.6068	—	3.5167	3.5094	3.4205	3B	3.4450	3.4646	3.5167	3.5262	3.6250
3⅝–8 UN	2A	0.0027	3.6223	3.6073	—	3.5411	3.5322	3.4699	2B	3.490	3.515	3.5438	3.5554	3.6250
	3A	0.0000	3.6250	3.6100	—	3.5438	3.5371	3.4716	3B	3.4900	3.5052	3.5438	3.5535	3.6250
3⅝–12 UN	2A	0.0019	3.6231	3.6117	—	3.5690	3.5626	3.5209	2B	3.535	3.553	3.5709	3.5793	3.6250
	3A	0.0000	3.6250	3.6136	—	3.5709	3.5661	3.5228	3B	3.5350	3.5448	3.5709	3.5772	3.6250
3⅝–16 UN	2A	0.0017	3.6233	3.6139	—	3.5827	3.5769	3.5466	2B	3.557	3.571	3.5844	3.5919	3.6250
	3A	0.0000	3.6250	3.6156	—	3.5844	3.5801	3.5483	3B	3.5570	3.5658	3.5844	3.5890	3.6250
3¾–4 UNC	1A	0.0034	3.7466	3.7109	3.7109	3.5842	3.5674	3.4399	1B	3.479	3.517	3.5876	3.6094	3.7500
	2A	0.0034	3.7466	3.7228	—	3.5842	3.5730	3.4399	2B	3.479	3.517	3.5876	3.6021	3.7500
	3A	0.0000	3.7500	3.7262	—	3.5876	3.5792	3.4433	3B	3.4790	3.5094	3.5876	3.5985	3.7500
3¾–6 UN	2A	0.0029	3.7471	3.7289	—	3.6388	3.6290	3.5426	2B	3.570	3.600	3.6417	3.6544	3.7500
	3A	0.0000	3.7500	3.7318	—	3.6417	3.6344	3.5455	3B	3.5700	3.5896	3.6417	3.6512	3.7500
3¾–8 UN	2A	0.0027	3.7473	3.7323	3.7248	3.6661	3.6571	3.5939	2B	3.615	3.640	3.6688	3.6805	3.7500
	3A	0.0000	3.7500	3.7350	—	3.6688	3.6621	3.5966	3B	3.6150	3.6297	3.6688	3.6776	3.7500
3¾–10 UNS	2A	0.0021	3.7479	3.7350	—	3.6829	3.6760	3.6252	2B	3.642	3.663	3.6850	3.6940	3.7500
3¾–12 UN	2A	0.0019	3.7481	3.7367	—	3.6940	3.6876	3.6459	2B	3.660	3.678	3.6959	3.7043	3.7500
	3A	0.0000	3.7500	3.7386	—	3.6959	3.6911	3.6478	3B	3.6600	3.6698	3.6959	3.7022	3.7500
3¾–14 UNS	2A	0.0018	3.7482	3.7379	—	3.7018	3.6957	3.6606	2B	3.673	3.688	3.7036	3.7115	3.7500
3¾–16 UN	2A	0.0017	3.7483	3.7389	—	3.7077	3.7019	3.6716	2B	3.682	3.696	3.7094	3.7169	3.7500
	3A	0.0000	3.7500	3.7406	—	3.7094	3.7051	3.6733	3B	3.6820	3.6908	3.7094	3.7150	3.7500
3¾–18 UNS	2A	0.0017	3.7483	7.7396	—	3.7122	3.7067	3.6801	2B	3.690	3.703	3.7139	3.7211	3.7500
3⅞–6 UN	2A	0.0030	3.8720	3.8538	—	3.7637	3.7538	3.6675	2B	3.695	3.725	3.7667	3.7795	3.8750
	3A	0.0000	3.8750	3.8568	—	3.7667	3.7593	3.6705	3B	3.6950	3.7146	3.7667	3.7763	3.8750

† Use UNS threads only if Standard Series do not meet requirements. See footnotes b, d, e, and f at end of table.

Table 5. (*Concluded*). Standard Series and Selected Combinations† — Unified Screw Threads

Nominal Size, Threads per Inch, and Series Designation^f	Class	External							Internal					
		Allowance	Major Diameter			Pitch Diameter		Minor Diameter	Class	Minor Diameter^e		Pitch Diameter		Major Diameter
			Max^b	Min	Min^d	Max^b	Min			Min	Max	Min	Max	Min
3⅞-8 UN	2A	0.0027	3.8723	3.8573	—	3.7911	3.7820	3.7189	2B	3.740	3.765	3.7938	3.8056	3.8750
	3A	0.0000	3.8750	3.8600	—	3.7938	3.7870	3.7216	3B	3.7400	3.7552	3.7938	3.8026	3.8750
3⅞-12 UN	2A	0.0020	3.8730	3.8616	—	3.8189	3.8124	3.7708	2B	3.785	3.803	3.8209	3.8294	3.8750
	3A	0.0000	3.8750	3.8636	—	3.8209	3.8160	3.7728	3B	3.7850	3.7948	3.8209	3.8273	3.8750
3⅞-16 UN	2A	0.0018	3.8732	3.8638	—	3.8326	3.8267	3.7965	2B	3.807	3.821	3.8344	3.8420	3.8750
	3A	0.0000	3.8750	3.8656	—	3.8344	3.8300	3.7983	3B	3.8070	3.8158	3.8344	3.8401	3.8750
4-4 UNC	1A	0.0034	3.9966	3.9669	3.9669	3.8342	3.8172	3.6899	1B	3.729	3.767	3.8376	3.8597	4.0000
	2A	0.0034	3.9966	3.9728	—	3.8342	3.8229	3.6899	2B	3.729	3.767	3.8376	3.8523	4.0000
	3A	0.0000	4.0000	3.9762	—	3.8376	3.8291	3.6933	3B	3.7594	3.7594	3.8376	3.8487	4.0000
4-6 UN	2A	0.0030	3.9970	3.9788	—	3.8887	3.8788	3.7925	2B	3.820	3.850	3.8917	3.9046	4.0000
	3A	0.0000	4.0000	3.9818	—	3.8917	3.8843	3.7955	3B	3.8200	3.8396	3.8917	3.9014	4.0000
4-8 UN	2A	0.0027	3.9973	3.9823	3.9748	3.9161	3.9070	3.8439	2B	3.865	3.890	3.9188	3.9307	4.0000
	3A	0.0000	4.0000	3.9850	—	3.9188	3.9120	3.8466	3B	3.8650	3.8797	3.9188	3.9277	4.0000
4-10 UNS	2A	0.0021	3.9979	3.9850	—	3.9329	3.9259	3.8752	2B	3.892	3.913	3.9350	3.9441	4.0000
4-12 UN	2A	0.0020	3.9980	3.9866	—	3.9439	3.9374	3.8958	2B	3.910	3.928	3.9459	3.9544	4.0000
	3A	0.0000	4.0000	3.9886	—	3.9459	3.9410	3.8978	3B	3.9100	3.9198	3.9459	3.9523	4.0000
4-14 UNS	2A	0.0018	3.9982	3.9879	—	3.9518	3.9456	3.9106	2B	3.923	3.938	3.9536	3.9616	4.0000
4-16 UN	2A	0.0018	3.9982	3.9888	—	3.9576	3.9517	3.9215	2B	3.932	3.946	3.9594	3.9670	4.0000
	3A	0.0000	4.0000	3.9906	—	3.9594	3.9550	3.9233	3B	3.9320	3.9408	3.9594	3.9651	4.0000

† Use UNS threads only if Standard Series do not meet requirements. For sizes above 4 inches see ANSI B1.1-1974.
b For Class 2A threads having an additive finish the maximum is increased, by the allowance, to the basic size, the value being the same as for Class 3A.
d For unfinished hot-rolled material.
e Revised minor diameter limits Classes 1B and 2B.
f Use UNR designations instead of UN wherever UNR thread form is desired for external use.

Measuring Screw Threads

Pitch and Lead of Screw Threads. — The *pitch* of a screw thread is the distance from the center of one thread to the center of the next thread. This applies no matter whether the screw has a single, double, triple or quadruple thread. The *lead* of a screw thread is the distance the nut will move forward on the screw if it is turned around one full revolution. In a single-threaded screw, the pitch and lead are equal, because the nut would move forward the distance from one thread to the next, if turned around once. In a double-threaded screw, the nut will move forward two threads, or twice the pitch, so that in this case the lead equals twice the pitch. In a triple-threaded screw, the lead equals three times the pitch, and so on.

The word "pitch" is often, although improperly, used to denote the *number of threads per inch*. Screws are spoken of as having a 12-pitch thread, when twelve threads per inch is what is really meant. The number of threads per inch equals 1 divided by the pitch, or expressed as a formula:

$$\text{Number of threads per inch} = \frac{1}{\text{pitch}}$$

The pitch of a screw equals 1 divided by the number of threads per inch, or:

$$\text{Pitch} = \frac{1}{\text{number of threads per inch}}$$

If the number of threads per inch equals 16, the pitch = $\frac{1}{16}$. If the pitch equals 0.05, the number of threads equals $1 \div 0.05 = 20$. If the pitch is $\frac{2}{5}$ inch, the number of threads per inch equals $1 \div \frac{2}{5} = 2\frac{1}{2}$.

Confusion is often caused by the indefinite designation of multiple-thread screws (double, triple, quadruple, etc.). The expression, "four threads per inch, triple," for example, is not to be recommended. It means that the screw is cut with four triple threads or with twelve threads per inch, if the threads are counted by placing a scale alongside the screw. To cut this screw, the lathe would be geared to cut four threads per inch, but they would be cut only to the depth required for twelve threads per inch. The best expression, when a multiple-thread is to be cut, is to say, in this case, "¼ inch lead, ¹⁄₁₂ inch pitch, triple thread." For single-threaded screws, only the number of threads per inch and the form of the thread are specified. The word "single" is not required.

Measuring Screw Thread Pitch Diameters by Thread Micrometers. — As the pitch or angle diameter of a tap or screw is the most important dimension, it is necessary that the pitch diameter of screw threads be measured, in addition to the outside diameter. One method of measuring the angle of a thread is by means of a special screw thread micrometer, as shown in the accompanying engraving, Fig. 1. The fixed anvil is V-shaped so as to fit over the thread, while the movable point is cone-shaped so as to enable it to enter the space

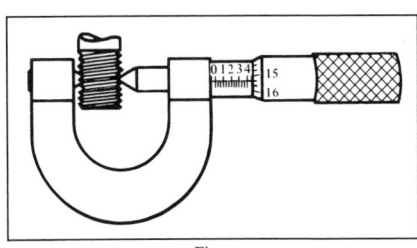

Fig. 1

between two threads, and at the same time be at liberty to revolve. The contact points are on the sides of the thread, as they necessarily must be in order that the

pitch diameter may be determined. The cone-shaped point of the measuring screw is slightly rounded so that it will not bear in the bottom of the thread. There is also sufficient clearance at the bottom of the V-shaped anvil to prevent it from bearing on the top of the thread. The movable point is adapted to measuring all pitches, but the fixed anvil is limited in its capacity. To cover the whole range of pitches, from the finest to the coarsest, a number of fixed anvils are therefore required.

To find the theoretical pitch diameter, which is measured by the micrometer, subtract twice the addendum of the thread from the standard outside diameter. The addendum of the thread for the American and other standard threads is given in the section on screw thread systems.

Measuring Screw Threads by Three-wire Method. — The *effective* or *pitch diameter* of a screw thread may be measured very accurately by means of some form of micrometer and three wires of equal diameter. This method is extensively used in checking the accuracy of threaded plug gages and other precision screw threads. Two of the wires are placed in contact with the thread on one side and the third wire in a position diametrically opposite as illustrated by the diagram, (see table "Formulas for Checking Pitch Diameters of Screw Threads") and the dimension over the wires is determined by means of a micrometer. An ordinary micrometer is commonly used but some form of "floating micrometer" is preferable, especially for measuring thread gages and other precision work. The floating micrometer is mounted upon a compound slide so that it can move freely in directions parallel or at right angles to the axis of the screw, which is held in a horizontal position between adjustable centers. With this arrangement the micrometer is held constantly at right angles to the axis of the screw so that only one wire on each side may be used instead of having two on one side and one on the other, as is necessary when using an ordinary micrometer. The accuracy of the pitch diameter may be determined provided the correct micrometer reading for wires of a given size is known.

Classes of Formulas for Three-wire Measurement. — Various formulas have been established for checking the pitch diameters of screw threads by measurement over wires of known size. These formulas differ in regard to their simplicity or complexity and resulting accuracy. They also differ in that some show what measurement M over the wires should be to obtain a given pitch diameter E, whereas others show the value of the pitch diameter E for a given measurement M.

Formulas for Finding Measurement M: In using a formula for finding the value of measurement M, the required pitch diameter E is inserted in the formula. Then, in cutting or grinding a screw thread, the *actual* measurement M is made to conform to the *calculated* value of M. Formulas for finding measurement M may be modified so that the basic major or outside diameter is inserted in the formula instead of the pitch diameter; however, the pitch diameter type of formula is preferable because this is a more important dimension than the major diameter.

Formulas for Finding Pitch Diameters E: Some formulas are arranged to show the value of the pitch diameter E when measurement M is known. Thus the value of M is first determined by actual measurement and then it is inserted in the formula for finding the corresponding pitch diameter E. This type of formula is useful for determining the pitch diameter of an existing thread gage or other screw thread in connection with inspection work. The formula for finding measurement M is more convenient to use in the shop or tool-room in cutting or grinding new threads, because the pitch diameter is specified on the drawing and the problem is to find the value of measurement M for obtaining this pitch diameter.

Formulas for Checking Pitch Diameters of Screw Threads

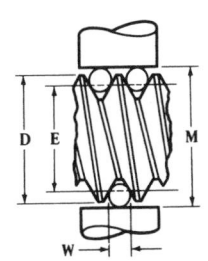

The formulas below do not compensate for the effect of the lead angle upon measurement M, but they are sufficiently accurate for checking standard single-thread screws unless exceptional accuracy is required. See accompanying information on effect of lead angle; also matter relating to measuring wire sizes, accuracy required for such wires, and contact or measuring pressure. The approximate best wire size for pitch-line contact may be obtained by the formula

$W = 0.5 \times \text{pitch} \times \sec \frac{1}{2}$ included thread angle

For 60-degree threads, $W = 0.57735 \times \text{pitch}$.

Form of Thread	Formulas for determining measurement M corresponding to correct pitch diameter and the pitch diameter E corresponding to a given measurement over wires.*
American National Unified Inch and Metric Threads	When measurement M is known. $E = M + 0.86603P - 3W$ When pitch diameter E is used in formula. $M = E - 0.86603P + 3W$ The American Standard formerly was known as U.S. Standard.
British Standard Whitworth	When measurement M is known. $E = M + 0.9605P - 3.1657W$ When pitch diameter E is used in formula. $M = E - 0.9605P + 3.1657W$
British Association Standard	When measurement M is known. $E = M + 1.1363P - 3.4829W$ When pitch diameter E is used in formula. $M = E - 1.1363P + 3.4829W$
Lowenherz Thread	When measurement M is known. $E = M + P - 3.2359W$ When pitch diameter E is used in formula. $M = E - P + 3.2359W$
Sharp V-Thread	When measurement M is known. $E = M + 0.86603P - 3W$ When pitch diameter E is used in formula. $M = E - 0.86603P + 3W$

* The wires must be lapped to a uniform diameter and it is very important to insert in the rule or formula the wire diameter as determined by precise means of measurement. Any error will be multiplied. See paragraph on Wire Sizes for Checking Pitch Diameters.

Wire Sizes for Checking Pitch Diameters of Screw Threads. — In checking screw threads by the 3-wire method, the general practice is to use measuring wires of the so-called "best size." The "best size" wire is one which contacts at the pitch line or mid-slope of the thread because then the measurement of the pitch diameter is least affected by an error in the thread angle. In the following formula for determining approximately the "best size" wire or the diameter for pitch-line contact, $A =$ one-half included angle of thread in axial plane.

$$\text{Best size wire} = \frac{0.5 \text{ pitch}}{\cos A} = 0.5 \text{ pitch} \times \sec A$$

For 60-degree threads this formula reduces to

$$\text{Best size wire} = 0.57735 \times \text{pitch}$$

These formulas are based upon a thread groove of zero lead angle because ordinary variations in the lead angle have little effect on the wire diameter and it is desirable to use one wire size for a given pitch regardless of the lead angle. A theoretically correct solution for finding the *exact* size for pitch-line contact involves the use of cumbersome indeterminate equations with solution by successive trials. The accompanying table gives the wire sizes for both American Standard (formerly U. S. Standard) and the Whitworth Standard Threads. The following formulas for determining wire diameters do not give the extreme theoretical limits but the smallest and largest sizes which are practicable. The diameters in the table are based upon these approximate formulas.

American Standard Unified and Metric Threads
$$\begin{cases} \text{Smallest wire diameter} = 0.56 \times \text{pitch} \\ \text{Largest wire diameter} = 0.90 \times \text{pitch} \\ \text{Diameter for pitch-line contact} = 0.57735 \times \text{pitch} \end{cases}$$

Whitworth
$$\begin{cases} \text{Smallest wire diameter} = 0.54 \times \text{pitch} \\ \text{Largest wire diameter} = 0.76 \times \text{pitch} \\ \text{Diameter for pitch-line contact} = 0.56368 \times \text{pitch} \end{cases}$$

Measuring Wire Accuracy. — A set of three measuring wires should have the same diameter within 0.00002 inch. In order to measure the pitch diameter of a screw-thread gage to an accuracy of 0.0001 inch by means of wires, it is necessary to know the wire diameters to 0.00002 inch. If the diameters of the wires are known only to an accuracy of 0.0001 inch, an accuracy better than 0.0003 inch in the measurement of pitch diameter cannot be expected. The wires should be accurately finished hardened steel cylinders of the maximum possible hardness without being brittle. The hardness should not be less than that corresponding to a Knoop indentation number of 630. A wire of this hardness can be cut with a file only with difficulty. The surface should not be rougher than the equivalent of one measuring 3 microinches deviation from a true cylindrical surface.

Diameters of Wires for Measuring American Standard, Unified, Metric, and British Standard Whitworth Screw Threads

Threads per Inch	Pitch, Inch	Wire Diameters for American Standard Threads			Wire Diameters for Whitworth Standard Threads		
		Max.	Min.	Pitch-line Contact	Max.	Min.	Pitch-line Contact
4	0.2500	0.2250	0.1400	0.1443	0.1900	0.1350	0.1409
4½	0.2222	0.2000	0.1244	0.1283	0.1689	0.1200	0.1253
5	0.2000	0.1800	0.1120	0.1155	0.1520	0.1080	0.1127
5½	0.1818	0.1636	0.1018	0.1050	0.1382	0.0982	0.1025
6	0.1667	0.1500	0.0933	0.0962	0.1267	0.0900	0.0939
7	0.1428	0.1286	0.0800	0.0825	0.1086	0.0771	0.0805
8	0.1250	0.1125	0.0700	0.0722	0.0950	0.0675	0.0705

Diameters of Wires for Measuring American Standard, Unified, Metric, and British Standard Whitworth Screw Threads *(Concluded)*

Threads per Inch	Pitch, Inch	Wire Diameters for American Standard Threads			Wire Diameters for Whitworth Standard Threads		
		Max.	Min.	Pitch-line Contact	Max.	Min.	Pitch-line Contact
9	0.1111	0.1000	0.0622	0.0641	0.0844	0.0600	0.0626
10	0.1000	0.0900	0.0560	0.0577	0.0760	0.0540	0.0564
11	0.0909	0.0818	0.0509	0.0525	0.0691	0.0491	0.0512
12	0.0833	0.0750	0.0467	0.0481	0.0633	0.0450	0.0470
13	0.0769	0.0692	0.0431	0.0444	0.0585	0.0415	0.0434
14	0.0714	0.0643	0.0400	0.0412	0.0543	0.0386	0.0403
16	0.0625	0.0562	0.0350	0.0361	0.0475	0.0337	0.0352
18	0.0555	0.0500	0.0311	0.0321	0.0422	0.0300	0.0313
20	0.0500	0.0450	0.0280	0.0289	0.0380	0.0270	0.0282
22	0.0454	0.0409	0.0254	0.0262	0.0345	0.0245	0.0256
24	0.0417	0.0375	0.0233	0.0240	0.0317	0.0225	0.0235
28	0.0357	0.0321	0.0200	0.0206	0.0271	0.0193	0.0201
32	0.0312	0.0281	0.0175	0.0180	0.0237	0.0169	0.0176
36	0.0278	0.0250	0.0156	0.0160	0.0211	0.0150	0.0156
40	0.0250	0.0225	0.0140	0.0144	0.0190	0.0135	0.0141

Diameters of Wires for Measuring Metric Threads

Pitch		Wire Diameters (in.)			Wire Diameters (mm)		
mm	in.	Max.	Min.	Pitch-line Contact	Max.	Min.	Pitch-line Contact
0.35	0.01378	0.01392	0.00696	0.00796	0.35357	0.17678	0.20218
0.40	0.01575	0.01591	0.00796	0.00909	0.40411	0.20218	0.23089
0.45	0.01772	0.01790	0.00895	0.01023	0.45466	0.22733	0.25984
0.50	0.01969	0.01989	0.00994	0.01137	0.50521	0.25278	0.28880
0.60	0.02362	0.02387	0.01193	0.01364	0.60630	0.30302	0.34646
0.70	0.02756	0.02784	0.01392	0.01591	0.70714	0.35357	0.40411
0.75	0.02953	0.02983	0.01492	0.01705	0.75769	0.37897	0.43307
0.8	0.03150	0.03182	0.01591	0.01818	0.80823	0.40411	0.46177
1.0	0.03937	0.03978	0.01989	0.02273	1.01041	0.50521	0.57734
1.25	0.04921	0.04972	0.02486	0.02841	1.26289	0.63144	0.72161
1.50	0.05906	0.05967	0.02983	0.03410	1.51562	0.75768	0.86614
1.75	0.06890	0.06961	0.03481	0.03978	1.76809	0.88417	1.01041
2.0	0.07874	0.07966	0.03978	0.04546	2.02082	1.01041	1.15468
2.5	0.09843	0.09945	0.04972	0.05683	2.52603	1.26289	1.44348
3.0	0.11811	0.11933	0.05967	0.06819	3.03098	1.51562	1.73203
3.5	0.13780	0.13922	0.06961	0.07956	3.53619	1.76809	2.02082
4.0	0.15748	0.15911	0.07955	0.09092	4.04139	2.02057	2.30937
4.5	0.17717	0.17900	0.08950	0.10229	4.54660	2.27330	2.59817
5.0	0.19685	0.19889	0.09945	0.11365	5.05181	2.52603	2.88671
5.5	0.21654	0.21878	0.10939	0.12502	5.55701	2.77851	3.17551
6.0	0.23622	0.23867	0.11933	0.13638	6.06222	3.03098	3.46405

This table may be used when measuring metric threads with either inch or millimeter size wires. Inch values are based on the pitch, in inches, converted from the millimeter sizes, multiplied by the wire size values for the maximum, minimum, and best wire sizes. Millimeter values are direct calculations for the appropriate wire sizes.

ASTM and SAE Grade Markings for Steel Bolts and Screws
(ANSI B18.2.1-1981, Appendix III)

Grade Marking	Specification	Material
NO MARK	SAE — Grade 1	Low or Medium Carbon Steel
	ASTM — A307	Low Carbon Steel
	SAE — Grade 2	Low or Medium Carbon Steel
	SAE — Grade 5	Medium Carbon Steel, Quenched and Tempered
	ASTM — A 449	
	SAE — Grade 5.2	Low Carbon Martensite Steel, Quenched and Tempered
A 325	ASTM — A 325 Type 1	Medium Carbon Steel, Quenched and Tempered Radial dashes optional
A 325	ASTM — A 325 Type 2	Low Carbon Martensite Steel, Quenched and Tempered
A 325	ASTM — A 325 Type 3	Atmospheric Corrosion (Weathering) Steel, Quenched and Tempered
BC	ASTM — A 354 Grade BC	Alloy Steel, Quenched and Tempered
	SAE — Grade 7	Medium Carbon Alloy Steel, Quenched and Tempered, Roll Threaded After Heat Treatment
	SAE — Grade 8	Medium Carbon Alloy Steel, Quenched and Tempered
	ASTM — A 354 Grade BD	Alloy Steel, Quenched and Tempered
	SAE — Grade 8.2	Low Carbon Martensite Steel, Quenched and Tempered
A 490	ASTM — A 490 Type 1	Alloy Steel, Quenched and Tempered
A 490	ASTM — A 490 Type 3	Atmospheric Corrosion (Weathering) Steel, Quenched and Tempered

Table 1. American National Standard Hexagon and Spline Socket Head Cap Screws (1960 Series) (ANSI B18.3-1982)

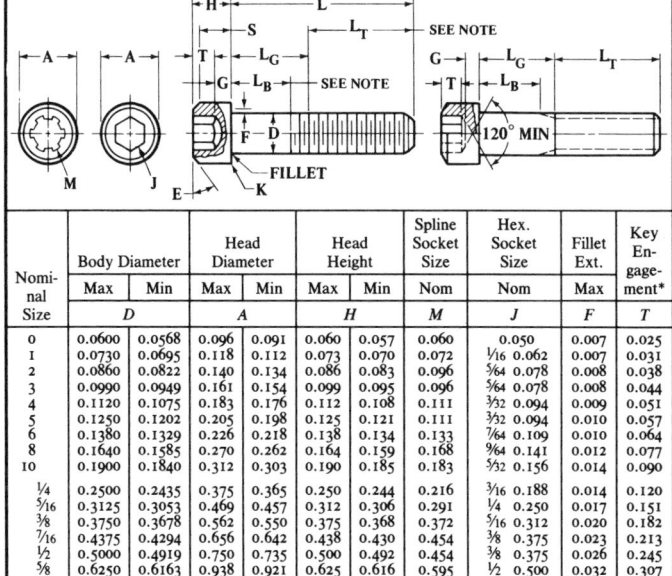

Nominal Size	Body Diameter Max	Body Diameter Min	Head Diameter Max	Head Diameter Min	Head Height Max	Head Height Min	Spline Socket Size Nom	Hex. Socket Size Nom	Fillet Ext. Max	Key Engagement*
	D		A		H		M	J	F	T
0	0.0600	0.0568	0.096	0.091	0.060	0.057	0.060	0.050	0.007	0.025
1	0.0730	0.0695	0.118	0.112	0.073	0.070	0.072	1/16 0.062	0.007	0.031
2	0.0860	0.0822	0.140	0.134	0.086	0.083	0.096	5/64 0.078	0.008	0.038
3	0.0990	0.0949	0.161	0.154	0.099	0.095	0.096	5/64 0.078	0.008	0.044
4	0.1120	0.1075	0.183	0.176	0.112	0.108	0.111	3/32 0.094	0.009	0.051
5	0.1250	0.1202	0.205	0.198	0.125	0.121	0.111	3/32 0.094	0.010	0.057
6	0.1380	0.1329	0.226	0.218	0.138	0.134	0.133	7/64 0.109	0.010	0.064
8	0.1640	0.1585	0.270	0.262	0.164	0.159	0.168	9/64 0.141	0.012	0.077
10	0.1900	0.1840	0.312	0.303	0.190	0.185	0.183	5/32 0.156	0.014	0.090
1/4	0.2500	0.2435	0.375	0.365	0.250	0.244	0.216	3/16 0.188	0.014	0.120
5/16	0.3125	0.3053	0.469	0.457	0.312	0.306	0.291	1/4 0.250	0.017	0.151
3/8	0.3750	0.3678	0.562	0.550	0.375	0.368	0.372	5/16 0.312	0.020	0.182
7/16	0.4375	0.4294	0.656	0.642	0.438	0.430	0.454	3/8 0.375	0.023	0.213
1/2	0.5000	0.4919	0.750	0.735	0.500	0.492	0.454	3/8 0.375	0.026	0.245
5/8	0.6250	0.6163	0.938	0.921	0.625	0.616	0.595	1/2 0.500	0.032	0.307
3/4	0.7500	0.7406	1.125	1.107	0.750	0.740	0.620	5/8 0.625	0.039	0.370
7/8	0.8750	0.8647	1.312	1.293	0.875	0.864	0.698	3/4 0.750	0.044	0.432
1	1.0000	0.9886	1.500	1.479	1.000	0.988	0.790	3/4 0.750	0.050	0.495
1 1/8	1.1250	1.1086	1.688	1.665	1.125	1.111	...	7/8 0.875	0.055	0.557
1 1/4	1.2500	1.2336	1.875	1.852	1.250	1.236	...	7/8 0.875	0.060	0.620
1 3/8	1.3750	1.3568	2.062	2.038	1.375	1.360	...	1 1.000	0.065	0.682
1 1/2	1.5000	1.4818	2.250	2.224	1.500	1.485	...	1 1.000	0.070	0.745
1 3/4	1.7500	1.7295	2.625	2.597	1.750	1.734	...	1 1/4 1.250	0.080	0.870
2	2.0000	1.9780	3.000	2.970	2.000	1.983	...	1 1/2 1.500	0.090	0.995
2 1/4	2.2500	2.2280	3.375	3.344	2.250	2.232	...	1 3/4 1.750	0.100	1.120
2 1/2	2.5000	2.4762	3.750	3.717	2.500	2.481	...	1 3/4 1.750	0.110	1.245
2 3/4	2.7500	2.7262	4.125	4.090	2.750	2.730	...	2 2.000	0.120	1.370
3	3.0000	2.9762	4.500	4.464	3.000	2.979	...	2 1/4 2.250	0.130	1.495
3 1/4	3.2500	3.2262	4.875	4.837	3.250	3.228	...	2 1/4 2.250	0.140	1.620
3 1/2	3.5000	3.4762	5.250	5.211	3.500	3.478	...	2 3/4 2.750	0.150	1.745
3 3/4	3.7500	3.7262	5.625	5.584	3.750	3.727	...	2 3/4 2.750	0.160	1.870
4	4.0000	3.9762	6.000	5.958	4.000	3.976	...	3 3.000	0.170	1.995

* Key engagement depths are minimum.

All dimensions in inches. The body length L_B of the screw is the length of the unthreaded cylindrical portion of the shank. The length of thread, L_T, is the distance from the extreme point to the last complete (full form) thread. Standard length increments for screw diameters up to 1 inch are 1/16 inch for lengths 1/8 through 1/4 inch, 1/8 inch for lengths 1/4 through 1 inch, 1/4 inch for lengths 1 through 3 1/2 inches, 1/2 inch for lengths 3 1/2 through 7 inches, 1 inch for lengths 7 through 10 inches and for diameters over 1 inch are 1/2 inch for lengths 1 through 7 inches, 1 inch for lengths 7 through 10 inches and 2 inches for lengths over 10 inches. Heads may be plain or knurled, and chamfered to an angle E of 30 to 45 degrees with the surface of the flat. The thread conforms to the Unified Standard with radius root, Class 3A, UNRC and UNRF for screw sizes No. 0 through 1 inch inclusive, Class 2A, UNRC and UNRF for over 1 inch through 1 1/2 inches inclusive, and Class 2A UNRC for sizes larger than 1 1/2 inches. Socket dimensions are given in Table 9. For manufacturing details not shown, including materials, see American National Standard ANSI B18.3-1982.

Table 2. Drill and Counterbore Sizes For Socket Head Cap Screws (1960 Series)

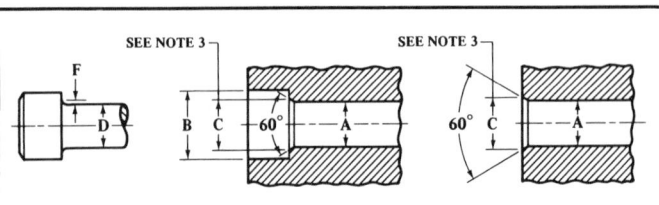

Nominal Size or Basic Screw Diameter	Nominal Drill Size				Counterbore Diameter	Countersink Diameter[3]
	Close Fit[1]		Normal Fit[2]			
	Number or Fractional Size	Decimal Size	Number or Fractional Size	Decimal Size		
	A				B	C
0 0.0600	51	0.067	49	0.073	1/8	0.074
1 0.0703	46	0.081	43	0.089	5/32	0.087
2 0.0860	3/32	0.094	36	0.106	3/16	0.102
3 0.0990	36	0.106	31	0.120	7/32	0.115
4 0.1120	1/8	0.125	29	0.136	7/32	0.130
5 0.1250	9/64	0.141	23	0.154	1/4	0.145
6 0.1380	23	0.154	18	0.170	9/32	0.158
8 0.1640	15	0.180	10	0.194	5/16	0.188
10 0.1900	5	0.206	2	0.221	3/8	0.218
1/4 0.2500	17/64	0.266	9/32	0.281	7/16	0.278
5/16 0.3125	21/64	0.328	11/32	0.344	17/32	0.346
3/8 0.3750	25/64	0.391	13/32	0.406	5/8	0.415
7/16 0.4375	29/64	0.453	15/32	0.469	23/32	0.483
1/2 0.5000	33/64	0.516	17/32	0.531	13/16	0.552
5/8 0.6250	41/64	0.641	21/32	0.656	1	0.689
3/4 0.7500	49/64	0.766	25/32	0.781	1 3/16	0.828
7/8 0.8750	57/64	0.891	29/32	0.906	1 3/8	0.963
1 1.0000	1 1/64	1.016	1 1/32	1.031	1 5/8	1.100
1 1/4 1.2500	1 9/32	1.281	1 5/16	1.312	2	1.370
1 1/2 1.5000	1 17/32	1.531	1 9/16	1.562	2 3/8	1.640
1 3/4 1.7500	1 25/32	1.781	1 13/16	1.812	2 3/4	1.910
2 2.0000	2 1/32	2.031	2 1/16	2.062	3 1/8	2.180

All dimensions in inches.

[1] *Close Fit:* The close fit is normally limited to holes for those lengths of screws which are threaded to the head in assemblies where only one screw is to be used or where two or more screws are to be used and the mating holes are to be produced either at assembly or by matched and coordinated tooling.

[2] *Normal Fit:* The normal fit is intended for screws of relatively long length or for assemblies involving two or more screws where the mating holes are to be produced by conventional tolerancing methods. It provides for the maximum allowable eccentricity of the longest standard screws and for certain variations in the parts to be fastened, such as: deviations in hole straightness, angularity between the axis of the tapped hole and that of the hole for the shank, differences in center distances of the mating holes, etc.

[3] *Countersink:* It is considered good practice to countersink or break the edges of holes which are smaller than (D Max + $2F$ Max) in parts having a hardness which approaches, equals or exceeds the screw hardness. If such holes are not countersunk, the heads of screws may not seat properly or the sharp edges on holes may deform the fillets on screws thereby making them susceptible to fatigue in applications involving dynamic loading. The countersink or corner relief, however, should not be larger than is necessary to insure that the fillet on the screw is cleared.

Source: Appendix to American National Standard ANSI B18.3-1982.

Table 3. Applicability of Hexagon and Spline Keys and Bits

Nominal Key or Bit Size	Cap Screws 1960 Series	Flat Countersunk Head Cap Screws	Button Head Cap Screws	Shoulder Screws	Set Screws
			Nominal Screw Sizes		
HEXAGON KEYS AND BITS					
0.028	...	...	...	...	o
0.035	...	o	o	...	1 & 2
0.050	o	1 & 2	1 & 2	...	3 & 4
1/16 0.062	1	3 & 4	3 & 4	...	5 & 6
5/64 0.078	2 & 3	5 & 6	5 & 6	...	8
3/32 0.094	4 & 5	8	8	...	10
7/64 0.109	6	...	...	...	...
1/8 0.125	...	10	10	1/4	1/4
9/64 0.141	8	...	...	...	...
5/32 0.156	10	1/4	1/4	5/16	5/16
3/16 0.188	1/4	5/16	5/16	3/8	3/8
7/32 0.219	...	3/8	3/8	...	7/16
1/4 0.250	5/16	7/16	...	1/2	1/2
5/16 0.312	3/8	1/2	1/2	5/8	5/8
3/8 0.375	7/16 & 1/2	5/8	5/8	3/4	3/4
7/16 0.438	...	...	...	...	...
1/2 0.500	5/8	3/4	...	1	7/8
9/16 0.562	...	7/8	...	...	1 & 1 1/8
5/8 0.625	3/4	1	...	1 1/4	1 1/4 & 1 3/8
3/4 0.750	7/8 & 1	1 1/8	...	...	1 1/2
7/8 0.875	1 1/8 & 1 1/4	1 1/4 & 1 3/8	...	1 1/2	...
1 1.000	1 3/8 & 1 1/2	1 1/2	...	1 3/4	1 3/4 & 2
1 1/4 1.250	1 3/4	...	...	2	...
1 1/2 1.500	2	...	...	...	...
1 3/4 1.750	2 1/4 & 2 1/2	...	...	...	...
2 2.000	2 3/4	...	...	...	...
2 1/4 2.250	3 & 3 1/4	...	...	...	...
2 3/4 2.750	3 1/2 & 3 3/4	...	...	...	...
3 3.000	4	...	...	...	...
SPLINE KEYS AND BITS					
0.033	...	...	...	...	o & 1
0.048	...	o	o	...	2 & 3
0.060	o	1 & 2	1 & 2	...	4
0.072	1	3 & 4	3 & 4	...	5 & 6
0.096	2 & 3	5 & 6	5 & 6	...	8
0.111	4 & 5	8	8	...	10
0.133	6	...	...	...	...
0.145	...	10	10	...	1/4
0.168	8	...	...	...	...
0.183	10	1/4	1/4	...	5/16
0.216	1/4	5/16	5/16	...	3/8
0.251	...	3/8	3/8	...	7/16
0.291	5/16	7/16	...	...	1/2
0.372	3/8	1/2	1/2	...	5/8
0.454	7/16 & 1/2	5/8 & 3/4	5/8	...	3/4
0.595	5/8	...	...	...	7/8
0.620	3/4	...	...	...	...
0.698	7/8	...	...	...	...
0.790	1	...	...	...	...

Source: Appendix to American National Standard ANSI B18.3-1982.

Table 4. American National Standard Socket Head Cap Screws—Metric Series
(ANSI B18.3.1M-1982)

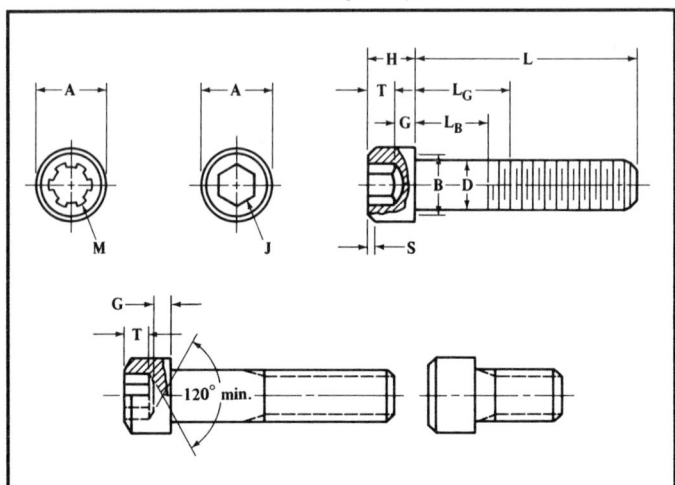

Nom. Size and Thread Pitch	Body Diameter, D		Head Diameter A		Head Height H		Chamfer or Radius S	Hexagon Socket Size‡ J	Spline Socket Size‡ M	Key Engagement T	Transition Diam. B†
	Max	Min	Max	Min	Max	Min	Max	Nom.	Nom.	Min	Max
M1.6 × 0.35	1.60	1.46	3.00	2.87	1.60	1.52	0.16	1.5	1.829	0.80	2.0
M2 × 0.4	2.00	1.86	3.80	3.65	2.00	1.91	0.20	1.5	1.829	1.00	2.6
M2.5 × 0.45	2.50	2.36	4.50	4.33	2.50	2.40	0.25	2.0	2.438	1.25	3.1
M3 × 0.5	3.00	2.86	5.50	5.32	3.00	2.89	0.30	2.5	2.819	1.50	3.6
M4 × 0.7	4.00	3.82	7.00	6.80	4.00	3.88	0.40	3.0	3.378	2.00	4.7
M5 × 0.8	5.00	4.82	8.50	8.27	5.00	4.86	0.50	4.0	4.648	2.50	5.7
M6 × 1	6.00	5.82	10.00	9.74	6.00	5.85	0.60	5.0	5.486	3.00	6.8
M8 × 1.25	8.00	7.78	13.00	12.70	8.00	7.83	0.80	6.0	7.391	4.00	9.2
M10 × 1.5	10.00	9.78	16.00	15.67	10.00	9.81	1.00	8.0	. . .	5.00	11.2
M12 × 1.75	12.00	11.73	18.00	17.63	12.00	11.79	1.20	10.0	. . .	6.00	14.2
M14 × 2*	14.00	13.73	21.00	20.60	14.00	13.77	1.40	12.0	. . .	7.00	16.2
M16 × 2	16.00	15.73	24.00	23.58	16.00	15.76	1.60	14.0	. . .	8.00	18.2
M20 × 2.5	20.00	19.67	30.00	29.53	20.00	19.73	2.00	17.0	. . .	10.00	22.4
M24 × 3	24.00	23.67	36.00	35.48	24.00	23.70	2.40	19.0	. . .	12.00	26.4
M30 × 3.5	30.00	29.67	45.00	44.42	30.00	29.67	3.00	22.0	. . .	15.00	33.4
M36 × 4	36.00	35.61	54.00	53.37	36.00	35.64	3.60	27.0	. . .	18.00	39.4
M42 × 4.5	42.00	41.61	63.00	62.31	42.00	41.61	4.20	32.0	. . .	21.00	45.6
M48 × 5	48.00	47.61	72.00	71.27	48.00	47.58	4.80	36.0	. . .	24.00	52.6

All dimensions are in millimeters
* The M14 × 2 size is not recommended for use in new designs.
† See Countersink footnote in Table 8. ‡ See also Table 6.
L_G is grip length and L_B is body length (see Table 5).
For length of complete thread, see Table 5.
For additional manufacturing and acceptance specifications, see Standard.

Table 5. Socket Head Cap Screws (Metric Series) — Length of Complete Thread
(ANSI B18.3.1M-1982)

Nominal Size	Length of Complete Thread, L_T	Nominal Size	Length of Complete Thread, L_T	Nominal Size	Length of Complete Thread, L_T
M1.6	15.2	M6	24.0	M20	52.0
M2	16.0	M8	28.0	M24	60.0
M2.5	17.0	M10	32.0	M30	72.0
M3	18.0	M12	36.0	M36	84.0
M4	20.0	M14	40.0	M42	96.0
M5	22.0	M16	44.0	M48	108.0

Grip length, L_G, equals screw length, L, minus L_T. Total length of thread L_{TT} equals L_T plus 5 times the pitch of the coarse thread for the respective screw size. Body length L_B equals L minus L_{TT}.

Table 6. American National Standard Hexagon and Spline Sockets for Socket Head Cap Screws — Metric Series (ANSI B18.3.1M-1982)

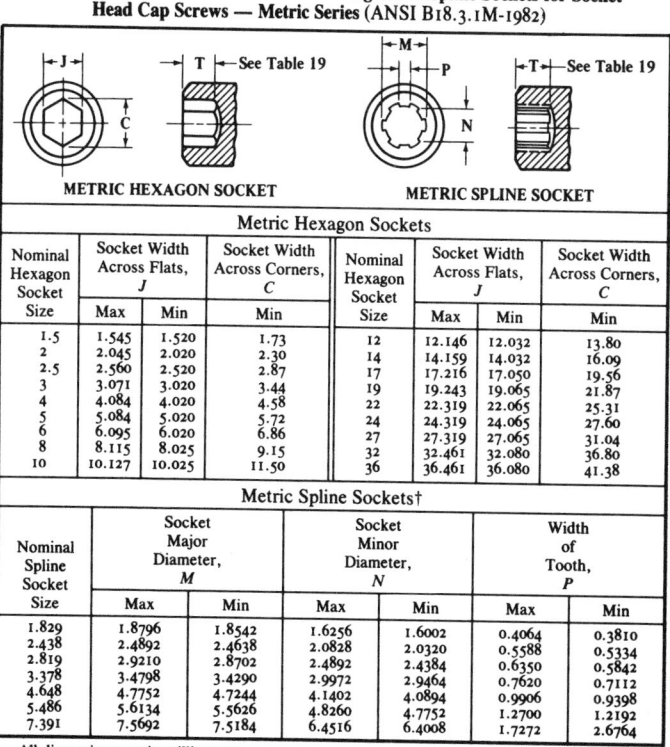

METRIC HEXAGON SOCKET   **METRIC SPLINE SOCKET**

Metric Hexagon Sockets

Nominal Hexagon Socket Size	Socket Width Across Flats, J		Socket Width Across Corners, C	Nominal Hexagon Socket Size	Socket Width Across Flats, J		Socket Width Across Corners, C
	Max	Min	Min		Max	Min	Min
1.5	1.545	1.520	1.73	12	12.146	12.032	13.80
2	2.045	2.020	2.30	14	14.159	14.032	16.09
2.5	2.560	2.520	2.87	17	17.216	17.050	19.56
3	3.071	3.020	3.44	19	19.243	19.065	21.87
4	4.084	4.020	4.58	22	22.319	22.065	25.31
5	5.084	5.020	5.72	24	24.319	24.065	27.60
6	6.095	6.020	6.86	27	27.319	27.065	31.04
8	8.115	8.025	9.15	32	32.461	32.080	36.80
10	10.127	10.025	11.50	36	36.461	36.080	41.38

Metric Spline Sockets†

Nominal Spline Socket Size	Socket Major Diameter, M		Socket Minor Diameter, N		Width of Tooth, P	
	Max	Min	Max	Min	Max	Min
1.829	1.8796	1.8542	1.6256	1.6002	0.4064	0.3810
2.438	2.4892	2.4638	2.0828	2.0320	0.5588	0.5334
2.819	2.9210	2.8702	2.4892	2.4384	0.6350	0.5842
3.378	3.4798	3.4290	2.9972	2.9464	0.7620	0.7112
4.648	4.7752	4.7244	4.1402	4.0894	0.9906	0.9398
5.486	5.6134	5.5626	4.8260	4.7752	1.2700	1.2192
7.391	7.5692	7.5184	6.4516	6.4008	1.7272	2.6764

All dimensions are in millimeters.

† The tabulated dimensions represent direct metric conversions of the equivalent inch size spline sockets shown in American National Standard Socket Cap, Shoulder and Set Screws — Inch Series (ANSI B18.3). Therefore, the spline keys and bits shown therein are applicable for wrenching the corresponding size metric spline sockets.

Table 7. Diameter-Length Combinations for Socket Head Cap Screws (Metric Series)

Nominal Length, L	Nominal Size													
	M1	M2	M2.5	M3	M4	M5	M6	M8	M10	M12	M14	M16	M20	M24
20	X	X												
25	X	X	X	X										
30	X	X	X	X	X	X	X							
35		X	X	X	X	X	X	X						
40		X	X	X	X	X	X	X	X					
45			X	X	X	X	X	X	X	X				
50				X	X	X	X	X	X	X	X			
55				X	X	X	X	X	X	X	X	X		
60				X	X	X	X	X	X	X	X	X	X	
65					X	X	X	X	X	X	X	X	X	X
70					X	X	X	X	X	X	X	X	X	X
80						X	X	X	X	X	X	X	X	X
90							X	X	X	X	X	X	X	X
100								X	X	X	X	X	X	X
110								X	X	X	X	X	X	X
120								X	X	X	X	X	X	X
130								X	X	X	X	X	X	X
140								X	X	X	X	X	X	X
150								X	X	X	X	X	X	X
160								X	X	X	X	X	X	X
180								X	X	X	X	X	X	X
200									X	X	X	X	X	X
220										X	X	X	X	X
240											X	X	X	X
260											X	X	X	X
300												X	X	X

All dimensions are in millimeters.

Screws with lengths above heavy cross lines are threaded full length.

In addition to the lengths shown, the following lengths are standard: 2.5, 3, 4, 5, 6, 8, 10, 12, AND 16 mm. No diameter-length combinations are given in the Standard for these lengths.

Diameter-length combinations are indicated by the symbol X.

Screws larger than M24 with lengths equal to or shorter than L_{TT} (see Table 5) are threaded full-length.

Table 8. Drill and Counterbore Sizes for Metric Socket Head Cap Screws

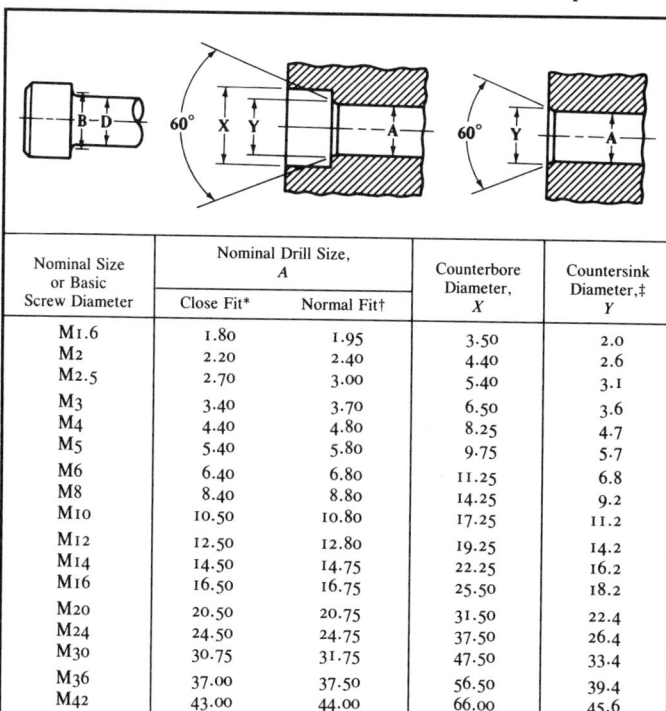

Nominal Size or Basic Screw Diameter	Nominal Drill Size, A		Counterbore Diameter, X	Countersink Diameter,‡ Y
	Close Fit*	Normal Fit†		
M1.6	1.80	1.95	3.50	2.0
M2	2.20	2.40	4.40	2.6
M2.5	2.70	3.00	5.40	3.1
M3	3.40	3.70	6.50	3.6
M4	4.40	4.80	8.25	4.7
M5	5.40	5.80	9.75	5.7
M6	6.40	6.80	11.25	6.8
M8	8.40	8.80	14.25	9.2
M10	10.50	10.80	17.25	11.2
M12	12.50	12.80	19.25	14.2
M14	14.50	14.75	22.25	16.2
M16	16.50	16.75	25.50	18.2
M20	20.50	20.75	31.50	22.4
M24	24.50	24.75	37.50	26.4
M30	30.75	31.75	47.50	33.4
M36	37.00	37.50	56.50	39.4
M42	43.00	44.00	66.00	45.6
M48	49.00	50.00	75.00	52.6

All dimensions are in millimeters.

* *Close Fit:* The close fit is normally limited to holes for those lengths of screws which are threaded to the head in assemblies where only one screw is to be used or where two or more screws are to be used and the mating holes are to be produced either at assembly or by matched and coordinated tooling.

† *Normal Fit:* The normal fit is intended for screws of relatively long length or for assemblies involving two or more screws where the mating holes are to be produced by conventional tolerancing methods. It provides for the maximum allowable eccentricity of the longest standard screws and for certain variations in the parts to be fastened, such as: deviations in hole straightness, angularity between the axis of the tapped hole and that of the hole for shank, differences in center distances of the mating holes, etc.

‡ *Countersink:* It is considered good practice to countersink or break the edges of holes which are smaller than *B* Max. (see Table 4) in parts having a hardness which approaches, equals, or exceeds the screw hardness. If such holes are not countersunk, the heads of screws may not seat properly or the sharp edges on holes may deform the fillets on screws, thereby making them susceptible to fatigue in applications involving dynamic loading. The countersink or corner relief, however, should not be larger than is necessary to ensure that the fillet on the screw is cleared. Normally, the diameter of countersink does not have to exceed *B* Max. Countersinks or corner relief in excess of this diameter reduce the effective bearing area and introduce the possibility of embedment where the parts to be fastened are harder than the screws.

Table 9. American National Standard Hardened Ground Machine Dowel Pins (ANSI B18.8.2-1978)

Nominal Size[1] or Nominal Pin Diameter	Pin Diameter, A						Point Diameter, B		Crown Height or Radius, C		Range of Preferred Lengths,[2] L	Double Shear Load, Min, lb for Carbon or Alloy Steel	Suggested Hole Diameter[3]	
	Standard Series Pins			Oversize Series Pins										
	Basic	Max	Min	Basic	Max	Min	Max	Min	Max	Min			Max	Min
1/16 0.0625	0.0627	0.0628	0.0626	0.0635	0.0636	0.0634	0.058	0.048	0.020	0.008	3/16-3/4	800	0.0625	0.0620
*5/64 0.0781	0.0783	0.0784	0.0782	0.0791	0.0792	0.0790	0.074	0.064	0.026	0.010	⋯	1,240	0.0781	0.0776
3/32 0.0938	0.0940	0.0941	0.0939	0.0948	0.0949	0.0947	0.089	0.079	0.031	0.012	5/16-1	1,800	0.0937	0.0932
1/8 0.1250	0.1252	0.1253	0.1251	0.1260	0.1261	0.1259	0.120	0.110	0.041	0.016	3/8-1	3,200	0.1250	0.1245
*5/32 0.1562	0.1564	0.1565	0.1563	0.1572	0.1573	0.1571	0.150	0.140	0.052	0.020	⋯	5,000	0.1562	0.1557
3/16 0.1875	0.1877	0.1878	0.1876	0.1885	0.1886	0.1884	0.180	0.170	0.062	0.023	⋯	7,200	0.1875	0.1870
1/4 0.2500	0.2502	0.2503	0.2501	0.2510	0.2511	0.2509	0.240	0.230	0.083	0.031	1/2-2 1/2	12,800	0.2500	0.2495
5/16 0.3125	0.3127	0.3128	0.3126	0.3135	0.3136	0.3134	0.302	0.290	0.104	0.039	1/2-2 1/2	20,000	0.3125	0.3120
3/8 0.3750	0.3752	0.3753	0.3751	0.3760	0.3761	0.3759	0.365	0.350	0.125	0.047	1/2-3	28,700	0.3750	0.3745
7/16 0.4375	0.4377	0.4378	0.4376	0.4385	0.4386	0.4384	0.424	0.409	0.146	0.055	7/8-3	39,100	0.4375	0.4370
1/2 0.5000	0.5002	0.5003	0.5001	0.5010	0.5011	0.5009	0.486	0.471	0.167	0.063	3/4,1-4	51,000	0.5000	0.4995
5/8 0.6250	0.6252	0.6253	0.6251	0.6260	0.6261	0.6259	0.611	0.595	0.208	0.078	1 1/4-5	79,800	0.6250	0.6245
3/4 0.7500	0.7502	0.7503	0.7501	0.7510	0.7511	0.7509	0.735	0.715	0.250	0.094	1 1/2-6	114,000	0.7500	0.7495
7/8 0.8750	0.8752	0.8753	0.8751	0.8760	0.8761	0.8759	0.860	0.840	0.293	0.109	2,2 1/2-6	156,000	0.8750	0.8745
1 1.0000	1.0002	1.0003	1.0001	1.0010	1.0011	1.0009	0.980	0.960	0.333	0.125	2,2 1/2-5,6	204,000	1.0000	0.9995

All dimensions are in inches.

* Nonpreferred sizes, not recommended for use in new designs.

[1] Where specifying nominal size as basic diameter, zeros preceding decimal and in the fourth decimal place are omitted.

[2] Lengths increase in 1/16-inch steps up to 3/8 inch, in 1/8-inch steps from 3/8 inch to 1 inch, in 1/4-inch steps from 1 inch to 2 1/2 inches, and in 1/2-inch steps above 2 1/2 inches. Tolerance on length is ±0.010 inch.

[3] These hole sizes have been commonly used for press fitting Standard Series machine dowel pins into materials such as mild steels and cast iron. In soft materials such as aluminum or zinc die castings, hole size limits are usually decreased by 0.0005 inch to increase the press fit.

Table 10. American National Standard Taper Pins (ANSI B18.8.2-1978)

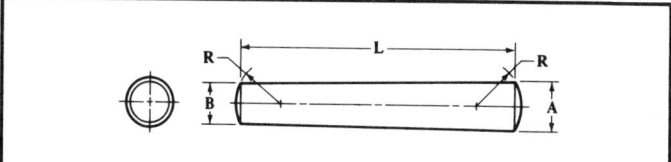

Pin Size Number and Basic Pin Diameter[1]	Major Diameter (Large End), A				End Crown Radius, R		Range of Lengths,[2] L	
	Commercial Class		Precision Class					
	Max	Min	Max	Min	Max	Min	Stand. Reamer Avail.[3]	Other
7/0 0.0625	0.0638	0.0618	0.0635	0.0625	0.072	0.052	...	¼–1
6/0 0.0780	0.0793	0.0773	0.0790	0.0780	0.088	0.068	...	¼–1½
5/0 0.0940	0.0953	0.0933	0.0950	0.0940	0.104	0.084	¼–1	1¼, 1½
4/0 0.1090	0.1103	0.1083	0.1100	0.1090	0.119	0.099	¼–1	1¼–2
3/0 0.1250	0.1263	0.1243	0.1260	0.1250	0.135	0.115	¼–1	1¼–2
2/0 0.1410	0.1423	0.1403	0.1420	0.1410	0.151	0.131	½–1¼	1½–2½
0 0.1560	0.1573	0.1553	0.1570	0.1560	0.166	0.146	½–1¼	1½–3
1 0.1720	0.1733	0.1713	0.1730	0.1720	0.182	0.162	¾–1¼	1½–3
2 0.1930	0.1943	0.1923	0.1940	0.1930	0.203	0.183	¾–1½	1¾–3
3 0.2190	0.2203	0.2183	0.2200	0.2190	0.229	0.209	¾–1¾	2–4
4 0.2500	0.2513	0.2493	0.2510	0.2500	0.260	0.240	¾–2	2¼–4
5 0.2890	0.2903	0.2883	0.2900	0.2890	0.299	0.279	1–2½	2¾–6
6 0.3410	0.3423	0.3403	0.3420	0.3410	0.351	0.331	1¼–3	3¼–6
7 0.4090	0.4103	0.4083	0.4100	0.4090	0.419	0.399	1¼–3¾	4–8
8 0.4920	0.4933	0.4913	0.4930	0.4920	0.502	0.482	1¼–4½	4¾–8
9 0.5910	0.5923	0.5903	0.5920	0.5910	0.601	0.581	1¼–5¼	5½–8
10 0.7060	0.7073	0.7053	0.7070	0.7060	0.716	0.696	1½–6	6¼–8
11 0.8600	0.8613	0.8593	...	...	0.870	0.850	...	2–8
12 1.0320	1.0333	1.0313	...	...	1.042	1.022	...	2–9
13 1.2410	1.2423	1.2403	...	...	1.251	1.231	...	3–11
14 1.5210	1.5223	1.5203	...	...	1.531	1.511	...	3–13

All dimensions are in inches.
[1] When specifying nominal pin size in decimals, zeros preceding the decimal and in the fourth decimal place are omitted.
[2] Lengths increase in ⅛-inch steps up to 1 inch and in ¼-inch steps above 1 inch.
[3] Standard reamers are available for pin lengths in this column.

To determine the number and sizes of step drills required: (1) find the length of pin to be used at the top of the chart on page 288 and follow this length down to the intersection with that heavy line which represents the size of taper pin (see taper pin numbers at the right-hand end of each heavy line). (2) If the length of pin falls between the first and second dots, counting from the left, only one drill is required. Its size is indicated by following the nearest horizontal line from the point of intersection (of the pin length) on the heavy line over to the drill diameter values at the left. (3) If the intersection of pin length comes between the second and third dots, then two drills are required. In this case the smaller has a size corresponding to the intersection of the pin length and heavy line and the larger is the corresponding drill diameter for the intersection of one-half this length with the heavy line. (4) Should the pin length fall between the third and fourth dots, then three drills are required. The smallest will have a diameter corresponding to the intersection of the total pin length with the heavy line, the next in size will have a diameter corresponding to the intersection of two-thirds of this length with the heavy line and the largest will have a diameter corresponding to the intersection of one-third of this length with the heavy line. Where the intersection falls between two drill sizes, use the smaller.

Chart to Facilitate Selection of Number and Sizes of Drills for Step-Drilling Prior to Taper Reaming

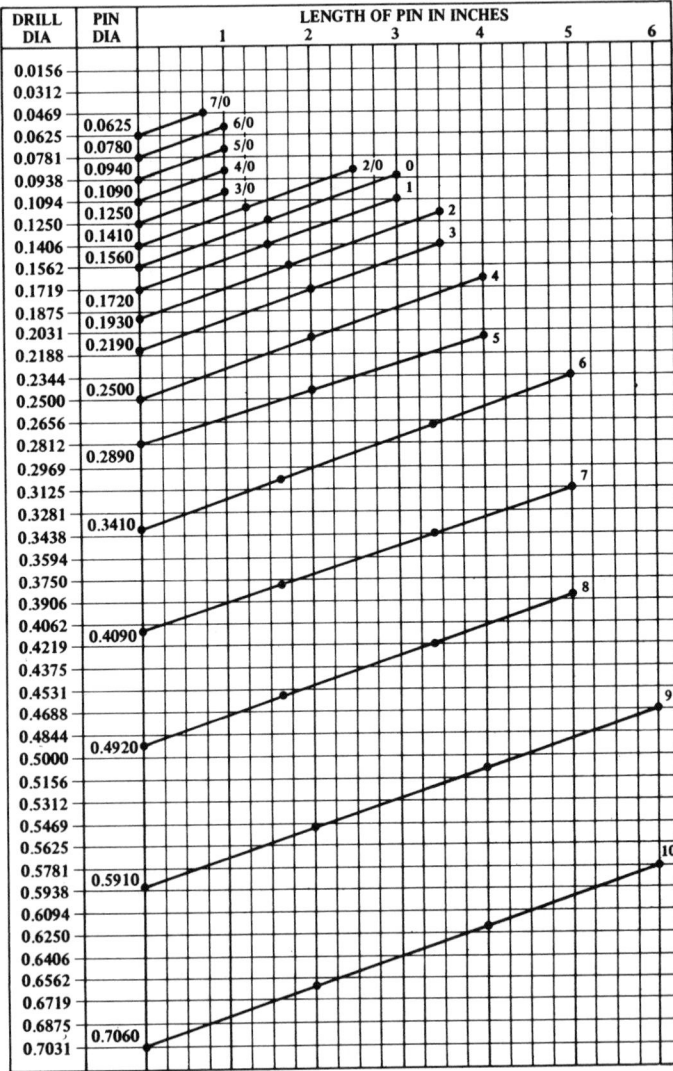

Table 11. Nominal Diameter at Small Ends of Standard Taper Pins

Pin Length in Inches	Pin Number and Small End Diameter for Given Length										
	0	1	2	3	4	5	6	7	8	9	10
¾	0.140	0.156	0.177	0.203	0.235	0.273	0.325	0.393	0.476	0.575	0.690
1	0.135	0.151	0.172	0.198	0.230	0.268	0.320	0.388	0.471	0.570	0.685
1¼	0.130	0.146	0.167	0.192	0.224	0.263	0.315	0.382	0.466	0.565	0.680
1½	0.125	0.141	0.162	0.187	0.219	0.258	0.310	0.377	0.460	0.560	0.675
1¾	0.120	0.136	0.157	0.182	0.214	0.252	0.305	0.372	0.455	0.554	0.669
2	0.114	0.130	0.151	0.177	0.209	0.247	0.299	0.367	0.450	0.549	0.664
2¼	0.109	0.125	0.146	0.172	0.204	0.242	0.294	0.362	0.445	0.544	0.659
2½	0.104	0.120	0.141	0.166	0.198	0.237	0.289	0.356	0.440	0.539	0.654
2¾	0.099	0.115	0.136	0.161	0.193	0.232	0.284	0.351	0.434	0.534	0.649
3	0.094	0.110	0.131	0.156	0.188	0.227	0.279	0.346	0.429	0.528	0.643
3¼	...	...	...	0.151	0.182	0.221	0.273	0.340	0.424	0.523	0.638
3½	...	...	...	0.146	0.177	0.216	0.268	0.335	0.419	0.518	0.633
3¾	...	...	...	0.141	0.172	0.211	0.263	0.330	0.414	0.513	0.628
4	...	...	...	0.136	0.167	0.206	0.258	0.326	0.409	0.508	0.623
4¼	...	...	...	0.131	0.162	0.201	0.253	0.321	0.403	0.502	0.617
4½	...	...	...	0.125	0.156	0.195	0.247	0.315	0.398	0.497	0.612
5	...	...	...	...	0.146	0.185	0.237	0.305	0.389	0.487	0.602
5½	...	...	...	...	...	...	...	0.294	0.377	0.476	0.591
6	...	...	...	...	...	...	...	0.284	0.367	0.466	0.581

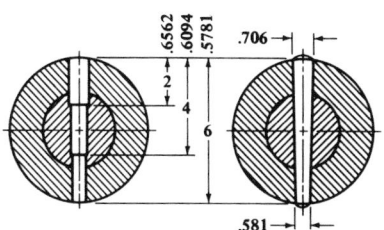

Examples; For a No. 10 taper pin 6-inches long, three drills would be used, of the sizes and for the depths shown in the accompanying diagram.

For a No. 10 taper pin 3-inches long, two drills would be used since the 3-inch length falls between the second and third dots. The first or through drill will be 0.6406 inch and the second drill, 0.6719 inch for a depth of 1½ inches.

GEARS

Gears are a very important part of many mechanical assemblies. They transmit power and torque, increase or decrease speed, and provide a constant slip-free rate of speed. The most common types of gears in use today are spur gears, helical gears, bevel gears, and worm gears, Fig. 1. In addition to these basic gear forms there are also several variations, or modified forms, of these basic gear types. These include herringbone, hypoid, spiral bevel, and elliptical gears.

Spur Gears. — Spur gears are the simplest and most common type of gears used in industry today. These gears have straight teeth cut parallel with the axis of rotation of the gear body. Spur gear dimensions, like other gear dimensions on a shop print, are generally divided into two groups: gear blank dimensions and gear cutting dimensions. Gear blank dimensions deal directly with the gear blank and are usually shown on the drawn detail. The gear cutting dimensions refer to the actual cutting of the gear and are generally shown in a separate data block on the print. Many times the gear cutting dimensions will include the note (REF). In these cases the note is used to indicate a calculated dimension that cannot be measured directly.

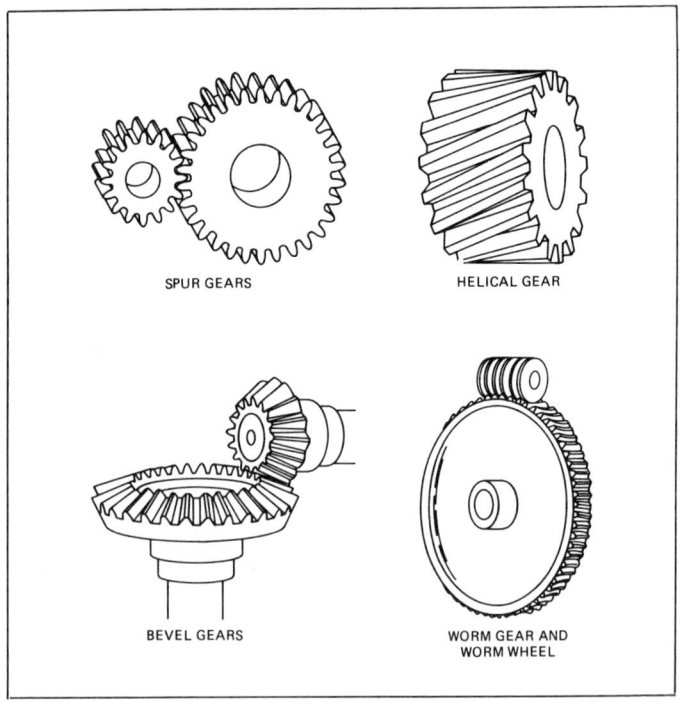

SPUR GEARS

HELICAL GEAR

BEVEL GEARS

WORM GEAR AND
WORM WHEEL

Courtesy of South-Western Publishing Co.

Fig. 1. Type of Gears

Spur gears are usually shown in prints by a front view and a full section side view. Rarely are complete gear profiles shown on prints. Generally the addendum circle and the dedendum circle are shown with phantom lines and the pitch circle is shown with a center line. Occasionally a few gear teeth may be drawn for clarity and to avoid confusion. In addition to the external spur gears, there are also two other spur gear forms: the internal gear and the rack. An internal spur gear is basically the same as an external gear with the exception of the location and the shape of the teeth. Internal spur gear teeth have the same shape as the spaces between the teeth of the mating external spur gear. A rack gear is simply a flat gear. A rack and spur gear are often used to transfer rotary motion into linear motion or vice versa.

Helical Gears. — Helical gears are a variation of the spur gear that have teeth cut at an angle to the axis of rotation. Helical gears are cut by advancing a cutter across the gear blank while the blank slowly revolves. This method produces a tooth form which is smoother running and quieter than normal spur gears. Spur gears engage across the entire face of the tooth while helical gears engage on a small area which moves across the face of the tooth.

Helical gears may be used for parallel gear shafts, or shafts that are at an angle to each other. Helical gears are made as either right hand or left hand. The hand of a gear may be determined by looking at the tooth. If the tooth runs down on the right side it is a right-hand helix, likewise, if it runs down on the left side it is a left-hand helical gear. Helical gears are usually specified in much the same way as spur gears. The only major difference is in the additions to the gear cutting data block. In most cases the drawn view of the helical gear is the same as that used for a spur gear.

Bevel Gears. — Bevel gears are characterized by a variable pitch diameter, tooth depth, and tooth thickness. Like helical gears, bevel gears are also used for applications where the gears must be positioned at an angle to each other. The most common type of bevel gears run at a 90° angle. These gears are called miter gears. Bevel gears can best be compared to cones that have teeth. Unlike spur gears and helical gears which engage on their outer surfaces, bevel gears engage, or mesh, along the angular side of their conical form. Bevel gears are normally shown as full sections on most prints. When two mating bevel gears are required, they may both be shown in their operational position. In this case the gear cutting data block is divided into two sections, one for the driver and one for the driven gear. Depending on their construction they may also be called a pinion (driver) and ring (driven) gear.

Worm Gears. — Worm gearing is primarily used to obtain great reductions in speed and an increase in power. The two basic parts that are included in a worm drive are the worm and the worm gear. Owing to their basic design, the worm is always the driver and the worm gear is always the driven gear in this sytem. Worm gearing can be compared to an endless screw. As the worm rotates the worm gear revolves, much the same as the movement of a nut along a bolt. The helix angle of the thread is what causes the worm to drive the worm gear. So the greater the helix angle the faster the movement. Worms and worm gears are sometimes shown as individual details in a print. In some cases only the worm gear will be shown and the cutting data of the worm will be shown in the data block along with the data necessary to make the worm gear. As is the case with other forms of gears, the drawn views are normally intended to show the dimensions needed to make the blank while all relevant gear cutting data is contained in a special data block on the print.

Metric Gears. — Metric gears use basically the same values as inch type gears. The only major difference is the use of the term module in place of diametral pitch and circular pitch. Where the term diametral pitch is defined as the ratio of the number of teeth per inch of pitch diameter, the module is equal to the length of pitch diameter per tooth expressed in millimeters. So, rather than a ratio, the module is a dimension.

Definitions of Gear Terms. — The terms which follow are commonly applied to various classes of gearing.

Addendum: Height of tooth above pitch circle or the radial distance between the pitch circle and the top of the tooth (see illustration).

Approach Ratio: The ratio of the arc of approach to the arc of action.

Arc of Action: Arc of the pitch circle through which a tooth travels from the first point of contact with the mating tooth to the point where contact ceases.

Arc of Approach: Arc of the pitch circle through which a tooth travels from the first point of contact with the mating tooth to the pitch point.

Arc of Recess: Arc of the pitch circle through which a tooth travels from its contact with the mating tooth at the pitch point to the point where its contact ceases.

Axial Plane: In a pair of gears it is the plane that contains the two axes. In a single gear, it may be any plane containing the axis and a given point.

Backlash: The amount by which the width of a tooth space exceeds the thickness of the engaging tooth on the pitch circles. As actually indicated by measuring devices, backlash may be determined variously in the transverse, normal or axial planes, and either in the direction of the pitch circles or on the line of action. Such measurements should be converted to corresponding values on transverse pitch circles for general comparison.

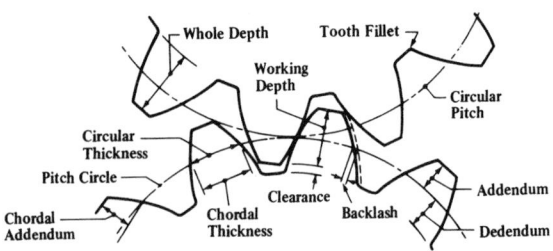

Gear Tooth Parts

Base Circle: The circle from which an involute tooth curve is generated or developed.

Base Helix Angle: The angle, at the base cylinder of an involute gear, that the tooth makes with the gear axis.

Base Pitch: In an involute gear it is the pitch on the base circle or along the line of action. Corresponding sides of involute teeth are parallel curves, and the base pitch is the constant and fundamental distance between them along a common normal in a plane of rotation. The *Normal Base Pitch* is the base pitch in the normal plane, and the *Axial Base Pitch* is the base pitch in the axial plane.

Center Distance: The distance between the parallel axes of spur gears and parallel helical gears, or between the crossed axes of crossed helical gears and worm gears. Also, it is the distance between the centers of the pitch circles.

Central Plane: In a worm gear this is the plane perpendicular to the gear axis and contains the common perpendicular of the gear and worm axes. In the usual case with the axes at right angles, it contains the worm axis.

Chordal Addendum: The height from the top of the tooth to the chord subtending the circular-thickness arc.

Chordal Thickness: Length of the chord subtended by the circular thickness arc (the dimension obtained when a gear-tooth caliper is used to measure the thickness at the pitch circle).

Circular Pitch: Length of the arc of the pitch circle between the centers or other corresponding points of adjacent teeth (see illustration). *Normal Circular Pitch* is the circular pitch in the normal plane.

Circular Thickness: The length of arc between the two sides of a gear tooth, on the pitch circle unless otherwise specified. *Normal Circular Thickness* is the circular thickness in the normal plane.

Clearance: The amount by which the dedendum in a given gear exceeds the addendum of its mating gear. It is also the radial distance between the top of a tooth and the bottom of the mating tooth space.

Contact Diameter: The smallest diameter on a gear tooth with which the mating gear makes contact.

Contact Ratio: The ratio of the arc of action to the circular pitch. It is sometimes thought of as the average number of teeth in contact. For involute gears, the contact ratio is obtained most directly as the ratio of the length of action to the base pitch.

Contact Stress: The maximum compressive stress within the contact area between mating gear tooth profiles. It is also called Hertz stress.

Cycloid: The curve formed by the path of a point on a circle as it rolls along a straight line. When this circle rolls along the outer side of another circle, the curve is called an *Epicycloid*; when it rolls along the inner side of another circle it is called a *Hypocycloid*. These curves are used in defining the former American Standard composite tooth form.

Dedendum: The depth of tooth space below the pitch circle or the radial dimension between the pitch circle and the bottom of the tooth space (see illustration).

Diametral Pitch: The ratio of the number of teeth to the number of inches of pitch diameter — equals number of gear teeth to each inch of pitch diameter. *Normal Diametral Pitch* is the diametral pitch as calculated in the normal plane and is equal to the diametral pitch divided by the cosine of the helix angle.

Effective Face Width: That portion of the face width that actually comes into contact with mating teeth, as occasionally one member of a pair of gears may have a greater face width than the other.

Efficiency: The actual torque ratio of a gear set divided by its gear ratio.

External Gear: A gear with teeth on the outer cylindrical surface.

Face of Tooth: That surface of the tooth which is between the pitch circle and the top of the tooth.

Fillet Curve: The concave portion of the tooth profile where it joins the bottom of the tooth space. The approximate radius of this curve is called the *Fillet Radius*.

Fillet Stress: The maximum tensile stress in the gear tooth fillet.

Flank of Tooth: That surface which is between the pitch circle and the bottom land. The flank includes the fillet.

Helical Overlap: The effective face width of a helical gear divided by the gear axial pitch; also called the *Face Overlap*.

Helix Angle: The angle that a helical gear tooth makes with the gear axis at the pitch circle unless otherwise specified.

Hertz Stress: See Contact Stress.

Highest Point of Single Tooth Contact: The largest diameter on a spur gear at which a single tooth is in contact with the mating gear. Often referred to as HPSTC.

Internal Diameter: The diameter of a circle coinciding with the tops of the teeth of an internal gear.

Internal Gear: A gear with teeth on the inner cylindrical surface.

Involute: The curve formed by the path of a point on a straight line, called the generatrix, as it rolls along a convex base curve. (The base curve is usually a circle.) This curve is generally used as the profile of gear teeth.

Land: The *Top Land* is the top surface of a tooth, and the *Bottom Land* is the surface of the gear between the fillets of adjacent teeth.

Lead: The distance a helical gear or worm would thread along its axis in one revolution if it were free to move axially.

Length of Action: The distance on an involute line of action through which the point of contact moves during the action of the tooth profile.

Line of Action: The path of contact in involute gears. It is the straight line passing through the pitch point and tangent to the base circles.

Lowest Point of Single Tooth Contact: The smallest diameter on a spur gear at which a single tooth of one gear is in contact with its mating gear. Often referred to as LPSTC. Gear set contact stress is determined with a load placed at this point on the pinion.

Module: Ratio of the pitch diameter to the number of teeth. Ordinarily, module is understood to mean ratio of pitch diameter in *millimeters* to the number of teeth. The *English Module* is a ratio of the pitch diameter in inches to the number of teeth.

Normal Plane: A plane normal to the tooth surfaces at a point of contact, and perpendicular to the pitch plane.

Pitch: The distance between similar, equally-spaced tooth surfaces, in a given direction and along a given curve or line. The single word "pitch" without qualification has been used to designate circular pitch, axial pitch, and diametral pitch, but such confusing usage should be avoided.

Pitch Circle: A circle the radius of which is equal to the distance from the gear axis to the pitch point (see illustration).

Pitch Diameter: The diameter of the pitch circle. In parallel shaft gears the pitch diameters can be determined directly from the center distance and the numbers of teeth by proportionality. *Operating Pitch Diameter* is the pitch diameter at which the gears operate. *Generating Pitch Diameter* is the pitch diameter at which the gear is generated. In a bevel gear the pitch diameter is understood to be at the outer ends of the teeth unless otherwise specified. (See also reference to standard pitch diameter under *Pressure Angle*.)

Pitch Plane: In a pair of gears it is the plane perpendicular to the axial plane and tangent to the pitch surfaces. In a single gear it may be any plane tangent to its pitch surface.

Pitch Point: This is the point of tangency of two pitch circles (or of a pitch circle and a pitch line) and is on the line of centers. The pitch point of a tooth profile is at its intersection with the pitch circle.

Plane of Rotation: Any plane perpendicular to a gear axis.

Pressure Angle: The angle between a tooth profile and a radial line at its pitch point. In involute teeth, pressure angle is often described as the angle between the line of action and the line tangent to the pitch circle. *Standard Pressure Angles* are established in connection with standard gear-tooth proportions. A given pair of involute profiles will transmit smooth motion at the same velocity ratio even when the center distance is changed. Changes in center distance, however, in gear design and gear manufacturing operations, are accompanied by changes in pitch diameter, pitch, and pressure angle. Different values of pitch diameter and pressure angle therefore may occur in the same gear under different conditions. Usually in gear design, and unless otherwise specified, the pressure angle is the *standard pressure angle* at the *standard pitch diameter,* and is standard for the hob or cutter used to generate the teeth. The *Operating Pressure Angle* is determined by the center distance at which a pair of gears operates. The *Generating Pressure Angle* is the angle at the pitch diameter in effect when the gear is generated. Other pressure angles may be considered in gear calculations. In gear cutting tools and cutters, the pressure angle indicates the direction of the cutting edge as referred to some principal direction. In oblique teeth, that is helical, spiral, etc., the pressure angle

may be specified in the *transverse, normal,* or *axial* plane. For a spur gear or a straight bevel gear, in which only one direction of cross-section needs to be considered, the general term pressure angle may be used without qualification to indicate transverse pressure angle. In spiral bevel gears, unless otherwise specified, pressure angle means normal pressure angle at the mean cone distance.

Principal Reference Planes: These are a pitch plane, axial plane, and transverse plane, all intersecting at a point and mutually perpendicular.

Rack: A gear with teeth spaced along a straight line, and suitable for straightline motion. A *Basic Rack* is one that is adopted as the basis of a system of interchangeable gears. Standard gear-tooth proportions are often illustrated on an outline of the basic rack (see diagrams on page 621). A *Generating Rack* is a rack outline used to indicate tooth details and dimensions for the design of a required generating tool, such as a hob or gear-shaper cutter.

Ratio of Gearing: Ratio of the numbers of teeth on mating gears. Ordinarily the ratio is found by dividing the number of teeth on the larger gear by the number of teeth on the smaller gear or pinion. For example, if the ratio is 2 or "2 to 1," this usually means that the smaller gear or pinion makes two revolutions to one revolution of the larger mating gear.

Roll Angle: The angle subtended at the center of a base circle from the origin of an involute to the point of tangency of the generatrix from any point on the same involute. The radian measure of this angle is the tangent of the pressure angle of the point on the involute.

Root Circle: A circle coinciding with or tangent to the bottoms of the tooth spaces.

Root Diameter: Diameter of the root circle.

Tangent Plane: A plane tangent to the tooth surfaces at a point or line of contact.

Tip Relief: An arbitrary modification of a tooth profile whereby a small amount of material is removed near the tip of the gear tooth.

Total Face Width: The actual width dimension of a gear blank. It may exceed the effective face width, as in the case of double-helical gears where the total face width includes any distance separating the right-hand and left-hand helical teeth.

Transverse Plane: A plane perpendicular to the axial plane and to the pitch plane. In gears with parallel axes, the transverse plane and the plane of rotation coincide.

Trochoid: The curve formed by the path of a point on the extension of a radius of a circle as it rolls along a curve or line. It is also the curve formed by the path of a point on a perpendicular to a straight line as the straight line rolls along the convex side of a base curve. By the first definition the trochoid is a derivative of the cycloid; by the second definition it is a derivative of the involute.

True Involute Form Diameter: The smallest diameter on the tooth at which the involute exists. Usually this is the point of tangency of the involute tooth profile and the fillet curve. This is usually referred to as the *TIF diameter*.

Undercut: A condition in generated gear teeth when any part of the fillet curve lies inside of a line drawn tangent to the working profile at its lowest point. Undercut may be deliberately introduced to facilitate finishing operations, as in preshaving.

Whole Depth: The total depth of a tooth space, equal to addendum plus dedendum, also equal to working depth plus clearance.

Working Depth: The depth of engagement of two gears, that is, the sum of their addendums. The standard working distance is the depth to which a tooth extends into the tooth space of a mating gear when the center distance is standard.

Definitions of gear terms are given in AGMA Standards 112.05, 115.01, and 116.01 entitled "Terms, Definitions, Symbols and Abbreviations," "Reference Information — Basic Gear Geometry," and "Glossary — Terms Used in Gearing," respectively; obtainable from American Gear Manufacturers Assn., 1901 No. Ft. Myer Dr., Arlington, Va. 22209.

Gear Teeth of Different Diametral Pitch, Full Size

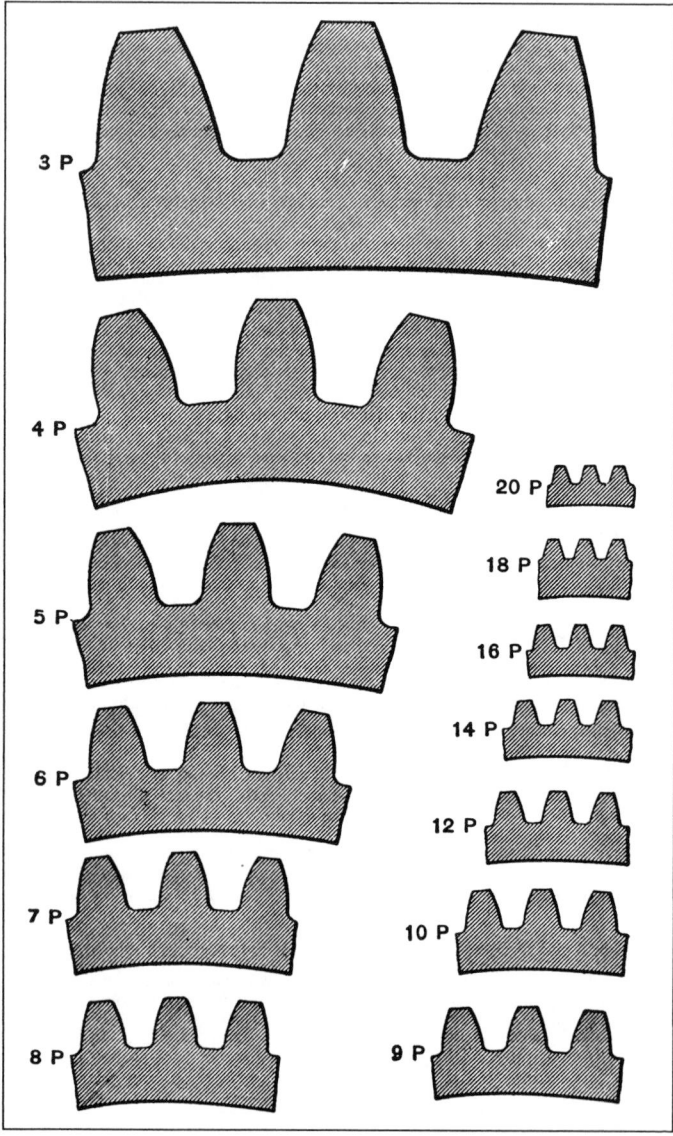

Properties of the Involute Curve. — The involute curve is used almost exclusively for gear-tooth profiles, because of the following important properties.

1. The form or shape of an involute curve depends upon the diameter of the base circle from which it is derived. (If a taut line were unwound from the circumference of a circle — the *base circle* of the involute — the end of that line or any point on the unwound portion, would describe an involute curve.)

2. If a gear tooth of involute curvature acts against the involute tooth of a mating gear while rotating at a uniform rate, the angular motion of the driven gear will also be uniform, even though the center-to-center distance is varied.

3. The relative rate of motion between driving and driven gears having involute tooth curves, is established by the diameters of their base circles.

4. Contact between intermeshing involute teeth on a driving and driven gear is along a straight line that is tangent to the two base circles of these gears. This is the *line of action*.

5. The point where the line of action intersects the common center-line of the mating involute gears, establishes the radii of the pitch circles of these gears; hence true pitch circle diameters are affected by a change in the center distance. (Pitch diameters obtained by dividing the number of teeth by the diametral pitch applies when the center distance equals the total number of teeth on both gears divided by twice the diametral pitch.)

6. The pitch diameters of mating involute gears are directly proportional to the diameters of their respective base circles; thus, if the base circle of one mating gear is three times as large as the other, the pitch circle diameters will be in the same ratio.

7. The angle between the line of action and a line perpendicular to the common center-line of mating gears, is the *pressure angle;* hence the pressure angle is affected by any change in the center distance.

8. When an involute curve acts against a straight line (as in the case of an involute pinion acting against straight-sided rack teeth), the straight line is tangent to the involute and perpendicular to its line of action.

9. The pressure angle, in the case of an involute pinion acting against straight-sided rack teeth, is the angle between the line of action and the line of the rack's motion. If the involute pinion rotates at a uniform rate, movement of the rack will also be uniform.

Diametral and Circular Pitch Systems. — Gear tooth system standards are established by specifying the tooth proportions of the basic rack. The diametral pitch system is applied to most of the gearing produced in the United States. If gear teeth are larger than about one diametral pitch, it is common practice to use the circular pitch system. The circular pitch system is also applied to cast gearing and it is commonly used in connection with the design and manufacture of worm gearing.

Pitch Diameters Obtained with Diametral Pitch System. — The diametral pitch system is arranged to provide a series of standard tooth sizes, the principle being similar to the standardization of screw thread pitches. Inasmuch as there must be a whole number of teeth on each gear, the increase in pitch diameter per tooth varies according to the pitch. For example, the pitch diameter of a gear having, say, 20 teeth of 4 diametral pitch, will be 5 inches; 21 teeth, 5¼ inches; and so on, the increase in diameter for each additional tooth being equal to ¼ inch for 4 diametral pitch. Similarly, for 2 diametral pitch the variations for successive numbers of teeth would equal ½ inch, and for 10 diametral pitch the variations would equal 1/10 inch, etc.

Table for Chordal Thicknesses and Chordal Addenda of Full-depth Teeth. — The table on page 300 gives values for chordal thickness and chordal addendum of full-depth spur gear teeth of 1 diametral pitch and from 10 to 156 teeth for gears of standard outside diameter. For any other diametral pitch the values are to be divided by the required pitch.

Helical Gears: In applying this table to helical gears, especially when the number of teeth is small and the helix angle large, the equivalent number of teeth N_e for entering the table is found by the formula, $N_e = N \div \cos^3 \psi$, where N is the actual number of teeth in the helical gear and ψ is the helix angle. The values obtained from the table should be divided by the *normal* diametral pitch of the helical gear to get the normal chordal thickness and the normal chordal addendum.

Example: Find the normal chordal thickness and the normal addendum of a helical gear having 54 teeth of 6 normal diametral pitch and a helix angle of 45 degrees.

$$N_e = \frac{54}{\cos^3 45°} = \frac{54}{(0.70711)^3} = 153 \text{ teeth}$$

$$\text{Normal chordal thickness} = \frac{1.57077}{6} = 0.26180 \text{ inch}$$

$$\text{Normal chordal addendum} = \frac{1.00405}{6} = 0.16734 \text{ inch}$$

Tables for Chordal Thicknesses and Chordal Addenda of Milled, Full-depth Teeth. — Two convenient tables for checking gears with milled, full-depth teeth are given on pages 302 and 303. The first shows chordal thicknesses and chordal addenda for the lowest number of teeth cut by gear cutters Nos. 1 through 8, and for the commonly used diametral pitches. The second gives similar data for commonly used circular pitches. In each case the data shown are accurate for the number of gear teeth indicated, but are approximate for other numbers of teeth within the range of the cutter under which they appear in the table. For the higher diametral pitches and lower circular pitches, the error introduced by using the data for any tooth number within the range of the cutter under which it appears is comparatively small. The chordal thicknesses and chordal addenda for gear cutters Nos. 1 through 8 of the more commonly used diametral and circular pitches can be obtained from the table and formulas on pages 302 and 303.

Caliper Measurement of Gear Tooth. — In cutting gear teeth, the general practice is to adjust the cutter or hob until it grazes the outside diameter of the blank; the cutter is then sunk to the total depth of the tooth space plus whatever slight additional amount may be required to provide the necessary play or backlash between the teeth. If the outside diameter of the gear blank is correct, the tooth thickness should also be correct after the cutter has been sunk to the depth required for a given pitch and backlash. However, it is advisable to check the tooth thickness by measuring it, and the vernier gear-tooth caliper (see accompanying illustration) is commonly used in measuring the thickness.

The vertical scale of this caliper is set so that when it rests upon the top of the tooth as shown, the lower ends of the caliper jaws will be at the height of the pitch circle; the horizontal scale then shows the chordal thickness of the tooth at this point. If the gear is being cut on a milling machine or with the type of gear-cutting machine employing a formed milling cutter, the tooth thickness is checked by first taking a trial cut for a short distance at one side of the blank; then the gear blank is indexed for the next space and another cut is taken far enough to mill the full outline of the tooth. The tooth thickness is then measured.

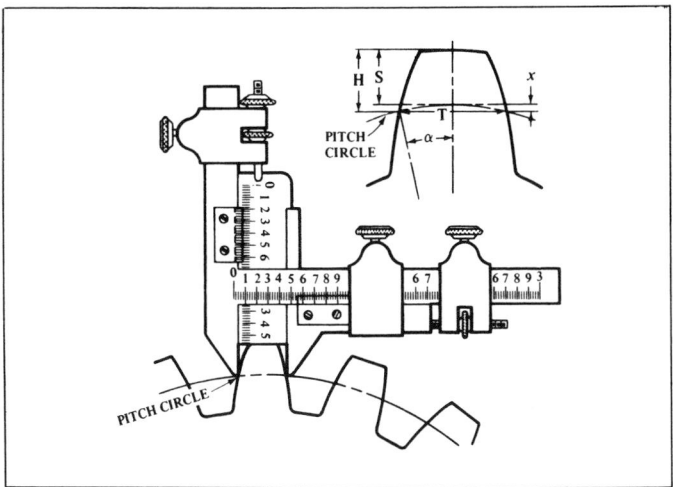

Before the gear-tooth caliper can be used, it is necessary to determine the correct chordal thickness and also the chordal addendum (or "corrected addendum" as it is sometimes called). The vertical scale is set to the chordal addendum, thus locating the ends of the jaws at the height of the pitch circle. The rules or formulas to use in determining the chordal thickness and chordal addendum will depend upon the outside diameter of the gear; for example, if the outside diameter of a small pinion is enlarged to avoid undercut and improve the tooth action, this must be taken into account in figuring the chordal thickness and chordal addendum as shown by the accompanying rules. The detail of a gear tooth included with the gear-tooth caliper illustration, represents the chordal thickness T, the addendum S, and the chordal addendum H.

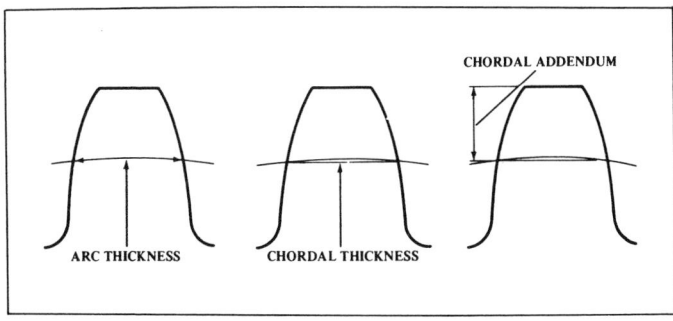

Chordal Thicknesses and Chordal Addenda of Full-depth Gear Teeth

This table is for spur gears of one diametral pitch. For any other diametral pitch, divide the given value by the required pitch. Table gives the chordal thickness and chordal addendum at the pitch circle when addendum is standard for full-depth teeth. Table is applicable to helical gears as explained on page 298.

No. of Teeth	Chordal Thickness	Chordal Addend.	No. of Teeth	Chordal Thickness	Chordal Addend.	No. of Teeth	Chordal Thickness	Chordal Addend.
10	1.56435	1.06156	59	1.57061	1.01046	108	1.57074	1.00570
11	1.56546	1.05598	60	1.57062	1.01029	109	1.57075	1.00565
12	1.56631	1.05133	61	1.57062	1.01011	110	1.57075	1.00560
13	1.56698	1.04739	62	1.57063	1.00994	111	1.57075	1.00556
14	1.56752	1.04401	63	1.57063	1.00978	112	1.57075	1.00551
15	1.56794	1.04109	64	1.57064	1.00963	113	1.57075	1.00546
16	1.56827	1.03852	65	1.57064	1.00947	114	1.57075	1.00541
17	1.56856	1.03625	66	1.57065	1.00933	115	1.57075	1.00537
18	1.56880	1.03425	67	1.57065	1.00920	116	1.57075	1.00533
19	1.56899	1.03244	68	1.57066	1.00907	117	1.57075	1.00529
20	1.56918	1.03083	69	1.57066	1.00893	118	1.57075	1.00524
21	1.56933	1.02936	70	1.57067	1.00880	119	1.57075	1.00519
22	1.56948	1.02803	71	1.57067	1.00867	120	1.57075	1.00515
23	1.56956	1.02681	72	1.57067	1.00855	121	1.57075	1.00511
24	1.56967	1.02569	73	1.57068	1.00843	122	1.57075	1.00507
25	1.56977	1.02466	74	1.57068	1.00832	123	1.57076	1.00503
26	1.56986	1.02371	75	1.57068	1.00821	124	1.57076	1.00499
27	1.56991	1.02284	76	1.57069	1.00810	125	1.57076	1.00495
28	1.56998	1.02202	77	1.57069	1.00799	126	1.57076	1.00491
29	1.57003	1.02127	78	1.57069	1.00789	127	1.57076	1.00487
30	1.57008	1.02055	79	1.57069	1.00780	128	1.57076	1.00483
31	1.57012	1.01990	80	1.57070	1.00772	129	1.57076	1.00479
32	1.57016	1.01926	81	1.57070	1.00762	130	1.57076	1.00475
33	1.57019	1.01869	82	1.57070	1.00752	131	1.57076	1.00472
34	1.57021	1.01813	83	1.57070	1.00743	132	1.57076	1.00469
35	1.57025	1.01762	84	1.57071	1.00734	133	1.57076	1.00466
36	1.57028	1.01714	85	1.57071	1.00725	134	1.57076	1.00462
37	1.57032	1.01667	86	1.57071	1.00716	135	1.57076	1.00457
38	1.57035	1.01623	87	1.57071	1.00708	136	1.57076	1.00454
39	1.57037	1.01582	88	1.57071	1.00700	137	1.57076	1.00451
40	1.57039	1.01542	89	1.57072	1.00693	138	1.57076	1.00447
41	1.57041	1.01504	90	1.57072	1.00686	139	1.57076	1.00444
42	1.57043	1.01471	91	1.57072	1.00679	140	1.57076	1.00441
43	1.57045	1.01434	92	1.57072	1.00672	141	1.57076	1.00439
44	1.57047	1.01404	93	1.57072	1.00665	142	1.57076	1.00435
45	1.57048	1.01370	94	1.57072	1.00658	143	1.57076	1.00432
46	1.57050	1.01341	95	1.57073	1.00651	144	1.57076	1.00429
47	1.57051	1.01311	96	1.57073	1.00644	145	1.57077	1.00425
48	1.57052	1.01285	97	1.57073	1.00637	146	1.57077	1.00422
49	1.57053	1.01258	98	1.57073	1.00630	147	1.57077	1.00419
50	1.57054	1.01233	99	1.57073	1.00623	148	1.57077	1.00416
51	1.57055	1.01209	100	1.57073	1.00617	149	1.57077	1.00413
52	1.57056	1.01187	101	1.57074	1.00611	150	1.57077	1.00411
53	1.57057	1.01165	102	1.57074	1.00605	151	1.57077	1.00409
54	1.57058	1.01143	103	1.57074	1.00599	152	1.57077	1.00407
55	1.57058	1.01121	104	1.57074	1.00593	153	1.57077	1.00405
56	1.57059	1.01102	105	1.57074	1.00587	154	1.57077	1.00402
57	1.57060	1.01083	106	1.57074	1.00581	155	1.57077	1.00400
58	1.57061	1.01064	107	1.57074	1.00575	156	1.57077	1.00397

Selection of Involute Gear Milling Cutter for a Given Diametral Pitch and Number of Teeth. — When gear teeth are cut by using formed milling cutters, the cutter must be selected to suit both the pitch and the number of teeth, because the shapes of the tooth spaces vary according to the number of teeth. For instance, the tooth spaces of a small pinion are not of the same shape as the spaces of a large gear of equal pitch. Theoretically, there should be a different formed cutter for every tooth number, but such refinement is unnecessary in practice. The involute formed cutters commonly used are made in series of eight cutters for each diametral pitch (see accompanying table). The shape of each cutter in this series is correct for a certain number of teeth only, but it can be used for other numbers within the limits given. For instance, a No. 6 cutter may be used for gears having from 17 to 20 teeth, but the tooth outline is correct only for 17 teeth or the lowest number in the range, which is also true of the other cutters listed. When this cutter is used for a gear having, say, 19 teeth, too much material is removed from the upper surfaces of the teeth, although the gear meets ordinary requirements. When greater accuracy of tooth shape is desired to ensure smoother or quieter operation, an intermediate series of cutters having half-numbers may be used provided the number of gear teeth is between the number listed for the regular cutters (see table).

Involute gear milling cutters are designed to cut a composite tooth form, the center portion being a true involute while the top and bottom portions are cycloidal. This composite form is necessary to prevent tooth interference when milled mating gears are meshed with each other. Because of their composite form, milled gears will not mate satisfactorily enough for high grade work with those of generated, full-involute form. Composite form hobs are available, however, which will produce generated gears that mesh with those cut by gear milling cutters.

Metric Module Gear Cutters: The accompanying table for selecting the cutter number to be used to cut a given number of teeth may be used also to select metric module gear cutters except that the numbers are designated in reverse order. For example, cutter No. 1, in the metric module system, is used for 12–13 teeth, cutter No. 2 for 14–16 teeth, etc.

Series of Involute, Finishing Gear Milling Cutters for Each Pitch*

Number of Cutter	Will cut Gears from	Number of Cutter	Will cut Gears from
1	135 teeth to a rack	5	21 to 25 teeth
2	55 to 134 teeth	6	17 to 20 teeth
3	35 to 54 teeth	7	14 to 16 teeth
4	26 to 34 teeth	8	12 to 13 teeth

The regular cutters listed above are used ordinarily. The cutters listed below (an intermediate series having half numbers) may be used when greater accuracy of tooth shape is essential in cases where the number of teeth is between the numbers for which the regular cutters are intended.

Number of Cutter	Will cut Gears from	Number of Cutter	Will cut Gears from
1½	80 to 134 teeth	5½	19 to 20 teeth
2½	42 to 54 teeth	6½	15 to 16 teeth
3½	30 to 34 teeth	7½	13 teeth
4½	23 to 25 teeth	. . .	. . .

* Roughing cutters are made with No. 1 form only.

Chordal Thicknesses and Chordal Addenda of Milled, Full-depth Gear Teeth and of Gear Milling Cutters

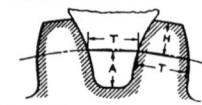

T = chordal thickness of gear tooth and cutter tooth at pitch line;
H = chordal addendum for full-depth gear tooth;
A = chordal addendum of cutter = $(2.157 \div$ diametral pitch$) - H = (0.6866 \times$ circular pitch$) - H$.

Diametral Pitch	Dimension	Number of Gear Cutter, and Corresponding Number of Teeth							
		No. 1 135 Teeth	No. 2 55 Teeth	No. 3 35 Teeth	No. 4 26 Teeth	No. 5 21 Teeth	No. 6 17 Teeth	No. 7 14 Teeth	No. 8 12 Teeth
1	T	1.5707	1.5706	1.5702	1.5698	1.5694	1.5686	1.5675	1.5663
	H	1.0047	1.0112	1.0176	1.0237	1.0294	1.0362	1.0440	1.0514
1½	T	1.0471	1.0470	1.0468	1.0465	1.0462	1.0457	1.0450	1.0442
	H	0.6698	0.6741	0.6784	0.6824	0.6862	0.6908	0.6960	0.7009
2	T	0.7853	0.7853	0.7851	0.7849	0.7847	0.7843	0.7837	0.7831
	H	0.5023	0.5056	0.5088	0.5118	0.5147	0.5181	0.5220	0.5257
2½	T	0.6283	0.6282	0.6281	0.6279	0.6277	0.6274	0.6270	0.6265
	H	0.4018	0.4044	0.4070	0.4094	0.4117	0.4144	0.4176	0.4205
3	T	0.5235	0.5235	0.5234	0.5232	0.5231	0.5228	0.5225	0.5221
	H	0.3349	0.3370	0.3392	0.3412	0.3431	0.3454	0.3480	0.3504
3½	T	0.4487	0.4487	0.4486	0.4485	0.4484	0.4481	0.4478	0.4475
	H	0.2870	0.2889	0.2907	0.2919	0.2935	0.2954	0.2977	0.3004
4	T	0.3926	0.3926	0.3926	0.3924	0.3923	0.3921	0.3919	0.3915
	H	0.2511	0.2528	0.2544	0.2559	0.2573	0.2590	0.2610	0.2628
5	T	0.3141	0.3141	0.3140	0.3139	0.3138	0.3137	0.3135	0.3132
	H	0.2009	0.2022	0.2035	0.2047	0.2058	0.2072	0.2088	0.2102
6	T	0.2618	0.2617	0.2617	0.2616	0.2615	0.2614	0.2612	0.2610
	H	0.1674	0.1685	0.1696	0.1706	0.1715	0.1727	0.1740	0.1752
7	T	0.2244	0.2243	0.2243	0.2242	0.2242	0.2240	0.2239	0.2237
	H	0.1435	0.1444	0.1453	0.1462	0.1470	0.1480	0.1491	0.1502
8	T	0.1963	0.1963	0.1962	0.1962	0.1961	0.1960	0.1959	0.1958
	H	0.1255	0.1264	0.1272	0.1279	0.1286	0.1295	0.1305	0.1314
9	T	0.1745	0.1745	0.1744	0.1744	0.1743	0.1743	0.1741	0.1740
	H	0.1116	0.1123	0.1130	0.1137	0.1143	0.1151	0.1160	0.1168
10	T	0.1570	0.1570	0.1570	0.1569	0.1569	0.1568	0.1567	0.1566
	H	0.1004	0.1011	0.1017	0.1023	0.1029	0.1036	0.1044	0.1051
11	T	0.1428	0.1428	0.1427	0.1427	0.1426	0.1426	0.1425	0.1424
	H	0.0913	0.0919	0.0925	0.0930	0.0935	0.0942	0.0949	0.0955
12	T	0.1309	0.1309	0.1308	0.1308	0.1308	0.1307	0.1306	0.1305
	H	0.0837	0.0842	0.0848	0.0853	0.0857	0.0863	0.0870	0.0876
14	T	0.1122	0.1122	0.1121	0.1121	0.1121	0.1120	0.1119	0.1118
	H	0.0717	0.0722	0.0726	0.0731	0.0735	0.0740	0.0745	0.0751
16	T	0.0981	0.0981	0.0981	0.0981	0.0980	0.0980	0.0979	0.0979
	H	0.0628	0.0632	0.0636	0.0639	0.0643	0.0647	0.0652	0.0657
18	T	0.0872	0.0872	0.0872	0.0872	0.0872	0.0871	0.0870	0.0870
	H	0.0558	0.0561	0.0565	0.0568	0.0571	0.0575	0.0580	0.0584
20	T	0.0785	0.0785	0.0785	0.0785	0.0784	0.0784	0.0783	0.0783
	H	0.0502	0.0505	0.0508	0.0511	0.0514	0.0518	0.0522	0.0525

Chordal Thicknesses and Chordal Addenda of Milled, Full-depth Gear Teeth and of Gear Milling Cutters

Circular Pitch	Dimension	Number of Gear Cutter, and Corresponding Number of Teeth							
		No. 1 135 Teeth	No. 2 55 Teeth	No. 3 35 Teeth	No. 4 26 Teeth	No. 5 21 Teeth	No. 6 17 Teeth	No. 7 14 Teeth	No. 8 12 Teeth
¼	T	0.1250	0.1250	0.1249	0.1249	0.1249	0.1248	0.1247	0.1246
	H	0.0799	0.0804	0.0809	0.0814	0.0819	0.0824	0.0830	0.0836
⁵⁄₁₆	T	0.1562	0.1562	0.1562	0.1561	0.1561	0.1560	0.1559	0.1558
	H	0.0999	0.1006	0.1012	0.1018	0.1023	0.1030	0.1038	0.1045
⅜	T	0.1875	0.1875	0.1874	0.1873	0.1873	0.1872	0.1871	0.1870
	H	0.1199	0.1207	0.1214	0.1221	0.1228	0.1236	0.1245	0.1254
⁷⁄₁₆	T	0.2187	0.2187	0.2186	0.2186	0.2185	0.2184	0.2183	0.2181
	H	0.1399	0.1408	0.1416	0.1425	0.1433	0.1443	0.1453	0.1464
½	T	0.2500	0.2500	0.2499	0.2498	0.2498	0.2496	0.2495	0.2493
	H	0.1599	0.1609	0.1619	0.1629	0.1638	0.1649	0.1661	0.1673
⁹⁄₁₆	T	0.2812	0.2812	0.2811	0.2810	0.2810	0.2808	0.2806	0.2804
	H	0.1799	0.1810	0.1821	0.1832	0.1842	0.1855	0.1868	0.1882
⅝	T	0.3125	0.3125	0.3123	0.3123	0.3122	0.3120	0.3118	0.3116
	H	0.1998	0.2012	0.2023	0.2036	0.2047	0.2061	0.2076	0.2091
11⁄16	T	0.3437	0.3437	0.3436	0.3435	0.3434	0.3432	0.3430	0.3427
	H	0.2198	0.2213	0.2226	0.2239	0.2252	0.2267	0.2283	0.2300
¾	T	0.3750	0.3750	0.3748	0.3747	0.3747	0.3744	0.3742	0.3740
	H	0.2398	0.2414	0.2428	0.2443	0.2457	0.2473	0.2491	0.2509
13⁄16	T	0.4062	0.4062	0.4060	0.4059	0.4059	0.4056	0.4054	0.4050
	H	0.2598	0.2615	0.2631	0.2647	0.2661	0.2679	0.2699	0.2718
⅞	T	0.4375	0.4375	0.4373	0.4372	0.4371	0.4368	0.4366	0.4362
	H	0.2798	0.2816	0.2833	0.2850	0.2866	0.2885	0.2906	0.2927
15⁄16	T	0.4687	0.4687	0.4685	0.4684	0.4683	0.4680	0.4678	0.4674
	H	0.2998	0.3018	0.3035	0.3054	0.3071	0.3092	0.3114	0.3137
1	T	0.5000	0.5000	0.4998	0.4997	0.4996	0.4993	0.4990	0.4986
	H	0.3198	0.3219	0.3238	0.3258	0.3276	0.3298	0.3322	0.3346
1⅛	T	0.5625	0.5625	0.5623	0.5621	0.5620	0.5617	0.5613	0.5610
	H	0.3597	0.3621	0.3642	0.3665	0.3685	0.3710	0.3737	0.3764
1¼	T	0.6250	0.6250	0.6247	0.6246	0.6245	0.6241	0.6237	0.6232
	H	0.3997	0.4023	0.4047	0.4072	0.4095	0.4122	0.4152	0.4182
1⅜	T	0.6875	0.6875	0.6872	0.6870	0.6869	0.6865	0.6861	0.6856
	H	0.4397	0.4426	0.4452	0.4479	0.4504	0.4534	0.4567	0.4600
1½	T	0.7500	0.7500	0.7497	0.7495	0.7494	0.7489	0.7485	0.7480
	H	0.4797	0.4828	0.4857	0.4887	0.4914	0.4947	0.4983	0.5019
1¾	T	0.8750	0.8750	0.8746	0.8744	0.8743	0.8737	0.8732	0.8726
	H	0.5596	0.5633	0.5666	0.5701	0.5733	0.5771	0.5813	0.5855
2	T	1.0000	1.0000	0.9996	0.9994	0.9992	0.9986	0.9980	0.9972
	H	0.6396	0.6438	0.6476	0.6516	0.6552	0.6596	0.6644	0.6692
2¼	T	1.1250	1.1250	1.1246	1.1242	1.1240	1.1234	1.1226	1.1220
	H	0.7195	0.7242	0.7285	0.7330	0.7371	0.7420	0.7474	0.7528
2½	T	1.2500	1.2500	1.2494	1.2492	1.2490	1.2482	1.2474	1.2464
	H	0.7995	0.8047	0.8095	0.8145	0.8190	0.8245	0.8305	0.8365
3	T	1.5000	1.5000	1.4994	1.4990	1.4990	1.4978	1.4970	1.4960
	H	0.9594	0.9657	0.9714	0.9774	0.9828	0.9894	0.9966	1.0038

Gear Design Based upon Module System. — The *module* of a gear equals the pitch diameter divided by the number of teeth, whereas *diametral pitch* equals the number of teeth divided by the pitch diameter. The module system is in general use in countries which have adopted the metric system; hence the term module is usually understood to mean the pitch diameter *in millimeters* divided by the number of teeth. The module system may, however, also be based upon inch measurements and then it is known as English module to avoid confusion with the metric module. Module is an actual dimension, whereas diametral pitch is only a ratio. Thus, if the pitch diameter of a gear is 50 millimeters and the number of teeth 25, the module is 2 which means that there are 2 millimeters of pitch diameter for each tooth. The table "Tooth Dimensions Based Upon Module System" shows the relation between module, diametral pitch, and circular pitch.

German Standard Tooth Form for Spur and Bevel Gears (DIN — 867)

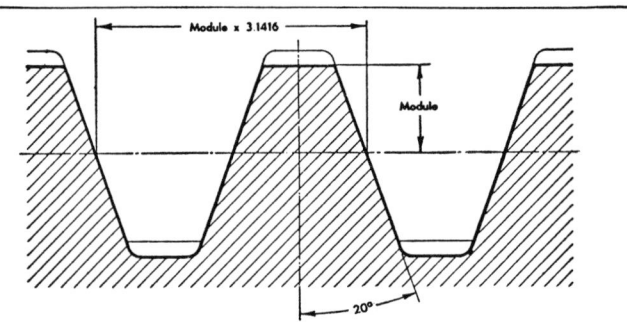

The flanks or sides are straight (involute system) and the pressure angle is 20 degrees. The shape of the root clearance space and the amount of clearance depend upon the method of cutting and special requirements. The amount of clearance may vary from 0.1 × module to 0.3 × module.

To Find	Module Known	Circular Pitch Known
Addendum	Equals module	0.3183 × Circular pitch
Dedendum	1.157 × module* 1.167 × module**	0.3683 × Circular pitch* 0.3714 × Circular pitch**
Working Depth	2 × module	0.6366 × Circular pitch
Total Depth	2.157 × module* 2.167 × module**	0.6866 × Circular pitch* 0.6898 × Circular pitch**
Tooth Thickness on Pitch Line	1.5708 × module	0.5 × Circular pitch

Formulas for dedendum and total depth, marked (*) are used when clearance equals 0.157 × module. Formulas marked (**) are used when clearance equals one-sixth module. It is the common practice among American cutter manufacturers to make the clearance of metric or module cutters equal to 0.157 × module.

Tooth Dimensions Based Upon Module System

Module, DIN Standard Series	Equivalent Diametral Pitch	Circular Pitch Millimeters	Inches	Addendum, Millimeters	Dedendum, Millimeters*	Whole Depth,* Millimeters	Whole Depth,† Millimeters
0.3	84.667	0.943	0.0371	0.30	0.35	0.650	0.647
0.4	63.500	1.257	0.0495	0.40	0.467	0.867	0.863
0.5	50.800	1.571	0.0618	0.50	0.583	1.083	1.079
0.6	42.333	1.885	0.0742	0.60	0.700	1.300	1.294
0.7	36.286	2.199	0.0865	0.70	0.817	1.517	1.510
0.8	31.750	2.513	0.0989	0.80	0.933	1.733	1.726
0.9	28.222	2.827	0.1113	0.90	1.050	1.950	1.941
1	25.400	3.142	0.1237	1.00	1.167	2.167	2.157
1.25	20.320	3.927	0.1546	1.25	1.458	2.708	2.697
1.5	16.933	4.712	0.1855	1.50	1.750	3.250	3.236
1.75	14.514	5.498	0.2164	1.75	2.042	3.792	3.774
2	12.700	6.283	0.2474	2.00	2.333	4.333	4.314
2.25	11.289	7.069	0.2783	2.25	2.625	4.875	4.853
2.5	10.160	7.854	0.3092	2.50	2.917	5.417	5.392
2.75	9.236	8.639	0.3401	2.75	3.208	5.958	5.932
3	8.466	9.425	0.3711	3.00	3.500	6.500	6.471
3.25	7.815	10.210	0.4020	3.25	3.791	7.041	7.010
3.5	7.257	10.996	0.4329	3.50	4.083	7.583	7.550
3.75	6.773	11.781	0.4638	3.75	4.375	8.125	8.089
4	6.350	12.566	0.4947	4.00	4.666	8.666	8.628
4.5	5.644	14.137	0.5566	4.50	5.25	9.750	9.707
5	5.080	15.708	0.6184	5.00	5.833	10.833	10.785
5.5	4.618	17.279	0.6803	5.50	6.416	11.916	11.864
6	4.233	18.850	0.7421	6.00	7.000	13.000	12.942
6.5	3.908	20.420	0.8035	6.50	7.583	14.083	14.021
7	3.628	21.991	0.8658	7.	8.166	15.166	15.099
8	3.175	25.132	0.9895	8.	9.333	17.333	17.256
9	2.822	28.274	1.1132	9.	10.499	19.499	19.413
10	2.540	31.416	1.2368	10.	11.666	21.666	21.571
11	2.309	34.558	1.3606	11.	12.833	23.833	23.728
12	2.117	37.699	1.4843	12.	14.000	26.000	25.884
13	1.954	40.841	1.6079	13.	15.166	28.166	28.041
14	1.814	43.982	1.7317	14.	16.332	30.332	30.198
15	1.693	47.124	1.8541	15.	17.499	32.499	32.355
16	1.587	50.266	1.9790	16.	18.666	34.666	34.512
18	1.411	56.549	2.2263	18.	21.000	39.000	38.826
20	1.270	62.832	2.4737	20.	23.332	43.332	43.142
22	1.155	69.115	2.7210	22.	25.665	47.665	47.454
24	1.058	75.398	2.9685	24.	28.000	52.000	51.768
27	0.941	84.823	3.339	27.	31.498	58.498	58.239
30	0.847	94.248	3.711	30.	35.000	65.000	64.713
33	0.770	103.673	4.082	33.	38.498	71.498	71.181
36	0.706	113.097	4.453	36.	41.998	77.998	77.652
39	0.651	122.522	4.824	39.	45.497	84.497	84.123
42	0.605	131.947	5.195	42.	48.997	90.997	90.594
45	0.564	141.372	5.566	45.	52.497	97.497	97.065
50	0.508	157.080	6.184	50.	58.330	108.330	107.855
55	0.462	172.788	6.803	55.	64.163	119.163	118.635
60	0.423	188.496	7.421	60.	69.996	129.996	129.426
65	0.391	204.204	8.040	65.	75.829	140.829	140.205
70	0.363	219.911	8.658	70.	81.662	151.662	150.997
75	0.339	235.619	9.276	75.	87.495	162.495	161.775

* Dedendum and total depth when clearance = 0.1666 × module, or one-sixth module.
† Total depth equivalent to American standard full-depth teeth. (Clearance = 0.157 × module.)

Rules for Module System of Gearing

To Find	Rule
Metric Module	*Rule* 1: To find the metric module, divide the pitch diameter in millimeters by the number of teeth. *Example* 1: The pitch diameter of a gear is 200 millimeters and the number of teeth, 40; then $$\text{module} = \frac{200}{40} = 5$$ *Rule* 2: Multiply circular pitch in millimeters by 0.3183. *Example* 2: (Same as Example 1. Circular pitch of this gear equals 15.708 millimeters). $$\text{module} = 15.708 \times 0.3183 = 5$$ *Rule* 3: Divide outside diameter in millimeters by the number of teeth plus 2.
English Module	*Note:* The module system is usually applied when gear dimensions are expressed in millimeters, but module may also be based upon inch measurements. *Rule:* To find the English module, divide pitch diameter in inches by number of teeth. *Example:* A gear has 48 teeth and a pitch diameter of 12 inches. $$\text{module} = \frac{12}{48} = \frac{1}{4} \text{ module or 4 diametral pitch}$$
Metric Module Equivalent to Diametral Pitch	*Rule:* To find the metric module equivalent to a given diametral pitch, divide 25.4 by the diametral pitch. *Example:* Determine metric module equivalent to 10 diametral pitch. $$\text{Equivalent module} = \frac{25.4}{10} = 2.54$$ *Note:* The nearest standard module is 2.5.
Diametral Pitch Equivalent to Metric Module	*Rule:* To find the diametral pitch equivalent to a given module, divide 25.4 by the module. (25.4 = number of millimeters per inch.) *Example:* The module is 12; determine equivalent diametral pitch. $$\text{Equivalent diametral pitch} = \frac{25.4}{12} = 2.117$$ *Note:* A diametral pitch of 2 is the nearest *standard* equivalent.
Pitch Diameter	*Rule:* Multiply number of teeth by module. *Example:* The metric module is 8 and gear has 40 teeth; then $$D = 40 \times 8 = 320 \text{ millimeters} = 12.598 \text{ inches}$$
Outside Diameter	*Rule:* Add 2 to the number of teeth and multiply sum by the module. *Example:* A gear has 40 teeth and module is 6. Find outside or blank diameter. $$\text{Outside diameter} = (40 + 2) \times 6 = 252 \text{ millimeters}$$

For tooth dimensions, see table Tooth Dimensions Based Upon Module System; also formulas below German Standard Tooth Form.

Equivalent Diametral Pitches, Circular Pitches, and Metric Modules

Commonly Used Pitches and Modules in Bold Type

Diametral Pitch	Circular Pitch, Inches	Module Milli-meters	Diametral Pitch	Circular Pitch, Inches	Module Milli-meters	Diametral Pitch	Circular Pitch, Inches	Module Milli-meters
½	6.2832	50.8000	2.2848	1⅜	11.1170	10.0531	5⁄16	2.5266
0.5080	6.1842	50	2.3091	1.3605	11	10.1600	0.3092	2½
0.5236	6	48.5104	2½	1.2566	10.1600	11	0.2856	2.3091
0.5644	5.5658	45	2.5133	1¼	10.1063	12	0.2618	2.1167
0.5712	5½	44.4679	2.5400	1.2368	10	12.5664	¼	2.0213
0.6283	5	40.4253	2¾	1.1424	9.2364	12.7000	0.2474	2
0.6350	4.9474	40	2.7925	1⅛	9.0957	13	0.2417	1.9538
0.6981	4½	36.3828	2.8222	1.1132	9	14	0.2244	1.8143
0.7257	4.3290	35	3	1.0472	8.4667	15	0.2094	1.6933
¾	4.1888	33.8667	3.1416	1	8.0851	16	0.1963	1.5875
0.7854	4	32.3403	3.1750	0.9895	8	16.7552	3⁄16	1.5160
0.8378	3¾	30.3190	3.3510	15⁄16	7.5797	16.9333	0.1855	1½
0.8467	3.7105	30	3½	0.8976	7.2571	17	0.1848	1.4941
0.8976	3½	28.2977	3.5904	⅞	7.0744	18	0.1745	1.4111
0.9666	3¼	26.2765	3.6286	0.8658	7	19	0.1653	1.3368
1	3.1416	25.4000	3.8666	13⁄16	6.5691	20	0.1571	1.2700
1.0160	3.0921	25	3.9078	0.8040	6½	22	0.1428	1.1545
1.0472	3	24.2552	4	0.7854	6.3500	24	0.1309	1.0583
1.1424	2¾	22.2339	4.1888	¾	6.0638	25	0.1257	1.0160
1¼	2.5133	20.3200	4.2333	0.7421	6	25.1328	⅛	1.0106
1.2566	2½	20.2127	4.5696	11⁄16	5.5585	25.4000	0.1237	1
1.2700	2.4737	20	4.6182	0.6803	5½	26	0.1208	0.9769
1.3963	2¼	18.1914	5	0.6283	5.0800	28	0.1122	0.9071
1.4111	2.2263	18	5.0265	⅝	5.0532	30	0.1047	0.8467
1½	2.0944	16.9333	5.0800	0.6184	5	32	0.0982	0.7937
1.5708	2	16.1701	5.5851	9⁄16	4.5478	34	0.0924	0.7470
1.5875	1.9790	16	5.6443	0.5566	4½	36	0.0873	0.7056
1.6755	1⅞	15.1595	6	0.5236	4.2333	38	0.0827	0.6684
1.6933	1.8553	15	6.2832	½	4.0425	40	0.0785	0.6350
1¾	1.7952	14.5143	6.3500	0.4947	4	42	0.0748	0.6048
1.7952	1¾	14.1489	7	0.4488	3.6286	44	0.0714	0.5773
1.8143	1.7316	14	7.1808	7⁄16	3.5372	46	0.0683	0.5522
1.9333	1⅝	13.1382	7.2571	0.4329	3½	48	0.0654	0.5292
1.9538	1.6079	13	8	0.3927	3.1750	50	0.0628	0.5080
2	1.5708	12.7000	8.3776	⅜	3.0319	50.2656	1⁄16	0.5053
2.0944	1½	12.1276	8.4667	0.3711	3	50.8000	0.0618	½
2.1167	1.4842	12	9	0.3491	2.8222	56	0.0561	0.4536
2¼	1.3963	11.2889	10	0.3142	2.5400	60	0.0524	0.4233

The module of a gear is the pitch diameter divided by the number of teeth. The module may be expressed in any units; but when no units are stated, it is understood to be in millimeters. The metric module, therefore, equals the pitch diameter in millimeters divided by the number of teeth. To find the metric module equivalent to a given diametral pitch, divide 25.4 by the diametral pitch. To find the diametral pitch equivalent to a given module, divide 25.4 by the module. (25.4 = number of millimeters per inch.)

Rules and Formulas for Spur Gear Calculations
(Circular Pitch)

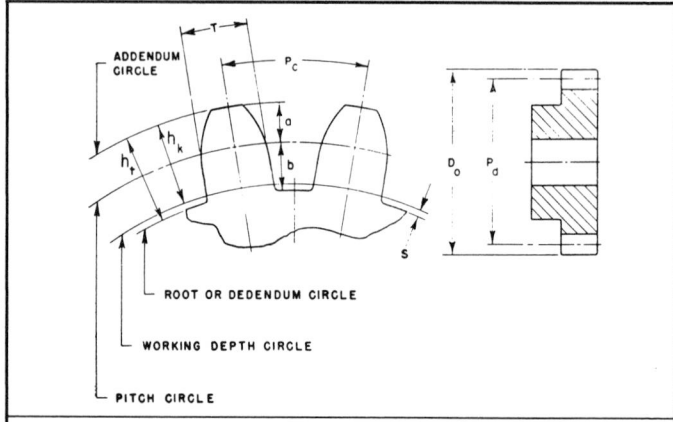

The following symbols are used in conjunction with the formulas for determining the proportions of spur gear teeth.

P = Diametral pitch.

P_c = Circular pitch.

P_d = Pitch diameter.

D_o = Outside diameter.

N = Number of teeth in the gear.

T = Tooth thickness.

a = Addendum.

b = Dedendum.

h_k = Working depth.

h_t = Whole depth.

S = Clearance.

C = Center distance.

L = Length of rack.

Courtesy of Cincinnatti Milacron

Rules and Formulas for Spur Gear Calculations—Continued
(Circular Pitch)

TO FIND	RULE	FORMULA
Diametral pitch P	Divide 3.1416 by the circular pitch.	$P = \dfrac{3.1416}{P_c}$
Circular pitch P_c	Divide 3.1416 by the diametral pitch.	$P_c = \dfrac{3.1416}{P}$
Pitch diameter P_d	Divide the number of teeth by the diametral pitch.	$P_d = \dfrac{N}{P}$
Outside diameter D_o	Add 2 to the number of teeth and divide the sum by the diametral pitch.	$D_o = \dfrac{N + 2}{P}$
Number of teeth N	Multiply the pitch diameter by the diametral pitch.	$N = P_d\,P$
Tooth circular thickness T	Divide 1.5708 by the diametral pitch.	$T = \dfrac{1.5708}{P}$
Addendum a	Divide 1.0 by the diametral pitch.	$a = \dfrac{1.0}{P}$
Dedendum b	Divide 1.157 by the diametral pitch.	$b = \dfrac{1.157}{P}$
Working depth h_k	Divide 2 by the diametral pitch.	$h_k = \dfrac{2}{P}$
Whole depth h_t	Divide 2.157 by the diametral pitch.	$h_t = \dfrac{2.157}{P}$
Clearance S	Divide 0.157 by the diametral pitch.	$S = \dfrac{0.157}{P}$
Center distance C	Add the number of teeth in both gears and divide the sum by two times the diametral pitch.	$C = \dfrac{N_1 + N_2}{2P}$
Length of rack L	Multiply the number of teeth in the rack by the circular pitch.	$L = N\,P_c$

Courtesy of Cincinnati Milacron

Rules and Formulas for Helical Gear Calculations

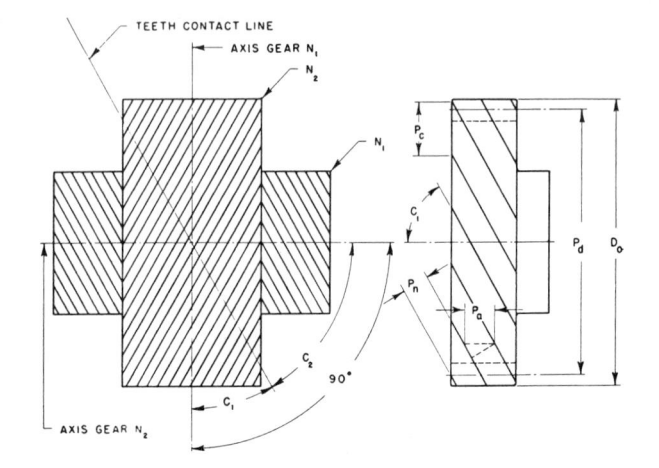

The following symbols are used in conjunction with the formulas for determining the proportions of helical gear teeth.

P_{nd} = Normal diametral pitch (pitch of cutter).

P_c = Circular pitch.

P_a = Axial pitch.

P_n = Normal pitch.

P_d = Pitch diameter.

S = Center distance.

C, C_1, C_2 = Helix angle of the gears.

L = Lead of tooth helix.

T_n = Normal tooth thickness at pitch line.

a = Addendum.

h_t = Whole depth of tooth.

N, N_1, N_2 = Number of teeth in the gears.

D_o = Outside diameter.

N_c = Hypothetical number of teeth for which the gear cutter should be selected.

Courtesy of Cincinnati Milacron

Rules and Formulas for Helical Gear Calculations—Continued

TO FIND	RULE	FORMULA
Normal diametral pitch P_{nd}	Divide the number of teeth by the product of the pitch diameter and the cosine of the helix angle.	$P_{nd} = \dfrac{N}{P_d \cos C_1}$
Circular pitch P_c	Multiply the pitch diameter of the gear by 3.1416, and divide the product by the number of teeth in the gear.	$P_c = \dfrac{3.1416\ P_d}{N}$
Axial pitch P_a	Multiply the circular pitch by the cotangent of the helix angle.	$P_a = P_c \cot C_1$
Normal pitch P_n	Divide 3.1416 by the normal diametral pitch.	$P_n = \dfrac{3.1416}{P_{nd}}$
Pitch diameter P_d	Divide the number of teeth by the product of the normal pitch and the cosine of the helix angle.	$P_d = \dfrac{N}{P_{nd} \cos C_1}$
Center distance S	Divide the sum of the pitch diameters of the mating gears by 2.	$S = \dfrac{P_{d_1} + P_{d_2}}{2}$
Checking Formulas (shafts at right angles)	Multiply the number of teeth in the first gear by the tangent of the tooth angle of that gear, and add the number of teeth in the second gear to the product. The sum should equal twice the product of the center distance multiplied by the normal diametral pitch, multiplied by the sine of the helix angle.	$N_1 + (N_2 \tan C_2) =$ $2\ S\ P_{nd} \sin C_1$
Lead of tooth helix L	Multiply the pitch diameter by 3.1416 times the cotangent of the helix angle.	$L = 3.1416\ P_d \cot C_1$
Normal circular tooth thickness at pitch line T_n	Divide 1.571 by the normal diametral pitch.	$T_n = \dfrac{1.571}{P_{nd}}$
Addendum a	Divide the normal pitch by 3.1416.	$a = \dfrac{P_n}{3.1416}$
Whole depth of tooth h_t	Divide 2.157 by the normal diametral pitch.	$h_t = \dfrac{2.157}{P_{nd}}$
Outside diameter D_o	Add twice the addendum to the pitch diameter.	$D_o = P_d + 2\ a$
Hypothetical number of teeth for which gear cutter should be selected N_c	Divide the number of teeth in the gear by the cube of the cosine of the helix angle.	$N_c = \dfrac{N_1}{(\cos C_1)^3}$

Courtesy of Cincinnati Milacron

Rules and Formulas for Bevel Gear Calculations
(Shafts at Right Angles)

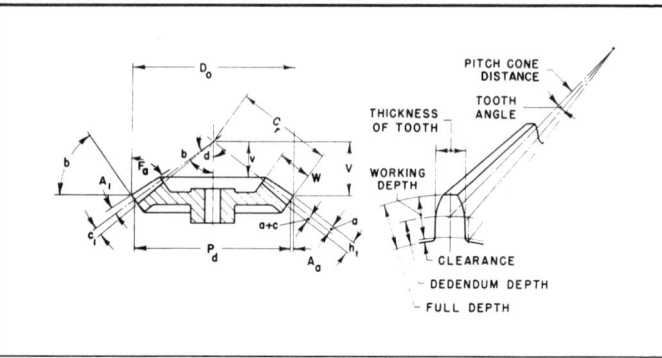

The following symbols are used in conjunction with the formulas for determining the proportions of bevel gear teeth.

P = Diametral pitch.

P_c = Circular pitch.

P_d = Pitch diameter.

b = Pitch angle.

C_r = Pitch cone distance.

a = Addendum.

A_1 = Addendum angle.

A_a = Angular addendum.

D_o = Outside diameter.

c_1 = Dedendum angle.

$a+c$ = Addendum plus clearance.

a_s = Addendum of small end of tooth.

T_L = Thickness of tooth at pitch line.

T_s = Thickness of tooth at pitch line at small end of gear.

F_a = Face angle.

h_t = Whole depth of tooth space.

V = Apex distance at large end of tooth.

v = Apex distance at small end of tooth.

m_g = Gear ratio.

N = Number of teeth.

N_g = Number of teeth in gear.

N_p = Number of teeth in pinion.

d = Root angle.

W = Width of gear tooth face.

N_c = Number of teeth of imaginary spur gear for which cutter is selected.

Courtesy of Cincinnati Milacron

Rules and Formulas for Bevel Gear Calculations—Continued
(Shafts at Right Angles)

TO FIND	RULE	FORMULA
Diametral pitch P	Divide the number of teeth by the pitch diameter.	$P = \dfrac{N}{P_d}$
Circular pitch P_c	Divide 3.1416 by the diametral pitch.	$P_c = \dfrac{3.1416}{P}$
Pitch diameter P_d	Divide the number of teeth by the diametral pitch.	$P_d = \dfrac{N}{P}$
Pitch angle of pinion $\tan b_p$	Divide the number of teeth in the pinion by the number of teeth in the gear to obtain the tangent.	$\tan b_p = \dfrac{N_p}{N_g}$
Pitch angle of gear $\tan b_g$	Divide the number of teeth in the gear by the number of teeth in the pinion to obtain the tangent.	$\tan b_g = \dfrac{N_g}{N_p}$
Pitch cone distance C_r	Divide the pitch diameter by twice the sine of the pitch angle.	$C_r = \dfrac{P_d}{2\,(\sin b)}$
Addendum a	Divide 1.0 by the diametral pitch.	$a = \dfrac{1.0}{P}$
Addendum angle $\tan A_1$	Divide the addendum by the pitch cone distance to obtain the tangent.	$\tan A_1 = \dfrac{a}{C_r}$
Angular addendum A_a	Multiply the addendum by the cosine of the pitch angle.	$A_a = a \cos b$
Outside diameter D_o	Add twice the angular addendum to the pitch diameter.	$D_o = P_d + 2A_a$
Dedendum angle $\tan c_1$	Divide the dedendum by the pitch cone distance to obtain the tangent.	$\tan c_1 = \dfrac{a + c}{C_r}$
Addendum of small end of tooth a_s	Subtract the width of face from the pitch cone distance, divide the remainder by the pitch cone distance and multiply by the addendum.	$a_s = a\left(\dfrac{C_r - \mathrm{W}}{C_r}\right)$

Courtesy of Cincinnati Milacron.

Rules and Formulas for Bevel Gear Calculations—Continued
(Shafts at Right Angles)

TO FIND	RULE	FORMULA
Thickness of tooth at pitch line T_L	Divide the circular pitch by 2.	$T_L = \dfrac{P_c}{2}$
Thickness of tooth at pitch line at small end of gear T_s	Subtract the width of face from the pitch cone distance, divide the remainder by the pitch cone distance and multiply by the thickness of the tooth at the pitch line.	$T_s = T_L \left(\dfrac{C_r - W}{C_r} \right)$
Face angle F_a	Face cone of blank turned parallel to root cone of mating gear.	$F_a = b + c_l$
Whole depth of tooth space h_t	Divide 2.157 by the diametral pitch.	$h_t = \dfrac{2.157}{P}$
Apex distance at large end of tooth V	Multiply one-half the outside diameter by the tangent of the face angle.	$V = \left(\dfrac{D_o}{2} \right) \tan F_a$
Apex distance at small end of tooth v	Subtract the width of face from the pitch cone distance, divide the remainder by the pitch cone distance and multiply by the apex distance.	$v = V \left(\dfrac{C_r - W}{C_r} \right)$
Gear ratio m_g	Divide the number of teeth in the gear by the number of teeth in the pinion.	$m_g = \dfrac{N_g}{N_p}$
Number of teeth in gear and/or pinion N_g, N_p	Multiply the pitch diameter by the diametral pitch.	$N_g = P_d P$ $N_p = P_d P$
Cutting angle d	Subtract the addendum plus clearance angle from the pitch angle.	$d = b - c_l$
Number of teeth of imaginary spur gear for which cutter is selected N_o	Divide the number of teeth in actual gear by the cosine of the pitch angle.	$N_c = \dfrac{N}{\cos b}$

Courtesy of Cincinnati Milacron.

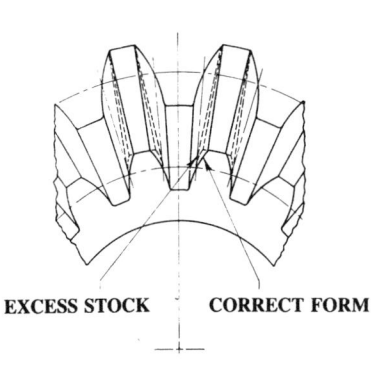

EXCESS STOCK   **CORRECT FORM**

Figure 2

Milling Bevel Gears. — Once the teeth of a bevel gear have been milled to their proper depth, only the back, or lage end of the teeth, have the proper size and form. To achieve the proper form on the small end of the teeth, the gear blank must be adjusted and the excess material, shown in the Fig. 2, must be removed. To determine the proper position of the blank for this cutting operation, the following formulas can be used to calculate the set over and roll of the blank.

Calculating The Angle Of Roll: The roll of the blank needed to produce the proper tooth form is calculated using the roll formula shown below. Once calculated, the dividing head is adjusted the required angular amount to produce the required tooth form.

Calculating The Set Over: In addition to roll, the gear blank must also be moved to produce the required form. The amount of this lateral movement is called the set over and is found using the set over formula shown on next page.

Rules and Formulas for Worm Gear Calculations (Solid Type)
(Single and Double Thread—14½° Pressure Angle)

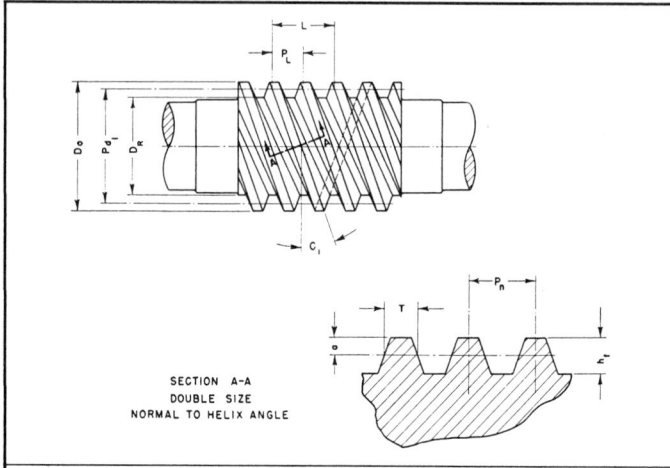

SECTION A-A
DOUBLE SIZE
NORMAL TO HELIX ANGLE

The following symbols are used in conjunction with the formulas for determining the proportions of worm gear teeth.

P_L = Linear pitch.
P_{d_1} = Pitch diameter.
D_o = Outside diameter.
N_W = Number of threads.
D_R = Root diameter.
h_t = Whole depth of tooth.

C_1 = Helix angle.
P_n = Normal pitch
a = Addendum.
L = Lead.
T = Normal tooth thickness.
t = Width of thread tool at end.

Roll

$$C = \frac{57.3}{P_d}\left(\frac{P_c}{2} - \frac{C_r}{W}(T_L - T_s)\right)$$

Set Over

$$n = \frac{T_L}{2} - \frac{T_L - T_s}{2}\frac{C_r}{W}$$

where:

C = angle of roll, degrees.
P_d = pitch diameter at large end of gear, inches.
P_c = circular pitch at large end of gear, inches.
C_r = pitch cone distance at large end of gear, inches.
T_s, T_L = chordal thickness of gear cutter tooth corresponding to pitch line at small and large ends of gear, respectively, inches.
57.3 = degrees per radian.
W = width of gear tooth face, inches.

Courtesy of Cincinnati Milacron

Rules and Formulas for Worm Gear Calculations (Solid Type)—Continued
(Single and Double Thread—$14\frac{1}{2}°$ Pressure Angle)

TO FIND	RULE	FORMULA
Linear pitch P_L	Divide the lead by the number of threads in the whole worm: i.e., one if single-threaded or four if quadrupled threaded.	$P_L = \dfrac{L}{N_W}$
Pitch diameter P_{d_1}	Subtract twice the addendum from the outside diameter.	$P_{d_1} = D_o - 2\,a$
Outside diameter D_o	Add twice the addendum of the worm to the pitch diameter of the worm wheel.	$D_o = P_{d_1} + 2\,a$
Root diameter D_R	Subtract twice the whole depth of the tooth from the outside diameter.	$D_R = D_o - 2\,h_t$
Whole depth of tooth h_t	Multiply the linear pitch by 0.6866.	$h_t = 0.6866 P_L$
Helix angle C_1	Multiply the pitch diameter of the worm by 3.1416, and divide the product by the lead. The quotient is the cotangent of the helix angle.	$\cot C_1 = \dfrac{3.1416 P_{d_2}}{L}$
Normal pitch P_n	Multiply the linear pitch by the cosine of the helix angle of the worm.	$P_n = P_L \cos C_1$
Addendum a	Multiply the linear pitch by 0.3183.	$a = 0.3183\,P_L$
Lead L	Multiply the linear pitch by the number of threads.	$L = P_L\,N_W$
Normal tooth thickness T	Multiply one-half the linear pitch by the cosine of the helix angle.	$T = \dfrac{P_L}{2} \cos C_1$
Width of thread tool at end t	Multiply the linear pitch by 0.31.	$t = 0.31\,P_L$

Courtesy of Cincinnati Milacron

Rules and Formulas for Worm Wheel Calculations
(Single and Double Thread—14½° Pressure Angle)

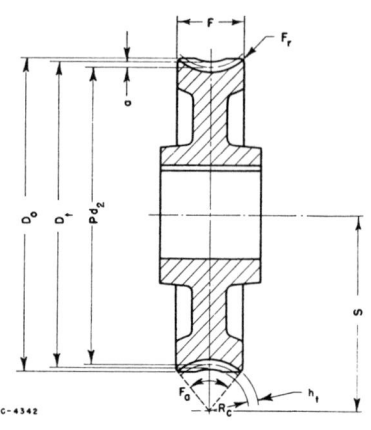

PC-4342

The following symbols are used in conjunction with the formulas for determining the proportions of worm wheel teeth.

$$P_c = \text{Circular pitch.}$$

$$P_{d_2} = \text{Pitch diameter.}$$

$$N = \text{Number of teeth.}$$

$$D_o = \text{Outside diameter.}$$

$$D_t = \text{Throat diameter.}$$

$$R_c = \text{Radius of curvature of worm wheel throat.}$$

$$D = \text{Diameter to sharp corners.}$$

$$F_a = \text{Face angle.}$$

$$F = \text{Face width of rim.}$$

$$F_r = \text{Radius at edge of face.}$$

$$a = \text{Addendum.}$$

$$h_t = \text{Whole depth of tooth.}$$

$$S = \text{Center distance between worm and worm wheel.}$$

$$G = \text{Gashing angle.}$$

Rules and Formulas for Worm Wheel Calculations—Continued
(Single and Double Thread—14½° Pressure Angle)

TO FIND	RULE	FORMULA
Circular pitch P_c	Divide the pitch diameter by the product of 0.3183 and the number of teeth.	$P_c = \dfrac{P_{d_2}}{0.3183\,N}$
Pitch diameter P_{d_2}	Multiply the number of teeth in the worm wheel by the linear pitch of the worm, and divide the product by 3.1416.	$P_{d_2} = \dfrac{NP_L}{3.1416}$
Outside diameter D_o	Multiply the circular pitch by 0.4775 and add the product to the throat diameter.	$D_o = D_t + 0.4775\,P_c$
Throat diameter D_t	Add twice the addendum of the worm tooth to the pitch diameter of the worm wheel.	$D_t = P_{d_2} + 2\,a$
Radius of curvature of worm wheel throat R_c	Subtract twice the addendum of the worm tooth from half the outside diameter of the worm.	$R_c = \dfrac{D_o}{2} - 2\,a$
Diameter to sharp corners D	Multiply the radius of curvature of the worm-wheel throat by the cosine of half the face angle, subtract this quantity from the radius of curvature. Multiply the remainder by 2, and add the product to the throat diameter of the worm wheel.	$D = 2\,(R_c - R_c \times \cos\dfrac{F_a}{2}) + D_t$
Face width of rim F	Multiply the circular pitch by 2.38 and add 0.25 to the product.	$F = 2.38\,P_c + 0.25$
Radius at edge of face F_r	Divide the circular pitch by 4.	$F_r = \dfrac{P_c}{4}$
Addendum a	Multiply the circular pitch by 0.3183.	$a = 0.3183\,P_c$
Whole depth of tooth h_t	Multiply the circular pitch by 0.6866.	$h_t = 0.6866\,P_c$
Center distance between worm and worm wheel S	Add the pitch diameter of the worm to the pitch diameter of the worm wheel and divide the sum by 2.	$S = \dfrac{P_{d_1} + P_{d_2}}{2}$
Gashing angle G	Divide the lead of the worm by the circumference of the pitch circle. The result will be the cotangent of the gashing angle.	$\cot G = \dfrac{L}{3.1416d}$

Courtesy of Cincinnati Milacron

INDEXING

Simple Indexing. — A general rule for determining the number of turns the crank of a dividing head must make, to obtain a given number of divisions, is as follows: Divide the number of turns required for one revolution of the dividing-head spindle by the number of divisions into which the periphery of the work is to be divided.

Example:—If 40 turns of the index crank are required for one revolution of the spindle, and 12 divisions are required, the number of turns of the index crank for each indexing would equal $40 \div 12 = 3\frac{1}{3}$ turns.

Angular Indexing. — With the ordinary indexing head, in which 40 turns of the index crank are required for one revolution of the work, one turn of the index crank equals 9 degrees. Hence, when one complete turn of the index crank equals 9 degrees, two holes in the 18-hole circle, or 3 holes in the 27-hole circle, must correspond to one degree. The first principle or rule for indexing for angles is therefore that two holes in the 18-hole circle or 3 holes in the 27-hole circle equals a movement of one degree of the index head spindle and the work.

Assume that an indexing movement of 35 degrees is required. One complete turn of the index crank equals 9 degrees; therefore, first divide the number of degrees for which to index, by 9, in order to find how many complete turns the index crank should make. The number of degrees left to turn after having completed the full turns are indexed by taking two holes in the 18-hole circle for each degree. In this case, $\frac{35}{9} = 3\frac{8}{9}$, which indicates that the index crank must be turned three full revolutions, and then 8 degrees more are indexed by moving 16 holes in the 18-hole circle.

To index for $11\frac{1}{2}$ degrees, for example, first turn the index crank one revolution, this being a 9-degree movement. Then to index $2\frac{1}{2}$ degrees, move the index crank 5 holes in the 18-hole circle (4 holes for the two whole degrees and one hole for the $\frac{1}{2}$ degree equals the total movement of 5 holes).

Below is shown how this calculation may be carried out to plainly indicate the movement required for this angle:

$11\frac{1}{2}$ deg. = 9 deg. + 2 deg. + $\frac{1}{2}$ deg.

1 turn + 4 holes + 1 hole in the 18-hole circle.

Should it be required to index only $\frac{1}{3}$ degree, this may be done by using the 27-hole circle. In this circle a three-hole movement equals one degree, and a one-hole movement in that circle thus equals $\frac{1}{3}$ degree, or 20 minutes. Assume that it is required to index the work through an angle of 48 degrees 40 minutes. Below is plainly shown how this calculation may be carried out:

48 deg. 40 min. = 45 deg. + 3 deg. + 40 min.

5 turns + 9 holes + 2 holes in the 27-hole circle.

Indexing Movements for Standard Index Plate — Cincinnati Milling Machine

The standard index plate indexes all numbers up to and including 60; all even numbers and those divisible by 5 up to 120; and all divisions listed below up to 400. This plate is drilled on both sides, and has holes as follows:

First side: 24, 25, 28, 30, 34, 37, 38, 39, 41, 42, 43.
Second side: 46, 47, 49, 51, 53, 54, 57, 58, 59, 62, 66.

No. of Divisions	Circle	Turns	Holes	No. of Divisions	Circle	Holes	No. of Divisions	Circle	Holes	No. of Divisions	Circle	Holes
2	Any	20	...	44	66	60	104	39	15	205	41	8
3	24	13	8	45	54	48	105	42	16	210	42	8
4	Any	10	...	46	46	40	106	53	20	212	53	10
5	Any	8	...	47	47	40	108	54	20	215	43	8
6	24	6	16	48	24	20	110	66	24	216	54	10
7	28	5	20	49	49	40	112	28	10	220	66	12
8	Any	5	...	50	25	20	114	57	20	224	28	5
9	54	4	24	51	51	40	115	46	16	228	57	10
10	Any	4	...	52	39	30	116	58	20	230	46	8
11	66	3	42	53	53	40	118	59	20	232	58	10
12	24	3	8	54	54	40	120	66	22	235	47	8
13	39	3	3	55	66	48	124	62	20	236	59	10
14	49	2	42	56	28	20	125	25	8	240	66	11
15	24	2	16	57	57	40	130	39	12	245	49	8
16	24	2	12	58	58	40	132	66	20	248	62	10
17	34	2	12	59	59	40	135	54	16	250	25	4
18	54	2	12	60	42	28	136	34	10	255	51	8
19	38	2	4	62	62	40	140	28	8	260	39	6
20	Any	2	...	64	24	15	144	54	15	264	66	10
21	42	1	38	65	39	24	145	58	16	270	54	8
22	66	1	54	66	66	40	148	37	10	272	34	5
23	46	1	34	68	34	20	150	30	8	280	28	4
24	24	1	16	70	28	16	152	38	10	290	58	8
25	25	1	15	72	54	30	155	62	16	296	37	5
26	39	1	21	74	37	20	156	39	10	300	30	4
27	54	1	26	75	30	16	160	28	7	304	38	5
28	42	1	18	76	38	20	164	41	10	310	62	8
29	58	1	22	78	39	20	165	66	16	312	39	5
30	24	1	8	80	34	17	168	42	10	320	24	3
31	62	1	18	82	41	20	170	34	8	328	41	5
32	28	1	7	84	42	20	172	43	10	330	66	8
33	66	1	14	85	34	20	176	66	15	336	42	5
34	34	1	6	86	43	20	180	54	12	340	34	4
35	28	1	4	88	66	30	184	46	10	344	43	5
36	54	1	6	90	54	24	185	37	8	360	54	6
37	37	1	3	92	46	20	188	47	10	368	46	5
38	38	1	2	94	47	20	190	38	8	370	37	4
39	39	1	1	95	38	16	192	24	5	376	47	5
40	Any	1	...	96	24	10	195	39	8	380	38	4
41	41	...	40	98	49	20	196	49	10	390	39	4
42	42	...	40	100	25	10	200	30	6	392	49	5
43	43	...	40	102	51	20	204	51	10	400	30	3

Tables for Angular Indexing. — The table, "Angular Indexing," gives the number of turns of the index crank for indexing various angles. In the column headed, "Turns of Index Crank," the whole number (where given) indicates the number of full revolutions; the numerator of the fraction, the number of holes additional; and the denominator, the number of holes in the index circle to be used. The angular movement obtained for a movement of one hole, in various index plates is given in the table, "Angular Values of One-Hole Moves."

Angular Indexing

Angle in Degs.	Turns of Index Crank	Angle in Degs.	Turns of Index Crank	Angle in Degs.	Turns of Index Crank	Angle in Degs.	Turns of Index Crank	Angle in Degs.	Turns of Index Crank
1	$\frac{2}{18}$	10	$1\frac{2}{18}$	19	$2\frac{2}{18}$	28	$3\frac{2}{18}$	37	$4\frac{2}{18}$
$1\frac{1}{3}$	$\frac{4}{27}$	$10\frac{1}{3}$	$1\frac{4}{27}$	$19\frac{1}{3}$	$2\frac{4}{27}$	$28\frac{1}{3}$	$3\frac{4}{27}$	$37\frac{1}{3}$	$4\frac{4}{27}$
$1\frac{1}{2}$	$\frac{3}{18}$	$10\frac{1}{2}$	$1\frac{3}{18}$	$19\frac{1}{2}$	$2\frac{3}{18}$	$28\frac{1}{2}$	$3\frac{3}{18}$	$37\frac{1}{2}$	$4\frac{3}{18}$
$1\frac{2}{3}$	$\frac{5}{27}$	$10\frac{2}{3}$	$1\frac{5}{27}$	$19\frac{2}{3}$	$2\frac{5}{27}$	$28\frac{2}{3}$	$3\frac{5}{27}$	$37\frac{2}{3}$	$4\frac{5}{27}$
2	$\frac{4}{18}$	11	$1\frac{4}{18}$	20	$2\frac{4}{18}$	29	$3\frac{4}{18}$	38	$4\frac{4}{18}$
$2\frac{1}{3}$	$\frac{7}{27}$	$11\frac{1}{3}$	$1\frac{7}{27}$	$20\frac{1}{3}$	$2\frac{7}{27}$	$29\frac{1}{3}$	$3\frac{7}{27}$	$38\frac{1}{3}$	$4\frac{7}{27}$
$2\frac{1}{2}$	$\frac{5}{18}$	$11\frac{1}{2}$	$1\frac{5}{18}$	$20\frac{1}{2}$	$2\frac{5}{18}$	$29\frac{1}{2}$	$3\frac{5}{18}$	$38\frac{1}{2}$	$4\frac{5}{18}$
$2\frac{2}{3}$	$\frac{8}{27}$	$11\frac{2}{3}$	$1\frac{8}{27}$	$20\frac{2}{3}$	$2\frac{8}{27}$	$29\frac{2}{3}$	$3\frac{8}{27}$	$38\frac{2}{3}$	$4\frac{8}{27}$
3	$\frac{6}{18}$	12	$1\frac{6}{18}$	21	$2\frac{6}{18}$	30	$3\frac{6}{18}$	39	$4\frac{6}{18}$
$3\frac{1}{3}$	$\frac{10}{27}$	$12\frac{1}{3}$	$1\frac{10}{27}$	$21\frac{1}{3}$	$2\frac{10}{27}$	$30\frac{1}{3}$	$3\frac{10}{27}$	$39\frac{1}{3}$	$4\frac{10}{27}$
$3\frac{1}{2}$	$\frac{7}{18}$	$12\frac{1}{2}$	$1\frac{7}{18}$	$21\frac{1}{2}$	$2\frac{7}{18}$	$30\frac{1}{2}$	$3\frac{7}{18}$	$39\frac{1}{2}$	$4\frac{7}{18}$
$3\frac{2}{3}$	$\frac{11}{27}$	$12\frac{2}{3}$	$1\frac{11}{27}$	$21\frac{2}{3}$	$2\frac{11}{27}$	$30\frac{2}{3}$	$3\frac{11}{27}$	$39\frac{2}{3}$	$4\frac{11}{27}$
4	$\frac{8}{18}$	13	$1\frac{8}{18}$	22	$2\frac{8}{18}$	31	$3\frac{8}{18}$	40	$4\frac{8}{18}$
$4\frac{1}{3}$	$\frac{13}{27}$	$13\frac{1}{3}$	$1\frac{13}{27}$	$22\frac{1}{3}$	$2\frac{13}{27}$	$31\frac{1}{3}$	$3\frac{13}{27}$	$40\frac{1}{3}$	$4\frac{13}{27}$
$4\frac{1}{2}$	$\frac{9}{18}$	$13\frac{1}{2}$	$1\frac{9}{18}$	$22\frac{1}{2}$	$2\frac{9}{18}$	$31\frac{1}{2}$	$3\frac{9}{18}$	$40\frac{1}{2}$	$4\frac{9}{18}$
$4\frac{2}{3}$	$\frac{14}{27}$	$13\frac{2}{3}$	$1\frac{14}{27}$	$22\frac{2}{3}$	$2\frac{14}{27}$	$31\frac{2}{3}$	$3\frac{14}{27}$	$40\frac{2}{3}$	$4\frac{14}{27}$
5	$\frac{10}{18}$	14	$1\frac{10}{18}$	23	$2\frac{10}{18}$	32	$3\frac{10}{18}$	41	$4\frac{10}{18}$
$5\frac{1}{3}$	$\frac{16}{27}$	$14\frac{1}{3}$	$1\frac{16}{27}$	$23\frac{1}{3}$	$2\frac{16}{27}$	$32\frac{1}{3}$	$3\frac{16}{27}$	$41\frac{1}{3}$	$4\frac{16}{27}$
$5\frac{1}{2}$	$\frac{11}{18}$	$14\frac{1}{2}$	$1\frac{11}{18}$	$23\frac{1}{2}$	$2\frac{11}{18}$	$32\frac{1}{2}$	$3\frac{11}{18}$	$41\frac{1}{2}$	$4\frac{11}{18}$
$5\frac{2}{3}$	$\frac{17}{27}$	$14\frac{2}{3}$	$1\frac{17}{27}$	$23\frac{2}{3}$	$2\frac{17}{27}$	$32\frac{2}{3}$	$3\frac{17}{27}$	$41\frac{2}{3}$	$4\frac{17}{27}$
6	$\frac{12}{18}$	15	$1\frac{12}{18}$	24	$2\frac{12}{18}$	33	$3\frac{12}{18}$	42	$4\frac{12}{18}$
$6\frac{1}{3}$	$\frac{19}{27}$	$15\frac{1}{3}$	$1\frac{19}{27}$	$24\frac{1}{3}$	$2\frac{19}{27}$	$33\frac{1}{3}$	$3\frac{19}{27}$	$42\frac{1}{3}$	$4\frac{19}{27}$
$6\frac{1}{2}$	$\frac{13}{18}$	$15\frac{1}{2}$	$1\frac{13}{18}$	$24\frac{1}{2}$	$2\frac{13}{18}$	$33\frac{1}{2}$	$3\frac{13}{18}$	$42\frac{1}{2}$	$4\frac{13}{18}$
$6\frac{2}{3}$	$\frac{20}{27}$	$15\frac{2}{3}$	$1\frac{20}{27}$	$24\frac{2}{3}$	$2\frac{20}{27}$	$33\frac{2}{3}$	$3\frac{20}{27}$	$42\frac{2}{3}$	$4\frac{20}{27}$
7	$\frac{14}{18}$	16	$1\frac{14}{18}$	25	$2\frac{14}{18}$	34	$3\frac{14}{18}$	43	$4\frac{14}{18}$
$7\frac{1}{3}$	$\frac{22}{27}$	$16\frac{1}{3}$	$1\frac{22}{27}$	$25\frac{1}{3}$	$2\frac{22}{27}$	$34\frac{1}{3}$	$3\frac{22}{27}$	$43\frac{1}{3}$	$4\frac{22}{27}$
$7\frac{1}{2}$	$\frac{15}{18}$	$16\frac{1}{2}$	$1\frac{15}{18}$	$25\frac{1}{2}$	$2\frac{15}{18}$	$34\frac{1}{2}$	$3\frac{15}{18}$	$43\frac{1}{2}$	$4\frac{15}{18}$
$7\frac{2}{3}$	$\frac{23}{27}$	$16\frac{2}{3}$	$1\frac{23}{27}$	$25\frac{2}{3}$	$2\frac{23}{27}$	$34\frac{2}{3}$	$3\frac{23}{27}$	$43\frac{2}{3}$	$4\frac{23}{27}$
8	$\frac{16}{18}$	17	$1\frac{16}{18}$	26	$2\frac{16}{18}$	35	$3\frac{16}{18}$	44	$4\frac{16}{18}$
$8\frac{1}{3}$	$\frac{25}{27}$	$17\frac{1}{3}$	$1\frac{25}{27}$	$26\frac{1}{3}$	$2\frac{25}{27}$	$35\frac{1}{3}$	$3\frac{25}{27}$	$44\frac{1}{3}$	$4\frac{25}{27}$
$8\frac{1}{2}$	$\frac{17}{18}$	$17\frac{1}{2}$	$1\frac{17}{18}$	$26\frac{1}{2}$	$2\frac{17}{18}$	$35\frac{1}{2}$	$3\frac{17}{18}$	$44\frac{1}{2}$	$4\frac{17}{18}$
$8\frac{2}{3}$	$\frac{26}{27}$	$17\frac{2}{3}$	$1\frac{26}{27}$	$26\frac{2}{3}$	$2\frac{26}{27}$	$35\frac{2}{3}$	$3\frac{26}{27}$	$44\frac{2}{3}$	$4\frac{26}{27}$
9	1	18	2	27	3	36	4	45	5
$9\frac{1}{3}$	$1\frac{1}{27}$	$18\frac{1}{3}$	$2\frac{1}{27}$	$27\frac{1}{3}$	$3\frac{1}{27}$	$36\frac{1}{3}$	$4\frac{1}{27}$	$45\frac{1}{3}$	$5\frac{1}{27}$
$9\frac{1}{2}$	$1\frac{1}{18}$	$18\frac{1}{2}$	$2\frac{1}{18}$	$27\frac{1}{2}$	$3\frac{1}{18}$	$36\frac{1}{2}$	$4\frac{1}{18}$	$45\frac{1}{2}$	$5\frac{1}{18}$
$9\frac{2}{3}$	$1\frac{2}{27}$	$18\frac{2}{3}$	$2\frac{2}{27}$	$27\frac{2}{3}$	$3\frac{2}{27}$	$36\frac{2}{3}$	$4\frac{2}{27}$	$45\frac{2}{3}$	$5\frac{2}{27}$

Angular Values of One-Hole Moves—B. & S. Index Plates

15-hole circle = 36   minutes

16-hole circle = 33.750 minutes

17-hole circle = 31.765 minutes

18-hole circle = 30   minutes

19-hole circle = 28.421 minutes

20-hole circle = 27   minutes

21-hole circle = 25.714 minutes

23-hole circle = 23.478 minutes

27-hole circle = 20   minutes

29-hole circle = 18.621 minutes

31-hole circle = 17.419 minutes

33-hole circle = 16.364 minutes

37-hole circle = 14.595 minutes

39-hole circle = 13.846 minutes

41-hole circle = 13.171 minutes

43-hole circle = 12.558 minutes

47-hole circle = 11.489 minutes

49-hole circle = 11.020 minutes

Approximate Indexing for Angles.—The following general rule for *approximate* indexing of small angles is applicable to any index head requiring 40 revolutions of the index crank for one revolution of the work.

Rule: Divide 540 by the total number of minutes to be indexed. If the quotient is approximately equal to the number of holes in any index circle available, the angular movement is obtained by moving the crank one hole in this index circle; but if the quotient is not approximately equal, multiply it by any trial number which will give a product equal to the number of holes in an available index circle and move the index crank as many holes as are indicated by the trial number. (If the quotient of 540 divided by the total number of minutes is greater than the number of holes in any of the index circles, it is not possible to obtain the required movement for the angle by simple indexing.)

Example:—Assume that it is required to index to an angle of 2 degrees 46 minutes. Changing this to minutes gives a total of 166 minutes. Dividing 540 by 166 we have 540 ÷ 166 = 3.253. This quotient is next multiplied by some trial number to obtain a product which equals the number of holes in an available index circle. Multiplying by 12, we have 3.253 × 12 = 39.036. Therefore, for indexing 2 degrees 46 minutes, the 39-hole circle can be used and the index crank would be moved 12 holes.

MACHINING METHODS*

1. Good machining practice requires a *rigid setup* in addition to the selection of proper cutting speed, feed, tool material, tool geometry and cutting fluid.

2. The machine tool must be capable of providing the rigidity required for the machining conditions used. If the size of the machine tool is not adequate or if looseness exists in the moving parts, such as spindle bearings or gibs, chatter will occur and poor tool life will result. When a rigid setup cannot be made, the feed and/or depth of cut must be reduced accordingly.

3. Excessive tool overhang is a source of trouble in a machining operation. When this condition exists, poor tool life and surface finish result, and dimensional accuracy is difficult to maintain. Stub-length drills should be used instead of jobbers-length drills where the depth of hole permits. Milling cutters should be mounted as close to the spindle as the job will allow. The length of end mills should be kept at a minimum. Climb milling usually gives better tool life and surface finish than does conventional milling if the machine tool and setup have sufficient rigidity and the feed mechanism is free from backlash.

4. Misalignment and tool runout cause other machining problems, such as oversize and bellmouthed holes when reaming and rough and torn threads when tapping and die threading.

5. Maintenance of cutting tools must be given careful consideration in the development of good machining practice.

6. When dimensional accuracy and surface integrity are not critical, high speed steel tools should be removed when the wearland on the flank of the tool reaches approximately 0.060 in. (1.5 mm) width. In the case of carbide tools, the maximum width should not be allowed to exceed 0.030 in. (0.75 mm), otherwise complete tool failure may occur. On components where dimensional accuracy and surface integrity are critical, the tool wear must be carefully limited.

7. Wearland measurement is not always possible on the tool; however, it is practical to instruct the operator to change tools after a predetermined number of pieces have been machined. The number of parts to be machined per tool should be set conservatively so that the cutter will not fail. Occasionally by this procedure, a cutter may be removed before it is dull; therefore, the resharpening time will be short, but catastrophic failure will have been avoided.

8. The cutting fluid system should provide a copious flow of cutting fluid to the area where the chip is being formed. In the case of machining operations where cutting fluids are used with carbide tools, a continuous flow of the cutting fluid is imperative. Interrupted or intermittent flow can cause thermal shock and breakage of the carbide tool.

9. The concept of good machining practice involves consideration of all factors associated with the machining operation. Each detail—workpiece, fixturing, speed, feed, tool material, tool geometry, cutting fluid, and the machine tool itself—must be given careful attention to ensure success for the machining operation under consideration.

10. The *machinability* of a work material must be determined in order to select the proper machining conditions. The machinability of a material can be defined in terms of three major factors: surface integrity, tool life, and power or force requirements.

*Reprinted from the MACHINING DATA HANDBOOK, 3rd Edition, by permission of the Machinability Data Center. © 1980 by Metcut Research Associates Inc.

CUTTING SPEEDS AND FEEDS

Work Materials. — The large number of work materials that are commonly machined vary greatly in their basic structure and the ease with which they can be machined. Yet it is possible to group together certain materials having similar machining characteristics, for the purpose of recommending the cutting speed at which they can be cut. Most materials that are machined are metals and it has been found that the most important single factor influencing the ease with which a metal can be cut is its microstructure, followed by any cold work that may have been done to the metal, which increases its hardness. Metals that have a similar, but not necessarily the same microstructure, will tend to have similar machining characteristics. Thus, the grouping of the metals in the accompanying tables has been done on the basis of their microstructure.

With the exception of a few soft and gummy metals, experience has shown that harder metals are more difficult to cut than softer metals. Furthermore, any given metal is more difficult to cut when it is in a harder form than when it is softer. It is more difficult to penetrate the harder metal and more power is required to cut it. This in turn will generate a higher cutting temperature at any given cutting speed thereby making it necessary to use a slower speed, for the cutting temperature must always be kept within the limits that can be sustained by the cutting tool without failure. Hardness, then, is an important property that must be considered when machining a given metal. Hardness alone, however, cannot be used as a measure of cutting speed. For example, if pieces of AISI 11L17 and AISI 1117 steel both have a hardness of 150 Bhn, their recommended cutting speed for high speed steel tools will be 140 fpm and 130 fpm respectively. In some metals two entirely different microstructures can produce the same hardness. As an example, a fine pearlite microstructure and a tempered martensite microstructure can result in the same hardness in a steel. These microstructures will not machine alike. For practical purposes, however, information on hardness is usually easier to obtain than information on microstructure; thus, hardness alone is usually used to differentiate between different cutting speeds for machining a metal.

In some situations the hardness of a metal to be machined is not known. When this occurs, the material condition, which is listed separately in the tables, can be used as a guide. Lacking more specific information on the hardness of the metal to be machined, a knowledge of its material condition is helpful in selecting the cutting speed.

The surface of ferrous metal castings has a scale that is more difficult to machine than the metal below. Some scale is more difficult to machine than others, depending on the foundry sand used, the casting process, the method of cleaning the casting, and the type of metal cast. Special electrochemical treatments can be used in some cases that almost entirely eliminate the effect of the scale on machining, although castings so treated are not frequently encountered. Usually when casting scale is encountered, the cutting speed is reduced approximately 5 or 10 per cent. Difficult-to-machine surface scale can also be encountered when machining hot rolled or forged steel bars.

Metallurgical differences that affect machining characteristics are often found within a single piece of metal. The occurrence of hard spots in castings is an example. Different microstructures and hardness levels may occur within a casting as a result of variations in the cooling rate in different parts of the casting. Such variations are less severe in castings that have been heat treated. Steel bar stock is usually harder toward the outside than toward the center of the bar. Sometimes there are slight metallurgical differences along the length of a bar which can affect its cutting characteristics.

Cutting Speed, Feed, Depth of Cut. — The cutting conditions that determine the rate of metal removal are the cutting speed, the feed rate, and the depth of cut. These cutting conditions and the nature of the material to be cut determine the power required to take the cut. The cutting conditions must be adjusted to stay within the power available on the machine tool to be used.

The cutting conditions must also be considered in relation to the tool life. Tool life can be defined as the length of time that a cutting tool will cut before becoming dull or before it must be replaced. The end of tool life is defined as a given amount of wear on the flank of the tool by the ANSI Standard Specification For Tool Life Testing With Single-Point Tools-ANSI B94.34-1946(R1971) and B94.36-1956(R1971). These standards are followed when making scientific machinability tests with single-point cutting tools in order to achieve uniformity in testing procedure so that results from different machinability laboratories can readily be compared. It is not practicable or necessary to follow this standard in the shop; however, it should be understood that the cutting conditions and tool life are related. For example, the cutting speed and the feed may be increased if a shorter tool life is accepted. Furthermore, the decrease in the tool life will be proportionately greater than the increase in the cutting speed or the feed. Conversely, if the cutting speed or the feed is decreased, the increase in the tool life will be proportionately greater than the decrease in the cutting speed or the feed.

Tool life is influenced most by cutting speed, then by the feed rate, and least by the depth of cut. After the depth of cut is about 10 times greater than the feed rate, a further increase in the depth of cut will have no significant effect on the tool life. This characteristic of the performance of cutting tools is very important in determining the operating or cutting conditions for machining metals.

The first step in selecting the cutting conditions is to select the depth of cut. The depth of cut will be limited by the amount of metal that is to be machined from the workpiece, by the power available on the machine tool, by the rigidity of the workpiece and the cutting tool, and by the rigidity of the setup. Since the depth of cut has the least effect upon the tool life, always use the heaviest depth of cut that is possible.

The second step is to select the feed rate. In selecting the feed rate, consideration must be given to the power available on the machine tool, to the rigidity of the workpiece and the cutting tool, to the rigidity of the setup, and to the surface finish required on the finished workpiece. The available power must be considered in relation to the depth of cut previously selected in considering the feed. Select the maximum feed possible; however, it must not be greater than that which will produce an acceptable surface finish.

The third step is to select the cutting speed. The accompanying tables provide recommended cutting speeds. If previous experience has been had in machining a certain material, this may form the basis for selecting the cutting speed. In either case, however, the depth of cut should be selected first, followed by the feed, and the last to be selected should be the cutting speed.

Use of Cutting Speed Tables. — On the following pages tables of recommended cutting speeds are provided. The values in these tables are for average conditions and serve as a basis from which to start. In many cases they will be found to be the most satisfactory cutting speeds, while in others, modifications may offer advantages as experience is gained on a particular job. It is not possible to specify a single optimum cutting speed for a material that will fit every situation. Many factors unique to each job may make a modification desirable. These factors include: the size, type, model, and make of the machine tool; its power, rigidity, and the foundation on which it is standing; the workpiece configuration; the rigidity of the workpiece setup, or fixturing; safety aspects of the setup; the particular grade of cemented carbide used; the influence of a cutting fluid; and, the tool life desired. Except for certain difficult-to-machine materials, most materials can be cut successfully over a rather wide range of cutting speeds with a particular type of cutting tool material; however, not just any speed within this range should be used. A cutting speed that is too slow will result in a loss of production and in an increase in the cost of the part. Likewise, a cutting speed that is too fast can have the same result because the tool life will be too short and the production must frequently be interrupted to change tools. Moreover, the cost of sharpening or replacing the cutting tool must be considered. There is usually a narrow range of cutting speeds within which the most economical results will be obtained.

When making a modification to the cutting speed the fundamental behaviour of cutting tools must be kept in mind; i.e., increasing the cutting speed will result in a proportionately larger reduction in the tool life; reducing the cutting speed will result in a proportionately larger increase in the tool life.

Feed and Depth of Cut Factors: Factors used to modify the cutting speeds in compensation for different feed rates and depths of cut are given in Table 5. These factors should be used only with the cutting speeds listed in Tables 1 through 4. Moreover, they do not apply when the hardness of a material exceeds approximately 350 to 400 HB. They should, however, be used in the other hardness ranges, at which most materials are machined. Experience in production and scientific investigations conducted in machinability laboratories have shown that a change in the feed or in the depth of cut will require a compensating change in the cutting speed if the tool life is to remain unchanged. For this reason a modification in the cutting speed should be made if the feed or depth of cut is changed significantly when machining the materials listed in the other tables.

Cutting Tool Material: The cutting speeds in the tables listed under cemented carbide are based on the assumption that a correct grade of uncoated straight tungsten carbide or an uncoated crater resisting carbide is used. Most carbide producers will be able to recommend a grade of carbide that can be used at the cutting speed given. An incorrect grade of carbide may, however, require a modification to the cutting speed, and in some instances it may result in an extremely short tool life. High speed steels are less sensitive in this respect; i.e., any type of high speed steel can cut most of the materials listed in the tables at the recommended cutting speed for high speed steel. It is true, however, that certain types of high speed steels do offer advantages in some applications, such as a longer tool life. Very hard and tough materials, such as the superalloys, should be cut with T15 high speed steel, or with one of the M-40 types, as for example M42. These premium high speed steels should not be used to cut most other materials, since they can be cut as well by other types of high speed steel.

Coated Carbides: A faster cutting speed can be used with coated carbide cutting tools than with uncoated tools when machining materials that can be cut by the coated grades. When using titanium carbide or titanium nitride coated cutting tools, the recommended cutting speed in the tables should be increased by 20 to 30 per cent· in some cases an increase up to 50 per cent is possible. The cutting speed should be increased even more when using aluminum oxide coated carbides.

Titanium Carbides: The cutting speed should be increased by 30 to 60 per cent over the values given in the tables when using titanium carbide cutting tools.

Milling: The recommended cutting speed for milling given in the tables can be used for all face milling operations, slab milling operations, slab milling type cuts taken with end milling cutters, and for shallow slotting cuts taken with either end milling cutters or with side milling cutters. When milling deep slots with end milling cutters or with side milling cutters, the cutting speed should be reduced approximately 10 per cent. Likewise, the cutting speed should be reduced about 10 per cent when taking wide side milling cuts with side milling cutters; no reduction in the cutting speed is required if the width of the side milling cut is small.

Planing and Shaping: The cutting speeds in Tables 1 through 4, and Tables 6 through 9 can be used for planing and shaping. The feed and depth of cut factors in Table 5 should also be used as explained previously. Very often other factors relating to the machine or the setup will require a reduction in the actual cutting speed used on the job.

Surface Scale: Certain heavy and abrasive surface scales encountered on castings and on some wrought metals require a reduction in the cutting speed, especially when drilling.

Cutting Fluids: Many cutting fluids permit a somewhat higher cutting speed to be used. It is not possible, however, to provide specific recommendations because each proprietary cutting fluid exhibits its own characteristics.

Metric Units: In metric practice the units used for the cutting speed is meters per minute, abbreviated m/min. or mpm. The cutting speeds in the tables are given in inch units, or feet per minute, which is abbreviated fpm. These units can be converted as follows: to obtain meters per minute, multiply feet per minute by 0.3; to obtain feet per minute, multiply meters per minute by 3.28.

Cutting Speed Formulas: Most machining operations are conducted on machine tools having a rotating spindle, and the cutting speed in feet or meters per minute must be converted to a spindle speed, or to revolutions per minute; this is accomplished by use of the following formulas:

For inch units only:

$$N = \frac{12V}{\pi D}$$

For metric units only:

$$N = \frac{1000V}{\pi D}$$

Where: N = Spindle speed; rpm

V = Cutting speed; fpm. or m/min

D = Diameter; in., or mm (For turning, D is the outside diameter of the workpiece. For milling, drilling and reaming, D is the diameter of the cutter.)

$\pi = 3.14$

Summary of Principal Tables in the Speeds and Feeds Section

Recommended Cutting Speeds for Turning:
Table 1.	Plain carbon and alloy steels
Table 2.	Tool steels
Table 3.	Stainless steels
Table 4.	Ferrous cast metals
Table 5.	Cutting speed feed and depth of cut factors for use with Tables 1-4
Table 6.	Light metals (For milling also)
Table 7.	Copper alloys (For milling also)
Table 8.	Titanium and titanium alloys (For milling and drilling also)
Table 9.	Superalloys (For milling and drilling also)
Table 10.	Hard-to-machine alloys with cubic boron nitride cutting tools

Recommended Cutting Speeds for Milling:
Table 11.	Plain carbon and alloy steels
Table 12.	Tool steels
Table 13.	Stainless steels
Table 14.	Ferrous cast metals
Table 15.	Feed rates using high speed steel cutters
Table 16.	Feed rates using cemented carbide cutters

Recommended Cutting Speeds for Drilling and Reaming:
Table 17.	Plain carbon and alloy steels
Table 18.	Tool steels
Table 19.	Stainless steels
Table 20.	Ferrous cast metals
Table 21.	Light metals
Table 22.	Copper alloys

The feed and depth-of-cut factors in Table 5 are only to be used together with Tables 1 through 4. While the values in Table 5 are for inch units, they can be used with metric units by converting the metric feed and depth of cut into inch units. First select the cutting speed from the appropriate table and then apply the factors from Table 5, using the formula given below:

$$V = V_o F_f F_d$$

Where: V = Cutting speed to be used; fpm, or m/min.

           V_o = Cutting speed from tables; fpm, or m/min.

           F_f = Feed Factor   (From Table 5)

           F_d = Depth-of-cut factor   (From Table 5)

Example: Using both the inch and the metric formulas, calculate the spindle speed for turning a 1¼ inch (31.75 mm) bar of 200-220 HB AISI 1040 steel using depth of cut of .100 in. (2.54 mm) and a feed rate of .015 in. (0.38 mm/rev.). (A small insignificant difference in the answers may occur, which is caused by the conversion of the units and the rounding off of numbers.)

From Table 1: V_o = 85 fpm.

From Table 5: F_f = .91;

              F_d = 1.03

$$V = V_o F_f F_d = 85 \times .91 \times 1.03 = 80 \text{ fpm}$$

$$N = \frac{12V}{\pi D} = \frac{12 \times 80}{\pi \times 1.250} = 244 \text{ rpm}$$

$$V_o = 85 \times .3 = 25.5 \text{ m/min}$$

$$V = V_o F_f F_d = 25.5 \times .91 \times 1.03 = 24 \text{ m/min}$$

$$N = \frac{1000V}{\pi D} = \frac{1000 \times 24}{\pi \times 31.75} = 241 \text{ rpm}$$

It is often necessary to calculate the cutting speed in feet per minute or in meters per minute, when the diameter of the workpiece or of the cutting tool and the spindle speed is known. In this event, the following formulas are used.

For inch units only:

$$V = \frac{\pi D N}{12} \text{ , where } D \text{ is in inches}$$

For metric units only:

$$V = \frac{\pi D N}{1000} \text{ , where } D \text{ is in millimeters}$$

Example: Calculate the cutting speed in feet per minute and in meters per minute when the spindle speed of a ¾ inch (19.05 mm) drill is 400 rpm.

$$V = \frac{\pi D N}{12} = \frac{\pi \times .75 \times 400}{12} = 78.5 \text{ fpm}$$

$$V = \frac{\pi D N}{1000} = \frac{\pi \times 19.05 \times 400}{1000} = 24 \text{ m/min}$$

Table 1. Recommended Cutting Speeds in Feet per Minute for Turning Plain Carbon and Alloy Steels

Material AISI and SAE Steels	Hardness, HB*	Material Condition*	Cutting Speed, fpm	
			HSS	Carbide
Free Machining Plain Carbon Steels (Resulphurized), 1212, 1213, 1215	100-150	HR, A	150	600
	150-200	CD	160	625
1108, 1109, 1115, 1117, 1118, 1120 1126, 1211	100-150	HR, A	130	500
	150-200	CD	120	525
1132, 1137, 1139, 1140, 1144, 1146, 1151	175-225	HR, A, N, CD	120	400
	275-325	Q and T	75	300
	325-375	Q and T	50	225
	375-425	Q and T	40	200
Free Machining Plain Carbon Steels (Leaded), 11L17, 11L18, 12L13, 12L14	100-150	HR, A, N, CD	140	550
	150-200	HR, A, N, CD	145	560
	200-250	N, CD	110	400
Plain Carbon Steels, 1006, 1008, 1009, 1010, 1012, 1015, 1016, 1017, 1018, 1019, 1020, 1021, 1022, 1023, 1024, 1025, 1026, 1513, 1514	100-125	HR, A, N, CD	120	450
	125-175	HR, A, N, CD	110	400
	175-225	HR, N, CD	90	350
	225-275	CD	70	300
1027, 1030, 1033, 1035, 1036, 1037, 1038, 1039, 1040, 1041, 1042, 1043, 1045, 1046, 1048, 1049, 1050, 1052, 1524, 1526, 1527, 1541	125-175	HR, A, N, CD	100	375
	175-225	HR, A, N, CD	85	325
	225-275	N, CD, Q and T	70	225
	275-325	Q and T	60	200
	325-375	Q and T	40	160
	375-425	Q and T	30	140
1055, 1060, 1064, 1065, 1070, 1074, 1078, 1080, 1084, 1086, 1090, 1095, 1548, 1551, 1552, 1561, 1566	125-175	HR, A, N, CD	100	370
	175-225	HR, A, N, CD	80	320
	225-275	N, CD, Q and T	65	220
	275-325	Q and T	50	180
	325-375	Q and T	35	150
	375-425	Q and T	30	130
Free Machining Alloy Steels (Resulphurized), 4140, 4150	175-200	HR, A, N, CD	110	400
	200-250	HR, N, CD	90	350
	250-300	Q and T	65	300
	300-375	Q and T	50	225
	375-425	Q and T	40	165
Free Machining Alloy Steels (Leaded), 41L30, 41L40, 41L47, 41L50, 43L47, 51L32, 52L100, 86L20, 86L40	150-200	HR, A, N, CD	120	430
	200-250	HR, N, CD	100	380
	250-300	Q and T	75	275
	300-375	Q and T	55	220
	375-425	Q and T	50	200
Alloy Steels, 4012, 4023, 4024, 4028, 4118, 4320, 4419, 4422, 4427, 4615, 4620, 4621, 4626, 4718, 4720, 4815, 4817, 4820, 5015, 5117, 5120, 6118, 8115, 8615, 8617, 8620, 8622, 8625, 8627, 8720, 8822, 94B17	125-175	HR, A, N, CD	100	400
	175-225	HR, N, CD	90	350
	225-275	CD, N, Q and T	70	300
	275-325	Q and T	60	250
	325-375	Q and T	50	200
	375-425	Q and T	35	175

Based on a feed rate of .012 in. per rev. and a depth of cut of .125 in.
* Abbreviations designate: HR, hot rolled; CD, cold drawn; A, annealed; N, normalized; Q and T, quenched and tempered; and HB, Brinell hardness number.

Table 1 (*Concluded*). **Recommended Cutting Speeds in Feet per Minute for Turning Plain Carbon and Alloy Steels**

Material AISI and SAE Steels	Hardness, HB*	Material Condition*	Cutting Speed, fpm	
			HSS	Carbide
Alloy Steels, 1330, 1335, 1340, 1345 4032, 4037, 4042, 4047, 4130, 4135, 4137, 4140, 4142, 4145, 4147, 4150, 4161, 4337, 4340, 50B44, 50B46, 50B50, 50B60, 5130, 5132, 5140, 5145, 5147, 5150, 5160, 51B60, 6150, 81B45, 8630, 8635, 8637, 8640, 8642, 8645, 8650, 8655, 8660, 8740, 9254, 9255, 9260, 9262, 94B30	175-225 225-275 275-325 325-375 375-425	HR, A, N, CD N, CD, Q and T N, Q and T N, Q and T Q and T	85 70 60 40 30	325 275 230 200 150
Alloy Steels, E51100, E52100	175-225 225-275 275-325 325-375 375-425	HR, A, CD N, CD, Q and T N, Q and T N, Q and T Q and T	70 65 50 30 20	310 260 220 180 140
Ultra High Strength Steels (Not AISI) AMS 6421 (98B37 Mod.), AMS 6422 (98BV40), AMS 6424, AMS 6427, AMS 6428, AMS 6430, AMS 6432, AMS 6433, AMS 6434, AMS 6436, AMS 6442, 300M, D6ac	220-300 300-350 350-400 43-48 HRC 48-52 HRC	A N N Q and T Q and T	65 50 35 25 10	270 200 150 120 80
Maraging Steels (Not AISI) 18% Ni Grade 200 18% Ni Grade 250 18% Ni Grade 300 18% Ni Grade 350	250-325 50-52 HRC	A Maraged	60 10	300 80
Nitriding Steels (Not AISI) Nitralloy 125 Nitralloy 135 Nitralloy 135 Mod. Nitralloy 225 Nitralloy 230 Nitralloy N Nitralloy EZ Nitrex 1	200-250 300-350	A N, Q and T	70 30	300 225

Based on a feed rate of .012 in. per rev. and a depth of cut of .125 in.

* Abbreviations designate: HR, hot rolled; CD, cold drawn; A, annealed; N, normalized; Q and T, quenched and tempered; HB, Brinell hardness number; and HRC, Rockwell C scale hardness number.

Cutting Time for Turning, Boring, and Facing. — The time required to turn a length of metal can be determined by the following formula in which T = time in minutes, L = length of cut in inches, f = feed in inches per revolution, and N = lathe spindle speed in revolutions per minute.

$$T = \frac{L}{fN}$$

When making job estimates, the time required to load and to unload the workpiece on the machine, and the machine handling time must be added to the cutting time for each length cut to obtain the floor-to-floor time.

Table 2. Recommended Cutting Speeds in Feet per Minute for Turning
Tool Steels

Material Tool Steels (AISI Types)	Hardness, HB*	Material Condition*	Cutting Speed, fpm HSS	Cutting Speed, fpm Carbide
Water Hardening W_1, W_2, W_5	150-200	A	100	325
Shock Resisting S_1, S_2, S_5, S_6, S_7	175-225	A	70	300
Cold Work, Oil Hardening O_1, O_2, O_6, O_7	175-225	A	70	250
Cold Work, High Carbon High Chromium D_2, D_3, D_4, D_5, D_7	200-250	A	45	175
Cold Work, Air Hardening $A_2, A_3, A_8, A_9, A_{10}$	200-250	A	70	250
A_4, A_6	200-250	A	55	200
A_7	225-275	A	45	175
Hot Work, Chromium Type $H_{10}, H_{11}, H_{12}, H_{13}, H_{14}, H_{19}$	150-200	A	80	300
	200-250	A	65	225
	325-375	Q and T	50	175
	48-50 HRC	Q and T	20	95
	50-52 HRC	Q and T	10	80
	52-54 HRC	Q and T	—	60
	54-56 HRC	Q and T	—	40
Hot Work, Tungsten Type $H_{21}, H_{22}, H_{23}, H_{24}, H_{25}, H_{26}$	150-200	A	60	250
	200-250	A	50	200
Hot Work, Molybdenum Type H_{41}, H_{42}, H_{43}	150-200	A	55	225
	200-250	A	45	175
Special Purpose, Low Alloy L_2, L_3, L_6	150-200	A	75	325
Mold P_2, P_3, P_4, P_5, P_6	100-150	A	90	400
P_{20}, P_{21}	150-200	A	80	350
High Speed Steel $M_1, M_2, M_6, M_{10}, T_1, T_2, T_6$	200-250	A	65	225
$M_{3}-1, M_4, M_7, M_{30}, M_{33}, M_{34}, M_{36}, M_{41}, M_{42}, M_{43}, M_{44}, M_{46}, M_{47}, T_5, T_8$	225-275	A	55	200
$T_{15}, M_{3}-2$	225-275	A	45	170

Based on a feed rate of .012 in. per rev. and a depth of cut of .125 in.
* Abbreviations designate: A, annealed; Q and T, quenched and tempered; HB, Brinell hardness number; and HRC, Rockwell C scale hardness number.

Cutting Speed for Tapping. — A table of cutting speeds for tapping is not given. Several factors, singly or in combination, can cause very great differences in the permissible tapping speed. The principal factors affecting the tapping speed are the pitch of the thread, the chamfer length on the tap, the percentage of full thread to be cut, the length of the hole to be tapped, the cutting fluid used, whether the threads are straight or tapered, the machine tool used to perform the operation, and the material to be tapped.

The cutting speed for coarse pitch taps must be slower than for fine pitch taps

Table 3. Recommended Cutting Speeds in Feet per Minute for Turning
Stainless Steels

Material Stainless Steels	Hardness, HB*	Material Condition*	Cutting Speed, fpm	
			HSS	Carbide
Free Machining Stainless Steels				
(Ferritic), 430F, 430F Se	135-185	A	110	400
(Austenitic), 203EZ, 303, 303Se, 303MA, 303Pb, 303Cu, 303 Plus X	135-185	A	100	350
	225-275	CD	80	325
(Martensitic); 416, 416Se, 416 Plus X, 420F, 420F Se, 440F, 440F Se	135-185	A	110	400
	185-240	A, CD	100	350
	275-325	Q and T	60	250
	375-425	Q and T	30	125
Stainless Steels				
(Ferritic), 405, 409, 429, 430, 434, 436, 442, 446, 502	135-185	A	90	300
(Austenitic), 201, 202, 301, 302, 304, 304L, 305, 308, 321, 347, 348	135-185	A	75	225
	225-275	CD	65	200
(Austenitic), 302B, 309, 309S, 310, 310S, 314, 316, 316L, 317, 330	135-185	A	70	225
(Martensitic), 403, 410, 420, 501	135-175	A	95	350
	175-225	A	85	300
	275-325	Q and T	55	200
	375-425	Q and T	35	125
(Martensitic), 414, 431, Greek Ascoloy	225-275	A	60	250
	275-325	Q and T	50	200
	375-425	Q and T	30	125
(Martensitic), 440A, 440B, 440C	225-275	A	55	200
	275-325	Q and T	45	150
	375-425	Q and T	30	125
(Precipitation Hardening) 15-5PH, 17-4PH, 17-7PH, AF-71, 17-14CuMo, AFC-77, AM-350, AM-355, AM-362, Custom 455, HNM, PH13-8, PH14-8Mo, PH15-7Mo, Stainless W	150-200	A	60	225
	275-325	H	50	200
	325-375	H	40	130
	375-450	H	25	90

Based on a feed rate of .012 in. per rev. and a depth of cut of .125 in.
* Abbreviations designate: A, annealed; CD, cold drawn; Q and T, quenched and tempered; H, precipitation hardened; and HB, Brinell hardness number.

with the same diameter. Usually the difference in pitch becomes more pronounced as the diameter of the tap becomes larger and slight differences in the pitch of smaller diameter taps have little significant effect on the cutting speed. Unlike all other cutting tools, the feed per revolution of a tap cannot be independently adjusted—it is always equal to the lead of the thread and is always greater for coarse pitches than for fine pitches. Furthermore, the thread form of a coarse pitch thread is larger than that of a fine pitch thread; therefore, it is necessary to remove more metal when cutting a coarse pitch thread.

Taps with a long chamfer, such as starting or taper taps, can cut faster in a short hole than short chamfer taps, such as plug taps. In deep holes, however, short chamfer or plug taps can run faster than long chamfer taps. Bottoming taps must be run more slowly than either starting or plug taps. The chamfer helps to start the tap in the hole. It also functions to involve more threads, or thread form cutting edges, on the tap in cutting the thread in the hole. This reduces the cutting load on

Table 4. **Recommended Cutting Speeds in Feet per Minute for Turning Ferrous Cast Metals**

Material Ferrous Cast Metals	Hardness, HB*	Material Condition*	Cutting Speed, fpm	
			HSS	Carbide
Gray Cast Iron				
ASTM Class 20	120-150	A	120	450
ASTM Class 25	160-200	AC	90	350
ASTM Class 30, 35, and 40	190-220	AC	80	275
ASTM Class 45 and 50	220-260	AC	60	200
ASTM Class 55 and 60	250-320	AC, HT	35	125
ASTM Type 1, 1b, 5 (Ni Resist)	100-215	AC	70	225
ASTM Type 2, 3, 6 (Ni Resist)	120-175	AC	65	210
ASTM Type 2b, 4 (Ni Resist)	150-250	AC	50	200
Malleable Iron				
(Ferritic), 32510, 35018	110-160	MHT	130	500
(Pearlitic), 40010, 43010, 45006, 45008, 48005, 50005	160-200 200-240	MHT MHT	95 75	400 275
(Martensitic), 53004, 60003, 60004	200-255	MHT	70	250
(Martensitic), 70002, 70003	220-260	MHT	60	225
(Martensitic), 80002	240-280	MHT	50	140
(Martensitic), 90001	250-320	MHT	30	125
Nodular (Ductile) Iron				
(Ferritic), 60-40-18, 65-45-12	140-190	A	100	450
(Ferritic-Pearlitic), 80-55-06	190-225 225-260	AC AC	80 65	350 210
(Pearlitic-Martensitic), 100-70-03	240-300	HT	45	175
(Martensitic), 120-90-02	270-330 330-400	HT HT	30 15	100 50
Cast Steels				
(Low Carbon), 1010, 1020	100-150	AC, A, N	110	400
(Medium Carbon), 1030, 1040, 1050	125-175 175-225 225-300	AC, A, N AC, A, N AC, HT	100 90 70	400 350 300
(Low Carbon Alloy), 1320, 2315, 2320 4110, 4120, 4320, 8020, 8620	150-200 200-250 250-300	AC, A, N AC, A, N AC, HT	90 80 60	350 325 250
(Medium Carbon Alloy), 1330, 1340, 2325, 2330, 4125, 4130, 4140, 4330, 4340, 8030, 80B30, 8040, 8430, 8440, 8630, 8640, 9525, 9530, 9535	175-225 225-250 250-300 300-350 350-400	AC, A, N AC, A, N AC, HT AC, HT HT	80 70 55 45 30	325 280 250 220 150

Based on a feed rate of .012 inch per revolution and a depth of cut of .125 inch.
* Abbreviations designate: A, annealed; AC, as cast; N, normalized; HT, heat treated; MHT, malleablizing heat treatment; and HB, Brinell hardness number.

Table 5. Cutting Speed Feed and Depth of Cut Factors for Turning*

Feed, in./rev.	Feed Factor, F_f	Depth of Cut, in.	Depth of Cut Factor, F_d
.002	1.50	.005	1.50
.003	1.50	.010	1.42
.004	1.50	.016	1.33
.005	1.44	.031	1.21
.006	1.34	.047	1.15
.007	1.25	.062	1.10
.008	1.18	.078	1.07
.009	1.12	.094	1.04
.010	1.08	.100	1.03
.011	1.04	.125	1.00
.012	1.00	.150	.97
.013	.97	.188	.94
.014	.94	.200	.93
.015	.91	.250	.91
.016	.88	.312	.88
.018	.84	.375	.86
.020	.80	.438	.84
.022	.77	.500	.82
.025	.73	.625	.80
.028	.70	.688	.78
.030	.68	.750	.77
.032	.66	.812	.76
.035	.64	.938	.75
.040	.60	1.000	.74
.045	.57	1.125	.73
.050	.55	1.250	.72
.060	.50	1.375	.71

* For use in conjunction with Tables 1, 2, 3, and 4 only.

any one set of thread form cutting edges. In so doing, more chips and thinner chips are produced which are difficult to remove from deeper holes. Shortening the chamfer length causes fewer thread form cutting edges to cut, thereby producing fewer and thicker chips that can be easily disposed of. Only one or two sets of thread form cutting edges are cut on bottoming taps which cause these cutting edges to assume a heavy cutting load and produce very thick chips.

Spiral pointed taps can operate at a faster cutting speed than taps with normal flutes. These taps are made with supplementary angular flutes on the end which push the chips ahead of the tap and prevent the tapped hole from becoming clogged with chips. They are used primarily to tap open or through holes although some are made with shorter supplementary flutes for tapping blind holes.

The tapping speed must be reduced as the percentage of full thread to be cut is increased. Experiments have shown that the torque required to cut a 100 per cent thread form is more than twice that required to cut a 50 per cent thread form. An increase in the percentage of full thread will also produce a greater volume of chips.

The tapping speed must be lowered as the length of the hole to be tapped is increased. More friction must be overcome in turning the tap and more chips accumulate in the hole. It will be more difficult to apply the cutting fluid at the cutting edges and to lubricate the tap in order to reduce friction. This is especially true when the hole is being tapped in a horizontal position.

Cutting fluids have a very great effect on the cutting speed for tapping. While other operating conditions when tapping frequently cannot be changed, a free selection of the cutting fluid can usually be made. When planning the tapping operation, the selection of a cutting fluid warrants a very careful consideration and perhaps an investigation.

Taper threaded taps, such as pipe taps, must be operated at a slower speed than

Table 6. Recommended Cutting Speeds in Feet per Minute for Turning and Milling
Light Metals

Material, Light Metals	Material Condition*	Cutting Speed, fpm	
		HSS	Carbide
All Wrought Aluminum Alloys	CD	600	1200
	ST and A	500	1100
All Aluminum Sand and Permanent Mold Casting Alloys	AC	750	1400
	ST and A	600	1200
All Aluminum Die Casting Alloys†	AC	125	550
	ST and A	100	450
†except Alloys 390.0 and 392.0	AC	80	500
	ST and A	60	425
All Wrought Magnesium Alloys	A, CD, ST, and A	800	2000
All Cast Magnesium Alloys	A, AC, ST and A	800	2000

* Abbreviations designate: A, annealed; AC, as cast; CD, cold drawn; and ST and A, solution treated and aged.

straight thread taps with a comparable diameter. All of the thread form cutting edges of a taper threaded tap that are engaged in the work cut and produce a chip, while only those cutting edges along the chamfer length cut on straight thread taps. In many cases the pipe tap is required to cut the tapered thread from a straight hole which adds to the cutting burden.

The machine tool on which the tapping operation is performed must be considered in selecting the tapping speed. Tapping machines and other machines that are able to feed the tap at a rate of advance equal to the lead of the tap and which have provisions for quickly reversing the spindle can be operated at high cutting speeds. On machines where the feed of the tap is controlled manually—such as on drill presses and turret lathes—the tapping speed must be reduced to allow the operator to maintain a safe control of the operation.

There are other special considerations in selecting the tapping speed. Very accurate threads are usually tapped more slowly than threads with a commercial grade of accuracy. Thread forms that require deep threads for which a large amount of metal must be removed, producing a large volume of chips, require special techniques and slower cutting speeds. Acme, buttress, and square threads are, therefore, generally cut at lower speeds.

Not the least important consideration is the material to be tapped. Like other metal cutting operations, tapping is affected by the basic structure of the metal being tapped and its hardness. Because of all of the factors mentioned above, it is not practical to tabulate cutting speeds for tapping as has been done for other operations. On a relative basis, the cutting speeds for tapping can be compared to the cutting speed for drilling with high speed steel drills. The speeds may be used as a starting point but normally must be reduced, and must not be applied as such without weighing every factor that affects the tapping operation.

Cutting Speed for Broaching. — Broaching offers many advantages in manufacturing metal parts, including high production rates, excellent surface finishes, and close dimensional tolerances. These advantages are not derived from the use of high cutting speeds; they are derived from the large number of cutting teeth that can be applied in consecutive order in a given period of time, from their configuration and precise dimensions, and from the width or diameter of the surface that can be machined in a single stroke. Most broaching cutters are expensive in their initial

Table 7.   Recommended Cutting Speeds in Feet per Minute for Turning
and Milling Copper Alloys

Material Copper Alloys (Copper Alloy Nos. as per the Copper Development Assn. Inc.)		Material Condition*	Cutting Speed, fpm	
			HSS	Carbide
314	Leaded Commercial Bronze	A CD	300 350	650 600
332	High Leaded Brass			
340	Medium Leaded Brass			
342	High Leaded Brass			
353	High Leaded Brass			
356	Extra High Leaded Brass			
360	Free Cutting Brass			
370	Free Cutting Muntz Metal			
377	Forging Brass			
385	Architectural Bronze			
485	Leaded Naval Brass			
544	Free Cutting Phosphor Bronze			
226	Jewelry Bronze	A CD	200 250	500 550
230	Red Brass			
240	Low Brass			
260	Cartridge Brass 70%			
268	Yellow Brass			
280	Muntz Metal			
335	Low Leaded Brass			
365	Leaded Muntz Metal			
368	Leaded Muntz Metal			
443	Admiralty Brass (inhibited)			
445	Admiralty Brass (inhibited)			
651	Low Silicon Bronze			
655	High Silicon Bronze			
675	Manganese Bronze			
687	Aluminum Brass			
770	Nickel Silver			
796	Leaded Nickel Silver			
102	Oxygen Free Copper	A CD	100 110	200 225
110	Electrolytic Tough Pitch Copper			
122	Phosphorus Deoxidized Copper			
170	Beryllium Copper			
172	Beryllium Copper			
175	Beryllium Copper			
210	Gilding, 95%			
220	Commercial Bronze			
502	Phosphor Bronze 1.25%			
510	Phosphor Bronze 5%			
521	Phosphor Bronze 8%			
524	Phosphor Bronze 10%			
614	Aluminum Bronze			
706	Copper Nickel 10%			
715	Copper Nickel 30%			
745	Nickel Silver			
752	Nickel Silver			
754	Nickel Silver			
757	Nickel Silver			

* Abbreviations used in this column are as follows: A, annealed; CD, cold drawn.

Table 8. Recommended Cutting Speeds in Feet per Minute for Turning, Milling, and Drilling Titanium and Titanium Alloys

Material Titanium and Titanium Alloys	Hardness, HB*	Material Condition*	Cutting Speed, fpm	
			HSS	Carbide
Commercially Pure				
99.5 Ti	110-150	A	110	400
99.1 Ti, 99.2 Ti	180-240	A	90	300
99.0% Ti	250-275	A	70	250
Low Alloyed				
99.5 Ti-.15 Pd	110-150	A	100	350
99.2 Ti-.15 Pd, 98.9 Ti-.8 Ni-.3 Mo	180-250	A	85	280
Alpha Alloys and Alpha-Beta Alloys 5Al-2.5Sn, 8Mn, 2Al-11Sn-5Zr-1Mo, 4Al-3Mo-1V, 5Al-6Sn-2Zr-1Mo, 6Al-2Sn-4Zr-2Mo, 6Al-2Sn-4Zr-6Mo, 6Al-2Sn-4Zr-2Mo-.25Si	300-350	A	50	200
6Al-4V	310-350	A	40	125
6Al-6V-2Sn, 7Al-4Mo, 8Al-1Mo-1V	320-370	A	30	100
8V-5Fe-1Al	320-380	A	20	90
6Al-4V, 6Al-2Sn-4Zr-2Mo, 6Al-2Sn-4Zr-6Mo, 6Al-2Sn-4Zr-2Mo-.25Sn	320-380	ST and A	40	100
4Al-3Mo-1V, 6Al-6V-2Sn, 7Al-4Mo	375-420	ST and A	20	80
1Al-8V-5Fe	375-440	ST and A	20	75
Beta Alloys 13V-11Cr-3Al, 8Mo-8V-2Fe-3Al, 3Al-8V-6Cr-4Mo-4Zr, 11.5Mo-6Zr-4.5Sn	275-350 / 350-440	A, ST / ST and A	25 / 20	100 / 60

* Abbreviations designate: A, annealed; ST, solution treated; ST and A, solution treated and aged; and HB, Brinell hardness number.

cost and are expensive to sharpen. For these reasons a long tool life is desirable, and to obtain a long tool life relatively slow cutting speeds are used. In many instances slower cutting speeds are used because of the limitations of the machine in accelerating and stopping heavy broaching cutters. At other times the available power on the machine places a limit on the cutting speed that can be used; i.e., the cubic inches of metal removed per minute must be within the power capacity of the machine.

The cutting speeds for high speed steel broaches range from 3 to 50 feet per minute, although faster speeds have been used. In general, the harder and more difficult to machine materials are cut at a slower cutting speed and those that are easier to machine are cut at a faster speed. Some typical recommendations for high speed steel broaches are: AISI 1040, 10 to 30 fpm; AISI 1060, 10 to 25 fpm; AISI 4140, 10 to 25 fpm; AISI 41L40, 20 to 30 fpm; 201 austenitic stainless steel, 10 to 20 fpm; Class 20 gray cast iron, 20 to 30 fpm; Class 40 gray cast iron, 15 to 25 fpm; aluminum and magnesium alloys, 30 to 50 fpm; copper alloys, 20 to 30 fpm; commercially pure titanium, 20 to 25 fpm; alpha and beta titanium alloys, 5 fpm; and, the superalloys, 3 to 10 fpm. Surface broaching operations on gray iron castings have been conducted at a cutting speed of 150 fpm, using indexable insert cemented carbide broaching cutters. In selecting the speed for broaching, the cardinal principle of the performance of all metal cutting tools should be kept in mind;

Table 9. Recommended Cutting Speeds in Feet per Minute for Turning, Milling, and Drilling* Superalloys

Material	Cutting Speed, fpm			
	Roughing		Finishing	
	HSS	Carbide	HSS	Carbide
A-286	30-35	120-145	35-40	145-155
AF2-1DA	8-10	35-45	10-15	40-50
Air Resist 213	15-20	55-65	20-25	70-85
Air Resist 13, and 215	10-12	35-40	10-15	45-55
Astroloy	5-10	25-50	5-15	50-75
B-1900	8-10	30-35	8-10	35-50
CW-12M	8-12	55-65	10-15	65-85
Discalloy	15-35	100-150	35-40	140-180
FSX-H14	10-12	35-40	10-15	45-55
GMR-235, and 235D	8-10	30-35	8-10	40-50
Hastelloy B, C, G, and X (wrought)	15-20	60-90	20-25	80-100
Hastelloy B, and C (cast)	8-12	55-65	10-15	75-85
Haynes 25, and 188	15-20	55-65	20-25	70-95
Haynes 36, and 151	10-12	35-40	10-15	45-55
HS 6, 21, 25, 31(X40), 36, and 151	10-12	35-40	10-15	45-55
IN 100, and 738	8-10	30-35	8-10	35-50
Incoloy 800, 801, and 802	30-35	120-160	35-40	145-180
Incoloy 804, and 825	15-20	60-90	20-25	80-100
Incoloy 901	10-20	30-60	20-35	40-80
Inconel 625, 702, 706, 718 (wrought), 721, 722, X750, 751, 901, 600, and 604	15-20	35-60	20-25	60-90
Inconel 700, and 702	10-12	40-65	12-15	65-70
Inconel 713C, and 718 (cast)	8-10	30-35	8-10	40-50
J1300	15-25	80-100	20-30	100-125
J1570	15-20	55-65	20-25	70-85
M252 (wrought)	15-20	65-75	20-25	75-85
M252 (cast)	8-10	30-35	8-10	40-50
Mar-M200, M246, M421, and M432	8-10	30-35	10-12	35-50
Mar-M905, and M918	15-20	55-65	20-25	70-85
Mar-M302, M322, and M509	10-12	35-40	10-15	45-55
N-12M	8-12	55-65	10-15	75-85
N-155	15-20	50-70	15-25	55-75
Nasa Co-W-Re	10-12	35-40	10-15	45-55
Nimonic 75, and 80	15-20	65-75	20-25	75-85
Nimonic 90 and 95	10-12	55-65	12-15	65-75
Refractaloy 26	15-20	60-90	20-25	80-100
René 41	10-15	35-60	12-20	55-80
René 80, and 95	8-10	30-45	10-15	40-50
S-590	10-20	60-90	15-30	80-100
S-816	10-15	45-65	15-20	50-75
TD-Nickel	70-80	250-290	80-100	300-350
Udimet 500, 700, and 710	10-15	30-50	12-20	40-60
Udimet 630	10-20	30-80	20-25	80-100
Unitemp 1753	8-10	35-45	10-15	40-50
V-36	10-15	45-65	15-20	50-75
V-57	30-35	120-160	35-40	145-180
W-545	25-35	110-155	30-40	140-175
WI-52	10-12	35-40	10-15	45-55
Waspaloy	10-30	30-60	25-35	50-95
X-45	10-12	35-40	10-15	45-55
16-25-6	30-35	120-160	35-40	145-180
19-9DL	25-35	110-150	30-40	140-180

* For milling and drilling, use the cutting speeds recommended under roughing.

i.e., increasing the cutting speed results in a proportionately larger reduction in tool life, and conversely, reducing the cutting speed results in a proportionately larger increase in the tool life. When broaching most materials, a suitable cutting fluid should be used to obtain a good surface finish and a better tool life. Gray cast iron can be broached without using a cutting fluid although some shops prefer to use a soluble oil.

Table 10.   Representative Cutting Conditions for Rough Turning Hard-to-Machine
Materials with Single Point Cubic Boron Nitride (CBN) Cutting Tools.

Material	Hardness, HRC	Cutting Speed, fpm	Feed, in./rev.	Depth of Cut, in.
Hardened Ferrous Alloys				
AISI 8620	63	250	.005	.060
AISI 52100	70	270	.020	.050
A2, A6, Cold Work Tool Steel	58	250	.008	.060
D2, Cold Work Tool Steel	54	250	.008	.060
H10, Hot Work Tool Steel	56	200	.005	.060
S5, Shock Resisting Tool Steel	60	400	.008	.060
O1, Oil Hardening Tool Steel	58	250	.008	.060
M2, High Speed Steel	62	250	.008	.060
Chilled Gray Iron	60	400	.010	.200
Meehanite Cast Iron	56	600	.008	.250
Superalloys				
Colmonoy	. . .	600	.006	.125
Incoloy 901	. . .	800	.006	.125
Inconel 600	. . .	600	.006	.125
Inconel 718	. . .	600	.006	.125
K-Monel	. . .	600	.006	.125
René 41	. . .	600	.006	.125
René 77	. . .	500	.006	.015
René 95 (Hot Isostatic Pressed)	. . .	900	.005	.125
René 95 (Forged)	. . .	450	.005	.125
Stellite	. . .	600	.006	.125
Waspaloy	. . .	600	.003	.060

Thread Cutting with Single Point Cutting Tools. — Whenever possible the cutting speed recommended for turning should be used to cut internal and external threads with single point thread cutting tools. This cutting speed can frequently be used on numerically controlled lathes, using either cemented carbide or high speed steel tools. There are occasions, however, when a slower cutting speed must be used as a result of the workpiece configuration, the setup, or when cutting certain difficult-to-machine threads, such as coarse pitch Acme threads. A slightly reduced cutting speed is sometimes used on numerically controlled lathes to obtain a longer tool life.

Thread cutting on an engine lathe is not necessarily a slow speed operation, although on these machines the operation is controlled manually. However, there must never be a compromise with safety; the operator must always be sure that he has control over the machine to the extent that he or others will not be injured, and that the machine or the workpiece will not be damaged. On some jobs a skilled operator can safely manipulate a lathe with such skill that a fast spindle speed can be used to cut the thread, in which case the cutting speed may be equal to that recommended for turning with high speed steel, or with cemented carbide in the case of the more difficult-to-machine materials. Other jobs require using a slower cutting speed, even when the thread cutting operation is performed by a highly skilled operator. Some of the reasons for cutting threads at a slower speed have been given in the previous paragraph. Other reasons involve the ability to safely manipulate the machine such as when cutting a thread close to a large shoulder, or when cutting an internal thread with the cutting tool feeding into the bore.

Cutting Speed for Thread Chasing. — Cutting threads with a self-opening die head is called thread chasing. The die head contains a set of thread chasers that cut the thread and feed the die head in a nut-and-screw-like action. The feed rate is determined entirely by the lead or pitch of the thread. Since the feed of the die head must not be

Tool Trouble-Shooting Check List

Problem	Tool Material	Remedy
Excessive flank wear — Tool life too short	Carbide	1. Change to harder, more wear-resistant grade 2. Reduce the cutting speed 3. Reduce the cutting speed and increase the feed to maintain production 4. Reduce the feed 5. For work hardenable materials — increase the feed 6. Increase the lead angle 7. Increase the relief angles
	HSS	1. Use a coolant 2. Reduce the cutting speed 3. Reduce the cutting speed and increase the feed to maintain production 4. Reduce the feed 5. For work hardenable materials — increase the feed 6. Increase the lead angle 7. Increase the relief angle
Excessive cratering	Carbide	1. Use a crater-resistant grade 2. Use a harder, more wear-resistant grade 3. Reduce the cutting speed 4. Reduce the feed 5. Widen the chip breaker groove
	HSS	1. Use a coolant 2. Reduce the cutting speed 3. Reduce the feed 4. Widen the chip breaker groove
Cutting edge chipping	Carbide	1. Increase the cutting speed 2. Lightly hone the cutting edge 3. Change to a tougher grade 4. Use negative rake tools 5. Increase the lead angle 6. Reduce the feed 7. Reduce the depth of cut 8. Reduce the relief angles 9. If low cutting speed must be used — use a high additive EP cutting fluid
	HSS	1. Use a high additive EP cutting fluid 2. Lightly hone the cutting edge before using 3. Increase the lead angle 4. Reduce the feed 5. Reduce the depth of cut 6. Use a negative rake angle 7. Reduce the relief angles
	Carbide and HSS	1. Check the setup for cause if chatter occurs 2. Check the grinding procedure for tool overheating 3. Reduce the tool overhang
Cutting edge deformation	Carbide	1. Change to a grade containing more tantalum 2. Reduce the cutting speed 3. Reduce the feed

Tool Trouble-Shooting Check List (*Continued*).

Problem	Tool Material	Remedy
Poor surface finish	Carbide	1. Increase the cutting speed 2. If low cutting speed must be used — use a high additive EP cutting fluid 3. For light cuts — use straight titanium carbide grade 4. Increase the nose radius 5. Reduce the feed 6. Increase the relief angles 7. Use positive rake tools
	HSS	1. Use high additive EP cutting fluid 2. Increase the nose radius 3. Reduce the feed 4. Increase the relief angles 5. Increase the rake angles
	Diamond	1. Use diamond tool for soft materials
Notching at the depth of cut line	Carbide and HSS	1. Increase the lead angle 2. Reduce the feed

too rapid, the thread lead and pitch place a limit on the spindle speed, and thereby on the cutting speed. Other factors affecting the cutting speed are the work material, the type and size of the thread, the thread tolerance, and the finish required. A cutting fluid should be used in most cases, which may also have an effect on the cutting speed. Much slower cutting speeds are recommended for thread chasing, as compared to turning. A cutting speed that is too fast will reduce the life of the thread chasers and may cause the threads cut to be rough or torn. Some typical cutting speeds recommended by one manufacturer of self-opening die heads are given below. These cutting speeds may have to be modified somewhat to suit existing conditions on each job.

Material	Threads per Inch			
	3-7½	8-15	16-24	25-Up
	Cutting Speed (fpm) for Threads per Inch			
AISI 1010-1035 Steel	20	30	40	50
AISI 1112-1340 Steel	20	30	40	50
AISI 1040-1095 Steel	15	20	25	30
AISI 4130-4820 Steel	8	10	15	20
AISI 5120-52100 Steel	8	10	15	20
Stainless Steel	8	10	15	20
Gray Cast Iron	25	40	50	80
Aluminum Alloys	50	100	150	200
Brass Bar Stock	50	100	150	200
Phosphor Bronze	40	80	100	150
Zinc Die Castings	50	100	150	200

Feed Rate for Milling. — Whenever the power feed is to be used to perform a milling operation, the table feed rate, in inches per minute, should always be calculated in order to achieve the best results. The table feed rate governs the production rate. Failure to calculate the table feed rate may result in overloading the milling cutter which can have serious consequences. Aside from the possible

Table 11. Cutting Speeds in Feet per Minute for Milling
Plain Carbon and Alloy Steels

Material AISI and SAE Steels	Hardness, HB*	Material Condition*	Cutting Speed, fpm HSS	Cutting Speed, fpm Carbide
Free Machining Plain Carbon Steels (Resulphurized), 1212, 1213, 1215	100-150	HR, A	140	600
	150-200	CD	130	550
1108, 1109, 1115, 1117, 1118, 11120, 1126, 1211	100-150	HR, A	130	550
	150-200	CD	115	500
1132, 1137, 1139, 1140, 1144, 1146 1151	175-225	HR, A, N, CD	115	450
	275-325	Q and T	70	290
	325-375	Q and T	45	200
	375-425	Q and T	35	170
Free Machining Plain Carbon Steels (Leaded), 11L17, 11L18, 12L13, 12L14	100-150	HR, A, N, CD	140	600
	150-200	HR, A, N, CD	130	625
	200-250	N, CD	110	400
Plain Carbon Steels, 1006, 1008, 1009, 1010, 1012, 1015, 1016, 1017, 1018, 1019, 1020, 1021, 1022, 1023, 1024, 1025, 1026, 1513, 1514	100-125	HR, A, N, CD	110	425
	125-175	HR, A, N, CD	110	400
	175-225	HR, N, CD	90	350
	225-275	CD	65	250
1027, 1030, 1033, 1035, 1036, 1037, 1038, 1039, 1040, 1041, 1042, 1043, 1045, 1046, 1048, 1049, 1050, 1052, 1524, 1526, 1527, 1541	125-175	HR, A, N, CD	100	375
	175-225	HR, A, N, CD	85	325
	225-275	N, CD, Q and T	70	225
	275-325	Q and T	55	200
	325-375	Q and T	35	160
	375-425	Q and T	25	140
1055, 1060, 1064, 1065, 1070, 1074, 1078, 1080, 1084, 1086, 1090, 1095, 1548, 1551, 1552, 1561, 1566	125-175	HR, A, N, CD	90	350
	175-225	HR, A, N, CD	75	300
	225-275	N, CD, Q and T	60	200
	275-325	Q and T	45	160
	325-375	Q and T	30	145
	375-425	Q and T	15	125
Free Machining Alloy Steels (Resulphurized), 4140, 4150	175-200	HR, A, N, CD	100	400
	200-250	HR, N, CD	90	350
	250-300	Q and T	60	280
	300-375	Q and T	45	220
	375-425	Q and T	35	160
Free Machining Alloy Steels (Leaded), 41L30, 41L40, 41L47, 41L50, 43L47, 51L32, 52L100, 86L20, 86L40	150-200	HR, A, N, CD	115	425
	200-250	HR, N, CD	95	375
	250-300	Q and T	70	260
	300-375	Q and T	50	210
	375-425	Q and T	40	180
Alloy Steels, 4012, 4023, 4024, 4028, 4118, 4320, 4419, 4422, 4427, 4615, 4620, 4621, 4626, 4718, 4720, 4815, 4817, 4820, 5015, 5117, 5120, 6118, 8115, 8615, 8617, 8620, 8622, 8625, 8627, 8720, 8822, 94B17	125-175	HR, A, N, CD	100	400
	175-225	HR, N, CD	90	350
	225-275	CD, N, Q and T	60	250
	275-325	Q and T	50	200
	325-375	Q and T	40	175
	375-425	Q and T	25	150

* Abbreviations designate: HR, hot rolled; CD, cold drawn; A, annealed; N, normalized; Q and T, quenched and tempered; HB, Brinell hardness number; and HRC, Rockwell C scale hardness number.

Table 11 (*Concluded*). **Cutting Speeds in Feet per Minute for Milling Plain Carbon and Alloy Steels**

Material AISI and SAE Steels	Hardness, HB*	Material Condition*	Cutting Speed, fpm	
			HSS	Carbide
Alloy Steels, 1330, 1335, 1340, 1345, 4032, 4037, 4042, 4047, 4130, 4135, 4137, 4140, 4142, 4145, 4147, 4150, 4161, 4337, 4340, 50B44, 50B46, 50B50, 50B60, 5130, 5132, 5140, 5145, 5147, 5150, 5160, 51B60, 6150, 81B45, 8630, 8635, 8637, 8640, 8642, 8645, 8650, 8655, 8660, 8740, 9254, 9255, 9260, 9262, 94B30	175-225 225-275 275-325 325-375 375-425	HR, A, N, CD N, CD, Q and T N, Q and T N, Q and T Q and T	75 60 50 35 20	310 260 210 180 140
Alloy Steels, E51100, E52100	175-225 225-275 275-325 325-375 375-425	HR, A, CD N, CD, Q and T N, Q and T N, Q and T Q and T	65 60 40 30 20	300 250 130 100 60
Ultra High Strength Steels (Not AISI) AMS 6421 (98B37 Mod.), AMS 6422 (98BV40), AMS 6424, AMS 6427, AMS 6428, AMS 6430, AMS 6432, AMS 6433, AMS 6434, AMS 6436, AMS 6442, 300M, D6 ac	220-300 300-350 350-400 43-48 HRC 48-52 HRC	A N N Q and T Q and T	60 45 20	250 180 130 100 60
Maraging Steels (Not AISI) 18% Ni Grade 200 18% Ni Grade 250 18% Ni Grade 300 18% Ni Grade 350	250-325 50-52 HRC	A Maraged	50 ..	250 60
Nitriding Steels (Not AISI) Nitralloy 125 Nitralloy 135 Nitralloy 135 (Mod.) Nitralloy 225 Nitralloy 230 Nitralloy N Nitralloy EZ Nitrex 1	200-250 300-350	A N, Q and T	60 25	280 200

* Abbreviations designate: HR, hot rolled; CD, cold drawn; A, annealed; N, normalized; Q and T, quenched and tempered; HB, Brinell hardness number; and HRC, Rockwell C scale hardness number.

breakage of equipment, overloading the cutter will cause the tool life of the cutter to decrease. The tool life can also be decreased if the feed rate is too slow, which, in addition, certainly leads to a loss of production.

The basic feed rate for milling cutters is the feed per tooth (f) which is expressed in inches per tooth. There are many factors to consider in selecting the feed per tooth and no formula is available to resolve these factors. Among the factors to consider are: 1) the cutting tool material; 2) the work material and its hardness; 3) the width and the depth of the cut to be taken; 4) the type of milling cutter to be used and its size; 5) the surface finish to be produced; 6) the power available on the milling machine; 7) the rigidity of the milling machine, the workpiece, the setup of the workpiece, the milling cutter, and the cutter mounting.

As a guide to help in the selection of the feed rate, two tables are given; Table 15 is for high-speed steel cutters, and Table 16 is for cemented carbide cutters.

As a cardinal principle, always use the maximum feed rate that conditions will

Table 12. Cutting Speed in Feet per Minute for Milling Tool Steels

Material Tool Steels (AISI Types)	Hardness, HB*	Material Condition*	Cutting Speed, fpm	
			HSS	Carbide
Water Hardening W1, W2, W5	150-200	A	85	250
Shock Resisting S1, S2, S5, S6, S7	175-225	A	55	215
Cold Work, Oil Hardening O1, O2, O6, O7	175-225	A	50	200
Cold Work, High Carbon High Chromium D2, D3, D4, D5, D7	200-250	A	40	150
Cold Work, Air Hardening				
A2, A3, A8, A9, A10	200-250	A	50	200
A4, A6	200-250	A	45	160
A7	225-275	A	40	140
Hot Work, Chromium Type H10, H11, H12, H13, H14, H19	150-200	A	60	250
	200-250	A	50	200
	325-375	Q and T	30	150
	48-50 HRC	Q and T	—	80
	50-52 HRC	Q and T	—	60
	52-54 HRC	Q and T	—	40
	54-56 HRC	Q and T	—	20
Hot Work, Tungsten Type H21, H22, H23, H24, H25, H26	150-200	A	55	200
	200-250	A	45	170
Hot Work, Molybdenum Type H41, H42, H43	150-200	A	55	180
	200-250	A	45	140
Special Purpose, Low Alloy L2, L3, L6	150-200	A	65	300
Mold				
P2, P3, P4, P5, P6	100-150	A	75	350
P20, P21	150-200	A	60	300
High-Speed Steel				
M1, M2, M6, M10, T1, T2, T6	200-250	A	50	175
M3-1, M4, M7, M30, M33, M34, M36, M41, M42, M43, M44, M46, M47, T5, T8	225-275	A	40	150
T15, M3-2	225-275	A	30	130

* Abbreviations designate: A, annealed; Q and T, quenched and tempered; and HB, Brinell hardness number.

permit. Avoid, if possible, using a feed rate that is less than .001 inch per tooth because this will result in a decrease in the tool life of the cutter. When milling hard materials with small diameter end mills, such small feed rates may be necessary, but otherwise use as much feed as possible. Harder materials in general will require lower feed rates than softer materials. The width and the depth of cut also affect the feed rate; wider and deeper cuts must be fed somewhat more slowly than narrow and shallow cuts. A slower feed rate will result in a better surface finish; however, always use the heaviest feed rate that will produce the surface finish

Table 13.   Recommended Cutting Speeds in Feet per Minute for Milling
Stainless Steels

Material Stainless Steels	Hardness, HB*	Material Condition*	Cutting Speed, fpm	
			HSS	Carbide
Free Machining Stainless Steels				
(Ferritic), 430F, 430F Se	135-185	A	95	375
(Austenitic), 203EZ, 303, 303 Se, 303MA, 303Pb, 303Cu, 303 Plus X	135-185	A	90	325
	225-275	CD	75	300
(Martensitic), 416, 416 Se, 416 Plus X, 420F, 420F Se, 440F, 440F Se	135-185	A	95	375
	185-240	CD	80	325
	275-325	Q and T	50	225
	375-425	Q and T	20	100
Stainless Steels				
(Ferritic), 405, 409, 429, 430, 434, 436, 442, 446, 502	135-185	A	75	275
(Austenitic), 201, 202, 301, 302, 304, 304L, 305, 308, 321, 347, 348	135-185	A	60	200
	225-275	CD	50	180
(Austenitic), 302B, 309, 309S, 310, 310S, 314, 316, 316L, 317, 330	135-185	A	50	200
(Martensitic), 403, 410, 420, 501	135-175	A	75	325
	175-225	A	65	275
	275-325	Q and T	40	175
	375-425	Q and T	25	100
(Martensitic), 414, 431, Greek Ascoloy	225-275	A	55	225
	275-325	Q and T	45	180
	375-425	Q and T	25	100
(Martensitic), 440A, 440B, 440C	225-275	A	50	180
	275-325	Q and T	40	140
	375-425	Q and T	20	100
(Precipitation Hardening) 15-5PH, 17-4PH, 17-7PH, AF-71, 17-14Cu Mo, AFC-77, AM-350, AM-355, AM-362, Custom 455, HNM, PH13-8, PH14-8Mo, PH15-7Mo, Stainless W	150-200	A	60	200
	275-325	H	50	180
	325-375	H	40	110
	375-450	H	25	75

* Abbreviations designate: A, annealed; CD, cold drawn; Q and T, quenched and tempered; H, precipitation hardened; and HB, Brinell hardness number.

desired. Fine chips produced by fine feeds are dangerous when milling magnesium because spontaneous combustion can occur. Thus, when milling magnesium, a fast feed that will produce a relatively thick chip should be used. Cutting stainless steel produces a work hardened layer on the surface that has been cut. When milling this material, the feed should be large enough to allow each cutting edge on the cutter to penetrate below the work hardened layer produced by the previous cutting edge. The heavy feeds recommended for face milling cutters are to be used primarily by larger cutters on milling machines having an adequate amount of power. For smaller face milling cutters, start with the slower feeds and increase the feed as indicated by the performance of the cutter and the machine.

When planning a milling operation that is to entail the use of a high cutting speed and a fast feed rate, always check to determine if the power required to take the cut is within the capacity of the milling machine. Such cutting conditions are often encountered when milling with cemented carbide cutters. The large metal removal

(Continued on page 351)

Table 14. Recommended Cutting Speeds in Feet per Minute for Milling Ferrous Cast Metals

Material Ferrous Cast Metals	Hardness, HB*	Material Condition*	Cutting Speed, fpm	
			HSS	Carbide
Gray Cast Iron				
ASTM Class 20	120-150	A	100	425
ASTM Class 25	160-200	AC	80	325
ASTM Class 30, 35, and 40	190-220	AC	70	250
ASTM Class 45 and 50	220-260	AC	50	190
ASTM Class 55 and 60	250-260	AC, HT	30	110
ASTM Type 1, 1b, 5 (Ni-Resist)	100-215	AC	50	200
ASTM Type 2, 3, 6 (Ni-Resist)	120-175	AC	40	190
ASTM Type 2b, 4 (Ni-Resist)	150-250	AC	30	180
Malleable Iron				
(Ferritic), 32510, 35018	110-160	MHT	110	475
(Pearlitic), 40010, 43010, 45006, 45008, 48005,50005	160-200	MHT	80	375
	200-240	MHT	65	250
(Martensitic), 53004, 60003, 60004	200-255	MHT	55	225
(Martensitic), 70002, 70003	220-260	MHT	50	200
(Martensitic), 80002	240-280	MHT	45	130
(Martensitic), 90001	250-320	MHT	25	110
Nodular (Ductile) Iron				
(Ferritic), 60-40-18, 65-45-12	140-190	A	75	425
(Ferritic-Pearlitic), 80-55-06	190-225	AC	60	325
	225-260	AC	50	200
(Pearlitic-Martensitic), 100-70-03	240-300	HT	40	160
(Martensitic), 120-90-02	270-330	HT	25	90
	330-400	HT	—	30
Cast Steels				
(Low Carbon), 1010, 1020	100-150	AC, A, N	100	375
(Medium Carbon), 1030, 1040, 1050	125-175	AC, A, N	95	375
	175-225	AC, A, N	80	325
	225-300	AC, HT	60	250
(Low Carbon Alloy), 1320, 2315, 2320, 4110, 4120, 4320, 8020, 8620	150-200	AC, A, N	85	325
	200-250	AC, A, N	75	300
	250-300	AC, HT	50	225
(Medium Carbon Alloy), 1330, 1340, 2325, 2330, 4125, 4130, 4140, 4330, 4340, 8030, 80B30, 8040, 8430, 8440, 8630, 8640, 9525, 9530, 9535	175-225	AC, A, N	70	300
	225-250	AC, A, N	65	250
	250-300	AC, HT	50	200
	300-350	AC, HT	30	180
	350-400	HT	..	125

* Abbreviations designate: A, annealed; AC, as cast; N, normalized; HT, heat treated; MHT, malleablizing heat treatment; and HB, Brinell hardness number.

Table 15. Recommended Feed in Inches per Tooth (f_t) for Milling with High Speed Steel Cutters

Material	Hardness, HB	End Mills — Depth of Cut, .250 in, Cutter Diam., in			End Mills — Depth of Cut, .050 in, Cutter Diam., in				Plain or Slab Mills	Form Relieved Cutters	Face Mills and Shell End Mills	Slotting and Side Mills
		1/2	3/4	1 and up	1/4	1/2	3/4	1 and up				
					Feed per Tooth, inch							
Free Machining Plain Carbon Steels	100-185	.001	.003	.004	.001	.002	.003	.004	.003-.008	.005	.004-.012	.002-.008
Plain Carbon Steels, AISI 1006 to 1030; 1513 to 1522	100-150	.001	.003	.003	.001	.002	.003	.004	.003-.008	.004	.004-.012	.002-.008
	150-200	.001	.003	.003	.001	.002	.002	.003	.003-.008	.004	.003-.012	.002-.008
AISI 1033 to 1095; 1524 to 1566	120-180	.001	.003	.003	.001	.002	.003	.004	.003-.008	.004	.004-.012	.002-.008
	180-220	.001	.002	.003	.001	.002	.002	.003	.003-.008	.004	.003-.012	.002-.008
	220-300	.001	.002	.002	.001	.001	.002	.003	.002-.006	.003	.002-.008	.002-.006
Alloy Steels having less than 3% Carbon. Typical examples: AISI 4012, 4023, 4027, 4118, 4320, 4422, 4427, 4615, 4620, 4626, 4720, 4820, 5015, 5120, 6118, 8115, 8620, 8627, 8720, 8822, 9310, 93B17	125-175	.001	.003	.003	.001	.002	.003	.004	.003-.008	.004	.004-.012	.002-.008
	175-225	.001	.002	.003	.001	.002	.003	.003	.003-.008	.004	.003-.012	.002-.008
	225-275	.001	.002	.003	.001	.001	.002	.003	.002-.006	.003	.003-.008	.002-.006
	275-335	.001	.002	.002	.001	.001	.002	.002	.002-.005	.003	.002-.008	.002-.005
Alloy Steels have 3% Carbon or more. Typical examples: AISI 1330, 1340, 4032, 4037, 4130, 4140, 4150, 4340, 50B40, 50B60, 5130, 51B60, 6150, 81B45, 8630, 8640, 86B45, 8660, 8740, 94B30	175-225	.001	.002	.003	.001	.002	.003	.004	.003-.008	.004	.003-.012	.002-.008
	225-275	.001	.002	.003	.001	.001	.002	.003	.002-.006	.003	.003-.010	.002-.006
	275-335	.001	.002	.002	.001	.001	.002	.003	.002-.005	.003	.002-.008	.002-.005
	325-375	.001	.002	.002	.001	.001	.002	.002	.002-.004	.002	.002-.008	.002-.005
Tool Steel	150-200	.001	.002	.002	.001	.002	.003	.003	.003-.008	.004	.003-.010	.002-.006
	200-250	.001	.002	.002	.001	.002	.002	.003	.002-.006	.003	.003-.008	.002-.005
Gray Cast Iron	120-180	.001	.003	.004	.002	.003	.004	.004	.004-.012	.005	.005-.016	.002-.010
	180-225	.001	.002	.003	.001	.002	.003	.003	.003-.010	.004	.004-.012	.002-.008
	225-300	.001	.002	.002	.001	.001	.002	.002	.002-.006	.003	.002-.008	.002-.005
Ferritic Malleable Iron	110-160	.001	.003	.004	.002	.003	.004	.004	.003-.010	.005	.005-.016	.002-.010
Pearlitic-Martensitic Malleable Iron	160-200	.001	.003	.004	.001	.002	.003	.004	.003-.010	.004	.004-.012	.002-.008
	200-240	.001	.002	.003	.001	.002	.003	.003	.003-.007	.004	.003-.010	.002-.006
	240-300	.001	.002	.002	.001	.001	.002	.002	.002-.006	.003	.002-.008	.002-.005

Table 15 (Concluded). Recommended Feed in Inches per Tooth (f_t) for Milling with High Speed Steel Cutters

Material	Hardness, HB	End Mills							Plain or Slab Mills	Form Relieved Cutters	Face Mills and Shell End Mills	Slotting and Side Mills
		Depth of Cut, .250 in.			Depth of Cut, .050 in.							
		Cutter Diam., in.			Cutter Diam., in.							
		½	¾	1 and up	¼	½	¾	1 and up				
		Feed per Tooth, inch										
Cast Steel	100-180	.001	.003	.003	.001	.002	.003	.004	.003-.008	.004	.003-.012	.002-.008
	180-240	.001	.002	.003	.001	.002	.003	.003	.003-.008	.004	.003-.010	.002-.006
	240-300	.001	.002	.002	.0005	.002	.002	.002	.002-.006	.003	.003-.008	.002-.005
Zinc Alloys (Die Castings)	…	.002	.003	.004	.001	.003	.004	.006	.003-.010	.005	.004-.015	.002-.012
Copper Alloys (Brasses & Bronzes)	100-150	.002	.004	.005	.002	.003	.005	.006	.003-.015	.004	.004-.020	.002-.010
	150-250	.002	.003	.004	.001	.003	.004	.005	.003-.015	.004	.003-.012	.002-.008
Free Cutting Brasses & Bronzes	80-100	.002	.004	.005	.002	.003	.005	.006	.003-.015	.004	.004-.015	.002-.010
Cast Aluminum Alloys—As Cast	…	.003	.004	.005	.002	.004	.005	.006	.005-.016	.006	.005-.020	.004-.012
Cast Aluminum Alloys—Hardened	…	.003	.004	.005	.002	.003	.004	.005	.004-.012	.005	.005-.020	.004-.012
Wrought Aluminum Alloys—Cold Drawn	…	.003	.004	.005	.002	.003	.004	.005	.004-.014	.005	.005-.020	.004-.012
Wrought Aluminum Alloys—Hardened	…	.002	.003	.004	.001	.002	.003	.004	.003-.012	.004	.005-.020	.004-.012
Magnesium Alloys	…	.003	.004	.005	.003	.004	.005	.007	.005-.016	.006	.005-.020	.005-.012
Ferritic Stainless Steel	135-185	.001	.002	.003	.001	.002	.003	.003	.002-.006	.004	.004-.008	.002-.007
Austenitic Stainless Steel	135-185	.001	.002	.003	.001	.002	.003	.003	.003-.007	.004	.005-.008	.002-.007
	185-275	.001	.002	.003	.001	.002	.002	.002	.003-.006	.003	.004-.006	.002-.007
Martensitic Stainless Steel	135-185	.001	.002	.002	.001	.002	.003	.003	.003-.006	.004	.004-.010	.002-.007
	185-225	.001	.002	.002	.001	.002	.002	.003	.003-.006	.004	.003-.008	.002-.007
	225-300	.0005	.002	.002	.0005	.001	.002	.002	.002-.005	.003	.002-.006	.002-.005
Monel	100-160	.001	.003	.004	.001	.002	.003	.003	.002-.006	.004	.002-.008	.002-.006

Table 16. Recommended Feed in Inch per Tooth (f_t) for Milling with
Cemented Carbide Cutters

Material	Hardness, HB	Face Mills	Slotting and Side Mills
		Feed per Tooth, inch	
Free Machining Plain Carbon Steels	100-185	.008-.020	.003-.010
Plain Carbon Steels, AISI 1006 to 1030, 1513 to 1522	100-150	.008-.020	.003-.010
	150-200	.008-.020	.003-.010
Plain Carbon Steels, AISI 1033 to 1095, 1524 to 1566	120-180	.005-.020	.003-.010
	180-220	.005-.020	.003-.010
	220-300	.003-.012	.003-.008
Alloy Steels having less than .3% Carbon content. Typical examples: AISI 4012, 4023, 4027, 4118, 4320, 4422, 4427, 4615, 4620, 4626, 4720, 4820, 5015, 5120, 6118, 8115, 8620, 8627, 8720, 8822, 9310, 93B17	125-175	.006-.020	.003-.010
	175-225	.006-.020	.003-.010
	225-275	.006-.016	.003-.010
	275-325	.004-.012	.003-.008
	325-375	.003-.008	.003-.007
Alloy Steels having .3% Carbon content, or more. Typical examples: AISI 1330, 1340, 4032, 4037, 4130, 4140, 4150, 4340, 50B40, 50B60, 5130, 51B60, 6150, 81B45, 8630, 8640, 86B45, 8660, 8740, 94B30	175-225	.005-.020	.003-.010
	225-275	.004-.012	.003-.008
	275-325	.003-.010	.003-.008
	325-375	.003-.008	.003-.007
Tool Steels	200-275	.004-.012	.003-.007
	275-325	.003-.010	.003-.006
	36-45 HRC	.003-.006	.002-.005
	45-55 HRC	.003-.005	.002-.003
Ferritic Stainless Steels	110-160	.005-.015	.003-.010
Austenitic Stainless Steels	135-185	.005-.012	.003-.010
	185-275	.005-.010	.003-.008
Martensitic Stainless Steel	135-185	.005-.015	.003-.010
	185-225	.005-.010	.003-.008
	225-300	.004-.008	.003-.007
Precipitation Hardening Stainless Steels	Annealed	.004-.012	.003-.010
	275-350	.003-.008	.002-.005
	350-450	.002-.005	.002-.004
Cast Steel	100-180	.008-.020	.003-.010
	180-240	.005-.016	.003-.010
	240-300	.004-.012	.003-.008
Gray Cast Iron	140-185	.008-.020	.005-.012
	185-225	.008-.016	.005-.010
	225-300	.005-.012	.004-.008
Ferritic Malleable Iron	110-160	.005-.020	.004-.012
Pearlitic-Martensitic Malleable Iron	160-200	.005-.020	.003-.010
	200-240	.005-.016	.003-.010
	240-300	.004-.010	.003-.008
Nodular (Ductile) Iron	140-200	.008-.020	.003-.010
	200-275	.006-.014	.003-.008
	275-325	.005-.012	.003-.007
	325-400	.003-.008	.002-.004
Copper Alloys (Brasses and Bronzes)	100-150	.005-.020	.003-.012
	150-250	.004-.014	.003-.010

Table 16 (*Concluded*). Recommended Feed in Inch per Tooth (f_t) for Milling with Cemented Carbide Cutters

Material	Hardness, HB	Face Mills	Slotting and Side Mills
		Feed per Tooth, inch	
Wrought and Cast Aluminum Alloys	....	.005-.020	.005-.020
Wrought and Cast Magnesium Alloys	....	.005-.020	.005-.020
Superalloys	....	.003-.010	.002-.006
Titanium Alloys	....	.003-.010	.002-.006
Nickel Alloys	....	.003-.010	.002-.006
Monel	....	.003-.010	.002-.006
Plastics, Hard Rubber, etc.	....	.003-.015	.003-.012

rates that can be attained require a high horsepower output. An example of this type of calculation is given in the section on milling under "Horsepower for Machining." If the size of the cut must be reduced in order to stay within the power capacity of the machine, start by reducing the cutting speed rather than the feed rate in inches per tooth.

The formula for calculating the table feed rate, when the feed in inches per tooth is known, is given below:

$$f_m = f_t \, n_t \, N \qquad (4)$$

Where: f_m = Milling machine table feed rate in inches per minute (ipm)
f_t = Feed rate in inch per tooth (ipt)
n_t = Number of teeth in the milling cutter
N = Spindle speed of the milling machine in revolutions per minute (rpm)

Example: Calculate the feed rate for milling a piece of AISI 1040 steel having a hardness of 160 Bhn. The cutter is a 3-inch diameter high speed steel plain or slab milling cutter with 8 teeth. The width of the cut is 2 inches, the depth of cut is .062 inch and the cutting speed is 100 fpm. From the Table 16, the feed rate selected is .008 inch per tooth.

$$N = \frac{12\,V}{\pi\,D} = \frac{12 \times 100}{3.14 \times 3} = 127 \text{ rpm} \qquad (1)$$
$$f_m = f_t \, n_t \, N = .008 \times 8 \times 127$$
$$= 8 \text{ ipm (approximately)}$$

Feed Rates for Drilling. — The feed rate for drilling is governed primarily by the size of the drill and by the material to be drilled. Other factors that also affect the feed rate that can be used are the workpiece configuration, the rigidity of the machine tool and the workpiece setup, and the length of the chisel edge. A chisel edge that is too long will result in a very significant increase in the thrust force, which may cause large deflections to occur on the machine tool and drill breakage. For ordinary twist drills the feed rate used is .001 to .003 in./rev. for drills smaller than ⅛ in.; .002 to .006 in./rev. for ⅛ to ¼ in. drills; .004 to .010 in./rev. for ¼ to ½ in. drills; .007 to .015 in./rev. for ½ to 1 in. drills; and, .010 to .025 in./rev. for drills larger than 1 inch. The lower values in the feed ranges should be used for hard materials such as tool steels, superalloys, and work hardening stainless steels; the higher values in the feed ranges should be used to drill soft materials such as aluminum and brass.

Table 17. Recommended Cutting Speeds in Feet per Minute for Drilling and Reaming Plain Carbon and Alloy Steels

Material AISI and SAE Steels	Hardness, HB*	Material Condition*	Drilling HSS	Reaming HSS	Reaming Carbide
Free Machining Plain Carbon Steels (Resulphurized) 1212, 1213, 1214	100-150	HR, A	120	80	400
	150-200	CD	125	80	350
1108, 1109, 1115, 1117, 1118, 1120, 1126, 1211	100-150	HR, A	110	75	375
	150-200	CD	120	80	350
1132, 1137, 1139, 1140, 1144, 1146, 1151	175-225	HR, A, N, CD	100	65	350
	275-325	Q and T	70	45	250
	325-375	Q and T	45	30	175
	375-425	Q and T	35	20	100
Free Machining Plain Carbon Steels (Leaded) 11L17, 11L18, 12L13, 12L14	100-150	HR, A, N, CD	130	85	400
	150-200	HR, A, N, CD	120	80	375
	200-250	N, CD	90	60	275
Plain Carbon Steels, 1006, 1008, 1009, 1010, 1012, 1015, 1016, 1017, 1018, 1019, 1020, 1021, 1022, 1023, 1024, 1025, 1026, 1513, 1514	100-125	HR, A, N, CD	100	65	300
	125-175	HR, A, N, CD	90	60	275
	175-225	HR, N, CD	70	45	200
	225-275	CD	60	40	175
1027, 1030, 1033, 1035, 1036, 1037, 1038, 1039, 1040, 1041, 1042, 1043, 1045, 1046, 1048, 1049, 1050, 1052, 1524, 1526, 1527, 1541	125-175	HR, A, N, CD	90	60	250
	175-225	HR, A, N, CD	75	50	200
	225-275	N, CD, Q and T	60	40	150
	275-325	Q and T	50	30	120
	325-375	Q and T	35	20	100
	375-425	Q and T	25	15	80
1055, 1060, 1064, 1065, 1070, 1074, 1078, 1080, 1084, 1086, 1090, 1095, 1548, 1551, 1552, 1561, 1566	125-175	HR, A, N, CD	85	55	250
	175-225	HR, A, N, CD	70	45	200
	225-275	N, CD, Q and T	50	30	140
	275-325	Q and T	40	25	110
	325-375	Q and T	30	20	90
	375-425	Q and T	15	10	70
Free Machining Alloy Steels (Resulphurized), 4140, 4150	175-200	HR, A, N, CD	90	60	250
	200-250	HR, N, CD	80	50	225
	250-300	Q and T	55	30	200
	300-375	Q and T	40	25	150
	375-425	Q and T	30	15	100
Free Machining Alloy Steels (Leaded), 41L30, 41L40, 41L47, 41L50, 43L47, 51L32, 52L100, 86L20, 86L40	150-200	HR, A, N, CD	100	65	285
	200-250	HR, N, CD	90	60	250
	250-300	Q and T	65	40	200
	300-375	Q and T	45	30	150
	375-425	Q and T	30	15	110
Alloy Steels, 4012, 4023, 4024, 4028, 4118, 4120, 4419, 4422, 4427, 4615, 4620, 4621, 4626, 4718, 4720, 4815, 4817, 4820, 5015, 5017, 5020, 6118, 8115, 8615, 8617, 8620, 8622, 8625, 8627, 8620, 8822, 94B17	125-175	HR, A, N, CD	85	55	250
	175-225	HR, N, CD	70	45	225
	225-275	CD, N, Q and T	55	35	200
	275-325	Q and T	50	30	150
	325-375	Q and T	35	25	125
	375-425	Q and T	25	15	90

* Abbreviations designate: A, annealed; HR, hot rolled; CD, cold drawn; N, normalized; Q and T, quenched and tempered; and HB, Brinell hardness number.

Table 17 (*Concluded*).   Recommended Cutting Speeds in Feet per Minute for Drilling
and Reaming Plain Carbon and Alloy Steels

Material AISI and SAE Steels	Hardness, HB*	Material Condition*	Cutting Speed, fpm		
			Drilling	Reaming	
			HSS	HSS	Carbide
Alloy Steels, 1330, 1335, 1340, 1345, 4032, 4037, 4042, 4047, 4130, 4135, 4137, 4140, 4142, 4145, 4147, 4150, 4160, 4337, 4340, 50B44, 50B46, 50B50, 50B60, 5130, 5132, 5140, 5145, 5147, 5150, 5160, 51B60, 6150, 81B45, 8630, 8635, 8637, 8640, 8642, 8645, 8650, 8655, 8660, 8740, 9254, 9255, 9260, 9262, 94B30	175-225 225-275 275-325 325-375 375-425	HR, A, N, CD N, CD, Q and T N, Q and T N, Q and T Q and T	75 60 45 30 20	50 40 30 15 15	200 175 150 100 80
Alloy Steels, E51100, E52100	175-225 225-275 275-325 325-375 375-425	HR, A, CD N, CD, Q and T N, Q and T N, Q and T Q and T	60 50 35 30 20	40 30 25 20 10	200 125 100 80 50
Ultra High Strength Steels (Not AISI), AMS6424, AMS6421 (98B37 Mod.), AMS6422 (98BV40), AMS6427, AMS6428, AMS6430, AMS6432, AMS6433, AMS6434, AMS6436, AMS6442, 300M, D6ac	220-300 300-350 350-400	A N N	50 35 20	30 20 10	180 125 90
Maraging Steels (Not AISI) 18% Nickel Grade 200 18% Nickel Grade 250 18% Nickel Grade 300 18% Nickel Grade 350	250-225	A	50	30	175
Nitriding Steels Nitralloy 125 Nitralloy 135 Nitralloy 135 (Mod.) Nitralloy 225 Nitralloy 230 Nitralloy N Nitralloy EZ Nitrex 1	200-250 250-300	A N, Q and T	60 35	40 20	175 125

* Abbreviations designate: A, annealed; HR, hot rolled; CD, cold drawn; N, normalized; Q and T, quenched and tempered; and HB, Brinell hardness number.

Drilling Difficulties. — A drill split up the web is evidence of too much feed or insufficient lip clearance at the center due to improper grinding. The rapid wearing away of the extreme outer corners of the cutting edges indicates that the speed is too high. A drill chipping or breaking out at the cutting edges indicates that either the feed is too heavy or the drill has been ground with too much lip clearance. Nothing will "check" a high-speed drill quicker than to turn a stream of cold water on it after it has been heated while in use. It is equally bad to plunge it in cold water after the point has been heated in grinding. the small checks or cracks resulting from this practice will eventually chip out and cause rapid wear or breakage. Insufficient speed in drilling small holes with hand feed greatly increases the risk of breakage, especially at the moment the drill is breaking through the farther side of the work. This is due to the operator's inability to gage the feed when the drill is running too slowly.

Table 18. **Recommended Cutting Speeds in Feet per Minute for Drilling and Reaming Tool Steels**

Material Tool Steels	Hardness, HB*	Material Condition*	Cutting Speed, fpm		
			Drilling	Reaming	
			HSS	HSS	Carbide
Water Hardening W1, W2, W5	150-200	A	85	55	200
Shock Resisting S1, S2, S5, S6, S7	175-225	A	50	35	175
Cold Work, Oil Hardening O1, O2, O6, O7	175-225	A	45	30	150
Cold Work, High Carbon High Chromium, D2, D3, D4, D5, D7	200-250	A	30	20	80
Cold Work, Air Hardening A2, A3, A8, A9, A10	200-250	A	50	35	150
A4, A6	200-250	A	45	30	125
A7	225-275	A	30	20	100
Hot Work, Chromium Type H10, H11, H12, H13, H14, H19	150-200 200-250 325-375	A A Q and T	60 50 30	40 30 20	200 150 100
Hot Work, Tungsten Type H21, H22, H23, H24, H25, H26	150-200 200-250	A A	55 40	35 25	150 125
Hot Work, Molybdenum Type H41, H42, H43	150-200 200-250	A A	45 35	30 20	150 125
Special Purpose, Low Alloy L2, L3, L6	150-200	A	60	40	200
Mold P2, P3, P4, P5, P6	100-150	A	75	50	225
P20, P21	150-200	A	60	40	200
High-Speed Steel M1, M2, M6, M10, T1, T2, T6	200-250	A	45	30	150
M3-1, M4, M7, M30, M33, M34, M36, M41, M42, M43, M44, M46, M47, T5, T8	225-275	A	35	20	100
T15, M3-2	225-275	A	25	15	80

* Abbreviations designate: A, annealed; Q and T, quenched and tempered; and HB, Brinell Hardness Number.

Small drills have heavier webs and smaller flutes in proportion to their size than do larger drills, and breakage due to clogging of chips in the flutes is more likely to occur. When drilling holes more than three times the diameter of the drill, it is advisable to withdraw the drill at intervals to remove the chips and permit coolant to reach the tip of the drill.

Drilling Holes in Glass. — There are several methods of drilling holes in glass. For holes of medium and large size, use brass or copper tubing, having an outside diameter equal to the size of hole required. Revolve the tube at a peripheral speed of about 100 feet per minute, and use carborundum (80 to 100 grit) and light machine oil between the

Table 19. Recommended Cutting Speeds in Feet per Minute for Drilling
and Reaming Stainless Steels

Material Stainless Steels	Hardness, HB*	Material Condition*	Cutting Speed, fpm		
			Drilling	Reaming	
			HSS	HSS	Carbide
Free Machining Stainless Steels					
(Ferritic), 430F, 430F Se	135-185	A	90	60	250
(Austenitic), 203EZ, 303, 303 Se, 303 MA, 303 Pb, 303 Cu, 303 Plus X	135-185 / 225-275	A / CD	85 / 70	55 / 45	225 / 200
(Martensitic), 416, 416 Se, 416 Plus X, 420F, 420F Se, 440F, 440F Se	135-185 / 185-240 / 275-325 / 375-425	A / CD / Q and T / Q and T	90 / 70 / 40 / 20	60 / 45 / 25 / 10	250 / 200 / 150 / 80
Stainless Steels					
(Ferritic), 405, 409, 429, 430, 434, 436, 442, 446, 502	135-185	A	65	45	200
(Austenitic), 201, 202, 301, 302, 304, 304L, 305, 308, 321, 347, 348	135-185 / 225-275	A / CD	55 / 50	35 / 30	150 / 125
(Austenitic), 302B, 309, 309S, 310, 310S, 314, 316, 316L, 317, 330	135-185	A	50	30	150
(Martensitic), 403, 410, 420, 501	135-175 / 175-225 / 275-325 / 375-425	A / A / Q and T / Q and T	75 / 65 / 40 / 25	50 / 45 / 25 / 15	225 / 200 / 125 / 80
(Martensitic), 414, 431 Greek Ascoloy	225-275 / 275-325 / 375-425	A / Q and T / Q and T	50 / 40 / 25	30 / 25 / 15	150 / 125 / 80
(Martensitic), 440A, 440B, 440C	225-275 / 275-325 / 375-425	A / Q and T / Q and T	45 / 40 / 20	30 / 25 / 10	125 / 100 / 75
(Precipitation Hardening) 15-5PH, 17-4PH, 17-7Ph, 17-14Cu Mo, AF-71, AFC-77, AM-350, AM-355, AM-362, Custom 455, HNM, PH13-8, PH14-8Mo, PH15-7Mo, Stainless W	150-200 / 275-325 / 325-375 / 375-425	A / H / H / H	50 / 45 / 35 / 20	30 / 25 / 20 / 10	150 / 125 / 75 / 50

* Abbreviations designate: A, annealed; CD, cold drawn; Q and T, quenched and tempered; H, precipitation hardened; and HB, Brinell hardness number.

end of the pipe and the glass. Insert the abrasive under the drill with a thin piece of soft wood, to avoid scratching the glass. The glass should be supported by a felt or rubber cushion, not much larger than the hole to be drilled. If practicable, it is well to drill about halfway through and then turn the glass over and drill down to meet the first cut. Any fin that may be left in the hole can be removed with a round second-cut file wet with turpentine.

Smaller diameter holes are generally drilled with triangular shaped cemented carbide drills that can be purchased in standard sizes. The end of the drill is shaped into a long tapering triangular point. The other end of the cemented carbide bit is brazed on to a

Table 20. Recommended Cutting Speeds in Feet per Minute for Drilling and Reaming Ferrous Cast Metals

Material Ferrous Cast Metals	Hardness, HB*	Material Condition*	Cutting Speed, fpm		
			Drilling	Reaming	
			HSS	HSS	Carbide
Gray Cast Iron					
ASTM Class 20	120-150	A	100	65	300
ASTM Class 25	160-200	AC	90	60	225
ASTM Class 30, 35, and 40	190-220	AC	80	55	180
ASTM Class 45 and 50	220-260	AC	60	40	125
ASTM Class 55 and 60	250-320	AC, HT	30	20	80
ASTM Type 1, 1b, 5 (Ni-Resist)	100-215	AC	50	30	150
ASTM Type 2, 3, 6 (Ni-Resist)	120-175	AC	40	25	140
ASTM Type 2b, 4 (Ni-Resist)	150-250	AC	30	20	125
Malleable Iron					
(Ferritic), 32510, 35018	110-160	MHT	110	75	325
(Pearlitic), 40010, 43010, 45006, 45008, 48005, 50005	160-200	MHT	80	55	250
	200-240	MHT	70	45	180
(Martensitic), 53004, 60003, 60004	200-255	MHT	55	35	160
(Martensitic), 70002, 70003	220-260	MHT	50	30	150
(Martensitic), 80002	240-280	MHT	45	30	100
(Martensitic), 90001	250-320	MHT	25	15	80
Nodular (Ductile) Iron					
(Ferritic), 60-40-18, 65-45-12	140-190	A	100	65	300
(Ferritic-Pearlitic), 80-55-06	190-225	AC	70	45	225
	225-260	AC	50	30	140
(Pearlitic-Martensitic), 100-70-03	240-300	HT	40	25	115
(Martensitic), 120-90-02	270-330	HT	25	15	65
	330-400	HT	10	5	30
Cast Steels					
(Low Carbon), 1010, 1020	100-150	AC, A, N	100	65	250
(Medium Carbon), 1030, 1040, 1050	125-175	AC, A, N	90	60	240
	175-225	AC, A, N	70	45	220
	225-300	AC, HT	55	35	200
(Low Carbon Alloy), 1320, 2315, 2320, 4110, 4120, 4320, 8020, 8620	150-200	AC, A, N	75	50	220
	200-250	AC, A, N	65	40	200
	250-300	AC, HT	50	30	150
(Medium Carbon Alloy), 1330, 1340, 2325, 2330, 4125, 4130, 4140, 4330, 4340, 8030, 80B30, 8040, 8430, 8440, 8630, 8640, 9525, 9530, 9535	175-225	AC, A, N	70	45	200
	225-250	AC, A, N	60	35	180
	250-300	AC, HT	45	30	150
	300-350	AC, HT	30	20	140
	350-400	HT	20	10	100

* Abbreviations designate: A, annealed; AC, as cast; N, normalized; HT, heat treated; MHT, Malleablizing heat treatment; and HB, Brinell hardness number.

Table 21. Recommended Cutting Speeds in Feet per Minute for Drilling
and Reaming Light Metals

Material Light Metals	Material Condition*	Cutting Speed, fpm		
		Drilling	Reaming	
		HSS	HSS	Carbide
All Wrought Aluminum Alloys	CD	400	400	800
	ST and A	350	350	750
All Aluminum Sand and Permanent Mold Casting Alloys	AC	500	500	900
	ST and A	350	350	750
All Aluminum Die Casting Alloys†	AC	300	300	500
	ST and A	70	70	200
†Except Alloys 390.0 and 392.0	AC	125	100	250
	ST and A	45	40	200
All Wrought Magnesium Alloys	A, CD, ST and A	500	500	1000
All Cast Magnesium Alloys	A, AC, ST and A	450	450	1000

* Abbreviations designate: A, annealed; AC, as cast; CD, cold drawn; and ST and A, solution treated and aged.

steel shank. A glass drill can be made to the same shape from hardened drill rod or an old three-cornered file. The location at which the hole is to be drilled is marked on the workpiece. A dam of putty or glazing compound is built up on the work surface to contain the cutting fluid, which can be either kerosene or turpentine mixed with camphor. Chipping on the back edge of the hole can be prevented by placing a scrap plate of glass behind the area to be drilled and drilling into the back-up glass. This procedure also provides additional support to the workpiece and is essential for drilling very thin plates. The hole is usually drilled with an electric hand drill. When the hole is being produced the drill should be given a small circular motion using the point as a fulcrum, thereby providing a clearance for the drill in the hole.

Very small round or intricately shaped holes and narrow slots can be cut in glass by the ultrasonic machining process or by the abrasive jet cutting process.

Estimating Planer Cutting Speeds. — While most planers of modern design have a means of indicating the speed at which the table is traveling, or cutting, many older planers do not. Thus, the following formulas are useful for planers that do not have a means of indicating the table or cutting speed. It is not practicable to provide a formula for calculating the exact cutting speed at which a planer is operating because the time to stop and start the table when reversing varies greatly. The formulas below will, however, provide a reasonable estimate.

$$V_c \cong S_c\,L$$

$$S_c \cong \frac{V_c}{L}$$

Where: V_c = Cutting speed; fpm, or m/min

S_c = Number of cutting strokes per minute of planer table

L = Length of table cutting stroke; ft, or m

Table 22. Recommended Cutting Speeds in Feet per Minute for
Drilling and Reaming Copper Alloys

Material Copper Alloys (Copper Alloy Nos. as per the Copper Development Assn. Inc.)	Material Condition*	Cutting Speed, fpm		
		Drilling	Reaming	
		HSS	HSS	Carbide
314 Leaded Commercial Bronze 332 High Leaded Brass 340 Medium Leaded Brass 342 High Leaded Brass 353 High Leaded Brass 356 Extra High Leaded Brass 360 Free Cutting Brass 370 Free Cutting Muntz Metal 377 Forging Brass 385 Architectural Bronze 485 Leaded Naval Brass 544 Free Cutting Phosphor Bronze	A CD	160 175	160 175	320 360
226 Jewelry Bronze 230 Red Brass 240 Low Brass 260 Cartridge Brass 70% 268 Yellow Brass 280 Muntz Metal 335 Low Leaded Brass 365 Leaded Muntz Metal 368 Leaded Muntz Metal 443 Admiralty Brass (inhibited) 445 Admiralty Brass (inhibited) 651 Low Silicon Bronze 655 High Silicon Bronze 675 Manganese Bronze 687 Aluminum Brass 770 Nickel Silver 796 Leaded Nickel Silver	A CD	120 140	110 120	250 275
102 Oxygen Free Copper 110 Electrolytic Tough Pitch Copper 122 Phosphorus Deoxidized Copper 170 Beryllium Copper 172 Beryllium Copper 175 Beryllium Copper 210 Guilding, 95% 220 Commercial Bronze 502 Phosphor Bronze 1.25% 510 Phosphor Bronze 5% 521 Phosphor Bronze 8% 524 Phosphor Bronze 10% 614 Aluminum Bronze 706 Copper Nickel 10% 715 Copper Nickel 30% 745 Nickel Silver 752 Nickel Silver 754 Nickel Silver 757 Nickel Silver	A CD	60 65	50 60	180 200

* Abbreviations used in this column are as follows: A, annealed; CD, cold drawn

Cutting Speeds and Equivalent R.P.M. for Drills of Number and Letter Sizes

Size No.	Cutting Speed, Feet per Minute										
	30'	40'	50'	60'	70'	80'	90'	100'	110'	130'	150'
	Revolutions per Minute for Number Sizes										
1	503	670	838	1005	1173	1340	1508	1675	1843	2179	2513
2	518	691	864	1037	1210	1382	1555	1728	1901	2247	2593
4	548	731	914	1097	1280	1462	1645	1828	2010	2376	2741
6	562	749	936	1123	1310	1498	1685	1872	2060	2434	2809
8	576	768	960	1151	1343	1535	1727	1919	2111	2495	2879
10	592	790	987	1184	1382	1579	1777	1974	2171	2566	2961
12	606	808	1010	1213	1415	1617	1819	2021	2223	2627	3032
14	630	840	1050	1259	1469	1679	1889	2099	2309	2728	3148
16	647	863	1079	1295	1511	1726	1942	2158	2374	2806	3237
18	678	904	1130	1356	1582	1808	2034	2260	2479	2930	3380
20	712	949	1186	1423	1660	1898	2135	2372	2610	3084	3559
22	730	973	1217	1460	1703	1946	2190	2433	2676	3164	3649
24	754	1005	1257	1508	1759	2010	2262	2513	2764	3267	3769
26	779	1039	1299	1559	1819	2078	2338	2598	2858	3378	3898
28	816	1088	1360	1631	1903	2175	2447	2719	2990	3534	4078
30	892	1189	1487	1784	2081	2378	2676	2973	3270	3864	4459
32	988	1317	1647	1976	2305	2634	2964	3293	3622	4281	4939
34	1032	1376	1721	2065	2409	2753	3097	3442	3785	4474	5162
36	1076	1435	1794	2152	2511	2870	3228	3587	3945	4663	5380
38	1129	1505	1882	2258	2634	3010	3387	3763	4140	4892	5645
40	1169	1559	1949	2339	2729	3118	3508	3898	4287	5067	5846
42	1226	1634	2043	2451	2860	3268	3677	4085	4494	5311	6128
44	1333	1777	2221	2665	3109	3554	3999	4442	4886	5774	6662
46	1415	1886	2358	2830	3301	3773	4244	4716	5187	6130	7074
48	1508	2010	2513	3016	3518	4021	4523	5026	5528	6534	7539
50	1637	2183	2729	3274	3820	4366	4911	5457	6002	7094	8185
52	1805	2406	3008	3609	4211	4812	5414	6015	6619	7820	9023
54	2084	2778	3473	4167	4862	5556	6251	6945	7639	9028	10417
Size	Revolutions per Minute for Letter Sizes										
A	491	654	818	982	1145	1309	1472	1636	1796	2122	2448
B	482	642	803	963	1124	1284	1445	1605	1765	2086	2407
C	473	631	789	947	1105	1262	1420	1578	1736	2052	2368
D	467	622	778	934	1089	1245	1400	1556	1708	2018	2329
E	458	611	764	917	1070	1222	1375	1528	1681	1968	2292
F	446	594	743	892	1040	1189	1337	1486	1635	1932	2229
G	440	585	732	878	1024	1170	1317	1463	1610	1903	2195
H	430	574	718	862	1005	1149	1292	1436	1580	1867	2154
I	421	562	702	842	983	1123	1264	1404	1545	1826	2106
J	414	552	690	827	965	1103	1241	1379	1517	1793	2068
K	408	544	680	815	951	1087	1223	1359	1495	1767	2039
L	395	527	659	790	922	1054	1185	1317	1449	1712	1976
M	389	518	648	777	907	1036	1166	1295	1424	1683	1942
N	380	506	633	759	886	1012	1139	1265	1391	1644	1897
O	363	484	605	725	846	967	1088	1209	1330	1571	1813
P	355	473	592	710	828	946	1065	1183	1301	1537	1774
Q	345	460	575	690	805	920	1035	1150	1266	1496	1726
R	338	451	564	676	789	902	1014	1127	1239	1465	1690
S	329	439	549	659	769	878	988	1098	1207	1427	1646
T	320	426	533	640	746	853	959	1066	1173	1387	1600
U	311	415	519	623	727	830	934	1038	1142	1349	1557
V	304	405	507	608	709	810	912	1013	1114	1317	1520
W	297	396	495	594	693	792	891	989	1088	1286	1484
X	289	385	481	576	672	769	865	962	1058	1251	1443
Y	284	378	473	567	662	756	851	945	1040	1229	1418
Z	277	370	462	555	647	740	832	925	1017	1202	1387

For fractional drill sizes, use table on page 361 and 362.

Length of Point on Twist Drills and Centering Tools

Size or Diam. of Drill	Decimal Equivalent	Length of Point when Included Angle = 90°	Length of Point when Included Angle = 118°
60	0.0400	0.020	0.012
59	0.0410	0.021	0.012
58	0.0420	0.021	0.013
57	0.0430	0.022	0.013
56	0.0465	0.023	0.014
55	0.0520	0.026	0.016
54	0.0550	0.028	0.017
53	0.0595	0.030	0.018
52	0.0635	0.032	0.019
51	0.0670	0.034	0.020
50	0.0700	0.035	0.021
49	0.0730	0.037	0.022
48	0.0760	0.038	0.023
47	0.0785	0.040	0.024
46	0.0810	0.041	0.024
45	0.0820	0.041	0.025
44	0.0860	0.043	0.026
43	0.0890	0.045	0.027
42	0.0935	0.047	0.028
41	0.0960	0.048	0.029
40	0.0980	0.049	0.029
39	0.0995	0.050	0.030
38	0.1015	0.051	0.030
37	0.1040	0.052	0.031
36	0.1065	0.054	0.032
35	0.1100	0.055	0.033
34	0.1110	0.056	0.033
33	0.1130	0.057	0.034
32	0.1160	0.058	0.035
31	0.1200	0.060	0.036
30	0.1285	0.065	0.039
29	0.1360	0.068	0.041
28	0.1405	0.070	0.042
27	0.1440	0.072	0.043
26	0.1470	0.074	0.044
25	0.1495	0.075	0.045
24	0.1520	0.076	0.046
23	0.1540	0.077	0.046
22	0.1570	0.079	0.047
21	0.1590	0.080	0.048
20	0.1610	0.081	0.048
19	0.1660	0.083	0.050
18	0.1695	0.085	0.051
17	0.1730	0.087	0.052
16	0.1770	0.089	0.053
15	0.1800	0.090	0.054
14	0.1820	0.091	0.055
13	0.1850	0.093	0.056
12	0.1890	0.095	0.057
11	0.1910	0.096	0.057
10	0.1935	0.097	0.058
9	0.1960	0.098	0.059
8	0.1990	0.100	0.060
7	0.2010	0.101	0.060
6	0.2040	0.102	0.061
5	0.2055	0.103	0.062
4	0.2090	0.105	0.063
3	0.2130	0.107	0.064
2	0.2210	0.111	0.067
1	0.2280	0.114	0.068
15/64	0.2344	0.117	0.070
1/4	0.2500	0.125	0.075
17/64	0.2656	0.133	0.080
9/32	0.2813	0.141	0.084
19/64	0.2969	0.148	0.089
5/16	0.3125	0.156	0.094
21/64	0.3281	0.164	0.098
11/32	0.3438	0.171	0.103
23/64	0.3594	0.180	0.108
3/8	0.3750	0.188	0.113
25/64	0.3906	0.195	0.117
13/32	0.4063	0.203	0.122
27/64	0.4219	0.211	0.127
7/16	0.4375	0.219	0.131
29/64	0.4531	0.227	0.136
15/32	0.4688	0.234	0.141
31/64	0.4844	0.242	0.145
1/2	0.5000	0.250	0.150
33/64	0.5156	0.258	0.155
17/32	0.5313	0.266	0.159
35/64	0.5469	0.273	0.164
9/16	0.5625	0.281	0.169
37/64	0.5781	0.289	0.173
19/32	0.5938	0.297	0.178
39/64	0.6094	0.305	0.183
5/8	0.6250	0.313	0.188
41/64	0.6406	0.320	0.192
21/32	0.6563	0.328	0.197
43/64	0.6719	0.336	0.202
11/16	0.6875	0.344	0.206
23/32	0.7188	0.359	0.216
3/4	0.7500	0.375	0.225

Revolutions per Minute for Various Cutting Speeds and Diameters

Diam-eter, Inches	Cutting Speed, Feet per Minute											
	40	50	60	70	80	90	100	120	140	160	180	200
	Revolutions per Minute											
¼	611	764	917	1070	1222	1376	1528	1834	2139	2445	2750	3056
5/16	489	611	733	856	978	1100	1222	1466	1711	1955	2200	2444
⅜	408	509	611	713	815	916	1018	1222	1425	1629	1832	2036
7/16	349	437	524	611	699	786	874	1049	1224	1398	1573	1748
½	306	382	459	535	611	688	764	917	1070	1222	1375	1528
9/16	272	340	407	475	543	611	679	813	951	1086	1222	1358
⅝	245	306	367	428	489	552	612	736	857	979	1102	1224
11/16	222	273	333	389	444	500	555	666	770	888	999	1101
¾	203	254	306	357	408	458	508	610	711	813	914	1016
13/16	190	237	284	332	379	427	474	569	664	758	853	948
⅞	175	219	262	306	349	392	438	526	613	701	788	876
15/16	163	204	244	285	326	366	407	488	570	651	733	814
1	153	191	229	267	306	344	382	458	535	611	688	764
1 1/16	144	180	215	251	287	323	359	431	503	575	646	718
1⅛	136	170	204	238	272	306	340	408	476	544	612	680
1 3/16	129	161	193	225	258	290	322	386	451	515	580	644
1¼	123	153	183	214	245	274	306	367	428	490	551	612
1 5/16	116	146	175	204	233	262	291	349	407	466	524	582
1⅜	111	139	167	195	222	250	278	334	389	445	500	556
1 7/16	106	133	159	186	212	239	265	318	371	424	477	530
1½	102	127	153	178	204	230	254	305	356	406	457	508
1 9/16	97.6	122	146	171	195	220	244	293	342	390	439	488
1⅝	93.9	117	141	165.	188	212	234	281	328	374	421	468
1 11/16	90.4	113	136	158	181	203	226	271	316	362	407	452
1¾	87.3	109	131	153	175	196	218	262	305	349	392	436
1⅞	81.5	102	122	143	163	184	204	244	286	326	367	408
2	76.4	95.5	115	134	153	172	191	229	267	306	344	382
2⅛	72.0	90.0	108	126	144	162	180	216	252	288	324	360
2¼	68.0	85.5	102	119	136	153	170	204	238	272	306	340
2⅜	64.4	80.5	96.6	113	129	145	161	193	225	258	290	322
2½	61.2	76.3	91.7	107	122	138	153	184	213	245	275	306
2⅝	58.0	72.5	87.0	102	116	131	145	174	203	232	261	290
2¾	55.6	69.5	83.4	97.2	111	125	139	167	195	222	250	278
2⅞	52.8	66.0	79.2	92.4	106	119	132	158	185	211	238	264
3	51.0	63.7	76.4	89.1	102	114	127	152	178	203	228	254
3⅛	48.8	61.0	73.2	85.4	97.6	110	122	146	171	195	219	244
3¼	46.8	58.5	70.2	81.9	93.6	105	117	140	164	188	211	234
3⅜	45.2	56.5	67.8	79.1	90.4	102	113	136	158	181	203	226
3½	43.6	54.5	65.5	76.4	87.4	98.1	109	131	153	174	196	218
3⅝	42.0	52.5	63.0	73.5	84.0	94.5	105	126	147	168	189	210
3¾	40.8	51.0	61.2	71.4	81.6	91.8	102	122	143	163	184	205
3⅞	39.4	49.3	59.1	69.0	78.8	88.6	98.5	118	138	158	177	197
4	38.2	47.8	57.3	66.9	76.4	86.0	95.6	115	134	153	172	191
4¼	35.9	44.9	53.9	62.9	71.8	80.8	89.8	108	126	144	162	180
4½	34.0	42.4	51.0	59.4	67.9	76.3	84.8	102	119	136	153	170
4¾	32.2	40.2	48.2	56.3	64.3	72.4	80.4	96.9	113	129	145	161
5	30.6	38.2	45.9	53.5	61.1	68.8	76.4	91.7	107	122	138	153
5¼	29.1	36.4	43.6	50.9	58.2	65.4	72.7	87.2	102	116	131	145
5½	27.8	34.7	41.7	48.6	55.6	62.5	69.4	83.3	97.2	111	125	139
5¾	26.6	33.2	39.8	46.5	53.1	59.8	66.4	80.0	93.0	106	120	133
6	25.5	31.8	38.2	44.6	51.0	57.2	63.6	76.3	89.0	102	114	127
6¼	24.4	30.6	36.7	42.8	48.9	55.0	61.1	73.3	85.5	97.7	110	122
6½	23.5	29.4	35.2	41.1	47.0	52.8	58.7	70.4	82.2	93.9	106	117
6¾	22.6	28.3	34.0	39.6	45.3	50.9	56.6	67.9	79.2	90.6	102	113
7	21.8	27.3	32.7	38.2	43.7	49.1	54.6	65.5	76.4	87.4	98.3	109
7¼	21.1	26.4	31.6	36.9	42.2	47.4	52.7	63.2	73.8	84.3	94.9	105
7½	20.4	25.4	30.5	35.6	40.7	45.8	50.9	61.1	71.0	81.4	91.6	102
7¾	19.7	24.6	29.5	34.4	39.4	44.3	49.2	59.0	68.9	78.7	88.6	98.4
8	19.1	23.9	28.7	33.4	38.2	43.0	47.8	57.4	66.9	76.5	86.0	95.6

Revolutions per Minute for Various Cutting Speeds and Diameters

Diameter, Inches	Cutting Speed, Feet per Minute											
	225	250	275	300	325	350	375	400	425	450	500	550
	Revolutions per Minute											
¼	3438	3820	4202	4584	4966	5348	5730	6112	6493	6875	7639	8403
5/16	2750	3056	3362	3667	3973	4278	4584	4889	5195	5501	6112	6723
⅜	2292	2546	2801	3056	3310	3565	3820	4074	4329	4584	5093	5602
7/16	1964	2182	2401	2619	2837	3056	3274	3492	3710	3929	4365	4802
½	1719	1910	2101	2292	2483	2675	2866	3057	3248	3439	3821	4203
9/16	1528	1698	1868	2037	2207	2377	2547	2717	2887	3056	3396	3736
⅝	1375	1528	1681	1834	1987	2139	2292	2445	2598	2751	3057	3362
11/16	1250	1389	1528	1667	1806	1941	2084	2223	2362	2501	2779	3056
¾	1146	1273	1401	1528	1655	1783	1910	2038	2165	2292	2547	2802
13/16	1058	1175	1293	1410	1528	1646	1763	1881	1998	2116	2351	2586
⅞	982	1091	1200	1310	1419	1528	1637	1746	1855	1965	2183	2401
15/16	917	1019	1120	1222	1324	1426	1528	1630	1732	1834	2038	2241
1	859	955	1050	1146	1241	1337	1432	1528	1623	1719	1910	2101
1 1/16	809	899	988	1078	1168	1258	1348	1438	1528	1618	1798	1977
1⅛	764	849	933	1018	1103	1188	1273	1358	1443	1528	1698	1867
1 3/16	724	804	884	965	1045	1126	1206	1287	1367	1448	1609	1769
1¼	687	764	840	917	993	1069	1146	1222	1299	1375	1528	1681
1 5/16	654	727	800	873	946	1018	1091	1164	1237	1309	1455	1601
1⅜	625	694	764	833	903	972	1042	1111	1181	1250	1389	1528
1 7/16	598	664	730	797	863	930	996	1063	1129	1196	1329	1461
1½	573	636	700	764	827	891	955	1018	1082	1146	1273	1400
1 9/16	550	611	672	733	794	855	916	978	1039	1100	1222	1344
1⅝	528	587	646	705	764	822	881	940	999	1057	1175	1293
1 11/16	509	566	622	679	735	792	849	905	962	1018	1132	1245
1¾	491	545	600	654	709	764	818	873	927	982	1091	1200
1 13/16	474	527	579	632	685	737	790	843	895	948	1054	1159
1⅞	458	509	560	611	662	713	764	815	866	917	1019	1120
1 15/16	443	493	542	591	640	690	739	788	838	887	986	1084
2	429	477	525	573	620	668	716	764	811	859	955	1050
2⅛	404	449	494	539	584	629	674	719	764	809	899	988
2¼	382	424	468	509	551	594	636	679	721	764	849	933
2⅜	362	402	442	482	522	563	603	643	683	724	804	884
2½	343	382	420	458	496	534	573	611	649	687	764	840
2⅝	327	363	400	436	472	509	545	582	618	654	727	800
2¾	312	347	381	416	451	486	520	555	590	625	694	763
2⅞	299	332	365	398	431	465	498	531	564	598	664	730
3	286	318	350	381	413	445	477	509	541	572	636	700
3⅛	274	305	336	366	397	427	458	488	519	549	611	672
3¼	264	293	323	352	381	411	440	470	499	528	587	646
3⅜	254	283	311	339	367	396	424	452	481	509	566	622
3½	245	272	300	327	354	381	409	436	463	490	545	600
3⅝	237	263	289	316	342	368	395	421	447	474	527	579
3¾	229	254	280	305	331	356	382	407	433	458	509	560
3⅞	221	246	271	295	320	345	369	394	419	443	493	542
4	214	238	262	286	310	334	358	382	405	429	477	525
4¼	202	224	247	269	292	314	337	359	383	404	449	494
4½	191	212	233	254	275	297	318	339	360	382	424	466
4¾	180	201	221	241	261	281	301	321	341	361	402	442
5	171	191	210	229	248	267	286	305	324	343	382	420
5¼	163	181	199	218	236	254	272	290	308	327	363	399
5½	156	173	190	208	225	242	260	277	294	312	347	381
5¾	149	166	182	199	215	232	249	265	282	298	332	365
6	143	159	174	190	206	222	238	254	270	286	318	349
6¼	137	152	168	183	198	213	229	244	259	274	305	336
6½	132	146	161	176	190	205	220	234	249	264	293	322
6¾	127	141	155	169	183	198	212	226	240	254	283	311
7	122	136	149	163	177	190	204	218	231	245	272	299
7¼	118	131	144	158	171	184	197	210	223	237	263	289
7½	114	127	139	152	165	178	190	203	216	229	254	279
7¾	111	123	135	148	160	172	185	197	209	222	246	271
8	107	119	131	143	155	167	179	191	203	215	238	262

Revolutions per Minute for Various Cutting Speeds and Diameters
(Metric units)

Diam., mm	Cutting Speed, Metres per Minute											
	5	6	8	10	12	16	20	25	30	35	40	45
	Revolutions per Minute											
5	318	382	509	637	764	1019	1273	1592	1910	2228	2546	2865
6	265	318	424	530	637	849	1061	1326	1592	1857	2122	2387
8	199	239	318	398	477	637	796	995	1194	1393	1592	1790
10	159	191	255	318	382	509	637	796	955	1114	1273	1432
12	133	159	212	265	318	424	531	663	796	928	1061	1194
16	99.5	119	159	199	239	318	398	497	597	696	796	895
20	79.6	95.5	127	159	191	255	318	398	477	557	637	716
25	63.7	76.4	102	127	153	204	255	318	382	446	509	573
30	53.1	63.7	84.9	106	127	170	212	265	318	371	424	477
35	45.5	54.6	72.8	90.9	109	145	182	227	273	318	364	409
40	39.8	47.7	63.7	79.6	95.5	127	159	199	239	279	318	358
45	35.4	42.4	56.6	70.7	84.9	113	141	177	212	248	283	318
50	31.8	38.2	51	63.7	76.4	102	127	159	191	223	255	286
55	28.9	34.7	46.3	57.9	69.4	92.6	116	145	174	203	231	260
60	26.6	31.8	42.4	53.1	63.7	84.9	106	133	159	186	212	239
65	24.5	29.4	39.2	49	58.8	78.4	98	122	147	171	196	220
70	22.7	27.3	36.4	45.5	54.6	72.8	90.9	114	136	159	182	205
75	21.2	25.5	34	42.4	51	68	84.9	106	127	149	170	191
80	19.9	23.9	31.8	39.8	47.7	63.7	79.6	99.5	119	139	159	179
90	17.7	21.2	28.3	35.4	42.4	56.6	70.7	88.4	106	124	141	159
100	15.9	19.1	25.5	31.8	38.2	51	63.7	79.6	95.5	111	127	143
110	14.5	17.4	23.1	28.9	34.7	46.2	57.9	72.3	86.8	101	116	130
120	13.3	15.9	21.2	26.5	31.8	42.4	53.1	66.3	79.6	92.8	106	119
130	12.2	14.7	19.6	24.5	29.4	39.2	49	61.2	73.4	85.7	97.9	110
140	11.4	13.6	18.2	22.7	27.3	36.4	45.5	56.8	68.2	79.6	90.9	102
150	10.6	12.7	17	21.2	25.5	34	42.4	53.1	63.7	74.3	84.9	95.5
160	9.9	11.9	15.9	19.9	23.9	31.8	39.8	49.7	59.7	69.6	79.6	89.5
170	9.4	11.2	15	18.7	22.5	30	37.4	46.8	56.2	65.5	74.9	84.2
180	8.8	10.6	14.1	17.7	21.2	28.3	35.4	44.2	53.1	61.9	70.7	79.6
190	8.3	10	13.4	16.8	20.1	26.8	33.5	41.9	50.3	58.6	67	75.4
200	8	9.5	12.7	15.9	19.1	25.5	31.8	39.8	47.7	55.7	63.7	71.6
220	7.2	8.7	11.6	14.5	17.4	23.1	28.9	36.2	43.4	50.6	57.9	65.1
240	6.6	8	10.6	13.3	15.9	21.2	26.5	33.2	39.8	46.4	53.1	59.7
260	6.1	7.3	9.8	12.2	14.7	19.6	24.5	30.6	36.7	42.8	49	55.1
280	5.7	6.8	9.1	11.4	13.6	18.2	22.7	28.4	34.1	39.8	45.5	51.1
300	5.3	6.4	8.5	10.6	12.7	17	21.2	26.5	31.8	37.1	42.4	47.7
350	4.5	5.4	7.3	9.1	10.9	14.6	18.2	22.7	27.3	31.8	36.4	40.9
400	4	4.8	6.4	8	9.5	12.7	15.9	19.9	23.9	27.9	31.8	35.8
450	3.5	4.2	5.7	7.1	8.5	11.3	14.1	17.7	21.2	24.8	28.3	31.8
500	3.2	3.8	5.1	6.4	7.6	10.2	12.7	15.9	19.1	22.3	25.5	28.6

Revolutions per Minute for Various Cutting Speeds and Diameters
(Metric units)

Diam., mm	Cutting Speed, Metres per Minute											
	50	55	60	65	70	75	80	85	90	95	100	200
	Revolutions per Minute											
5	3183	3501	3820	4138	4456	4775	5093	5411	5730	6048	6366	12,732
6	2653	2918	3183	3448	3714	3979	4244	4509	4775	5039	5305	10,610
8	1989	2188	2387	2586	2785	2984	3183	3382	3581	3780	3979	7958
10	1592	1751	1910	2069	2228	2387	2546	2706	2865	3024	3183	6366
12	1326	1459	1592	1724	1857	1989	2122	2255	2387	2520	2653	5305
16	995	1094	1194	1293	1393	1492	1591	1691	1790	1890	1989	3979
20	796	875	955	1034	1114	1194	1273	1353	1432	1512	1592	3183
25	637	700	764	828	891	955	1019	1082	1146	1210	1273	2546
30	530	584	637	690	743	796	849	902	955	1008	1061	2122
35	455	500	546	591	637	682	728	773	819	864	909	1818
40	398	438	477	517	557	597	637	676	716	756	796	1592
45	354	389	424	460	495	531	566	601	637	672	707	1415
50	318	350	382	414	446	477	509	541	573	605	637	1273
55	289	318	347	376	405	434	463	492	521	550	579	1157
60	265	292	318	345	371	398	424	451	477	504	530	1061
65	245	269	294	318	343	367	392	416	441	465	490	979
70	227	250	273	296	318	341	364	387	409	432	455	909
75	212	233	255	276	297	318	340	361	382	403	424	849
80	199	219	239	259	279	298	318	338	358	378	398	796
90	177	195	212	230	248	265	283	301	318	336	354	707
100	159	175	191	207	223	239	255	271	286	302	318	637
110	145	159	174	188	203	217	231	246	260	275	289	579
120	133	146	159	172	186	199	212	225	239	252	265	530
130	122	135	147	159	171	184	196	208	220	233	245	490
140	114	125	136	148	159	171	182	193	205	216	227	455
150	106	117	127	138	149	159	170	180	191	202	212	424
160	99.5	109	119	129	139	149	159	169	179	189	199	398
170	93.6	103	112	122	131	140	150	159	169	178	187	374
180	88.4	97.3	106	115	124	133	141	150	159	168	177	354
190	83.8	92.1	101	109	117	126	134	142	151	159	167	335
200	79.6	87.5	95.5	103	111	119	127	135	143	151	159	318
220	72.3	79.6	86.8	94	101	109	116	123	130	137	145	289
240	66.3	72.9	79.6	86.2	92.8	99.5	106	113	119	126	132	265
260	61.2	67.3	73.4	79.6	85.7	91.8	97.9	104	110	116	122	245
280	56.8	62.5	68.2	73.9	79.6	85.3	90.9	96.6	102	108	114	227
300	53.1	58.3	63.7	69	74.3	79.6	84.9	90.2	95.5	101	106	212
350	45.5	50	54.6	59.1	63.7	68.2	72.8	77.3	81.8	99.1	91	182
400	39.8	43.8	47.7	51.7	55.7	59.7	63.7	67.6	71.6	75.6	79.6	159
450	35.4	38.9	42.4	46	49.5	53.1	56.6	60.1	63.6	67.2	70.7	141
500	31.8	35	38.2	41.4	44.6	47.7	50.9	54.1	57.3	60.5	63.6	127

Planing Time. — The approximate time required to plane a surface can be determined from the following formula in which T = time in minutes, L = length of stroke in feet, V_c = cutting speed in feet per minute, V_r = return speed in feet per minute, W = width of surface to be planed in inches, F = feed in inches, and 0.025 = approximate reversal time factor per stroke in minutes for most planers.

$$T = \frac{W}{F}\left[L \times \left(\frac{1}{V_c} + \frac{1}{V_r} \right) + 0.025 \right]$$

Speeds for Metal-cutting Saws. — The following speeds and feeds for metal-cutting saws are recommended by Henry Disston & Sons, Inc. Speeds and feeds vary over a wide range depending upon the kind of machine as well as the size, shape, and kind of metal to be cut; hence, the following speeds are intended as a general guide only.

Solid-tooth Circular Saws. — Low-carbon steel: Speed, 60–90 feet per minute; feed, 3–6 inches per minute. Tool and alloy steel: Speed, 40–60 feet per minute; feed, $\frac{1}{32}$–$2\frac{1}{2}$ inches per minute. Cast iron: Speed, 60–70 feet per minute; feed, 5 inches per minute. Brass: Speed, 500–800 feet per minute; feed, 80 inches per minute. Copper: Speed, 600–700 feet per minute; feed, 60 inches per minute. Aluminum: Speed, 800–1000 feet per minute; feed, 90 inches per minute.

"Hot saws" for sawing hot metal (structural shapes, rails, billets, etc., in rolling mills) should operate at speeds of 22,000 to 25,000 feet per minute.

Inserted-tooth Metal-cutting Saws. — Generally speaking, an inserted-tooth saw is not recommended smaller than 18 inches in diameter, although smaller sizes are made. Small saws usually are applied to work requiring fine or relatively fine teeth which are impossible when the teeth are of the inserted type. The following speeds are not suitable for all compositions but represent general practice. Steel: Speed, 40 feet per minute; feed, $2\frac{1}{2}$–10 inches per minute. Cast iron: Speed, 40 feet per minute; feed, 5 inches per minute. Brass: Speed, 500 feet per minute; feed, 80 inches per minute. Copper: Speed, 750 feet per minute; feed, 60 inches per minute. Aluminum: Speed, 1200 feet per minute; feed, 90 inches per minute.

Metal-cutting Band Saws. — The speeds which follow apply to Disston "hard-edge" saws which have milled teeth. Aluminum sheets: Speed of blade, 1000–3000 feet per minute. Bakelite: Speed, 800–1000 feet per minute. Brass sheets and tubing: Speed, 700–1500 feet per minute. Carbon tool steel: Speed, 100–150 feet per minute. Cast iron: Speed, 100–150 feet per minute. Cold-rolled steel: Speed, 150–200 feet per minute. High-speed steel: Speed, 90–125 feet per minute. Malleable iron: Speed, 150–200 feet per minute. Hard rubber: Speed, 150–200 feet per minute. Slate: Speed, 100–150 feet per minute. The number of saw teeth per inch recommended for these different materials ranges from 8 to 14, and the saw user should be guided by the recommendations of the manufacturer.

Hacksaw Speeds. — The following hacksaw speeds are given on the authority of a leading manufacturer: The average rate of travel of the cutting blade in feet per minute, including forward and return strokes, should be as follows: For mild steel, 130; for annealed tool steel, 90; and for unannealed tool steel, 60 feet per minute. Thus in the case of a 6-inch stroke, for example, the revolutions per minute of the driving crank should be 130 for mild steel, 90 for annealed tool steel, and 60 for unannealed tool steel. All of these steels are cut with the use of a cutting compound. Bronze can ordinarily be cut at the same speed as mild steel when a suitable compound is used. Brass heats the blade very rapidly if cut dry, and must be cut with a cooling compound adapted to brass. It also fills up the teeth of the saw if not used with the right kind of compound, but with a suitable compound, it may be cut at the same speed as machine steel.

Speeds for Turning Unusual Materials. — *Slate,* on account of its peculiarly stratified formation, is rather difficult to turn, but if handled carefully, can be machined in an ordinary lathe. The cutting speed should be about the same as for cast iron. A sheet of fiber or pressed paper should be interposed between the chuck or steadyrest jaws and the slate, to protect the latter. Slate rolls must not be centered and run on the tailstock. A satisfactory method of supporting a slate roll having journals at the ends is to bore a piece of lignum vitae to receive the turned end of the roll, and center it for the tailstock spindle.

Rubber can be turned at a peripheral speed of 200 feet per minute, although it is much easier to grind it with an abrasive wheel that is porous and soft. For cutting a rubber roll in two, the ordinary parting tool should not be used, but a tool shaped like a knife; such a tool severs the rubber without removing any material.

Gutta percha can be turned as easily as wood, but the tools must be sharp and a good soap-and-water lubricant used.

Copper can be turned easily at 200 feet per minute.

Lime-stone such as is used in the construction of pillars for balconies, etc., can be turned at 150 feet per minute, and the formation of ornamental contours is quite easy. *Marble* is a treacherous material to turn. It should be cut with a tool such as would be used for brass, but at a speed suitable for cast iron. It must be handled very carefully to prevent flaws in the surface.

The foregoing speeds are for high-speed steel tools. Tools tipped with tungsten carbide are adapted for cutting various non-metallic products which cannot be machined readily with steel tools, such as slate, marble, synthetic plastic materials, etc. In drilling slate and marble, use flat drills; and for plastic materials, tungsten-carbide-tipped twist drills. Cutting speeds ranging from 75 to 150 feet per minute have been used for drilling slate (without coolant) and a feed of 0.025 inch per revolution for drills ¾ and 1 inch in diameter.

CUTTING TOOLS

Tool Contour. — Tools for turning, planing, etc., are made in straight, bent, offset, and other forms to locate the cutting edges in convenient positions for operating on differently located surfaces. The contour or shape of the cutting edge may also be varied to suit different classes of work. Tool shapes, however, are not only related to the kind of operation, but, in the case of roughing tools particularly, the contour may have a decided effect upon the cutting efficiency of the tool. To illustrate, an increase in the side cutting-edge angle of a roughing tool, or in the nose radius, tends to permit higher cutting speeds because the chip will be thinner for a given feed rate. Such changes, however, may result in chattering or vibrations unless the work and the machine are rigid; hence, the most desirable contour may be a compromise between the ideal form and one that is needed to meet practical requirements.

Terms and Definitions. — The terms and definitions relating to single-point tools vary somewhat in different plants, but the following are in general use.

Single-point Tool: This term is applied to tools for turning, planing, boring, etc., which have a cutting edge at one end. This cutting edge may be formed on one end of a solid piece of steel, or the cutting part of the tool may consist of an insert or tip which is held to the body of the tool either by brazing, welding, or by mechanical means. Figure 1 shows the more common shapes of high-speed-steel single-point tools used for turning operations.

Shank: The shank is the main body of the tool. If the tool is an inserted cutter type, the shank supports the cutter or bit. (See diagram, Fig. 2.)

Nose: This is a general term sometimes used to designate the cutting end but usually it relates more particularly to the rounded tip of the cutting end.

Face: The surface against which the chips bear, as they are severed in turning or planing operations, is called the face.

Flank: The flank is that end surface that is adjacent to the cutting edge and below it when the tool is in a horizontal position as for turning.

Base: The base is that surface of the tool shank which bears against the supporting tool-holder or block.

Side Cutting Edge: The side cutting edge is the cutting edge located on the side of the tool. Tools, such as shown in Fig. 1, do the bulk of the cutting with this cutting edge and are, therefore, sometimes called side cutting edge tools.

End Cutting Edge: The end cutting edge is the cutting edge located at the end of the tool. On side cutting edge tools, the end cutting edge can be used for light plunging and facing cuts. Cut-off tools and similar tools have only one cutting edge located on the end. These tools and other tools that are intended to cut primarily with the end cutting edge are sometimes called end cutting edge tools.

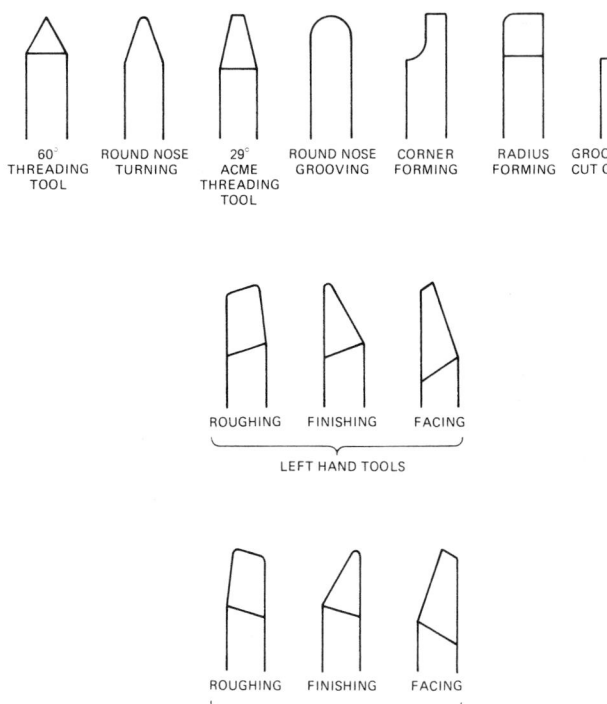

Fig. 1. Lathe Tools—Standard Shapes

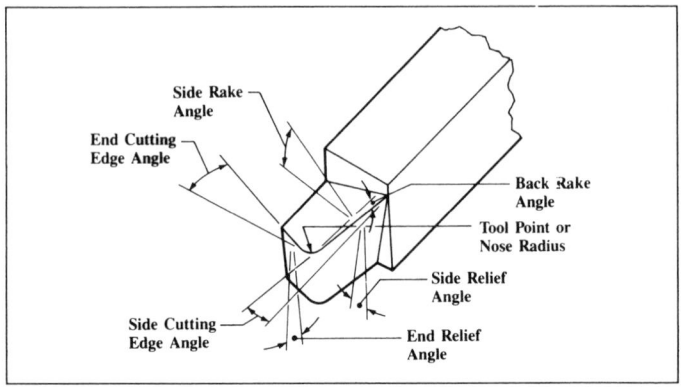

Fig. 2. Terms Applied to Single-point Turning Tools

Rake: A metal cutting tool is said to have rake when the tool face or surface against which the chips bear as they are being severed, is inclined for the purpose of either increasing or diminishing the keenness or bluntness of the edge. The magnitude of the rake is most conveniently measured by two angles called the back rake angle and the side rake angle. The tool shown in Fig. 2 has rake. If the face of the tool did not incline but was parallel to the base, there would be no rake; the rake angles would be zero.

Positive Rake: If the inclination of the tool face is such as to make the cutting edge keener or more acute than when the rake angle is zero, the rake angle is defined as positive.

Negative Rake: If the inclination of the tool face makes the cutting edge less keen or more blunt than when the rake angle is zero, the rake is defined as negative.

Back Rake: The back rake is the inclination of the face toward or away from the end or the end cutting edge of the tool. When the inclination is away from the end cutting edge, as shown in Fig. 2, the back rake is positive. If the inclination is downward toward the end cutting edge it is negative.

Side Rake: The side rake is the inclination of the face toward or away from the side cutting edge. When the inclination is away from the side cutting edge, as shown in Fig. 1, the side rake is positive. If the inclination is toward the side cutting edge the side rake is negative.

Relief: The flanks below the side cutting edge and the end cutting edge must be relieved to allow these cutting edges to penetrate into the workpiece when taking a cut. If the flanks are not provided with relief, the cutting edges will rub against the workpiece and be unable to penetrate in order to form the chip. Relief is also provided below the nose of the tool to allow it to penetrate into the workpiece. The relief at the nose is usually a blend of the side relief and the end relief.

End Relief Angle: The end relief angle is a measure of the relief below the end cutting edge.

Side Relief Angle: The side relief angle is a measure of the relief below the side cutting edge.

Back Rake Angle: The back rake angle is a measure of the back rake. It is

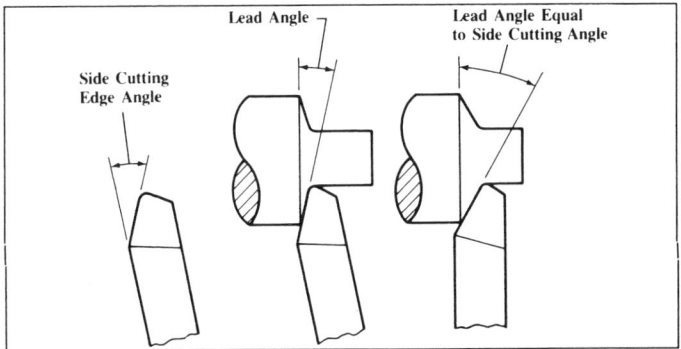

Fig. 3. Lead Angle on Single-point Turning Tool

measured in a plane that passes through the side cutting edge and is perpendicular to the base. Thus, the back rake angle can be measured by measuring the inclination of the side cutting edge with respect to a line or plane that is parallel to the base. The back rake angle may be positive, negative, or zero depending upon the magnitude and direction of the back rake.

Side Rake Angle: The side rake angle is a measure of the side rake. It is always measured in a plane that is perpendicular to the side cutting edge and perpendicular to the base. Thus, the side rake angle is the angle of the inclination of the face perpendicular to the side cutting edge with reference to a line or a plane that is parallel to the base.

End Cutting Edge Angle: The end cutting edge angle is the angle made by the end cutting edge with respect to a plane perpendicular to the axis of the tool shank. It is provided to allow the end cutting edge to clear the finish machined surface on the workpiece.

Side Cutting Edge Angle: The side cutting edge angle is the angle made by the side cutting edge and a plane that is parallel to the side of the shank.

Nose Radius: The nose radius is the radius of the nose of the tool. The performance of the tool, in part, is influenced by nose radius and, for this reason, it must be carefully controlled.

Lead Angle: The lead angle, shown in Fig. 3, is not ground on the tool. It is a tool setting angle which has a great influence on the performance of the tool. The lead angle is bounded by the side cutting edge and a plane perpendicular to the workpiece surface when the tool is in position to cut; or, more correctly, the lead angle is the angle between the side cutting edge and a plane perpendicular to the direction of the feed travel.

Solid Tool: A solid tool is a cutting tool made from one piece of tool material.

Brazed Tool: A brazed tool is a cutting tool having a blank of cutting-tool material permanently brazed to a steel shank.

Blank: A blank is an unground piece of cutting-tool material from which a brazed tool is made.

Tool Bit: A tool bit is a relatively small cutting tool that is clamped in a holder in such a way that it can readily be removed and replaced. It is intended primarily to be reground when dull and not indexed.

Tool-bit Blank: The tool-bit blank is an unground piece of cutting-tool material from which a tool bit can be made by grinding. It is available in standard sizes and shapes.

Tool-bit Holder: Made from forged steel, the tool-bit holder is used to hold the tool bit, to act as an extended shank for the tool bit, and to provide a means for clamping in the tool post.

Straight-shank Tool-bit Holder: A straight-shank tool-bit holder has a straight shank when viewed from the top. The axis of the tool bit is held parallel to the axis of the shank.

Offset-shank Tool-bit Holder: An offset-shank tool-bit holder has the shank bent to the right or left, as seen in Fig. 4. The axis of the tool bit is held at an angle with respect to the axis of the shank.

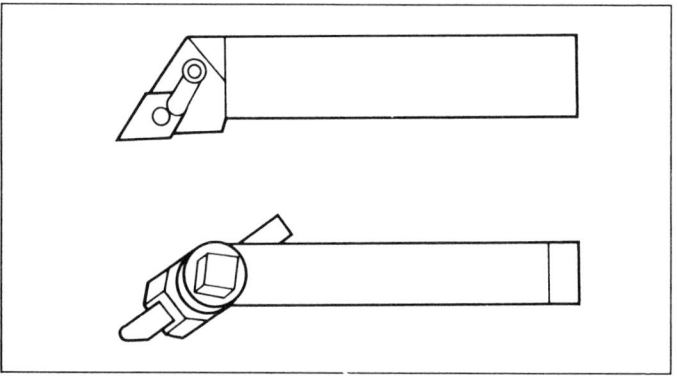

Fig. 4. Top: Right-hand Offset-shank, Indexable Insert Holder
Bottom: Right-hand Offset-shank Tool-bit Holder

Indexable Inserts: An indexable insert is a relatively small piece of cutting-tool material that is geometrically shaped to have two or several cutting edges that are used until dull. It is then indexed on the holder to apply a sharp cutting edge. When all of the cutting edges have been dulled, the insert is discarded. The insert is held in a pocket or against other seating surfaces on an indexable insert holder by means of a mechanical clamping device that can be tightened or loosened easily.

Indexable Insert Holder: Made of steel, an indexable insert holder is used to hold indexable inserts. It is equipped with a mechanical clamping device that holds the inserts firmly in a pocket or against other seating surfaces.

Straight-shank Indexable Insert Holder: A straight-shank indexable insert toolholder is essentially straight when viewed from the top, although the cutting edge of the insert may be oriented parallel, or at an angle, to the axis of the holder.

Offset-shank indexable Insert Holder: An offset-shank indexable insert holder has the head end, or the end containing the insert pocket, offset to the right or left, as shown in Fig. 4.

Side cutting Tool: A side cutting tool has its major cutting edge on the side of the cutting part of the tool. The major cutting edge may be parallel or at an angle with respect to the axis of the tool.

End cutting Tool: An end cutting tool has its major cutting edge on the end of the cutting part of the tool. The major cutting edge may be perpendicular or at an angle, with respect to the axis of the tool.

Curved Cutting-edge Tool: A curved cutting-edge tool has a continuously variable side cutting edge angle. The cutting edge is usually in the form of a smooth, continuous curve along its entire length, or along a large portion of its length.

Right-hand Tool: A right-hand tool has the major, or working, cutting edge on the right-hand side when viewed from the cutting end with the face up. As used in a lathe, it is usually fed into the work from right to left, when viewed from the shank end.

Left-hand Tool: A left-hand tool has the major or working cutting edge on the left-hand side when viewed from the cutting end with the face up. As used in a lathe, it is usually fed into the work from left to right, when viewed from the shank end.

Neutral-hand Tool: A neutral-hand tool is a tool to cut either left to right or right to left; or the cut may be parallel to the axis of the shank as when plunge cutting.

Chipbreaker: A groove formed in or on a shoulder formed on the face of a turning tool back of the cutting edge, for the purpose of breaking up the chips and thus preventing the formation of long, continuous chips which would be dangerous to the operator and also bulky and cumbersome to handle. A chipbreaker of the shoulder type may be formed directly on the tool face or it may consist of a separate piece that is either held by brazing or by clamping.

Relief Angles. — The end relief angle and the side relief angle on single-point cutting tools are usually made equal to each other, although this is not always the case. The relief angle under the nose of the tool is a blend of the side and end relief angles.

The size of the relief angles has a pronounced effect on the performance of the cutting tool. If the relief angles are too large, the cutting edge will be weakened and in danger of breaking when a heavy cutting load is placed on it by a hard and tough material. On finish cuts, rapid wear of the cutting edge may cause problems with size control on the part. Relief angles that are too small will cause the rate of wear on the flank of the tool below the cutting edge to increase, thereby significantly reducing the tool life. In general, when cutting hard and tough materials, the relief angles should be 6 to 8 degrees for high-speed steel tools and 5 to 7 degrees for carbide tools. For medium steels, mild steels, cast iron, and other average work the recommended values of the relief angles are 8 to 12 degrees for high-speed steel tools and 5 to 10 degrees for carbides. Ductile materials having a relatively low modulus of elasticity should be cut using larger relief angles. For example, the relief angles recommended for turning copper, brass, bronze, aluminum, ferritic malleable iron, and similar metals are 12 to 16 degrees for high-speed steel tools and 8 to 14 degrees for carbides.

Larger relief angles generally tend to produce a better surface finish on the finish machined surface because less surface of the worn flank of the tool rubs against the workpiece. For this reason, single-point thread-cutting tools should be provided with relief angles that are as large as circumstances will permit. Problems encountered when machining stainless steel may be overcome by increasing the size of the relief angle. The relief angles used should never be smaller than necessary.

Rake Angles. — Machinability tests have confirmed that when the rake angle along which the chip slides, called the true rake angle, is made larger in the positive

direction, the cutting force and the cutting temperature will decrease. Also, the tool life for a given cutting speed will increase with increases in the true rake angle up to an optimum value, after which it will decrease again. For turning tools which cut primarily with the side cutting edge, the true rake angle corresponds rather closely with the side rake angle except when taking shallow cuts. It can safely be stated that increasing the side rake angle in the positive direction lowers the cutting temperature and the cutting force, while at the same time it results in a longer tool life or a higher permissible cutting speed up to an optimum value of the side rake angle. After the optimum value is exceeded, the cutting temperature and the cutting force will continue to drop; however, the tool life and the permissible cutting speed will decrease. As an approximation, the magnitude of the cutting force will decrease about one per cent per degree increase in the side rake angle. While not exact, this rule of thumb does correspond approximately to test results and can be used to make rough estimates. Of course, the cutting force also increases about one per cent per degree decrease in the side rake angle. The limiting value of the side rake angle for optimum tool life or cutting speed depends upon the work material and the cutting tool material. In general, it will occur at lower values for hard and tough work materials. Cemented carbides are harder and more brittle than high-speed steel; therefore, the rake angles usually used for cemented carbides are larger than for high-speed steel.

Negative rake angles cause the face of the tool to slope in the opposite direction from positive rake angles and, as might be expected, they have an opposite effect. For side cutting edge tools, increasing the side rake angle in a negative direction will result in an increase in the cutting temperature and an increase in the cutting force of approximately one per cent per degree change in rake angle. For example, if the side rake angle is changed from 5 degrees positive to 5 degrees negative, the cutting force will be about 10 per cent larger. Usually the tool life will also decrease when negative side rake angles are used, although the tool life will sometimes increase when the negative rake angle is not too large and when a fast cutting speed is used.

Negative side rake angles are usually used in combination with negative back rake angles on single-point cutting tools. The negative rake angles strengthen the cutting edges enabling them to sustain heavier cutting loads and shock loads. They are recommended for turning very hard materials and for heavy interrupted cuts. There is also an economic advantage in favor of using negative rake indexable inserts and tool holders inasmuch as the cutting edges on both the top and bottom of the insert can be used.

On turning tools that cut primarily with the side cutting edge, the effect of the back rake angle alone is much less than the effect of the side rake angle although the direction of the change in cutting temperature, cutting force, and tool life is the same. The effect that the back rake angle has can be ignored unless, of course, extremely large changes in this angle are made. A positive back rake angle does improve the performance of the nose of the tool somewhat and is helpful in taking light finishing cuts. A negative back rake angle strengthens the nose of the tool and is helpful when interrupted cuts are taken. The back rake angle does have a very significant effect on the performance of end cutting edge tools, such as cut-off tools. For these tools, the effect of the back rake angle is very similar to the effect of the side rake angle on side cutting edge tools.

Side Cutting Edge and Lead Angles. — These angles are considered together because the side cutting edge angle is usually designed to provide the desired lead angle when the tool is being used. The side cutting edge angle and the lead angle will be equal when the shank of the cutting tool is positioned perpendicular to the workpiece, or, more correctly, perpendicular to the direction of the feed. When the shank is not perpendicular, the lead angle is determined by the side cutting edge and an imaginary line perpendicular to the feed direction.

The flow of the chips over the face of the tool is approximately perpendicular to the side cutting edge except when shallow cuts are taken. The thickness of the undeformed chip is measured perpendicular to the side cutting edge. As the lead angle is increased, the length of chip in contact with the side cutting edge is increased, and the chip will become longer and thinner. This effect is the same as increasing the depth of cut and decreasing the feed, although the actual depth of cut and feed remain the same and the same amount of metal is removed. The effect of lengthening and thinning the chip by increasing the lead angle is very beneficial as it increases the tool life for a given cutting speed or the cutting speed can be increased. Increasing the cutting speed while the feed and the tool life remain the same leads to faster production.

There is, however, an adverse effect that must be considered. Chatter can be caused by a cutting edge that is oriented at a lead angle when turning and in some cases, such as when turning long and slender shafts, only a small lead angle can cause chatter. In fact, the lead angle of the side cutting edge is one of the principal causes of chatter. When chatter occurs, often simply reducing the lead angle will be the cure. Sometimes, very long and slender shafts can be turned successfully with a tool having a zero degree lead angle (and having a small nose radius). Boring bars, being usually somewhat long and slender, are also susceptible to chatter if a large lead angle is used. The lead angle for boring bars should be kept small, and for very long and slender boring bars a zero degree lead angle is recommended. It is impossible to provide a rule that will determine when chatter caused by a lead angle will occur and when it will not occur. In making a judgment, the first consideration is the length to diameter ratio of the part to be turned, or of the boring bar. Then the method of holding the workpiece must be considered — a part that is firmly held is less apt to chatter. Finally, the overall condition and rigidity of the machine must be considered because, in some cases, this may be the real cause of chatter.

Although chatter can be a problem, the advantages gained are such that the lead angle should be as large as possible at all times.

End Cutting Edge Angle. — The size of the end cutting edge angle is important when tool wear by cratering occurs. Frequently, the crater will enlarge until it breaks through the end cutting edge just behind the nose, and tool failure follows shortly. Reducing the size of the end cutting edge angle tends to delay the time of crater breakthrough. When cratering takes place, the recommended end cutting edge angle is 8 to 15 degrees. When cratering does not take place it can be made larger. Larger end cutting edge angles may be required for profile turning tools to enable these tools to plunge into the work without interference from the end cutting edge.

Nose Radius. — The nose is a very critical part of the cutting edge since it cuts the finished surface on the workpiece. If the nose is made to a sharp point, the finish machined surface will usually be unacceptable and the life of the tool will be short. Thus, a nose radius is required to obtain an acceptable surface finish and tool life. The surface finish obtained is determined by the feed rate and by the nose radius if extraneous factors such as the work material, the cutting speed, and cutting fluids are not considered. A large nose radius will give a better surface finish and will permit a faster feed rate to be used.

Machinability tests have demonstrated that increasing the nose radius will also improve the tool life or allow a faster cutting speed to be used. For example, high-speed steel tools were used to turn an alloy steel in one series of tests where complete or catastrophic tool failure was used as a criterion for the end of tool life. The cutting speed for a 60-minute tool life was found to be 125 fpm when the nose radius was $\frac{1}{16}$ inch and 160 fpm when the nose was $\frac{1}{4}$ inch.

While a very large nose radius can often be used, a limit is sometimes imposed because the tendency for chatter to occur is increased as the nose radius is made larger. A nose radius that is too large can cause chatter and when this occurs a smaller nose radius must be used on the tool. It is always good practice, however, to make the nose radius as large as is compatible with the operation being performed.

Chipbreakers. — Many steel turning tools are equipped with chipbreaking devices to prevent the formation of long continuous chips in connection with the turning of steel at the high speeds made possible by high-speed steel and especially sintered carbide tools. Long steel chips are dangerous to the operator, cumbersome to handle, and they may twist around the tool and cause damage. Broken chips not only occupy less space, but permit a better flow of coolant to the cutting edge. Several different forms of chipbreakers are illustrated in Fig. 5.

Angular Shoulder Type: The angular shoulder type shown at *A* is one of the commonly used forms. As the enlarged sectional view shows, the chipbreaking shoulder is located back of the cutting edge. The angle *a* between the shoulder and cutting edge may vary from 6 to 15 degrees or more, 8 degrees being a fair average. The ideal angle, width *W* and depth *G*, depends upon the speed and feed, the depth of cut, and the material. As a general rule, width *W* at the end of the tool, varies from ⅜₂ to ⁷⁄₃₂ inch, and the depth *G* may range from ⅟₆₄ to ⅟₁₆ inch. The shoulder radius equals depth *D*. If the tool has a large nose radius, the corner of the shoulder at the nose end may be beveled off, as illustrated at *B*, to prevent it from coming into contact with the work. The width *K* for type *B* should equal approximately 1.5 times the nose radius.

Parallel Shoulder Type: Diagram C illustrates the type which has a chipbreaking shoulder that is parallel with the cutting edge. With this form, the chips are likely to come off in short curled sections. The parallel form may also be applied to straight tools which do not have a side cutting-edge angle. The tendency with this parallel shoulder form is to force the chips against the work and thus cause breakage.

Groove Type: This type (diagram D) consists of a groove that is ground into the face of the tool. Between the groove and the cutting edge, there is a land *L*. Under ideal conditions, this width *L*, the groove width *W*, and the groove depth *G*, would be varied according to the feed, depth of cut and material. For average use, *L* is about ⅟₃₂ inch; *G*, ⅟₃₂ inch; and *W*, ⅟₁₆ inch. There are differences of opinion concerning the relative merits of the groove type and the shoulder type. Both types have proved satisfactory when properly proportioned for a given class of work.

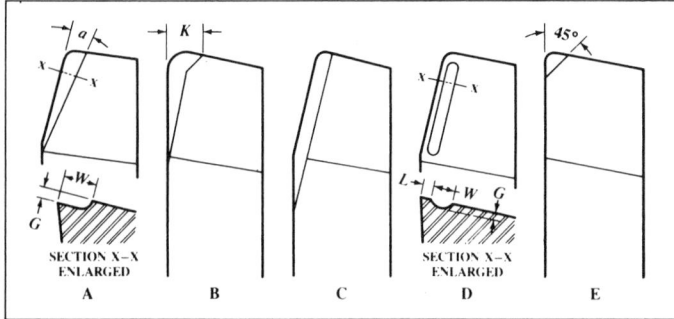

Fig. 5. Different Forms of Chipbreakers for Turning Tools.

Chipbreaker for Light Cuts: Diagram E illustrates a form of chipbreaker that is sometimes used on tools for finishing cuts having a maximum depth of about ⅟32 inch. This is a shoulder type having an angle of 45 degrees and a maximum width of about ⅟16 inch. It is important in grinding all chipbreakers to give the chip-bearing surfaces a fine finish, such as would be obtained by honing. This greatly increases the life of the tool.

Planing Tools. — Many of the principles which govern the shape of turning tools also apply in the grinding of tools for planing. The amount of rake depends upon the hardness of the material, and the direction of the rake should be away from the *working part* of the cutting edge. The angle of clearance should be about 4 or 5 degrees for planer tools, which is less than for lathe tools. This small clearance is allowable because a planer tool is held about square with the platen, whereas a lathe tool, the height and inclination of which can be varied, may not always be clamped in the same position.

Carbide Tools: Carbide tools for planing usually have negative rake. Round-nose and square-nose end-cutting tools should have a "negative back rake" (or front rake) of 2 or 3 degrees. Side cutting tools may have a negative back rake of 10 degrees, a negative side rake of 5 degrees, and a side cutting-edge angle of 8 degrees.

Indexable Inserts. — These inserts are also called throwaway inserts and disposable inserts. A large proportion of the cemented carbide, single-point cutting tools are indexable inserts and indexable insert tool holders. Samples of the many insert shapes are shown in Table 3. Most modern, cemented carbide, face milling cutters are of the indexable insert type. Larger size end milling cutters, side milling or slotting cutters, boring tools, and a wide variety of special tools are made to use indexable inserts. These inserts are primarily made from cemented carbide although most of the cemented oxide cutting tools are also indexable inserts.

The objective of this type of tooling is to provide an insert with several cutting edges. When one edge is worn the insert is indexed in the tool holder until all of the cutting edges are used up, after which it is discarded. The insert is not intended to be reground. The advantages are that the cutting edges on the tool can be rapidly changed without moving the tool holder from the machine, tool-grinding costs are eliminated, and the cost of the insert is less than the cost of a similar, brazed carbide tool. Of course, the cost of the tool holder must be added to the cost of the insert; however, one tool holder will usually last for a long time before it, too, must be replaced.

Indexable inserts and tool holders are made with a negative rake or with a positive rake. Negative rake inserts have the advantage of having twice as many cutting edges available as comparable positive rake inserts, because the cutting edges on both the top and bottom of negative rake inserts can be used, while only the top cutting edges can be used on positive rake inserts. When machining long and slender parts, thin-walled parts, or other parts that are subject to bending or chatter when the cutting load is applied to them, positive rake inserts have a distinct advantage because the cutting force is significantly lower, when they are used, as compared to that for negative rake inserts. Indexable inserts can be obtained in the following forms: utility ground, or ground on top and bottom only; precision ground, or ground on all surfaces; prehoned to produce a slight rounding of the cutting edge; and, precision molded, in which case they are unground. Positive-negative rake inserts are available. These inserts are held on a negative-rake tool holder and have a chipbreaker groove that is formed to produce an effective positive-rake angle while cutting. They may have cutting

edges available on the top surface only, or on both top and bottom surfaces. The positive-rake chipbreaker surface may be ground or precision molded on the insert.

Many materials, such as gray cast iron, form a discontinuous chip. For these materials an insert that has plain faces without chipbreaker grooves should always be used. Steels and other ductile materials form a continuous chip that must be broken into small segments when machined on lathes and planers having single-point, cemented-carbide and cemented-oxide cutting tools; otherwise, the chips can cause injury to the operator. In this case a chipbreaker must be used. Some inserts are made with chipbreaker grooves molded or ground directly on the insert. When inserts with plain faces are used, a cemented-carbide plate-type chipbreaker is clamped on top of the insert.

Identification System for Indexable Inserts. — The size of indexable inserts is determined by the diameter of an inscribed circle (I.C.), except for rectangular and parallelogram inserts where the length and width dimensions are used. To describe an insert in its entirety, a standard (ANSI B94.4-1976) identification system is used where each position number designates a feature of the insert. The ANSI Standard includes items now commonly used and facilitates identification of items not in common use. Identification consists of up to ten positions; each position defines a characteristic of the insert as shown below:

$$1 \quad 2 \quad 3 \quad 4 \quad 5 \quad 6 \quad 7 \quad 8^* \quad 9 \quad 10^*$$

$$\textbf{T} \ \textbf{N} \ \textbf{M} \ \textbf{G} \ \textbf{5} \ \textbf{4} \ \textbf{3} \quad\ \textbf{A}$$

* Eighth and Tenth Position only used when required.

1. *Shape:* The shape of an insert is designated by a letter: **R** for round; **S**, square; **T**, triangle; **A**, 85° parallelogram; **B**, 82° parallelogram; **C**, 80° diamond; **D**, 55° diamond; **E**, 75° diamond; **H**, hexagon; **K**, 55° parallelogram; **L**, rectangle; **M**, 86° diamond; **O**, octagon; **P**, pentagon; **V**, 35° diamond; and **W**, 80° trigon.

2. *Relief Angle:* The second position is a letter denoting the relief angles; **N** for 0°; **A**, 3°; **B**, 5°; **C**, 7°; **P**, 11°; **D**, 15°; **E**, 20°; **F**, 25°; **G**, 30°; **H**, 0° & 11°*; **J**, 0° & 14°*; **K**, 0° & 17°*; **L**, 0° & 20°*; **M**, 11° & 14°*; **R**, 11° & 17°*; **S**, 11° & 20°*. When mounted on a holder, the actual relief angle may be different from that on the insert.

* Second angle is secondary facet angle, which may vary by ±1°.

3. *Tolerances:* The third position is a letter and indicates the tolerances which control the indexability of the insert. Tolerances specified do not imply the method of manufacture.

Symbol	Tolerance (± from nominal)		Symbol	Tolerance (± from nominal)	
	Inscribed Circle, Inch	Thickness, Inch		Inscribed Circle, Inch	Thickness, Inch
A	0.001	0.001	**H**	0.0005	0.001
B	0.001	0.005	**J**	0.002–0.005	0.001
C	0.001	0.001	**K**	0.002–0.005	0.001
D	0.001	0.005	**L**	0.002–0.005	0.001
E	0.001	0.001	**M**	0.002–0.004†	0.005
F	0.0005	0.001	**U**	0.005–0.010†	0.005
G	0.001	0.005	**N**	0.002–0.004†	0.001

† Exact tolerance is determined by size of insert. (See ANSI B94.25).

4. *Type:* The type of insert is designated by a letter. **A**, with hole; **B**, with hole and countersink; **C**, with hole and two countersinks; **D**, smaller than ¼ inch I.C. with hole; **E**, smaller than ¼ inch I.C.; **F**, with 0° top rake land, chip grooves both surfaces, no hole; **G**, Same as **F** but with hole; **H**, with hole, one countersink and chip groove on one top rake surface; **J**, with hole, two countersinks and chip grooves on both top rake surfaces; **K**, smaller than ¼ inch I.C. with hole and chip grooves on both top rake surfaces; **L**, smaller than ¼ inch I.C. without hole and with chip

grooves on both top rake surfaces; **M**, with hole and chip groove on one top rake surface; **P**, with 10° top rake chip grooves, both surfaces with hole; **R**, without hole but with chip groove on one top rake surface; and **S**, with 20° top rake chip groove on one surface and with hole. *Note:* a dash may be used after position 4 to separate the shape describing portion from the following dimensional description of the insert and is not to be considered a position in the standard description.

5. *Size:* The size of the insert is designated by a number having the following meaning: For inserts less than ¼ inch I.C., the number represents the I.C. diameter in ⅟32nds of an inch. For inserts ¼ inch I.C., and over, the number represents the I.C. diameter in ⅛ths of an inch. Rectangular and parallelogram inserts require two digits: the first digit indicates the number of ⅛ths of an inch width and the second digit, the number of ¼ths of an inch length.

6. *Thickness:* The thickness is designated by a number: For inserts less than ¼ inch I.C., the number represents ⅟32nds inch of thickness. For inserts ¼ inch I.C. and over, the number represents the ⅟16ths inch of thickness. For rectangular and parallelogram inserts, use width dimension in place of I.C. to designate thickness.

7. *Cutting Point Configuration:* The cutting point, or nose radius, is designated by a number representing ⅟64ths of an inch; a flat at the cutting point or nose, is designated by a letter: **0** for sharp corner; **1**, ⅟64 inch radius; **2**, ⅟32 inch radius; **3**, ³⁄64 inch radius; **4**, ⅟16 inch radius; **6**, ³⁄32 inch radius; **8**, ⅛ inch radius; **A**, square insert with 45° chamfer; **D**, square insert with 30° chamfer; **E**, square insert with 15° chamfer; **F**, square insert with 3° chamfer; **K**, square insert with 30° double chamfer; **L**, square insert with 15° double chamfer; **M**, square insert with 3° double chamfer; **N**, truncated triangle insert; and **P**, flatted corner triangle insert.

8. *Special Cutting Point Definition:* The eighth position, if it follows a letter in the 7th position, is a number indicating the number of ⅟32nds of an inch measured parallel to the edge of the facet.

9. *Other Conditions:* The ninth position defines special conditions (such as edge treatment, surface finish) as follows: **A**, honed, 0.001″ to less than 0.003″; **B**, honed, 0.003″ to less than 0.005″; **C**, honed, 0.005″ to less than 0.007″; **J**, polished, 4 microinch arithmetic average (AA) on rake surfaces only.

10. *Hand:* **R**, right; **L**, left; to be used when required in tenth position.

Indexable Insert Tool Holders. — Indexable insert tool holders are made from a good grade of steel which is heat treated to a hardness of 44 to 48 Rc for most normal applications. Accurate pockets are machined in the end of tool holders which serve to locate the insert in position and to provide surfaces against which the insert can be clamped. In almost all cases, a cemented carbide seat is provided which is held in the bottom of the pocket by a screw or by the clamping pin, if one is used. The seat is necessary to provide a flat bearing surface upon which the insert can rest and, in so doing, it adds materially to the ability of the insert to withstand the cutting load. The seating surface of the holder may provide a positive-, negative-, or a neutral-rake orientation to the insert when it is in position on the holder. Holders, therefore, are classified as positive, negative, or neutral rake.

There are four basic methods used to clamp the insert on the holder: 1. Top clamping; 2. Pin-lock clamping; 3. Multiple clamping using a top clamp and a pin lock; and, 4. Clamping the insert with a machine screw. All top clamps are actuated by a screw that forces the clamp directly against the insert. When required, a cemented-carbide, plate type chipbreaker is placed between the clamp and the insert. Pin-lock clamps require an insert having a hole: the pin acts against

Table 1.   Standard Shank Sizes for Indexable Insert Holders

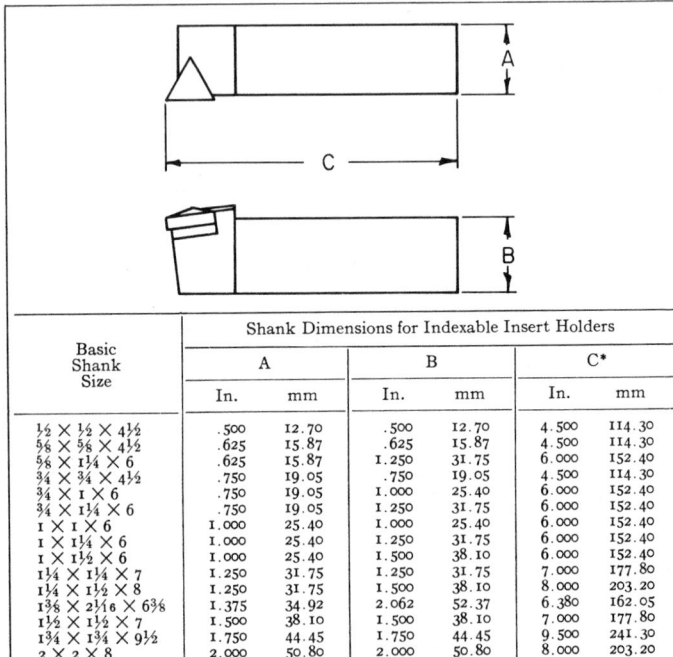

Basic Shank Size	Shank Dimensions for Indexable Insert Holders					
	A		B		C*	
	In.	mm	In.	mm	In.	mm
½ × ½ × 4½	.500	12.70	.500	12.70	4.500	114.30
⅝ × ⅝ × 4½	.625	15.87	.625	15.87	4.500	114.30
⅝ × 1¼ × 6	.625	15.87	1.250	31.75	6.000	152.40
¾ × ¾ × 4½	.750	19.05	.750	19.05	4.500	114.30
¾ × 1 × 6	.750	19.05	1.000	25.40	6.000	152.40
¾ × 1¼ × 6	.750	19.05	1.250	31.75	6.000	152.40
1 × 1 × 6	1.000	25.40	1.000	25.40	6.000	152.40
1 × 1¼ × 6	1.000	25.40	1.250	31.75	6.000	152.40
1 × 1½ × 6	1.000	25.40	1.500	38.10	6.000	152.40
1¼ × 1¼ × 7	1.250	31.75	1.250	31.75	7.000	177.80
1¼ × 1½ × 8	1.250	31.75	1.500	38.10	8.000	203.20
1⅜ × 2¹⁄₁₆ × 6⅜	1.375	34.92	2.062	52.37	6.380	162.05
1½ × 1½ × 7	1.500	38.10	1.500	38.10	7.000	177.80
1¾ × 1¾ × 9½	1.750	44.45	1.750	44.45	9.500	241.30
2 × 2 × 8	2.000	50.80	2.000	50.80	8.000	203.20

* Holder length; may vary by manufacturer.   Actual shank length depends on holder style.

the walls of the hole to clamp the insert firmly against the seating surfaces of the holder.   Multiple, or combination clamping, simultaneously using both a pin-lock and a top clamp, is recommended when taking heavier or interrupted cuts.   Holders are available on which all of the above-mentioned methods of clamping may be used. Others are made with only a top clamp or a pin lock.   Screw-on type holders use a machine screw to hold the insert in the pocket.   Most standard indexable insert holders are either straight-shank or offset-shank, although special holders can be made having a wide variety of configurations.

The common shank sizes of indexable insert tool holders are shown in Table 1. Not all styles are available in every shank size.   Positive- and negative-rake tools are also not available in every style or shank size.   Some manufacturers provide additional shank sizes for certain tool holder styles.   For more complete details the manufacturers' catalogs must be consulted.

Identification System for Indexable Insert Holders. — The following identification system generally conforms to the American National Standard, ANSI B94.26-1969, R1977.   In a few instances some additions are included which conform

with current usage; these additions are marked with an asterisk. Each position in the system designates a feature of the holder in the following order:

$$
\begin{array}{cccccccc}
\text{I} & 2 & 3 & 4 & 5 & 6 & 7 & 8 \\
\text{C} & \text{T} & \text{N} & \text{A} & \text{R} - & \text{8} \;\; \text{5} - & \text{5} & \text{D}
\end{array}
$$

1. *Method of Holding Horizontally Mounted Insert:* The method of holding or clamping is designated by a letter: **C** for clamp (normally, top clamping); **P**, pin lock (insert with hole in center); **M*** or **D***, multiple or combination pin lock and top clamp; and **K***, alternate for clamp (normally, top clamp).

2. *Insert Shape:* The insert shape is designated by a letter: **S** for square; **R**, round; **T**, triangle; **C**, 80° diamond; **D**, 55° diamond; **V**, 35° diamond; **F***, 70° parallelogram; **L***, rectangular; and **G***, deep grooving.

3. *Holder Style:* The holder style designates the shank style and the side cutting edge angle, or end cutting edge angle, or the purpose for which the holder is used. It is designated by a letter: **A** for straight shank with 0° side cutting edge angle; **B**, straight shank with 15° side cutting edge angle; **C***, straight-shank end cutting tool with 0° end cutting edge angle; **D**, straight shank with 45° side cutting edge angle; **E**, straight shank with 30° side cutting edge angle; **F**, offset shank with 0° end cutting edge angle; **G**, offset shank with 0° side cutting edge angle; **H**, straight shank with 38° side cutting edge angle; **J**, offset shank with negative 3° side cutting edge angle; **K**, offset shank with 15° end cutting edge angle; **L**, offset shank with negative 5° side cutting edge angle; **M**, straight shank with 40° side cutting edge angle; **N**, straight shank with 27° side cutting edge angle; **P**, straight shank with 27½° side cutting edge angle; **S**, offset shank with 45° side cutting edge angle;

Table 2. Qualified Indexable Insert Tool Holders: (left) Letter Designations for Shank Qualification and (right) Insert Nose Radius for Corresponding Inscribed Circle of Gage Insert

Length and Width Control		Length and Side Control		Insert Nose Radius, Inch	Insert Inscribed Circle (I.C.) Diam., Inch
Letter	Length, Inch	Letter	Length, Inch		
A	4.000	M	4.000	.015	¼ or ⁵⁄₁₆
B	4.500	N	4.500		
C	5.000	P	5.000	.031	⅜ or ½
D	6.000	R	6.000		
E	7.000	S	7.000		
F	8.000	T	8.000	.047	⅝ or ¾
G	5.500	U	5.500		
H	9.000	V	9.000		
J	Unassigned	W	Unassigned		
K	12.000	X	12.000	.094	I
L	14.000	Y	14.000		
		Z	Length Only		

* Common usage, but not standard.

U*, straight-shank deep grooving tool; **V***, straight-shank end cutting tool for threading and grooving; and **W***, offset shank with 10° side cutting edge angle.

4. *Rake Angle:* The rake angle of the insert seat orientation on the holder is designated by a letter: **N** for negative rake; **P**, positive rake; and **H*** or **O***, neutral rake.

5. *Hand of tool:* The hand of the tool is designated by a letter: **R** for right-hand; **L**, left-hand; and **N**, neutral, or either hand.

6. *Holder Size:* The holder size or shank size is given by a significant two-digit number which indicates the holder or shank cross section. For shanks ⅝-inch square, and over, the numbers represent the number of sixteenths-inch of width and height. For shanks under ⅝-inch square the number of sixteenths-inch of cross section will be preceded by a zero. For rectangular holders the first digit represents the number of eighth inches of width and the second digit the number of quarter inches of height, except the following holder: 1¼ × 1½, which is given the number 91.

7. *Insert Size:* For triangular, square, diamond, and round inserts, the insert size shall be specified in eighths of an inch of inscribed circle (I.C.) radius. For rectangle and parallelogram inserts two digits are required: first digit denotes number of eighth inches, in width; second digit denotes number of one-quarter inches, in height.

8. *Position Accuracy (Optional):* Used to designate the tolerance or qualified accuracy of the cutting-tool point, or nose, with respect to the shank, as determined by a measurement over the cutting point, or nose radius, of a master insert. These tolerances are designated by a letter symbol, which is used in the eighth position when appropriate. The letter symbols are shown in Table 2.

9. *Shank Modification (Not shown, and optional):* Modifications to the shank for mounting the holder in NC machines are designated by a letter symbol supplied by the holder manufacturer and used only when appropriate.

Selecting Indexable Insert Holders. — A guide for selecting indexable insert holders is provided by Table 3. Some operations such as deep grooving, cut-off, and threading are not given in this table. However, tool holders designed specifically for these operations are available. The boring operations listed in Table 3 refer primarily to larger holes, into which the holders will fit. Smaller holes are bored using boring bars. An examination of this table shows that several tool-holder styles can be used and frequently are used for each operation. However, it is often necessary to select the best holder for a given job. This depends largely on the job and there are certain basic facts that should be considered in making the selection.

Rake Angle: A negative-rake insert has twice as many cutting edges available as a comparable positive-rake insert. In some cases the tool life obtained when using the second face may be less than that obtained on the first face because the tool wear on the cutting edges of the first face may reduce the insert strength. Nevertheless, the advantage of negative-rake inserts and holders is such that they should be considered first in making any choice. Positive-rake holders should be used where lower cutting faces are required, as when machining slender or small-diameter parts, when chatter may occur, and for machining some materials, such as aluminum, copper, and certain grades of stainless steel. In these cases positive-negative rake inserts can sometimes be used to advantage. These are inserts held on negative-rake holders that have their rake surfaces ground or molded to form a positive-rake angle.

Insert Shape: The configuration of the workpiece, the operation to be performed, and the lead angle required often determine the insert shape. When these factors need not be considered, the insert shape should be selected on the basis of insert

* Common usage, but not standard.

Table 3. Indexable Insert Holder Application Guide

Tool	Tool Holder Style	Insert Shape	Rake N-Negative P-Positive	Application								
				Turn	Face	Turn and Face	Turn and Backface	Trace	Groove	Chamfer	Bore	Plane
0°	A	T	N	●	●						●	
			P	●	●						●	
0°	A	T	N	●	●			●				
			P	●	●			●				
	A	R	N	●	●	●						●
	A	R	N	●	●	●		●				●
15°	B	T	N	●	●						●	
			P	●	●						●	
15°	B	T	N	●	●			●			●	
			P	●	●			●			●	
15°	B	S	N	●	●						●	
			P	●	●						●	
5°, 15°	B	C	N	●	●	●					●	●
	C	T	N	●	●				●	●		
			P	●	●				●	●		
45°	D	S	N	●	●	●		●		●	●	●
			P	●	●	●		●		●	●	●
30°	E	T	N	●	●			●	●	●		
			P	●	●			●	●	●		
	F	T	N	●	●						●	
			P	●	●						●	
0°	G	T	N	●	●						●	
			P	●	●						●	

Table 3 (*Continued*). **Indexable Insert Holder Application Guide**

Tool	Tool Holder Style	Insert Shape	Rake N-Negative P-Positive	Turn	Face	Turn and Face	Turn and Backface	Trace	Groove	Chamfer	Bore	Plane
	G	R	N	●	●	●						
	G	C	N	●	●	●						
			P	●	●	●						
	H	D	N	●	●			●				
	J	T	N				●	●				
			P				●	●				
	J	D	N				●	●				
	J	V	N				●	●				
	K	S	N	●	●						●	
			P	●	●						●	
	K	C	N	●	●						●	
	L	C	N			●	●					
	N	T	N	●	●			●				
			P	●	●			●				
	N	D	N	●	●			●				
	S	S	N	●	●	●		●		●	●	●
			P	●	●	●		●		●	●	●
	W	S	N	●	●							

strength and the maximum number of cutting edges available. Thus, a round insert is the strongest and has a maximum number of available cutting edges. It can be used with heavier feeds while producing a good surface finish. Round inserts are limited by their tendency to cause chatter, which may preclude using them. The square insert is the next most effective shape, providing good corner strength and more cutting edges than all other inserts except the round insert. The only limitation of this insert shape is that it must be used with a lead angle. Therefore, the square insert cannot be used for turning square shoulders or for back facing. Triangle inserts are the most versatile and can be used to perform more operations than any other insert shape. The 80-degree diamond insert is designed primarily for heavy turn and face operations, using the 100-degree corners, and for turning and back-facing square shoulders using the 80-degree corners. The 55- and 35-degree diamond inserts are intended primarily for tracing.

Lead Angle: Tool holders should be selected to provide the largest possible lead angle, although limitations are sometimes imposed by the nature of the job. For example, when turning and back-facing a shoulder, a negative lead angle must be used. Slender or small-diameter parts may deflect, causing difficulties in holding size or chatter when the lead angle is too large.

End Cutting Edge Angle: When tracing or contour turning, the plunge angle is determined by the end cutting edge angle. A two-degree minimum clearance angle should be provided between the workpiece surface and the end cutting edge of the insert. Table 4 provides the maximum plunge angle for holders commonly used to plunge when tracing. When severe cratering cannot be avoided, an insert having a small, end cutting edge angle is desirable to delay the crater break-through behind the nose. For very heavy cuts a small, end cutting edge angle will strengthen the corner of the tool.

Table 4. Maximum Plunge Angle for Tracing or Contour Turning

Tool Holder Style	Insert* Shape	Maximum Plunge Angle	Tool Holder Style	Insert* Shape	Maximum Plunge Angle
E	T	58°	J	D	30°
D and S	S	43°	J	V	50°
H	D	71°	N	T	55°
J	T	25°	N	D	58°–60°

* Insert Shape: S = square; T = triangle; D = 55° diamond; V = 35° diamond.

Sharpening Of Twist Drills. — Twist drills are cutting tools designed to perform concurrently several functions, such as penetrating directly into solid material, ejecting the removed chips outside the cutting area, maintaining the essentially straight direction of the advance movement and controlling the size of the drilled hole. The geometry needed for these multiple functions is incorporated into the design of the twist drill in such a manner that it can be retained even after repeated sharpening operations. Twist drills are actually resharpened many times during their service life, with the practically complete restitution of their original operational characteristics. However, in order to assure all the benefits which the design of the twist drill is capable of providing, the surface generated in the sharpening process must agree with the original form of the tool's operating surface, unless a change of shape is deliberately sought because the drill will be used for a different work material.

The principal elements of the tool geometry which are essential for the adequate cutting performance of twist drills are shown in Fig. 1. The generally used values for these dimensions are the following:

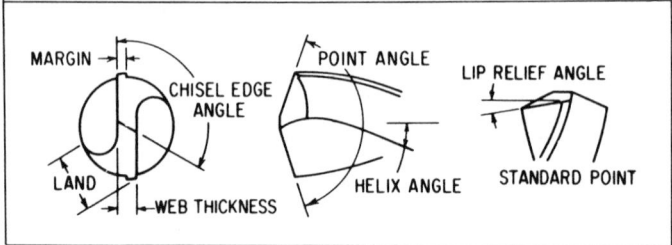

Fig. 1. The principal elements of tool geometry on twist drills.

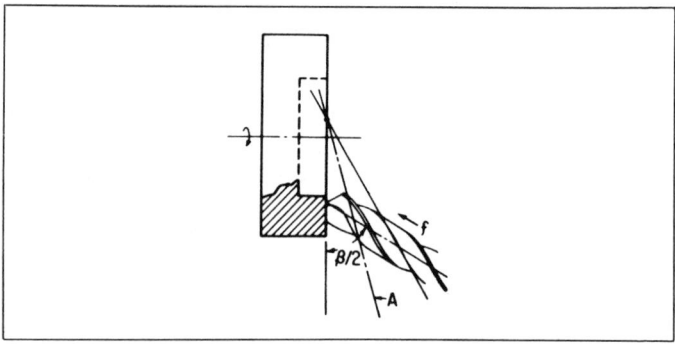

Fig. 2. In grinding the face of the twist drill the tool is swung around the axis A of an imaginary cone, while resting in a support tilted by half of the point angle β with respect to the face of the grinding wheel. Feed f for stock removal is in the direction of the drill axis.

Point angle: Commonly 118°, except for high strength steels, 118° to 135°; aluminum alloys, 90° to 140°; and magnesium alloys, 70° to 118°.

Helix angle: Commonly 24° to 32°, except for magnesium and copper alloys, 10° to 30°.

Lip relief angle: Commonly 10° to 15°, except for high strength or tough steels, 7° to 12°. The lower values of these angle ranges are used for drills of larger diameter, the higher values for the smaller diameters. For drills of diameters less than ¼ inch, the lip relief angles are increased beyond the listed max. values up to 24°. For soft and free machining materials, 12° to 18° except for diameters less than ¼ inch, 20° to 26°.

Relief Grinding of the Tool Flanks. — In the sharpening of the twist drill the tool flanks, containing the two cutting edges are ground. Each of the flanks consists of a curved surface which provides the relief needed for the easy penetration and free cutting of the tool edges. In grinding the flanks, Fig. 2, the drill is swung around the axis A of an imaginary cone while resting in a support which holds the drill at one-half the point angle B with respect to the face of the grinding wheel. Feed f for stock removal is in the direction of the drill axis. That relief angle is

usually measured at the periphery of the twist drill and is also specified by that value. It is not a constant but should increase toward the center of the drill.

The relief grinding of the flank surfaces will generate the chisel angle on the web of the twist drill. The value of that angle, typically 55°, which can be measured, for example, with the protractor of an optical projector, is indicative of the correctness of the relief grinding.

Drill Point Thinning. — The chisel edge is the least efficient operating surface element of the twist drill because it does not cut, but actually squeezes or extrudes the work material. In order to improve the inefficient cutting conditions caused by the chisel edge, its width is often reduced in a drill-point thinning operation, resulting in a condition such as that shown in Fig. 3. Point thinning is particularly desirable on larger size drills and also on those which become shorter in usage, because the thickness of the web increases toward the shaft of the twist drill, thereby adding to the length of the chisel edge. The extent of point thinning is limited by the minimum strength of the web needed to avoid splitting of the drill point under the influence of cutting forces.

Both sharpening operations — the relieved face grinding and the point thinning — should be carried out in special drill grinding machines or with twist drill grinding fixtures mounted on general-purpose tool grinding machines, in order to assure the essential accuracy of the produced tool geometry. Off-hand grinding may be used

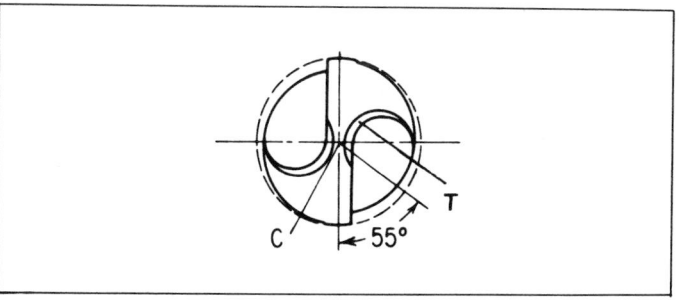

Fig. 3. The chisel edge *C* after thinning the web by grinding off area *T*.

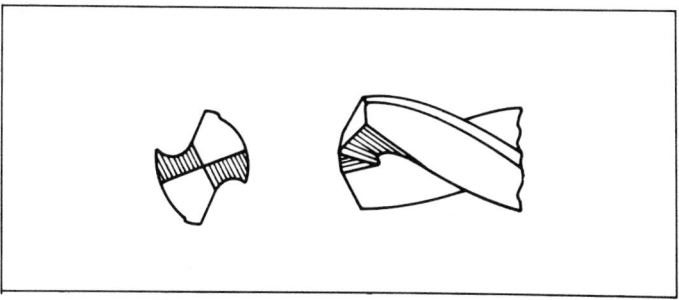

Fig. 4. Split point or "crankshaft" type web thinning.

for the important web thinning when a special machine is not available; however, such operation requires skill and experience.

Improperly sharpened twist drills, e.g. those with unequal edge length or asymmetrical point angle, will tend to produce holes with poor diameter and directional control.

For deep holes and also drilling into stainless steel, titanium alloys, high temperature alloys, nickel alloys, very high strength materials and in some cases tool steels, split point grinding, resulting in a "crankshaft" type drill point, is recommended. In this type of pointing, see Fig. 4, the chisel edge is entirely eliminated, extending the positive rake cutting edges to the center of the drill, thereby greatly reducing the required thrust in drilling.

Sharpening Carbide Tools. — Cemented carbide indexable inserts are usually not resharpened but sometimes they require a special grind in order to form a contour on the cutting edge to suit a special purpose. Brazed type carbide cutting tools are resharpened after the cutting edge has become worn. On brazed carbide tools the cutting-edge wear should not be allowed to become excessive before the tool is resharpened. One method of determining when brazed carbide tools need resharpening is by periodic inspection of the flank wear and the condition of the face. Another method is to determine the amount of production which is normally obtained before excessive wear has taken place, or to determine the equivalent period of time. One disadvantage of this method is that slight variations in the work material will often cause the wear rate not to be uniform and the number of parts machined before regrinding will not be the same each time. Usually, sharpening should not require the removal of more than .005 to .010 inch of carbide.

General Procedure in Carbide Tool Grinding: The general procedure depends upon the kind of grinding operation required. If the operation is to resharpen a dull tool, a diamond wheel of 100 to 120 grain size is recommended although a finer wheel—up to 150 grain size—is sometimes used to obtain a better finish. If the tool is new or is a "standard" design and changes in shape are necessary, a 100-grit diamond wheel is recommended for roughing and a finer grit diamond wheel can be used for finishing. Some shops prefer to rough grind the carbide with a vitrified silicon carbide wheel, the finish grinding being done with a diamond wheel. A final operation commonly designated as lapping may or may not be employed for obtaining an extra-fine finish.

Wheel Speeds: — The speed of silicon carbide wheels usually is about 5000 feet per minute. The speeds of diamond wheels generally range from 5000 to 6000 feet per minute; yet lower speeds (550 to 3000 fpm) can be effective.

Offhand Grinding. — In grinding single-point tools (excepting chip breakers) the common practice is to hold the tool by hand, press it against the wheel face and traverse it continuously across the wheel face while the tool is supported on the machine rest or table which is adjusted to the required angle. This is known as "offhand grinding" to distinguish it from the machine grinding of cutters as in regular cutter grinding practice. The selection of wheels adapted to carbide tool grinding is very important.

Silicon Carbide Wheels. — The green colored silicon carbide wheels generally are preferred to the dark gray or gray-black variety, although the latter are sometimes used.

Grain or Grit Sizes: — For roughing, a grain size of 60 is very generally used. For finish grinding with silicon carbide wheels, a finer grain size of 100 or 120 is common. A silicon carbide wheel such as C60-I-7V may be used for grinding both the steel shank and carbide tip. However, for under-cutting steel shanks up to the carbide tip, it may be advantageous to use an aluminum oxide wheel suitable for high speed steel grinding.

Grade: — According to the standard system of marking, different grades from soft to hard are indicated by letters from A to Z. For carbide tool grinding fairly soft grades such as G, H, I and J are used. The usual grades for roughing are I or J and for finishing H, I and J. The grade should be such that a sharp free-cutting wheel will be maintained without excessive grinding pressure. Harder grades than those indicated tend to overheat and crack the carbide.

Structure: — The common structure numbers for carbide tool grinding are 7 and 8. The larger cup-wheels (10 to 14 inches) may be of the porous type and be designated as 12P. The standard structure numbers range from 1 to 15 with progressively higher numbers indicating less density and more open wheel structure.

Diamond Wheels. — Wheels with diamond-impregnated grinding faces, are fast and cool cutting and have a very low rate of wear. They are used extensively both for resharpening and for finish grinding of carbide tools when preliminary roughing is required. Diamond wheels are also adapted for sharpening multi-tooth cutters such as milling cutters, reamers, etc., which are ground in a cutter grinding machine.

Resinoid bonded wheels are commonly used for grinding chip breakers, milling cutters, reamers or other multi-tooth cutters. They are also applicable to precision grinding of carbide dies, gages, and various external, internal and surface grinding operations. Fast, cool cutting action is characteristic of these wheels.

Metal bonded wheels are often used for offhand grinding of single-point tools especially when durability or long life and resistance to grooving of the cutting face, are considered more important than the rate of cutting. *Vitrified bonded* wheels are used both for roughing of chipped or very dull tools and for ordinary resharpening and finishing. They provide rigidity for precision grinding, a porous structure for fast cool cutting, sharp cutting action and durability.

Diamond Wheel Grit Sizes. — For roughing with diamond wheels a grit size of 100 is the most common both for offhand and machine grinding. Grit sizes of 120 and 150 are frequently used in offhand grinding of single point tools (1) for resharpening, (2) for a combination roughing and finishing wheel and (3) for chip-breaker grinding. Grit sizes of 220 or 240 are used for ordinary finish grinding all types of tools (offhand and machine) and also for cylindrical, internal and surface finish grinding. Grits of 320 and 400 are used for "lapping" to obtain very fine finishes, and for hand hones. A grit of 500 is for lapping to a mirror finish on such work as carbide gages and boring or other tools for exceptionally fine finishes.

Diamond Wheel Grades. — Diamond wheels are made in several different grades to better adapt them to different classes of work. The grades vary for different types and shapes of wheels. Standard Norton grades are H, J and L for resinoid bonded wheels, grade N for metal bonded wheels and grades J, L, N and P for vitrified wheels. Harder and softer grades than standard may at times be used to advantage.

Diamond Concentration. — The relative amount (by carat weight) of diamond in the diamond section of the wheel is known as the "diamond concentration." Concentrations of 100 (high), 50 (medium) and 25 (low) ordinarily are supplied. A concentration of 50 represents one-half the diamond content of 100 (if the depth of the diamond is the same in each case) and 25 equals one-fourth the content of 100 or one-half the content of 50 concentration.

100 Concentration: — Recommended (especially in grit sizes up to about 220) for general machine grinding of carbides, and for grinding cutters and chip breakers. Vitrified and metal bonded wheels usually have 100 concentration.

50 Concentration: — In the finer grit sizes of 220, 240, 320, 400 and 500, a 50 concentration is recommended for offhand grinding with resinoid bonded cup-wheels.

25 Concentration: — A low concentration of 25 is recommended for offhand grinding with resinoid bonded cup-wheels with grit sizes of 100, 120 and 150.

Depth of Diamond Section: — The radial depth of the diamond section usually varies from 1/16 to 1/4 inch. The depth varies somewhat according to the wheel size and type of bond.

Dry Versus Wet Grinding of Carbide Tools. — In using silicon carbide wheels, grinding should be done either absolutely dry or with enough coolant to flood the wheel and tool. Satisfactory results may be obtained either by the wet or dry method. However, dry grinding is the most prevalent usually because, in wet grinding, operators tend to use an inadequate supply of coolant to obtain better visibility of the grinding operation and avoid getting wet; hence checking or cracking in many cases is more likely to occur in wet grinding than in dry grinding.

Wet Grinding with Silicon Carbide Wheels: — One advantage commonly cited in connection with wet grinding is that an ample supply of coolant permits using wheels about one grade harder than in dry grinding thus increasing the wheel life. Plenty of coolant also prevents thermal stresses and the resulting cracks, and there is less tendency for the wheel to load. A dust exhaust system also is unnecessary.

Wet Grinding with Diamond Wheels: — In grinding with diamond wheels the general practice is to use a coolant to keep the wheel face clean and promote free cutting. The amount of coolant may vary from a small stream to a coating applied to the wheel face by a felt pad.

Coolants for Carbide Tool Grinding. — In grinding either with silicon carbide or diamond wheels a coolant that is used extensively consists of water plus a small amount either of soluble oil, sal soda, or soda ash to prevent corrosion. One prominent manufacturer recommends for silicon carbide wheels about 1 ounce of soda ash per gallon of water and for diamond wheels kerosene. The use of kerosene is quite general for diamond wheels and usually it is applied to the wheel face by a felt pad. Another coolant recommended for diamond wheels consists of 80 per cent water and 20 per cent soluble oil.

Peripheral Versus Flat Side Grinding. — In grinding single point carbide tools with silicon carbide wheels, the roughing preparatory to finishing with diamond wheels may be done either by using the flat face of a cup-shaped wheel (side grinding) or the periphery of a "straight" or disk-shaped wheel. Even where side grinding is preferred, the periphery of a straight wheel may be used for heavy roughing

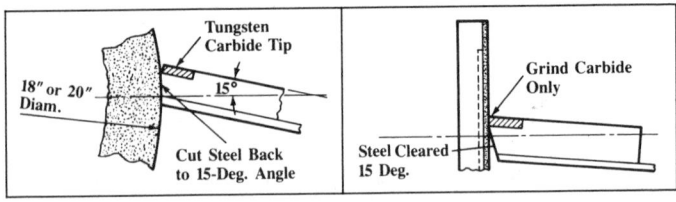

as in grinding back chipped or broken tools (see left-hand diagram). Reasons for preferring peripheral grinding include faster cutting with less danger of localized heating and checking especially in grinding broad surfaces. The advantages usually claimed for side grinding are that proper rake or relief angles are easier to obtain and

the relief or land is ground flat. The diamond wheels used for tool sharpening are designed for side grinding. (See right-hand diagram.)

Lapping Carbide Tools. — Carbide tools may be finished by lapping, especially if an exceptionally fine finish is required on the work as, for example, tools used for precision boring or turning non-ferrous metals. If the finishing is done by using a diamond wheel of very fine grit (such as 240, 320 or 400), the operation is often called "lapping." A second lapping method is by means of a power-driven lapping disk charged with diamond dust, Norbide powder, or silicon carbide finishing compound. A third method is by using a hand lap or hone usually of 320 or 400 grit. In many plants the finishes obtained with carbide tools meet requirements without a special lapping operation. In all cases any feather edge which may be left on tools should be removed and it is good practice to bevel the edges of roughing tools at 45 degrees to leave a chamfer 0.005 to 0.010 inch wide. This is done by hand honing and the object is to prevent crumbling or flaking off at the edges when hard scale or heavy chip pressure is encountered.

Hand Honing: The cutting edge of carbide tools, and tools made from other tool materials, is sometimes hand honed before it is used in order to strengthen the cutting edge. When interrupted cuts or heavy roughing cuts are to be taken, or when the grade of carbide is slightly too hard, hand honing is beneficial because it will prevent chipping, or even possibly, breakage of the cutting edge. Whenever chipping is encountered, hand honing the cutting edge before use will be helpful. It is important, however, to hone the edge lightly and only when necessary. Heavy honing will always cause a reduction in tool life. Normally, removing .002 to .004 inch from the cutting edge is sufficient. When indexable inserts are used, the use of pre-honed inserts is preferred to hand honing although sometimes an additional amount of honing is required. Hand honing of carbide tools in between cuts is sometimes done to defer grinding or to increase the life of a cutting edge on an indexable insert. If correctly done, so as not to change the relief angle, this procedure is sometimes helpful. If improperly done, it can result in a reduction in tool life.

Chip Breaker Grinding. — For this operation a straight diamond wheel is used on a universal tool and cutter grinder, a small surface grinder, or a special chipbreaker grinder. A resinoid bonded wheel of the grade J or N commonly is used and the tool is held rigidly in an adjustable holder or vise. The width of the diamond wheel usually varies from ⅛ to ¼ inch. A vitrified bond may be used for wheels as thick as ¼ inch, and a resinoid bond for relatively narrow wheels.

Summary of Miscellaneous Points. — In grinding a single-point carbide tool, traverse it across the wheel face continuously to avoid localized heating. This traverse movement should be quite rapid in using silicon carbide wheels and comparatively slow with diamond wheels. A hand traversing and feeding movement, whenever practicable, is generally recommended because of greater sensitivity. In grinding, maintain a constant, moderate pressure. Never cool a hot tool by dipping it in a liquid, as this may crack the tip. Wheel rotation should preferably be *against* the cutting edge or from the front face toward the back. If the grinder is driven by a reversing motor, opposite sides of a cup wheel can be used for grinding right- and left-hand tools and with rotation against the cutting edge. If it is necessary to grind the top face of a single-point tool, this should precede the grinding of the side and front relief, and top-face grinding should be minimized to maintain the tip thickness. In machine grinding with a diamond wheel, limit the feed per traverse to 0.001 inch for 100 to 120 grit; 0.0005 inch for 150 to 240 grit; and 0.0002 inch for 320 grit and finer.

Single-Point Turning and Boring Tools — Rake and Relief Angles-1

Note: All rake and relief angles are measured in normal direction.

*Use the largest nose radius and the largest side cutting edge angle or end cutting edge angle that are consistent with part requirements.

Material	Hardness Bhn	High Speed Steel				Carbide					
						Brazed			Indexable		
		Back Rake Angle degrees	Side Rake Angle degrees	End Relief Angle degrees	Side Relief Angle degrees	Back Rake Angle degrees	Side Rake Angle degrees	Relief Angles degrees	Back Rake Angle degrees	Side Rake Angle degrees	Relief Angles degrees
Free Machining Carbon Steels-Wrought	85-225	10	12	5	5	0	6	7	0	5	5
Carbon Steels-Wrought and Cast	225-325	8	10	5	5	0	6	7	0	5	5
Free Machining Alloy Steels-Wrought	325-52Rc	0	10	5	5	0	6	7	-5	-5	5
Alloy Steels-Wrought and Cast; High Strength Steels-Wrought; Maraging Steels-Wrought; Tool Steels-Wrought; Nitriding Steels-Wrought; Armor Plate-Wrought; Structural Steels-Wrought	52Rc-58Rc	—	—	—	—	—	—	—	-5	-5	5
Free Machining Stainless Steels-Wrought	135-275	5	8	5	5	0	6	7	-5	-5	5
	275-425	0	10	5	5	0	6	7	-5	-5	5
Stainless Steels, Ferritic-Wrought and Cast	135-185	5	8	5	5	0	6	7	0	5	5
Stainless Steels, Austenitic-Wrought and Cast	135-275	0	10	5	5	0	6	7	0	5	5
Stainless Steels, Martensitic-Wrought and Cast	325-425 48Rc-52Rc	0	10	5	5	0	6	7	-5	5	5

Reprinted from the MACHINING DATA HANDBOOK, 3rd Edition, by permission of the Machinability Data Center. © 1980 by Metcut Research Associates Inc.

Single-Point Turning and Boring Tools—Rake and Relief Angles-2

| | | High Speed Steel | | | | Carbide | | | | | |
| | | | | | | Brazed | | | Indexable | | |
Material	Hardness Bhn	Back Rake Angle degrees	Side Rake Angle degrees	End Relief Angle degrees	Side Relief Angle degrees	Back Rake Angle degrees	Side Rake Angle degrees	Relief Angles degrees	Back Rake Angle degrees	Side Rake Angle degrees	Relief Angles degrees
Precipitation Hardening Stainless Steels-Wrought and Cast	150-450	0	10	5	5	0	6	7	-5	-5	5
Gray Cast Irons Ductile Cast Irons Malleable Cast Irons Compacted Graphite Cast Irons	100-200 200-300	5 5	10 8	5 5	5 5	0 0	6 6	7 7	-5 -5	-5 -5	5 5
White Cast Irons	300-400	5	5	5	5	-5	-5	7	-5	-5	5
Aluminum Alloys-Wrought and Cast	30-150 500 kg	20	15	12	10	3	15	7	0	5	5
Magnesium Alloys-Wrought and Cast	40-90 500 kg	20	15	12	10	3	15	7	0	5	5
Titanium Alloys-Wrought and Cast	110-440	0	5	5	5	0	6	7	-5	-5	5
Copper Alloys-Wrought and Cast	40-200 500 kg	5	10	8	8	0	8	7	0	5	5
Nickel Alloys-Wrought and Cast Chrome-Nickel Alloys Beryllium-Nickel Alloys	80-360	8	10	12	12	0	6	7	-5	-5	5
Nitinol Alloys-Wrought	210-340 48Rc-52Rc	—	—	—	—	0	5	7	0	5	5
High Temperature Alloys-Wrought and Cast	140-475	0	10	5	5	0	6	7	5	0	5

Single-Point Turning and Boring Tools—Rake and Relief Angles-3

Material	Hardness									
Columbium Alloys-Wrought, Cast, P/M; Molybdenum Alloys-Wrought, Cast, P/M; Tantalum Alloys-Wrought, Cast, P/M	170-290	0	20	5	5	0	20	7	—	—	5
Tungsten Alloys-Wrought, Cast, P/M	180-320	—	—	—	—	-15	0	7	—	—	—
Zinc Alloys-Cast	80-100	10	10	12	4	5	5	7	0	5	5
Uranium Alloys-Wrought	190-210	—	—	—	—	0	0	7	-5	0	7
Zirconium Alloys-Wrought	140-280	15	10	10	10	5	5	7	5	5	6
Thermoplastics	All	0	0	20 to 30	15 to 20	0	0	20 to 30	0	0	20 to 30
Thermosetting Plastics	All	0	0	20 to 30	15 to 20	0	15	7	0	15	5
Magnetic Alloys, Nickel- and Cobalt-Base Controlled Expansion Alloys	125-250	10	8	8	8	—	—	—	—	—	—
Powder Metal Alloys — Copper	All	10	8	8	8	6	12	7	0	0	—
Powder Metal Alloys — Iron	All	0	0	8	8	6	16	7	0	5	5
Magnetic Core Iron	185-240	15	30	20	20	20	0	20	5	5	15
Carbon and Graphite	All	0	0	5	5	0	0	7	0	0	5
Machinable Carbide (Ferro-Tic)	40Rc-51Rc	-5	-5	5	5	-5	-5	7	-5	-5	5
Machinable Glass Ceramic	250 Knoop 100 g	0	15	5	5	1	0	7	0	0	5

Reprinted from the MACHINING DATA HANDBOOK, 3rd Edition, by permission of the Machinability Data Center. © 1980 by Metcut Research Associates Inc.

Arbor-Mounted Side and Slot Mills—Rake and Relief Angles

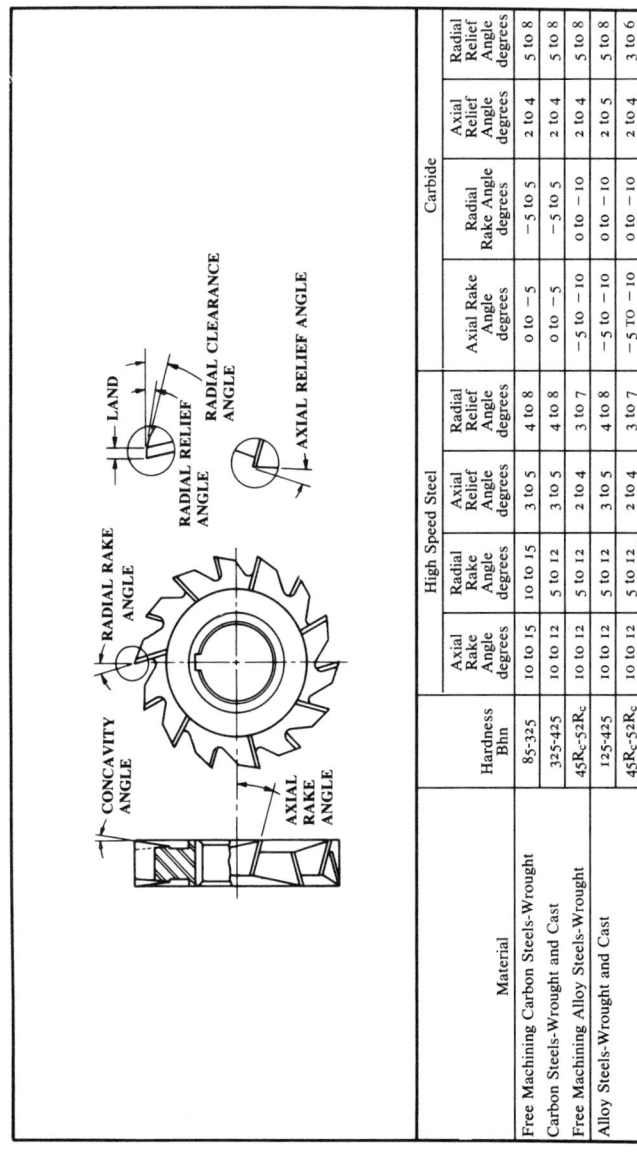

Material	Hardness Bhn	High Speed Steel				Carbide			
		Axial Rake Angle degrees	Radial Rake Angle degrees	Axial Relief Angle degrees	Radial Relief Angle degrees	Axial Rake Angle degrees	Radial Rake Angle degrees	Axial Relief Angle degrees	Radial Relief Angle degrees
Free Machining Carbon Steels-Wrought	85-325	10 to 15	10 to 15	3 to 5	4 to 8	0 to −5	−5 to 5	2 to 4	5 to 8
Carbon Steels-Wrought and Cast	325-425	10 to 12	5 to 12	3 to 5	4 to 8	0 to −5	−5 to 5	2 to 4	5 to 8
Free Machining Alloy Steels-Wrought	$45R_c$–$52R_c$	10 to 12	5 to 12	2 to 4	3 to 7	−5 to −10	0 to −10	2 to 4	5 to 8
Alloy Steels-Wrought and Cast	125-425	10 to 12	5 to 12	3 to 5	4 to 8	−5 to −10	0 to −10	2 to 5	5 to 8
	$45R_c$–$52R_c$	10 to 12	5 to 12	2 to 4	3 to 7	−5 TO −10	0 to −10	2 to 4	3 to 6

Reprinted from the MACHINING DATA HANDBOOK, 3rd Edition, by permission of the Machinability Data Center. © 1980 by Metcut Research Associates Inc.

Material	Hardness								
High Strength Steels, Maraging Steels and Tool Steels-Wrought	100-52R$_c$	10 to 12	5 to 12	2 to 4	3 to 7	-5 to -10	0 to -10	2 to 4	5 to 8
Nitriding Steels-Wrought	200-350	10 to 12	5 to 12	2 to 4	3 to 7	-5 to -10	0 to -10	2 to 4	3 to 6
Armor Plate-Wrought	250-320	0 to 5	0 to 5	2 to 4	3 to 7	-5 to -10	-5 to -10	2 to 4	3 to 6
Structural Steels-Wrought	100-50R$_c$	10 to 12	5 to 12	3 to 5	4 to 8	0 to -5	0 to -10	2 to 4	5 to 8
Free Machining Stainless Steels-Wrought	135-425	10 to 12	5 to 12	3 to 5	4 to 8	0 to 5	-5 to 5	2 to 4	5 to 8
Stainless Steels, Ferritic-Wrought and Cast / Stainless Steels, Austenitic-Wrought and Cast	135-52R$_c$	10 to 12	5 to 12	3 to 5	4 to 8	0 to 5	-5 to 5	2 to 4	5 to 8
Stainless Steels, Martensitic-Wrought and Cast	135-52R$_c$	10 to 12	5 to 12	2 to 4	3 to 7	-5 to -10	0 to -10	2 to 4	5 to 8
Precipitation Hardening Stainless Steels-Wrought and Cast	150-450	10 to 12	5 to 12	2 to 4	4 to 8	0 to -5	0 to -10	2 to 4	5 to 8
Gray Cast Irons / Ductile Cast Irons / Malleable Cast Irons	100-400	10 to 12	10 to 20	2 to 4	3 to 7	0 to -10	5 to -10	3 to 5	5 to 8
Aluminum Alloys-Wrought and Cast	30-150 500 kg	12 to 25	10 to 20	5 to 7	5 to 11	10 to 20	5 to 15	5 to 7	7 to 10
Magnesium Alloys-Wrought and Cast	40-90 500 kg	12 to 25	10 to 20	5 to 7	5 to 11	10 to 20	5 to 15	5 to 7	7 to 10
Titanium Alloys-Wrought	110-440	10 to 15	5 to 10	5 to 7	5 to 11	0 to -10	0 to -10	5 to 7	5 to 8
Copper Alloys-Wrought and Cast	40-200 500 kg	12 to 25	10 to 20	5 to 7	5 to 11	10 to 20	5 to 10	4 to 7	5 to 8
Nickel Alloys-Wrought and Cast	80-360	10 to 20	10 to 15	3 to 5	4 to 8	-5 to -10	0 to -10	3 to 5	5 to 8
High Temperature Alloys-Wrought and Cast	140-300 / 300 to 475	10 to 15	10 to 15	1 to 5	5 to 10	-5 to -10	0 to -10	3 to 5	5 to 8
Columbium, Molybdenum Alloys-Wrought, Cast, P/M	170-290	0	15 to 20	3 to 5	5 to 10	0	5 to 15	7 to 10	7 to 10
Tantalum Alloys-Wrought, Cast, P/M	200-250	0	15 to 20	3 to 5	5 to 10	0	5 to 15	7 to 10	7 to 10
Tungsten Alloys-Wrought, Cast, P/M	180-320	—	—	—	—	-10 to -15	5 to 15	10 to 15	10 to 15
Zinc Alloys-Cast	80-100	10 to 20	10 to 20	5 to 7	8 to 11	10 to 15	10 to 15	7 to 10	7 to 10

Reprinted from the MACHINING DATA HANDBOOK, 3rd Edition, by permission of the Machinability Data Center. © 1980 by Metcut Research Associates Inc.

High-Speed-Steel Peripheral and Slotting End Mills—Tool Geometry

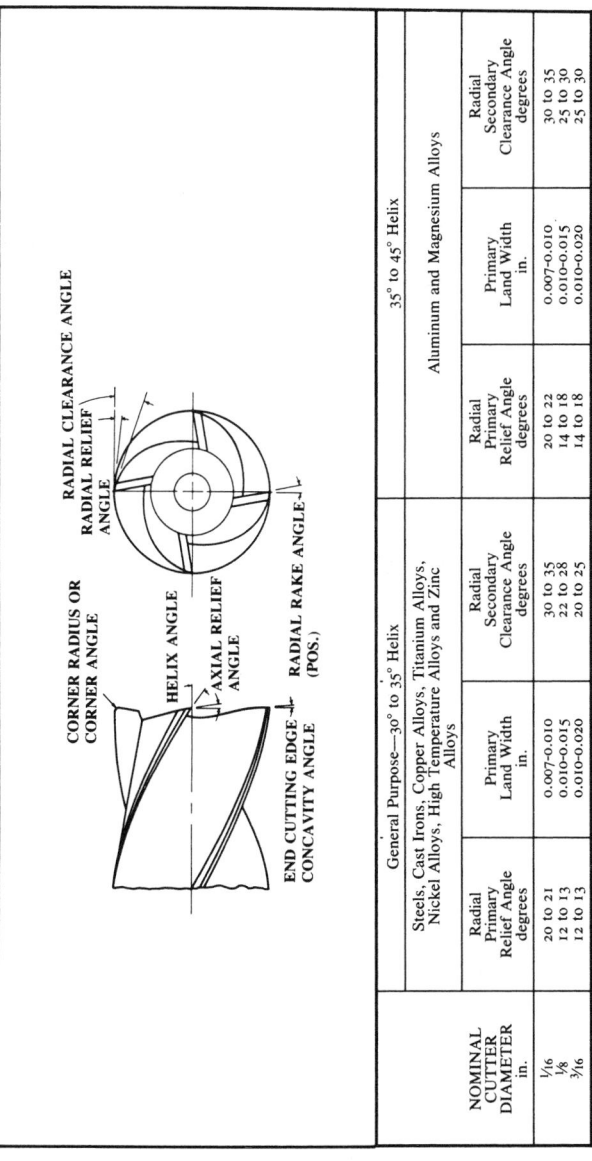

CORNER RADIUS OR
CORNER ANGLE

RADIAL CLEARANCE ANGLE

RADIAL RELIEF
ANGLE

HELIX ANGLE

AXIAL RELIEF
ANGLE

END CUTTING EDGE
CONCAVITY ANGLE

RADIAL RAKE ANGLE
(POS.)

NOMINAL CUTTER DIAMETER in.	General Purpose—30° to 35° Helix				35° to 45° Helix			
	Steels, Cast Irons, Copper Alloys, Titanium Alloys, Nickel Alloys, High Temperature Alloys and Zinc Alloys				Aluminum and Magnesium Alloys			
	Radial Primary Relief Angle degrees	Primary Land Width in.	Radial Secondary Clearance Angle degrees		Radial Primary Relief Angle degrees	Primary Land Width in.	Radial Secondary Clearance Angle degrees	
1/16	20 to 21	0.007-0.010	30 to 35		20 to 22	0.007-0.010	30 to 35	
1/8	12 to 13	0.010-0.015	22 to 28		14 to 18	0.010-0.015	25 to 30	
3/16	12 to 13	0.010-0.020	20 to 25		14 to 18	0.010-0.020	25 to 30	

Reprinted from the MACHINING DATA HANDBOOK, 3rd Edition, by permission of the Machinability Data Center. © 1980 by Metcut Research Associates Inc.

High-Speed-Steel Peripheral and Slotting End Mills—Tool Geometry

	degrees	mm	degrees	degrees	mm	degrees
1/4	10 to 11	0.010-0.020	20 to 25	12 to 15	0.010-0.020	22 to 28
5/16	10 to 11	0.015-0.025	20 to 25	12 to 14	0.015-0.025	21 to 28
3/8	10 to 11	0.015-0.025	17 to 20	12 to 14	0.015-0.025	19 to 26
7/16	9 to 10	0.020-0.030	17 to 20	11 to 13	0.020-0.030	18 to 25
1/2	9 to 10	0.020-0.030	17 to 20	11 to 13	0.020-0.030	18 to 25
5/8	9 to 10	0.025-0.035	17 to 20	11 to 13	0.025-0.035	18 to 25
3/4	8 to 9	0.030-0.040	15 to 18	10 to 12	0.030-0.040	17 to 24
7/8	8 to 9	0.030-0.040	15 to 18	10 to 12	0.030-0.040	17 to 24
1	8 to 9	0.035-0.050	15 to 18	10 to 12	0.035-0.050	16 to 23
1 1/4	7 to 8	0.040-0.060	13 to 18	9 to 11	0.040-0.060	14 to 22
1 1/2	7 to 8	0.040-0.060	11 to 17	9 to 11	0.040-0.060	13 to 21
1 3/4	7 to 8	0.040-0.060	10 to 16	8 to 10	0.040-0.060	12 to 20
2	6 to 7	0.040-0.060	9 to 15	8 to 10	0.040-0.060	12 to 20
mm	**degrees**	**mm**	**degrees**	**degrees**	**mm**	**degrees**
M1.6	20 to 21	0.200-0.250	30 to 35	20 to 22	0.200-0.250	30 to 35
M3	14 to 15	0.250-0.350	24 to 30	15 to 20	0.250-0.350	25 to 30
M4	12 to 13	0.250-0.400	22 to 28	14 to 18	0.250-0.400	25 to 30
M6	10 to 11	0.250-0.500	20 to 25	12 to 15	0.250-0.500	22 to 28
M7	10 to 11	0.250-0.500	20 to 25	12 to 15	0.250-0.500	22 to 28
M8	10 to 11	0.400-0.650	20 to 25	12 to 14	0.400-0.650	19 to 26
M10	10 to 11	0.400-0.650	17 to 20	12 to 14	0.400-0.650	19 to 26
M12	9 to 10	0.500-0.750	17 to 20	11 to 13	0.500-0.750	18 to 25
M14	9 to 10	0.500-0.750	17 to 20	11 to 13	0.500-0.750	18 to 25
M16	9 to 10	0.650-0.900	17 to 20	11 to 13	0.650-0.900	18 to 25
M20	8 to 9	0.750-1.000	15 to 18	10 to 12	0.750-1.000	17 to 24
M22	8 to 9	0.75-1.000	15 to 18	10 to 12	0.75-1.000	17 to 24
M25	8 to 9	0.900-1.250	15 to 18	10 to 12	0.900-1.250	16 to 23
M32	7 to 8	1.000-1.500	13 to 18	9 to 11	1.000-1.500	14 to 22
M40	7 to 8	1.000-1.500	11 to 17	9 to 11	1.000-1.500	13 to 21
M45	7 to 8	1.000-1.500	10 to 16	8 to 10	1.000-1.500	12 to 20

Reprinted from the MACHINING DATA HANDBOOK, 3rd Edition, by permission of the Machinability Data Center. © 1980 by Metcut Research Associates Inc.

High-Speed-Steel Twist Drills—Tool Geometry

STANDARD POINT

CRANKSHAFT POINT

Material	Hardness Bhn	Drill Type	Point Angle* degrees	Lip Relief Angle‡ degrees	Helix Angle	Point Grind
Free Machining Carbon Steels-Wrought Carbon Steels-Wrought, Cast and P/M Free Machining Alloy Steels-Wrought Alloy Steels-Wrought, Cast and P/M Maraging Steels-Wrought	85-225	General Purpose	118	A	Standard	Standard
Tool Steels-Wrought, Cast and P/M	225-325	General Purpose	118	A	Standard	Standard
Nitriding Steels-Wrought	325-425	General Purpose	118 to 135	B	Standard	Crankshaft
Armor Plate-Wrought Structural Steels-Wrought	45R$_c$-52R$_c$	Heavy Web	118 to 135	B	Standard	Crankshaft
High Strength Steels-Wrought	175-325 325-52R$_c$	Heavy Web	118 118 to 135	A B	Standard Standard	Crankshaft

Reprinted from the MACHINING DATA HANDBOOK, 3rd Edition, by permission of the Machinability Data Center. © 1980 by Metcut Research Associates Inc.

Material							
Austenitic Manganese Steels-Cast	150-220	Rail Drill	135	B	135	Low	Split
Free Machining Stainless Steels-Wrought	135-425	General Purpose	118	A	118	Standard	Crankshaft
Stainless Steels, Ferritic-Wrought and Cast	135-200	General Purpose	118 to 135	A	118 to 135	Standard	Standard
Stainless Steels, Austenitic-Wrought, Cast and P/M	200-325	General Purpose	118 to 135	A	118 to 135	Standard	Crankshaft
Stainless Steels, Martensitic-Wrought, Cast and P/M Precipitation Hardening Stainless Steels-Wrought and Cast	325-425 48Rc-52Rc	Heavy Web	118 to 135	B	118 to 135	Standard	Crankshaft
Gray Cast Irons Ductile Cast Irons	110-220	General Purpose	118	A	118	Standard	Standard
Malleable Cast Irons Compacted Graphite Cast Iron White Cast Iron	220-400	Heavy Web	118	A	118	Standard	Standard
Aluminum Alloys-Wrought and Cast	30-150 500 kg	Polished Flutes	90 to 118	C	90 to 118	High	Standard
Magnesium Alloys-Wrought and Cast	40-90 500 kg	Polished Flutes	70 to 118	C	70 to 118	High	Standard
Titanium Alloys-Wrought and Cast	110-275	General Purpose	118 to 135	B	118 to 135	Standard	Crankshaft
	275-440	Heavy Web	118 to 135	B	118 to 135	Standard	Crankshaft
Copper Alloys-Wrought, Cast and P/M	40-200 500 kg	Polished Flutes	118	C	118	Low	Standard
Nickel Alloys-Wrought, Cast and P/M Chromium-Nickel Alloys-Cast	80-360	General Purpose	118	B	118	Standard	Crankshaft
Nitinol Alloys-Wrought	210-360 48Rc-52Rc	General Purpose	118	B	118	Standard	Crankshaft
High Temperature Alloys-Wrought and Cast	140-475	Heavy Web	118 to 135	B	118 to 135	Standard	Crankshaft
Columbium, Molybdenum and Tantalum Alloys-Wrought, Cast, P/M	170-290	General Purpose	118	B	118	Standard	Standard

Tungsten Alloys (Anviloy)†	290-320	General Purpose	118	B	Standard	Standard
Zinc Alloys-Cast	80-100	General Purpose	118	C	Standard	Standard
Uranium Alloys-Wrought§	190-210	Special Carbide	118	5-8	20°	−5° Land on Drill Lip
Zirconium Alloys-Wrought	140-280	General Purpose	118	A	Standard	Crankshaft
Thermoplastics and Thermosetting Plastics†	All	Special Polished	60 to 90	C	Low	Standard
Magnetic Core Iron	185-240	General Purpose	100 to 118	A	High	Standard
Controlled Expansion Alloys	125-250	General Purpose	118	A	Standard	Crankshaft
Magnetic Alloys (Hi Perm 49, HyMu 80)	185-240	General Purpose	118	A	Standard	Crankshaft
Carbon and Graphite	8-100 Shore	General Purpose	90 to 118	C	Standard	Crankshaft
Machinable Carbide (Ferro-Tic)	40Rc-51Rc	General Purpose	118	B	Standard	Crankshaft

NOTE: Use stub-length drills whenever possible on high strength materials.

*Chisel edge angle: 115° to 135°

†For both hss and carbide drills.

§For carbide drills.

‡See following chart.

Lip Relief Angles at Periphery	Drill Size							
	#80 to #61	#60 to #41	#40 to #31	1/8 to 1/4	1/4 to 3/8	3/8 to 1/2	1/2 to 3/4	1 inch up
A	24°	21°	18°	16°	14°	12°	10°	8°
B	20°	18°	16°	14°	12°	10°	8°	7°
C	26°	24°	22°	20°	18°	16°	14°	12°

Length of Point on Twist Drills and Centering Tools

Size of Drill	Decimal Equiva-lent	Length of Point when Included Angle = 90°	Length of Point when Included Angle = 118°	Size of Drill	Decimal Equiva-lent	Length of Point when Included Angle = 90°	Length of Point when Included Angle = 118°	Size or Diam. of Drill	Decimal Equiva-lent	Length of Point when Included Angle = 90°	Length of Point when Included Angle = 118°	Diam. of Drill	Decimal Equiva-lent	Length of Point when Included Angle = 90°	Length of Point when Included Angle = 118°
60	0.0400	0.020	0.012	37	0.1040	0.052	0.031	14	0.1820	0.091	0.055	3/8	0.3750	0.188	0.113
59	0.0410	0.021	0.012	36	0.1065	0.054	0.032	13	0.1850	0.093	0.056	25/64	0.3906	0.195	0.117
58	0.0420	0.021	0.013	35	0.1100	0.055	0.033	12	0.1890	0.095	0.057	13/32	0.4063	0.203	0.122
57	0.0430	0.022	0.013	34	0.1110	0.056	0.033	11	0.1910	0.096	0.057	27/64	0.4219	0.211	0.127
56	0.0465	0.023	0.014	33	0.1130	0.057	0.034	10	0.1935	0.097	0.058	7/16	0.4375	0.219	0.131
55	0.0520	0.026	0.016	32	0.1160	0.058	0.035	9	0.1960	0.098	0.059	29/64	0.4531	0.227	0.136
54	0.0550	0.028	0.017	31	0.1200	0.060	0.036	8	0.1990	0.100	0.060	15/32	0.4688	0.234	0.141
53	0.0595	0.030	0.018	30	0.1285	0.065	0.039	7	0.2010	0.101	0.060	31/64	0.4844	0.242	0.145
52	0.0635	0.032	0.019	29	0.1360	0.068	0.041	6	0.2040	0.102	0.061	1/2	0.5000	0.250	0.150
51	0.0670	0.034	0.020	28	0.1405	0.070	0.042	5	0.2055	0.103	0.062	33/64	0.5156	0.258	0.155
50	0.0700	0.035	0.021	27	0.1440	0.072	0.043	4	0.2090	0.105	0.063	17/32	0.5313	0.266	0.159
49	0.0730	0.037	0.022	26	0.1470	0.074	0.044	3	0.2130	0.107	0.064	35/64	0.5469	0.273	0.164
48	0.0760	0.038	0.023	25	0.1495	0.075	0.045	2	0.2210	0.111	0.067	9/16	0.5625	0.281	0.169
47	0.0785	0.040	0.024	24	0.1520	0.076	0.046	1	0.2280	0.114	0.068	37/64	0.5781	0.289	0.173
46	0.0810	0.041	0.024	23	0.1540	0.077	0.046	15/64	0.2344	0.117	0.070	19/32	0.5938	0.297	0.178
45	0.0820	0.041	0.025	22	0.1570	0.079	0.047	1/4	0.2500	0.125	0.075	39/64	0.6094	0.305	0.183
44	0.0860	0.043	0.026	21	0.1590	0.080	0.048	17/64	0.2656	0.133	0.080	5/8	0.6250	0.313	0.188
43	0.0890	0.045	0.027	20	0.1610	0.081	0.048	9/32	0.2813	0.141	0.084	41/64	0.6406	0.320	0.192
42	0.0935	0.047	0.028	19	0.1660	0.083	0.050	19/64	0.2969	0.148	0.089	21/32	0.6563	0.328	0.197
41	0.0960	0.048	0.029	18	0.1695	0.085	0.051	5/16	0.3125	0.156	0.094	43/64	0.6719	0.336	0.202
40	0.0980	0.049	0.029	17	0.1730	0.087	0.052	21/64	0.3281	0.164	0.098	11/16	0.6875	0.344	0.206
39	0.0995	0.050	0.030	16	0.1770	0.089	0.053	11/32	0.3438	0.171	0.103	23/32	0.7188	0.359	0.216
38	0.1015	0.051	0.030	15	0.1800	0.090	0.054	23/64	0.3594	0.180	0.108	3/4	0.7500	0.375	0.225

Cutting Fluids for Machining

Cutting fluids can improve the performance of cutting tools by cooling the cutting tool and the workpiece, by lubricating the cutting and noncutting surfaces on the tool, by inhibiting seizure between the chip and the tool, and by flushing the chips from the work area. The results of the action of the cutting fluid can be a lowered tool force, improved tool life or a faster cutting speed, improved surface finish on the machined surface, and better control of workpiece accuracy. Cutting fluids are also used on grinding operations to cool the workpiece and to improve the surface finish.

Some cutting fluids are sometimes called coolants because of their cooling action. Cooling the cutting edge of the tool reduces the rate of tool wear and improves the tool life obtained at a given cutting speed. A faster cutting speed can be used if the tool life can be the same as for cutting dry. Sometimes difficulties can occur when applying a coolant to carbides. This is discussed further on. Cutting fluids also keep the workpiece cool and prevent dimensional changes that would affect precision measurements. On high-speed cutting operations the cooling action may be required to allow the operator to handle the workpiece. Friction always occurs between the sliding chip and the face of the tool, and between the worn flank of the tool and the workpiece. It also occurs on non-cutting tool surfaces which rub against the workpiece or over which the chip passes, when drilling, reaming, tapping, and broaching. Reducing the friction in any of these areas by using the lubricating qualities of the cutting fluid improves the surface finish on the work and lowers the tool force. In some applications the cutting fluid is selected to perform the usual function of a cutting fluid and, in addition, serve as a lubricant for the machine tool slides located near the cutting tools. Rust prevention on machined surfaces is another important function of the cutting fluid. The anti-seizure or anit-welding property of some cutting fluids is very important to attain a good finish on the machined surface. This property is generally only effective when cutting at the lower range (less than 200 fpm for steel) of cutting speed and, therefore, cutting fluids having anti-seizure properties usually are not used at higher speeds except, in some cases, for grinding. The anti-seizure properties inhibit the formation of the built-up edge and other adhesions to the cutting tool which are the principal cause of surface roughness on the machined surface. Other properties of cutting fluids that must also be considered are: the hygienic properties to prevent dermatitis on the hands of machine operators; the toxicity of the cutting fluid; the possibility of bacterial growth which causes unpleasant odors; the ease of handling, preparation, and concentration; the ability to resist foaming, smoking, and misting; and, the service life and cost.

There are four basic types of cutting fluids: 1) water-base soluble or emulsifying oils; 2) oil-base cutting fluids; 3) chemical or synthetic fluids; 4) gaseous fluids. Water-base soluble or emulsifying oils are sometimes classified as aqueous cutting fluids. Depending upon the application, they are mixed with water to a concentration of 1 part oil to 70 parts water up to as much as 1 part oil to 5 parts water. Soluble or emulsifying oils are very effective coolants and are used extensively as general-purpose cutting fluids for almost all machining operations and for grinding. Heavy-duty soluble oils and emulsions are available that contain certain additives which provide lubricating qualities and surface chemical effects in addition to their ability to act as a coolant. Improvements in tool life and surface finish are obtained through the use of heavy-duty soluble oils and emulsions. Chemical cutting fluids are also classified as synthetic cutting fluids. They are available to suit a wide range of applications. Some have excellent coolant properties and others combine cooling with very good properties as a lubricant. The most common gaseous cutting fluid is compressed air. Other gaseous fluids that have been used are carbon dioxide (CO_2), liquid argon, and liquid nitrogen. Oil base cutting fluids have been combined

with compressed air to form a mist that is sprayed on the work and the tool and acts as a cutting fluid. In some cases this has been very effective. Care must be exercised with this type of cooling to prevent injurious effects to the machine operator. Sprayed or brushed-on cutting fluids used to perform difficult operations such as tapping have also proven to be effective.

There are no fixed rules or formulas for selecting cutting fluids, except the rule of an actual test with a given tool or operation and material. The following paragraphs treat the more widely used types of cutting fluids used in machine shop practice.

Emulsifying or Soluble Oils. — Soluble oils are extensively used for machining operations, particularly when an inexpensive cooling medium meets practical requirements. When an emulsifying oil is mixed with water, an emulsion is formed rather than a solution, but the name "soluble oil" has been generally accepted because the oil apparently dissolves or goes into solution with the water. To obtain a mixture of oil and water, an emulsifying agent is required and soap has proved to be very effective. The emulsion formed contains an infinite number of minute and invisible oil particles which give the mixture a milky or creamy white color. The proportions of the mixture should be determined by test. If the mixture is too weak, it may cause corrosion of both work and machine. The mixture may range from 1 part soluble oil to 5 parts water up to 1 part soluble oil and 50 or even 100 parts water. Soluble oils are easily mixed; but since the emulsion is of the oil-in-water type, it should always be prepared by pouring the *oil* into the water instead of pouring the water into the oil. The emulsification is aided at temperatures somewhat above 40 degrees F., but the use of boiling water may have a reverse tendency. Agitation or stirring increases the emulsifying tendency unless too violent, when there may be a reverse tendency. Even though a soluble oil may serve readily with hard water, it may be desirable to soften the water for obtaining one of the leaner mixtures or to eliminate scum formation. To prevent corrosion when a soluble oil is used, there should be sufficient air circulation to evaporate the water from the machined surface, thus leaving the oil to form a protective coating. In other words, if rapid drying is prevented either by lack of air circulation or by high humidity, rusting is likely to occur.

Regular Soluble Oils: These are primarily considered to be coolants; however, heavy-duty soluble oils are available which contain additives that provide lubrication and other surface effects in addition. Because of their superior cooling properties when compared to oil-base cutting fluids, they drastically reduce or entirely eliminate the problem of smoking and fogging when heavy cuts are taken and, they have replaced oil-base fluids on some applications. True solution type cutting fluids and temperature-soluble dispersions have been developed which are superior to emulsions for machining and grinding titanium. These fluids have a strong detergent action and the coolant system of the machine must be thoroughly cleaned before they are used.

Paste Compounds: Emulsions may be formed by mixing paste compounds with water. A paste made of saponified mineral oil with a high soap content has the consistency of grease. Paste compounds are not used as extensively as soluble oils but they are often applied particularly in connection with grinding operations. These paste compounds do not emulsify with water as readily as the soluble oils and this is particularly true with hard water. Hot water and steam may be required in some cases to obtain the best mixing results.

Straight Cutting Oils. — Straight cutting oils may be either in the mineral class or in the fatty or animal class such as the lard oil.

Straight Mineral Oils: The chief use of mineral oils in connection with machining processes is for blending with base cutting oils for obtaining whatever properties are

required for different machining operations. Straight mineral oils are applicable to light machining operations especially on non-ferrous metals or in machining free-cutting steel, particularly if both cooling and lubricating effects are required. The low viscosity grades are preferable because of their cooling and penetrating properties. Straight mineral oils are sometimes used on automatics in preference to an emulsion which might interfere with proper lubrication of adjacent machine slides or other parts.

Straight Fatty Oils: Straight fatty oils have in the past been used quite extensively, especially for the heavier machining operations, but certain objectionable features have resulted in the substitution of other cutting fluids. The cost of a fatty oil such as lard oil is not only high but a straight animal oil tends to become rancid and produce an objectionable odor. Bacteria breed in such oils and cause skin troubles among machine operators. Sulphurized cutting oils have supplanted the fatty oils to a considerable degree, for machining operations requiring chip lubrication. Fatty oils, however, may be used in cases where a sulphurized mineral oil would tarnish the machined parts unless some preventive treatment is applied.

Mineral Lard Oil. — Mixtures of mineral oil and lard oil may be used to obtain a cutting fluid having greater lubricating value than a straight mineral oil and much lower cost than a straight lard oil. They may be used when the chip pressures are moderate. The proportion of lard oil and mineral oil depends upon the character of the machining operation and the need for both cooling and lubricating effects. For light machining, straight mineral oil may be satisfactory; for heavier duty or where the metal is removed at comparatively high rate, the mineral oil may contain from 10 to 40 per cent of lard oil. As a general rule, the percentage of lard oil is increased with the hardness of the stock to be machined. The No. 1 or prime lard oil is the most commonly used animal fatty oil. There are, however, synthetic fatty oils which are used in conjunction with mineral oils to obtain the so-called mineral lard oils. Mineral lard oils are frequently used in connection with automatic screw machine practice. The Saybolt viscosity (which is controlled by the mineral oil) generally varies from 150 to 225 seconds at 100 degrees F. for the blended oil.

Sulfurized and Chlorinated Cutting Oils. — The oils in these two general classes are especially useful when there is high chip-bearing pressure as in machining alloy or other tough steels. The sulfurized and chlorinated base oils are blended with mineral oils and produce metallic oxides or a *metallic*-film lubrication instead of fluid-film lubrication, as a result of the heat generated by the cutting tool. While mineral oils may contain natural sulfur, it has little or no value in cutting oils. There are two methods of adding sulfur. The sulfur may be cooked in and bonded with a fatty oil such as lard or sperm oil, thus obtaining a compound or "sulfur base" which is added to the mineral oil. The transparent oils are produced by this method, and the required viscosity is obtained by blending the base with a mineral oil. A second method is to add the sulfur directly to the mineral oil by the application of heat. The result is a dark or opaque oil which is suitable where work visibility is not required as in pipe threading and miscellaneous roughing operations. Chlorine, like sulfur, is added to mineral oils to obtain cutting fluids suitable for high chip-bearing pressures or to assist in obtaining extremely high pressure (EP) lubrication.

Chemical or Synthetic Cutting Fluids. — These cutting and grinding fluids are true solutions of organic and inorganic materials in water. They contain little or no petroleum products. They may be classified into two types: 1) those with added lubricants and wetting agents, and 2) those without.

Chemical cutting fluids containing lubricants and wetting agents are recommended for use on tough machining operations such as tapping, threading, sawing, broaching, gear shaving, and gear cutting. They are also recommended for turning,

boring, drilling, and milling. They are excellent rust inhibitors, have a high heat conductivity, and their excellent lubricity keeps the machine tool slides moving smoothly. They tend to produce good finishes on the machined surfaces and they do not require degreasing of the parts which is an advantage in production. They do not combine readily with hard water and have a tendency to foam which makes them unsuitable for use on disc-wheel type surface grinders.

Chemical cutting fluids that do not contain lubricants or wetting agents do not foam readily and are excellent coolants and rust inhibitors. They are clear solutions and are recommended for surface grinding operations. They are not recommended for the severe machining operations where their lack of lubricity is a disadvantage. Evaporation of the water will leave behind deposits that can accumulate and interfere with the operation of machine slides. They do resist the growth of undesirable bacteria and do not form a scum in hard water.

Aqueous Solutions. — When the function of a cutting fluid is merely to cool the work and possibly wash away chips, water containing some alkali has often been used, although these aqueous solutions have been replaced quite largely by modern cutting fluids. They do not, of course, provide the lubricating film that is important for many classes of work and they also cause corrosion of both work and the machine. These aqueous solutions may contain carbonate of soda, borax, caustic soda, etc. Lard oil and soft soap may also be used to improve the properties. An inexpensive mixture for turning, milling, etc., is made in the following proportions: 1 pound of sal-soda (carbonate of soda), 1 quart of lard oil, 1 quart of soft soap, and enough water to make 10 or 12 gallons. This mixture is boiled for one-half hour, preferably by passing a steam coil through it. If the solution should have an objectionable smell, this can be eliminated by adding about 2 pounds of unslaked lime.

Cutting Fluids for Different Materials and Operations. — The selection of the cutting fluid depends upon the machinability of the metal, the severity of the operation, the cutting tool material, and the overall cost. Other factors, too, must be considered. Some shops standardize on a few cutting fluids which are to serve all purposes. In other cases, one cutting fluid must be used for all of the operations performed on a machine. Sometimes, in such cases, a very severe operating condition is alleviated by applying the "right" cutting fluid manually while the machine supplies the cutting fluid for the other operations through its coolant system. There are many excellent proprietary cutting fluids to select from and it is not possible to specify from among these. While the following recommendations represent good practice, they are to serve as a guide only, and it is not intended to say that other cutting fluids will not, in certain specific cases, also be effective.

Steels: Caution should be used when using a cutting fluid on steel that is being turned at a high cutting speed with cemented carbide cutting tools. See "The Application of Cutting Fluids to Carbides" later. Frequently this operation is performed dry. If a cutting fluid is used, it should be a soluble oil mixed to a consistency of about 1 part oil to 20 to 30 parts water. A sulphurized mineral oil is recommended for reaming with carbide tipped reamers although a heavy-duty soluble oil has also been used successfully.

The cutting fluid recommended for machining steel with high speed cutting tools depends largely on the severity of the operation. For ordinary turning, boring, drilling, and milling on medium and low strength steels, use a soluble oil having a consistency of 1 part oil to 10 to 20 parts water. For tool steels and tough alloy steels, a heavy-duty soluble oil having a consistency of 1 part oil to 10 parts water is recommended for turning and milling, while for drilling and reaming a light sul-

Cutting Fluids Recommended for Machining Operations — 1

Soluble Oils. — Types of oils or paste compounds which form emulsions when mixed with water: Soluble oils are used extensively in machining both ferrous and non-ferrous metals when the cooling quality is paramount and the chip-bearing pressure is not excessive. Care should be taken in selecting the proper soluble oil for precision grinding operations. Grinding coolants should be free from fatty materials that tend to load the wheel, thus affecting the finish on the machined part. Soluble coolants should contain rust preventive constituents to prevent corrosion.

Base Oils. — Various types of highly sulfurized and chlorinated oils containing inorganic, animal, or fatty materials. This "base stock" usually is "cut back" or blended with a lighter oil, unless the chip-bearing pressures are high, as in cutting alloy steel, in which case the base stock may be used straight. Base oils usually have a viscosity range of from 300 to 900 seconds at 100 degrees F.

Mineral Oils. — This group includes all types of oils extracted from petroleum such as paraffin oil, mineral seal oil, and kerosene. Mineral oils are often blended with base stocks, but they are generally used in the original form for light machining operations on both free-machining steels and non-ferrous metals. The coolants in this class should be of a type that has a relatively high flash point. Care should be taken to see that they are nontoxic, so that they will not be injurious to the operator. The heavier mineral oils (paraffin oils) usually have a viscosity of about 100 seconds at 100 degrees F. Mineral seal oil and kerosene have a viscosity of 35 to 60 seconds at 100 degrees F.

Material to be Cut	Turning	Milling
Aluminum (Note 1–See notes on next page)	Mineral Oil with 10 Percent Fat (or) Soluble Oil	Soluble Oil (96 Per Cent Water) (or) Mineral Seal Oil (or) Mineral Oil
Alloy Steels (Note 2)	25 Per Cent Sulfur base Oil* with 75 Per Cent Mineral Oil	10 Per Cent Lard Oil with 90 Per Cent Mineral Oil
Brass	Mineral Oil with 10 Per Cent Fat	Soluble Oil (96 Per Cent Water)
Tool Steels and Low-carbon Steels	25 Per Cent Lard Oil with 75 Per Cent Mineral Oil	Soluble Oil
Copper	Soluble Oil	Soluble Oil
Monel Metal	Soluble Oil	Soluble Oil
Cast Iron (Note 3)	Dry	Dry
Malleable Iron	Soluble Oil	Soluble Oil
Bronze	Soluble Oil	Soluble Oil
Magnesium (Note 5)	10 Per Cent Lard Oil with 90 Per Cent Mineral Oil	Mineral Seal Oil

Cutting Fluids Recommended for Machining Operations — 2

Material to be Cut	Drilling	Tapping
Aluminum (Note 4)	Soluble Oil (75 to 90 Per Cent Water) (or) 10 Per Cent Lard Oil with 90 Per Cent Mineral Oil	Lard Oil (or) Sperm Oil (or) Wool Grease (or) 25 Per Cent Sulfur-base Oil* Mixed with Mineral Oil
Alloy Steels (Note 2)	Soluble Oil	30 Per Cent Lard Oil with 70 Per Cent Mineral Oil
Brass	Soluble Oil (75 to 90 Per Cent Water) (or) 30 Per Cent Lard Oil with 70 Per Cent Mineral Oil	10 to 20 Per Cent Lard Oil with Mineral Oil
Tool Steels and Low-carbon Steels	Soluble Oil	25 to 40 Per Cent Lard Oil with Mineral Oil (or) 25 Per Cent Sulfur-base Oil* with 75 Per Cent Mineral Oil
Copper	Soluble Oil	Soluble Oil
Monel Metal	Soluble Oil	25 to 40 Per Cent Lard Oil Mixed with Mineral Oil (or) Sulfur-base Oil* Mixed with Mineral Oil
Cast Iron (Note 3)	Dry	Dry (or) 25 Per Cent Lard Oil with 75 Per Cent Mineral Oil
Malleable Iron	Soluble Oil	Soluble Oil
Bronze	Soluble Oil	20 Per Cent Lard Oil with 80 Per Cent Mineral Oil
Magnesium (Note 5)	60-second Mineral Oil	20 Per Cent Lard Oil with 80 Per Cent Mineral Oil

Note 1. In machining aluminum, several varieties of coolants may be used. For rough machining, where the stock removal is sufficient to produce heat, water soluble mixtures can be used with good results to dissipate the heat. Other oils that may be recommended are straight mineral seal oil; a 50–50 mixture of mineral seal oil and kerosene; a mixture of 10 per cent lard oil with 90 per cent kerosene; and a 100-second mineral oil cut back with mineral seal oil or kerosene.

* *Note* 2. The sulfur-base oil referred to contains 4½ per cent sulfur compound. Base oils are usually dark in color. As a rule, they contain sulfur compounds resulting from a thermal or catalytic refinery process. When so processed, they are more suitable for industrial coolants than when they have had such compounds as flowers of sulfur added by hand. The adding of sulfur compounds by hand to the coolant reservoir is of temporary value only, and the non-uniformity of the solution may affect the machining operation.

Note 3. A soluble oil or low-viscosity mineral oil may be used in machining cast iron to prevent excessive metal dust.

Note 4. Sulfurized oils ordinarily are not recommended for tapping aluminum; however, for some tapping operations they have proved very satisfactory, although the work should be slushed in a solvent right after machining to prevent discoloration.

Note 5. When machining magnesium, use an anhydrous non-acid oil. Water solubles or emulsions should never be used because water would intensify an accidental chip fire.

furized mineral-fatty oil is used. For tough operations such as tapping, threading, and broaching, a sulfochlorinated mineral-fatty oil is recommended for tool steels and high-strength steels, and a heavy sulfurized mineral-fatty oil or a sulfochlorinated mineral oil can be used for medium- and low-strength steels. Straight sulfurized mineral oils are often recommended for machining tough, stringy low carbon steels in order to reduce tearing and rough surface finishes.

Stainless Steel: For ordinary turning and milling a heavy-duty soluble oil mixed to a consistency of 1 part oil to 5 parts water is recommended. Broaching, threading, drilling, and reaming can be done with best results by using a sulfochlorinated mineral-fatty oil.

Copper Alloys: Most brasses, bronzes, and copper are stained when exposed to cutting oils containing active sulfur and chlorine; thus, sulfurized and sulfochlorinated oils should not be used. For most operations a straight soluble oil, mixed to 1 part oil and 20 to 25 parts water is satisfactory. For very severe operations and for automatic screw machine work a mineral-fatty oil is used. A typical mineral-fatty oil might contain 5 to 10 per cent lard oil with the remainder mineral oil.

Monel Metal: When turning this material, an emulsion gives a slightly longer tool life than a sulfurized mineral oil, but the latter aids in chip breakage, which is frequently desirable.

Aluminum Alloys: Aluminum and aluminum alloys are frequently machined dry. When a cutting fluid is used it should be selected for its ability to act as a coolant. Soluble oils mixed to a consistency of 1 part oil to 20 to 30 parts water can be used. Mineral oil-base cutting fluids, when used to machine aluminum alloys, are frequently cut back to increase their viscosity in order to obtain good cooling characteristics and to flow easily to cover the tool and the work. For example, a mineral-fatty oil or a mineral plus a sulfurized fatty oil can be cut back by the addition of as much as 50 per cent kerosene.

Cast Iron: Ordinarily, cast iron is machined dry. Some increase in tool life can be obtained or a faster cutting speed can be used with a chemical cutting fluid or a soluble oil mixed to consistency of 1 part oil and 20 to 40 parts water. A soluble oil is sometimes used to reduce the amount of dust around the machine.

Magnesium: A water soluble oil must never be used to machine magnesium. Thin magnesium chips are subject to spontaneous combustion and, in the event of a fire, the water will release hydrogen which will intensify the fire. Usually, magnesium is machined dry. Light mineral oil or mineral-fatty oils are sometimes used to cool the work and to reduce the fire hazard.

Grinding: Soluble oil emulsions or emulsions made from paste compounds are used extensively in precision grinding operations. For cylindrical grinding use 1 part oil to 40 to 50 parts water. Solution type fluids and translucent grinding emulsions are particularly suited for many fine-finish grinding applications. Mineral oil-base grinding fluids are recommended for many applications where a fine surface finish is required on the ground surface. Mineral oils are used with vitrified wheels but are not recommended for wheels with rubber or shellac bonds. Under certain conditions the oil vapor mist caused by the action of the grinding wheel can be ignited by the grinding sparks and explode. To quench the grinding spark a secondary coolant line to direct a flow of grinding oil below the grinding wheel is recommended.

Broaching: For steel, use a heavy mineral oil such as sulfurized oil of 300 to 500 Saybolt viscosity at 100 degrees F. to provide both adequate lubricating effect and a dampening of the shock loads. Soluble oil emulsions may be used for the lighter broaching operations.

The Application of Cutting Fluids to Carbides. — Carbide turning, boring, and similar operations on lathes are performed dry or with the help of soluble oil or chemical cutting fluids. The effectiveness of cutting fluids in improving tool life or by permitting higher cutting speeds to be used, is less with carbides than with high-speed steel tools. Furthermore, the effectiveness of the cutting fluid is lessened as the cutting speed is increased. Cemented carbides are very sensitive to sudden changes in temperature and to temperature gradients within the carbide. Thermal shocks to the carbide will cause thermal cracks to form near the cutting edge which are a prelude to tool failure. An unsteady or interrupted flow of the coolant reaching the cutting edge will generally cause thermal cracks to form near the cutting edge which results in a rapid failure of the tool. The flow of the chip over the face of the tool can cause an interruption to the flow of the coolant reaching the cutting edge even though a steady stream of coolant is directed at the tool. When a cutting fluid is used and frequent tool breakage is encountered, it is often best to cut dry. When a cutting fluid must be used to keep the workpiece cool for size control or to allow it to be handled by the operator, special precautions must be used. Sometimes applying the coolant from the front and the side of the tool simultaneously is helpful. On lathes equipped with overhead shields, it is very effective to apply the coolant from below the tool into the space between the shoulder of the work and the tool flank, in addition to applying the coolant from the top. Another method is not to direct the coolant stream at the cutting tool at all but to direct it at the workpiece above or behind the cutting tool.

The danger of thermal cracking is great when milling with carbide cutters. The nature of the milling operation itself tends to promote thermal cracking because the cutting edge is constantly heated to a high temperature and rapidly cooled as it enters and leaves the workpiece. For this reason, carbide milling operations should be performed dry.

Lower cutting-edge temperatures diminish the danger of thermal cracking. The cutting-edge temperatures usually encountered when reaming with solid carbide or carbide tipped reamers are generally such that thermal cracking is not apt to occur except when reaming certain difficult-to-machine metals. Therefore, cutting fluids are very effective when used on carbide reamers. Practically every kind of cutting fluid has been used, depending on the job material encountered. For difficult surface-finish problems in holes, heavy duty soluble oils, sulfurized mineral-fatty oils, and sulfo-chlorinated mineral-fatty oils have been used successfully. In some cases, the grade and the hardness of the carbide also have an effect on the surface finish of the hole.

Application of Cutting Fluids. — Cutting fluids should be applied where the cutting action is taking place and at the highest possible velocity without causing splashing. As a general rule, it is preferable to supply from 3 to 5 gallons per minute for each single-point tool on a machine such as a turret lathe or automatic. The temperature of the cutting fluid should be kept below 110 degrees F. If the volume of fluid used is not sufficient to maintain the proper temperature, means of cooling it should be provided.

Cutting Fluids for Machining Magnesium. — In machining magnesium, it is the general but not invariable practice in the United States to use a cutting fluid, whereas, in England, magnesium usually is machined dry except in cases where heat generated by high cutting speeds would not be dissipated rapidly enough without a cutting fluid. This condition may exist when, for example, small tools without much heat-conducting capacity are employed on automatics.

The cutting fluid for magnesium should be an anhydrous oil having, at most, a very low acid content. Various mineral-oil cutting fluids are used for magnesium.

To secure adequate cooling, the supply of fluid should be large (4 to 5 gallons per minute) and the viscosity low; however, to avoid too low a flash point, a compromise between cooling capacity and flash point is necessary. *Soluble oils or emulsions should never be used for machining magnesium.* Compressed air may be preferable to a fluid because it leaves the chips or swarf clean and dry.

A cutting fluid serves primarily to cool the work and also eliminate a possible fire hazard, especially when dull tools are operated at high speeds with fine feeds. Even when using sharp tools, the cut should not be less than 0.001 inch because fine chips are more likely to become ignited at high speeds. While a variety of mineral oils may be used, the following properties are recommended: Specific gravity 0.79 to 0.86; viscosity (Saybolt) at 100 degrees F., up to 55 seconds; flash point, minimum value (closed cup), 160 degrees F.; saponification No., 16 (max.); free acid (max.) 0.2 per cent. Oil-water emulsions, while good coolants, are objectionable because water will greatly intensify any accidental chip fire.

Machining Magnesium. — Magnesium alloys are readily machined and with relatively low power consumption per cubic inch of metal removed. The usual practice is to employ high cutting speeds with relatively coarse feeds and deep cuts. Exceptionally fine finishes can be obtained so that grinding to improve the finish usually is unnecessary. The horsepower normally required in machining magnesium varies from 0.15 to 0.30 per cubic inch per minute. While this value is low, especially in comparison with power required for cast iron and steel, the total amount of power for machining magnesium usually is high because of the exceptionally rapid rate at which metal is removed.

Carbide tools are recommended for maximum efficiency, although high-speed steel frequently is employed. Tools should be designed so as to dispose of chips readily or without excessive friction, by employing polished chip-bearing surfaces, ample chip spaces, large clearances, and small contact areas. *Keen-edged tools should always be used.*

Feeds and Speeds for Magnesium: Speeds ordinarily range up to 5000 feet per minute for rough- and finish-turning, up to 3000 feet per minute for rough-milling, and up to 9000 feet per minute for finish-milling. For rough-turning, the following combinations of speed in feet per minute, feed per revolution, and depth of cut are recommended: Speed 300 to 600 feet per minute — feed 0.030 to 0.100 inch, depth of cut 0.5 inch; speed 600 to 1000 — feed 0.020 to 0.080, depth of cut 0.4; speed 1000 to 1500 — feed 0.010 to 0.060, depth of cut 0.3; speed 1500 to 2000 — feed 0.010 to 0.040, depth of cut 0.2; speed 2000 to 5000 — feed 0.010 to 0.030, depth of cut 0.15.

Lathe Tool Angles for Magnesium: The true or actual rake angle resulting from back and side rakes usually varies from 10 to 15 degrees. Back rake varies from 10 to 20, and side rake from 0 to 10 degrees. Reduced back rake may be employed to obtain better chip breakage. The back rake may also be reduced to from 2 to 8 degrees on form tools or other broad tools to prevent chatter.

Parting Tools: For parting tools, the back rake varies from 15 to 20 degrees, the front end relief 8 to 10 degrees, the side relief measured perpendicular to the top face 8 degrees, the side relief measured in the plane of the top face from 3 to 5 degrees.

Milling Magnesium: In general, the coarse-tooth type of cutter is recommended. The number of teeth or cutting blades may be one-third to one-half the number normally used; however, the two-blade fly-cutter has proved to be very satisfactory. As a rule, the land relief or primary peripheral clearance is 10 degrees followed by secondary clearance of 20 degrees. The lands should be narrow, the width being about 3/64 to 1/16 inch. The rake, which is positive, is about 15 degrees.

For rough-milling and speeds in feet per minute up to 900 — feed, inch per tooth, 0.005 to 0.025, depth of cut up to 0.5; for speeds 900 to 1500 — feed

0.005 to 0.020, depth of cut up to 0.375; for speeds 1500 to 3000 — feed 0.005 to 0.010, depth of cut up to 0.2.

Drilling Magnesium: If the depth of a hole is less than five times the drill diameter, an ordinary twist drill with highly polished flutes may be used. The included angle of the point may vary from 70 degrees to the usual angle of 118 degrees. The relief angle is about 12 degrees. The drill should be kept sharp and the outer corners rounded to produce a smooth finish and prevent burr formation. For deep hole drilling, use a drill having a helix angle of 40 to 45 degrees with large polished flutes of uniform cross-section throughout the drill length to facilitate the flow of chips. A pyramid-shaped "spur" or "pilot point" at the tip of the drill will reduce the "spiraling or run-off."

Drilling speeds vary from 300 to 2000 feet per minute with feeds per revolution ranging from 0.015 to 0.050 inch.

Reaming Magnesium: Reamers up to 1 inch in diameter should have four flutes; larger sizes, six flutes. These flutes may be either parallel with the axis or have a negative helix angle of 10 degrees. The positive rake angle varies from 5 to 8 degrees, the relief angle from 4 to 7 degrees, and the clearance angle from 15 to 20 degrees.

Tapping Magnesium: Standard taps may be used unless Class 3B tolerances are required, in which case the tap should be designed for use in magnesium. A high-speed steel concentric type with a ground thread is recommended. The concentric form, which eliminates the radial thread relief, prevents jamming of chips while the tap is being backed out of the hole. The positive rake angle at the front may vary from 10 to 25 degrees and the "heel rake angle" at the back of the tooth from 3 to 5 degrees. The chamfer extends over two to three threads. For holes up to ¼ inch in diameter, two-fluted taps are recommended; for sizes from ½ to ¾ inch, three flutes; and for larger holes, four flutes. Tapping speeds ordinarily range from 75 to 200 feet per minute, and mineral oil cutting fluid should be used.

Threading Dies for Magnesium: Threading dies for use on magnesium should have about the same cutting angles as taps. Narrow lands should be used to provide ample chip space. Either solid or self-opening dies may be used. The latter type is recommended when maximum smoothness is required. Threads may be cut at speeds up to 1000 feet per minute.

Grinding Magnesium: As a general rule, magnesium is ground dry. The highly inflammable dust should be formed into a sludge by means of a spray of water or low-viscosity mineral oil. Accumulations of dust or sludge should be avoided. For surface grinding, when a fine finish is desirable, a low-viscosity mineral oil may be used.

Machining Aluminum. — Some of the alloys of aluminum have been machined successfully without any lubricant or cutting compound, but in order to obtain the best results, some form of lubricant is desirable. For many purposes, a soluble cutting oil is good.

Tools for aluminum and aluminum alloys should have larger relief and rake angles than tools for cutting steel. For high-speed steel turning tools the following angles are recommended: relief angles, 14 to 16 degrees; back rake angle, 5 to 20 degrees; side rake angle, 15 to 35 degrees. For very soft alloys even larger side rake angles are sometimes used. High silicon aluminum alloys and some others have a very abrasive effect on the cutting tool. While these alloys can be cut successfully with high-speed-steel tools, cemented carbides are recommended because of their superior abrasion resistance. The tool angles recommended for cemented carbide turning tools are: relief angles, 12 to 14 degrees; back rake angle, 0 to 15 degrees; side rake angle, 8 to 30 degrees.

Cut-off tools and necking tools for machining aluminum and its alloys should have from 12 to 20 degrees back rake angle and the end relief angle should be from 8 to 12

degrees. Excellent threads can be cut with single-point tools in even the softest aluminum. Experience seems to vary somewhat regarding the rake angle for single-point thread cutting tools. Some prefer to use a rather large back and side rake angle although this requires a modification in the included angle of the tool to produce the correct thread contour. When both rake angles are zero, the included angle of the tool is ground equal to the included angle of the thread. Excellent threads have been cut in aluminum with zero rake angle thread-cutting tools using large relief angles, which are 16 to 18 degrees opposite the front side of the thread and 12 to 14 degrees opposite the back side of the thread. In either case, the cutting edges should be ground and honed to a keen edge. It is sometimes advisable to give the face of the tool a few strokes with a hone between cuts when chasing the thread in order to remove any built-up edge on the cutting edge.

Fine surface finishes are often difficult to obtain on aluminum and aluminum alloys, particularly the softer metals. When a fine finish is required, the cutting tool should be honed to a keen edge and the surfaces of the face and the flank will also benefit by being honed smooth. Tool wear is inevitable; however, it should not be allowed to progress too far before the tool is changed or sharpened. A sulphurized mineral oil or a heavy-duty soluble oil will sometimes be helpful in obtaining a satisfactory surface finish. For best results, however, a diamond cutting tool is recommended. Excellent surface finishes can be obtained on even the softest aluminum and aluminum alloys with these tools.

Although ordinary milling cutters can be used successfully in shops where aluminum parts are only machined occasionally, the best results are obtained with coarse-tooth, large helix-angle cutters having large rake and clearance angles. Clearance angles up to 10 to 12 degrees are recommended. When slab milling and end milling a profile, using the peripheral teeth on the end mill, climb milling (also called down milling) will generally produce a better finish on the machined surface than conventional (or up) milling. Face milling cutters should have a large axial rake angle. Standard twist drills can be used without difficulty in drilling aluminum and aluminum alloys although high helix-angle drills are preferred. The wide flutes and high helix-angle in these drills helps to clear the chips. In some cases the use of split-point drills is preferred. Carbide tipped twist drills can be used for drilling aluminum and its alloys which may afford advantages in some production applications. Ordinary hand and machine taps can be used to tap aluminum and its alloys although spiral-fluted ground thread taps give superior results. Experience has shown that such taps should have a right-hand ground flute when intended to cut right-hand threads and the helix angle should be similar to that used in an ordinary twist drill.

Machining Plastics. — Molded plastic parts do not require machining, as a general rule, unless the tolerances are exceptionally small or there are undercuts, angular holes, or other openings difficult or impracticable to reproduce in a mold. It is common practice, however, to machine laminated phenolic plastics and also cast phenolic plastics, as well as sheet, bar, and tube stock such as is commonly used for making parts such as pinions or gears. The machining characteristics of different plastics vary somewhat so that the general recommendations given may require some modification to obtain the best results for a given class of work. Although plastics are poor conductors of heat, they usually are machined dry or without a cutting fluid. In some cases, either an air jet, water, or a soap solution is used. The maximum speed at which a cutting tool can operate without excessive heating should be determined by actual test.

Turning and Boring Plastics: Tools of high-speed steel, Stellite and carbide are commonly used in machining plastics. The general practice in turning and boring

is similar to that for brass so far as the feed and speed are concerned. Speeds usually vary from 250 to 500 feet per minute for high-speed steel tools and from 500 to 1500 for Stellite and carbide tools. According to the Haynes Stellite Co., hot-set molded parts require tools having less clearance and more rake than those used for steel or other metals. The cold-set acetate and polystyrene molded parts should be turned with tools having no rake and plenty of clearance. Parts molded of "Vinylite" plastic can usually be machined satisfactorily with ordinary metal-cutting tools, provided the front and side clearances are about double those required for machine steel.

Cutting Off Plastics: Tools for cutting off should have greater front and side clearances than for steel. The cutting speed should be about half that employed for a turning operation.

Drilling Plastics: Standard twist drills may be used for small holes up to ⅛ or 3⁄16 inch, but for larger sizes particularly, it is preferable to use the commercial high-speed steel drills designed especially for plastics. These drills are made both in wire gage and fractional sizes. They have relatively large flutes to provide greater space for chips, and polished flutes are preferable. For large-quantity production, carbide-tipped drills are recommended. Extra clearance back of the edges of the flutes tends to reduce friction and heating. Frequent removal of the drill may be necessary, especially in drilling comparatively deep holes. A feed of 0.007 to 0.015 inch per revolution is a common range. Drilling speeds usually vary from 150 to 300 feet per minute. With carbide-tipped drills, the speeds may be as high as 12,000 to 15,000 rpm. The drill point should have an included angle of 55 to 60 degrees for thin sections and 90 degrees for the thicker sections. The clearance angle usually is 15 degrees. To avoid excessive heating and aid in chip removal, an air jet may be directed into the hole. In some cases, a soap solution is effective as a cutting fluid. If the drill-spindle movement is cam-operated, design the cam to advance the drill tip slowly at the beginning or for about 0.010 inch, then continue at the full rate of speed and allow a dwell of about 2 to 5 revolutions at the bottom of the hole; then withdraw the drill rapidly.

Tapping and Threading Plastics: For tapping small holes in hot- or cold-set molded parts, a high-speed steel nitrided chromium-plated tap is recommended. The rake may vary from 0 to 5 degrees negative. The size of the hole should allow for about three-fourths of the standard thread depth. Small holes generally are tapped dry. If a coolant is used, water is preferable to oil. Tapping speeds usually vary from 40 to 55 per minute. Threaded brass inserts or bushings are often used, in which case they may be tapped either before or after insertion. In cutting a 60-degree thread, such as the Unified Standard, use a tool that is ground to cut on one side only and feed it in at an angle of 30 degrees by setting the compound rest at this angle.

Drill Jigs for Plastics: In the design of jigs for drilling, close-fitting drill bushings should be avoided. They may increase not only the friction on the drill but also the tendency of the chips to plug up the drill flutes. If the operation is such that a drill bushing is absolutely essential, a floating leaf or templet should be employed. When using a templet, the hole should be spotted with the templet in place, using the drill size corresponding to the final hole size; then the templet should be removed and the hole completed. Pilot holes should be avoided, except in special instances when the hole is to be reamed or counterbored.

Sawing Hot-set Molded Parts: The sawing of molded hot-set parts is done chiefly on circular and band saws. Band saws are to be recommended at times for straight cutting because they run cooler than circular saws. Band saw manufacturers advocate saw teeth set to clear, some advocating one-half the thickness of the blade on each side so that saws give a width of cut double their thickness. Narrower saws

and more set are needed for cutting curves than for straight cuts. Band saws just soft enough to permit filing are recommended, but saws must be kept sharp. Dull saws cause chipping and might result in saw breakage. Sawing is usually done dry but some recommend water for cooling. Saw teeth should have little set and eight to nine teeth per inch. The speed is 1800 to 2500 feet per minute.

Sawing Cold-set Molded Parts: Sawing of cold-set polystyrene or cellulose acetate can be done with circular saws having nine to twelve teeth per inch for thin sheets and six teeth per inch for thickness over ¼ inch. Saws six to nine inches in diameter are run at speeds of 3000 to 3600 rpm and should be hollow ground. They usually are ¹⁄₃₂ to ¹⁄₁₆ inch thick. The use of a water spray gives a cleaner cut. One large saw manufacturer recommends that pieces be cut with a stream of water running in the kerf while the saw is cutting. This applies to both circular and band saws; otherwise, the cold-set type of material will fuse. The circular saws recommended are 14 inches, 12 and 9 gage, 130 teeth, 10 degrees rake to be operated at 3000 rpm and made of a special alloy steel stock.

For band sawing, manufacturers suggest a band saw which is 19 to 20 gage, having twenty points to the inch and hardened and tempered. The saw should be operated at 4000 to 4500 feet per minute.

Milling Hot-set and Cold-set Molded Parts: Milling of molded parts is not as a rule feasible, but where it is required, milling speeds and feeds of the range used for brass are recommended. A speed of 400 fpm with carbon steel cutters and 1200–1600 fpm is recommended with carbide cutters. Single- and double-bladed fly cutters are sometimes used at high speed with fine cuts. Where little material has to be removed, a high-speed woodworking shaper with a carbide-tipped tool can be used to advantage. It is desirable and may be necessary to use an air blast to assure proper chip removal from the milling cutter. Wherever possible, it is recommended that spiral milling cutters be used, and that the number of teeth in the cutter head be such that at least two of them are in contact with the work at all times.

The same general rules that apply to turning, facing, and boring operations also hold for milling Vinyl molded parts. Standard cutters can be used, but higher speeds are feasible if extra clearances are ground on the cutter blades.

Machining Non-metallic Gear Blanks. — Laminated phenolic plastics are extensively used for non-metallic gears and pinions. A non-metallic pinion should preferably be used in conjunction with either a hardened steel or a cast-iron gear, thus providing a durable and comparatively noiseless drive. Small- or medium-sized gears may be formed by molding, provided the quantity is large enough to warrant the cost of a mold. Most non-metallic gears, however, are machined from blanks cut from laminated phenolic plastics which have physical properties superior to the molded gears. These blanks may be cut either by punching in a die, by shearing, or by sawing. An efficient method of producing small blanks is to punch them from thin sheet stock. If necessary, the face width of the gear or pinion may be increased by riveting together two or more of these blanks. Gear blanks may also be cut either from a laminated plastic bar or thick tube. Before cutting the gear teeth, the blanks are machined, as by grinding, to obtain a concentric blank of the required diameter. The larger gears are sawed from sheet stock and then turned to size. Gear blanks which are ready for the gear-cutting operation are supplied by manufacturers of laminated plastic materials. The gear teeth are cut with standard gear-cutting equipment. Metal reinforcing end plates and bushings are commonly employed in connection with laminated gears.

Punching Operations: Most grades of laminated phenolics can be punched either hot or cold. The die must be kept sharp, however, in order to produce good results. The minimum clearance between individual punchings, and also between punchings and the edge of the material strips, should be about three times the thickness of the

material. For hot punching or shearing, the material is heated in a steam or electric oven, designed to give a uniform temperature throughout the heating chamber. The material is left in the oven just long enough to be uniformly heated to oven temperature. Further heating causes brittleness. Temperatures of 100 to 120 degrees C (212 to 248 degrees F) are recommended. The heating time ranges from five minutes for 1/16-inch material to thirty minutes for 1/4-inch material.

Dies for punching laminated plastics are designed the same as for punching metal, except that smaller clearances are allowed between punch and die. In cold punching, this clearance is small, approaching a "sliding fit." The strippers are close fitting and backed with strong springs.

Because these plastic materials expand after being compressed in a die, blanks will be larger than the die diameter and holes will be smaller. On hot punchings, allowance should be made for shrinkage of the material after punching. This shrinkage varies with the grade of material, thickness of piece, and the temperature of the material during the punching operation. For very small holes and blanks, allowances for shrinkage are often neglected, while for large pieces and accurate work, they must be carefully considered. As an example, suppose a 1-inch diameter hole is to be punched hot in 3/32-inch thick stock; this would require a die of 1.009 inches and a punch of 1.007 inches in diameter. If this piece is punched cold, however, the die should be 1.005 inches in diameter and the punch 1.003 inches in diameter.

Shearing Laminated Plastics: Shears suitable for thin metal sheet are used for cutting laminated plastics. The knife must be kept sharp. In trimming paper-base grades cold, clean-cut edges are obtained with thicknesses of 1/32 inch and under, and when trimmed hot, up to 1/8 inch. Fabric-base grades are trimmed cold up to 1/16 inch, and hot up to 1/8 inch. Greater thicknesses can be sheared if the condition of the edge is not important.

Sawing Plastics: Material up to 1 inch thick may be cut with a 12- to 16-inch circular saw at about 3000 rpm, and material 1 inch thick and over, with a 16-inch saw at about 2400 rpm. A saw used for roughing cuts has bevel teeth, seven to the inch, while a smooth saw, with no set — similar to that for metal — is used for finishing cuts. For use on all thick material, the saws should be hollow-ground to prevent binding. The smaller the projection of the saw above the material on the sawing table, the better will be the sawed edges. A thin sheet of plastic or other material placed under the piece to be sawed, is of advantage when extreme smoothness of cut is desired.

A band saw is used for sawing round blanks from plate stock. The usual band saw is of the bevel-tooth type, with some set, and has three to seven teeth per inch. It is run at 3000 feet per minute.

Machining Zinc Alloy Die-Castings. — Machining of zinc alloy die-castings is mostly done without a lubricant. For particular work, especially deep drilling and tapping, a lubricant such as lard oil and kerosene (about half and half) or a 50–50 mixture of kerosene and machine oil may be used to advantage. A mixture of turpentine and kerosene has been been found effective on certain difficult jobs.

In drilling, standard carbon steel drills are used for shallow holes and high-speed drills of the high-spiral type are recommended for deep holes. The standard 118-degree angle of point between cutting edges is recommended. A lip clearance of 12 degrees is satisfactory for most drilling, but in some cases may be increased up to 15 degrees. Flutes that are larger than normal offer the advantage of providing plenty of chip clearance. Straight flute drills have been found useful in enlarging existing holes. Peripheral speeds of from 200 to 300 feet per minute are generally found satisfactory for high-speed steel drills and about half this speed for carbon steel drills.

Threading: Button or acorn dies are satisfactory for threading small diameters of work. Either radial or tangent type chasers may be employed for the larger diameters. For radial type chasers, one manufacturer recommends a 10-degree radial hook for straight threads, a 7-degree radial hook for tapered threads; and a surface speed of 50 feet per minute for 3½ to 7½ threads per inch, 100 feet per minute for 8 to 11 threads per inch, and 200 feet per minute for 12 to 32 threads per inch. In using tangent type chasers with zinc alloys, the cutting edge must be on or very near the center to avoid rapid wearing of the chasers just behind the cutting edge. A 5-degree positive rake is recommended.

Reaming: In reaming, tools with six straight flutes are commonly used, although tools with eight flutes irregularly spaced have been found to yield better results by one manufacturer. Many standard reamers have a land that is too wide for best results. A land about 0.015 inch wide is recommended but this may often be ground down to around 0.007 or even 0.005 inch to obtain freer cutting, less tendency to loading, and reduced heating.

Turning: Tools of high-speed steel are commonly employed although the application of Stellite and carbide tools, even on short runs, is feasible. For steel or Stellite, a positive top rake of from 0 to 20 degrees and an end clearance of about 15 degrees is commonly recommended. Where side cutting is involved, a side clearance of about 4 degrees minimum is recommended. With carbide tools, the end clearance should not exceed 6 to 8 degrees and the top rake should be from 5 to 10 degrees positive. For boring, facing, and other lathe operations, rake and clearance angles are about the same as for tools used in turning.

Machining Monel and Nickel Alloys. — These alloys are machined with high-speed steel and with cemented carbide cutting tools. High-speed steel lathe tools usually have a back rake of 6 to 8 degrees, a side rake of 10 to 15 degrees, and relief angles of 8 to 12 degrees. Broad-nose finishing tools have a back rake of 20 to 25 degrees and an end relief angle of 12 to 15 degrees. In most instances, standard commercial cemented-carbide tool holders and tool shanks can be used which provide an acceptable tool geometry. Honing the cutting edge lightly will help if chipping is encountered.

The most satisfactory tool materials for machining Monel and the softer nickel alloys, such as Nickel 200 and Nickel 230, are M2 and T5 for high-speed steel and crater resistant grades of cemented carbides. For the harder nickel alloys such as K Monel, Permanickel, Duranickel, and Nitinol alloys, the recommended tool materials are T15, M41, M42, M43, and for high-speed steel, M42. For carbides, a grade of crater resistant carbide is recommended when the hardness is less than 300 Bhn, and when the hardness is more than 300 Bhn, a grade of straight tungsten carbide will often work best, although some crater resistant grades will also work well.

A sulfurized oil or a water-soluble oil is recommended for rough and finish turning. A sulfurized oil is also recommended for milling, threading, tapping, reaming, and broaching. Recommended cutting speeds for Monel and the softer nickel alloys are 70 to 100 fpm for high-speed steel tools and 200 to 300 fpm for cemented carbide tools. For the harder nickel alloys, the recommended speed for high-speed steel is 40 to 70 fpm for a hardness up to 300 Bhn and for a higher hardness, 10 to 20 fpm; for cemented carbides, 175 to 225 fpm when the hardness is less than 300 Bhn and for a higher hardness, 30 to 70 fpm.

Nickel alloys have a high tendency to work harden. To minimize work hardening caused by machining, the cutting tools should be provided with adequate relief angles and positive rake angles. Furthermore, the cutting edges should be kept sharp and replaced when dull to prevent burnishing of the work surface. The depth of cut and feed should be sufficiently large to ensure that the tool penetrates the work without rubbing.

In all phases of steel production, various practices are employed which determine the quality and types of the finished material.

Quality Classifications: The term "quality" as it technically relates to steel products may be indicative of many conditions such as the degree of internal soundness, relative uniformity of composition, relative freedom from injurious surface imperfections, and finish. Steel quality also relates to general suitability for particular applications. Sheet steel surface requirements may be broadly identified as to the end use by the suffix E for exposed parts requiring a good painted surface and suffix U for unexposed parts for which surface finish is unimportant.

Carbon Steel may be obtained in a number of fundamental qualities which reflect various degrees of the quality conditions mentioned above. Some of those qualities may be modified by such requirements as limited Austenitic Grain Size, Specified Discard, Macroetch Test, Special Heat Treating, Maximum Incidental Alloy Elements, Restricted Chemical Composition and Nonmetallic Inclusions. In addition, several of the products have special qualities which are intended for specific end uses or fabricating practices.

Alloy Steels also may be obtained in special qualities with such requirements as Extensometer Test, Fracture Test, Impact Test, Macroetch Test, Nonmetallic Inclusion Test, Special Hardenability Test, and Grain Size Test.

For complete descriptions of the qualities and supplementary requirements for carbon and alloy steels, reference should be made to the latest applicable American Iron and Steel Institute (AISI) Steel Products Manual Section.

High Strength, Low Alloy Steel, SAE 950. — High strength, low alloy steel represents a specific type of steel in which enhanced mechanical properties and, in most cases, good resistance to atmospheric corrosion are obtained by the addition of moderate amounts of one or more alloying elements other than carbon.

Steels of this type are normally furnished in the hot rolled or annealed condition to minimum mechanical properties. They are not intended for quenching and tempering. The user should not subject them to such treatment without assuming responsibility for the ensuing mechanical properties. Where these steels are used for fabrication by welding, no preheat or postheat is required. In certain complex structures, stress relieving may be desirable. These steels may be obtained in the standard shapes or forms normally available in carbon steel.

Application: These steels, because of their enhanced strength, corrosion and erosion resistance, and their high strength-to-weight ratio and service life, are adapted particularly for use in mobile equipment and other structures where substantial weight savings are generally desirable. Typical applications are automotive bumper face bars, truck bodies, frames and structural members, scrapers, dump wagons, cranes, shovels, booms, chutes, conveyors, railroad and industrial cars.

Certain Minimum Properties of SAE 950 Steel as Furnished by the Mill

Property *	Thickness or Diameter, inches				
	Up to 0.0709, inclusive	0.0710 to 0.2299, inclusive	0.2300 to ½, inclusive	Over ½ to 1, inclusive	Over 1 to 2, inclusive
Minimum yield point, psi......	50,000	50,000	50,000	47,000	45,000
Minimum tensile strength, psi..	70,000	70,000	70,000	67,000	65,000
Elongation in 2 in., %.........	20	22	22	22	22

* For severe cold forming operations requiring greater ductility, relaxation of the yield point and tensile strength requirements is commonly negotiated between producer and consumer.

Carbon Steels. — *SAE steels 1006, 1008, 1010, 1015:* These steels are the lowest carbon steels of the plain carbon type, and are selected where cold formability is the primary requisite of the user. They are produced both as rimmed and killed steels. Rimmed steel is used for sheet, strip, rod, and wire where excellent surface finish or good drawing qualities are required, such as body and fender stock, hoods, lamps, oil pans, and other deep drawn and formed products. It is also used for cold heading wire for tacks, and rivets and low carbon wire products. Killed steel (usually aluminum killed or special killed) is used for difficult stampings or where non-aging properties are needed. Killed steels (usually silicon killed) should be used in preference to rimmed steel for forging or heat treating applications.

These steels have relatively low tensile values and should not be selected where much strength is desired. Within the carbon range of the group, strength and hardness will increase with increase in carbon and/or with cold work, but such increases in strength are at the sacrifice of ductility or the ability to withstand cold deformation. Where cold rolled strip is used the proper temper designation should be specified to obtain the desired properties.

When under 0.15 carbon, the steels are susceptible to serious grain growth, causing brittleness, which may occur as the result of a combination of critical strain (from cold work) followed by heating to certain elevated temperatures. If cold worked parts formed from these steels are to be later heated to temperatures in excess of 1100 degrees F., the user should exercise care to avoid trouble from this cause. When this condition develops it can be overcome by heating the parts to a temperature well in excess of the upper critical point, or at least 1750 degrees F.

Steels in this group, being nearly pure iron or ferritic in structure, do not machine freely and should be avoided for cut screws and operations requiring broaching or smooth finish on turning. The machinability of bar, rod and wire products is improved by cold drawing. Steels in this group are readily welded.

SAE 1016, 1017, 1018, 1019, 1020, 1021, 1022, 1023, 1024, 1025, 1026, 1027, 1030: Steels in this group, due to the carbon range covered, have increased strength and hardness, and reduced cold formability compared to the lowest carbon group. For heat treating purposes they are known as carburizing or case hardening grades. When uniform response to heat treatment is required, or for forgings, killed steel is preferred; for other uses, semi-killed or rimmed steel may be indicated, depending on the combination of properties desired. Rimmed steels can ordinarily be supplied up to 0.25 carbon.

Selection of one of these steels for carburizing applications depends on the nature of the part, the properties desired, and the processing practice preferred. Increase in carbon gives greater core hardness with a given quench, or permits the use of thicker sections. Increase in manganese improves the hardenability of both the core and case; in carbon steels this is the only change in composition that will increase case hardenability. The higher manganese variants also machine much better. For carburizing applications SAE 1016, 1018, and 1019 are widely used for thin sections or water quenched parts. SAE 1022 and 1024 are used for heavier sections or where oil quenching is desired, and SAE 1024 is sometimes used for such parts as transmission and rear axle gears. SAE 1027 is used for parts given a light case to obtain satisfactory core properties without drastic quenching. SAE 1025 and 1030, while not usually regarded as carburizing types, are sometimes used in this manner for larger sections or where greater core hardness is needed.

For cold formed or headed parts the lowest manganese grades (SAE 1017, 1020, and 1025) offer the best formability at their carbon level. SAE 1020 is used for fan blades and some frame members, and SAE 1020 and 1025 are widely used for low strength bolts. The next higher manganese types (SAE 1018, 1021, and 1026) provide increased strength.

All of these steels may be readily welded or brazed by the common commercial methods. SAE 1020 is frequently used for welded tubing. These steels are used for numerous forged parts, the lower carbon grades where high strength is not essential. Forgings from the lower carbon steels usually machine better in the as forged condition without annealing, or after normalizing.

SAE 1030, 1033, 1034, 1035, 1036, 1038, 1039, 1040, 1041, 1042, 1043, 1045, 1046, 1049, 1050, 1052: These steels, of the medium carbon type, are selected for uses where higher mechanical properties are needed and are frequently further hardened and strengthened by heat treatment or by cold work. These grades are ordinarily produced as killed steels.

Steels in this group are suitable for a wide variety of automotive type applications. The particular carbon and manganese level selected is affected by a number of factors. Increase in the mechanical properties required in section thickness, or in depth of hardening, ordinarily indicates either higher carbon or manganese or both. The heat treating practice preferred, particularly the quenching medium, has a great effect on the steel selected. In general, any of the grades over 0.30 carbon may be selectively hardened by induction or flame methods.

The lower carbon and manganese steels in this group find usage for certain types of cold formed parts. SAE 1030 is used for shift and brake levers, SAE 1034 and 1035 are used in the form of wire and rod for cold upsetting such as bolts, and SAE 1038 for bolts and studs. In practically all cases the parts cold formed from these steels are heat treated prior to use. Stampings are usually limited to flat parts or simple bends. The higher carbon SAE 1038, 1040, and 1042 are frequently cold drawn to specified physical properties for use without heat treatment for some applications, such as cylinder head studs.

All of this group of steels are used for forgings, the selection being governed by the section size and the physical properties desired after heat treatment. Thus SAE 1030 and 1035 are used for shifter forks and many small forgings where moderate properties are desired, but the deeper hardening SAE 1036 is used for more critical parts where a higher strength level and more uniformity is essential, such as some front suspension parts. Forgings such as connecting rods, steering arms, truck front axles, axle shafts, and tractor wheels are commonly made from the SAE 1038 to 1045 group. Larger forgings at similar strength levels need more carbon and perhaps more manganese. Examples are crankshafts from SAE 1046 and 1052. These steels are also used for small forgings where high hardness after oil quenching is desired. Suitable heat treatment is necessary on forgings from this group to provide machinability. These steels are also widely used for parts machined from bar stock, the selection following an identical pattern to that described for forgings. They are used both with and without heat treatment, depending on the application and the level of properties needed. As a class they are considered good for normal machining operations. It is also possible to weld these steels by most commercial methods, but precautions should be taken to avoid cracking from too rapid cooling.

SAE 1055, 1060, 1062, 1064, 1065, 1066, 1070, 1074, 1078, 1080, 1085, 1086, 1090, 1095: Steels in this group are of the high carbon type, having more carbon than is required to achieve maximum as quenched hardness. They are used for applications where the higher carbon is needed to improve wear characteristics for cutting edges, to make springs, and for special purposes. Selection of a particular grade is affected by the nature of the part, its end use, and the manufacturing methods available.

In general, cold forming methods are not practical on this group of steels, being limited to flat stampings and springs coiled from small diameter wire. Practically all parts from these steels are heat treated before use, with some variations in heat treating methods to obtain optimum properties for the particular use to which the steel is to be put.

Uses in the spring industry include SAE 1065 for pretempered wire and SAE 1066 for cushion springs of hard drawn wire, SAE 1064 may be used for small washers and thin stamped parts, SAE 1074 for light flat springs formed from annealed stock, and SAE 1080 and 1085 for thicker flat springs. SAE 1085 is also used for heavier coil springs. Valve spring wire and music wire are special products.

Due to good wear properties when properly heat treated, the high carbon steels find wide usage in the farm implement industry. SAE 1070 has been used for plow beams, SAE 1074 for plow shares, and SAE 1078 for such parts as rake teeth, scrapers, cultivator shovels and plow shares. SAE 1085 has been used for scraper blades, disks, and for spring tooth harrows. SAE 1086 and 1090 find use as mower and binder sections, twine holders, and knotter disks.

Free Cutting Steels. — *SAE 1111, 1112, 1113:* This class of steels is intended for those uses where easy machining is the primary requirement. They are character-ized by a higher sulphur content than comparable carbon steels. This results in some sacrifice of cold forming properties, weldability, and forging characteristics. In general the uses are similar to those for carbon steels of similar carbon and manganese content.

These steels are commonly known as Bessemer screw stock, and are considered the best machining steels available, machinability improving within the group as sulphur increases. They are used for a wide variety of machined parts. While of excellent strength in the cold drawn condition, they have an unfavorable property of cold shortness and are not commonly used for vital parts. These steels may be cyanided or carburized but when uniform response to heat treating is necessary, open hearth steels are recommended.

SAE 1109, 1114, 1115, 1116, 1117, 1118, 1119, 1120, 1126: Steels in this group are used where a combination of good machinability and more uniform response to heat treatment is needed. The lower carbon varieties are used for small parts which are to be cyanided or carbonitrided. SAE 1116, 1117, 1118, and 1119 carry more manganese for better hardenability, permitting oil quenching after case hardening heat treatments in many instances. The higher carbon SAE 1120 and 1126 provide more core hardness when this is needed.

SAE 1132, 1137, 1138, 1140, 1141, 1144, 1145, 1146, 1151: This group of steels has characteristics comparable to carbon steels of the same carbon level, except for changes due to higher sulphur as noted previously.

They are widely used for parts where a large amount of machining is necessary, or where threads, splines or other operations offer special tooling problems. SAE 1137, for example, is widely used for nuts and bolts and studs with machined threads. The higher manganese SAE 1132, 1137, 1141, and 1144 offer greater hardenability, the higher carbon types being suitable for oil quenching for many parts. All of these steels may be selectively hardened by induction or flame heating if desired.

Carburizing Grades of Alloy Steels. *Properties of the Case:* The properties of carburized and hardened cases depend upon the carbon and alloy content, the structure of the case, and the degree and distribution of residual stresses. The carbon content of the case depends upon the details of the carburizing process, and the response of iron and the alloying elements present, to carburization. The original carbon content of the steel has little or no effect upon the carbon content produced in the case. The hardenability of the case therefore depends upon the alloy content of the steel and the final carbon content produced by carburizing, but not upon the initial carbon content of the steel.

With complete carbide solution the effect of alloying elements upon the harden-ability of the case, will in general be the same as the effect of these elements upon the hardenability of the core. As an exception to this, any element which inhibits

carburizing may reduce the hardenability of the case. It is also true that some elements which raise the hardenability of the core may tend to produce more retained austenite and consequently somewhat lower hardness in the case.

Alloy steels are frequently used for case hardening because the required surface hardness can be obtained by moderate speeds of quenching. This may mean less distortion than would be encountered with water quenching. It is usually desirable to select a steel which will attain a minimum surface hardness of 58 or 60 Rockwell C after carburizing and oil quenching. Where section sizes are large, a high hardenability alloy steel may be necessary, while for medium and light sections, low hardenability steels will suffice.

In general, the case hardening alloy steels may be divided into two classes so far as the hardenability of the case is concerned. Only the general type of steel (SAE 3300–4100, etc.) is given. As the original carbon content of the steel has no effect upon the carbon content of the case, the last two digits in the specification numbers are not meaningful so far as the case is concerned.

(a) — *High Hardenability Case.* — *SAE 2500, 3300, 4300, 4800, 9300*

As these are high alloy steels, both the case and the core have high hardenability. These types of steel are used particularly for carburized parts having thick sections, such as bevel drive pinions and heavy gears. Good case properties can be obtained by oil quenching. These steels are likely to have retained austenite in the case after carburizing and quenching, consequently special precautions or treatments, such as refrigeration, may be required.

(b) — *Medium Hardenability Case.* — *SAE 1300, 2300, 4000, 4100, 4600, 5100, 8600, 8700*

Carburized cases of these steels have medium hardenability which means that their hardenability is intermediate between that of plain carbon steel and the higher alloy carburizing steels just described. In general, these steels can be used for average size case hardened automotive parts such as gears, pinions, piston pins, ball studs, universal crosses, crankshafts, etc. Satisfactory case hardness should be produced in most cases by oil quenching.

Core Properties: The core properties of case hardened steels depend upon both carbon and alloy content of the steel. Each of the general types of alloy case hardening steel is usually made with two or more carbon contents so as to produce different hardenability in the core.

The most desirable hardness for the core depends upon the design and functioning of the individual part. In general, where high compressive loads are encountered, relatively high core hardness is beneficial in supporting the case. Low core hardnesses may be desirable where great toughness is essential.

The case hardening steels may be divided into three general classes depending upon hardenability of the core.

(a) — *Low Hardenability Core.* — *SAE 4017, 4023, 4024, 4027[1], 4028[1], 4608, 4615, 4617[1], 8615[1], 8617[1]*

(b) — *Medium Hardenability Core.* — *SAE 1320, 2317, 2512, 2515[1], 3115, 3120, 4032, 4119, 4317, 4620, 4621, 4812, 4815[1], 5115, 5120, 8620, 8622, 8720, 9420*

(c) — *High Hardenability Core.* — *SAE 2517, 3310, 3316, 4320, 4817, 4820, 9310, 9315, 9317*

Heat Treatments: In general, all of the alloy carburizing steels are made fine grain and most are suitable for direct quenching from the carburizing temperature. Several other types of heat treatment involving single and double quenching are also used for most of these steels. (See tables of Typical Heat Treatments for SAE Steels.)

---

[1] Borderline classifications might be considered in the next higher hardenability group.

Directly Hardenable Grades of Alloy Steels. — These steels may be considered in five groups on the basis of approximate mean carbon content of the SAE specification. In general, the last two figures of the specification agree with the mean carbon content. Consequently the heading " .30–.37 Mean Carbon Content of SAE Specification " includes steels such as SAE 1330, 3135, and 4137.

Mean Carbon Content of SAE Specification	Common Applications
(a) — .30–.37 per cent	Heat treated parts requiring moderate strength and great toughness.
(b) — .40–.42 per cent	Heat treated parts requiring higher strength and good toughness.
(c) — .45–.50 per cent	Heat treated parts requiring fairly high hardness and strength with moderate toughness.
(d) — .50–.62 per cent	Springs and hand tools.
(e) — 1.02 per cent	Ball and roller bearings.

It is necessary to deviate from the above plan in the classification of the carbon molybdenum steels. When carbon molybdenum steels are used, it is customary to specify higher carbon content for any given application than would be specified for other alloy steels, due to the low alloy content of these steels. For example, SAE 4063 is used for the same applications as SAE 4140, 4145 and 5150. Consequently in the following discussion, the carbon molybdenum steels have been shown in the groups where they belong on the basis of applications rather than carbon content.

For the present discussion, steels of each carbon content are divided into two or three groups on the basis of hardenability. Transformation ranges and consequently heat treating practices vary somewhat with different alloying elements even though the hardenability is not changed.

.30–.37 Mean Carbon Content of SAE Specification. — These steels are frequently used for water quenched parts of moderate section size and for oil quenched parts of small section size. Typical applications of these steels are connecting rods, steering arms and steering knuckles, axle shafts, bolts, studs, screws, and other parts requiring strength and toughness where section size is small enough to permit obtaining the desired physical properties with the customary heat treatment.

Steels falling in this classification may be subdivided into two groups on the basis of hardenability:

(a) — Low Hardenability: SAE 1330, 1335, 4037, 4042, 4130, 5130, 5132, 8630

(b) — Medium Hardenability: SAE 2330, 3130, 3135, 4137, 5135, 8632, 8635, 8637, 8735, 9437

.40–.42 Mean Carbon Content of SAE Specification. — In general, these steels are used for medium and large size parts requiring high degree of strength and toughness. The choice of the proper steel depends upon the section size and the mechanical properties which must be produced. The low and medium hardenability steels are used for average size automotive parts such as steering knuckles, axle shafts, propeller shafts, etc. The high hardenability steels are used particularly for large axles and shafts for large aircraft parts.

These steels are usually considered as oil quenching steels, although some large parts made of the low and medium hardenability classifications may be quenched in water under properly controlled conditions.

These steels may be divided into three groups on the basis of hardenability:

(a) — Low Hardenability: SAE 1340, 4047, 5140, 9440

(b) — Medium Hardenability: SAE 2340, 3140, 3141, 4053, 4063, 4140, 4640, 8640, 8641, 8642, 8740, 8742, 9442

(c) — High Hardenability: SAE 4340, 9840

.45–.50 Mean Carbon Content of SAE Specification. — These steels are used primarily for gears and other parts requiring fairly high hardness as well as strength and toughness. Such parts are usually oil quenched and a minimum of 90 per cent martensite in the as quenched condition is desirable.

(a) — Low Hardenability: SAE 5045, 5046, 5145, 9747, 9763

(b) — Medium Hardenability: SAE 2345, 3145, 3150, 4145, 5147, 5150, 8645, 8647, 8650, 8745, 8747, 8750, 9445, 9845

(c) — High Hardenability: SAE 4150, 9850

.50–.62 Mean Carbon Content of SAE Specification. — These steels are used primarily for springs and hand tools. The hardenability necessary depends upon the thickness of the material and the quenching practice.

(a) — Medium Hardenability: SAE 4068, 5150, 5152, 6150, 8650, 9254, 9255, 9260, 9261

(b) — High Hardenability: SAE 8653, 8655, 8660, 9262

1.02 Mean Carbon Content of SAE Specification. — SAE 50100, 51100, 52100

These are straight chromium electric furnace steels used primarily for the races and balls or rollers of anti-friction bearings. They are also used for other parts requiring high hardness and wear resistance. The compositions of the three steels are identical, except for a variation in chromium, with a corresponding variation in hardenability.

(a) — Low Hardenability: SAE 50100

(b) — Medium Hardenability: SAE 51100, 52100

Resulphurized Steel. — Some of the alloy steels, SAE 4024, 4028 and 8641, are made resulphurized so as to give better machinability at a relatively high hardness. In general, increased sulphur results in decreased transverse ductility, notched impact toughness, and weldability.

Chromium Nickel Austenitic Steels *(Not capable of heat treatment).* —

SAE 30301: This steel is capable of attaining high tensile strength and ductility by moderate or severe cold working. It is used largely in the cold rolled or cold drawn condition in the form of sheet, strip and wire. Its corrosion resistance is good but not equal to SAE 30302.

SAE 30302: This is the most widely used of the general purpose austenitic chromium nickel stainless steels. It is used for deep drawing largely in the annealed condition. It can be worked to high tensile strengths but with slightly lower ductility than SAE 30301.

SAE 30303F: This is a free machining type recommended for the manufacture of parts produced on automatic machines. Caution must be used in forging this steel.

SAE 30304: This is similar to SAE 30302 but somewhat superior in corrosion resistance and having superior welding properties for certain types of equipment.

SAE 30305: Similar to SAE 30304 but capable of lower hardness. Has greater ductility with slower work hardening tendency.

SAE 30309: This steel has high heat resisting qualities and is resistant to oxidation at temperatures up to about 1800 deg. F.

SAE 30310: This steel has the highest heat resisting properties of any of the chromium nickel steels listed herewith and is used to resist oxidation at temperatures up to about 1900 deg. F.

SAE 30316: This steel is recommended for use in parts where unusual resistance to chemical or salt water corrosion is necessary. It has superior creep strength at elevated temperatures.

SAE 30317: This steel is similar to SAE 30316 but has the highest corrosion resistance of all these alloys in many environments.

SAE 30321: This steel is recommended for use in the manufacture of welded structures where heat treatment after welding is not feasible. It is also recommended for use where temperatures up to 1600 deg. F. are encountered in service.

SAE 30325: Used for such parts as heat control shafts.

SAE 30347: This steel is similar to SAE 30321 with the following additional statement. This columbium alloy is sometimes preferred to titanium because less columbium is lost in the welding operation.

Stainless Chromium Irons and Steels. — *SAE 51410:* This is a general purpose stainless steel capable of heat treatment to show good physical properties. It is used for general stainless applications, both in the heat treated and annealed condition but it is not as resistant to corrosion as SAE 51430 in either the annealed or heat treated condition.

SAE 51414: This is a corrosion and heat resisting nickel-bearing chromium steel with somewhat better corrosion resistance than SAE 51410. It will attain slightly higher mechanical properties when heat treated than SAE 51410. It is used in the form of tempered strip or wire, and in bars and forgings for heat treated parts.

SAE 51416F: This is a free machining grade for the manufacture of parts produced in automatic screw machines.

SAE 51420: This steel is capable of heat treating to a relatively high hardness. It will harden to a maximum of approximately 500 Brinell. It has its maximum corrosion resisting qualities only in the fully hardened condition. It is used for cutlery, hardened pump shafts, etc.

SAE 51420F: This is similar to SAE 51420 except for its free machining properties.

SAE 51430: This is a steel of a high chromium type not capable of heat treatment and is recommended for use in parts of moderate draw. Corrosion and heat resistance are superior to SAE 51410.

SAE 51430F: This is similar to SAE 51430 except for its free machining properties.

SAE 51431: This is a nickel bearing chromium steel designed for heat treatment to high mechanical properties. Its corrosion resistance is superior to other hardenable steels.

SAE 51440A: A hardenable chromium steel with greater quenched hardness than SAE 51420 and greater toughness than SAE 51440B and 51440C. Maximum corrosion resistance is obtained in the fully hardened and polished condition.

SAE 51440B: A hardenable chromium steel with greater quenched hardness than SAE 51440A. Maximum corrosion resistance is obtained in the fully hardened and polished condition. Capable of hardening to 50–60 Rockwell C depending upon carbon content.

SAE 51440C: This steel has the greatest quenched hardness and wear resistance upon heat treatment of any corrosion or heat resistant steel.

SAE 51440F: The same as SAE 51440C, except for its free machining characteristics.

SAE 51442: A corrosion and heat resisting chromium steel with corrosion resisting properties slightly better than SAE 51430 and with good scale resistance up to 1600 deg. F.

SAE 51446: A corrosion and heat resisting steel with maximum amount of chromium consistent with commercial malleability. Used principally for parts which must resist high temperatures in service without scaling. Resists oxidation up to 2000 deg. F.

SAE 51501: Used for its heat and corrosion resistance and good mechanical properties at temperatures up to approximately 1000 deg. F.

General Applications of SAE Steels

These applications are intended as a general guide only since the selection may depend upon the exact character of the service, cost of material, machinability when machining is required, or other factors. When more than one steel is recommended for a given application, information on the characteristics of each steel listed will be found in the section beginning on page 417.

Application	SAE No.	Application	SAE No.
Adapters...............	1145	Chain pins, transmission ..	4320
Agricultural steel........	1070	" " " ..	4815
" "	1080	" " " ..	4820
Aircraft forgings.........	4140	Chains, transmission.....	3135
Axles, front or rear.......	1040	" "	3140
" " "........	4140	Clutch disks.............	1060
Axle shafts..............	1045	" "	1070
" "	2340	" "	1085
" "	2345	Clutch springs...........	1060
" "	3135	Coil springs.............	4063
" "	3140	Cold-headed bolts........	4042
" "	3141	Cold-heading steel........	30905
" "	4063	Cold-heading wire or rod..	rimmed*
" "	4340	" " " ..	1035
Ball-bearing races........	52100	Cold-rolled steel.........	1070
Balls for ball bearings.....	52100	Connecting-rods.........	1040
Body stock for cars.......	rimmed*	" "	3141
Bolts, anchor...........	1040	Connecting-rod bolts.....	3130
Bolts and screws.........	1035	Corrosion resisting.......	51710
Bolts, cold-headed........	4042	" "	30805
Bolts, connecting-rod.....	3130	Covers, transmission......	rimmed*
Bolts, heat-treated.......	2330	Crankshafts.............	1045
Bolts, heavy-duty........	4815	"	1145
" " "	4820	"	3135
Bolts, steering-arm.......	3130	"	3140
Brake levers.............	1030	"	3141
" "	1040	Crankshafts, Diesel engine.	4340
Bumper bars............	1085	Cushion springs.........	1060
Cams, free-wheeling......	4615	Cutlery, stainless.........	51335
" "	4620	Cylinder studs...........	3130
Camshafts..............	1020	Deep-drawing steel.......	rimmed*
"	1040	" " "	30905
Carburized parts.........	1020	Differential gears.........	4023
" "	1022	Disks, clutch............	1070
" "	1024	" "	1060
" "	1320	Ductile steel.............	30905
" "	2317	Fan blades..............	1020
" "	2515	Fatigue resisting.........	4340
" "	3310	" "	4640
" "	3115	Fender stock for cars.....	rimmed*
" "	3120	Forgings, aircraft........	4140
" "	4023	Forgings, carbon steel.....	1040
" "	4032	" " "	1045
" "	1117	Forgings, heat-treated.....	3240
" "	1118	" " "	5140

* The "rimmed" and "killed" steels listed are in the SAE 1008, 1010 and 1015 group. See general description of these steels.

General Applications of SAE Steels

These applications are intended as a general guide only since the selection may depend upon the exact character of the service, cost of material, machinability when machining is required, or other factors. When more than one steel is recommended for a given application, information on the characteristics of each steel listed will be found in the section beginning on page 417.

Application	SAE No.	Application	SAE No.
Forgings, heat-treated....	6150	Key stock..............	1030
Forgings, high-duty.......	6150	" "	2330
Forgings, small or medium.	1035	" "	3130
Forgings, large..........	1036	Leaf springs............	1085
Free-cutting carbon steel..	1111	" "	9260
" " " " ..	1113	Levers, brake...........	1030
Free-cutting chro.-ni. steel.	30615	" "	1040
Free-cutting mang. steel...	1132	Levers, gear shift........	1030
" " " " ...	1137	Levers, heat-treated......	2330
Gears, carburized........	1320	Lock-washers...........	1060
" "	2317	Mower knives...........	1085
" "	3115	Mower sections..........	1070
" "	3120	Music wire..............	1085
" "	3310	Nuts...................	3130
" "	4119	Nuts, heat-treated.......	2330
" "	4125	Oil-pans, automobile......	rimmed*
" "	4320	Pinions, carburized......	3115
" "	4615	" "	3120
" "	4620	" "	4320
" "	4815	Piston-pins.............	3115
" "	4820	" "	3120
Gears, heat-treated.......	2345	Plow beams............	1070
Gears, car and truck.....	4027	Plow disks.............	1080
" " " "	4032	Plow shares............	1080
Gears, cyanide-hardening..	5140	Propeller shafts.........	2340
Gears, differential........	4023	" "	2345
Gears, high duty.........	4640	" "	4140
" " "	6150	Races, ball-bearing.......	52100
Gears, oil-hardening......	3145	Ring gears..............	3115
" " "	3150	" "	3120
" " " "	4340	" "	4119
" " " "	5150	Rings, snap............	1060
Gears, ring.............	1045	Rivets.................	rimmed*
" "	3115	Rod and wire	killed*
" "	3120	Rod, cold-heading.......	1035
" "	4119	Roller bearings..........	4815
Gears, transmission.......	3115	Rollers for bearings.......	52100
" "	3120	Screws and bolts.........	1035
" "	4119	Screw stock, Bessemer....	1111
Gears, truck and bus.....	3310	" " "	1112
" " " "	4320	" " "	1113
Gear shift levers.........	1030	Screw stock, open hearth...	1115
Harrow disks...........	1080	Screws, heat-treated......	2330
" "	1095	Seat springs............	1095
Hay-rake teeth..........	1095	Shafts, axle............	1045

* The "rimmed" and "killed" steels listed are in the SAE 1008, 1010 and 1015 group. See general description of these steels.

General Applications of SAE Steels

These applications are intended as a general guide only since the selection may depend upon the exact character of the service, cost of material, machinability when machining is required, or other factors. When more than one steel is recommended for a given application, information on the characteristics of each steel listed will be found in the section beginning on page 417.

Application	SAE No.	Application	SAE No.
Shafts, cyanide-hardening..	5140	Steel, cold-heading.......	30905
Shafts, heavy-duty.......	4340	Steel, free-cutting carbon..	11111
" " "	6150	" " " " ..	1113
" " "	4615	Steel, free-cutting chro.-ni..	30615
" " "	4620	Steel, free-cutting mang...	1132
Shafts, oil-hardening......	5150	" " " " ..	0000
Shafts, propeller.........	2340	Steel, minimum distortion .	4615
" "	2345	" " " .	4620
" "	4140	" " " .	4640
Shafts, transmission......	4140	Steel, soft ductile	30905
Sheets and strips.........	rimmed*	Steering arms............	4042
Snap rings.............	1060	Steering-arm bolts........	3130
Spline shafts............	1045	Steering knuckles........	3141
" "	1320	Steering-knuckle pins.....	4815
" "	2340	" " "	4820
" "	2345	Studs...................	1040
" "	3115	"	1111
" "	3120	Studs, cold-headed.......	4042
" "	3135	Studs, cylinder..........	3130
" "	3140	Studs, heat-treated.......	2330
" "	4023	Studs, heavy-duty........	4815
Spring clips.............	1060	" " "	4820
Springs, coil.............	1095	Tacks...................	rimmed*
" "	4063	Thrust washers..........	1060
" "	6150	Thrust washers, oil-harden.	5150
Springs, clutch...........	1060	Transmission shafts......	4140
Springs, cushion.........	1060	Tubing.................	1040
Springs, leaf.............	1085	Tubing, front axle........	4140
" "	1095	Tubing, seamless........	1030
" "	4063	Tubing, welded..........	1020
" "	4068	Universal joints..........	1145
" "	9260	Valve springs............	1060
" "	6150	Washers, lock............	1060
Springs, hard-drawn coiled .	1066	Welded structures........	30705
Springs, oil-hardening.....	5150	Wire and rod............	killed*
Springs, oil-tempered wire.	1066	Wire, cold-heading.......	rimmed*
Springs, seat............	1095	" " "	1035
Springs, valve...........	1060	Wire, hard-drawn spring ..	1045
Spring wire..............	1045	" " " " ...	1055
Spring wire, hard-drawn...	1055	Wire, music.............	1085
Spring wire, oil-tempered..	1055	Wire, oil-tempered spring..	1055
Stainless irons...........	51210	Wrist-pins, automobile....	1020
" "	51710	Yokes..................	1145
Steel, cold-rolled........	1070		

* The "rimmed" and "killed" steels listed are in the SAE 1008, 1010 and 1015 group. See general description of these steels.

Numbering Systems for Metals and Alloys. — Several different numbering systems have been developed for metals and alloys by various trade associations, professional engineering societies, standards organizations, and by private industries for their own use. The numerical code used to identify the metal or alloy may or may not be related to a specification, which is a statement of the technical and commercial requirements that the product must meet. Numbering systems in use include those developed by the American Iron and Steel Institute (AISI), Society of Automotive Engineers (SAE), American Society for Testing and Materials (ASTM), American National Standards Institute (ANSI), Steel Founders Society of America, American Society of Mechanical Engineers (ASME), American Welding Society (AWS), Aluminum Association, Copper Development Association, U.S. Department of Defense (Military Specifications), and the General Accounting Office (Federal Specifications).

The Unified Numbering System (UNS) was developed through a joint effort of the ASTM and the SAE to provide a means of correlating the different numbering systems for metals and alloys that have a commercial standing. This system avoids the confusion caused when more than one identification number is used to specify the same material, or when the same number is assigned to two entirely different materials. It is important to understand that a UNS number is not a specification; it is an identification number for metals and alloys for which detailed specifications are provided elsewhere. There are seventeen series of UNS numbers, which are shown in Table 1. Each UNS number consists of a letter prefix followed by five digits. In some cases the letter is suggestive of the family of metals identified by the series, such as A for aluminum and C for copper. Whenever possible, the numbers in the UNS number groups contain numbering sequences taken directly from other systems in order to facilitate the identification of the material; e.g., the corresponding UNS number for AISI 1020 steel is G10200. The UNS numbers corresponding to the commonly used AISI-SAE numbers that are used to identify plain carbon alloy and tool steels are given in Table 2.

Table 1. Unified Numbering System (UNS) for Metals and Alloys

UNS Series	Metal
Nonferrous Metals and Alloys	
A00001 to A99999	Aluminum and aluminum alloys
C00001 to C99999	Copper and copper alloys
E00001 to E99999	Rare earth and rare earth-like metals and alloys
L00001 to L99999	Low melting metals and alloys
M00001 to M99999	Miscellaneous nonferrous metals and alloys
P00001 to P99999	Precious metals and alloys
R00001 to R99999	Reactive and refractory metals and alloys
Z00001 to Z99999	Zinc and zinc alloys
Ferrous Metals and Alloys	
D00001 to D99999	Specified mechanical property steels
F00001 to F99999	Cast irons
G00001 to G99999	AISI and SAE carbon and alloy steels (except tool steels)
H00001 to H99999	AISI H-steels
J00001 to J99999	Cast steels (except tool steels)
K00001 to K99999	Miscellaneous steels and ferrous alloys
S00001 to S99999	Heat and corrosion resistant (stainless) steels
T00001 to T99999	Tool steels

Table 2. AISI and SAE Numbers and Their Corresponding UNS Numbers for Plain Carbon, Alloy, and Tool Steels

AISI-SAE Numbers	UNS Numbers	AISI-SAE Numbers	UNS Numbers	AISI-SAE Numbers	UNS Numbers	AISI-SAE Numbers	UNS Numbers
Plain Carbon Steels							
1005	G10050	1030	G10300	1070	G10700	1566	G15660
1006	G10060	1035	G10350	1078	G10780	1110	G11100
1008	G10080	1037	G10370	1080	G10800	1117	G11170
1010	G10100	1038	G10380	1084	G10840	1118	G11180
1012	G10120	1039	G10390	1086	G10860	1137	G11370
1015	G10150	1040	G10400	1090	G10900	1139	G11390
1016	G10160	1042	G10420	1095	G10950	1140	G11400
1017	G10170	1043	G10430	1513	G15130	1141	G11410
1018	G10180	1044	G10440	1522	G15220	1144	G11440
1019	G10190	1045	G10450	1524	G15240	1146	G11460
1020	G10200	1046	G10460	1526	G15260	1151	G11510
1021	G10210	1049	G10490	1527	G15270	1211	G12110
1022	G10220	1050	G10500	1541	G15410	1212	G12120
1023	G10230	1053	G10530	1548	G15480	1213	G12130
1025	G10250	1055	G10550	1551	G15510	1215	G12150
1026	G10260	1059	G10590	1552	G15520	12L14	G12144
1029	G10290	1060	G10600	1561	G15610	…	…
Alloy Steels							
1330	G13300	4150	G41500	5140	G51400	8642	G86420
1335	G13350	4161	G41610	5150	G51500	8645	G86450
1340	G13400	4320	G43200	5155	G51550	8655	G86550
1345	G13450	4340	G43400	5160	G51600	8720	G87200
4023	G40230	E4340	G43406	E51100	G51986	8740	G87400
4024	G40240	4615	G46150	E52100	G52986	8822	G88220
4027	G40270	4620	G46200	6118	G61180	9260	G92600
4028	G40280	4626	G46260	6150	G61500	50B44	G50441
4037	G40370	4720	G47200	8615	G86150	50B46	G50461
4047	G40470	4815	G48150	8617	G86170	50B50	G50501
4118	G41180	4817	G48170	8620	G86200	50B60	G50601
4130	G41300	4820	G48200	8622	G86220	51B60	G51601
4137	G41370	5117	G51170	8625	G86250	81B45	G81451
4140	G41400	5120	G51200	8627	G86270	94B17	G94171
4142	G41420	5130	G51300	8630	G86300	94B30	G94301
4145	G41450	5132	G51320	8637	G86370	…	…
4147	G41470	5135	G51350	8640	G86400	…	…
Tool Steels (AISI and UNS Only)							
M1	T11301	T6	T12006	A6	T30106	P4	T51604
M2	T11302	T8	T12008	A7	T30107	P5	T51605
M4	T11304	T15	T12015	A8	T30108	P6	T51606
M6	T11306	H10	T20810	A9	T30109	P20	T51620
M7	T11307	H11	T20811	A10	T30110	P21	T51621
M10	T11310	H12	T20812	D2	T30402	F1	T60601
M3-1	T11313	H13	T20813	D3	T30403	F2	T60602
M3-2	T11323	H14	T20814	D4	T30404	L2	T61202
M30	T11330	H19	T20819	D5	T30405	L3	T61203
M33	T11333	H21	T20821	D7	T30407	L6	T61206
M34	T11334	H22	T20822	O1	T31501	W1	T72301
M36	T11336	H23	T20823	O2	T31502	W2	T72302
M41	T11341	H24	T20824	O6	T31506	W5	T72305
M42	T11342	H25	T20825	O7	T31507	CA2	T90102
M43	T11343	H26	T20826	S1	T41901	CD2	T90402
M44	T11344	H41	T20841	S2	T41902	CD5	T90405
M46	T11346	H42	T20842	S4	T41904	CH12	T90812
M47	T11347	H43	T20843	S5	T41905	CH13	T90813
T1	T12001	A2	T30102	S6	T41906	CO1	T91501
T2	T12002	A3	T30103	S7	T41907	CS5	T91905
T4	T12004	A4	T30104	P2	T51602	…	…
T5	T12005	A5	T30105	P3	T51603	…	…

Standard Steels — Compositions, Applications, and Heat-treatments

The standard steel compositions of the Society of Automotive Engineers (SAE), Inc., given in the accompanying table, are considered adequate for practically all parts made of ferrous materials that are necessary for the production of automotive apparatus, and include grades that have been found commercially available and technically adequate for the service required of such parts. Definite applications of SAE steels are not specified as the selection of a proper steel for a given part must depend upon an intimate knowledge of a number of important factors, such as the availability and price of the material, the detailed design of the part and the severity of the service to be imposed, whether the part is to be forged or machined and its machineability; hence only general applications are indicated. (See following text and tables.)

AISI-SAE System of Designating Carbon and Alloy Steels

AISI-SAE Designation	Type of Steel and Nominal Alloy Content
	Carbon Steels
10xx	Plain Carbon (Mn 1.00% max.)
11xx	Resulfurized
12xx	Resulfurized and Rephosphorized
15xx	Plain Carbon (Max. Mn range 1.00 to 1.65%)
	Manganese Steels
13xx	Mn 1.75
	Nickel Steels
23xx	Ni 3.50
25xx	Ni 5.00
	Nickel-Chromium Steels
31xx	Ni 1.25; Cr 0.65 and 0.80
32xx	Ni 1.75; Cr 1.07
33xx	Ni 3.50; Cr 1.50 and 1.57
34xx	Ni 3.00; Cr 0.77
	Molybdenum Steels
40xx	Mo 0.20 and 0.25
44xx	Mo 0.40 and 0.52
	Chromium-Molybdenum Steels
41xx	Cr 0.50, 0.80, and 0.95; Mo 0.12, 0.20, 0.25, and 0.30
	Nickel-Chromium-Molybdenum Steels
43xx	Ni 1.82; Cr 0.50 and 0.80; Mo 0.25
43BVxx	Ni 1.82; Cr 0.50; Mo 0.12 and 0.35; V 0.03 min.
47xx	Ni 1.05; Cr 0.45; Mo 0.20 and 0.35
81xx	Ni 0.30; Cr 0.40; Mo 0.12
86xx	Ni 0.55; Cr 0.50; Mo 0.20
87xx	Ni 0.55; Cr 0.50; Mo 0.25
88xx	Ni 0.55; Cr 0.50; Mo 0.35
93xx	Ni 3.25; Cr 1.20; Mo 0.12
94xx	Ni 0.45; Cr 0.40; Mo 0.12
97xx	Ni 0.55; Cr 0.20; Mo 0.20
98xx	Ni 1.00; Cr 0.80; Mo 0.25
	Nickel-Molybdenum Steels
46xx	Ni 0.85 and 1.82; Mo 0.20 and 0.25
48xx	Ni 3.50; Mo 0.25
	Chromium Steels
50xx	Cr 0.27, 0.40, 0.50, and 0.65
51xx	Cr 0.80, 0.87, 0.92, 0.95, 1.00, and 1.05
50xxx	Cr 0.50; C 1.00 min.
51xxx	Cr 1.02; C 1.00 min.
52xxx	Cr 1.45; C 1.00 min.
	Chromium-Vanadium Steels
61xx	Cr 0.60, 0.80, and 0.95; V 0.10 and 0.15 min
	Tungsten-Chromium Steels
72xx	W 1.75; Cr 0.75
	Silicon-Manganese Steels
92xx	Si 1.40 and 2.00; Mn 0.65, 0.82, and 0.85; Cr 0.00 and 0.65
	High-Strength Low-Alloy Steels
9xx	Various SAE grades

Numbering Systems for Steel. — The primary numbering systems for identifying steels are the American Iron and Steel Institute (AISI) and the Society of Automotive Engineers (SAE) systems. Although they are entirely separate systems, they are closely coordinated and are nearly identical. The basic AISI-SAE numbering system for plain carbon and alloy steels is shown in the table on page 429. All steels are identified by four numbers, except certain chromium steels which have five numbers. The first two numbers identify the type of steel and the last two numbers indicate the approximate amount of carbon in hundredths of a per cent. Placed between the first and second pair of numbers, the letter L indicates that the steel contains lead to improve its machinability; when placed in this position, the letter B indicates a boron steel. The prefix E, as in E52100, indicates a steel made by the basic electric furnace method; the suffix H, as in 4150H, indicates a steel that is produced to specific hardenability limits. The Unified Number System (UNS) for metals and alloys is also used to designate steels (see pages 427 and 428).

Classification of Tool Steels. — Steels for tools must satisfy a number of different, often conflicting requirements. The need for specific steel properties arising from widely varying applications has led to the development of many compositions of tool steels, each intended to meet a particular combination of applicational requirements. The resultant diversity of tool steels, their number being continually expanded by the addition of new developments, made it extremely difficult for the user to select the type best suited to his needs, or to find equivalent alternatives for specific types available from particular sources.

As a cooperative industrial effort under the sponsorship of AISI and SAE, a tool classification system has been developed in which the commonly used tool steels are grouped into seven major categories. These categories, several of which contain more than a single group, are listed in the following with the letter symbols used for their identification. The individual types of tool steels within each category are identified by suffix numbers following the letter symbols.

Category Designation	Letter Symbol	Group Designation
High Speed Tool Steels	M T	Molybdenum types Tungsten types
Hot Work Tool Steels	H1–H19 H20–H39 H40–H59	Chromium types Tungsten types Molybdenum types
Cold Work Tool Steels	D A O	High carbon, high chromium types Medium alloy, air hardening types Oil hardening types
Shock Resisting Tool Steels	S	. . . . .
Mold Steels	P	. . . . .
Special Purpose Tool Steels	L F	Low alloy types Carbon tungsten types
Water Hardening Tool Steels	W	. . . . .

Table 3. Quick Reference Guide for Tool Steel Selection — 1

Application Areas	High Speed Tool Steels, M and T	Hot Work Tool Steels H	Cold Work Tool Steels, D, A and O	Shock Resisting Tool Steels, S	Mold Steels, P	Special Purpose Tool Steels, L and F	Water Hardening Tool Steels, W
			Examples of Typical Application				
Cutting Tools Single point types (lathe, planer, boring) Milling cutters Drills Reamers Taps Threading dies Form cutters	General purpose production tools: M2, T1 For increased abrasion resistance: M3, M4, and M10 Heavy-duty work calling for high hot hardness: T5, T15 Heavy-duty work calling for high abrasion resistance: M42, M44		Tools with keen edges (knives, razors) Tools for operations where no high speed is involved, yet stability in heat treatment and substantial abrasion resistance are needed	Pipe cutter wheels			Uses which do not require hot hardness or high abrasion resistance. Examples with carbon content of applicable group: Taps (1.05/1.10%C) Reamers (1.10/1.15%C) Twist drills (1.20/1.25%C) Files (1.35/1.40%C)
Hot Forging Tools and Dies Dies and inserts Forging machine plungers and pierces	For combining hot hardness with high abrasion resistance: M2, T1	Dies for presses and hammers: H20, H21 For severe conditions over extended service periods: H22 to H26, also H43	Hot trimming dies: D2	Hot trimming dies Blacksmith tools Hot swaging dies			Smith's tools (.65/.70%C) Hot chisels (.70/.75%C) Drop forging dies (.90/1.00%C) Applications limited to short-run production
Hot Extrusion Tools and Dies Extrusion dies and mandrels, Dummy blocks Valve extrusion tools	Brass extrusion dies: T1	Extrusion dies and dummy blocks: H20 to H26 For tools which are exposed to less heat: H10 to H19		Compression molding: S1			

Table 3. Quick Reference Guide for Tool Steel Selection — 2

Application Areas	High Speed Tool Steels, M and T	Hot Work Tool Steels, H	Cold Work Tool Steels, D, A and O	Shock Resisting Tool Steels, S	Mold Steels, P	Special Purpose Tool Steels, L and F	Water Hardening Tool Steels, W
			Tool Steel Categories and AISI Letter Symbols				
			Examples of Typical Application				
Cold Forming Dies Bending, forming, drawing, and deep drawing dies and punches	Burnishing tools: M1, T1	Cold heading die casings: H13	Drawing dies: or Coining tools: O1, D2 Forming and bending dies: A2 Thread rolling dies: D2	Hobbing and short-run applications: S1, S7 Rivet sets and rivet busters		Blanking, forming, and trimmer dies when toughness has precedence over abrasion resistance: L6	Cold heading dies: W1 or W2 (C = 1.00%). Bending dies: W1 (C = 1.00%)
Shearing Tools Dies for piercing, punching, and trimming Shear blades	Special dies for cold and hot work: T1 For work requiring high abrasion resistance: M2, M3	For shearing knives: H11, H12 For severe hot shearing applications: M21, M25	Dies for medium runs: A2, A6 also O1 and O4 Dies for long runs: D2, D3 Trimming dies (also for hot trimming): A2	Cold and hot shear blades Hot punching and piercing tools Boilermaker's tools		Knives for work requiring high toughness: L6	Trimming dies (.90/.95%C) Cold blanking and punching dies (1.00%C)
Die Casting and Molding Dies		For aluminum and lead: H11 and H13 For brass: H21	A2 and A6 O1		Plastic molds: P2 to P4, and P20		
Structural Parts for Severe Service Conditions	Roller bearings for high temperature environment: T1 Lathe centers: M2 and T1	For aircraft components (landing gears, arrester hooks, rocket cases): H11	Lathe centers: D2, D3 Arbors: O1 Bushings: A4 Gages: D2	Pawls Clutch parts		Spindles, clutch parts (where high toughness is needed): L6	Spring steel (1.10/1.15%C)
Battering Tools for Hand and Power Tool Use				Pneumatic chisels for cold work: S5 For higher performance: S7			For intermittent use: W1 (.80%C)

Table 4. Molybdenum High Speed Steels

Identifying Chemical Composition and Typical Heat Treatment Data

		M1	M2	M3 Cl.1	M3 Cl.2	M4	M6	M7	M10	M30	M33	M34	M36	M41	M42	M43	M44	M46	M47
Identifying Chemical Composition Elements in Per Cent	C	.80	.85; 1.00	1.05	1.20	1.30	.80	1.00	.85; 1.00	.80	.90	.90	.80	1.10	1.10	1.20	1.15	1.25	1.10
	W	1.50	6.00	6.00	6.00	5.50	4.00	1.75	…	2.00	1.50	2.00	6.00	6.75	1.50	2.75	5.25	2.00	1.50
	Mo	8.00	5.00	5.00	5.00	4.50	5.00	8.75	8.00	8.00	9.50	8.00	5.00	3.75	9.50	8.00	6.25	8.25	9.50
	Cr	4.00	4.00	4.00	4.00	4.00	4.00	4.00	4.00	4.00	4.00	4.00	4.00	4.25	3.75	3.75	4.25	4.00	3.75
	V	1.00	2.00	2.40	3.00	4.00	1.50	2.00	2.00	1.25	1.15	2.00	2.00	2.00	1.15	1.60	2.25	3.20	1.25
	Co	…	…	…	…	…	12.00	…	…	5.00	8.00	8.00	8.00	5.00	8.00	8.25	12.00	8.25	5.00
Heat Treat Data	Hardening Temperature Range, °F	2150-2225	2175-2225	2200-2250	2200-2250	2200-2250	2150-2200	2150-2225	2150-2225	2200-2250	2200-2250	2200-2250	2225-2275	2175-2220	2175-2210	2175-2220	2190-2240	2175-2225	2150-2200
	Tempering Temperature Range, °F	1000-1100	1000-1160	1000-1100	1000-1100	1000-1100	1000-1100	1000-1100	1000-1100	1000-1100	1000-1100	1000-1100	1000-1100	1000-1100	950-1100	950-1100	1000-1160	975-1050	975-1100
	Approx. Tempered Hardness, R_c	65-66	65-66	66-61	66-61	66-61	66-61	66-61	65-66	65-66	65-66	65-66	65-66	70-65	70-65	70-65	70-62	69-67	70-65

Relative Ratings of Properties (A = greatest to E = least):

		M1	M2	M3 Cl.1	M3 Cl.2	M4	M6	M7	M10	M30	M33	M34	M36	M41	M42	M43	M44	M46	M47
Characteristics in Heat Treatment	Safety in Hardening	D	D	D	D	D	D	D	D	D	D	D	D	D	D	D	D	D	D
	Depth of Hardening	A	A	A	A	A	A	A	A	A	A	A	A	A	A	A	A	A	A
	Resistance to Decarburization	C	B	B	B	B	C	C	C	C	C	C	C	C	C	C	C	C	C
	Stability of Shape in Heat Treatment (Quenching Medium: Air or Salt / Oil)	C	C	C	C	C	C	C	C	C	C	C	C	C	C	C	C	C	C
Service Properties	Machinability	D	D	D	D/E	D	D	D	D	D	D	D	D	D	D	D	D	D	D
	Hot Hardness	B	B	B	B	B	B	B	B	B	B	B	B	B	B	B	B	B	B
	Wear Resistance	B	B	B	A	A	B	B	B	B	B	B	B	B	B	B	B	B	B
	Toughness	E	E	E	E	E	E	E	E	E	E	E	E	E	E	E	E	E	E

Table 6. Hot Work Tool Steels

Identifying Chemical Composition and Typical Heat Treatment Data

Relative Ratings of Properties (A = greatest to D = least)

		Chromium Types						Tungsten Types						Molybdenum Types		
	Type	H10	H11	H12	H13	H14	H19	H21	H22	H23	H24	H25	H26	H41	H42	H43
Identifying Elements in %	C	.40	.35	.35	.35	.40	.40	.35	.35	.35	.45	.25	.50	.65	.60	.55
	W	..	..	1.50	..	5.00	4.25	9.00	11.00	12.00	15.00	15.00	18.00	1.50	6.00	..
	Mo	2.50	1.50	1.50	1.50	..	..	..	..	..	..	..	..	8.00	5.00	8.00
	Cr	3.25	5.00	5.00	5.00	5.00	4.25	3.50	2.00	12.00	3.00	4.00	4.00	4.00	4.00	4.00
	V	.40	.40	.40	1.00	..	2.00	..	..	..	..	..	1.00	1.00	2.00	2.00
	Co	..	..	..	..	..	4.25	..	..	..	..	..	..	..	..	..
Heat Treat Data	Hardening Temperature Range °F	1850-1900	1825-1875	1825-1875	1825-1900	1850-1950	2000-2200	2000-2200	2000-2200	2000-2300	2000-2250	2100-2300	2150-2300	2000-2175	2050-2225	2000-2175
	Tempering Temperature Range °F	1000-1200	1000-1200	1000-1200	1000-1200	1100-1200	1000-1300	1100-1250	1100-1250	1200-1500	1050-1200	1050-1250	1050-1250	1050-1200	1050-1200	1050-1200
	Approx. Tempered Hardness, R_c	56-39	54-38	55-38	53-38	47-40	59-40	54-36	52-39	47-30	55-45	44-35	58-43	60-50	60-50	58-45
Characteristics in Heat Treatment	Safety in Hardening	A	A	A	A	A	B	B	B	B	B	B	B	C	C	C
	Depth of Hardening	A	A	A	A	A	A	A	A	A	A	A	A	A	A	A
	Resistance to Decarburization	B	B	B	B	B	B	B	B	B	B	B	B	C	B	C
	Stability of shape in heat treatment — Quenching medium — Air or Salt	B	B	B	B	C	C	C	C	..	C	C	C	C	C	C
	Stability of shape in heat treatment — Quenching medium — Oil	..	..	..	..	..	..	..	..	..	..	..	..	..	..	..
Properties in Service	Machinability	C/D	C/D	C/D	C/D	D	D	D	D	D	D	D	D	D	D	D
	Hot Hardness	C	C	C	C	C	C	C	C	C	C	C	C	D	D	D
	Wear Resistance	D	D	D	D	D	C/D	C	C/D	C/D	C	D	C	C	C	C
	Toughness	C	B	B	B	C	C	C/D	C	D	D	C	D	D	D	D

Table 5. Tungsten High Speed Tool Steels

Identifying Chemical Composition Heat and Treatment Data

	AISI Type	T1	T2	T4	T5	T6	T8	T15
Identifying Chemical Elements in Per Cent	C	.75	.80	.75	.80	.80	.75	1.50
	W	18.00	18.00	18.00	18.00	20.00	14.00	12.00
	Cr	4.00	4.00	4.00	4.00	4.50	4.00	4.00
	V	1.00	2.00	1.00	2.00	1.50	2.00	5.00
	Co	...	...	5.00	...	...	5.00	5.00
Heat Treat Data	Hardening Temperature Range, °F	2300-2375	2300-2375	2300-2375	2325-2375	2325-2375	2300-2375	2200-2300
	Tempering Temperature Range, °F	1000-1100	1000-1100	1000-1100	1000-1100	1000-1100	1000-1100	1000-1200
	Approx. Tempered Hardness, R_c	65-60	66-61	66-62	65-60	65-60	65-60	68-63

Relative Ratings of Properties (A = greatest to E = least)

		T1	T2	T4	T5	T6	T8	T15
Characteristics in Heat Treatment	Safety in Hardening	C	C	D	D	D	D	D
	Depth of Hardening	A	A	A	A	A	A	A
	Resistance to Decarburization	A	A	B	C	C	B	B
	Stability of Shape in Heat Treatment — Quenching Medium Air or Salt	C	C	C	C	C	C	C
	Stability of Shape in Heat Treatment — Quenching Medium Oil	D	D	D	D	D	D	D
Service Properties	Machinability	D	D	D	D	D/E	D	D/E
	Hot Hardness	B	B	A	A	A	A	A
	Wear Resistance	B	B	B	B	B	B	A
	Toughness	E	E	E	E	E	E	E

Table 7. Cold Work Tool Steels

Identifying Chemical Composition and Typical Heat Treatment Data

AISI Group Types	High Carbon High Chromium Types					Medium Alloy Air Hardening Types								Oil Hardening Types			
	D2	D3	D4	D5	D7	A2	A3	A4	A6	A7	A8	A9	A10	O1	O2	O6	O7
Identifying Elements in %																	
C	1.50	2.25	2.25	1.50	2.35	1.00	1.25	1.00	.70	2.25	.55	.50	1.35	.90	.90	1.45	1.20
Mn						1.00		2.00	2.00				1.80	1.00	1.60		
Si													1.25			1.00	
W										1.00				.50			1.75
Mo	1.00		1.00	1.00	1.00	1.00	1.00	1.00	1.25	1.00	1.25	1.40	1.50			.25	
Cr	12.00	12.00	12.00	12.00	12.00	5.00	5.00	1.00	1.00	5.25	5.00	5.00	1.00	.50			.75
V	1.00		1.00		4.00		1.00			4.75							
Co				3.00													
Ni													1.50				
Heat Treatment Data																	
Hardening Temperature Range °F	1800-1875	1700-1800	1775-1850	1800-1875	1850-1950	1700-1800	1750-1850	1500-1600	1525-1600	1750-1800	1800-1850	1800-1875	1450-1500	1450-1500	1400-1475	1450-1500	1550-1525
Quenching Medium	Air	Oil	Air	Air	Air	Air	Air	Air	Air	Air	Air	Air	Air	Oil	Oil	Oil	Oil
Tempering Temperature Range °F	400-1000	400-1000	400-1000	400-1000	300-1000	350-1000	350-1000	350-800	300-800	300-1000	350-1100	950-1150	350-800	350-500	350-500	350-600	350-550
Approx. Tempered Hardness, R_c	61-54	61-54	61-54	61-54	65-58	62-57	65-57	62-54	60-54	67-57	60-50	56-35	62-55	62-57	62-57	63-58	64-58
Properties — Service Characteristics — Relative Ratings of Properties (A = greatest to E = least)																	
Safety in Hardening	A	C	A	A	A	A	A	A	A	A	A	A	A	B	B	B	B
Depth of Hardening	A	A	A	A	A	A	A	A	A	A	A	A	A	B	B	B	B
Resistance to Decarburization	B	B	B	B	B	B	B	A/B	A/B	B	B	B	A/B	A	A	A	A
Stability of Shape in Heat Treatment	A	B	A	A	A	A	A	A	A	A	A	A	A	B	B	B	B
Machinability	E	E	E	E	E	D	D	D/E	D	E	D	D	C/D	C	C	B	C
Hot Hardness	C	C	C	C	C	C	C	D	D	C	C	C	D	E	E	E	E
Wear Resistance	B/C	B	B	B/C	A	B	B	C/D	C/D	A	B	C/D	C	C	E	D	D
Toughness	E	E	E	E	E	D	D	D	D	E	B	C	D	C	D	D	C

Table 8. Shock Resisting, Mold, and Special Purpose Tool Steels

Identifying Chemical Compositions and Typical Heat Treatment Data

Category		Shock Resisting Tool Steels				Mold Steels							Special Purpose Tool Steels				
Types		S1	S2	S5	S7	P2	P3	P4	P5	P6	P20	P21[a]	L2[b]	L3[b]	L6	F1	F2
Identifying Elements in Per Cent	C	.50	.50	.55	.50	.07	.10	.07	.10	.10	.35	.20	.50/1.10	1.00	.70	1.00	1.25
	Mn			.80													
	Si		1.00	2.00													
	W	2.50														1.25	3.50
	Mo		.50	.40	1.40	.20		.75			.40				.25		
	Cr	1.50			3.25	2.00	.60	5.00	2.25	1.50	1.25		1.00	1.50	.75		
	V												.20	.20			
	Ni					.50	1.25			3.50		4.00			1.50		
Heat Treat Data	Hardening Temperature	1650-1750	1550-1650	1600-1700	1700-1750	1525-1550*	1475-1525*	1775-1825*	1550-1600*	1450-1500*	1500-1600*	So't'n treat.	1550-1700	1500-1600	1450-1550	1450-1600	1450-1600
	Tempering Temp. Range, °F	400-1200	350-800	350-800	400-1150	350-500	350-500	350-900	350-500	350-450	900-1100	Aged	350-1000	350-600	350-1000	350-500	350-500
	Approx. Tempered Hardness R_c	58-40	60-50	60-50	57-45	64-58‡	64-58‡	64-58‡	64-58‡	61-58‡	37-28‡	40-30	63-45	63-56	62-45	64-60	65-62

Relative Ratings of Properties (A = greatest to E = least)

Category		S1	S2	S5	S7	P2	P3	P4	P5	P6	P20	P21[a]	L2[b]	L3[b]	L6	F1	F2
Characteristics in Heat Treatment	Safety in Hardening	C	E	C	B/C	C	C	C	C	B	C	A	D	D	C	E	E
	Depth of Hardening	B	B	B	A	B‡	B‡	B‡	B‡	A‡	B	A	B	B	B	C	C
	Resist. to Decarb.	B	C	C	B	A	A	A	A	A	A	A	A	A	A	A	A
	Stability of Shape in Heat Treatment — Quench Med. Air				A			B				A					
	Quench Med. Oil	D		D		C/D	C		C	C	E		C	C	D/E		
	Quench Med. Water[c]		E													D	E
Service Properties	Machinability	D	C/D	C/D	D	C/D	D	D/E	D	D	C/D	D	C	E	D	C	D
	Hot Hardness	D	E	E	C	E	E	D	E	E	E	D	E	E	E	E	E
	Wear Resistance	D/E	D/E	D/E	D/E	D	D	C	D	D	D/E	D	D/E	D	D	B/C	B/C
	Toughness	B	A	A	B	C	C	C	C	C	C	D	D	E	C	E	E

*After carburizing   †Carburized case   ‡Core Hardenability
bQuenched in oil.   cSometimes brine is used.
aContains also about 1.20 per cent Al. Solution treated in hardening.

Trade Names of AISI Classified Tool Steels* — I

AISI Type	Al-Tech	Atlas Steels	Bethlehem	Braeburn	Carpenter	Columbia	Producer — Crucible	Jessop	Latrobe	Simonds	Teledyne Vasco	Universal-Cyclops
HIGH SPEED TOOL STEELS — TUNGSTEN TYPES												
T1	LXX	Spartan-7	Bethlehem T-1	Vinco	Star Zenith	Clarite	Rex AA	Supremus	Electrite No. 1 XL	Red Streak	Red-Cut Superior	…
T2	ML	…	…	Twinvan	…	Vanite	…	Supremus Extra	Electrite No. 19	Lock Port Special	E.V.M.	…
T4	Panther Special	…	…	Cobalt	…	Acmite	Rex AAA	Purple Label	Electrite Cobalt	Tunco	Red Cut Cobalt	…
T5	Super Panther	Nipigon	…	Bonded Carbide JR	…	Cobite	…	Purple Label Extra	Electrite Super Cobalt	Super Cobalt	Circle C	…
T6	…	…	…	…	…	Cobite II	…	King Cobalt	Electrite Ultra Cobalt	…	…	…
T8	…	…	…	…	…	Maxite	Rex 95	Jessop T8	…	…	…	…
T15	Panther 5	Sabre	…	Braeburn T15	…	Maxite 15	CPM Rex T-15	Jessop T15	Electrite Dynavan	…	Vasco Supreme	…
HIGH SPEED TOOL STEELS — MOLYBDENUM TYPES												
M1	LMW	Mohican-8	Bethlehem M-1	Mocut	Starmax, Starmax FM	Molite 1	Rex TMO	Mogul	Electrite Tatmo	STM	8-N-2	Motung
M2	DBL-2	Sixix	Bethlehem M-2	Braemow, Mocarb	Speed Star, Speed Star FM	Molite 2	Rex M-2, CPM Rex M-2	Mustang	Electrite Double Six M2 XL	Molva T	Vasco M-2	Motung 652
M3 Class 1	DBL-2½	Atlas M-3	…	Braevan	…	Molite 3 Class 1	Rex M-3-1	Jessop M3 Class 1	Electrite Corsair XL	Molva TC1	Van Cut	Unicut
M3 Class 2	DBL-3	…	…	Braevan 2	…	Molite 3 Class 2	Rex M-3-2, CPM Rex M-3-2	Jessop M3 Class 2	Electrite Crusader XL	Molva TC 2	Van Cut Type 2	Unicut 2

* *Source:* Committee of Tool Steel Producers, American Iron and Steel Institute, 1000 16th St., N. W., Washington, D.C. 20036

Trade Names of AISI Classified Tool Steels — 2

AISI Type	Al-Tech	Atlas Steels	Bethlehem	Braeburn	Carpenter	Columbia	Crucible	Jessop	Latrobe	Simonds	Teledyne Vasco	Universal-Cyclops
					Producer							
			HIGH SPEED TOOL STEELS — MOLYBDENUM TYPES (Continued)									
M4	DBL-4	Atlas M-4	...	Braefour	Four Star	Molite 4	CPM Rex M-4	Jessop M4	Electrite Stark	Molva HC	Neatro	Cyclops M4
M6	...	...	...	Congo	...	...	...	...	...	...	...	...
M7	LMW-V	Atlas M-7	Bethlehem M-7	Motuf	Seven Star	Molite 7	Rex M-7	Jessop M7	Electrite Tatmo V	Molva C	Vasco M-7	Motung CV
M10	VLM	Atlas M-10	Bethlehem M-10	Motemp	Ten Star	Molite 10	Rex VM	Jessop M10	Electrite TNW	Molva	Van Lom	...
M30	Super LMW	...	...	Como	...	Molite 30	...	...	Electrite Lacomo	...	8-N-2 Cobalt	Super Motung
M33	Super LMW Extra	...	...	Braeburn M33	...	Molite 33	Rex M-33	...	Electrite Kelvan	STMCO	8-N-2 Cobalt 8	Super Motung 33
M34	Super LMW Special	Atlas M-34	...	...	...	Molite 34	...	...	Electrite Tatmo Cobalt	...	...	...
M36	Super DBL	...	...	Moco	...	Molite 36	...	...	Electrite CO-6	...	Victory Cobalt	...
M41	...	...	...	...	...	Molite 41	Rex 49	Jessop RC 70	...	...	...	...
M42	Exocut	Atlas M-42	...	Braemax	Super Star	Molite 42	Rex M-42, CPM Rex M-42	...	Electrite Dynamax	...	Hypercut	Cyclops M-42
M43	...	...	...	...	...	...	...	...	Electrite Dynacut	...	...	...
M44	...	...	...	Braecut	...	...	...	...	...	...	...	...
M46	AL-46	...	...	...	...	...	Rex M-46	...	...	...	...	...
M47	Exohard	...	...	...	...	...	...	...	...	...	...	...

Trade Names of AISI Classified Tool Steels — 3

AISI Type						Producer						
	Al-Tech	Atlas Steels	Bethlehem	Braeburn	Carpenter	Columbia	Crucible	Jessop	Latrobe	Simonds	Teledyne Vasco	Universal-Cyclops
HIGH SPEED TOOL STEELS — MOLYBDENUM TYPES (Continued)												
M50	HTB-2	Atlas	…	…	Carpenter M50	…	…	…	MV-1	…	Vasco M50	BHT
M52	Oglala	…	…	Natrona 52	Carpenter M52	…	…	…	MV-2	…	Vasco M52	…
HOT WORK TOOL STEELS — CHROMIUM TYPES												
H10	AL-173	…	…	Pressurdie 6	…	Columbia H10	Peerless 56	…	Dart	…	…	…
H11	Potomac A	Atlas H-11	Cromo-V	Pressurdie 3L	882, 882 FM	Firedie	Nu-Die, Halcomb 218	Dica B (Modified)	Dycast No. 1	Howord A	Hot Form No. 2	Thermold H11
H12	Potomac	Crodi	Cromo-W	Pressurdie 2	345, 345 FM	Alcodie	Chro-Mow	Dica B	LPD	Howord B	Hot Form No. 1	Thermold H12
H13	Potomac M	Dievac	Cromo-High-V	Pressurdie 3	883, 883 FM	Firedie 13	Nu-Die V	Dica B Vanadium	VDC, Viscount 20, Viscount 44	Howord C	Hot Form V	Thermold H13
H14	…	Red Indian	…	Pressurdie 1	…	Firedie 14	…	…	Lumdie	…	…	…
H19	B-47	Atlas H-19	…	Pressurdie C	…	Firedie 19	Halcomb 425		Lesco 19	…	W.C.C.	…
HOT WORK TOOL STEELS — TUNGSTEN TYPES												
H21	Atlas A	Seneca	57 HW	T-Alloy A	TK	Formite 21	Peerless A	2-BLC	CLW	…	Marvel	Thermold H21
H22	Atlas B	…	…	T-Alloy	TK (Modified)	…	…	2-BMC	…	…	…	…
H23	…	…	…	HCA	…	Formite 23	…	…	Kalkos	…	W.W. Hot Work	…
H24	…	…	…	T-Alloy B	…	Formite 24	…	2-BHC	CHW	…	SC Special	…

Trade Names of AISI Classified Tool Steels — 4

AISI Type						Producer						
	Al-Tech	Atlas Steels	Bethlehem	Braeburn	Carpenter	Columbia	Crucible	Jessop	Latrobe	Simonds	Teledyne Vasco	Universal-Cyclops
HOT WORK TOOL STEELS — TUNGSTEN TYPES (Continued)												
H25	Mohawk	...	...	...	T-Alloy C	...	...	...	EHW No. 1	...	Forge-Die	...
H26	...	Spartan-5	Special HS-55	Vinco Hot Work	Carpenter H26	Clarite HW 26	Rex AA PX	...	Electrite No. 5	...	Red Cut Superior J	...
HOT WORK TOOL STEELS — MOLYBDENUM TYPE												
H42	...	...	...	...	...	...	...	Mustang L.C.	Electrite No. 7	...	...	...
COLD WORK TOOL STEELS — HIGH CARBON HIGH CHROMIUM TYPES												
D2	Ontario	FNS	Lehigh H	Superior 3	610, 610 FM	Atmodie	Airdi 150	CNS-1	Olympic FM	CCM	Ohio Die	Ultradie 3
D3	Huron	...	Lehigh S	Superior 1	Hampden	Superdie	...	CNS-2	GSN	HCCM	...	...
D4	...	NN	...	AT 2	...	Atmodie 4	HYCC	CNS-3	GSN + MO	...	Crocar	...
D5	AL-D-5	...	...	Superior 2	Carpenter D5	Atmodie 5	...	3 C Special	Cobalt Chrome FM	...	...	...
D7	Huron V	...	...	...	Carpenter D7	...	HYCV	Truwear	BR-4 FM	ARS	...	...
SPECIAL PURPOSE TOOL STEELS — LOW ALLOY TYPES												
L2	Albany Caroga	...	Tough M	...	...	Columbia L2	Halvan	ET-6	Crown, Superb	...	Vanadium Type H	Cyclops L2
L6	Tioga	Atlas L-6	Bethalloy	...	R.D.S.	Nicrodie	Champaloy	ET-4	NDS	...	Nikro M	Cyclops L6
COLD WORK TOOL STEELS — MEDIUM ALLOY AIR HARDENING TYPES												
A2	Sagamore	Cromoloy	A-H5	Airque	484, 484 FM	E-Z-Die	Airkool	Windsor	Select B FM	Airtrue	Air Hard	Sparta
A3	...	...	...	Airque V	...	...	...	...	...	...	...	...
A4	...	...	Air-4	...	...	...	...	...	...	...	...	...
A6	Apache	Nutherm	A-6	...	Vega	Uni-Die	CSM 6	Jess-Air	...	...	...	Lo-Air

Trade Names of AISI Classified Tool Steels — 5

AISI Type	Al-Tech	Atlas Steels	Bethlehem	Braeburn	Carpenter	Columbia	Crucible	Jessop	Latrobe	Simonds	Teledyne Vasco	Timken	Universal-Cyclops
COLD WORK TOOL STEELS — MEDIUM ALLOY AIR HARDENING TYPES (Continued)													
A7	Sagamore V	...	...	...	...	E-Z-Die V	Airkool V	BX 3	BR-3	A7W	Chrome-wear	...	...
A8	AL-158	...	Cromo-W55	Pressurdie 16	Carpenter A8	Columbia A-8	...	...	MGR	Airtrue LC	Hot Form No. 3	...	...
A9	...	...	...	...	Carpenter A9	Formdie	...	...	...	...	...	...	Thermold J
A10	...	...	...	...	...	...	...	...	...	...	...	Graph-Air	...
COLD WORK TOOL STEELS — OIL HARDENING TYPES													
O1	Saratoga	Keewatin	BTR	Kiski	Carpenter O-1	EXL-Die	Ketos	Truform	Badger	Teenax	Colonial No. 6	...	Wando
O2	Deward	...	...	...	Stentor	...	...	Special Oil Hardening	...	...	...	...	...
O6	Oilgraph	...	O-6 Graphitic	...	...	Col-Graph	Halgraph	Truglide	...	...	...	Graph-Mo	...
O7	Utica	...	...	...	...	Tapdie	...	...	W Tap	BFD	Red Star Tungsten	...	...
SHOCK RESISTING TOOL STEELS													
S1	Seminole	Falcon-6	67 Chisel	Vibro	Excelo	Buster Alloy	Atha Pneu	Top Notch	XL Chisel	Com-mando	Par-Exc	...	...
S2	...	...	Imperial	...	Solar	...	...	RTS	...	Havoc	...	...	Venango Special
S5	AL-602, AL-609	Monark-2	Omega	Alloy 10	481 Collet	Silico Alloy	La Belle Silicon #2	259 Grade	Lanark	Orleans	Mosil	...	Cyclops S5
S6	...	...	...	...	...	Columbia S6	La Belle HT	...	...	...	...	...	...
S7	AL-7	...	Bearcat	...	Carpenter S-7	Shock-Die	Crucible S-7	Super Shock 7	...	...	Simoch	...	...

Trade Names of AISI Classified Tool Steels — 6

AISI Type	Producer										
	Al-Tech	Atlas Steels	Bethlehem	Braeburn	Carpenter	Columbia	Crucible	Jessop	Latrobe	Simonds	Teledyne Vasco
MOLD STEELS											
P2	...	...	Duramold B	...	...	...	...	...	...	...	...
P4	...	...	Duramold A	...	Super Samson	...	...	...	...	...	...
P5	...	...	...	...	Samson Extra	...	...	...	...	...	Vasco Chromold VM
P6	...	Super Impacto PQ	Duramold N	...	No. 158	...	...	...	...	...	...
P20	Almold-20	Mold Special	Bethlehem P-20	...		...	CSM #2	P20	...	...	...
P21	...	...	...	...	...	...	...	...	Cascade	...	...
WATER HARDENING TOOL STEELS											
W1	Pompton	X-10, X-12, Alpha, XX-95	X, XCL, XX, Cold Header Die	Extra	Comet Green Label, H-9 Double Header, No. 11 Special, Titan, Reading Tap	Columbia Special, Extra, Extra Headerdie, Standard	Sanderson Extra, Labelle Cold Header, Black Diamond	Washington Lion, Lion Extra, New Process Cold Header	Carbon Types	Red Label, Blue Label, Diamond S, Green Label	Colonial No. 14
W2	Python	Asa-10	Best, Superior	Coldie	Nitro Special Vanadium	Vanadium Extra, Standard	Alva Extra	Lion Van., Lion Extra	Carbon Vanadium Types	Red Label Extra	Colonial No. 7
W5	Crow	Atlas "Q"	Bethlehem W-5	Braeburn W-5	U.D.R.	Waterdie Extra, Standard	...	W-5	CFS	...	...

BRASS, BRONZE, ALUMINUM AND OTHER NON-FERROUS ALLOYS

Cast Brass and Bronze. — The following information on S.A.E. Standard Brass and Bronze Castings includes typical applications of the different alloys in the automotive industry, the composition in percentage, and physical properties based upon standard test bars cast to size with only a minimum amount of machining to remove the fin gate. Standard specimens of wrought material are taken parallel to the direction of rolling and all rods, bars and shapes are tested in full size when practicable.

Red Brass Castings. — **S.A.E. Standard No. 40.** — Red brass is used for water-pump impellers, fittings for gasoline and oil lines, small bushings, small miscellaneous castings. This is a free-cutting brass with good casting and finished properties.

Composition of No. 40: Copper, 84 to 86; tin, 4 to 6; lead, 4 to 6; zinc, 4 to 6; iron, max., 0.25; nickel, max., 0.75; phosphorus, max., 0.05; aluminum, 0.00; sulphur, max., 0.05; antimony, max., 0.25; other impurities, max., 0.15 per cent.

Physical Properties: Tensile strength, 26,000 pounds per square inch; yield point, 12,000 pounds per square inch; elongation in 2 inches (or proportionate gage length), 15 per cent.

Yellow Brass Castings — **S.A.E. Standard No. 41.** — Yellow brass is used for radiator parts, fittings for water-cooling systems, battery terminals, miscellaneous castings. This alloy is intended for commercial castings when cheapness and good machining properties are essential.

Composition of No. 41: Copper, 62 to 67; lead, 1.50 to 3.50; tin, max., 1; iron, max., 0.75; nickel, max., 0.25; phosphorus, max., 0.03; aluminum, max., 0.30; sulphur, max., 0.05; antimony, max., 0.15; other impurities, max., 0.15 per cent; zinc, remainder.

Physical Properties: Tensile strength, 20,000 pounds per square inch; elongation in 2 inches (or proportionate gage length), 15 per cent.

Manganese Bronze Castings — **S.A.E. Standard No. 43.** — This alloy is intended for castings requiring strength and toughness. It is used for such automotive parts as gear-shifter forks; counters, spiders; brackets and similar fittings; parts for starting motors; landing-gear and tail-skid castings for airplanes.

Composition of No. 43: Copper, 55 to 60; zinc, 38 to 42; tin, max., 1.50; manganese, max., 3.50; aluminum, max., 1.50; iron, max., 2; lead, max., 0.40 per cent.

Physical Properties: Tensile strength, 65,000 pounds per square inch; elongation in 2 inches (or proportionate gage length), 25 per cent.

High Tensile Manganese Bronze Castings — **S.A.E. Standard No. 430.** — This alloy is intended for use in castings where high strength and toughness are required such as marine propellers, shafts and gears.

Composition of No. 430: Copper, 60 to 68; iron, 2 to 4; aluminum, 3 to 6; manganese, 2.5 to 5; tin, max., 0.50: lead, max., 0.20, and nickel, max., 0.50 per cent; zinc, remainder.

Physical Properties: This alloy is manufactured in two grades, distinguished by chemical composition: Grade A being in the lower, and Grade B in the higher range of manganese, aluminum and iron content. Tensile strength, Grade A, 90,000, and Grade B, 110,000 pounds per square inch; elongation in 2 inches, Grade A, 20, and Grade B, 12 per cent.

Cast Brass to be Brazed — **S.A.E. Standard No. 44.** — This brass is used for water-pipe fittings which are to be brazed. It begins to melt at about 1830

degrees F. and is entirely melted at approximately 1870 degrees F. The alloy or spelter used for brazing must have a lower melting temperature. Silver solder may be used.

Composition of No. 44: Copper, 83 to 86; zinc, 14 to 17; lead, max., 0.50; iron, max., 0.15 per cent.

Brazing Solder — S.A.E. Standard No. 45. — This solder begins to melt at approximately 1560 degrees F. and is entirely melted at about 1600 degrees F. It may be used by melting in a crucible under a flux of borax, with or without the addition of boric acid. The part to be brazed is dipped into the melted solder. When used in powdered form, this solder, mixed with a flux, is applied to the material and then melted either by means of a brazing torch or by using a furnace.

Composition of No. 45: Copper, 48 to 52; lead, max., 0.50; iron, max., 0.10 per cent; zinc, remainder.

Hard Bronze Castings — S.A.E. Standard No. 62.— This is a strong general utility bronze suitable for severe working conditions and heavy pressures. Typical applications include gears; bearings; bushings for severe service; valve guides; valve-tappet guides; camshaft bearings; fuel pump, timer and distributor parts; connecting-rod bushings; piston-pins; rocker lever; steering sector and hinge bushings; starting-motor parts.

Composition of No. 62: Copper, 86 to 89; tin, 9 to 11; lead, max., 0.20; iron, max., 0.06; zinc, 1 to 3 per cent.

Physical Properties: Tensile strength, 30,000 pounds per square inch; yield point, 15,000 pounds per square inch; elongation in 2 inches (or proportionate gage length), 14 per cent.

Leaded Gun Metal Castings — S.A.E. Standard No. 63. — This general-utility bronze combines strength with fair machining qualities. It is especially good for bushings subjected to heavy loads and severe working conditions. It is also used for fittings subjected to moderately high water or oil pressures.

Composition of No. 63: Copper, 86 to 89; tin, 9 to 11; phosphorus, max., 0.25; zinc and other impurities, max., 0.50; lead, 1 to 2.50 per cent.

Physical Properties: Tensile strength, 30,000 pounds per square inch; yield point, 12,000 pounds per square inch; elongation in 2 inches (or proportionate gage length), 10 per cent.

Phosphor Bronze Castings — S.A.E. Standard No. 64. — This alloy is excellent when anti-friction qualities are important and where resistance to wear and scuffing are desired. It is used for such parts as wrist-pins, piston-pins, valve rocker-arm bushings, fuel and water-pump bushings, steering-knuckle bushings, aircraft control bushings.

Properties of No. 64: Copper, 78.50 to 81.50; tin, 9 to 11; lead, 9 to 11; phosphorus, 0.05 to 0.25; zinc, max., 0.75; other impurities, max., 0.25 per cent.

Physical Properties: Tensile strength, 25,000 pounds per square inch; yield point, 12,000 pounds per square inch; elongation in 2 inches (or proportionate gage length), 8 per cent.

Phosphor Gear Bronze Castings — S.A.E. Standard No. 65. — This bronze is not used regularly but it may be employed for gears and worm wheels where the requirements are severe and a very hard bronze is necessary.

Properties of No. 65: Copper, 88 to 90; tin, 10 to 12; phosphorus, 0.10 to 0.30; nickel, max., 0.05; lead, zinc, and other impurities, max., 0.50 per cent.

Physical Properties: Tensile strength, 35,000 pounds per square inch; yield point, 20,000 pounds per square inch; elongation in 2 inches (or proportionate gage length), 10 per cent.

Bronze Backing for Lined Bearings — S.A.E. Standard No. 66. — This is an inexpensive but suitable alloy for bronze-backed bearings of connecting-rods or main engine bearings.

Composition: Copper, 83 to 86; tin, 4.50 to 6; lead, 8 to 10; zinc, max., 2; other impurities, max., 0.25 per cent.

Physical Properties: Tensile strength, 25,000 pounds per square inch; yield point, 12,000 pounds per square inch; elongation in 2 inches, 8 per cent.

Bronze Bearing Castings — S.A.E. Standard No. 660. — This composition is widely used for bronze bearings. Typical applications in the automotive industry include such parts as spring bushings, torque tube bushings, steering-knuckle bushings, piston-pin bushings, thrust washers, etc.

Composition of No. 660: Copper, 81 to 85; tin, 6.50 to 7.50; lead, 6 to 8; zinc, 2 to 4; iron, max., 0.20; antimony, max., 0.20; other impurities, max., 0.50 per cent.

Physical Properties: Tensile strength, 30,000 pounds per square inch; yield point, 14,000 pounds per square inch; elongation in 2 inches, 18 per cent.

Cast Aluminum Bronze — S.A.E. Standard No. 68. — This alloy has considerable strength, resistance to corrosion, hardness equal to manganese bronze, and good bearing qualities under certain conditions. It is used for worm-wheels, gears, valve guides, valve seats, and forgings.

Composition of No. 68: Copper, (Grade A) 87 to 89, (Grade B) 89.50 to 90.50; aluminum, (Grade A) 7 to 9, (Grade B) 9.50 to 10.50; iron, (Grade A) 2.50 to 4, (Grade B) not over 1; tin, max., (Grade A) 0.5, (Grade B) 0.2; total other impurities, (Grade A), 1, (Grade B) 0.5 per cent.

Physical Properties: Tensile strength, (Grades A and B) as cast, 65,000 pounds per square inch; tensile strength, (Grade B) as heat-treated, quenched and drawn, 80,000 pounds per square inch; yield point, (Grades A and B) as cast, 25,000 pounds per square inch; yield point, (Grade B) as heat-treated, 50,000 pounds per square inch. Elongation in 2 inches, (Grade A) as cast, 20 per cent; (Grade B) 15 per cent; (Grade B) as heat-treated, 4 per cent.

Wrought Copper and Copper Alloys

Brass Sheet and Strip — S.A.E. Standard No. 70. — There are two grades designated as 70A (Cartridge Brass) and 70C (Yellow Brass). Tempers range from quarter hard through extra spring. These are given in the accompanying table. The numbers following each temper designation in the table represent the amount of reduction in B. & S. gage numbers when the brass sheets are rolled. The greater the reduction, the harder the brass.

This alloy is used to make radiator cores and tanks in the automotive industry; bead chain, flashlight shells, socket and screw shells in the electrical industry; and eyelets, fasteners, springs and stampings in the hardware industry.

Composition of No. 70A: Copper, 68.5 to 71.5; lead, max., 0.07; iron, max., 0.05; zinc, remainder.

Composition of No. 70C: Copper, 64.0 to 68.5; lead, max., 0.15; iron, max., 0.05; zinc, remainder.

Mechanical Properties: Tensile strengths and Rockwell hardness numbers are given in the accompanying table.

Aluminum Bronze Rods, Bars, and Shapes — S.A.E. Standard No. 701. — This alloy is commonly used for bushings, gears, valve parts, bearings, sleeves, screws, pins, and fabricated sections. It is also used where strength at elevated temperatures, a low coefficient of friction against steel, or a combination of strength and corrosion resistance is required. Alloy grades are: 701B, 701C and 701D.

Composition of No. 701B: Copper, 80.0 to 93.0; aluminum, 6.5 to 11.0; iron, max., 4.00; nickel, max., 1.00; manganese, max., 1.50; silicon, max., 2.25; tin, max., 0.60; zinc, max., 1.0; tellurium, max., 0.6; other elements, max., 0.50.

Composition of No. 701C: Copper, 78.0; aluminum, 9.0 to 11.0; iron, 2.0 to 4.0; nickel, 4.0 to 5.5; manganese, max., 1.50; silicon, max., 0.25; tin, max., 0.20; other elements, max. 0.50.

Composition of No. 701D: Copper, 88.0 to 92.5; aluminum, 6.0 to 8.0; iron, 1.5 to 3.5; other elements, max., 0.50.

Mechanical Properties: Minimum tensile strengths of the No. 701B alloy grade range from 70,000 to 80,000 psi, minimum yield strengths from 30,000 to 40,000 psi and minimum elongations in 2 inches from 12 to 9 per cent depending on the shape or size of rod or bar. Minimum tensile strengths of the No. 701C alloy grade range

Hardness and Ultimate Strength of No. 70 Sheet Brass by Tempers

Temper of Brass Sheet (S.A.E. No. 70) and Equivalent Reduction in B. & S. Gage Numbers		Rockwell Hardness Numbers				Ultimate Strength Pounds per Square Inch	
		B Scale $\frac{1}{16}''$ Ball- 100 kg. Load		Superficial 30-T Scale $\frac{1}{16}''$ Ball- 30 kg. Load			
Temper	Gage Nos.	Min.	Max.	Min.	Max.	Min.	Max.
Grade A							
Quarter Hard	1	40	65	43	60	49,000	59,000
Half Hard	2	60	77	56	68	57,000	67,000
Three-Quarter Hard	3	72	82	65	72	64,000	74,000
Hard	4	79	86	70	74	71,000	81,000
Extra Hard	6	85	91	74	77	83,000	92,000
Spring	8	89	93	76	78	91,000	100,000
Extra Spring	10	91	95	77	79	95,000	104,000
Grade C							
Quarter Hard	1	40	65	43	60	49,000	59,000
Half Hard	2	57	74	54	66	55,000	65,000
Three-Quarter Hard	3	70	80	65	71	62,000	72,000
Hard	4	76	84	68	73	68,000	78,000
Extra Hard	6	83	89	73	76	79,000	89,000
Spring	8	87	92	75	78	86,000	95,000
Extra Spring	10	88	93	76	79	90,000	99,000

The hardness numbers equivalent to such temper designations as "quarter hard," "half hard," etc., vary over a wide range as shown by the table above. The hardness number represented by a given temper designation depends not only upon the kind of annealing and thickness of a given material, but may be affected decidedly by the composition or type of alloy. "Quarter hard" red brass sheet (S.A.E. No. 79), for example, may have a Rockwell hardness varying from 50 to 95 which differs considerably from the minimum and maximum numbers given in the table above opposite "quarter hard."

Hardness tests of the indentation type, such as Rockwell or Brinell, are generally used for thin materials; however, if the sheet is very thin, the test may be for comparison only with other sheets of the same composition and thickness. When the penetration is deep relative to the thickness, there may be an apparent decrease of hardness due to the flow or punching-through of the material because of lack of lateral support; however, when the penetration is even greater relative to thickness, there may be an apparent *increase* in hardness due to the pressure of the penetrator on the anvil of the instrument.

from 85,000 to 100,000 psi, minimum yield strengths from 42,500 to 50,000 psi and minimum elongations in 2 inches from 10 to 5 per cent depending on size.

This alloy must withstand cold bending without fracture through an angle of 120 degrees around a pin, the diameter of which is equal to twice the diameter of round rod or four times the thickness of bar or other shapes.

Copper Sheet and Strip — S.A.E. Standard No. 71. — This alloy is used for building fronts, roofing, radiators, chemical process equipment, rotating bands, and vats.

Composition of No. 71: Copper, min., 99.90 (plus silver). In one type of sheet used in the automotive industry 6 to 10 troy ounces of silver may be added to one ton (avoirdupois) of copper. This is sufficient to raise the recrystallization temperature appreciably.

Mechanical Properties: Minimum tensile strengths range from 30,000 to 52,000 psi depending on temper. Generally the higher the strength the harder the temper.

Free Cutting Brass Rod — S.A.E. Standard No. 72. — This alloy is used for small screw machine parts, pins, nuts, plugs, screws, valve discs and caps.

Composition of No. 72: Copper, 60.0 to 63.0; lead, 2.5 to 3.7; iron, max., 0.35; other elements, max., 0.50; zinc, remainder.

Mechanical Properties: In the soft temper the minimum tensile strength ranges from 40,000 to 48,000 psi, the minimum yield strength from 15,000 to 20,000 psi and the minimum elongation in 2 inches from 25 to 15 per cent as the thickness decreases down from over 2 inches. In the half hard temper the minimum tensile strength ranges from 45,000 to 57,000 psi, the minimum yield strength ranges from 15,000 to 25,000 psi and the minimum elongation in 2 inches from 20 to 7 per cent as the size decreases down from over 2 inches. In the hard temper the minimum tensile strength is 80,000 psi and the minimum yield strength 45,000 psi for thicknesses of ⅛ to ¾₆ inch. The minimum tensile strength is 70,000 psi, the minimum yield strength is 35,000 psi and the minimum elongation in 2 inches is 4 per cent for thicknesses over ¾₆ to ⁵⁄₆ inch.

Naval Brass Rods, Bars, Forgings, and Shapes — S.A.E. No. 73. — This material is intended for use where brass rod that is stronger, tougher, and more corrosion resistant than commercial brass rod is required. Uses include forgings, water pump and propeller shafts, studs and nuts, bushings, turnbuckle barrels, adjusting stud ends, and screw machine parts.

Composition of No. 73: Copper, 59.0 to 62.0; tin, 0.50 to 1.00; lead, max., 0.20; iron, max., 0.10; other elements, max., 0.10; zinc, remainder.

Mechanical Properties: Rods and bars in the soft temper have a minimum tensile strength ranging from 50,000 to 54,000 psi, a minimum yield strength of 20,000 psi and a minimum elongation in 4 times the diameter or thickness of 30 per cent as the size decreases. Rods and bars in the half hard temper have a minimum tensile strength ranging from 54,000 to 60,000 psi, a minimum yield strength ranging from 22,000 to 27,000 psi and a minimum elongation in 4 times the diameter or thickness ranging from 30 to 22 per cent as the size decreases. Rods and bars in the hard temper have a minimum tensile strength ranging from 54,000 to 67,000 psi, a minimum yield strength ranging from 22,000 to 45,000 psi and a minimum elongation in 4 times the diameter or thickness ranging from 30 to 13 per cent as the size decreases.

Seamless Brass Tubes — S.A.E. Standard No. 74. — The alloys comprising these tubes are identified by the letters *A, B, C,* and *D.* Nos. 74A and 74D are used for condenser and heat exchanger tubes and flexible hose. Nos. 74B and 74C

are general purpose materials used for water pipe radiator and ornamental work. The tubes may be formed, bent, upset, squeezed, swaged, flared, roll threaded and knurled.

Composition of No. 74A (Muntz Metal): Copper, 59.0 to 63.0; lead, max., 0.30; iron, max., 0.07; zinc, remainder.

Composition of No. 74B (Yellow Brass): Copper, 65.0 to 68.0; lead, 0.20 to 0.80; iron, max., 0.07; zinc, remainder.

Composition of No. 74C (Cartridge Brass): Copper, 68.5 to 71.5; lead, max., 0.07; iron, max., 0.05; zinc, remainder.

Composition of No. 74D (Red Brass, 85%): Copper, 84.0 to 86.0; lead, max., 0.06; iron, max., 0.05; zinc, remainder.

Mechanical Properties of No. 74A: This tube in drawn temper exhibits a minimum tensile strength of 54,000 psi. Common tempers of this tube include light annealed, drawn general purpose, and hard drawn.

Mechanical Properties of Nos. 74B and 74C: These tubes in drawn temper exhibit a minimum tensile strength of 54,000 psi. In hard temper they exhibit a minimum tensile strength of 66,000 psi. Common tempers of these tubes include drawn general purpose and hard drawn.

Mechanical Properties of No. 74D: This tube in light, drawn, and hard tempers exhibits minimum tensile strengths of 44,000, 44,000, and 57,000 psi, respectively. Common tempers of this tube include light, drawn general purpose, and hard.

Copper Tubes — S.A.E. Standard No. 75. — These tubes which contain a minimum of 99.90 per cent deoxidized copper are used for general engineering purposes, including gasoline, hydraulic and oil lines.

Mechanical Properties: In the light drawn temper the minimum tensile strength is 36,000 psi and the maximum tensile strength is 47,000 psi. In the drawn general purpose temper the minimum tensile strength is 36,000 psi and in the hard drawn temper (applying to tubes up to 1 inch outside diameter, inclusive, with wall thicknesses from 0.020 to 0.120 inch; tubes over 1 to 2 inches outside diameter, inclusive with wall thicknesses from 0.035 to 0.180 inch; and tubes over 2 to 4 inches outside diameter with wall thicknesses from 0.060 to 0.250 inch) the minimum tensile strength is 45,000 psi.

Phosphor Bronze Sheet and Strip — S.A.E. Standard No. 77. — Typical uses for this sheet and strip include springs, switch parts, sleeve bushings, clutch discs, diaphragms, fuse clips, and fasteners. There are two grades of this alloy, 77A and 77C. Six tempers are applied to this alloy, namely, soft, half hard, hard, extra hard, spring and extra spring.

Composition of No. 77A: Tin, 3.5 to 5.8; phosphorus, 0.03 to 0.35; lead, max., 0.05; iron, max., 0.10; zinc, max., 0.30; antimony, max., 0.01; copper, tin, and phosphorus, min., 99.50.

Composition of No. 77C: Tin, 7.0 to 9.0; phosphorus, 0.03 to 0.35; lead, max., 0.05; iron, max., 0.10; zinc, max., 0.20; antimony, max., 0.01; copper, tin, and phosphorus, min., 99.50.

Mechanical Properties: The minimum tensile strength of the No. 77A alloy ranges from 40,000 to 96,000 psi as the temper ranges from soft to extra spring. The minimum tensile strength of the No. 77C alloy ranges from 53,000 to 110,000 psi as the temper ranges from soft to extra spring.

Red Brass and Low Brass Sheet and Strip — S.A.E. Standard No. 79. — There are two grades designated as 79A (Red Brass, 85 per cent) and 79B (Low Brass, 80 per cent). Common tempers of No. 79A strip are quarter hard, half hard, extra hard, and spring. Common temper of No. 79A sheet is half hard. Common

tempers of No. 79B strip are quarter hard, half hard, hard, and spring. Typical uses include weather strip, trim, conduit, sockets, fasteners, radiator cores and costume jewelry.

Composition of No. 79A: Copper, 84.0 to 86.0; lead, 0.05; iron, 0.05; zinc, remainder.

Composition of No. 79B: Copper, 78.5 to 81.5; lead, 0.05; iron, 0.05; zinc, remainder.

Mechanical Properties: Minimum tensile strengths of the 79A alloy range from 44,000 to 82,000 psi as the temper ranges from quarter hard to extra spring and minimum tensile strengths of the 79B alloy range from 48,000 to 89,000 psi as the temper ranges from quarter hard to extra spring.

Brass Wire — S.A.E. Standard No. 80. — This wire is used for making springs, locking wire, rivets, screws, and for wrapping turnbuckles. There are two grades, 80A and 80B.

Composition of No. 80A: Copper, 68.5 to 71.5; lead, max., 0.07; iron, max., 0.05; zinc, remainder.

Composition of No. 80B: Copper, 63.0 to 68.5; lead, max., 0.10; iron, max., 0.05; zinc, remainder.

Mechanical Properties: Minimum tensile strengths of the 80A and 80B alloys range from 50,000 to 120,000 psi as tempers range from eighth hard to spring.

Phosphor Bronze Wire and Rod — S.A.E. Standard No. 81. — This alloy is used for springs, switch parts, fasteners, and cotter pins. It should withstand being bent cold through an angle of 120 degrees without fracture, around a pin with a diameter twice the diameter of the wire.

Composition of No. 81: Tin, 3.50 to 5.80; phosphorus, 0.03 to 0.35; lead, max., 0.05; iron, max., 0.10; zinc, max., 0.30; copper, tin, and phosphorus, min., 99.50.

Mechanical Properties: Minimum tensile strengths of hard drawn wire in coils range from 145,000 to 105,000 as the wire diameter ranges from 0.025 to 0.500 inch. Minimum tensile strengths of spring temper rods range from 125,000 to 90,000 psi as the rod diameter ranges from 0.025 to 0.500 inch.

Annealed Copper Wire — S.A.E. Standard No. 83. — This wire is used primarily for electrical purposes but it is also used for metal spraying and copper brazing. No composition limits are specified for this wire but the copper should be of such quality and purity that when drawn and annealed should exhibit the mechanical properties (maximum tensile strength and minimum elongation) and electrical characteristics called for in the standard. Its electrical resistivity should not exceed 875.20 ohms per mile-lb (100 per cent electrical conductivity IACS, International Annealed Copper Standard) at a temperature of 20 degrees C.

Mechanical Properties: Maximum tensile strengths of annealed wire range from 36,000 to 38,000 psi for wire diameters ranging from 0.4600 down to over 0.0201 inch. Minimum elongations in 10 inches of annealed wire range from 15 to 35 per cent as the wire diameter ranges from over 0.0030 to 0.4600 inch.

Brass Forgings — S.A.E. Standard No. 88. — Typical uses for this alloy are forgings and pressings of all kinds.

Composition of No. 88: Copper, 58.0 to 61.0; lead, 1.50 to 2.50; iron, max., 0.30; other elements, max., 0.50; zinc, remainder.

Mechanical Properties: Hot-pressed forgings made of this alloy should have tensile strengths ranging from 45,000 to 60,000 psi and elongations in 2 inches ranging from 25 to 60 per cent.

Aluminum and Aluminum Alloys

Pure aluminum is a silver-white metal characterized by a slightly bluish cast. It has a specific gravity of 2.70, resists the corrosive effects of many chemicals and has a malleability approaching that of gold. When alloyed with other metals numerous properties are obtained which make these alloys useful over a wide range of applications.

Aluminum alloys are light in weight compared to steel, brass, nickel or copper; can be fabricated by all common processes; are available in a wide range of sizes, shapes and forms; resist corrosion; readily accept a wide range of surface finishes; have good electrical and thermal conductivities; and are highly reflective to both heat and light.

Characteristics of Aluminum and Aluminum Alloys. — Aluminum and its alloys lose part of their strength at elevated temperatures, although some alloys retain good strength at temperatures from 400 to 500 degrees F. At subzero temperatures, however, their strength increases without loss of ductility so that aluminum is a particularly useful metal for low-temperature applications.

When aluminum surfaces are exposed to the atmosphere, a thin invisible oxide skin forms immediately which protects the metal from further oxidation. This self-protecting characteristic gives aluminum its high resistance to corrosion. Unless exposed to some substance or condition which destroys this protective oxide coating, the metal remains protected against corrosion. Aluminum is highly resistant to weathering, even in industrial atmospheres. It is also corrosion resistant to many acids. Alkalis are among the few substances that attack the oxide skin and therefore are corrosive to aluminum. Although the metal can safely be used in the presence of certain mild alkalis with the aid of inhibitors, in general, direct contact with alkaline substances should be avoided. Direct contact with certain other metals should be avoided in the presence of an electrolyte; otherwise galvanic corrosion of the aluminum may take place in the vicinity of the contact area. Where other metals must be fastened to aluminum, the use of a bituminous paint coating or insulating tape is recommended.

Aluminum is one of the two common metals having an electrical conductivity high enough for use as an electric conductor. The conductivity of electric-conductor (EC) grade is about 62 per cent that of the International Annealed Copper Standard. Because aluminum has less than one-third the specific gravity of copper, however, a pound of aluminum will go almost twice as far as a pound of copper when used for this purpose. Alloying lowers the conductivity somewhat so that wherever possible the EC grade is used in electric conductor applications.

Aluminum has nonsparking and nonmagnetic characteristics which make the metal useful for electrical shielding purposes such as in bus bar housings or enclosures for other electrical equipment and for use around inflammable or explosive substances.

Aluminum can be cast by any method known to foundrymen. It can be rolled to any desired thickness down to foil thinner than paper and in sheet form can be stamped, drawn, spun or rolled-formed. The metal also may be hammered or forged. Aluminum wire, drawn from rolled rod, may be stranded into cable of any desired size and type. The metal may be extruded into a variety of shapes. It may be turned, milled, bored, or machined in machines often operating at their maximum speeds. Aluminum rod and bar may readily be employed in the high-speed manufacture of automatic screw-machine parts.

Almost any method of joining is applicable to aluminum — riveting, welding or brazing. A wide variety of mechanical aluminum fasteners simplifies the assembly of many products. Resin bonding of aluminum parts has been successfully employed, particularly in aircraft components.

For the majority of applications, aluminum needs no protective coating. Mechanical finishes such as polishing, sand blasting or wire brushing meet the majority of needs. When additional protection is desired, chemical, electrochemical and paint finishes are all used. Vitreous enamels have recently been developed for aluminum, and the metal may also be electroplated.

Temper Designations for Aluminum Alloys. — The temper designation system adopted by The Aluminum Association and used in industry pertains to all forms of wrought and cast aluminum and aluminum alloys except ingot. It is based on the sequences of basic treatments used to produce the various tempers. The temper designation follows the alloy designation, being separated by a dash.

Basic temper designations consist of letters. Subdivisions of the basic tempers, where required, are indicated by one or more digits following the letter. These designate specific sequences of basic treatments, but only operations recognized as significantly influencing the characteristics of the product are indicated. Should some other variation of the same sequence of basic operations be applied to the same alloy, resulting in different characteristics, then additional digits are added.

The basic temper designations and subdivisions are as follows:

-F *as fabricated:* Applies to products which acquire some temper from shaping processes not having special control over the amount of strain-hardening or thermal treatment. For wrought products, there are no mechanical property limits.

-O *annealed, recrystallized (wrought products only):* Applies to the softest temper of wrought products.

-H *strain-hardened (wrought products only):* Applies to products which have their strength increased by strain-hardening with or without supplementary thermal treatments to produce partial softening.

The –H is always followed by two or more digits.
The first digit indicates the specific combination of basic operations, as follows:

-H1 *strain-hardened only:* Applies to products which are strain-hardened to obtain the desired mechanical properties without supplementary thermal treatment.

The number following this designation indicates the degree of strain-hardening.

-H2 *strain-hardened and then partially annealed:* Applies to products which are strain-hardened more than the desired final amount and then reduced in strength to the desired level by partial annealing. For alloys that age-soften at room temperature, the –H2 tempers have approximately the same ultimate strength as the corresponding –H3 tempers. For other alloys, the –H2 tempers have approximately the same ultimate strengths as the corresponding –H1 tempers and slightly higher elongations.

The number following this designation indicates the degree of strain-hardening remaining after the product has been partially annealed.

-H3 *strain-hardened and then stabilized:* Applies to products which are strain-hardened and then stabilized by a low temperature heating to slightly lower their strength and increase ductility. This designation applies only to the magnesium-containing alloys which, unless stabilized, gradually age-soften at room temperature.

The number following this designation indicates the degree of strain-hardening remaining after the product has been strain-hardened a specific amount and then stabilized.

The second digit following the designations $-H_1$, $-H_2$, and $-H_3$ indicates the final degree of strain-hardening. Numeral 8 has been assigned to indicate tempers having a final degree of strain-hardening equivalent to that resulting from approximately 75 per cent reduction of area. Tempers between $-O$ (annealed) and 8 (full hard) are designated by numerals 1 through 7. Material having an ultimate strength about midway between that of the $-O$ temper and that of the 8 temper is designated by the numeral 4 (half hard); between $-O$ and 4 by the numeral 2 (quarter hard); between 4 and 8 by the numeral 6 (three-quarter hard); etc. (NOTE. For two-digit $-H$ tempers whose second figure is odd, the standard limits for ultimate strength are exactly midway between those for the adjacent two-digit $-H$ tempers whose second figures are even). Numeral 9 designates extra hard tempers.

The third digit, when used, indicates a variation of a two-digit $-H$ temper. It is used when the degree of control of temper or the mechanical properties are different from but close to those for the two-digit $-H$ temper designation to which it is added. (NOTE. The minimum ultimate strength of a three-digit $-H$ temper is at least as close to that of the corresponding two-digit $-H$ temper as it is to the adjacent two-digit $-H$ tempers.) Numerals 1 through 9 may be arbitrarily assigned and registered with The Aluminum Association for an alloy and product to indicate a specific degree of control of temper or specific mechanical property limits. Zero has been assigned to indicate degrees of control of temper or mechanical property limits negotiated between the manufacturer and purchaser which are not used widely enough to justify registration with The Aluminum Association.

The following three-digit $-H$ temper designations have been assigned for wrought products in all alloys:

- $-H_{111}$ Applies to products which are strain-hardened less than the amount required for a controlled H_{11} temper.

- $-H_{112}$ Applies to products which acquire some temper from shaping processes not having special control over the amount of strain-hardening or thermal treatment, but for which there are mechanical property limits or mechanical property testing is required.

- $-H_{311}$ Applies to products which are strain-hardened less than the amount required for a controlled H_{31} temper.

The following three-digit $-H$ temper designations have been assigned for

Patterned or Embossed Sheet	Fabricated From
$-H_{114}$	$-O$ temper
$-H_{124}$, $-H_{224}$, $-H_{324}$	$-H_{11}$, $-H_{21}$, $-H_{31}$ temper, respectively
$-H_{134}$, $-H_{234}$, $-H_{334}$	$-H_{12}$, $-H_{22}$, $-H_{32}$ temper, respectively
$-H_{144}$, $-H_{244}$, $-H_{344}$	$-H_{13}$, $-H_{23}$, $-H_{33}$ temper, respectively
$-H_{154}$, $-H_{254}$, $-H_{354}$	$-H_{14}$, $-H_{24}$, $-H_{34}$ temper, respectively
$-H_{164}$, $-H_{264}$, $-H_{364}$	$-H_{15}$, $-H_{25}$, $-H_{35}$ temper, respectively
$-H_{174}$, $-H_{274}$, $-H_{374}$	$-H_{16}$, $-H_{26}$, $-H_{36}$ temper, respectively
$-H_{184}$, $-H_{284}$, $-H_{384}$	$-H_{17}$, $-H_{27}$, $-H_{37}$ temper, respectively
$-H_{194}$, $-H_{294}$, $-H_{394}$	$-H_{18}$, $-H_{28}$, $-H_{38}$ temper, respectively
$-H_{195}$, $-H_{395}$	$-H_{19}$, $-H_{39}$, temper, respectively

- $-W$ *solution heat-treated:* An unstable temper applicable only to alloys which spontaneously age at room temperature after solution heat-treatment. This designation is specific only when the period of natural aging is indicated: for example, $-W_{1/2}$ hour.

−T *thermally treated to produce stable tempers other than −F, −O, or −H:* Applies to products which are thermally treated, with or without supplementary strain-hardening, to produce stable tempers.

The −T is always followed by one or more digits. Numerals 2 through 10 have been assigned to indicate specific sequences of basic treatments, as follows:

−T2 *annealed (cast products only):* Designates a type of annealing treatment used to improve ductility and increase dimensional stability of castings.

−T3 *solution heat-treated and then cold worked:* Applies to products which are cold worked to improve strength, or in which the effect of cold work in flattening or straightening is recognized in applicable specifications.

−T4 *solution heat-treated and naturally aged to a substantially stable condition:* Applies to products which are not cold worked after solution heat-treatment, or in which the effect of cold work in flattening or straightening may not be recognized in applicable specifications.

−T5 *artificially aged only:* Applies to products which are artificially aged after an elevated-temperature rapid-cool fabrication process, such as casting or extrusion, to improve mechanical properties and/or dimensional stability.

−T6 *solution heat-treated and then artificially aged:* Applies to products which are not cold worked after solution heat-treatment, or in which the effect of cold work in flattening or straightening may not be recognized in applicable specifications.

−T7 *solution heat-treated and then stabilized:* Applies to products which are stabilized to carry them beyond the point of maximum hardness, providing control of growth and/or residual stress.

−T8 *solution heat-treated, cold worked, and then artificially aged:* Applies to products which are cold worked to improve strength, or in which the effect of cold work in flattening or straightening is recognized in applicable specifications.

−T9 *solution heat-treated, artificially aged, and then cold worked:* Applies to products which are cold worked to improve strength.

−T10 *artificially aged and then cold worked:* Applies to products which are artificially aged after an elevated-temperature rapid-cool fabrication process, such as casting or extrusion, and then cold worked to improve strength.

A period of natural aging at room temperature may occur between or after the operations listed for tempers −T3 through −T10. Control of this period is exercised when it is metallurgically important.

Additional digits may be added to designations −T2 through −T10 to indicate a variation in treatment which significantly alters the characteristics of the product. These may be arbitrarily assigned and registered with The Aluminum Association for an alloy and product to indicate a specific treatment or specific mechanical property limits.

These additional digits have been assigned for wrought products in all alloys:

−TX51 *stress-relieved by stretching:* Applies to products which are stress-relieved by stretching the following amounts after solution heat-treatment:

Plate	1½ to 3 per cent permanent set
Rod, Bar and Shapes	1 to 3 per cent permanent set

Applies directly to plate and rolled or cold-finished rod and bar.

These products receive no further straightening after stretching.

Applies to extruded rod, bar and shapes when designated as follows:

-TX510   Applies to extruded rod, bar and shapes which receive no further straightening after stretching.

-TX511   Applies to extruded rod, bar and shapes which receive minor straightening after stretching to comply with standard tolerances.

-TX52   *stress-relieved by compressing:* Applies to products which are stress-relieved by compressing after solution heat-treatment, to produce a nominal permanent set of 2½ per cent.

-TX53   *stress-relieved by thermal treatment*

The following two-digit -T temper designations have been assigned for wrought products in all alloys:

-T42   Applies to products solution heat-treated by the user which attain mechanical properties different from those of the -T4 temper.

-T62   Applies to products solution heat-treated and artificially aged by the user which attain mechanical properties different from those of the -T6 temper. (NOTE. Exceptions not conforming to the definitions given for -T42 and -T62 are 4032–T62, 6101–T62, 6061–T62, 6062–T62, 6063–T42 and 6463 -T42.)

Aluminum Alloy Designation Systems. — Aluminum casting alloys are listed in many specifications of various standardizing agencies. These include Federal Specifications, Military Specifications, ASTM Specifications and SAE Specifications, to mention some. The numbering systems used by each differ and are not always correlatable. Casting alloys are available from producers who use a commercial numbering system and this numbering system is the one used in the tables of aluminum casting alloys which are given further along in this section.

A system of four-digit numerical designations for wrought aluminum and wrought aluminum alloys was adopted by The Aluminum Association in 1954. This system is used by the commercial producers and is similar to the one used by the SAE; the difference being the addition of two prefix letters.

The first digit of the designation identifies the alloy type, 1, indicating an aluminum of 99.00 per cent or greater purity; 2, copper; 3, manganese; 4, silicon; 5, magnesium; 6, magnesium and silicon; 7, zinc; 8, some element other than those aforementioned; 9, unused (not assigned at present). If the second digit in the designation is zero, it indicates that there is no special control on individual impurities; while integers 1 through 9, indicate special control on one or more individual impurities.

In the 1000 series group for aluminum of 99.00 per cent or greater purity, the last two of the four digits indicate to the nearest hundredth the amount of aluminum above 99.00 per cent. Thus designation 1030 indicates 99.30 per cent minimum aluminum. In the 2000 to 8000 series groups the last two of the four digits have no significance but are used to identify different alloys in the group. At the time of adoption of this designation system most of the existing commercial designation numbers were used as these last two digits, as for example, 14S became 2014, 3S became 3003, and 75S became 7075. When new alloys are developed and are commercially used these last two digits are assigned consecutively beginning with -01, skipping any numbers previously assigned at the time of initial adoption.

Experimental alloys are also designated in accordance with this system but they are indicated by the prefix X. The prefix is dropped upon standardization.

Heat-treatability of Wrought Aluminum Alloys. — In high-purity form, aluminum is soft and ductile. Most commercial uses, however, require greater strength than pure aluminum affords. This is achieved in aluminum first by the addition of other elements to produce various alloys, which singly or in combination impart strength to the metal. Further strengthening is possible by means which classify the alloys roughly into two categories, non-heat-treatable and heat-treatable.

Non-heat-treatable alloys: The initial strength of alloys in this group depends upon the hardening effect of elements such as manganese, silicon, iron and magnesium, singly or in various combinations. The non-heat-treatable alloys are usually designated, therefore, in the 1000, 3000, 4000, or 5000 series. Since these alloys are work-hardenable, further strengthening is made possible by various degrees of cold working, denoted by the "H" series of tempers. Alloys containing appreciable amounts of magnesium when supplied in strain-hardened tempers are usually given a final elevated-temperature treatment called *stabilizing* for property stability.

Heat-treatable alloys: The initial strength of alloys in this group is enhanced by the addition of alloying elements such as copper, magnesium, zinc, and silicon. Since these elements singly or in various combinations show increasing solid solubility in aluminum with increasing temperature, it is possible to subject them to thermal treatments which will impart pronounced strengthening.

The first step, called *heat-treatment* or *solution heat-treatment*, is an elevated-temperature process designed to put the soluble element in solid solution. This is followed by rapid quenching, usually in water, which momentarily "freezes" the structure and for a short time renders the alloy very workable. It is at this stage that some fabricators retain this more workable structure by storing the alloys at below freezing temperatures until they are ready to form them. At room or elevated temperatures the alloys are not stable after quenching, however, and precipitation of the constituents from the supersaturated solution begins. After a period of several days at room temperature, termed *aging* or *room-temperature precipitation*, the alloy is considerably stronger. Many alloys approach a stable condition at room temperature, but some alloys, particularly those containing magnesium and silicon or magnesium and zinc, continue to age-harden for long periods of time at room temperature.

By heating for a controlled time at slightly elevated temperatures, even further strengthening is possible and properties are stabilized. This process is called *artificial aging* or *precipitation hardening*. By the proper combination of solution heat-treatment, quenching, cold working and artificial aging, the highest strengths are obtained.

Clad Aluminum Alloys. — The heat-treatable alloys in which copper or zinc are major alloying constituents, are less resistant to corrosive attack than the majority of non-heat-treatable alloys. To increase the corrosion resistance of these alloys in sheet and plate form they are often clad with high-purity aluminum, a low magnesium-silicon alloy, or an alloy containing 1 per cent zinc. The cladding, usually from 2½ to 5 per cent of the total thickness on each side, not only protects the composite due to its own inherently excellent corrosion resistance but also exerts a galvanic effect which further protects the core material.

Special composites may be obtained such as clad non-heat-treatable alloys for extra corrosion protection, for brazing purposes, or for special surface finishes. Some alloys in wire and tubular form are clad for similar reasons and on an experimental basis extrusions also have been clad.

Characteristics of Principal Aluminum Alloy Series Groups. — 1000 series: These alloys are characterized by high corrosion resistance, high thermal and electrical conductivity, low mechanical properties and good workability. Moderate

increases in strength may be obtained by strain-hardening. Iron and silicon are the major impurities.

2000 series: Copper is the principal alloying element in this group. These alloys require solution heat-treatment to obtain optimum properties; in the heat-treated condition mechanical properties are similar to, and sometimes exceed, those of mild steel. In some instances artificial aging is employed to further increase the mechanical properties. This treatment materially increases yield strength, with attendant loss in elongation; its effect on tensile (ultimate) strength is not as great. The alloys in the 2000 series do not have as good corrosion resistance as most other aluminum alloys and under certain conditions they may be subject to intergranular corrosion. Therefore, these alloys in the form of sheet are usually clad with a high-purity alloy or a magnesium-silicon alloy of the 6000 series which provides galvanic protection to the core material and thus greatly increases resistance to corrosion. Alloy 2024 is perhaps the best known and most widely used aircraft alloy.

3000 series: Manganese is the major alloying element of alloys in this group, which are generally non-heat-treatable. Because only a limited percentage of manganese, up to about 1.5 per cent, can be effectively added to aluminum, it is used as a major element in only a few instances. One of these, however, is the popular 3003, used for moderate-strength applications requiring good workability.

4000 series: The major alloying element of this group is silicon, which can be added in sufficient quantities to cause substantial lowering of the melting point without producing brittleness in the resulting alloys. For these reasons aluminum-silicon alloys are used in welding wire and as brazing alloys where a lower melting point than that of the parent metal is required. Most alloys in this series are non-heat-treatable, but when used in welding heat-treatable alloys they will pick up some of the alloying constituents of the latter and so respond to heat-treatment to a limited extent. The alloys containing appreciable amounts of silicon become dark gray when anodic oxide finishes are applied, and hence are in demand for architectural applications.

5000 series: Magnesium is one of the most effective and widely used alloying elements for aluminum. When it is used as the major alloying element or with manganese, the result is a moderate to high strength non-heat-treatable alloy. Magnesium is considerably more effective than manganese as a hardener, about 0.8 per cent magnesium being equal to 1.25 per cent manganese, and it can be added in considerably higher quantities. Alloys in this series possess good welding characteristics and good resistance to corrosion in marine atmospheres. However, certain limitations should be placed on the amount of cold work and the safe operating temperatures permissible for the higher magnesium content alloys (over about 3½ per cent for operating temperatures over about 150 deg. F.) to avoid susceptibility to stress corrosion.

6000 series: Alloys in this group contain silicon and magnesium in approximate proportions to form magnesium silicide, thus making them capable of being heat-treated. The major alloy in this series is 6061, one of the most versatile of the heat-treatable alloys. Though less strong than most of the 2000 or 7000 alloys, the magnesium-silicon (or magnesium-silicide) alloys possess good formability and corrosion resistance, with medium strength. Alloys in this heat-treatable group may be formed in the −T4 temper (solution heat-treated but not artificially aged) and then reach full −T6 properties by artificial aging.

7000 series: Zinc is the major alloying element in this group, and when coupled with a smaller percentage of magnesium results in heat-treatable alloys of very high strength. Usually other elements such as copper and chromium are also added in small quantities. Notable member of this group is 7075, which is among the highest strength aluminum alloys available and is used in air-frame structures and for highly stressed parts.

S.A.E. Cast Magnesium Alloys

S.A.E. Standard No. 50 Alloy. — This alloy is used for most commercial applications. It is used in the "as cast," "heat treated," or "heat treated and aged" condition as may be required.

Composition of No. 50: Aluminum, 5.3 to 6.7; manganese, min., 0.15; zinc, 2.5 to 3.5; silicon, max., 0.5; copper, max., 0.05; nickel, max., 0.03; other impurities, max., 0.3 per cent and the remainder, magnesium.

Physical Properties: For sand castings as in the "as cast," "heat treated" and "heat treated and aged" conditions the minimum tensile strengths are respectively: 24,000, 30,000 and 32,000 pounds per square inch; the minimum yield strengths are respectively: 10,000, 10,000 and 16,000 pounds per square inch and the elongations in 2 inches are respectively: 4, 6 and 2 per cent.

S.A.E. Standard No. 500 Alloy. — This is a sand casting alloy to be used particularly where maximum pressure tightness is required. It may be used in the "as cast," "heat treated" or "heat treated and aged" condition as may be required.

Composition of No. 500: Aluminum 8.3 to 9.7; manganese, min., 0.10; zinc, 1.7 to 2.3; silicon, max., 0.5; copper, max., 0.05; nickel, max., 0.03; other impurities, max., 0.3 per cent and the remainder, magnesium.

Physical Properties: For sand castings in the "as cast," "heat treated" and "heat treated and aged" conditions, the minimum tensile strengths are respectively: 20,000, 30,000 and 32,000 pounds per square inch; the yield strengths are respectively: 10,000, 10,000 and 17,000 pounds per square inch and the elongations in 2 inches are respectively: 1, 6 and 1 per cent.

S.A.E. Wrought Magnesium Alloys

S.A.E. Standard No. 51 Alloy. — This alloy is used where maximum salt water resistance and weldability are desired. It is used in the annealed temper for applications requiring maximum formability, such as aircraft tanks and wheel fairings.

Composition of No. 51: Manganese, min., 1.20; silicon, max., 0.3; copper, max., 0.05; nickel, max., 0.03; other impurities, max., 0.3 per cent and the remainder, magnesium.

Physical Properties: Standard tensile test specimens machined from plate or sheet stock in thicknesses between 0.016 inch and 0.025 inch have a minimum tensile strength of 32,000 pounds per square inch in the hard rolled temper, a maximum tensile strength of 35,000 pounds per square inch in the annealed temper and an elongation in 2 inches of 4 per cent in the hard rolled temper and 12 per cent in the annealed temper.

S.A.E. Standard No. 510 Alloy. — This alloy is generally used where moderate formability and mechanical properties are required.

Composition of No. 510: Aluminum, 3.3 to 4.7; manganese, min., 0.20; zinc, max., 0.3; silicon, max., 0.5; copper, max., 0.05; nickel, max., 0.03; other impurities, max., 0.3 per cent and the remainder, magnesium.

Physical Properties: Standard tensile test specimens machined from plate or sheet stock in thicknesses between 0.16 and 0.125 inch have a tensile strength of 36,000 pounds per square inch, minimum in the hard rolled temper and 38,000 pounds per square inch, maximum in the annealed temper; a yield strength of 25,000 pounds per square inch in the hard rolled temper and an elongation in 2

inches of 4 per cent in the hard rolled temper and 10 per cent in the annealed temper.

S.A.E. Standard No. 511 Alloy. — This alloy is used where high mechanical properties are required. It is available in the hard rolled and annealed tempers.

Composition of No. 511: Aluminum 5.8 to 7.2; manganese, min., 0.15; zinc, max., 0.3; silicon, max., 0.5; copper, max., 0.05; nickel, 0.03; other impurities, 0.3 per cent and the remainder, magnesium.

Physical Properties: Standard tension test specimens machined from plate or sheet stock in thicknesses between 0.016 inch and 0.125 inch have tensile strength of 39,000 pounds per square inch, minimum in the hard rolled temper and 42,000 pounds per square inch, maximum in the annealed temper; a yield strength of 28,000 pounds per square inch in the hard rolled temper and an elongation in 2 inches of 3 per cent in the hard rolled temper and 10 per cent in the annealed temper.

S.A.E. Standard No. 52 Alloy. — This is a general purpose alloy with moderate strength and fair weldability. It is especially suited for the production of thin wall tubing and other sections requiring good extrusion characteristics.

Composition of No. 52: Aluminum, 2.4 to 3.0; manganese, min., 0.20; zinc, 0.7 to 1.3; silicon, max., 0.5; copper, max., 0.05; nickel, max., 0.03; other impurities, max., 0.3 per cent and the remainder, magnesium.

Physical Properties: Standard test specimens machined from solid bar stock and structural shapes have a minimum tensile strength of 37,000 pounds per square inch in extruded bars up to 1½ inches and 34,000 pounds per square inch in structural shapes; a yield strength of 25,000 pounds per square inch in the former and 17,000 pounds per square inch in the latter and an elongation in 2 inches of 12 per cent in the former and 10 per cent in the latter.

S.A.E. Standard No. 520 Alloy. — This alloy is used for extruded bars, rods and shapes with good strength and fair weldability.

Composition of No. 520: Aluminum, 5.8 to 7.2; manganese, min., 0.15; zinc, 0.4 to 1.0; silicon, max., 0.5; iron, max., 0.05; nickel, max., 0.03; other impurities, max., 0.3 per cent and the remainder, magnesium.

Physical Properties: Standard tension test specimens machined from solid bar stock and structural shapes have a minimum tensile strength of 40,000 pounds per square inch in extruded bars up to 1½ inches and 38,000 pounds per square inch in structural shapes; a yield strength of 26,000 pounds per square inch in the former and 23,000 pounds per square inch in the latter and an elongation in 2 inches of 12 per cent in the former and 10 per cent in the latter.

S.A.E. Standard No. 522 Alloy. — This is an extrusion alloy used for applications requiring maximum weldability.

Composition of No. 522: Manganese, min. 1.2; silicon, max. 0.3; copper, max. 0.05; nickel, max. 0.03; and calcium, 0.3 per cent; remainder, magnesium.

Physical Properties: Tensile strength is 30,000 pounds per square inch for extruded bars, ¼ inch to 1½ inches; 29,000 pounds per square inch for structural shapes and 28,000 pounds per square inch for hollow shapes. Elongation in 2 inches is 3 per cent for extruded bars, ¼ inch to 1½ inches and 2 per cent for structural and hollow shapes.

S.A.E. Standard Nos. 53, 531, 532 and 533 Alloys. — These are forging alloys. Nos. 53 and 533 are suitable for hammer forging. The former has somewhat better physical properties but the latter may be readily welded and contains no tin. No.

533 may also be press forged. Hammer forgings are normally more economical than press forgings but can only be used for applications involving moderate stresses. Press forging alloys Nos. 531 and 532 are used in applications involving higher stresses. No. 532 is stronger than No. 531 but more difficult to forge and is usually employed only for comparatively simple forgings requiring highest physical properties.

Composition of No. 53: Aluminum, 3.0 to 4.0; manganese, min. 0.2; zinc, max. 0.3; silicon, max. 0.3; copper, max. 0.05; nickel, max. 0.005; iron, max. 0.005 and tin, 4.0 to 6.0 per cent; remainder, magnesium.

Composition of No. 531: Aluminum, 5.8 to 7.2; manganese, minimum 0.15; zinc, 0.4 to 1.5; silicon, maximum 0.3; copper, max. 0.05; nickel, maximum 0.005 and iron, maximum 0.005 per cent; remainder, magnesium.

Composition of No. 532: Aluminum, 7.8 to 9.2; manganese, minimum 0.12; zinc, 0.2 to 0.8; silicon, maximum 0.3; copper, maximum 0.05; nickel, maximum 0.005 and iron, maximum 0.005 per cent; remainder, magnesium.

Composition of No. 533: Manganese, minimum 1.2; silicon, maximum 0.3; copper, maximum 0.05 and nickel, maximum 0.03 per cent; remainder, magnesium.

Physical Properties: In the as forged condition, No. 53 has a minimum tensile strength of 36,000; No. 531, 38,000; No. 532, 42,000; and No. 533, 30,000 pounds per square inch. Nos. 53 and 531 have a yield strength of 22,000; No. 532, 26,000; and No. 533, 18,000 pounds per square inch. No. 53 has a minimum elongation in 2 inches of 7; No. 531, 6; No. 532, 5; and No. 533, 3 per cent.

Nickel and Nickel Alloys

Nickel. — Nickel is noted for its corrosion resistance, good electrical conductivity and high heat-transfer properties. It is used to fabricate process equipment for handling pure foods and drugs, electrical contact parts, and radio and X-ray tube elements.

Approximate Composition: (Commercially pure wrought nickel:) Nickel (including cobalt), 99.4; copper, 0.1; iron, 0.15; manganese, 0.25; silicon, 0.05; carbon, 0.05; and sulphur, 0.005. (Cast nickel:) Nickel, 97.0; copper, 0.3; iron, 0.25; manganese, 0.5; silicon, 1.6; and carbon, 0.5.

Average Physical Properties: Wrought nickel in the annealed, hot-rolled, cold-drawn, and hard temper cold-rolled conditions exhibits yield strengths (0.2 per cent offset) of 20,000, 25,000, 70,000, and 95,000 pounds per square inch, respectively; tensile strengths of 70,000, 75,000, 95,000, and 105,000 pounds per square inch, respectively; elongations in 2 inches of 40, 40, 25, and 5 per cent, respectively; and Brinell hardnesses of 100, 110, 170, and 210, respectively.

Low-Carbon Nickel. — A special type of nickel that is corrosion resistant and has a high ductility and heat resistance. It lends itself well to spinning and cold coining or forging and is used in the manufacture of tubing and molds for the beverage and food industries.

Approximate Composition: Nickel, 99.4; copper, 0.05; iron, 0.1; silicon, 0.15; manganese, 0.2; carbon, 0.01; and sulphur, 0.005.

Average Physical Properties: Annealed low-carbon nickel exhibits a yield strength (0.2 per cent offset) of 15,000 pounds per square inch, a tensile strength of 60,000 pounds per square inch, an elongation in 2 inches of 50 per cent and a Brinell hardness of 90.

Duranickel. — This age-hardenable alloy has good spring and low-sparking properties and is slightly magnetic after heat treatment. Items such as corrosion-resistant paper machine shaker springs, diaphragms, and extrusion dies for plastics are made from it.

Approximate Composition: Nickel, 93.7; copper, 0.05; iron, 0.35; aluminum, 4.4; silicon, 0.5; manganese, 0.3; carbon, 0.17; and sulphur, 0.005.

Average Physical Properties: In the hot-rolled, hot-rolled and age-hardened, cold-drawn, and cold-drawn and age-hardened conditions this alloy exhibits yield strengths (0.2 per cent offset) of 50,000, 130,000, 90,000, and 135,000 pounds per square inch, respectively; tensile strengths of 105,000, 170,000, 120,000, and 175,000 pounds per square inch, respectively; elongations in 2 inches of 35, 15, 25, and 15 per cent, respectively; and Brinell hardnesses of 180, 320, 220, and 340, respectively.

Monel. — This general purpose alloy is corrosion-resistant, strong, tough and has a silvery-white color. It is used for making abrasion- and heat-resistant valves and pump parts, propeller shafts, laundry machines, chemical processing equipment, etc.

Approximate Composition: Nickel, 67; copper, 30; iron, 1.4; silicon, 0.1; manganese, 1; carbon, 0.15; and sulphur 0.01.

Average Physical Properties: Wrought Monel in the annealed, hot-rolled, cold-drawn, and hard temper cold-rolled conditions exhibits yield strengths (0.2 per cent offset) of 35,000, 50,000, 80,000, and 100,000 pounds per square inch, respectively; tensile strengths of 75,000, 90,000, 100,000, and 110,000 pounds per square inch, respectively; elongations in 2 inches of 40, 35, 25, and 5 per cent, respectively; and Brinell hardnesses of 125, 150, 190, and 240, respectively.

"R" Monel. — This free-cutting, corrosion resistant alloy is used for automatic screw machine products such as bolts, screws and precision parts.

Approximate Composition: Nickel, 67; copper, 30; iron, 1.4; silicon, 0.05; manganese, 1; carbon, 0.15; and sulphur, 0.035.

Average Physical Properties: In the hot-rolled and cold-drawn conditions this alloy exhibits yield strengths (0.2 per cent offset) of 45,000 and 75,000 pounds per square inch, respectively; tensile strengths of 85,000 and 90,000 pounds per square inch, respectively; elongations in 2 inches of 35, and 25 per cent, respectively; and Brinell hardnesses of 145 and 180, respectively.

"K" Monel. — This strong and hard alloy, comparable to heat-treated alloy steel, is age-hardenable, non-magnetic and has low-sparking properties. It is used for corrosive applications where the material is to be machined or formed, then age hardened. Pump and valve parts, scrapers, and instrument parts are made from this alloy.

Approximate Composition: Nickel, 66; copper, 29; iron, 0.9; aluminum, 2.75; silicon, 0.5; manganese, 0.75; carbon, 0.15; and sulphur, 0.005.

Average Physical Properties: In the hot-rolled, hot-rolled and age-hardened, cold-drawn, and cold-drawn and age-hardened conditions the alloy exhibits yield strengths (0.2 per cent offset) of 45,000, 110,000, 85,000, and 115,000 pounds per square inch, respectively; tensile strengths of 100,000, 150,000, 115,000, and 155,000 pounds per square inch, respectively; elongations in 2 inches of 40, 25, 25, and 20 per cent, respectively; and Brinell hardnesses of 160, 280, 210, and 290, respectively.

"KR" Monel. — This strong, hard, age-hardenable and non-magnetic alloy is more readily machinable than "K" Monel. It is used for making valve stems, small parts for pumps, and screw machine products requiring an age-hardening material that is corrosion-resistant.

Approximate Composition: Nickel, 66; copper, 29; iron, 0.9; aluminum, 2.75; silicon, 0.5; manganese, 0.75; carbon, 0.28; and sulphur, 0.005.

Average Physical Properties: Essentially the same as "K" Monel.

"S" Monel. — This extra hard casting alloy is non-galling, corrosion-resisting, non-magnetic, age-hardenable and has low-sparking properties. It is used for gall-

resistant pump and valve parts which have to withstand high temperatures, corrosive chemicals and severe abrasion.

Approximate Composition: Nickel, 63; copper, 30; iron, 2; silicon, 4; manganese, 0.75; carbon, 0.1; and sulphur, 0.015.

Average Physical Properties: In the annealed sand-cast, as-cast sand-cast, and age-hardened sand-cast conditions it exhibits yield strengths (0.2 per cent offset) of 70,000, 100,000, and 100,000 pounds per square inch, respectively; tensile strengths of 90,000, 130,000, and 130,000 pounds per square inch, respectively; elongations in 2 inches of 3, 2, and 2 per cent, respectively; and Brinell hardnesses of 275, 320, and 350, respectively.

"H" Monel. — An extra hard casting alloy with good ductility, intermediate strength and hardness that is used for pumps, impellers and steam nozzles.

Approximate Composition: Nickel, 63; copper, 31; iron, 2; silicon, 3; manganese, 0.75; carbon, 0.1; and sulphur, 0.015.

Average Physical Properties: In the as-cast sand-cast condition this alloy exhibits a yield strength (0.2 per cent offset) of 60,000 pounds per square inch, a tensile strength of 100,000 pounds per square inch, an elongation in 2 inches of 15 per cent and a Brinell hardness of 210.

Inconel. — This heat resistant alloy retains its strength at high heats, resists oxidation and corrosion, has a high creep strength and is non-magnetic. It is used for high temperature applications (up to 2000 degrees F.) such as engine exhaust manifolds and furnace and heat treating equipment. Springs operating at temperatures up to 700 degrees F. are also made from it.

Approximate Composition: Nickel, 76; copper, 0.20; iron, 7.5; chromium, 15.5; silicon, 0.25; manganese, 0.25; carbon, 0.08; and sulphur, 0.007.

Physical Properties: Wrought Inconel in the annealed, hot-rolled, cold-drawn, and hard temper cold-rolled conditions exhibits yield strengths (0.2 per cent offset) of 35,000, 60,000, 90,000, and 110,000 pounds per square inch, respectively; tensile strengths of 85,000, 100,000, 115,000, and 135,000 pounds per square inch, respectively; elongations in 2 inches of 45, 35, 20, and 5 per cent, respectively; and Brinell hardnesses of 150, 180, 200, and 260, respectively.

Inconel "X". — This alloy has a low creep rate, is age-hardenable and non-magnetic, resists oxidation and exhibits a high strength at elevated temperatures. Uses include the making of bolts and turbine rotors used at temperatures up to 1500 degrees F., aviation brake drum springs and relief valve and turbine springs with low load-loss or relaxation for temperatures up to 1000 degrees F.

Approximate Composition: Nickel, 73; copper, 0.2 maximum; iron, 7; chromium, 15; aluminum, 0.7; silicon, 0.4; manganese, 0.5; carbon, 0.04; sulphur, 0.007; columbium, 1; and titanium, 2.5.

Average Physical Properties: Wrought Inconel " X " in the annealed and age-hardened hot-rolled conditions exhibits yield strengths (0.2 per cent offset) of 50,000 and 120,000 pounds per square inch, respectively; tensile strengths of 115,000 and 180,000 pounds per square inch, respectively; elongations in 2 inches of 50 and 25 per cent, respectively; and Brinell hardnesses of 200 and 360, respectively.

Titanium and Titanium Alloys

Titanium. — This metal is used in its commercially pure state and in alloy form (being alloyed with manganese or ferrochromium) for applications requiring a metal with properties of light weight, high strength, and good temperature- and corrosion-resistance. Titanium and its alloys weigh approximately 44 per cent less

Properties of Titanium and Titanium Alloys

Type*	Yield Strength, Pounds per Sq. In.	Ult. Tensile Strength, Pounds per Sq. In.	Per Cent Elongation in 2 In.
Titanium, pure	40,000 to 85,000	60,000 to 110,000	30 to 20
3% Ferrochromium or 4% Manganese	75,000 to 110,000	100,000 to 125,000	20 to 15
6% Ferrochromium	110,000 to 125,000	120,000 to 155,000	18 to 10
7% Manganese	120,000 to 160,000	130,000 to 170,000	18 to 7
* The percentage of chief alloying elements is given, except for the first entry which is commercially pure titanium.			

than stainless or alloy steels, are equal or greater in yield and ultimate tensile strength than structural alloys in common use, withstand temperatures up to 800 degrees F. and higher temperatures up to 2000 degrees F. for short periods and are resistant to the corrosive effects of salt water and many acids, alkalis and other chemicals. It is available in the form of plates, sheets, strip, forgings, ingots, bars, rods, and wire.

Composition and Properties: The accompanying table gives the nominal compositions, yield strengths, tensile strengths and elongations of titanium and some of its alloys.

Copper-Silicon and Beryllium Copper Alloys

Everdur. — This copper-silicon alloy is available in five slightly different nominal compositions for applications which require high strength, good fabricating and fusing qualities, immunity to rust, free-machining and a corrosion resistance equivalent to copper. The following table gives the nominal compositions and tensile strengths, yield strengths, and per cent elongations for various tempers and forms.

Nominal Composition and Properties of Everdur

Desig. No.	Nominal Composition[1]					Temper[2]	Strength, Thousands of Pounds per Square Inch		Per Cent Elongation
	Cu	Si	Mn	Pb	Al		Tensile	Yield	
1010	95.80	3.10	1.10	...	...	A HRA CRA CRHH CRH H	52 50 52 71 87 70 to 85	15 18 18 40 60 38 to 50	35* 40 35 10 3 17 to 8*
1015	98.25	1.50	0.25	...	...	AP HP XHB	38 50 75 to 85	10 40 45 to 55	35 7 8 to 6*
1012	95.60	3.00	1.00	0.40	...	A H	52 85	15 50	35* 13 to 8*
1000	94.90	4.00	1.10	...	...	AC	45	....	15
1014	90.75	2.00	...	...	7.25	A	75 to 90	37.5 to 45	12 to 9*

Designation numbers are those of The American Brass Co.

The values given for the tensile strength, yield strength and elongation are all minimum values. Where ranges are shown, the first values given are for the largest diameter or largest size specimens. Yield strength values were determined at 0.50 per cent elongation under load.

* Per cent elongation in 4 times the diameter or thickness of the specimen. All other values are per cent elongation in 2 inches.

[1] The following chemical symbols are used: Cu for copper, Si for silicon, Mn for manganese, Pb for lead, and Al for aluminum.

[2] Symbols used are: HRA for hot-rolled and annealed tank plates; CRA for cold-rolled sheets and strips; CRHH for cold-rolled half hard strips; and CRH for cold-rolled hard strips. For round, square, hexagonal, and octagonal rods: A for annealed; H for hard; and XHB for extra-hard bolt temper (in coils for cold-heading). For pipe and tube: AP for annealed; and HP for hard. For castings: AC for as cast.

Uses: (1010) Hot-rolled-and-annealed plates for unfired pressure vessels, and rods for hot forging, hot upsetting, and machining. (1015) Cold-headed-and-roll-threaded bolts and cold-drawn seamless tubes for electrical metallic tubing and rigid conduit. (1012) Screw machine products. (1000) Castings. (1014) Hot forgings and for free machining applications; not for cold working or welding.

Beryllium Copper. — These alloys which contain copper, beryllium, cobalt and in the case of one alloy, silver, fall into two groups. One group whose beryllium content is greater than one per cent is characterized by its high strength and hardness and the other, whose beryllium content is less than 1 per cent, by its high electrical and thermal conductivity. The alloys have many applications in the electrical and aircraft industries or wherever strength, corrosion resistance, conductivity, non-magnetic and non-sparking properties are essential. Beryllium copper is obtainable in the form of strips, rods and bars, wire, platers bars, billets, tubes, and casting ingots.

Composition and Physical Properties: The accompanying table lists some of the more common wrought and casting alloys and gives some of their physical properties.

Nominal Composition and Properties of Beryllium Copper Alloys

Alloy[1]	Composition[2] (Per Cent)				Form	Temper[3]	Tensile Strength, Thousands of Lbs. per Sq. In.	% Elong. in 2 In.	Rockwell Hardness, Scale-Range
	Be	Co	Ag	Cu					
25	1.80 to 2.05	0.18 to 0.30	...	Bal.	Strip	A / H / HT	60 to 78 / 100 to 120 / 190 to 215	35 to 60 / 2 to 7 / 1 to 3	B-45 to 78 / B-96 to 102 / C-40 to 45
					Rod	A / AT	60 to 85 / 165 to 190	35 to 60 / 4 to 10	B-45 to 85 / C-36 to 41
					Wire	A / AT	58 to 78 / 165 to 190	35 to 55 / 3 to 8	..... / .....
165	1.60 to 1.80	0.18 to 0.30	...	Bal.	Strip	A / H / HT	60 to 78 / 100 to 120 / 180 to 200	35 to 60 / 2 to 7 / 1 to 3	B-45 to 78 / B-96 to 102 / C-39 to 41
10	0.40 to 0.70	2.35 to 2.70	...	Bal.	Strip	A / H / HT	38 to 55 / 70 to 85 / 110 to 130	20 to 35 / 5 to 8 / 5 to 12	B-20 to 45 / B-70 to 80 / B-95 to 102
					Rod	A / AT	38 to 55 / 100 to 120	20 to 35 / 10 to 25	B-20 to 45 / B-92 to 100
50	0.25 to 0.50	1.40 to 1.70	0.90 to 1.10	Bal.	Rod	A / AT	38 to 55 / 100 to 120	20 to 35 / 10 to 25	B-20 to 45 / B-92 to 100
20C	1.90 to 2.15	0.35 to 0.65	...	Bal.	...	AT	150 to 175	1 to 3	C-38 to 45
275C	2.50 to 2.75	0.35 to 0.65	...	Bal.	...	AT	140 to 165	1 to 2	C-42 to 48
10C	0.45 to 0.75	2.35 to 2.70	...	Bal.	...	AT	100 to 120	5 to 12	B-92 to 103

[1] Alloys with number designations are wrought alloys and those with number and letter designations are casting alloys. Designations are those of The Beryllium Corp.

[2] Chemical symbols are used to designate the constituent metals: Be, beryllium; Co, cobalt; Ag, silver; Cu, copper.

[3] Temper and condition symbol designations: A, solution annealed; H, hard; HT, heat treated from hard; AT, heat treated from solution annealed.

Powdered Metal Process

This is a process by means of which metal parts in large quantities can be made by the compressing and sintering of various powdered metals such as brass, bronze and iron. The compressing of the metal powder into the shape of the part to be made is done by accurately formed dies and punches in special types of presses known as briquetting machines. The "green" compressed pieces are then sintered in an atmosphere controlled furnace at high temperatures, causing the metal powder to be bonded together into a solid mass. A subsequent sizing or pressing operation and supplementary heat treatments may also be employed in some cases. The physical properties of the final product are usually comparable to those of cast or wrought products of the same composition. Using closely controlled conditions, steel of high hardness and tensile strength has also been made by this process.

Any desired porosity from 5 to 50 per cent can be obtained in the final product. Large quantities of porous bronze and iron bearings which are impregnated with oil for self-lubrication, have been made by this process. Other porous powder metal products are being used for the filtering of liquids and gases. Where continuous porosity is desired in the final product, the voids between particles are kept connected or open by mixing one per cent of zinc stearate or other finely powdered metallic soap throughout the metal powder before briquetting and then boiling this out in a low temperature baking before the piece is sintered.

The dense type of powdered metal products include refractory metal wire and sheet, cemented carbide tools, and electrical contact materials (products which could not be made as satisfactorily by other processes) and gears or other complex shapes which might also have been made by die-casting or the precise machining of wrought or cast metal.

Advantages of Powdered Metal Process. — This process is advantageous when irregular curves, eccentrics, radial projections, or recesses are required. Where a part has irregular holes, keyways, flat sides, splines or square holes that are not easily machined, powdered metal parts may solve the problem. Tapered holes and counterbores are easily produced. Axial projections can be formed but the permissible size depends on the extent to which the powder will flow into the die recesses. Projections not more than one-quarter the length of the part are practicable. Slots, grooves, blind holes, and recesses of varied depths are also obtainable.

Limiting Factors in Powdered Metal Process. — The number and variety of shapes which may be obtained are limited by the lack of plastic flow of powders, i.e., the difficulty with which they can be made to flow around corners. Tolerances in diameter usually cannot be held closer than 0.001 inch and tolerances in length are limited to 0.005 inch. This difference in diameter and length tolerances may be due to the elasticity of the powder and spring of the press.

Factors Affecting Design of Briquetting Tools. — High-speed steel is recommended for dies and punches and oil-hardening steel for strippers and knock-outs. One manufacturer specifies dimensional tolerances of 0.0002 inch and super-finished surfaces for these tools. Because of the high pressures employed and the abrasive character of certain refractory materials used in some powdered metal compositions, there is frequently a tendency toward severe wear of dies and punches. In such cases, carbide inserts, chrome plating, or highly resistant die steels are employed. With regard to the shape of the die, corner radii, fillets, and bevels should be used to avoid sharp corners. Feather edges, threads, and reentrant angles are usually impracticable. The making of punches and dies is particularly exacting because allowances must be made for change in dimensions due to growth after briquetting and shrinkage or growth during sintering.

Classification, Approximate Compositions, and Properties Affecting Selection of Tool and Die Steels (From SAE Recommended Practice)

Type of Tool Steel	Chemical Composition†								Non-warping. Prop.	Safety in Hardening	Toughness	Depth of Hardening	Wear Resistance
	C	Mn	Si	Cr	V	W	Mo	Co					
Water Hardening													
.80 Carbon	.70– .85	*	*	*						Poor	Fair	Good[4]	Shallow	Fair
.90 Carbon	.85– .95	*	*	*						Poor	Fair	Good[4]	Shallow	Fair
1.00 Carbon	.95–1.10	*	*	*						Poor	Fair	Good[4]	Shallow	Good
1.20 Carbon	1.10–1.30	*	*	*						Poor	Fair	Good[4]	Shallow	Good
.90 Carbon-V	.85– .95	*	*	*	.15– .35					Poor	Fair	Good	Shallow	Fair
1.00 Carbon-V	.95–1.10	*	*	*	.15– .35					Poor	Fair	Good	Shallow	Good
1.00 Carbon-VV	.90–1.10	*	*	*	.35– .50					Poor	Fair	Good	Shallow	Good
Oil Hardening													
Low Manganese	.90	1.20	.25	.50	.20[1]	.50				Good	Good	Fair	Deep	Good
High Manganese	.90	1.60	.25	.35[1]	.20[1]	.75[1]	.30[1]			Good	Good	Fair	Deep	Good
High Carbon-High Chromium‡	2.15	.35	.35	12.00	.80[1]		.80[1]			Good	Good	Poor	Through	Best
Chromium	1.00	.35	.25	1.40			.40			Fair	Good	Fair	Deep	Good
Molybdenum Graphitic	1.45	.75	1.00				.25[1]			Fair	Good	Fair	Deep	Good
Nickel-Chromium**	.75	.70	.25	.85	.25[1]		.50[1]			Fair	Good	Fair	Deep	Fair
Air Hardening													
High Carbon-High Chromium	1.50	.40	.40	12.00	.80[1]		.90	.60[1]		Best	Best	Fair	Through	Best
5 Per Cent Chromium Air Hard.	1.00	.60	.25	5.25	.40[1]		1.10			Best	Best	Fair	Through	Good
High Carbon-High Chromium-Cobalt.	1.50	.40	.40	12.00	.80[1]		.90	3.10		Best	Best	Fair	Through	Best
Shock Resisting													
Chromium-Tungsten	.50	.25	.35	1.40	.20	2.25	.40[1]			Fair	Good	Good	Deep	Fair
Silicon-Molybdenum	.50	.40	1.00		.25[1]		.50			Poor[2]	Poor[3]	Best	Deep	Fair
Silicon-Manganese	.55	.80	2.00		.25[1]		.40[1]			Poor[2]	Poor[3]	Best	Deep	Fair
Hot Work													
Chromium-Molybdenum-Tungsten	.35	.30	1.00	5.00	.25[1]	1.25	1.50			Good	Good	Good	Through	Fair
Chromium-Molybdenum-V	.35	.30	1.00	5.00	.40		1.50			Good	Good	Good	Through	Fair
Chromium-Molybdenum-VV	.35	.30	1.00	5.00	.90		1.50			Good	Good	Good	Through	Fair
Tungsten	.32	.30	.20	3.25	.40	9.00				Good	Good	Good	Through	Fair

See footnotes at end of table.

Classification, Approximate Compositions, and Properties Affecting Selection of Tool and Die Steels (Concluded)

Type of Tool Steel	Chemical Composition†								Non-Warping Prop.	Safety in Hardening	Toughness	Depth of Hardening	Wear Resistance
	C	Mn	Si	Cr	V	W	Mo	Co					
High Speed													
Tungsten, 18-4-1	.70	.30	.30	4.10	1.10	18.00	...	...	Good	Good	Poor	Through	Good
Tungsten, 18-4-2	.80	.30	.30	4.10	2.10	18.50	.80	...	Good	Good	Poor	Through	Good
Tungsten, 18-4-3	1.05	.30	.30	4.10	3.25	18.50	.70	...	Good	Good	Poor	Through	Best
Cobalt-Tungsten, 14-4-2-5	.80	.30	.30	4.10	2.00	14.00	.80	5.00	Good	Fair	Poor	Through	Good
Cobalt-Tungsten, 18-4-1-5	.75	.30	.30	4.10	1.00	18.00	.80	5.00	Good	Fair	Poor	Through	Good
Cobalt-Tungsten, 18-4-2-8	.80	.30	.30	4.10	1.75	18.50	.80	8.00	Good	Fair	Poor	Through	Good
Cobalt-Tungsten, 18-4-2-12	.80	.30	.30	4.10	1.75	20.00	.80	12.00	Good	Fair	Poor	Through	Good
Molybdenum, 8-2-1	.80	.30	.30	4.00	1.15	1.50	8.50	...	Good	Fair	Poor	Through	Good
Molybdenum-Tungsten, 6-6-2	.83	.30	.30	4.10	1.90	6.25	5.00	...	Good	Fair	Poor	Through	Good
Molybdenum-Tungsten, 6-6-3	1.15	.30	.30	4.10	3.25	5.75	5.25	...	Good	Fair	Poor	Through	Best
Molybdenum-Tungsten, 6-6-4	1.30	.30	.30	4.25	4.25	5.75	5.25	...	Good	Fair	Poor	Through	Best
Cobalt-Molybdenum-Tungsten, 6-6-2-8	.85	.30	.30	4.10	2.00	6.00	5.00	8.00	Good	Fair	Poor	Through	Good

† C = carbon; Mn = manganese; Si = silicon; Cr = chromium; V = vanadium; W = tungsten; Mo = molybdenum; Co = cobalt.
‡ This steel may have 0.50 per cent nickel as an optional element. The steel has been found to give satisfactory application either with or without the element present.

* Carbon tool steels are usually available in four grades or qualities: Special (Grade 1) — The highest quality water-hardening carbon tool steel, controlled for hardenability, chemistry held to closest limits, and subject to rigid tests to insure maximum uniformity in performance; Extra (Grade 2) — A high quality water-hardening carbon tool steel, controlled for hardenability, subject to tests to insure good service; Standard (Grade 3) — A good quality water-hardening carbon tool steel, not controlled for hardenability, recommended for application where some latitude with respect to uniformity is permissible; Commercial (Grade 4) — A commercial quality water-hardening carbon tool steel, not controlled for hardenability, not subject to special tests. On special and extra grades, limits on manganese, silicon, and chromium are not generally required if Shepherd hardenability limits are specified. For standard and commercial grades, limits are 0.35 max. each for Mn and Si; 0.15 max. Cr for standard; 0.20 max. Cr for commercial.

** Approximate nickel content of this steel is 1.50%.

¹ Optional element. Steels have found satisfactory application either with or without the element present. In the case of silicon manganese steel listed under Shock Resisting Steels, if chromium, vanadium and molybdenum are not present, then hardenability will be affected.
² Poor when water quenched, fair when oil quenched. ³ Poor when water quenched, good when oil quenched.
⁴ Toughness decreases somewhat with increasing depth of hardening.

468   MATERIALS

Table 1. Typical Heat Treatments for SAE Carbon Steels

				Carburizing Grades				
SAE No.	Normalize Deg. F.	Carburize Deg. F.	Cool*	Reheat Deg. F.	Cool*	2nd Reheat Deg. F.	Cool*	Temper[3] Deg. F.
1010 to 1022	....	1650–1700	A	....	..	....	..	250–400
	....	1650–1700	B	1400–1450	A	....	..	250–400
	....	1650–1700	C	1400–1450	A	....	..	250–400
	....	1650–1700	C	1650–1700	B	1400–1450	A	250–400
	....	1500–1650[1]	B	....	..	....	..	Optional
	....	1350–1575[2]	D	....	..	....	..	Optional
1024	1650–1750[4]	1650–1700	E	....	..	....	..	250–400
	....	1350–1575[2]	D	....	..	....	..	Optional
1025	....	1650–1700	A	....	..	....	..	250–400
1026	....	1500–1650[1]	B	....	..	....	..	Optional
1027	....	1350–1575[2]	D	....	..	....	..	Optional
1030	....	1500–1650[1]	B	....	..	....	..	Optional
	....	1350–1575[2]	D	....	..	....	..	Optional
1111 1112 1113	....	1500–1650[1]	B	....	..	....	..	Optional
	....	1350–1575[2]	D	....	..	....	..	Optional
1109 to 1120	....	1650–1700	A	....	..	....	..	250–400
	....	1650–1700	B	1400–1450	A	....	..	250–400
	....	1650–1700	C	1400–1450	A	....	..	250–400
	....	1650–1700	C	1650–1700	B	1400–1450	A	250–400
	....	1500–1650[1]	B	....	..	....	..	Optional
	....	1350–1575[2]	D	....	..	....	..	Optional
1126	....	1500–1650[1]	B	....	..	....	..	Optional
	....	1350–1575[2]	D	....	..	....	..	Optional

		Heat Treating Grades			
SAE Number	Normalize Deg. F.	Anneal	Harden Deg. F.	Quench*	Temper Deg. F.
1025 & 1030	....	....	1575–1650	A	
1033 to 1035	....	....	1525–1575	B	
1036	1600–1700	....	1525–1575	B	
	....	....	1525–1575	B	
1038 to 1040	1600–1700	....	1525–1575	B	
	....	....	1525–1575	B	
1041	1600–1700 and/or	1400–1500	1475–1550	E	
1042 to 1050	1600–1700	....	1475–1550	B	
1052 & 1055	1550–1650 and/or	1400–1500	1475–1550	E	To Desired Hardness
1060 to 1074	1550–1650 and/or	1400–1500	1475–1550	E	
1078	....	1400–1500[5]	1450–1500	A	
1080 to 1090	1550–1650 and/or	1400–1500[5]	1450–1500	E[6]	
1095	....	1400–1500[5]	1450–1500	F	
	....	1400–1500[5]	1500–1600	E	
1132 & 1137	1600–1700 and/or	1400–1500	1525–1575	B	
1138 & 1140	....	....	1500–1550	B	
	1600–1700	....	1500–1550	B	
1141 & 1144	....	1400–1500	1475–1550	E	
	1600–1700	1400–1500	1475–1550	E	
1145 to 1151	....	....	1475–1550	B	
	1600–1700	....	1475–1550	B	

* Symbols: A = Water or Brine; B = Water or oil; C = Cool slowly; D = Air or Oil; E = Oil; F = Water, Brine or Oil.

[1] Activated or cyanide baths; may be given refining heat as in other processes.

[2] Carbonitriding atmospheres; may be given refining heat as in other processes.

[3] Even where tempering temperatures are shown, tempering is not mandatory on many applications. It is usually employed for partial stress relief, and improves resistance to grinding cracks.

[4] Normalizing temperatures at least 50 deg. F. above the carburizing temperature are sometimes recommended where minimum heat treat distortion is of vital importance.

[5] Slow cooling produces a spheroidal structure in these high carbon steels which is sometimes required for machining purposes.

[6] May be water or brine quenched by special techniques such as partial immersion or time quenched, otherwise they are subject to quench cracking.

Table 2. Typical Heat Treatments for SAE Alloy Steels

			Carburizing Grades				
SAE No.	Normalize[1]	Cycle Anneal[3]	Carburized Deg. F.	Cool*	Reheat Deg. F.	Cool*	Temper[8] Deg. F
1320	yes	...	1650-1700	E	1400-1450[6]	E	250-350
	yes	...	1650-1700	E	1475-1525[7]	E	250-350
	yes	...	1650-1700	C	1400-1450[6]	E	250-350
	yes	...	1650-1700	C	1500-1550[7]	E	250-350
	yes	...	1650-1700	E[5]	....	..	250-350
	yes	...	1500-1650[4]	E	....	..	250-350
2317	yes	yes	1650-1700	E	1375-1425[6]	E	250-350
	yes	yes	1650-1700	E	1450-1500[7]	E	250-350
	yes	yes	1650-1700	C	1375-1425[6]	E	250-350
	yes	yes	1650-1700	C	1475-1525[7]	E	250-350
	yes	yes	1650-1700	E[5]	....	..	250-350
	yes	yes	1450-1650[4]	E	....	..	250-350
2512 to 2517	yes[2]	...	1650-1700	C	1325-1375[6]	E	250-350
	yes[2]	...	1650-1700	C	1425-1475[7]	E	250-350
	yes	...	1650-1700	E	1400-1450[6]	E	250-350
	yes	...	1650-1700	E	1475-1525[7]	E	250-350
3115 & 3120	yes	...	1650-1700	C	1400-1450[6]	E	250-350
	yes	...	1650-1700	C	1500-1550[7]	E	250-350
	yes	...	1650-1700	E[5]	....	..	250-350
	yes	...	1500-1650[4]	E	....	..	250-350
3310 & 3316	yes[2]	...	1650-1700	E	1400-1450[6]	E	250-350
	yes[2]	...	1650-1700	C	1475-1500[7]	E	250-350
4017 to 4032	yes	yes	1650-1700	E[5]	....	..	250-350
4119 & 4125	yes	...	1650-1700	E[5]	....	..	250-350
4317 & 4320 4608 to 4621	yes	yes	1650-1700	E	1425-1475[6]	E	250-350
	yes	yes	1650-1700	E	1475-1525[7]	E	250-350
	yes	yes	1650-1700	C	1425-1475[6]	E	250-350
	yes	yes	1650-1700	C	1475-1525[7]	E	250-350
	yes	yes	1650-1700	E[5]	....	..	250-350
	yes	yes	1650-1700	E[5]	....	..	250-350
	yes	...	1500-1650[4]	E	....	..	250-350
4812 to 4820	yes[2]	yes	1650-1700	E	1375-1425[6]	E	250-350
	yes[2]	yes	1650-1700	E	1450-1500[7]	E	250-35c
	yes[2]	yes	1650-1700	C	1375-1425[6]	E	250-350
	yes[2]	...	1650-1700	C	1450-1500[7]	E	250-350
	...	...	1650-1700	E[5]	....	..	250-350
5115 & 5120	yes	...	1650-1700	E	1425-1475[6]	E	250-350
	yes	...	1650-1700	E	1500-1550[7]	E	250-350
	yes	...	1650-1700	C	1425-1475[6]	E	250-350
	yes	...	1650-1700	C	1500-1550[7]	E	250-350
	yes	...	1500-1650[4]	E	....	..	250-350
8615 to 8625 8720	yes	yes	1650-1700	E	1475-1525[6]	E	250-350
	yes	yes	1650-1700	E	1525-1575[7]	E	250-350
	yes	yes	1650-1700	C	1475-1525[6]	E	250-350
	yes	yes	1650-1700	C	1525-1575[7]	E	250-350
	yes	yes	1650-1700	E[5]	....	..	250-350
	yes	yes	1500-1650[4]	E	....	..	250-350
9310 to 9317	yes[2]	...	1650-1700	E	1400-1450[6]	E	250-350
	yes[2]	...	1650-1700	C	1500-1525	E	250-350

* Symbols: C = Cool slowly; E = Oil.

[1] Normalizing temperatures should not be less than 50 degrees F. higher than the carburizing temperature. Follow by air cooling.

[2] After normalizing, reheat to temperatures of 1000-1200 degrees F. and hold approximately 4 hours.

[3] Where cycle annealing is desired, heat to normalizing temperature — hold for uniformity — cool rapidly to 1000-1250 degrees F.; hold 1 to 3 hours, then air or furnace cool to obtain a structure suitable for machining and finish.

[4] This treatment is for activated or cyanide baths, and parts may be given refining heats as indicated for other heat treating processes.

[5] This treatment applicable to fine-grained steels only. When fine-grained steels are employed, a second reheat is often unnecessary.

[6] This treatment when case hardness only is paramount.

[7] This treatment when higher core hardness is desired.

[8] Tempering treatment is optional. Tempering is generally employed for partial stress relief and improved resistance to cracking from grinding operations.

Table 2 (*Continued*).   Typical Heat Treatments for SAE Alloy Steels

Directly Hardenable Grades					
SAE No.	Normalize Deg. F.	Anneal Deg. F.	Harden Deg. F.	Quench *	Temper Deg. F.
1330	.... 1600–1700 and/or 1500–1600	....	1525–1575 1525–1575	B B	To desired hardness To desired hardness
1335 & 1340	.... 1600–1700 and/or 1500–1600	....	1500–1550 1525–1575	E E	To desired hardness To desired hardness
2330	.... 1600–1700 and/or 1400–1500	....	1450–1500 1450–1500	E E	To desired hardness To desired hardness
2340 & 2345	.... 1600–1700 and/or 1400–1500	....	1425–1475 1425–1475	E E	To desired hardness To desired hardness
3130	1600–1700	....	1500–1550	B	To desired hardness
3135 to 3141	.... 1600–1700 and/or 1450–1550	....	1500–1550 1500–1550	E E	To desired hardness To desired hardness
3145 & 3150	.... 1600–1700 and/or 1400–1500	....	1500–1550 1500–1550	E E	To desired hardness To desired hardness
4037 & 4042	....	1525–1575	1500–1575	E	Gears, 350–450 To desired hardness
4047 & 4053	....	1450–1550	1500–1575	E	To desired hardness
4063 & 4068	....	1450–1550	1475–1550	E	To desired hardness
4130	1600–1700 and/or 1450–1550		1600–1650	B	To desired hardness
4137 & 4140	1600–1700 and/or 1450–1550		1550–1600	E	To desired hardness
4145 & 4150	1600–1700 and/or 1450–1550		1500–1600	E	To desired hardness
4340	1600–1700 and draw 1100–1225		1475–1525	E	To desired hardness
4640	1600–1700 and/or 1450–1550 1600–1700 and/or 1450–1550		1450–1500 1450–1500	E E	To desired hardness Gears, 350–450
5045 & 5046	1600–1700 and/or 1450–1550		1475–1500	E	250–300
5130 & 5132	1650–1750 and/or 1450–1550		1500–1550	G	To desired hardness
5135 to 5145	1650–1750 and/or 1450–1550		1500–1550	E	To desired hardness Gears, 350–400
5147 to 5152	1650–1750 and/or 1450–1550		1475–1550	E	To desired hardness Gears, 350–400
50100 51100 52100	....	1350–1450 1350–1450	1425–1475 1500–1600	H E	To desired hardness To desired hardness
6150	1650–1750 and/or 1550–1650		1600–1650	E	To desired hardness
9254 to 9262	....	....	1500–1650	E	To desired hardness
8627 to 8632	1600–1700 and/or 1450–1550		1550–1650	B	To desired hardness
8635 to 8641	1600–1700 and/or 1450–1550		1525–1575	E	To desired hardness
8642 to 8653	1600–1700 and/or 1450–1550		1500–1550	E	To desired hardness
8655 & 8660	1650–1750 and/or 1450–1550		1475–1550	E	To desired hardness
8735 & 8740	1600–1700 and/or 1450–1550		1525–1575	E	To desired hardness
8745 & 8750	1600–1700 and/or 1450–1550		1500–1550	E	To desired hardness
9437 & 9440	1600–1700 and/or 1450–1550		1550–1600	E	To desired hardness
9442 to 9747	1600–1700 and/or 1450–1550		1500–1600	E	To desired hardness
9840	1600–1700 and/or 1450–1550		1500–1550	E	To desired hardness
9845 & 9850	1600–1700 and/or 1450–1550		1500–1550	E	To desired hardness

Heat Treating Grades — Chromium-Nickel Austenitic Steels

SAE No.	Normalize	Anneal[9]	Harden Deg. F.	Quenching Medium	Temper
30301 to 30347	....	1800–2100	....	Water or Air	....

* Symbols: B = Water or oil; E = Oil; G = Water, caustic solution or oil; H = Water.
[9] Quench to produce full austenitic structure using water or air in accordance with thickness of section.   Annealing temperatures given cover process and full annealing as now used by industry, the lower end of the range being used for process annealing.

Table 2 (*Continued*). Typical Heat Treatments for SAE Alloy Steels

			Heat Treating Grades — Stainless Chromium Irons and Steels			
SAE No.*	Nor- malize	Sub-critical Anneal, Deg. F.	Full Anneal Deg. F.	Harden Deg. F.	Quenching Medium	Temper Deg. F.
51410	....	1300–1350[10]	1550–1650[11]	..........} 1750–1850}	Oil or air	To desired hardness
51414	....	1200–1250[10]	....	..........} 1750–1850}	Oil or air	To desired hardness
51416	....	1300–1350[10]	1550–1650[11]	..........} 1750–1850}	Oil or air	To desired hardness
51420 51420F	.....} .....}	1350–1450[10]	1550–1650[11]	..........} 1800–1850}	Oil or air	To desired hardness
51430	.....	1400–1500[12]	....	....	....	....
51430F	.....	1250–1500[12]	....	....	....	....
51431	.....	1150–1225[10]	....	1800–1900	Oil or air	To desired hardness
51440A 51440B 51440C 51440F	.....	1350–1440[10]	1550–1650[11]	1850–1950	Oil or air	To desired hardness
51442	.....	1400–1500[12]	....	....	....	....
51446	.....	1500–1650[12]	....	....	....	....
51501	.....	1325–1375[10]	1525–1600[11]	1600–1700	Oil or air	To desired hardness

* Suffixes A. B and C denote three types of steel differing in carbon content only. Suffix F denotes a free machining steel.
[10] Usually air cooled, but may be furnace cooled.
[11] Cool slowly in furnace.
[12] Cool rapidly in air.

TESTING THE HARDNESS OF METALS

Brinell Hardness Test. — The Brinell test for determining the hardness of metallic materials consists in applying a known load to the surface of the material to be tested through a hardened steel ball of known diameter. The diameter of the resulting permanent impression in the metal is measured and the Brinell Hardness Number (BHN) is then calculated from the following formula in which $D =$ diameter of ball in millimeters, d = measured diameter at the rim of the impression in millimeters, and P = applied load in kilograms.

$$\text{BHN} = \frac{\text{load on indenting tool in kilograms}}{\text{surface area of indentation in sq. mm.}} = \frac{P}{\frac{\pi D}{2}(D - \sqrt{D^2 - d^2})}$$

If the steel ball were not deformed under the applied load and if the impression were truly spherical, then the above formula would be a general one, and any combination of applied load and size of ball could be used. The impression, however, is not quite a spherical surface since there must always be some deformation of the steel ball and some recovery of form of the metal in the impression; hence for a standard Brinell test, the size and characteristics of the ball and the magnitude of the applied load must be standardized. In the standard Brinell test, a ball 10 millimeters in diameter and a load of 3000, 1500, or 500 kilograms is used. It is desirable, although not mandatory, that the test load be of such magnitude that the diameter

of the impression be in the range of 2.50 to 4.75 millimeters. The following test loads and approximate Brinell numbers for this range of impression diameters are: 3000 kg., 160 to 600 BHN; 1500 kg., 80 to 300 BHN; 500 kg., 26 to 100 BHN. In making a Brinell test the load should be applied steadily and without a jerk for at least 15 seconds in the case of iron and steel, and at least 30 seconds in testing other metals. A minimum period of two minutes, for example, has been recommended for magnesium and magnesium alloys. (For the softer metals, loads of 250 kg., 125 kg., or 100 kg., are sometimes used.)

According to the American Society for Testing and Materials Standard E10-66, a steel ball may be used on material having a BHN not over 450, a Hultgren ball on material not over 500, or a carbide ball on material not over 630. The Brinell hardness test is not recommended for material having a BHN over 630.

Rockwell Hardness Test. — The Rockwell hardness tester is essentially a machine that measures hardness by determining the depth of penetration of a penetrator into the specimen under certain fixed conditions of test. The penetrator may be either a steel ball or a diamond sphero-conical penetrator. The hardness number is related to the depth of indentation and the number is higher the harder the material. A minor load of 10 kg. is first applied which causes an initial penetration; the dial is set at zero on the black-figure scale, and the major load is applied. This major load is customarily 60 kg. or 100 kg. when a steel ball is used as a penetrator, but other loads may be used when found necessary. The ball penetrator is $\frac{1}{16}$ inch in diameter normally; but other penetrators of larger diameter, such as $\frac{1}{8}$ inch, may be employed for soft metals. When a diamond sphero-conical penetrator is employed the load usually is 150 kg. Experience decides the best combination of load and penetrator for use. After the major load is applied and removed, according to standard procedure, the reading is taken while the minor load is still applied.

The Rockwell Hardness Scales. — The various Rockwell scales and their applications are shown in the table below. The type of penetrator and load used with each are shown in Tables 1 and 2 which give comparative hardness values for different hardness scales.

Scale	Testing Application
A	For tungsten carbide and other extremely hard materials. Also for thin, hard sheets.
B	For materials of medium hardness such as low and medium carbon steels in the annealed condition.
C	For materials harder than Rockwell B–100.
D	Where somewhat lighter load is desired than on C scale, as on case hardened pieces.
E	For very soft materials such as bearing metals.
F	Same as E scale but using $\frac{1}{16}$-inch ball.
G	For metals harder than tested on B scale.
H & K	For softer metals.
15–N; 30–N; 45–N	Where shallow impression or small area is desired. For hardened steel and hard alloys.
15–T; 30–T; 45–T	Where shallow impression or small area is desired for materials softer than hardened steel.

Shore's Scleroscope. — The scleroscope is an instrument which measures the hardness of the work in terms of elasticity. A diamond-tipped hammer is allowed to drop from a known height on the metal to be tested. As this hammer strikes the metal, it rebounds, and the harder the metal, the greater the rebound. The extreme height of the rebound is recorded, and an average of a number of readings taken on a single piece will give a good indication of the hardness of the work. The surface smoothness of the work affects the reading of the instrument. The readings are also affected by the contour and mass of the work and the depth of the case, in carburized work, the soft core of light-depth carburizing, pack-hardening, or cyanide hardening, absorbing the force of the hammer fall and decreasing the rebound. The hammer weighs about 40 grains, the height of the rebound of hardened steel is in the neighborhood of 100 on the scale, or about 6¼ inches, while the total fall is about 10 inches or 255 millimeters.

Vickers Hardness Test. — The Vickers test is similar in principle to the Brinell test. The standard Vickers penetrator is a square-based diamond pyramid having an included point angle of 136 degrees. The numerical value of the hardness number equals the applied load in kilograms divided by the area of the pyramidal impression. A smooth, firmly supported, flat surface is required. The load, which usually is applied for 30 seconds, may either be 5, 10, 20, 30, 50 or 120 kilograms. The 50-kilogram load is usually employed. The hardness number is based upon the diagonal length of the square impression. The Vickers test, which is considered very accurate, may, with proper load regulation, be applied to thin sheets as well as to larger sections.

Knoop Hardness Numbers. — The Knoop hardness test is applicable to extremely thin metal, plated surfaces, exceptionally hard and brittle materials, very shallow carburized or nitrided surfaces, or whenever the applied load must be kept below 3600 grams. The Knoop indentor is a diamond ground to an elongated pyramidal form and it produces an indentation having long and short diagonals with a ratio of approximately 7 to 1. The longitudinal angle of the indentor is 172 degrees 30 minutes and the transverse angle 130 degrees. The Tukon Tester in which the Knoop indentor is used is fully automatic under electronic control. The Knoop hardness number equals load in kilograms divided by the projected area of indentation in square millimeters. The indentation number corresponding to the long diagonal and for a given load, may be determined from a table computed for a theoretically perfect indentor. The load, which may be varied from 25 to 3600 grams, is applied for a definite period and always normal to the surface tested. Lapped plane surfaces free from scratches are required.

Monotron Hardness Indicator. — With this instrument, a diamond-ball impressor point ¾ mm. in diameter is forced into the material to a depth of 9/5000 inch and the pressure required to produce this constant impression indicates the hardness. One of two dials shows the pressure in kilograms and pounds, and the other shows the depth of the impression in millimeters and inches. Readings in Brinell numbers may be obtained by means of a scale designated as $M — 1$.

Keep's Test. — With this apparatus a standard steel drill is caused to make a definite number of revolutions, while it is pressed with standard force against the specimen to be tested. The hardness is automatically recorded on a diagram on which a dead soft material gives a horizontal line, while a material as hard as the drill itself gives a vertical line, intermediate hardness being represented by the corresponding angle between 0 and 90 degrees.

Table 1. Comparative Hardness Scales for Steel — 1

Rockwell C-Scale Hardness Number	Diamond Pyramid Hardness Number Vickers	Brinell Hardness Number 10-mm. Ball, 3000-kgf Load			Rockwell Hardness Number		Rockwell Superficial Hardness Number Superficial Diam. Penetrator			Shore Sclero-scope Hardness Number
		Standard Ball	Hultgren Ball	Tungsten Carbide Ball	A-Scale 60-kgf Load Diam. Penetrator	D-Scale 100-kgf Load Diam. Penetrator	15-N Scale 15-kgf Load	30-N Scale 30-kgf Load	45-N Scale 45-kgf Load	
68	940	...	...	...	85.6	76.9	93.2	84.4	75.4	97
67	900	...	...	...	85.0	76.1	92.9	83.6	74.2	95
66	865	...	...	...	84.5	75.4	92.5	82.8	73.3	92
65	832	...	...	739	83.9	74.5	92.2	81.9	72.0	91
64	800	...	...	722	83.4	73.8	91.8	81.1	71.0	88
63	772	...	...	705	82.8	73.0	91.4	80.1	69.9	87
62	746	...	...	688	82.3	72.2	91.1	79.3	68.8	85
61	720	...	...	670	81.8	71.5	90.7	78.4	67.7	83
60	697	...	613	654	81.2	70.7	90.2	77.5	66.6	81
59	674	...	599	634	80.7	69.9	89.8	76.6	65.5	80
58	653	...	587	615	80.1	69.2	89.3	75.7	64.3	78
57	633	...	575	595	79.6	68.5	88.9	74.8	63.2	76
56	613	...	561	577	79.0	67.7	88.3	73.9	62.0	75
55	595	...	546	560	78.5	66.9	87.9	73.0	60.9	74
54	577	...	534	543	78.0	66.1	87.4	72.0	59.8	72
53	560	...	519	525	77.4	65.4	86.9	71.2	58.6	71
52	544	500	508	512	76.8	64.6	86.4	70.2	57.4	69
51	528	487	494	496	76.3	63.8	85.9	69.4	56.1	68
50	513	475	481	481	75.9	63.1	85.5	68.5	55.0	67
49	498	464	469	469	75.2	62.1	85.0	67.6	53.8	66
48	484	451	455	455	74.7	61.4	84.5	66.7	52.5	64
47	471	442	443	443	74.1	60.8	83.9	65.8	51.4	63
46	458	432	432	432	73.6	60.0	83.5	64.8	50.3	62
45	446	421	421	421	73.1	59.2	83.0	64.0	49.0	60
44	434	409	409	409	72.5	58.5	82.5	63.1	47.8	58
43	423	400	400	400	72.0	57.7	82.0	62.2	46.7	57
42	412	390	390	390	71.5	56.9	81.5	61.3	45.5	56
41	402	381	381	381	70.9	56.2	80.9	60.4	44.3	55
40	392	371	371	371	70.4	55.4	80.4	59.5	43.1	54
39	382	362	362	362	69.9	54.6	79.9	58.6	41.9	52
38	372	353	353	353	69.4	53.8	79.4	57.7	40.8	51
37	363	344	344	344	68.9	53.1	78.8	56.8	39.6	50
36	354	336	336	336	68.4	52.3	78.3	55.9	38.4	49
35	345	327	327	327	67.9	51.5	77.7	55.0	37.2	48
34	336	319	319	319	67.4	50.8	77.2	54.2	36.1	47
33	327	311	311	311	66.8	50.0	76.6	53.3	34.9	46
32	318	301	301	301	66.3	49.2	76.1	52.1	33.7	44
31	310	294	294	294	65.8	48.4	75.6	51.3	32.5	43
30	302	286	286	286	65.3	47.7	75.0	50.4	31.3	42
29	294	279	279	279	64.7	47.0	74.5	49.5	30.1	41
28	286	271	271	271	64.3	46.1	73.9	48.6	28.9	41
27	279	264	264	264	63.8	45.2	73.3	47.7	27.8	40

Note: The values in this table shown in **bold faced** type correspond to those shown in American Society for Testing and Materials Specification E140-67.

Table 1. Comparative Hardness Scales for Steel — 2

Rockwell C-Scale Hardness Number	Diamond Pyramid Hardness Number Vickers	Brinell Hardness Number 10-mm. Ball, 3000-kgf Load			Rockwell Hardness Number		Rockwell Superficial Hardness Number Superficial Brale Penetrator			Shore Sclero-scope Hardness Number
		Standard Ball	Hultgren Ball	Tungsten Carbide Ball	A-Scale 60-kgf Load Diam. Penetrator	D-Scale 100-kgf Load Diam. Penetrator	15-N Scale 15-kgf Load	30-N Scale 30-kgf Load	45-N Scale 45-kgf Load	
26	272	258	258	258	63.3	44.6	72.8	46.8	26.7	38
25	266	253	253	253	62.8	43.8	72.2	45.9	25.5	38
24	260	247	247	247	62.4	43.1	71.6	45.0	24.3	37
23	254	243	243	243	62.0	42.1	71.0	44.0	23.1	36
22	248	237	237	237	61.5	41.6	70.5	43.2	22.0	35
21	243	231	231	231	61.0	40.9	69.9	42.3	20.7	35
20	238	226	226	226	60.5	40.1	69.4	41.5	19.6	34
(18)	230	219	219	219	...	...	...	...	...	33
(16)	222	212	212	212	...	...	...	...	...	32
(14)	213	203	203	203	...	...	...	...	...	31
(12)	204	194	194	194	...	...	...	...	...	29
(10)	196	187	187	187	...	...	...	...	...	28
(8)	188	179	179	179	...	...	...	...	...	27
(6)	180	171	171	171	...	...	...	...	...	26
(4)	173	165	165	165	...	...	...	...	...	25
(2)	166	158	158	158	...	...	...	...	...	24
(0)	160	152	152	152	...	...	...	...	...	24

Note: The values in this table shown in **bold faced** type correspond to those shown in American Society for Testing and Materials Specification E140-67.
Values in ( ) are beyond the normal range and are given for information only.

Comparison of Hardness Scales. — Tables 1 and 2 show comparisons of various hardness scales. All such tables are based on the assumption that the metal tested is homogeneous to a depth several times that of the indentation. To the extent that the metal being tested is not homogeneous, errors are introduced because different loads and different shapes of penetrators meet the resistance of metal of varying hardness, depending on the depth of indentation. Another source of error is introduced in comparing the hardness of different materials as measured on different hardness scales. This arises from the fact that in any hardness test, metal that is severely cold-worked actually supports the penetrator and different metals, different alloys, and different analyses of the same type of alloy have different cold-working properties. In spite of the possible inaccuracies introduced by such factors, it is of considerable value to be able to compare hardness values in a general way.

The data shown in Table 1 are based upon extensive tests on carbon and alloy steels mostly in the heat-treated condition, but have been found to be reliable on constructional alloy steels and tool steels in the as-forged, annealed, normalized, quenched and tempered conditions, providing they are homogeneous. These hardness comparisons are not as accurate for special cases such as high manganese steel, 18–8 stainless steel and other austenitic steels, nickel base alloys, as well as constructional alloy steels and nickel base alloys in the cold-worked condition.

The data shown in Table 2 are for hardness measurements of unhardened steel, steel of soft temper, grey and malleable cast iron, and most non-ferrous metals. Again these hardness comparisons are not as accurate for annealed metals of high Rockwell B hardness such as austenitic stainless steel, nickel and high nickel alloys and cold-worked metals of low B-scale hardness such as aluminum and the softer alloys.

Table 2. Comparative Hardness Scales for Unhardened Steel, Soft-temper Steel, Grey and Malleable Cast Iron, and Non-ferrous Alloys* — 1

Rockwell Hardness Number			Rockwell Superficial Hardness Number			Rockwell Hardness Number			Brinell Hardness Number	
Rockwell B scale 1⁄16″ Ball Penetrator 100 kg. Load	Rockwell F scale 1⁄16″ Ball Penetrator 60 kg. Load	Rockwell G scale 1⁄16″ Ball Penetrator 150 kg. Load	Rockwell Superficial 15-T scale 1⁄16″ Ball Penetrator 15 kg. Load	Rockwell Superficial 30-T scale 1⁄16″ Ball Penetrator 30 kg. Load	Rockwell Superficial 45-T scale 1⁄16″ Ball Penetrator 45 kg. Load	Rockwell E scale 1⁄8″ Ball Penetrator 100 kg. Load	Rockwell K scale 1⁄8″ Ball Penetrator 150 kg. Load	Rockwell A scale "Brale" Penetrator 60 kg. Load	Brinell Scale 10 mm Standard Ball — 500 kg. Load	Brinell Scale 10 mm. Standard Ball — 3000 kg. Load
100	....	82.5	93.0	82.0	72.0	....	....	61.5	201	240
99	....	81.0	92.5	81.5	71.0	....	....	61.0	195	234
98	....	79.0	....	81.0	70.0	....	....	60.0	189	228
97	....	77.5	92.0	80.5	69.0	....	....	59.5	184	222
96	....	76.0	....	80.0	68.0	....	....	59.0	179	216
95	....	74.0	91.5	79.0	67.0	....	....	58.0	175	210
94	....	72.5	....	78.5	66.0	....	....	57.5	171	205
93	....	71.0	91.0	78.0	65.5	....	....	57.0	167	200
92	....	69.0	90.5	77.5	64.5	....	100	56.5	163	195
91	....	67.5	....	77.0	63.5	....	99.5	56.0	160	190
90	....	66.0	90.0	76.0	62.5	....	98.5	55.5	157	185
89	....	64.0	89.5	75.5	61.5	....	98.0	55.0	154	180
88	....	62.5	....	75.0	60.5	....	97.0	54.0	151	176
87	....	61.0	89.0	74.5	59.5	....	96.5	53.5	148	172
86	....	59.0	88.5	74.0	58.5	....	95.5	53.0	145	169
85	....	57.5	....	73.5	58.0	....	94.5	52.5	142	165
84	....	56.0	88.0	73.0	57.0	....	94.0	52.0	140	162
83	....	54.0	87.5	72.0	56.0	....	93.0	51.0	137	159
82	....	52.5	....	71.5	55.0	....	92.0	50.5	135	156
81	....	51.0	87.0	71.0	54.0	....	91.0	50.0	133	153
80	....	49.0	86.5	70.0	53.0	....	90.5	49.5	130	150
79	....	47.5	....	69.5	52.0	....	89.5	49.0	128	147
78	....	46.0	86.0	69.0	51.0	....	88.5	48.5	126	144
77	....	44.0	85.5	68.0	50.0	....	88.0	48.0	124	141
76	....	42.5	....	67.5	49.0	....	87.0	47.0	122	139
75	99.5	41.0	85.0	67.0	48.5	....	86.0	46.5	120	137
74	99.0	39.0	....	66.0	47.5	....	85.0	46.0	118	135
73	98.5	37.5	84.5	65.5	46.5	....	84.5	45.5	116	132
72	98.0	36.0	84.0	65.0	45.5	....	83.5	45.0	114	130
71	97.5	34.5	....	64.0	44.5	100	82.5	44.5	112	127
70	97.0	32.5	83.5	63.5	43.5	99.5	81.5	44.0	110	125
69	96.0	31.0	83.0	62.5	42.5	99.0	81.0	43.5	109	123
68	95.5	29.5	....	62.0	41.5	98.0	80.0	43.0	107	121
67	95.0	28.0	82.5	61.5	40.5	97.5	79.0	42.5	106	119
66	94.5	26.5	82.0	60.5	39.5	97.0	78.0	42.0	104	117
65	94.0	25.0	....	60.0	38.5	96.0	77.5	....	102	116
64	93.5	23.5	81.5	59.5	37.5	95.5	76.5	41.5	101	114
63	93.0	22.0	81.0	58.5	36.5	95.0	75.5	41.0	99	112
62	92.0	20.5	....	58.0	35.5	94.5	74.5	40.5	98	110
61	91.5	19.0	80.5	57.0	34.5	93.5	74.0	40.0	96	108
60	91.0	17.5	....	56.5	33.5	93.0	73.0	39.5	95	107
59	90.5	16.0	80.0	56.0	32.0	92.5	72.0	39.0	94	106

* See note at end of table.

Table 2. Comparative Hardness Scales for Unhardened Steel, Soft-temper Steel, Grey and Malleable Cast-iron, and Non-ferrous Alloys* — 2

Rockwell Hardness Number			Rockwell Superficial Hardness Number			Rockwell Hardness Number				Brinell
Rockwell B scale 1/16" Ball Penetrator 100 kg. Load	Rockwell F scale 1/16" Ball Penetrator 60 kg. Load	Rockwell G scale 1/16" Ball Penetrator 150 kg. Load	Rockwell Superficial 15-T scale 1/16" Ball Penetrator 15 kg. Load	Rockwell Superficial 30-T scale 1/16" Ball Penetrator 30 kg. Load	Rockwell Superficial 45-T scale 1/16" Ball Penetrator 45 kg. Load	Rockwell E scale 1/8" Ball Penetrator 100 kg. Load	Rockwell H scale 1/8" Ball Penetrator 60 kg. Load	Rockwell K scale 1/8" Ball Penetrator 150 kg. Load	Rockwell A scale "Brale" Penetrator 60 kg. Load	Brinell Scale 10-mm. Standard Ball — 500 kg. Load
58	90.0	14.5	79.5	55.0	31.0	92.0	....	71.0	38.5	92
57	89.5	13.0	....	54.5	30.0	91.0	....	70.5	38.0	91
56	89.0	11.5	79.0	54.0	29.0	90.5	....	69.5	....	90
55	88.0	10.0	78.5	53.0	28.0	90.0	....	68.5	37.5	89
54	87.5	8.5	....	52.5	27.0	89.5	....	68.0	37.0	87
53	87.0	7.0	78.0	51.5	26.0	89.0	....	67.0	36.5	86
52	86.5	5.5	77.5	51.0	25.0	88.0	....	66.0	36.0	85
51	86.0	4.0	....	50.5	24.0	87.5	....	65.0	35.5	84
50	85.5	2.5	77.0	49.5	23.0	87.0	....	64.5	35.0	83
50	85.5	2.5	77.0	49.5	23.0	87.0	....	64.5	35.0	83
49	85.0	1.0	76.5	49.0	22.0	86.5	....	63.5	....	82
48	84.5	....	....	48.5	20.5	85.5	....	62.5	34.5	81
47	84.0	....	76.0	47.5	19.5	85.0	....	61.5	34.0	80
46	83.0	....	75.5	47.0	18.5	84.5	....	61.0	33.5	..
45	82.5	....	....	46.0	17.5	84.0	....	60.0	33.0	79
44	82.0	....	75.0	45.5	16.5	83.5	....	59.0	32.5	78
43	81.5	....	74.5	45.0	15.5	82.5	....	58.0	32.0	77
42	81.0	....	....	44.0	14.5	82.0	....	57.5	31.5	76
41	80.5	....	74.0	43.5	13.5	81.5	....	56.5	31.0	75
40	79.5	....	73.5	43.0	12.5	81.0	....	55.5	....	..
39	79.0	....	....	42.0	11.0	80.0	....	54.5	30.5	74
38	78.5	....	73.0	41.5	10.0	79.5	....	54.0	30.0	73
37	78.0	....	72.5	40.5	9.0	79.0	....	53.0	29.5	72
36	77.5	....	....	40.0	8.0	78.5	100	52.0	29.0	..
35	77.0	....	72.0	39.5	7.0	78.0	99.5	51.5	28.5	71
34	76.5	....	71.5	38.5	6.0	77.0	99.0	50.5	28.0	70
33	75.5	....	....	38.0	5.0	76.5	....	49.5	....	69
32	75.0	....	71.0	37.5	4.0	76.0	98.5	48.5	27.5	..
31	74.5	....	....	36.5	3.0	75.5	98.0	48.0	27.0	68
30	74.0	....	70.5	36.0	2.0	75.0	....	47.0	26.5	67
29	73.5	....	70.0	35.5	1.0	74.0	97.5	46.0	26.0	..
28	73.0	....	....	34.5	....	73.5	97.0	45.0	25.5	66
27	72.5	....	69.5	34.0	....	73.0	96.5	44.5	25.0	..
26	72.0	....	69.0	33.0	....	72.5	....	43.5	24.5	65
25	71.0	....	....	32.5	....	72.0	96.0	42.5	....	64
24	70.5	....	68.5	32.0	....	71.0	95.5	41.5	24.0	..
23	70.0	....	68.0	31.0	....	70.5	....	41.0	23.5	63
22	69.5	....	....	30.5	....	70.0	95.0	40.0	23.0	..
21	69.0	....	67.5	29.5	....	69.5	94.5	39.0	22.5	62
20	68.5	....	....	29.0	....	68.5	....	38.0	22.0	..
19	68.0	....	67.0	28.5	....	68.0	94.0	37.5	21.5	61
18	67.0	....	66.5	27.5	....	67.5	93.5	36.5	....	..

* See note at end of table.

Table 2. Comparative Hardness Scales for Unhardened Steel, Soft-temper Steel, Grey and Malleable Cast Iron, and Non-ferrous Alloys* — 3

(Compiled by Wilson Mechanical Instrument Co.)

Rockwell Hardness Number			Rockwell Superficial Hardness Number			Rockwell Hardness Number				Brinell
Rockwell B scale 1/16" Ball Penetrator 100 kg. Load	Rockwell F scale 1/16" Ball Penetrator 60 kg. Load	Rockwell G scale 1/16" Ball Penetrator 150 kg. Load	Rockwell 15-T scale Superficial 1/16" Ball Penetrator 15 kg. Load	Rockwell 30-T scale Superficial 1/16" Ball Penetrator 30 kg. Load	Rockwell 45-T scale Superficial 1/16" Ball Penetrator 45 kg. Load	Rockwell E scale 1/8" Ball Penetrator 100 kg. Load	Rockwell H scale 1/8" Ball Penetrator 60 kg. Load	Rockwell K scale 1/8" Ball Penetrator 150 kg. Load	Rockwell A scale "Brale" Penetrator 60 kg. Load	Brinell Scale 10-mm. Standard Ball — 500 kg. Load
17	66.5	....	....	27.0	....	67.0	93.0	35.5	21.0	60
16	66.0	....	66.0	26.0	....	66.5	....	35.0	20.5	..
15	65.5	....	65.5	25.5	....	65.5	92.5	34.0	20.0	59
14	65.0	....	....	25.0	....	65.0	92.0	33.0	....	..
13	64.5	....	65.0	24.0	....	64.5	....	32.0	....	58
12	64.0	....	64.5	23.5	....	64.0	91.5	31.5	....	..
11	63.5	....	....	23.0	....	63.5	91.0	30.5	....	..
10	63.0	....	64.0	22.0	....	62.5	90.5	29.5	....	57
9	62.0	....	....	21.5	....	62.0	....	29.0	....	..
8	61.5	....	63.5	20.5	....	61.5	90.0	28.0	....	..
7	61.0	....	63.0	20.0	....	61.0	89.5	27.0	....	56
6	60.5	....	....	19.5	....	60.5	....	26.0	....	..
5	60.0	....	62.5	18.5	....	60.0	89.0	25.5	....	55
4	59.5	....	62.0	18.0	....	59.0	88.5	24.5	....	..
3	59.0	....	....	17.0	....	58.5	88.0	23.5	....	..
2	58.0	....	61.5	16.5	....	58.0	....	23.0	....	54
1	57.5	....	61.0	16.0	....	57.5	87.5	22.0	....	..
0	57.0	....	....	15.0	....	57.0	87.0	21.0	....	53

* Not applicable to annealed metals of high B-scale hardness such as austenitic stainless steels, nickel and high-nickel alloys nor to cold-worked metals of low B-scale hardness such as aluminum and the softer alloys.

Turner's Sclerometer. — In making this test a weighted diamond point is drawn, once forward and once backward, over the smooth surface of the material to be tested. The hardness number is the weight in grams required to produce a standard scratch.

Mohs's Hardness Scale. — Hardness, in general, is determined by what is known as Mohs's scale, a standard for hardness which is mainly applied to non-metallic elements and minerals. In this hardness scale there are ten degrees or steps, each designated by a mineral, the difference in hardness of the different steps being determined by the fact that any member in the series will scratch any of the preceding members. This scale is as follows:

1. Talc; 2. gypsum; 3. calcite; 4. fluor spar; 5. apatite; 6. orthoclase; 7. quartz; 8. topaz; 9. sapphire or corundum; 10. diamond.

These minerals, arbitrarily selected as standards, are successively harder, from talc, the softest of all minerals, to diamond, the hardest. This scale, which is now universally used for non-metallic minerals, is, however, not applied to metals.

Relation Between Hardness and Tensile Strength. — The approximate relationship between the hardness and tensile strength is shown by the following formula in which B = Brinell hardness number.

Tensile strength = $B \times 515$ (for Brinell numbers up to 175).

Tensile strength = $490 \times B$ (for Brinell numbers larger than 175).

These formulas give the tensile strength in pounds per square inch and apply to steels. This definite relationship between hardness and tensile strength does not apply to non-ferrous metals with the possible exception of certain aluminum alloys.

ENGINEERING DRAWINGS

The following information represents the standard methods and procedures used to prepare engineering drawings. [This section is taken from ANSI Y14.5M-1984 *Dimensioning and Tolerancing,* by permission of the American Society of Mechanical Engineers.]

FUNDAMENTAL RULES

Dimensioning and tolerancing shall clearly define engineering intent and shall conform to the following.

(*a*) Each dimension shall have a tolerance, except for those dimensions specifically identified as reference, maximum, minimum, or stock (commercial stock size). The tolerance may be applied directly to the dimension (or indirectly in the case of basic dimensions), indicated by a general note, or located in a supplementary block of the drawing format (see ANSI Y14.1).

(*b*) Dimensions for size, form, and location of features shall be complete to the extent that there is full understanding of the characteristics of each feature. Neither scaling (measuring the size of a feature directly from an engineering drawing) nor assumption of a distance or size is permitted.

NOTE: Undimensioned drawings—for example, loft, printed wiring, templates, master layouts, tooling layout—prepared on stable material are excluded, provided the necessary control dimensions are specifed.

(*c*) Each necessary dimension of an end product shall be shown. No more dimensions than those necessary for complete definition shall be given. The use of reference dimensions on a drawing should be minimized.

(*d*) Dimensions shall be selected and arranged to suit the function and mating relationship of a part and shall not be subject to more than one interpretation.

(*e*) The drawing should define a part without specifying manufacturing methods. Thus, only the diameter of a hole is given without indicating whether it is to be drilled, reamed, punched, or made by any other operation. However, in those instances where manufacturing, processing, quality assurance, or environmental information is essential to the definition of engineering requirements, it shall be specified on the drawing or in a document referenced on the drawing.

(*f*) It is permissible to identify as nonmandatory certain processing dimensions that provide for finish allowance, shrink allowance, and other requirements, provided the final dimensions are given on the drawing. Nonmandatory processing dimensions shall be identified by an appropriate note, such as NONMANDATORY (MFG DATA).

(*g*) Dimensions should be arranged to provide required information for optimum readability. Dimensions should be shown in true profile views and refer to visible outlines.

(*h*) Wires, cables, sheets, rods, and other materials manufactured to gage or code numbers shall be specified by linear dimensions indicating the diameter or thickness. Gage or code numbers may be shown in parentheses following the dimension.

(*i*) A 90° angle is implied where center lines and lines depicting features are shown on a drawing at right angles and no angle is specified.

(*j*) A 90° BASIC angle applies where center lines of features in a pattern or surfaces shown at right angles on the drawing are located or defined by basic dimensions and no angle is specified.

(k) Unless otherwise specified, all dimensions are applicable at 20°C (68°F). Compensation may be made for measurements made at other temperatures.

UNITS OF MEASUREMENT

SI (Metric) Linear Units. — The commonly used SI linear unit used on engineering drawings is the millimeter.

U.S. Customary Linear Units. — The commonly used U.S. customary linear unit used on engineering drawings is the decimal inch.

Identification of Linear Units. — On drawings where all dimensions are either in millimeters or inches, individual identification of linear units is not required. However, the drawing shall contain a note stating UNLESS OTHERWISE SPECIFIED, ALL DIMENSIONS ARE IN MILLIMETERS (or IN INCHES, as applicable).

Where some inch dimensions are shown on a millimeter-dimensioned drawing, the abbreviation IN. shall follow the inch values. Where some millimeter dimensions are shown on an inch-dimensioned drawing, the symbol mm shall follow the millimeter values.

Angular Units. — Angular dimensions are expressed in either degrees and decimal parts of a degree or in degrees, minutes, and seconds. These latter dimensions are expressed by symbols: for degrees °, for minutes ′, and for seconds ″. Where degrees are indicated alone, the numerical value shall be followed by the symbol °. Where only minutes or seconds are specified, the number of minutes or seconds shall be preceded by 0° or 0°0′, as applicable. See Fig. 1.

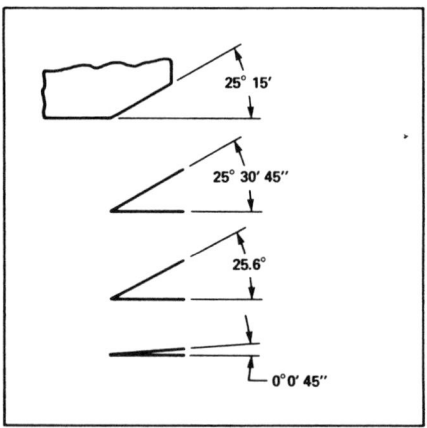

Figure 1. Angular Units

TYPES OF DIMENSIONING

Decimal dimensioning shall be used on drawings except where certain commercial commodities are identified by standardized nominal designations such as pipe and lumber sizes.

Millimeter Dimensioning. — The following shall be observed when specifying millimeter dimensions on drawings.

(*a*) Where the dimension is less than one millimeter, a zero precedes the decimal point. See Fig. 2.

(*b*) Where the dimension is a whole number, neither the decimal point nor a zero is shown. See Fig. 2.

(*c*) Where the dimension exceeds a whole number by a decimal fraction of one millimeter, the last digit to the right of the decimal point is not followed by a zero. See Fig. 2, except when used with a bilateral tolerance.

(*d*) Neither commas nor spaces shall be used to separate digits into groups in specifying millimeter dimensions on drawings.

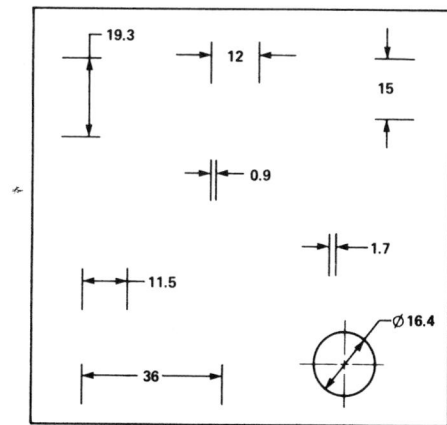

Figure 2. Millimeter Dimensions

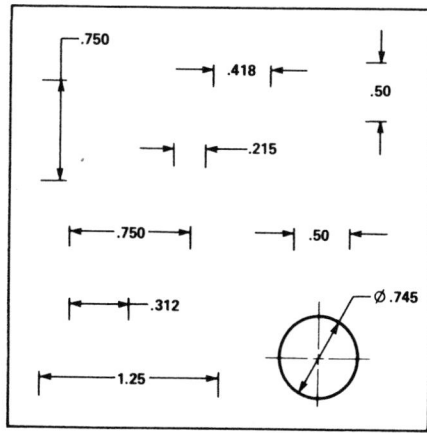

Figure 3. Decimal Inch Dimensions

Decimal Inch Dimensioning. — The decimal inch system is explained in ANSI B87.1. The following shall be observed when specifying decimal inch dimensions on drawings.
(a) A zero is not used before the decimal point for values less than one inch.
(b) A dimension is expressed to the same number of decimal places as its tolerance. Zeros are added to the right of the decimal point where necessary. See Fig. 3.

Decimal Points. — Decimal points must be uniform, dense, and large enough to be clearly visible and meet the reproduction requirements of ANSI Y14.2M. Decimal points are placed in line with the bottom of the associated digits.

APPLICATION OF DIMENSIONS

Dimensions are applied by means of dimension lines, extension lines, chain lines, or a leader from a dimension, note, or specification directed to the appropriate feature. See Fig. 4. General notes are used to convey additional information. For further information on dimension lines, extension lines, chain lines, and leaders, see ANSI Y14.2M.

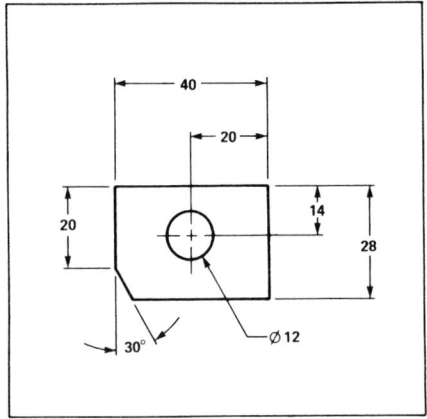

Figure 4. Application of Dimensions

Dimension Lines. — A dimension line, with its arrowheads, shows the direction and extent of a dimension. Numerals indicate the number of units of a measurement. Preferably, dimension lines should be broken for insertion of numerals as shown in Fig. 4. Where horizontal dimension lines are not broken, numerals are placed above and parallel to the dimension lines.

Dimension lines shall be aligned if practicable and grouped for uniform appearance. See Fig. 5.

Dimension lines are drawn parallel to the direction of measurement. The space between the first dimension line and the part outline should be not less than 10 mm; the space between succeeding parallel dimension lines should be not less than 6 mm. See Fig. 6.

NOTE: These spacings are intended as guides only. If the drawing meets the reproduction requirements of the accepted industry or military reproduction specification,

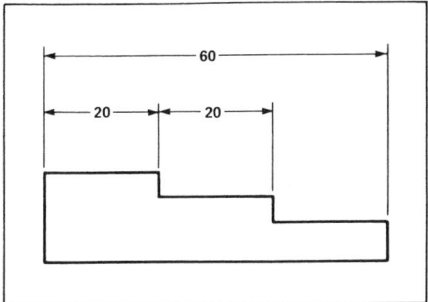

Figure 5. Grouping of Dimensions

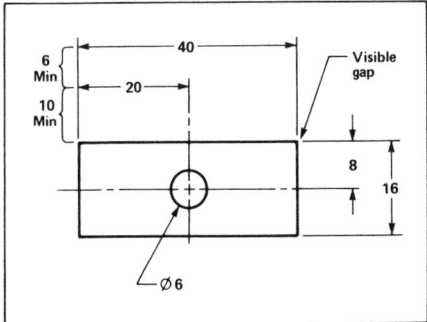

Figure 6. Spacing of Dimensions

nonconformance to these spacing requirements is not a basis for rejection of the drawing.

Where there are several parallel dimension lines, the numerals should be staggered for easier reading. See Fig. 7.

The following shall not be used as a dimension line: a center line, an extension line, a phantom line, a line that is part of the outline of the object, or a continuation of any of these lines. A dimension line is not used as an extension line, except where a simplified method of coordinate dimensioning is used to define curved outlines.

The dimension line of an angle is an arc drawn with its center at the apex of the angle. The arrowheads terminate at the extensions of the two sides. See Figs. 1 and 4.

Crossing dimension lines should be avoided. Where unavoidable, the dimension lines are unbroken.

Limited Length or Area Indication. — Where it is desired to indicate that a limited length or area of a surface is to receive additional treatment or consideration within limits specified on the drawing, the extent of these limits may be indicated by use of a chain line. See Fig. 8.

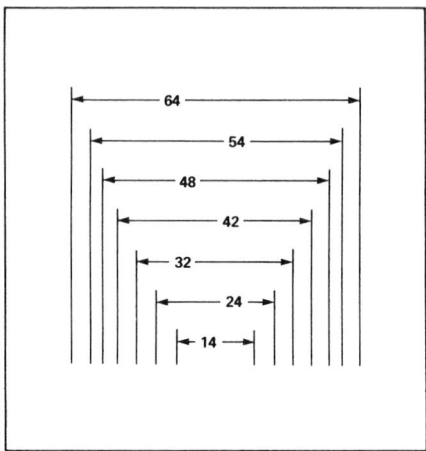

Figure 7. Staggered Dimensions

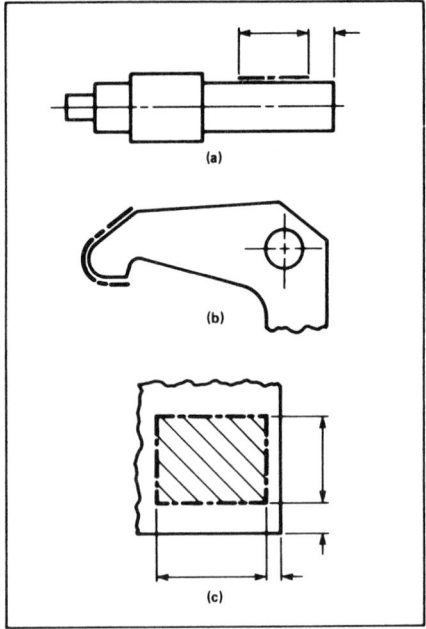

Figure 8. Limited Length or Area Indication

In an appropriate view or section, a chain line is drawn parallel to the surface profile at a short distance from it. Dimensions are added for length and location. If applied to a surface of revolution, the indication may be shown on one side only. See Fig. 8, part (a).

If the chain line clearly indicates the location and extent of the surface area, dimensions may be omitted. See Fig. 8, part (b).

Where the desired area is shown on a direct view of the surface, the area is section lined within the chain line boundary and appropriately dimensioned. See Fig. 8, part (c).

Reading Direction. — Dimensions and notes should be placed to be read from the bottom of the drawing with regard to orientation of the drawing format. See Fig. 9.

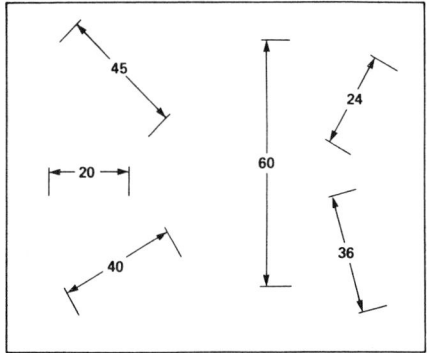

Figure 9. Reading Direction

Reference Dimensions. — The method for identifying a reference dimension (or reference data) on drawings is to enclose the dimension (or data) within parentheses. See Figs. 10 and 11.

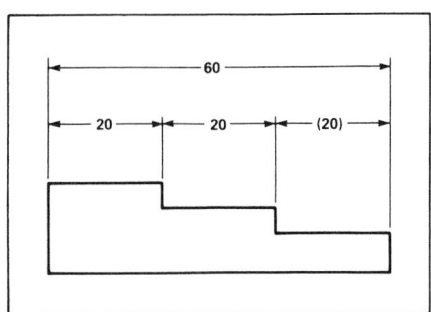

Figure 10. Intermediate Reference Dimension

Overall Dimensions. — Where an overall dimension is specified, one intermediate

dimension is omitted or identified as a reference dimension. See Fig. 10. Where the intermediate dimensions are more important than the overall dimension, the overall dimension, if used, is identified as a reference dimension. See Fig. 11.

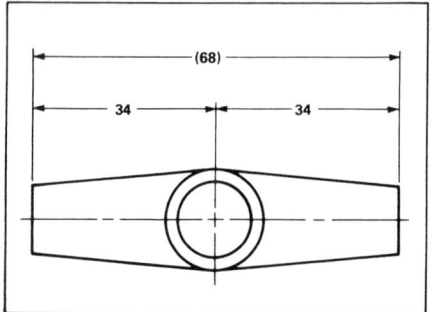

Figure 11. Overall Reference Dimension

Dimensioning Within the Outline of a View. — Dimensions are usually placed outside the outline of a view. Where directness of application makes it desirable, or where extension lines or leader lines would be excessively long, dimensions may be placed within the outline of a view.

Dimensions Not to Scale. — Where it is necessary or desirable to indicate that a particular feature is not to scale, the dimension should be underlined with a straight thick line.

DIRECT TOLERANCING METHODS

Limits and directly applied tolerance values are specified as follows.

(a) *Limit Dimensioning.* The high limit (maximum value) is placed above the low limit (minimum value). When expressed in a single line, the low limit precedes the high limit and a dash separates the two values. See Fig. 12.

(b) *Plus and Minus Tolerancing.* The dimension is given first and is followed by a plus and minus expression of tolerance. See Fig. 13.

TOLERANCE EXPRESSION

The conventions pertaining to the number of decimal places carried in the tolerance shown in the following paragraphs shall be observed.

Millimeter Tolerances. — Where millimeter dimensions are used on the drawings, the following applies.

(a) Where unilateral tolerancing is used and either the plus or minus value is nil, a single zero is shown without a plus or minus sign.

Example: $32_{-0.02}^{0}$ or $32_{0}^{+0.02}$

(b) Where bilateral tolerancing is used, both the plus and minus values have the same number of decimal places, using zeros where necessary.

22.5
22.0
Ø7.5
7.4
Ø7.6
7.5
Ø7.5 - 7.6
25°30'45"
25°30'15"
25.2°
25.1°

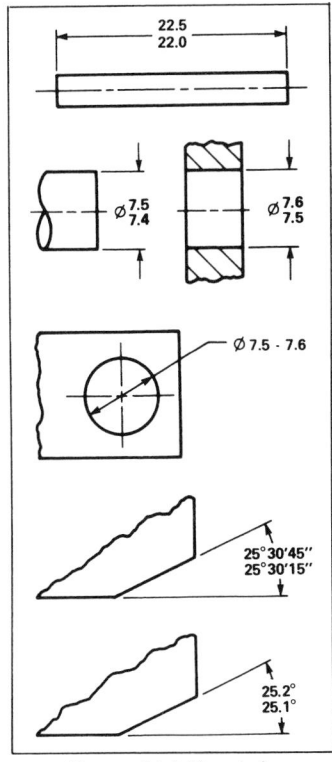

Figure 12. Limit Dimensioning

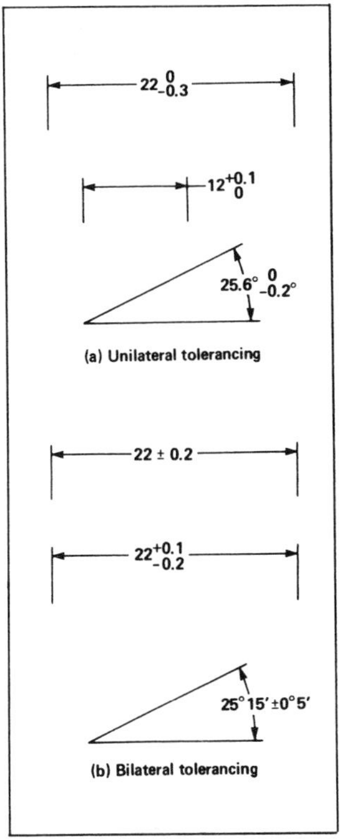

22 0 -0.3
12 +0.1 0
25.6° 0 -0.2°
(a) Unilateral tolerancing
22 ± 0.2
22 +0.1 -0.2
25°15'±0°5'
(b) Bilateral tolerancing

Figure 13. Plus and Minus Tolerancing

Example: $32^{+0.25}_{-0.10}$ not $32^{+0.25}_{-0.1}$

(*c*) Where limit dimensioning is used and either the maximum or minimum value has digits following a decimal point, the other value has zeros added for uniformity.

Example: $\begin{matrix}25{\cdot}45\\25{\cdot}00\end{matrix}$ not $\begin{matrix}25{\cdot}45\\25\end{matrix}$

Inch Tolerances. — Where inch dimensions are used on the drawing, both limit dimensions or the plus and minus tolerance and its dimension shall be expressed with the same number of decimal places.

Examples: $\begin{matrix}{\cdot}750\\{\cdot}748\end{matrix}$ not $\begin{matrix}{\cdot}75\\{\cdot}748\end{matrix}$

.500 ±.005   not   .50±.005

.500 $\begin{array}{l} +.005 \\ -.000 \end{array}$ not   .500 $\begin{array}{l} +.005 \\ 0 \end{array}$

25.0°±.2°   not   25° ±.2°

INTERPRETATION OF LIMITS

All limits are absolute. Dimensional limits, regardless of the number of decimal places, are used as if they were continued with zeros.

Examples: $\begin{array}{l} 12.2 \\ 12.0 \end{array}$ means $\begin{array}{l} 12.20{-}{-}0 \\ 12.00{-}{-}0 \end{array}$

$\begin{array}{l} 12.01 \\ 12.00 \end{array}$ means $\begin{array}{l} 12.010{-}{-}0 \\ 12.000{-}{-}0 \end{array}$

For the purpose of determining conformance within limits, the measured value is compared directly with the specified value and any deviation outside the specified limiting value signifies nonconformance with the limits.

Plated or Coated Parts. — Where a part is to be plated or coated, the drawing or referenced document shall specify whether the dimensions are before or after plating. Typical examples of notes are the following:

(*a*) DIMENSIONAL LIMITS APPLY AFTER PLATING.

(*b*) DIMENSIONAL LIMITS APPLY BEFORE PLATING.

(For coatings other than plating, substitute the appropriate term.)

SINGLE LIMITS

MIN or MAX is placed after a dimension where other elements of the design definitely determine the other unspecified limit. Features such as depth of holes, length of threads, corner radii, chamfers, etc., may be limited in this way. Single limits are used where the intent will be clear, and the unspecified limit can be zero or approach infinity and will not result in a condition detrimental to the design.

TOLERANCE ACCUMULATION

Figure 14 compares the tolerance values resulting from three methods of dimensioning.

(*a*) *Chain Dimensioning.*   The maximum variation between two features is equal to the sum of the tolerances on the intermediate distances; this results in the greatest tolerance accumulation. In Fig. 14, part (a), the tolerance accumulation between surfaces X and Y is ±0.15.

(*b*) *Base Line Dimensioning.*   The maximum variation between two features is equal to the sum of the tolerances on the two dimensions from their origin to the features; this results in a reduction of the tolerance accumulation. In Fig. 14, part (b), the tolerance accumulation between surfaces X and Y is ±0.1.

(*c*) *Direct Dimensioning.*   The maximum variation between two features is controlled by the tolerance on the dimension between the features; this results in the least tolerance. In Fig. 14, part (c), the tolerance between surfaces X and Y is ±0.05.

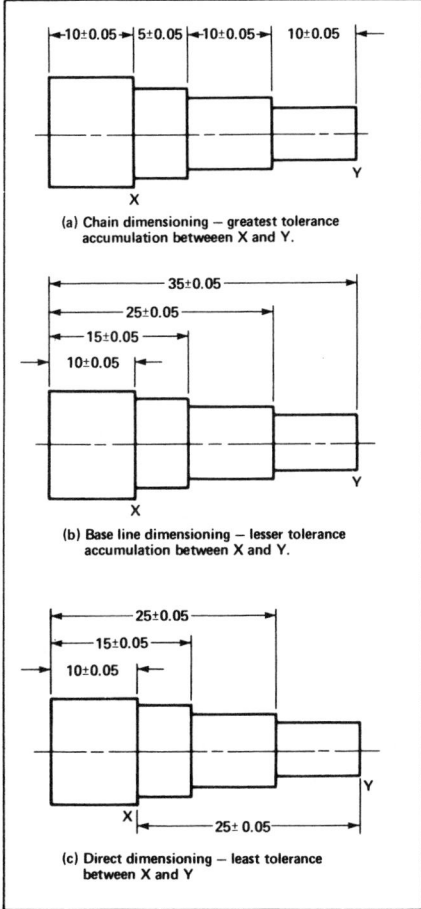

Figure 14. Tolerance Accumulation

Dimensional Limits Related to an Origin. — In certain cases it is necessary to indicate that a dimension between two features shall originate from one of these features and not the other. Such a case is illustrated in Fig. 15, where a part having two parallel surfaces of unequal length is to be mounted on the shorter surface. In this example, the dimension origin symbol signifies that the dimension originates from the shorter surface and dimensional limits apply to the other surface. Without such indication, the longer surface could have been selected as the origin, thus permitting a greater angular variation between surfaces.

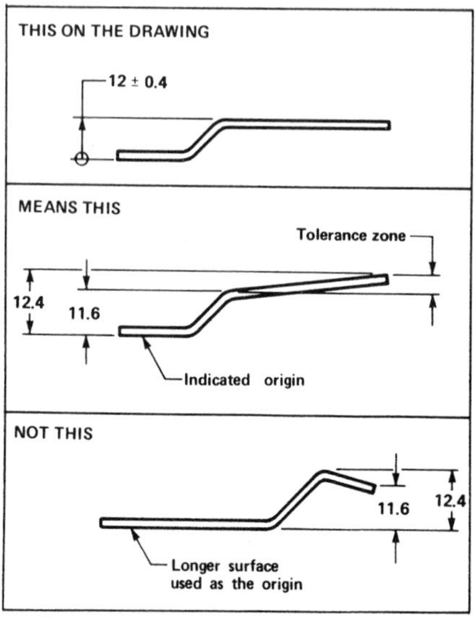

Figure 15. Relating Dimensional Limits to an Origin

	TYPE OF TOLERANCE	CHARACTERISTIC	SYMBOL	SEE:
FOR INDIVIDUAL FEATURES	FORM	STRAIGHTNESS	—	6.4.1
		FLATNESS	⏥	6.4.2
		CIRCULARITY (ROUNDNESS)	○	6.4.3
		CYLINDRICITY	⌭	6.4.4
FOR INDIVIDUAL OR RELATED FEATURES	PROFILE	PROFILE OF A LINE	⌒	6.5.2 (b)
		PROFILE OF A SURFACE	⌓	6.5.2 (a)
FOR RELATED FEATURES	ORIENTATION	ANGULARITY	∠	6.6.2
		PERPENDICULARITY	⊥	6.6.4
		PARALLELISM	//	6.6.3
	LOCATION	POSITION	⌖	5.2
		CONCENTRICITY	◎	5.11.3
	RUNOUT	CIRCULAR RUNOUT	↗*	6.7.2.1
		TOTAL RUNOUT	⌰*	6.7.2.2

*Arrowhead(s) may be filled in.

Figure 16. Geometric Characteristic Symbols

Geometric Characteristic Symbols. — The symbols denoting geometric characteristics are shown in Fig. 16.

Datum Feature Symbol. — The datum feature symbol consists of a frame containing the datum identifying letter preceded and followed by a dash. See Fig. 17.

Letters of the alphabet (excpet I, O, and Q) are used as datum identifying letters. Each datum feature requiring identification shall be assigned a different letter. When datum features requiring identification on a drawing are so numerous as to exhaust the single alpha series, the double alpha series shall be used—AA through AZ, BA through BZ, etc.

Where the same datum feature symbol is repeated to identify that same feature in other locations on a drawing, it need not be identified as reference.

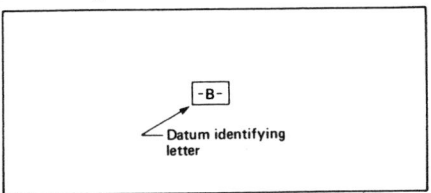

Figure 17. Datum Feature Symbol

Datum Target Symbol. — The datum target symbol is a circle divided horizontally into two halves. See Fig. 18. The lower half contains a letter identifying the associated datum, followed by the target number assigned sequentially starting with 1 for each datum. See Fig. 19. Where the datum target is an area, the area size may be entered in the upper half of the symbol; otherwise, the upper half is left blank. A radial line attached to the symbol is directed to a target point (indicated by an "X"), target line, or target area, as applicable.

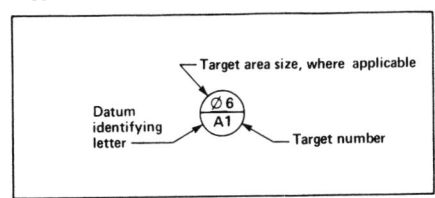

Figure 18. Datum Target Symbol

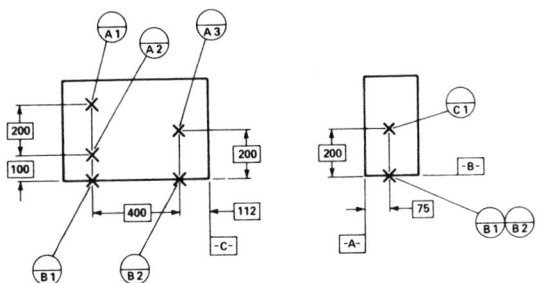

Figure 19. Dimensioning Datum Targets

Basic Dimension Symbol. — The symbol used to identify a basic dimension is shown in Fig. 20.

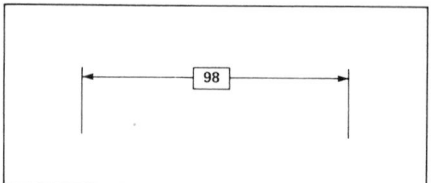

Figure 20. Basic Dimension Symbol

Material Condition Symbols. — The symbols used to indicate "at maximum material condition," "regardless of feature size," and "at least material condition" are shown in Fig. 21. The use of these symbols in local and general notes is prohibited.

Projected Tolerance Zone Symbol. — The symbol used to indicate a projected tolerance zone is shown in Fig. 21. The use of this symbol in local and general notes is prohibited.

Diameter and Radius Symbols. — The symbols used to indicate diameter, spherical diameter, radius, and spherical radius are shown in Fig. 21. These symbols precede the value of a dimension or tolerance given as a diameter or radius, as applicable.

Reference Symbol. — A reference dimension (or reference data) is identified by enclosing the dimension (or data) within parentheses. See Fig. 21.

Arc Length Symbol. — The symbol used to indicate that a linear dimension is an arc length measured on a curved outline is shown in Fig. 21. The symbol is placed above the dimension.

TERM	SYMBOL
AT MAXIMUM MATERIAL CONDITION	Ⓜ
REGARDLESS OF FEATURE SIZE	Ⓢ
AT LEAST MATERIAL CONDITION	Ⓛ
PROJECTED TOLERANCE ZONE	Ⓟ
DIAMETER	$\emptyset$
SPHERICAL DIAMETER	S$\emptyset$
RADIUS	R
SPHERICAL RADIUS	SR
REFERENCE	( )
ARC LENGTH	⌒

Figure 21. Modifying Symbols

Counterbore or Spotface Symbol. — The symbolic means of indicating a counterbore or spotface is shown in Fig. 22. The symbol precedes the dimension of the counterbore or spotface.

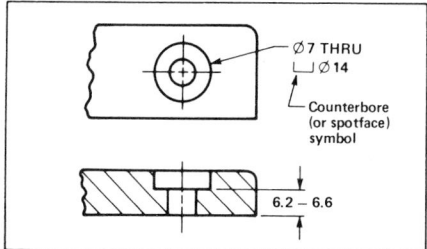

Figure 22. Counterbore or Spotface Symbol

Countersink Symbol. — The symbolic means of indicating a countersink is shown in Fig. 23. The symbol precedes the dimensions of the countersink.

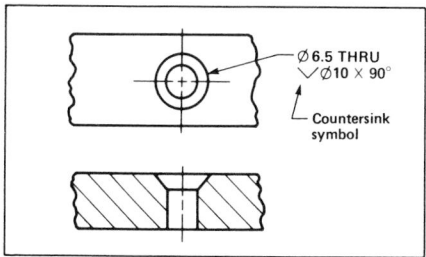

Figure 23. Countersink Symbol

Depth Symbol. — The symbolic means of indicating where a dimension applies to the depth of a feature is to precede that dimension with the depth symbol, as shown in Fig. 24.

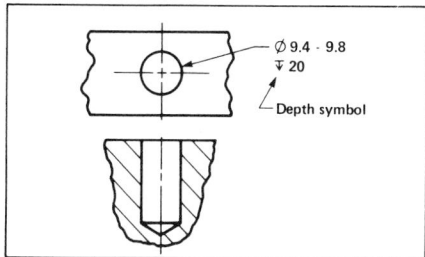

Figure 24. Depth Symbol

Square Symbol. — The symbol used to indicate that a single dimension applies to a square shape is to precede that dimension with the symbol for a square, as shown in Fig. 25.

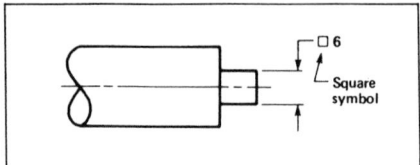

Figure 25. Square Symbol

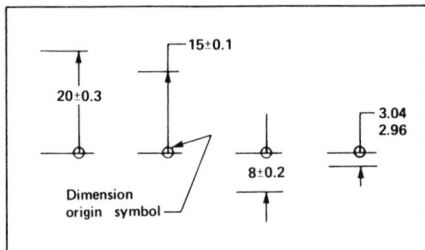

Figure 26. Dimension Origin Symbol

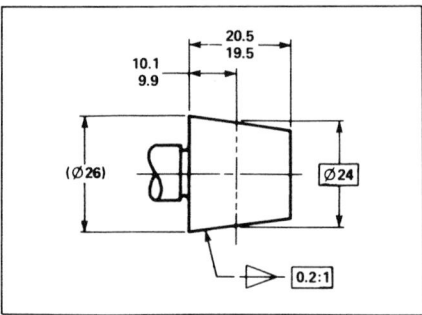

Figure 27. Taper Symbol

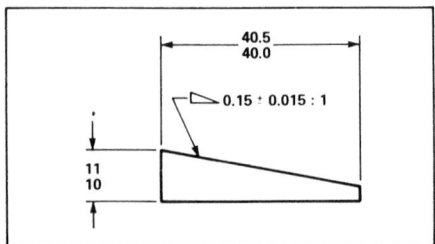

Figure 28. Slope Symbol

Dimension Origin Symbol. — The symbol used to indicate that a toleranced dimension between two features originates from one of these features is shown in Fig. 26.

Taper and Slope Symbols. — Symbols used for specifying taper and slope for conical and flat tapers are shown in Figs. 27 and 28. These symbols are always shown with the vertical leg to the left.

GEOMETRIC TOLERANCE SYMBOLS

Geometric characteristic symbols, the tolerance value, and datum reference letters, where applicable, are combined in a feature control frame to express a geometric tolerance.

Feature Control Frame. — A geometric tolerance for an individual feature is specified by means of a feature control frame divided into compartments containing the geometric characteristic symbol followed by the tolerance. See Fig. 29. Where applicable, the tolerance is preceded by the diameter symbol and followed by a material condition symbol.

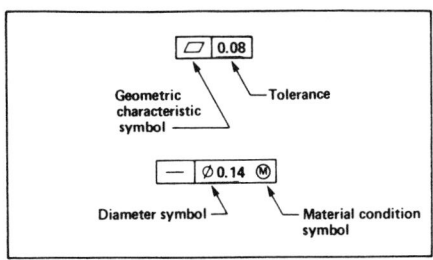

Figure 29. Feature Control Frame

Feature Control Frame Incorporating Datum References. — Where a geometric tolerance is related to a datum, this relationship is indicated by entering the datum reference letter in a compartment following the tolerance. See Fig. 30. Where applicable, the datum reference letter is followed by a material condition symbol.

Where a datum is established by two datum features—for example, an axis established by two datum diameters—both datum reference letters, separated by a dash, are entered in a single compartment. See Fig. 31, part (a). Where more than one datum is required, the datum reference letters (followed by a material condition symbol, where applicable) are entered in separate compartments in the desired order of prec-

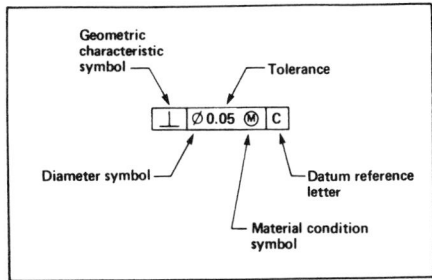

Figure 30. Feature Control Frame Incorporating a Datum Reference

edence, from left to right. See Fig. 31, parts (b) and (c). Datum reference letters need not be in alphabetical order in the feature control frame.

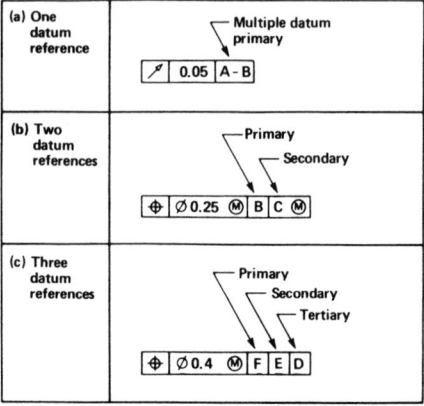

Figure 31. Order of Precedence of Datum References

A composite feature control frame is used where more than one tolerance is specified for the same geometric characteristic of a feature or features having different datum requirements. The composite frame contains a single entry of the geometric characteristic symbol followed by each tolerance and datum requirement, one above the other. See Fig. 32.

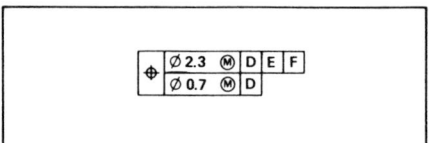

Figure 32. Composite Feature Control Frame

The symbol used to indicate that a profile tolerance applies to surfaces all around the part is a circle located at the junction of the leader from the feature control frame. See Fig. 33.

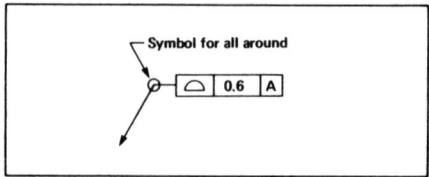

Figure 33. Symbol for All Around

Combined Feature Control Frame and Datum Feature Symbol. — Where a feature or pattern of features controlled by a geometric tolerance also serves as a datum feature, the feature control frame and datum feature symbol are combined. See Fig. 34.

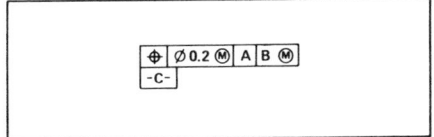

Figure 34. Combined Feature Control Frame and Datum Feature Symbol

Wherever a feature control frame and datum feature symbol are combined, datums referenced in the feature control frame are not considered part of the datum feature symbol. In the positional tolerance example, Fig. 34, a feature is controlled for position in relation to datums A and B, and identified as datum C. Whenever datum C is referenced elsewhere on the drawing, the reference applies to datum C, not to datums A and B.

Feature Control Frame With a Projected Tolerance Zone. — Where a positional or an orientation tolerance is specified as a projected tolerance zone, a frame containing the height dimension, followed by the projected tolerance zone symbol, is placed beneath the feature control frame. See Fig. 35. Where the projected tolerance zone is indicated with a chain line, as in Fig. 36, the height dimension is omitted from the frame.

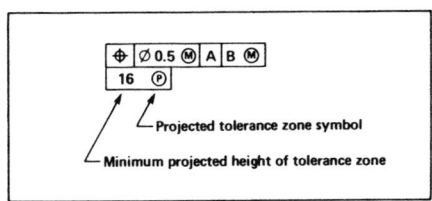

Figure 35. Feature Control Frame with a Projected Tolerance Zone

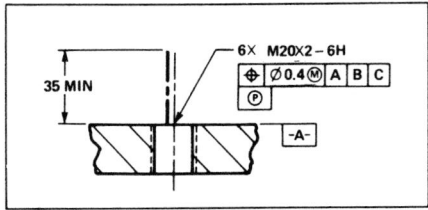

Figure 36. Showing Projected Tolerance Zone with a Chain Line

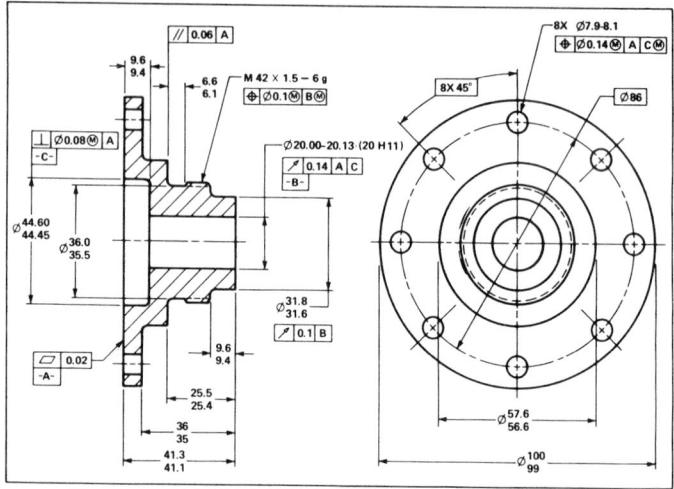

Figure 37. Feature Control Frame Placement

FEATURE CONTROL FRAME PLACEMENT

The feature control frame is related to the considered feature by one of the following methods as depicted in Fig. 37.

(*a*) locating the feature control frame below or attached to a leader-directed callout or dimension pertaining to the feature;

(*b*) running a leader from the frame to the feature;

(*c*) attaching a side or an end of the frame to an extension line from the feature, provided it is a plane surface;

(*d*) attaching a side or an end of the frame to an extension of the dimension line pertaining to a feature of size.

IDENTIFICATION OF THE TOLERANCE ZONE

Where the specified tolerance value represents the diameter of a cylindrical zone, the diameter symbol shall precede the tolerance value. Where the tolerance zone is other than a diameter, identification is unnecessary, and the specified tolerance value represents the distance between two parallel straight lines or planes, or the distance between two uniform boundaries, as the specific case may be.

TABULATED TOLERANCES

Where the tolerance in a feature control frame is to be tabulated, a letter representing the tolerance, preceded by the abbreviation TOL, is entered as shown in Fig. 38.

American National Standard Symbols for Section Lining (ANSI Y14.2M-1979)

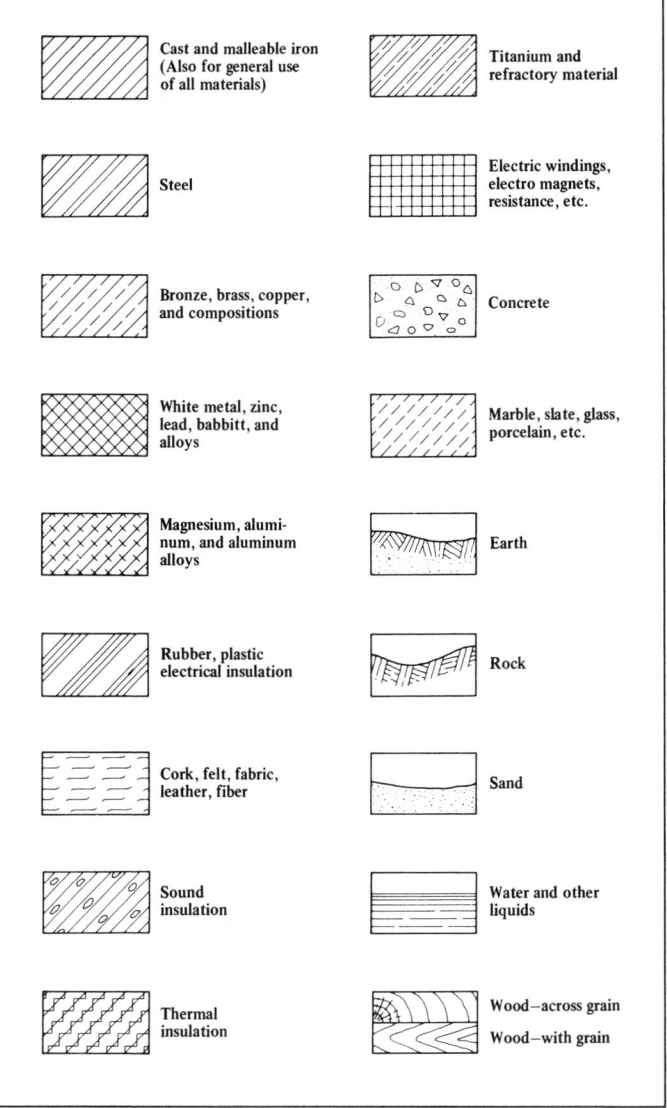

Cast and malleable iron (Also for general use of all materials)	Titanium and refractory material
Steel	Electric windings, electro magnets, resistance, etc.
Bronze, brass, copper, and compositions	Concrete
White metal, zinc, lead, babbitt, and alloys	Marble, slate, glass, porcelain, etc.
Magnesium, aluminum, and aluminum alloys	Earth
Rubber, plastic electrical insulation	Rock
Cork, felt, fabric, leather, fiber	Sand
Sound insulation	Water and other liquids
Thermal insulation	Wood—across grain / Wood—with grain

Control and Production of Surface Texture. — Surface characteristics should not be controlled on a drawing or specification unless such control is essential to functional performance or appearance of the product. Imposition of such restrictions when unnecessary may increase production costs and in any event will serve to lessen the emphasis on the control specified for important surfaces.

Smoothness and roughness are relative, i.e., surfaces may be either smooth or rough for the purpose intended; what is smooth for one purpose may be rough for another purpose. In the mechanical field comparatively few surfaces require any control of surface texture beyond that afforded by the processes required to obtain the necessary dimensional characteristics.

Working surfaces such as on bearings, pistons, and gears are typical of surfaces for which optimum performance may require control of the surface characteristics. Nonworking surfaces such as on the walls of transmission cases, crankcases, or housings seldom require any surface control. Experimentation or experience with surfaces performing similar functions is the best criterion on which to base selection of optimum surface characteristics.

Determination of required characteristics for working surfaces may involve consideration of such conditions as the area of contact, the load, speed, direction of motion, type and amount of lubricant, temperature, material and physical characteristics of component parts, and that variations in any one of the conditions may require a change in the specified surface characteristics.

Production: Surface texture is a result of the processing method, the surface obtained from casting, forging, or burnishing is the result of plastic deformation; if machined or ground, lapped or honed, the surface obtained is the result of action of cutting tools, abrasives or other forces. It is important to understand that surfaces with like roughness average ratings may not have the same performance, due to tempering, sub-surface effects, different profile waveforms, etc.

American National Standard Surface Texture (Surface Roughness, Waviness and Lay). — American National Standard ANSI B46.1-1978 is concerned with the geometric irregularities of surfaces of solid materials, physical specimens for gaging roughness, and the characteristics of stylus instrumentation for measuring roughness. It defines surface texture and its constituents: roughness, waviness, lay and flaws. A set of symbols for drawings, specifications and reports is established. In order to assure a uniform basis for measurements, it also provides specifications for Precision Reference Specimens, and Roughness Comparison Specimens, and establishes requirements for stylus type instruments. The standard is not concerned with luster, appearance, color, corrosion resistance, wear resistance, hardness, subsurface micro-structure, surface integrity and many other characteristics which may be governing considerations in specific applications.

The standard does not define the degrees of surface roughness and waviness or type of lay suitable for specific purposes, nor does it specify the means by which any degree of such irregularities may be obtained or produced. However, criteria for selection of surface qualities and information on instrument techniques and methods of producing, controlling and inspecting surfaces are included in Appendixes attached to the standard. The Appendix sections are not considered a part of the standard; they are included for clarification or information purposes only.

Surfaces, in general, are very complex in character. The standard deals only with the height, width, and direction of surface irregularities since these are of practical importance in specific applications. Surface texture designations as delineated in this standard may not be a sufficient index to performance. Other part characteristics such as dimensional and geometrical relationships, material, metallurgy, and stress must also be controlled.

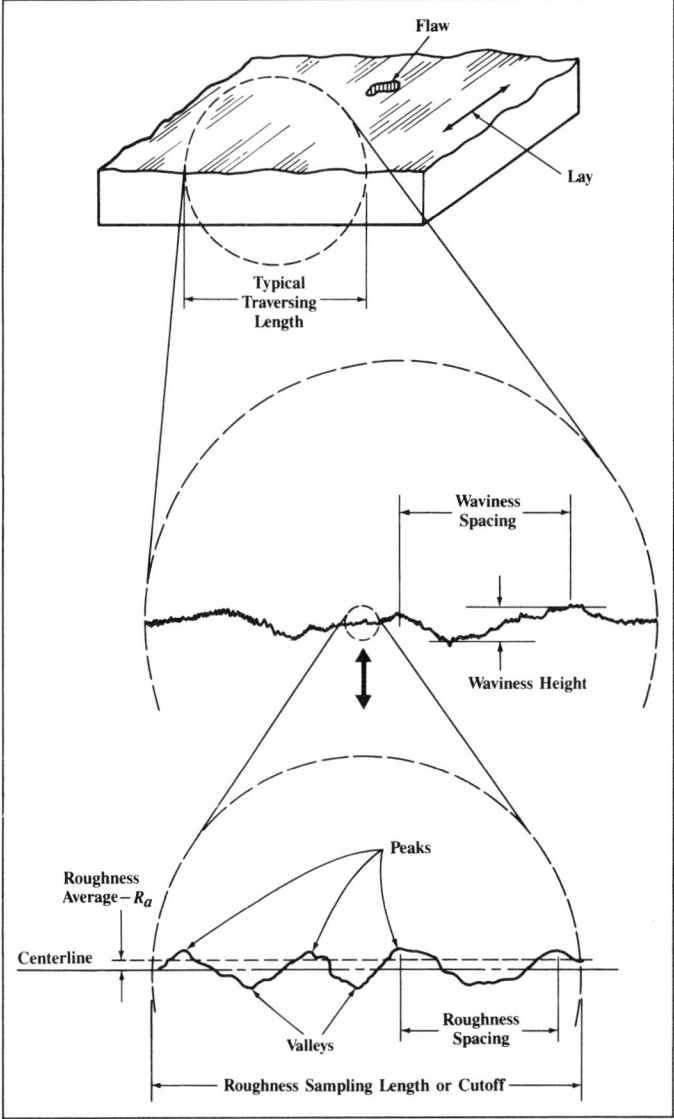

Fig. 1. Pictorial Display of Surface Characteristics

Definitions of Terms Relating to the Surfaces of Solid Materials. — The terms and ratings in the standard relate to surfaces produced by such means as abrading, casting, coating, cutting, etching, plastic deformation, sintering, wear, erosion, etc.

The *surface* of an object is the boundary which separates that object from another object, substance or space.

The *nominal surface* is the intended surface contour, the shape and extent of which is usually shown and dimensioned on a drawing or descriptive specification.

The *measured surface* is a representation of the surface obtained by instrumental or other means.

Surface texture is the repetitive or random deviations from the nominal surface which form the three-dimensional topography of the surface. Surface texture includes roughness, waviness, lay and flaws. Figure 1 is an example of a uni-directional lay surface. Roughness and waviness parallel to the lay are not represented in the expanded views.

Roughness consists of the finer irregularities of the surface texture, usually including those irregularities which result from the inherent action of the production process. These are considered to include traverse feed marks and other irregularities within the limits of the roughness sampling length.

Waviness is the more widely spaced component of surface texture. Unless otherwise noted, waviness is to include all irregularities whose spacing is greater than the roughness sampling length and less than the waviness sampling length. Waviness may result from such factors as machine or work deflections, vibration, chatter, heat-treatment or warping strains. Roughness may be considered superposed on a 'wavy' surface.

Lay is the direction of the predominant surface pattern, ordinarily determined by the production method used.

Flaws are unintentional irregularities which occur at one place or at relatively infrequent or widely varying intervals on the surface. Flaws include such defects as cracks, blow holes, inclusions, checks, ridges, scratches, etc. Unless otherwise specified, the effect of flaws shall not be included in the roughness average measurements. Where flaws are to be restricted or controlled, a special note as to the method of inspection should be included on the drawing or in the specifications.

The *error of form* is considered as being that deviation from the nominal surface which is not included in surface texture.

Definitions of Terms Relating to the Measurement of Surface Texture. — The *profile* is the contour of the surface in a plane perpendicular to the surface, unless some other angle is specified.

The *nominal profile* is a profile of the nominal surface; it is the intended profile. See Fig. 2.

The *measured profile* is a representation of the profile obtained by instrumental or other means. When the measured profile is a graphical representation, it will usually be distorted through the use of different vertical and horizontal magnifications but shall otherwise be as faithful to the profile as technically possible.

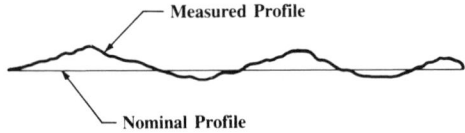

Fig. 2. Nominal and Measured Profiles

The *modified profile* is a measured profile where filter mechanisms (including the instrument datum) are used to minimize certain surface texture characteristics and emphasize others.

The *graphical centerline* is the line about which roughness is measured and is a line parallel to the general direction of the profile within the limits of the sampling length, such that the sums of the areas contained between it and those parts of the profile which lie on either side are equal. (See Fig. 3.)

The *electrical mean line* is the centerline established by the selected cutoff and its associated circuitry in an electric roughness average measuring instrument.

A *peak* is the point of maximum height on that portion of a profile which lies above the centerline and between the two intersections of the profile and the centerline.

A *valley* is the point of maximum depth on that portion of a profile which lies below the centerline and between two intersections of the profile and the centerline.

The *spacing* is the distance between specified points on the profile measured parallel to the nominal profile.

The *roughness spacing* is the average spacing between adjacent peaks of the measured profile within the roughness sampling length.

The *waviness spacing* is the average spacing between adjacent peaks of the measured profile within the waviness sampling length.

The *sampling length* is the nominal spacing within which a surface characteristic is determined.

The *roughness sampling length* is the sampling length within which the roughness average is determined. This length is chosen, or specified, to separate the profile irregularities which are designated as roughness from those irregularities designated as waviness.

The *cutoff* is the electrical response characteristic of the roughness average measuring instrument which is selected to limit the spacing of the surface irregularities to be included in the assessment of roughness average. The cutoff is rated in millimeters. In most electrical averaging instruments, the cutoff can be selected. It is a characteristic of the instrument rather than of the surface being measured. In specifying the cutoff, care must be taken to choose a value which will include all the surface irregularities which it is desired to assess.

The *waviness sampling length* is the sampling length within which the waviness height is determined.

The *traversing length* is the length of profile which is traversed by the stylus to establish a representative measurement.

Height is considered to be those measurements of the profile in a direction normal to the nominal profile.

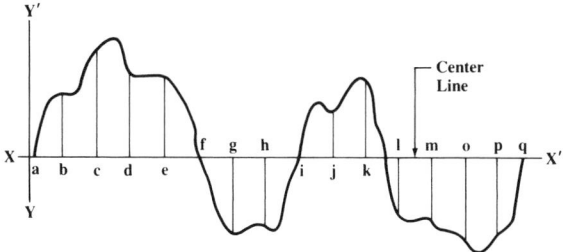

Fig. 3. Short Section of Hypothetical Profile Divided into Increments

Roughness average (R_a), also known as arithmetic average (AA) and centerline average (CLA), is the arithmetic average of the absolute values of the measured profile height deviations taken within the sampling length and measured from the graphical centerline. For graphical determinations of roughness average, the height deviations are measured normal to the chart centerline, as shown in Fig. 3. Roughness average is expressed in micrometers. A micrometer is one millionth of a meter, (0.000 001 meter). For written specifications or references as to surface requirements, micrometer can be abbreviated as μm. A microinch is one millionth of an inch, (0.000001 inch). For written specifications or reference to surface roughness requirements, microinch may be abbreviated as μin. One microinch equals 0.0254 micrometer (1 μin. = 0.0254 μm).

Roughness average value (R_a) from continuously averaging meter readings. So that uniform interpretation may be made of readings from stylus-type instruments of the continuously averaging type, it should be understood that the reading which is considered significant is the mean reading around which the needle tends to dwell or fluctuate under small amplitude.

The *peak-to-valley height* is the maximum excursion above the centerline plus the maximum excursion below the centerline within the sampling length. This value is typically 3 or more times the roughness average.

The *waviness height* is the peak-to-valley height of the modified profile from which roughness and flaws have been removed by filtering, smoothing, or other means. The measurement is to be taken normal to the nominal profile within the limits of the waviness sampling length and expressed in millimeters.

Relation of Surface Roughness to Tolerances. — Since the measurement of surface roughness involves the determination of the average linear deviation of the actual surface from the nominal surface, there is a direct relationship between the dimensional tolerance on a part and the permissible surface roughness. It is evident that a requirement for the accurate measurement of a dimension is that the variations introduced by surface roughness should not exceed the tolerance placed on the dimension. If this is not the case, the measurement of the dimension will be subject to an uncertainty greater than the required tolerance as illustrated in Fig. 4. The standard

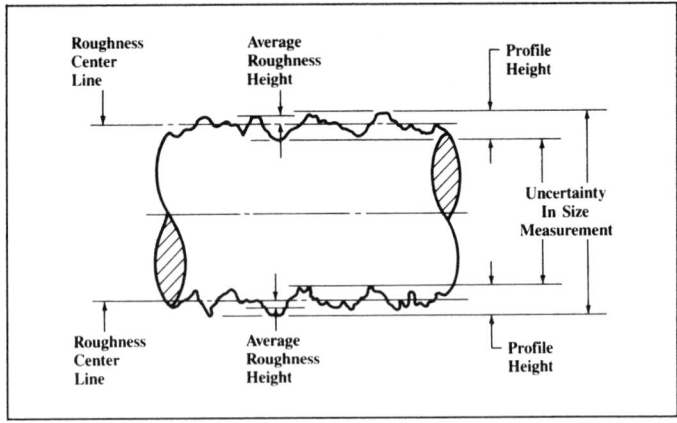

Fig. 4. Uncertainty in Dimensional Measurement due to Surface Roughness

method of measuring surface roughness involves the determination of the average deviation from the mean surface. On most surfaces the total profile height of the surface roughness (peak-to-valley height) will be approximately four times the measured average surface roughness in microinches. This factor will vary somewhat with the character of the surface under consideration, but the value of four may be used to establish approximate profile heights.

From these considerations it follows that if the arithmetical average value of surface roughness specified on a part exceeds one eighth of the dimensional tolerance, the whole tolerance will be taken up by the roughness height. In most cases, a smaller roughness specification than this will be found; but on parts where very small dimensional tolerances are given, it is necessary to specify a suitably small surface roughness in order that useful dimensional measurements can be made. Values for surface roughness produced by common processing methods are shown in Fig. 5. The

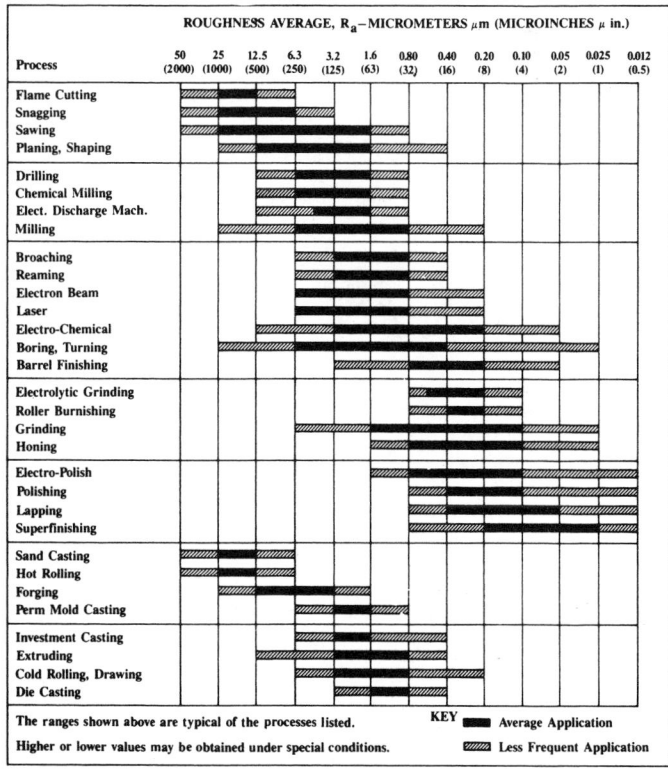

Fig. 5. Surface Roughness Produced by Common Production Methods

ability of a processing operation to produce a specific surface roughness depends on many factors. For example, in surface grinding, the final surface depends on the peripheral speed of the wheel, the speed of the traverse, the rate of feed, the grit size, bonding material and state of dress of the wheel, the amount and type of lubrication at the point of cutting, and the mechanical properties of the piece being ground. A small change in any of the above factors can have a marked effect on the surface produced.

Selecting Cutoff for Roughness Measurements. — In general, surfaces will contain irregularities with a large range of widths. Stylus-type instruments are designed to respond only to irregularity spacings less than a given value, called cutoff. In some cases, such as surfaces in which actual contact area with a mating surface is important, the largest convenient cutoff will be used. In other cases, such as surfaces subject to fatigue failure, only the irregularities of small width will be important, and more significant values will be obtained when a short cutoff is used. In still other cases, such as identifying chatter marks on machined surfaces, information is needed on only the widely spaced irregularities. For such measurements, a long cutoff instrument should be used.

The effect of variation in cutoff can be understood better by reference to Fig. 6. The profile at the top is the true movement of a stylus on a surface having a roughness spacing of about 1 mm and the profiles below are interpretations of the same surface with cutoff value settings of 0.8 mm, 0.25 mm and 0.08 mm, respectively. It can be seen that the trace based on 1 mm cutoff includes most of the coarse irregularities and all of the fine irregularities of the surface; that the trace based on 0.25 mm excludes the coarser irregularities but includes the fine and medium fine; and

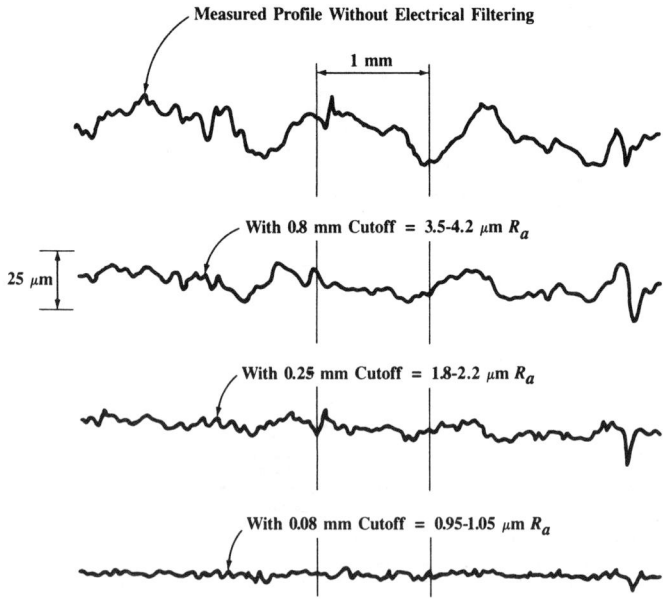

Fig. 6. Effects of Various Cutoff Values

that the trace based on 0.08 mm cutoff includes only the very fine irregularities. In this example the effect of reducing the cutoff has been to reduce the roughness height indication. However, had the surface been made up only of irregularities as fine as those of the bottom trace, the roughness height indications would have been the same for all three cutoff settings.

In other words, all irregularities having a spacing less than the value of the cutoff used are included in a measurement. Obviously, if the cutoff value is too small to include coarser irregularities of a surface, the measurements will not agree with those taken with a larger cutoff. For this reason, care must be taken to choose a cutoff value which will include all of the surface irregularities it is desired to assess.

To become proficient in the use of continuously averaging stylus-type instruments the inspector or machine operator must realize that for uniform interpretation, the reading which is considered significant is the mean reading around which the needle tends to dwell or fluctuate under small amplitude.

Drawing Practices for Surface Texture Symbols. — American National Standard ANSI Y14.36-1978 establishes the method to designate controls for surface texture of solid materials. It includes methods for controlling roughness, waviness, and lay, and provides a set of symbols for use on drawings, specifications, or other documents. The units (metric or nonmetric) shall be consistent with the other units used on the drawing or documents. The numerical values expressed in the standard are stated in metric units and are to be regarded as standard. Approximate nonmetric equivalents are shown for reference.

Surface Texture Symbol. — The symbol used to designate control of surface irregularities is shown in Fig. 1(a). Where surface texture values other than roughness average are specified, the symbol must be drawn with the horizontal extension as shown in Fig. 1(e).

Symbol	Meaning
(a) ∨	Basic Surface Texture Symbol. Surface may be produced by any method except when the bar or circle (Figure 1b or 1d) is specified.
(b) ▽	Material Removal By Machining Is Required. The horizontal bar indicates that material removal by machining is required to produce the surface and that material must be provided for that purpose.
(c) 3.5▽	Material Removal Allowance. The number indicates the amount of stock to be removed by machining in millimeters (or inches). Tolerances may be added to the basic value shown or in a general note.
(d) ⊘	Material Removal Prohibited. The circle in the vee indicates that the surface must be produced by processes such as casting, forging, hot finishing, cold finishing, die casting, powder metallurgy or injection molding without subsequent removal of material.
(e)  ∨—	Surface Texture Symbol. To be used when any surface characteristics are specified above the horizontal line or to the right of the symbol. Surface may be produced by any method except when the bar or circle (Figure 1b and 1d) is specified.
(f) 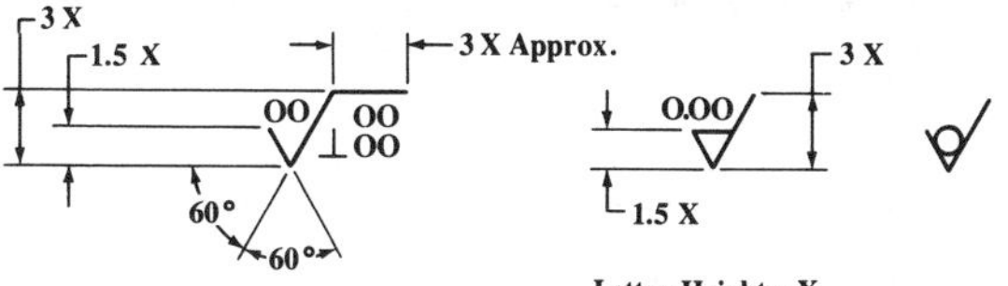	

Fig. 1. Surface Texture Symbols and Construction

Use of Surface Texture Symbols: When required from a functional standpoint, the desired surface characteristics should be specified. Where no surface texture control is specified, the surface produced by normal manufacturing methods is satisfactory provided it is within the limits of size (and form) specified in accordance with ANSI Y14.5-1973. Dimensioning and Tolerancing. This is not viewed as good practice; there should always be some maximum value, either specifically or by default (for example, in the manner of the note shown in Fig. 2).

Material Removal Required or Prohibited: The surface texture symbol is modified when necessary to require or prohibit removal of material. When it is necessary to indicate that a surface must be produced by removal of material by machining, specify the symbol shown in Fig. 1(b). When required, the amount of material to be removed is specified as shown in Fig. 1(c), in millimeters for metric drawings and in inches for nonmetric drawings. Tolerance for material removal may be added to the basic value shown or specified in a general note. When it is necessary to indicate that a surface must be produced without material removal, specify the machining prohibited symbol as shown in Fig. 1(d).

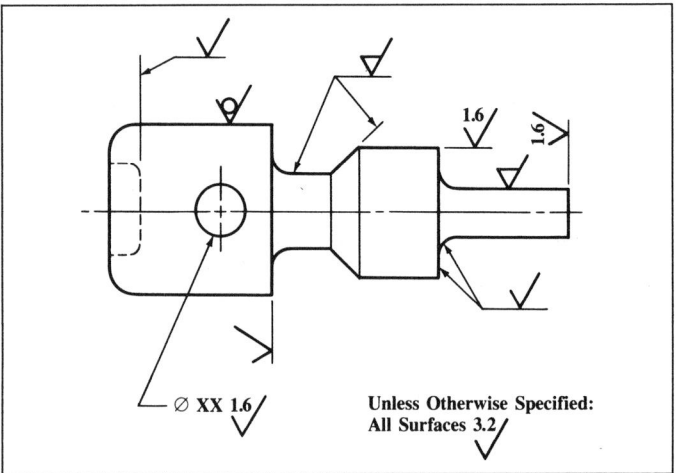

Fig. 2. Application of Surface Texture Symbols

Proportions of Surface Texture Symbols: The recommended proportions for drawing the surface texture symbol are shown in Fig. 1(f). The letter height and line width should be the same as that for dimensions and dimension lines.

Applying Surface Texture Symbols. — The point of the symbol should be on a line representing the surface, an extension line of the surface, or a leader line directed to the surface, or to an extension line. The symbol may be specified following a diameter dimension. The long leg (and extension) shall be to the right as the drawing is read. For parts requiring extensive and uniform surface roughness control, a general note may be added to the drawing which applies to each surface texture symbol specified without values as shown in Fig. 2.

When the symbol is used with a dimension, it affects the entire surface defined by the dimension. Areas of transition, such as chamfers and fillets, shall conform with the roughest adjacent finished area unless otherwise indicated.

Table 1. Preferred Series Roughness Average
Values (R$_a$)

μm	μin	μm	μin
0.012	0.5	1.25	50
0.025*	1*	1.60*	63*
0.050*	2*	2.0	80
0.075*	3	2.5	100
0.10*	4*	3.2*	125*
0.125	5	4.0	160
0.15	6	5.0	200
0.20*	8*	6.3*	250*
0.25	10	8.0	320
0.32	13	10.0	400
0.40*	16*	12.5*	500*
0.50	20	15	600
0.63	25	20	800
0.80*	32*	25*	1000*
1.00	40	. . .	. . .

* Recommended

Surface texture values, unless otherwise specified, apply to the complete surface. Drawings or specifications for plated or coated parts shall indicate whether the surface texture values apply before plating, after plating or both before and after plating.

Include in the symbol only those values required to specify and verify the required texture characteristics. Values should be in metric units for metric drawings and nonmetric units for nonmetric drawings.

Roughness and waviness measurements, unless otherwise specified, apply in a direction which gives the maximum reading; generally across the lay.

Table 2. Standard Roughness Sampling Length
(Cutoff) Values

mm	in.	mm	in.
0.08	0.003	2.5	0.1
0.25	0.010	8.0	0.3
0.80	0.030	25.0	1.0

Roughness Average (R$_a$): The preferred series of specified roughness average values is given in Table 1.

Cutoff or Roughness Sampling Length: Standard values are listed in Table 2. When no value is specified, the value 0.8 mm (0.030 in.) applies.

Waviness Height: The preferred series of maximum waviness height values is listed in Table 3. Waviness is not currently shown in ISO Standards. It is included here to follow present industry practice in the United States.

Table 3. Preferred Series Maximum Waviness
Height Values

mm	in.	mm	in.	mm	in.
0.0005	0.00002	0.008	0.0003	0.12	0.005
0.0008	0.00003	0.012	0.0005	0.20	0.008
0.0012	0.00005	0.020	0.0008	0.25	0.010
0.0020	0.00008	0.025	0.001	0.38	0.015
0.0025	0.0001	0.05	0.002	0.50	0.020
0.005	0.0002	0.08	0.003	0.80	0.030

Lay Symbol	Meaning	Example Showing Direction of Tool Marks
—	Lay approximately parallel to the line representing the surface to which the symbol is applied.	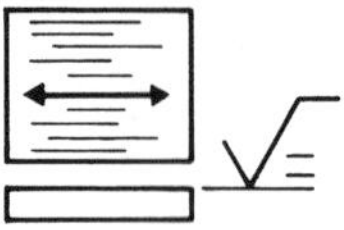
⊥	Lay approximately perpendicular to the line representing the surface to which the symbol is applied.	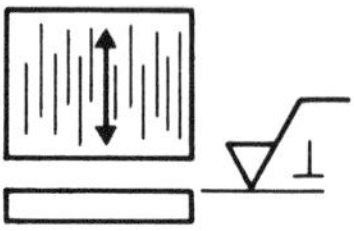
X	Lay angular in both directions to line representing the surface to which the symbol is applied.	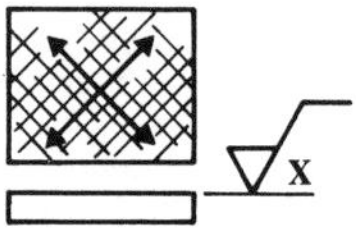
M	Lay multidirectional.	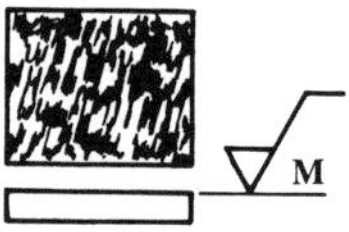
C	Lay approximately circular relative to the center of the surface to which the symbol is applied.	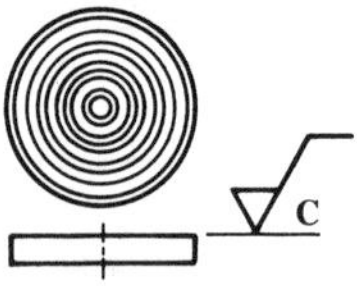
R	Lay approximately radial relative to the center of the surface to which the symbol is applied.	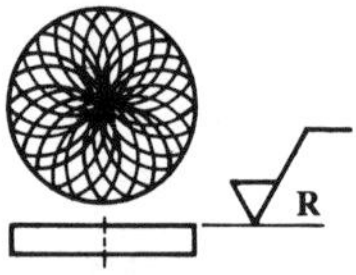
P	Lay particulate, non-directional, or protuberant.	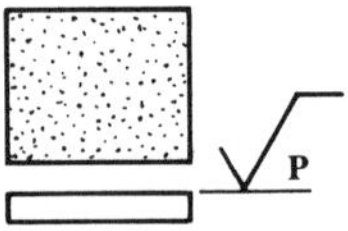

Fig. 3. Lay Symbols

Lay: Symbols for designating the direction of lay are shown and interpreted in Fig. 3.

Example Designations. — Figure 4 illustrates examples of designations of roughness, waviness, and lay by insertion of values in appropriate positions relative to the symbol. Where surface roughness control of several operations is required within a given area, or on a given surface, surface qualities may be designated, as in Fig. 5(a). If a surface must be produced by one particular process or a series of processes, they should be specified as shown in Fig. 5(b). Where special requirements are needed

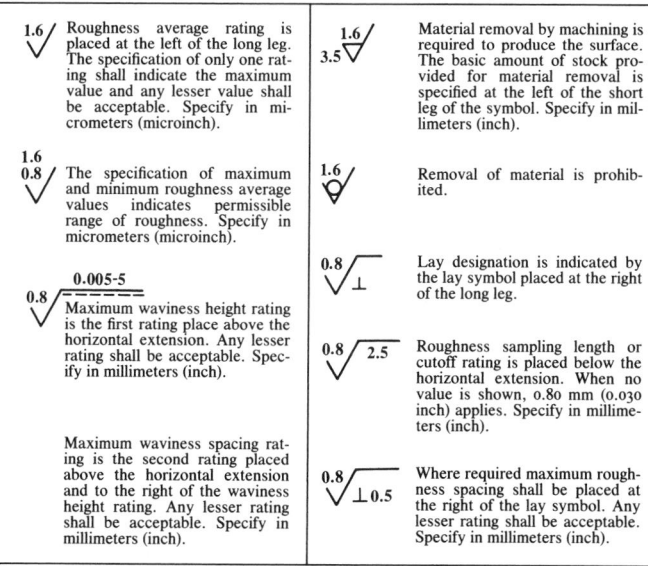

Fig. 4. Application of Surface Texture Values to Symbol

on a designated surface, a note should be added at the symbol giving the requirements and the area involved. An example is illustrated in Fig. 5(c).

Surface Texture of Castings. — Surface characteristics should not be controlled on a drawing or specification unless such control is essential to functional performance or appearance of the product. Imposition of such restrictions when unnecessary may increase production costs and in any event will serve to lessen the emphasis on the control specified for important surfaces. Surface characteristics of castings should never be considered on the same basis as machined surfaces. Castings are characterized by random distribution of non-directional deviations from the nominal surface.

Surfaces of castings rarely need control beyond that provided by the production method necessary to meet dimensional requirements. Comparison specimens are frequently used for evaluating surfaces having specific functional requirements. Surface texture control should not be specified unless required for appearance or function of the surface. Specification of such requirements may increase cost to the user.

Engineers should recognize that different areas of the same castings may have different surface textures. It is recommended that specifications of the surface be

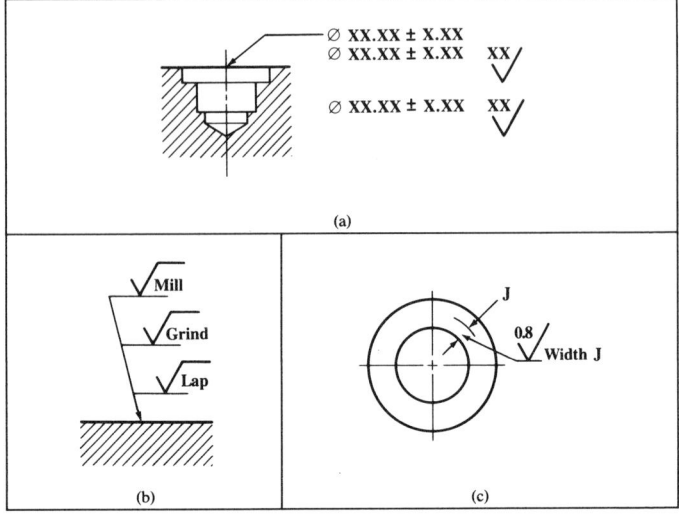

Fig. 5. Examples of Special Designations

limited to defined areas of the casting. Practicality of, and methods of determining that a casting's surface texture meets the specification shall be coordinated with the producer. The Society of Automotive Engineers standard J435 "Automotive Steel Castings" describes methods of evaluating steel casting surface texture used in the automotive and related industries.

Checking Drawings. — In order that the drawings may have a high standard of excellence, a set of instructions, as given in the following, has been issued to the checkers, and also to the draftsmen and tracers in the engineering department of a well-known machine-building company.

Inspecting a New Design: In case a new design is involved, first inspect the layouts carefully to see that the parts function correctly under all conditions, that they have the proper relative proportions, that the general design is correct in the matters of strength, rigidity, bearing areas, appearance, convenience of assembly, and direction of motion of the parts, and that there are no interferences. Consider the design as a whole to see if any improvements can be made. If the design appears to be unsatisfactory in any particular, or improvements appear to be possible, call the matter to the attention of the chief engineer.

Checking for Strength: Inspect the design of the part being checked for strength, rigidity, and appearance by comparing it with other parts for similar service whenever possible, giving preference to the later designs in such comparison, unless the later designs are known to be unsatisfactory. If there is any question regarding the matter, compute the stresses and deformations or find out whether the chief engineer has approved the stresses or deformations that will result from the forces applied to the part in service. In checking parts that are to go on a machine of increased size, be sure that standard parts used in similar machines and proposed for use on the larger

machine, have ample strength and rigidity under the new and more severe service to which they will be put.

Materials Specified. — Consider the kind of material required for the part and the various possibilities of molding, forging, welding, or otherwise forming the rough part from this material. Then consider the machining operations to see whether changes in form or design will reduce the number of operations or the cost of machining.

See that parts are designed with reference to the economical use of material, and whenever possible, utilize standard sizes of stock and material readily obtainable from local dealers. In the case of alloy steel, special bronze, and similar materials, be sure that the material can be obtained in the size required.

Method of Making Drawing. — Inspect the drawing to see that the projections and sections are made in such a way as to show most clearly the form of the piece and the work to be done on it. Make sure that any worker looking at the drawing will understand what the shape of the piece is and how it is to be molded or machined. Make sure that the delineation is correct in every particular, and that the information conveyed by the drawing as to the form of the piece is complete.

Checking Dimensions: — Check all dimensions to see that they are correct. Scale all dimensions and see that the drawing is to scale. See that the dimensions on the drawing agree with the dimensions scaled from the lay-out. Wherever any dimension is out of scale, see that the dimension is so marked. Investigate any case where the dimension, the scale of the drawing, and the scale of the lay-out do not agree. All dimensions not to scale must be underlined on the tracing. In checking dimensions, note particularly the following points:

See that all figures are correctly formed and that they will print clearly, so that the workers can easily read them correctly.

See that the over-all dimensions are given.

See that all witness lines go to the correct part of the drawing.

See that all arrow points go to the correct witness lines.

See that proper allowance is made for all fits.

See that the tolerances are correctly given where necessary.

See that all dimensions given agree with the corresponding dimensions of adjacent parts.

Be sure that the dimensions given on a drawing are those that the machinist will use, and that the worker will not be obliged to do addition or subtraction in order to obtain the necessary measurements for machining or checking his work.

Avoid strings of dimensions where errors can accumulate. It is generally better to give a number of dimensions from the same reference surface or center line.

When holes are to be located by boring on a horizontal spindle boring machine or other similar machine, give dimensions to centers of bored holes in rectangular coordinates and from the center lines of the first hole to be bored, so that the operator will not be obliged to add measurements or transfer gages.

Checking Assembly. — See that the part can readily be assembled with the adjacent parts. If necessary, provide tapped holes for eyebolts and cored holes for tongs, lugs, or other methods of handling.

Make sure that, in being assembled, the piece will not interfere with other pieces already in place and that the assembly can be taken apart without difficulty.

Check the sum of a number of tolerances; this sum must not be great enough to permit two pieces that should not be in contact to come together.

Checking Castings. — In the case of castings, study the form of the pattern, the methods of molding, the method of supporting and venting the cores, and the effect of draft and rough molding on clearances.

Avoid undue metal thickness, and especially avoid thick and thin sections in the same casting.

Indicate all metal thicknesses, so that the molder will know what chaplets to use for supporting the cores.

See that ample fillets are provided, and that they are properly dimensioned.

See that the cores can be assembled in the mold without crushing or interference.

See that swelling, shrinkage, or misalignment of cores will not make trouble in machining.

See that the amount of finish is indicated.

See that there is sufficient finish on large castings to permit them to be "cleaned up," even though they warp. In the case of such castings, make sure that the metal thickness will be sufficient after finishing, even though the castings do warp.

Make sure that sufficient sections are shown so that the patternmakers and molders will not be compelled to make assumptions about the form of any part of the casting. This is particularly important when a number of sections of the casting are similar in form, while others differ slightly.

Checking Machined Parts. — Study the sequences of operations in machining and see that all finish marks are indicated.

See that the finish marks are placed on the lines to which dimensions are given.

See that methods of machining are indicated where necessary.

Give all drill, reamer, tap, and rose bit sizes.

See that jig and gage numbers are indicated at the proper places.

See that all necessary bosses, lugs, and openings are provided for lifting, handling, clamping, and machining the piece.

See that adequate wrench room is provided for all nuts and bolt heads.

Avoid special tools, such as taps, drills, reamers, etc., unless such tools are specially authorized.

Where parts are right- and left-hand, be sure that the hand is correctly designated. When possible, mark parts symmetrical, so as to avoid having them right- and left-hand, but do not sacrifice correct design or satisfactory operation on this account.

When heat-treatment is required, the heat-treatment should be specified.

Check the title, size of machine, the scale, and the drawing number on both the drawing and the drawing record card.

Metric Dimensions on Drawings. — The length units of the metric system that are most generally used in connection with any work relating to mechanical engineering are the meter (39.37 inches) and the millimeter (0.03937 inch). One meter equals 1000 millimeters. On mechanical drawings, all dimensions are generally given in millimeters, no matter how large the dimensions may be. In fact, dimensions of such machines as locomotives and large electrical apparatus are given exclusively in millimeters. This practice is adopted to avoid mistakes due to misplacing decimal points, or mis-reading dimensions as when other units are used as well. When dimensions are given in millimeters, many of them can be given without resorting to decimal points, as a millimter is only a little more than $\frac{1}{32}$ inch. Only dimensions of precision need be given in decimals of a millimeter; such dimensions are generally given in hundredths of a millimeter — for example, 0.02 millimeter, which is equal to 0.0008 inch. As 0.01 millimeter is equal to 0.0004 inch, it is seldom that dimensions would be given with greater accuracy than to hundredths of a millimeter.

Scales of Metric Drawings. — Drawings made to the metric system are not made to scales of $\frac{1}{2}$, $\frac{1}{4}$, $\frac{1}{8}$, etc., as in the case of drawings made to the English system. If the object cannot be drawn full size, it may be drawn $\frac{1}{2.5}$, $\frac{1}{5}$, $\frac{1}{10}$, $\frac{1}{20}$, $\frac{1}{50}$, $\frac{1}{100}$, $\frac{1}{200}$, $\frac{1}{500}$, or $\frac{1}{1000}$ size. If the object is too small and has to be drawn larger it is drawn 2, 5, or 10 times its actual size.

INDEX

A

absolute temperature scales, 17
absolute zero, 17
algebraic identities, 26
allowances, preferred series of, 171
aluminum, 446, 451–458
  alloy designation system, 455
  characteristics of, 451–452
  clad alloys, 456
  heat treatability, 456
  series group, characteristics of, 456–457
  temper designations, 452–455
aluminum, machining, 410–411
angles, compound, 148–150

B

basic dimensions, 167
  preferred, 170
bolts and screws, steel, markings for, 278
brass, 444, 448–450
bronze, 444–446, 449, 450
Brown and Sharpe taper, 209, 213

C

cap screws, socket head
  drill and counterbore sizes for, 280
  hexagon and spline, 279
  metric, 283
  metric series, 282
    diameter-length combinations, 284
    drill and counterbore sizes for, 285
  length of complete thread, 283
centering tool, length of point, 400
chordal thicknesses and addenda of full depth gear teeth, 300
chords, lengths of, for spacing circumference of circles, 50–53
circle diameter-square side for equal area, 62
circles
  circumferences and areas of, 38–49
  properties of, 81
circular segment area, 50
compound angles, 148–150
conversion tables
  decimal equivalents, 8
  decimal of inch-millimeters, 10–11, 14
  fahrenheit-centigrade, 18–19
  inch-millimeter, 9
  metric-English measure, 1–7
  microinches-micrometers, 15–16
  millimeters-inches, 12–13
  miscellaneous English units, 7
copper, wrought and alloys, 446–448, 449
  beryllium, 464
  silicon (Everdur), 463–464
cutting fluids, 401–409
  recommendations, 405–406
cutting tools
  angles of end cutting edge, 373
    rake, 371–372
    relief, 371
    side cutting edge and head, 372
  chip breakers, 374–375
  contours of, 366
  definitions, 366–371
  dry versus wet grinding of carbide tools, 388
    coolants for, 388
  lapping of, 389
  nose radius, 373–374
  sharpening of carbide tools, 386
cubes of numbers from 1/32 to 100, 30–36

D

decimals of degree, into minutes and seconds, 147
degrees, minutes, seconds, into radians, 146
dial caliper (*see* vernier caliper)
die castings, zinc alloy, machining of, 414–415
die steels, selection of, 466–467
distance across corners
  of hexagons, 63
  of squares, 63
dovetail slides, measuring, 200
dowel pins, 286

E

engineering drawings, 479–516
  application of dimensions, 482–486
  check drawings, 514–516
  fundamental rules, 479–480

metric dimensions in drawings, 516
surface texture and roughness,
    501–514
symbols, 491–499
tolerances, 486–490
  geometric, symbols, 495–497
units of measurement, 480–482
equations, 26–29
cubic, 24
first degree, 26–27
quadractic, 27

F

fractions
common, 20–22
converting common-decimal, 25
decimal, 23–24

G

gear blank, nonmetallic, machining
    of, 413–414
gearing bevel
calculations, 312–314
milling of, 315
gearing, helical, calculations, 310–311
gearing, spur, calculations, 308–309
gearing, worm, calculations, 316–319
gears
definitions, 292–295
types of, 290–291
gear tooth caliper measurements,
    298–299
gear tooth milling, 301
chordal thicknesses and addenda
    for, 302–303
geometrical
constructions, 88–92
propositions, 82–87
grinding, chip breaker, 389
grinding wheels
diamond, 387
  concentration, 387–388
  grades, 387
  grit size, 387
dry vs. wet grinding of carbide
    tools, 388
  coolant for, 388
peripheral vs. flat side grinding,
    388–389
silicon carbide, 386–387

H

hardness testing, 471–478
Brinell hardness test, 471–472

comparative hardness scales, 474–
    478
Keep test, 473
Knoop hardness numbers, 473
Mohs's hardness scale, 478
Monotron hardness indicator, 473
Rockwell hardness test, 472
  scales, 472
Shore's Scleroscope, 473
Turner's Sclerometer, 478
Vickers hardness test, 473
heat treatment
of carbon steel, 468
of SAE alloy steels, 468–471
hexagons, distance across corners, 63

I

imaginary quantities, 29
indexable inserts, 375–376
identification of, 376
tool holders for, 377–378
  identification of, 378–380
  selecting, 380–383
indexing
angular, 320
  tables for, 322
approximate, for angles, 322
movements for standard index
    plate, 321
simple, 320
involute curve, properties of, 297
involute functions, 100
table of, 101–145

J

Jarno taper, 209, 214
jig boring
type "A" hole circles, 54–55
  central coordinates, 58–59
type "B" hole circles, 56–57
  central coordinates, 60–61

K

keys and bits, hex and spline,
    applicability of, 281
keys and keyseats
ANSI standard, 217, 218
chamfered, 222
depth control values S and T for
    shaft and hub, 219–220
depth of, 223
fillet, 222
parallel and taper, 221
size versus shaft diameter, 219
Woodruff, 217, 224–227

L

law of cosines, 94
law of sines, 94
limits and fits, 167, 169
    definitions of, 169–170
    description of, 172
    factors affecting selection of, 169
    preferred metric, 183–196
    standard, 172–173
        graphical presentation of, 173–174
        modified, 183
        parallel and taper keys, 221
        tables of, 175–182

M

machine tapers, American National Standard, 209–210
    steep, 215
machining methods, rules for, 324
magnesium alloys
    cast, 458
    wrought, 458–460
magnesium, machining, 409–410
    cutting fluids for, 408–409
mathematical signs and abbreviations, 75
micrometer, inch and millimeter, 151–153
    reading an inch, 152–153
    reading a millimeter, 153
minutes and seconds into decimals of a degree, 147
module system of gear design, 304–306
Monel, machining of, 415–416
Morse taper, 210–212
M profile screw threads
    external limiting dimensions, 243–246
    internal limiting dimensions, 241–242

N

nickel, 460–462
    Duranickel, 460–461
    Inconel, 462
    low-carbon, 460
    machining of, 415–416
    Monel, 461–462

P

pi (π), fraction of, 36
pipe threads
    railing point taper, NPTR, 248
    straight, NPSM and NPSL, 249
    taper, NPT, 247–248
plane figures, areas and dimensions, 65–72
planing tools, 375
plastics, machining of, 411–413
polygons, regular, formulas for, 64
powdered metal process, 465
powers of numbers, 78
powers of ten notation, 79–80
proportions, 76–78
protractor, 156–157

R

radians
    into degrees and decimals of degree, 147
    into degrees, minutes, and seconds, 147
Rankine temperature scale, 17
ratios, 76
rearrangement of formulas, 28–29
reciprocals, 78
roots, 78

S

scientific notation, 79–80
screw threads
    definitions, 229–230
    designations, 230–232
    measuring, 273–277
        three-wire method of, 274–277
screw thread systems, 229–277
segments of circles, 73–74
sevolute function, 100
side mills, arbor mounted
    rake angle, 393–394
    relief angle, 393–394
sine-bar, 158–159
    constants for setting 5-in., 160–166
slot mills, arbor mounted
    rake angle, 393–394
    relief angle, 393–394
speeds and feeds
    for broaching, 336, 338–339
    depth of cut, 325–326, 335
    equivalent rpm for drill sizes, 359, 361–364
    hard-to-machine materials, using CBN tools, 340
    drilling and reaming
        difficulties, 351
        feedrates for, 351
        glass, 354–355, 357

recommended speeds for
copper alloys, 358
ferrous cast materials, 356
light metals, 357
plain carbon and alloy steels,
352–353
stainless steels, 355
tool steels, 354
metal cutting saws, 365
milling
feedrates for, 342, 344–346, 351
recommended speeds for
cemented carbide cutters,
350–351
ferrous cast metals, 347–349
high-speed-steel cutters, 348
plain carbon and alloy steels,
343–344
stainless steels, 346
tool steels, 345
planers, 357
planing tools, 340
recommended cutting speeds for
copper alloys, 337
ferrous cast metals, 334
light metals, 336
plain carbon and alloy steels,
330–331
stainless steels, 333
superalloys, 339
titanium and titanium alloys, 338
tool steels, 332
for tapping, 332–333, 335
thread chasing, 340, 342
thread cutting, single-point tools,
340
troubleshooting, 341–342
use of cutting speed tables, 326–
329
unusual materials, 366
squares, distance across corners, 63
square side-circle diameters for equal
areas, 62
squares of numbers from 1/32 to 100,
30–36
steel
applications of SAE, 424–426
carbon, 417–419
carburizing grades, 419–420
chromium nickel austenitic, 422–423
free cutting, 419
high strength, low alloy, 416
molybdenum, high-speed, 433
stainless chromium, 423

T

tap drill sizes
general formulas for, 240
for metric threads, 238–239
for pipe threads, 240
for Unified threads, 233–237
tapers
for machine tool spindles, 214
measurement of angles and, 202
measuring with vee-block and sine-
bar, 199
per foot and corresponding angles,
201
pins, 205–208
rules for figuring, 200, 202–204
self-holding, 215, 216
standard, 209–216
taper pins, 287
titanium, 462–463
properties, 463
tolerance
application of, 168
bilateral, 167–168
gagemakers, 197
preferred series of, 171
standard, 171
unilateral, 167–168
tolerance dimensions, locating, 168
tolerance grades, relation to machine
processes, 171, 198
tool geometry
end mills, 395–396
twist drills, 397–399
tools, single point
rake angle, 390–392
relief angle, 390–392
tool steels
classification of, 430
cold work, 436
hot work, 434
mold, 437
shock resisting, 437
selection guide, 431–432, 466–467
special purpose, 437
trade names of, 438–443
tungsten, high speed, 435
triangles
oblique-angled, solution of, 98–99
right-angled, solution of, 96–97
trigonometric functions of angles, 93–
94
signs of, 94, 100
tables of, 100, 101–145
trigonometric identities, 95

transposition of formulas, 28–29
twist drills
   drill point thinning, 385–386
   length of point, 360, 400
   sharpening of, 383–384
   tool flanks, relief grinding of, 384–
     385

U

Unified Numbering System for metals
   and alloys, 427–430

Unified screw thread series, 250–
   272

V

vernier, 153–156
   caliper (dial caliper), 155–156
   reading a 25-division, 154
   reading a 50-division, 154
   reading a millimeter, 155
versed sine and cosine, 100